Work It Out

Like your instructors, authors Ratti and McWaters encourage you to practice this material frequently. Each chapter includes ample opportunities to master the material by solving problems and applying your understanding.

Procedure In Action

Procedure In Action boxes present important multi-step procedures in a two-column format, illustrating the numbered steps of a procedure with a worked example.

Practice Problems

All examples conclude with a **Practice Problem,** so you can try a new problem and make sure you grasp the concept before moving ahead.

End-of-Section Exercises

Each section ends with four levels of exercises for you to practice the math and apply your understanding:

- Basic Concepts and Skills
- Applying the Concepts
- Beyond the Basics
- Critical Thinking/Discussion/Writing

Preparation and Review

End-of-Chapter Material

Each chapter concludes with a **Summary of Definitions, Concepts, and Formulas; Review Exercises;** two **Practice Tests;** and a **Cumulative Review** to prove your mastery of the concepts presented in the chapter.

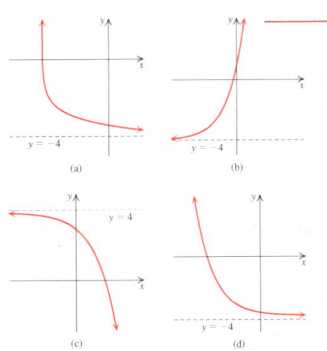

Student-friendly Support

The features in this book are designed to help you see what you are going to learn and why. In addition, integrated study aids give you hints and tips at strategic places in the text.

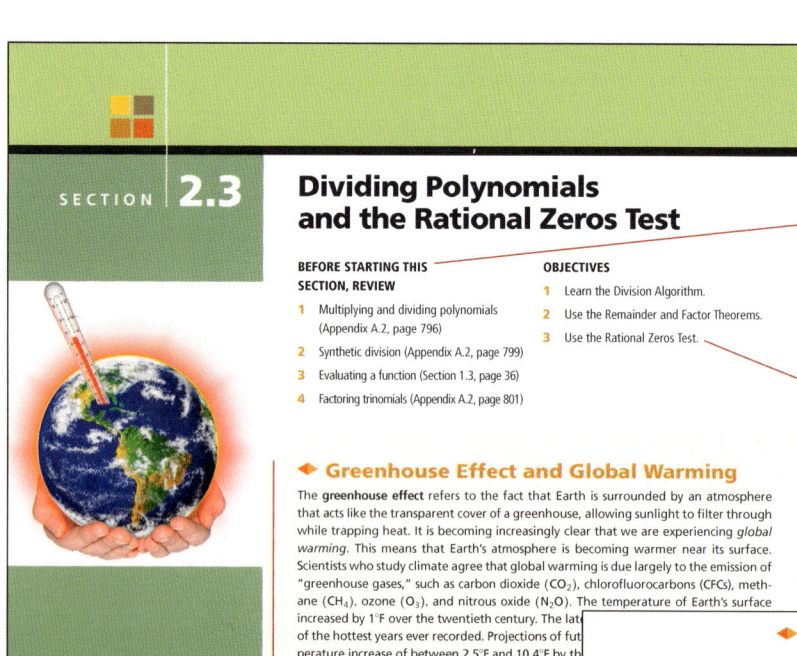

Topics for Review

Topics for Review help you refresh your memory on essential topics before beginning each chapter, listing specific page and section references to make it easy to find those concepts.

Objectives

Clear **Objectives** preview the key points of the section ahead, providing easy reference points when studying for exams. These **Objectives** appear again at the appropriate places in the section, putting each new topic into context.

Applications

Each section begins with an engaging **Application** that is also revisited later in the section through examples and exercises. The **Application** ties key ideas together and provides motivation for you to read and study.

This icon ◆ connects the section opening application to an example or exercise in the section.

Recall Notes

Located in the margins, **Recall** notes periodically remind you of a key idea learned earlier in the text that will help you work through the current problem.

Precalculus Essentials

Precalculus Essentials

J. S. Ratti
University of South Florida

Marcus McWaters
University of South Florida

PEARSON

Boston   Columbus   Indianapolis   New York   San Francisco   Upper Saddle River
Amsterdam   Cape Town   Dubai   London   Madrid   Milan   Munich   Paris   Montréal   Toronto
Delhi   Mexico City   São Paulo   Sydney   Hong Kong   Seoul   Singapore   Taipei   Tokyo

Editorial Director: Christine Hoag
Editor in Chief: Anne Kelly
Senior Development Editor: Elaine Page
Executive Content Editor: Christine O'Brien
Editorial Assistant: Judith Garber
Marketing Manager: Peggy Lucas
Marketing Assistant: Justine Goulart
Director of Production: Sarah McCracken
Senior Managing Editor: Karen Wernholm
Senior Production Project Manager: Beth Houston
Procurement Manager: Evelyn Beaton
Procurement Specialist: Linda Cox
Associate Design Director: Andrea Nix
Senior Designer: Barbara Atkinson
Cover and Text Designer: Ark Design
Cover Image: Franz Pritz/Picture Press/Getty Images
Executive Director of Development: Carol Trueheart
Digital Assets Manager: Marianne Groth
Supplements Production Coordinator: Kerri Consalvo
Math XL Content Supervisor: Kristina Evans
Senior Content Developer: Mary Durnwald
Senior Author Support/ Technology Specialist: Joe Vetere
Image Manager: Rachel Youdelman
Photo Researcher: Diahanne Lucas
Media Director: Ruth Berry
Media Producer: Tracy Menoza
Full-Service Project Management: PreMediaGlobal
Composition: PreMediaGlobal

Credits and acknowledgments borrowed from other sources and reproduced, with permission, in this textbook appear on the appropriate page within text or on page C-1.

Many of the designations by manufacturers and sellers to distinguish their products are claimed as trademarks. Where those designations appear in this book, and the publisher was aware of a trademark claim, the designations have been printed in initial caps or all caps.

Copyright © 2014 by Pearson Education, Inc. All rights reserved. Manufactured in the United States of America. This publication is protected by Copyright, and permission should be obtained from the publisher prior to any prohibited reproduction, storage in a retrieval system, or transmission in any form or by any means, electronic, mechanical, photocopying, recording, or likewise. To obtain permission(s) to use material from this work, please submit a written request to Pearson Education, Inc., Permissions Department, One Lake Street, Upper Saddle River, New Jersey 07458, or you may fax your request to 201-236-3290.

Library of Congress Cataloging-in-Publication Data
Ratti, J. S.
  Precalculus essentials / J.S. Ratti, University of South Florida,
Marcus McWaters, University of South Florida.
     pages cm
  Includes bibliographical references and index.
  ISBN 0-321-81696-X (student edition : alk. paper)

  1. Algebra. 2. Trigonometry. I. McWaters, Marcus M. II. Title.
  QA154.3.R38 2014
  515—dc23
                                                    2012039779

7 16

www.pearsonhighered.com

ISBN 13: 978-0-321-81696-2
ISBN 10: 0-321-81696-X

To Our Wives
Lata and Debra

FOREWORD

Our goal in writing this text is to provide material that will actively engage your students in the process of learning the mathematical skills necessary for success with calculus. As instructors ourselves, we are familiar with the challenges of making mathematics useful and interesting without sacrificing the solid mathematics that is necessary for conceptual understanding.

In this book, you will find a strong emphasis on both concept development and real-life applications. Topics such as functions, graphing, the difference quotient, and limiting processes provide thorough preparation for the study of calculus and improve students' comprehension of algebra. Just-in-time review throughout the text and MyMathLab course assures that all students are brought to the same level before being introduced to new concepts. Numerous applications are used to motivate students to apply the concepts and skills they learn in precalculus to other courses, including the physical and biological sciences, engineering, and economics, and to on-the-job and everyday problem solving. Students are given ample opportunities throughout this book to think about important mathematical ideas and to practice and apply algebraic skills.

Throughout the text, we emphasize why the material being covered is important and how it can be applied. By thoroughly developing mathematical concepts with clearly defined terminology, students see the "why" behind those concepts, paving the way for a deeper understanding, better retention, less reliance on rote memorization, and ultimately more success. The level of exposition was selected so that the material would be accessible to students and provide them with an opportunity to grow.

(Marcus McWaters)

(J. S. Ratti)

CONTENTS

Foreword   vii

Preface   xiv

Resources   xvi

Acknowledgments   xix

Chapter P

BASIC CONCEPTS OF ALGEBRA   1

P.1   The Real Numbers; Integer Exponents   2

Variables, Constants, and Operations ■ Classifying Sets of Numbers ■ Rational Numbers ■ Irrational Numbers ■ The Real Number Line ■ Inequality Symbols ■ Intervals ■ Absolute Value ■ Distance Between Two Points on a Real Number Line ■ Integer Exponents ■ Rules of Exponents

P.2   Radicals and Rational Exponents   12

Square Roots ■ Other Roots ■ Simplifying Radical Expressions ■ Like Radicals ■ Radicals with Different Indexes ■ Rationalizing ■ Conjugates ■ Rational Exponents

P.3   Solving Equations   21

Definitions ■ Equivalent Equations ■ Solving Linear Equations in One Variable ■ Solving Quadratic Equations ■ Factoring Method ■ Completing the Square ■ The Quadratic Formula ■ Solving Equations by Factoring ■ Rational Equations ■ Radical Equations ■ Equations That Are Quadratic in Form ■ Equations Involving Absolute Value

P.4   Inequalities   33

Inequalities ■ Linear Inequalities ■ Combining Two Inequalities ■ Using Test Points to Solve Inequalities ■ Inequalities Involving Absolute Value

P.5   Complex Numbers   45

Complex Numbers ■ Addition and Subtraction ■ Multiplying and Dividing Complex Numbers ■ Powers of i

Chapter 1

GRAPHS AND FUNCTIONS   51

1.1   Graphs of Equations   52

The Coordinate Plane ■ The Distance Formula ■ The Midpoint Formula ■ Graph of an Equation ■ Intercepts ■ Symmetry ■ Circles

viii

1.2 Lines 65
Slope of a Line ■ Point–Slope Form ■ Slope–Intercept Form ■ Equations of Horizontal and Vertical Lines ■ General Form of the Equation of a Line ■ Parallel and Perpendicular Lines

1.3 Functions 77
Functions ■ Function Notation ■ Representations of Functions ■ The Domain of a Function ■ The Range of a Function ■ Graphs of Functions ■ Function Information from Its Graph ■ Building Functions

1.4 A Library of Functions 94
Linear Functions ■ Increasing and Decreasing Functions ■ Relative Maximum and Minimum Values ■ Even–Odd Functions and Symmetry ■ Square Root and Cube Root Functions ■ Piecewise Functions ■ Graphing Piecewise Functions ■ Basic Functions ■ Average Rate of Change

1.5 Transformations of Functions 109
Transformations ■ Vertical and Horizontal Shifts ■ Reflections ■ Stretching or Compressing ■ Multiple Transformations in Sequence

1.6 Combining Functions; Composite Functions 124
Combining Functions ■ Composition of Functions ■ Domain of Composite Functions ■ Decomposition of a Function ■ Applications of Composite Functions

1.7 Inverse Functions 136
Inverses ■ Finding the Inverse Function ■ Finding the Range of a One-to-One Function ■ Applications

Chapter 1 Summary ■ Chapter 1 Review Exercises ■ Chapter 1 Exercises for Calculus ■ Chapter 1 Practice Test A ■ Chapter 1 Practice Test B

Chapter 2

POLYNOMIAL AND RATIONAL FUNCTIONS 154

2.1 Quadratic Functions 155
Quadratic Functions ■ Standard Form of a Quadratic Function ■ Graphing a Quadratic Function $f(x) = ax^2 + bx + c$ ■ Applications

2.2 Polynomial Functions 164
Polynomial Functions ■ Power Functions ■ End Behavior of Polynomial Functions ■ Zeros of a Function ■ Zeros and Turning Points ■ Graphing a Polynomial Function

2.3 Dividing Polynomials and the Rational Zeros Test 179

The Division Algorithm ▪ The Remainder and Factor Theorems ▪ The Rational Zeros Test

2.4 Zeros of a Polynomial Function 188

Descartes's Rule of Signs ▪ Bounds on the Real Zeros ▪ Complex Zeros of Polynomials ▪ Conjugate Pairs Theorem

2.5 Rational Functions 198

Rational Functions ▪ Vertical and Horizontal Asymptotes ▪ Translations of $f(x) = \dfrac{1}{x}$ ▪ Graphing Rational Functions ▪ Oblique Asymptotes ▪ Graph of a Revenue Curve

Chapter 2 Summary ▪ Chapter 2 Review Exercises ▪ Chapter 2 Exercises for Calculus ▪ Chapter 2 Practice Test A ▪ Chapter 2 Practice Test B

Chapter 3

EXPONENTIAL AND LOGARITHMIC FUNCTIONS 221

3.1 Exponential Functions 222

Exponential Functions ▪ Evaluate Exponential Functions ▪ Graphing Exponential Functions ▪ Simple Interest ▪ Compound Interest ▪ Continuous Compound Interest Formula ▪ The Natural Exponential Function ▪ Natural Exponential Growth and Decay

3.2 Logarithmic Functions 239

Logarithmic Functions ▪ Evaluating Logarithms ▪ Domain of Logarithmic Functions ▪ Graphs of Logarithmic Functions ▪ Common Logarithms ▪ Natural Logarithms ▪ Investments ▪ Newton's Law of Cooling

3.3 Rules of Logarithms 253

Rules of Logarithms ▪ Number of Digits ▪ Change of Base ▪ Growth and Decay ▪ Half-Life ▪ Radiocarbon Dating

3.4 Exponential and Logarithmic Equations and Inequalities 265

Solving Exponential Equations ▪ Applications of Exponential Equations ▪ Solving Logarithmic Equations ▪ Logarithmic and Exponential Inequalities

Chapter 3 Summary ▪ Chapter 3 Review Exercises ▪ Chapter 3 Exercises for Calculus ▪ Chapter 3 Practice Test A ▪ Chapter 3 Practice Test B

Chapter 4

TRIGONOMETRIC FUNCTIONS 282

4.1 Angles and Their Measure 283

Angles ■ Measuring Angles Using Degrees ■ Radian Measure ■ Relationship Between Degrees and Radians ■ Complements and Supplements ■ Length of an Arc of a Circle

4.2 The Unit Circle; Trigonometric Functions 293

The Unit Circle ■ Trigonometric Functions of Real Numbers ■ Finding Exact Trigonometric Function Values ■ Trigonometric Functions of an Angle ■ Trigonometric Function Values of an Angle θ ■ Evaluating Trigonometric Functions Using a Calculator ■ Signs of the Trigonometric Functions ■ Reference Angle

4.3 Graphs of the Sine and Cosine Functions 310

Properties of Sine and Cosine ■ Domain and Range of Sine and Cosine ■ Zeros of Sine and Cosine ■ Even–Odd Properties ■ Periodic Functions ■ Graphs of Sine and Cosine Functions ■ Graph of the Sine Function ■ Graph of the Cosine Function ■ Five Key Points ■ Amplitude and Period ■ Phase Shift ■ Vertical Shifts ■ Simple Harmonic Motion

4.4 Graphs of the Other Trigonometric Functions 327

Tangent Function ■ Graph of $y = \tan x$ ■ Graphs of the Reciprocal Functions ■ Graphing $y = a \csc(bx - k) + d$ and $y = a \sec(bx - k) + d$

4.5 Inverse Trigonometric Functions 337

The Inverse Sine Function ■ The Inverse Cosine Function ■ The Inverse Tangent Function ■ Other Inverse Trigonometric Functions ■ Evaluating Inverse Trigonometric Functions ■ Composition of Trigonometric and Inverse Trigonometric Functions

4.6 Right-Triangle Trigonometry 350

Trigonometric Ratios and Functions ■ Evaluating Trigonometric Functions ■ Complements ■ Solving Right Triangles ■ Applications

4.7 Trigonometric Identities 362

Fundamental Trigonometric Identities ■ Evaluating Trigonometric Functions ■ Trigonometric Equations and Identities ■ Verifying Trigonometric Identities ■ Methods of Verifying Trigonometric Identities

4.8 Sum and Difference Formulas 373

Sum and Difference Formulas for Cosine ■ Cofunction Identities ■ Sum and Difference Formulas for Sine ■ Sum and Difference Formulas for Tangent ■ Double-Angle Formulas ■ Power-Reducing Formulas ■ Half-Angle Formulas

Chapter 4 Summary ■ Chapter 4 Review Exercises ■ Chapter 4 Exercises for Calculus ■ Chapter 4 Practice Test A ■ Chapter 4 Practice Test B

Chapter 5

APPLICATIONS OF TRIGONOMETRIC FUNCTIONS 394

5.1 The Law of Sines and the Law of Cosines 395

Solving Oblique Triangles ▪ The Law of Sines ▪ Solving AAS and ASA Triangles ▪ Solving SSA Triangles—the Ambiguous Case ▪ Bearings ▪ The Law of Cosines ▪ Derivation of the Law of Cosines ▪ Solving SAS Triangles ▪ Solving SSS Triangles

5.2 Areas of Polygons Using Trigonometry 411

Geometry Formulas ▪ Area of SAS Triangles ▪ Area of AAS and ASA Triangles ▪ Area of SSS Triangles

5.3 Polar Coordinates 419

Polar Coordinates ▪ Converting Between Polar and Rectangular Forms ▪ Converting Equations Between Rectangular and Polar Forms ▪ The Graph of a Polar Equation

5.4 Parametric Equations 434

Parametric Equations ▪ Graphing a Plane Curve ▪ Eliminating the Parameter ▪ Finding Parametric Equations

Chapter 5 Summary ▪ Chapter 5 Review Exercises ▪ Chapter 5 Exercises for Calculus ▪ Chapter 5 Practice Test A ▪ Chapter 5 Practice Test B

Chapter 6

FURTHER TOPICS IN ALGEBRA 447

6.1 Sequences and Series 448

Sequences ▪ Recursive Formulas ▪ Factorial Notation ▪ Summation Notation ▪ Series

6.2 Arithmetic Sequences; Partial Sums 458

Arithmetic Sequence ▪ Sum of the First n Terms of an Arithmetic Sequence

6.3 Geometric Sequences and Series 465

Geometric Sequences ▪ Finding the Sum of a Finite Geometric Sequence ▪ Annuities ▪ Infinite Geometric Series

6.4 Systems of Equations in Two Variables 476

System of Equations ▪ Graphical Method ▪ Substitution Method ▪ Elimination Method ▪ Applications of Systems of Equations

6.5 Partial-Fraction Decomposition 488
Partial Fractions ▪ $Q(x)$ Has Only Distinct Linear Factors ▪ $Q(x)$ Has Repeated Linear Factors

Chapter 6 Summary ▪ Chapter 6 Review Exercises ▪ Chapter 6 Exercises for Calculus ▪ Chapter 6 Practice Test A ▪ Chapter 6 Practice Test B

Answers to Practice Problems   P-1
Answers to Selected Exercises   A-1
Credits   C-1
Index   I-1

PREFACE

Students begin precalculus classes with widely varying backgrounds. Some haven't taken a math course in several years and may need to spend time reviewing prerequisite topics, while others are ready to jump right in to new and challenging material. We have provided review material in Chapter P and in some of the early sections of other chapters in such a way that it can be used or omitted as is appropriate for your course. In addition, students may follow several different paths after completing a precalculus course. Many will continue their study of mathematics in courses such as finite mathematics, statistics, and calculus. For others, this course may be their last mathematics course. Responding to the current and future needs of all of these students was essential in creating this text. Overall, we present our content in a systematic way that illustrates how to study and what to review. We believe that if students use this textbook well, they will succeed in this course.

Features

CHAPTER OPENER Each chapter opener includes a description of applications (one of them illustrated) relevant to the content of the chapter, plus the list of topics that will be covered. In one page, students see what they are going to learn and why they are learning it.

SECTION OPENER with APPLICATION Each section opens with a list of prerequisite topics, complete with section and page references, which students can **review** prior to starting the section. The **Objectives** of the section are also clearly stated and numbered, and then referenced again in the margin of the lesson at the point where the objective's topic is taught. An **Application** then follows containing a motivating anecdote or interesting problem. An example later in the section, relating to this application and identified by the same icon ◆, is then solved using the mathematics covered in the section. These applications utilize material from a variety of fields: the physical and biological sciences (including health sciences), economics, art and architecture, history, and more.

EXAMPLES and PRACTICE PROBLEMS **Examples** include a wide range of computational, conceptual, and modern applied problems carefully selected to build confidence, competency, and understanding. Every example has a title indicating its purpose, and a detailed solution containing annotated steps. All examples are followed by a **Practice Problem** for students to try so that they can check their understanding of the concept covered. Answers to the Practice Problems are provided just before the section exercises.

ADDITIONAL PEDAGOGICAL FEATURES

Procedure In Action is a special feature which introduces procedure steps within the context of a work-out example. Important multistep procedures, such as the steps for doing synthetic division, are presented in a two-column format. The numbered steps of the procedure are given in the left column, and an example is worked out step-by-step, aligned with and numbered as the procedure steps, in the right column. This approach provides students with a clear solution model when encountering difficulty in their work.

Definitions, *Theorems*, *Properties*, and *Rules* are all boxed and titled for emphasis and ease of reference.

Warnings appear as appropriate throughout the text to let students know of common errors and pitfalls that can trip them up in their thinking or calculations.

Summary of Main Facts boxes summarize information related to equations and their graphs along with guidelines, such as those for verifying trigonometric identities.

MARGIN NOTES

Side Notes provide hints for handling newly introduced concepts.

Recall notes remind students of a key idea learned earlier in the text that will help them work through a current problem.

Technology Connections give students tips on using calculators to solve problems, check answers, and reinforce concepts. Note the use of graphing calculators is optional in this text.

Do You Know? provide students with additional interesting information on topics to keep them engaged in the mathematics presented.

Historical Notes give students information on key people or ideas in the history and development of mathematics.

EXERCISES The heart of any textbook is its exercises, so we have tried to ensure that the quantity, quality, and variety of exercises meet the needs of all students. Exercises are carefully graded to strengthen the skills developed in the section and are organized using the following categories. **Basic Concepts and Skills** develop fundamental skills—each odd-numbered exercise is closely paired with its consecutive even-numbered exercise. **Applying the Concepts** use the section's material to solve real-world problems—all are titled and relevant to the topics of the section. **Critical Thinking/Discussion/Writing**

exercises, appearing as appropriate, are designed to develop students' higher-level thinking skills. Calculator problems, identified by [icon] are included where needed. **Maintaining Skills** exercises help refresh important concepts learned in previous chapters and allow for practice of skills needed for more advanced concepts that appear in the upcoming chapter.

END-OF-CHAPTER The chapter-ending material begins with a **Summary of Definitions**, **Concepts**, **and Formulas** consisting of a brief description of key topics organized by section, encouraging students to reread sections rather than memorize definitions out of context.

Review Exercises provide students with an opportunity to practice what they have learned in the chapter. **Exercises for Calculus** allow students to practice skills that will be used in future calculus courses. Then students are given two chapter test options. They can take **Practice Test A** in the usual open-ended format and/or **Practice Test B**, covering the same topics, in multiple-choice format. All tests are designed to increase student comprehension and verify that students have mastered the skills and concepts in the chapter. Mastery of these materials should indicate a true comprehension of the chapter and the likelihood of success on the associated in-class examination.

RESOURCES

Media Resources

MYLABSPLUS®

MyMathLab®Plus/MyStatLab™Plus

MyLabsPlus combines proven results and engaging experiences from MyMathLab® and MyStatLab™ with convenient management tools and a dedicated services team. Designed to support growing math and statistics programs, it includes additional features such as:

- **Batch Enrollment:** Your school can create the login name and password for every student and instructor, so everyone can be ready to start class on the first day. Automation of this process is also possible through integration with your school's Student Information System.

- **Login from your campus portal:** You and your students can link directly from your campus portal into your MyLabsPlus courses. A Pearson service team works with your institution to create a single sign-on experience for instructors and students.

- **Advanced Reporting:** MyLabsPlus's advanced reporting allows instructors to review and analyze students' strengths and weaknesses by tracking their performance on tests, assignments, and tutorials. Administrators can review grades and assignments across all courses on your MyLabsPlus campus for a broad overview of program performance.

- **24/7 Support:** Students and instructors receive 24/7 support, 365 days a year, by email or online chat.

MyLabsPlus is available to qualified adopters. For more information, visit our website at HYPERLINK "http://www.mylabsplus.com/" www.mylabsplus.com or contact your Pearson representative.

MATHXL® ONLINE COURSE
(Access Code Required)

MathXL is the homework and assessment engine that runs MyMathLab. (MyMathLab is MathXL plus a learning management system.)

With MathXL, instructors can:

- Create, edit, and assign online homework and tests using algorithmically generated exercises correlated at the objective level to the textbook.

- Create and assign their own online exercises and import TestGen tests for added flexibility.

- Maintain records of all student work tracked in MathXL's online gradebook.

With MathXL, students can:

- Take chapter tests in MathXL and receive personalized study plans and/or personalized homework assignments based on their test results.

- Use the study plan and/or the homework to link directly to tutorial exercises for the objectives they need to study.

- Access supplemental animations and video clips directly from selected exercises.

MathXL is available to qualified adopters. For more information, visit our website at www.mathxl.com, or contact your Pearson representative.

MYMATHLAB® ONLINE COURSE
(Access Code Required)

MyMathLab from Pearson is the world's leading online resource in mathematics, integrating interactive homework, assessment, and media in a flexible, easy to use format.

MyMathLab delivers **proven results** in helping individual students succeed.

- MyMathLab has a consistently positive impact on the quality of learning in higher education math instruction. MyMathLab can be successfully implemented in any environment—lab-based, hybrid, fully online, traditional—and demonstrates the quantifiable difference that integrated usage has on student retention, subsequent success, and overall achievement.

- MyMathLab's comprehensive online gradebook automatically tracks your students' results on tests, quizzes, homework, and in the study plan. You can use the gradebook to quickly intervene if your students have trouble, or to provide positive feedback on a job well done. The data within MyMathLab is easily exported to a variety of spreadsheet programs, such as Microsoft Excel. You can determine which points of data you want to export, and then analyze the results to determine success.

MyMathLab provides **engaging experiences** that personalize, stimulate, and measure learning for each student.

- **Exercises:** The homework and practice exercises in MyMathLab are correlated to the exercises in the textbook, and they regenerate algorithmically to give students unlimited opportunity for practice and mastery. The software offers immediate, helpful feedback when students enter incorrect answers.

- **Multimedia Learning Aids:** Exercises include guided solutions, sample problems, animations, videos, and eText access for extra help at point-of-use.

- **Expert Tutoring:** Although many students describe the whole of MyMathLab as "like having your own personal tutor," students using MyMathLab do have access to live tutoring from Pearson, from qualified math and statistics instructors.

And, MyMathLab comes from an **experienced partner** with educational expertise and an eye on the future.

- Knowing that you are using a Pearson product means knowing that you are using quality content. That means that our eTexts are accurate and our assessment tools work. It means we are committed to making MyMathLab as accessible as possible.

- Whether you are just getting started with MyMathLab, or have a question along the way, we're here to help you learn about our technologies and how to incorporate them into your course.

Ratti/McWaters' MyMathLab course engages students and keeps them thinking.

- Author designated preassigned homework assignments are provided.

- Integrated Review provides optional quizzes throughout the course that test prerequisite knowledge. After taking each quiz, students receive a personalized, just-in-time review assignment to help them refresh forgotten skills.

- Interactive figures are available, enabling users to manipulate figures to bring hard-to-convey math concepts to life.

- Example Videos provide lectures for each section of the text to help students review important concepts and procedures 24/7. Assignable questions are available to check students' video comprehension.

To learn more about how MyMathLab combines proven learning applications with powerful assessment, visit **www.mymathlab.com** or contact your Pearson representative.

MYMATHLAB® READY TO GO COURSE
(Access Code Required)

These new Ready to Go courses provide students with all the same great MyMathLab features but make it easier for instructors to get started. Each course includes pre-assigned homework and quizzes to make creating your course even simpler. Ask your Pearson representative about the details for this particular course or to see a copy of this course.

Additional Student Resources
STUDENT'S SOLUTIONS MANUAL

- By Beverly Fusfield

- Provides detailed worked-out solutions to the odd numbered end-of-section and Chapter Review exercises and solutions to all of the Practice Problems, Practice Tests and Cumulative Review problems

- ISBN-13: 978-0-321-81699-3;
  ISBN-10: 0-321-81699-4

Additional Instructor Resources
ANNOTATED INSTRUCTOR'S EDITION

- Answers included on the same page beside the text exercises where possible for quick reference

- ISBN-13: 978-0-321-81697-9;
  ISBN-10: 0-321-81697-8

INSTRUCTOR'S SOLUTIONS MANUAL

- By Beverly Fusfield

- Complete solutions provided for all end-of-section exercises, including the Critical Thinking and Group Projects, Practice Problems, Chapter Review exercises, Practice Tests, and Cumulative Review problems

- ISBN-13: 978-0-321-81703-7;
  ISBN-10: 0-321-81703-6

TESTGEN®

TestGen® (www.pearsoned.com/testgen) enables instructors to build, edit, print, and administer tests using a computerized bank of questions developed to cover all the objectives of the text. TestGen is algorithmically based, allowing instructors to create multiple but equivalent versions of the same question or test with the click of a button. Instructors can also modify test bank questions or add new questions. The software and testbank are available for download from Pearson Education's online catalog.

INSTRUCTOR'S TESTING MANUAL

- By James Lapp

- Includes diagnostic pretests, chapter tests, and additional test items, grouped by section, with answers provided

- (Available online within MyMathLab or from the Instructor Resource Center at www.pearsonhighered.com/irc)

POWERPOINT LECTURE SLIDES

- Features presentations written and designed specifically for this text, including figures and examples from the text

- (Available online within MyMathLab or from the Instructor Resource Center at www.pearsonhighered.com/irc)

VIDEO LECTURES

- Video feature Section Summaries and Example Solutions. Section Summaries cover key definitions and procedures for most sections. Example Solutions walk students through the detailed solution process for many examples in the textbook.

- There are over 20 hours of video instruction specifically filmed for this book, making it ideal for distance learning or supplemental instruction on your home computer or in a campus computer lab.

- Videos include optional subtitles in English and Spanish.

ACKNOWLEDGMENTS

We would like to express our gratitude to the reviewers of this first edition, who provided such invaluable insights and comments. Their contributions helped shape the development of the text and carry out the vision stated in the preface.

Mario Barrientos, Angelo State University
Kate Bella, Manchester Community College
Sarah Bennet, University of Wisconsin–Barron County
Karen Briggs, North Georgia College & State University
Katina Davis, Wayne Community College
Nicole Dowd, Gainesville State College
Hussain Elaloui-Talibi, Tuskegee University
Cathy Famiglietti, University of California–Irvine
Sandi Fay, University of California Riverside
Thomas Fitzkee, Francis Marion University
Olivier Heubo-Kwegna, Saginaw Valley State University
Leif Jordan, College of the Desert
Diana Klimek, New Jersey Institute of Technology
Julie Kostka, Austin Community College
Dr. Carole King Krueger, University of Texas–Arlington
Lance Lana, University of Colorado Denver
Christy Schmidt, Northwest Vista
Comlan de Souza, California State University–Fresno
Linda Snellings Neal, Wright State University
Bob Strozak, Old Dominion University
German Vargas, College of Coastal Georgia
Marti Zimmerman, University of Louisville

Our sincerest thanks also go to the legion of dedicated individuals who worked tirelessly to make this book possible. We would also like to express our gratitude to our typist, Beverly DeVine-Hoffmeyer, for her amazing patience and skill. We must also thank Dr. Praveen Rohatgi, Dr. Nalini Rohatgi, and Dr. Bhupinder Bedi for the consulting they provided on all material relating to medicine. We particularly want to thank Professor Mile Krajcevski for many helpful discussions and suggestions, particularly for improving the exercise sets. Further gratitude is due to Irena Andreevska, Gokarna Aryal, Ferenc Tookos, and Christine Fitch for their assistance on the answers to the exercises in the text. In addition, we would like to thank Beverly Fusfield, Douglas Ewert, Viktor Maymeskul, Patricia Nelson, and Elka Block and Frank Purcell of Twin Prime Editorial for their meticulous accuracy in checking the text. Thanks are due as well to Laura Hakala and PreMediaGlobal for their excellent production work. Finally, our thanks are extended to the professional and remarkable staff at Pearson. In particular, we would like to thank Greg Tobin, Publisher; Anne Kelly, Editor in Chief; Elaine Page, Senior Development Editor; Joanne Dill, Senior Project Editor; Christine O'Brien, Senior Project Editor; Judith Garber, Editorial Assistant; Beth Houston, Senior Production Supervisor; Peggy Lucas, Marketing Manager; Justine Goulart, Marketing Assistant; Barbara Atkinson, Designer; Tracy Menoza, Media Producer; Karen Wernholm, Senior Managing Editor; and Joseph Vetere, Senior Author/Technical Art Support.

We invite all who use this book to send suggestions for improvements to Marcus McWaters at mmm@cas.usf.edu.

CHAPTER P

Basic Concepts of Algebra

TOPICS

P.1 The Real Numbers; Integer Exponents

P.2 Radicals and Rational Exponents

P.3 Solving Equations

P.4 Inequalities

P.5 Complex Numbers

SECTION P.1 The Real Numbers; Integer Exponents

BEFORE STARTING THIS SECTION, REVIEW FROM YOUR PREVIOUS MATHEMATICS TEXTS

1. Arithmetic of signed numbers
2. Arithmetic of fractions
3. Long division involving integers
4. Decimals
5. Arithmetic of real numbers

OBJECTIVES

1. Classify sets of real numbers and define the real number line.
2. Use interval notation.
3. Relate absolute value and distance on the real number line.
4. Define and use integer exponents.
5. Use the rules of exponents.

1 Classify sets of real numbers and define the real number line.

SIDE NOTE

Here is one difficulty with attempting to divide by 0: If, for example, $\frac{5}{0} = a$, then $5 = a \cdot 0$. However, $a \cdot 0 = 0$ for all numbers a. Because 5 does not equal 0, there is no appropriate solution for $\frac{5}{0}$.

Variables, Constants, and Operations

In algebra, we use letters such as a, b, x, and y to represent numbers. A letter that is used to represent one or more numbers is called a **variable**. A **constant** is a specific number such as 3 or $\frac{1}{2}$ or a letter that represents a fixed (but not necessarily specified) number. Physicists use the letter c as a constant to represent the speed of light ($c \approx 300{,}000{,}000$ meters per second).

We use two variables, a and b, to denote the results of the operations of addition $(a + b)$, subtraction $(a - b)$, multiplication $(a \cdot b \text{ or } ab)$, and division $\left(a \div b \text{ or } \frac{a}{b}\right)$.

These operations are called **binary operations** because each is performed on two numbers. We frequently omit the multiplication sign when writing a product involving two variables (or a constant and a variable) so that $a \cdot b$ and ab indicate the same product. Both a and b are called **factors** in the product $a \cdot b$. This is a good time to recall that we never divide by zero. For $\frac{a}{b}$ to represent a real number, b cannot be zero.

Classifying Sets of Numbers

The idea of a set is familiar to us. We regularly refer to "a set of baseball cards," a "set of rules," or "a set of dishes." In mathematics, as in everyday life, a **set** is a collection of objects. The objects in the set are called the **elements**, or **members**, of the set. In the study of algebra, we are interested primarily in sets of numbers.

In listing the elements of a set, it is customary to enclose the listed elements in braces, { }, and separate them with commas. We refer to a set with no elements as the **empty set**, denoted by the symbol $\emptyset$.

We can distinguish among various sets of numbers.

The numbers we use to count with constitute the set of **natural numbers**: $\{1, 2, 3, 4, \ldots\}$.

The three dots ... may be read as "and so on" to indicate that the pattern continues indefinitely.

The **whole numbers** are formed by adjoining the number 0 to the set of natural numbers to obtain $\{0, 1, 2, 3, 4, \ldots\}$.

2 Chapter P Basic Concepts of Algebra

The **integers** consist of the set of natural numbers together with their opposites and 0: $\{\ldots, -4, -3, -2, -1, 0, 1, 2, 3, 4, \ldots\}$.

Rational Numbers

The **rational numbers** consist of all numbers that *can* be expressed as the quotient or ratio, $\frac{a}{b}$ of two integers, where $b \neq 0$.

Examples of rational numbers are $\frac{1}{2}, \frac{5}{3}, -\frac{4}{17}$, and $0.07 = \frac{7}{100}$. Any integer a can be expressed as the quotient of two integers by writing $a = \frac{a}{1}$. Consequently, every integer is also a rational number. In particular, 0 is a rational number because $0 = \frac{0}{1}$.

The rational number $\frac{a}{b}$ can be written as a decimal by using long division. When any integer a is divided by an integer b, $b \neq 0$, the result is always a **terminating decimal** such as $\left(\frac{1}{2} = 0.5\right)$ or a **nonterminating repeating decimal** such as $\left(\frac{2}{3} = 0.666\ldots\right)$.

We sometimes place a bar over the repeating digits in a nonterminating repeating decimal. Thus, $\frac{2}{3} = 0.666\ldots = 0.\overline{6}$ and $\frac{13}{11} = 1.181818\ldots = 1.\overline{18}$.

EXAMPLE 1 Converting Decimal Rationals to a Quotient

Convert the rational number, $7.\overline{45}$, as the ratios of two integers in lowest terms.

Solution

Let $x = 7.454545\ldots$. Then

$$100x = 745.4545\ldots \quad \text{Multiply both sides by 100.}$$
$$\text{and} \quad x = 7.4545\ldots$$
$$\overline{}$$
$$99x = 738 \quad \text{Subtract.}$$
$$x = \frac{738}{99} \quad \text{Divide both sides by 99.}$$
$$= \frac{82 \times 9}{11 \times 9} \quad \text{Common factor}$$
$$x = \frac{82}{11} \quad \text{Reduce to lowest terms.}$$

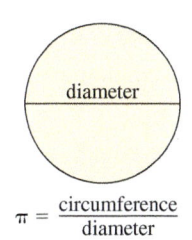

$\pi = \frac{\text{circumference}}{\text{diameter}}$

FIGURE P.1 Definition of π

Practice Problem 1 Repeat Example 1 for $2.132132132\ldots$.

Irrational Numbers

An **irrational number** is a number that cannot be written as a ratio of two integers. This means that its decimal representation must be nonrepeating and nonterminating. We can construct such a decimal using only the digits 0 and 1 such as: $0.01001000100001\ldots$. Because each group of zeros contains one more zero than the previous group, no group of digits repeats. Other numbers such as π (pi, see Figure P.1) and $\sqrt{2}$ (the square root of 2, see Figure P.2) can also be expressed as decimals that neither terminate nor repeat; so they are irrational numbers as well. We can obtain an approximation of an irrational number by using

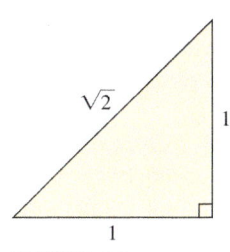

FIGURE P.2

an initial portion of its decimal representation. For example, we can write $\pi \approx 3.14159$ or $\sqrt{2} \approx 1.41421$, where the symbol $\approx$ is read "is approximately equal to."

No familiar process, such as long division, is available for obtaining the decimal representation of an irrational number. However, your calculator can provide a useful approximation for irrational numbers such as $\sqrt{2}$. (Try it!) Because a calculator displays a fixed number of decimal places, it gives a **rational approximation** of an irrational number.

It is usually not easy to determine whether a specific number is irrational. One helpful fact in this regard is that *the square root of any natural number that is not a perfect square is irrational.*

In the next example, we prove that $\sqrt{2}$ is irrational.

> **RECALL**
>
> An integer is a *perfect square* if it is a product $a \cdot a$, where a is an integer. For example, $9 = 3 \cdot 3$ is a perfect square.

EXAMPLE 2 Understanding the Meaning of Irrational Numbers

Prove that $\sqrt{2}$ is irrational.

Solution

We will show that it is impossible to write $\sqrt{2}$ as the ratio of two integers. We will assume that $\sqrt{2}$ is a rational number and show that this assumption leads to a contradiction. This process is called a *proof by contradiction*.

Suppose $\sqrt{2}$ is rational. Then we can write $\sqrt{2} = \dfrac{a}{b}$, such that a and b are integers and $b \neq 0$. We will assume that the ratio $\dfrac{a}{b}$ is in lowest terms. Then we have

$$\sqrt{2}b = a \qquad \text{Multiply both sides of } \sqrt{2} = \dfrac{a}{b} \text{ by } b.$$
$$(\sqrt{2}b)^2 = a^2 \qquad \text{Square both sides.}$$
$$2b^2 = a^2 \qquad \text{Simplify.}$$

This shows that a^2 is even. Hence, a is even because the product of an odd integer and itself is odd.

So $a = 2q$ for some integer q. Then we have

$$2b^2 = a^2$$
$$\text{or} \quad 2b^2 = (2q)^2 = 4q^2 \qquad \text{Replace } a \text{ with } 2q.$$
$$\text{or} \quad b^2 = 2q^2 \qquad \text{Simplify.}$$

So b^2 is even and hence b is even.

We have shown that both a and b are even, which contradicts our assumption that the ratio $\dfrac{a}{b}$ is in lowest terms. So our original assumption that $\sqrt{2}$ can be written as the ratio $\dfrac{a}{b}$ of two integers is false. Therefore, $\sqrt{2}$ is irrational.

Practice Problem 2 Prove that $\sqrt{8}$ is irrational.

Because rational numbers have decimal representations that either terminate or repeat, whereas irrational numbers do not have such representations, *no number is both rational and irrational.*

The rational numbers together with the irrational numbers form the **real numbers**.

The diagram in Figure P.3 shows how various sets of numbers are related. For example, every natural number is also a whole number, an integer, a rational number, and a real number.

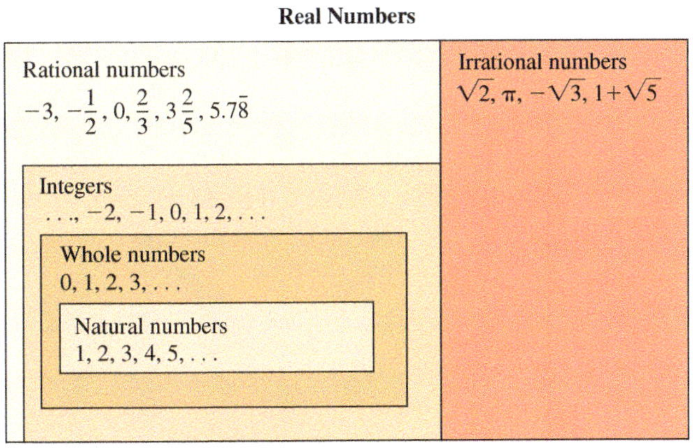

Real Numbers

Rational numbers	Irrational numbers
$-3, -\frac{1}{2}, 0, \frac{2}{3}, 3\frac{2}{5}, 5.7\bar{8}$	$\sqrt{2}, \pi, -\sqrt{3}, 1+\sqrt{5}$

Integers
..., −2, −1, 0, 1, 2, ...

Whole numbers
0, 1, 2, 3, ...

Natural numbers
1, 2, 3, 4, 5, ...

FIGURE P.3 Relationships among sets of real numbers

The Real Number Line

We associate the real numbers with points on a geometric line (imagined to be extended indefinitely in both directions) in such a way that each real number corresponds to exactly one point and each point corresponds to exactly one real number. The point is called the **graph** of the corresponding real number, and the real number is called the **coordinate** of the point. By agreement, the graphs of **positive numbers** lie to the right of the point corresponding to 0 and the graphs of **negative numbers** lie to the left of 0. See Figure P.4.

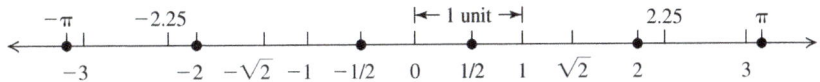

FIGURE P.4 The real number line

Notice that $\frac{1}{2}$ and $-\frac{1}{2}$, 2 and −2, and π and $-\pi$ correspond to pairs of points exactly the same distance from 0 but on opposite sides of 0.

When coordinates have been assigned to points on a line in the manner just described, the line is called a **real number line**, **coordinate line**, a **real line**, or simply a **number line**. The point corresponding to 0 is called the **origin**.

Inequality Symbols

The real numbers are **ordered** by their size. We say that *a is less than b* and write $a < b$, provided that $b = a + c$ for some *positive* number c. We also write $b > a$, meaning the same thing as $a < b$, and say that *b is greater than a*. On the real line, the numbers increase from left to right. Consequently, *a is to the left of b on the number line when $a < b$*. Similarly, a is to the right of b on the number line when $a > b$. We sometimes want to indicate that at least one of two conditions is correct: Either $a < b$ or $a = b$. In this case, we write $a \leq b$ or $b \geq a$. The four symbols, $<, >, \leq,$ and $\geq$ are called **inequality symbols**.

Frequently, we read $a > 0$ as "a is positive" instead of "a is greater than 0." We can also read $a < 0$ as "a is negative." If $a \geq 0$, then either $a > 0$ or $a = 0$ and we may say that "a is nonnegative."

2 Use interval notation.

Intervals

To specify a set, we do one of the following:

1. List the elements of the set (**roster method**)
2. Describe the elements of the set (often using **set-builder notation**)

Variables are helpful in describing sets when we use set-builder notation. The notation $\{x | x$ is a natural number less than $6\}$ is in set-builder notation and describes the set $\{1, 2, 3, 4, 5\}$ using the roster method. We read $\{x | x$ is a natural number less than $6\}$ as "the set of all x such that x is a natural number less than six." Generally, $\{x|x \text{ has property } P\}$ designates the set of all x such that (the vertical bar is read "such that") x has property P.

We now turn our attention to graphing certain sets of numbers. That is, we graph each number in a given set. We are particularly interested in sets of real numbers, called **intervals**, whose graphs correspond to special sections of the number line.

If $a < b$, then the set of real numbers between a and b, but not including either a or b, is called the **open interval** from a to b and is denoted by (a, b). See Figure P.5. Using set-builder notation, we can write

$$(a, b) = \{x | a < x < b\}.$$

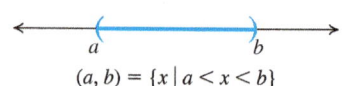
$(a, b) = \{x | a < x < b\}$
FIGURE P.5

We indicate graphically that the endpoints a and b are excluded from the open interval by drawing a left parenthesis at a and a right parenthesis at b. These parentheses enclose the numbers between a and b.

The **closed interval** from a to b is the set

$$[a, b] = \{x | a \leq x \leq b\}.$$

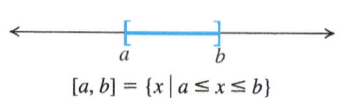
$[a, b] = \{x | a \leq x \leq b\}$
FIGURE P.6

The closed interval includes both endpoints a and b. We replace the parentheses with square brackets. See Figure P.6. This type of notation involving parentheses and/or square brackets is called **interval notation**.

We also want to graph all numbers to the left or right of a given number, such as

$$\{x | x > 2\}.$$

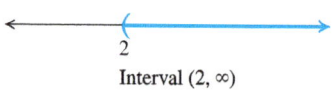
Interval $(2, \infty)$
FIGURE P.7

The number 2 is not included in this set, and we again indicate this fact graphically by drawing a left parenthesis at 2. We then graph all points to the right of 2, as shown in Figure P.7.

For any number a, we call $\{x | x > a\}$ an **unbounded** (or infinite) **interval** and denote this interval by (a, ∞). The symbol ∞ ("infinity") does not represent a number. The notation (a, ∞) is used to indicate the set of all real numbers that are greater than a, and the symbol ∞ is used to indicate that the interval extends indefinitely to the right of a. See Figure P.8.

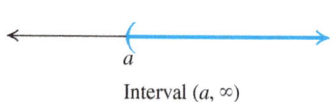
Interval (a, ∞)
FIGURE P.8

The symbol $-\infty$ is another symbol that does not represent a number. The notation $(-\infty, a)$ is used to indicate the set of all real numbers that are less than a. The notation $(-\infty, \infty)$ represents the set of all real numbers.

Table P.1 lists various types of intervals that we use in this book. In the table, when two points a and b are involved, we assume that $a < b$.

SIDE NOTE

The symbols ∞ and $-\infty$ are always used with parentheses, not square brackets. Also note that $<$ and $>$ correspond to parentheses and that $\leq$ and $\geq$ correspond to square brackets.

TABLE P.1

Interval Notation	Set-Builder Notation	Graph
(a, b)	$\{x | a < x < b\}$	
$[a, b]$	$\{x | a \leq x \leq b\}$	
$(a, b]$	$\{x | a < x \leq b\}$	
$[a, b)$	$\{x | a \leq x < b\}$	
(a, ∞)	$\{x | x > a\}$	
$[a, \infty)$	$\{x | x \geq a\}$	
$(-\infty, b)$	$\{x | x < b\}$	
$(-\infty, b]$	$\{x | x \leq b\}$	
$(-\infty, \infty)$	$\{x | x \text{ is a real number}\}$	

Section P.1 ■ The Real Numbers; Integer Exponents   7

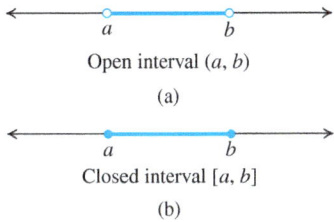

Open interval (a, b)
(a)

Closed interval $[a, b]$
(b)

FIGURE P.9

An alternative notation for indicating whether endpoints are included uses closed circles to show inclusion and open circles to show exclusion. See Figure P.9.

Sometimes it is useful to work with the union and intersection of sets.

Union and Intersection

Let A and B be two sets.

The **union** of A and B, denoted $A \cup B$, is the set of all elements in A or B (or both).

The **intersection** of A and B, denoted $A \cap B$, is the set of all elements in both A and B.

EXAMPLE 3 Union and Intersection of Intervals

Consider the two intervals $I_1 = (-3, 4)$ and $I_2 = [2, 6]$.

Find: **a.** $I_1 \cup I_2$ **b.** $I_1 \cap I_2$

Solution

a. From Figure P.10, we see that $I_1 \cup I_2 = (-3, 6]$. We note that every number in the interval $(-3, 6]$ is in either I_1 or I_2 or in both I_1 and I_2.

b. We see in Figure P.10 that $I_1 \cap I_2 = [2, 4)$. Every number in the interval $[2, 4)$ is in both I_1 and I_2.

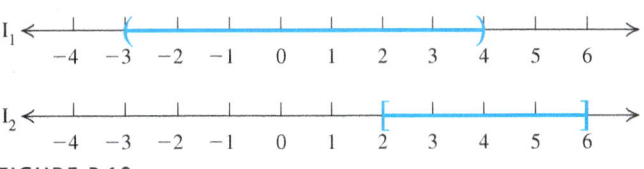

FIGURE P.10

Practice Problem 3 Let $I_1 = (-\infty, 5)$ and $I_2 = [2, \infty)$. Find the following.
a. $I_1 \cup I_2$ **b.** $I_1 \cap I_2$

3 Relate absolute value and distance on the real number line.

Absolute Value

The *absolute value* of a number a, denoted $|a|$, is the distance between the origin and the point on the number line with coordinate a. The point with coordinate -3 is three units from the origin, so we write $|-3| = 3$ and say that the absolute value of -3 is 3. See Figure P.11.

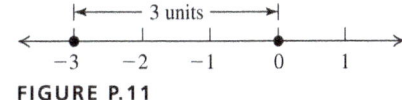

FIGURE P.11

Absolute Value

For any real number a, the **absolute value** of a, denoted $|a|$, is defined by

$$|a| = a \text{ if } a \geq 0 \quad \text{and} \quad |a| = -a \text{ if } a < 0.$$

EXAMPLE 4 Determining Absolute Value

Find the absolute value of each of the following expressions.

a. $|4|$ **b.** $|-4|$ **c.** $|0|$ **d.** $|(-3) + 1|$

8 Chapter P Basic Concepts of Algebra

Solution

a. $|4| = 4$ b. $|-4| = -(-4) = 4$
c. $|0| = 0$ d. $|(-3) + 1| = |-2| = -(-2) = 2$

Practice Problem 4 Find the absolute value of each of the following.

a. $|-10|$ b. $|3 - 4|$ c. $|2(-3) + 7|$

WARNING
The absolute value of a number represents a distance. Because distance can never be negative, the absolute value of a number is never negative; it is always positive or zero. However, if a is not 0, $-|a|$ is always negative. So, $-|5.3| = -5.3$, $-|-4| = -4$, and $-|1.18| = -1.18$.

Distance Between Two Points on a Real Number Line

Absolute value is used when finding the distance between two points on a number line.

> **DISTANCE FORMULA ON A NUMBER LINE**
>
> If a and b are the coordinates of two points on a number line, then the distance between a and b, denoted $d(a, b)$, is $|a - b|$. In symbols, $d(a, b) = |a - b|$.

EXAMPLE 5   **Finding the Distance Between Two Points**

Find the distance between -3 and 4 on the number line.

Solution

Figure P.12 shows that the distance between -3 and 4 is seven units. The distance formula gives the same answer.

$$d(-3, 4) = |-3 - 4| = |-7| = -(-7) = 7.$$

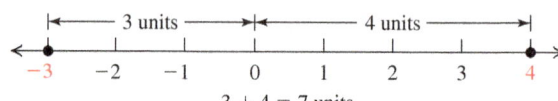

FIGURE P.12

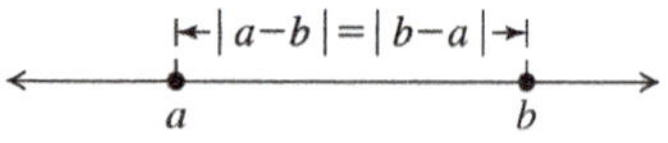

FIGURE P.13

Notice that reversing the order of -3 and 4 in this computation gives the same answer. That is, the distance between 4 and -3 is $|4 - (-3)| = |4 + 3| = |7| = 7$. It is always true that $|a - b| = |b - a|$. See Figure P.13.

Practice Problem 5 Find the distance between -7 and 2 on the number line.

4 Define and use integer exponents.

Integer Exponents

The area of a square with side 5 feet is $5 \cdot 5 = 25$ square feet. The volume of a cube whose sides are each 5 feet is $5 \cdot 5 \cdot 5 = 125$ cubic feet. A shorter notation for $5 \cdot 5$ is 5^2 and for $5 \cdot 5 \cdot 5$ is 5^3. The number 5 is called the *base* for both 5^2 and 5^3. The number 2 is called the *exponent* in the expression 5^2 and indicates that the base 5 appears as a factor twice. When an exponent on a particular base is an integer, we call that number an **integer exponent**.

Positive Integer Exponent

> If a is a real number and n is a positive integer, then
>
> $$a^n = \underbrace{a \cdot a \cdot \cdots \cdot a}_{n \text{ factors}},$$
>
> where a is the base and n is the exponent. We adopt the convention that $a^1 = a$.

Section P.1 ■ The Real Numbers; Integer Exponents 9

TECHNOLOGY CONNECTION

The notation for 5^3 on a graphing calculator is 5^3. Any expression on a calculator enclosed in parentheses and followed by ^n will be raised to the nth power. A common error is to forget parentheses when computing an expression such as $(-3)^2$, with the result being -3^2.

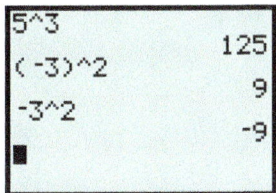

We can evaluate exponents as follows:

$$5^3 = 5 \cdot 5 \cdot 5 = 125$$
$$(-3)^2 = (-3)(-3) = 9$$
$$-3^2 = -(3 \cdot 3) = -9$$
$$(-2)^3 = (-2)(-2)(-2) = -8$$

Pay careful attention to the fact that $(-3)^2 \neq -3^2$.

Negative exponents indicate the reciprocal of a number.

Zero and Negative Integer Exponents

For any nonzero number a and any positive integer n,

$$a^0 = 1 \text{ and } a^{-n} = \frac{1}{a^n}.$$

EXAMPLE 6 Evaluating Zero or Negative Exponents

Evaluate.

a. $(-5)^{-2}$ b. -5^{-2} c. 8^0 d. $\left(\frac{2}{3}\right)^{-3}$

Solution

a. $(-5)^{-2} = \frac{1}{(-5)^2} = \frac{1}{25}$ b. $-5^{-2} = -\frac{1}{5^2} = -\frac{1}{25}$

c. $8^0 = 1$ d. $\left(\frac{2}{3}\right)^{-3} = \frac{1}{\left(\frac{2}{3}\right)^3} = \frac{1}{\frac{8}{27}} = \frac{27}{8}$

Practice Problem 6 Evaluate.

a. 2^{-1} b. $\left(\frac{4}{5}\right)^0$ c. $\left(\frac{3}{2}\right)^{-2}$

Rules of Exponents

5 Use the rules of exponents.

We now review the rules of exponents. All expressions are assumed to be defined.

RULES OF EXPONENTS

Product Rule: $a^m \cdot a^n = a^{m+n}$ **Quotient Rule**: $\frac{a^m}{a^n} = a^{m-n}$ **Power Rule**: $(a^m)^n = a^{mn}$

Power of a Product Rule: $(a \cdot b)^n = a^n \cdot b^n$ **Power of a Quotient Rule**: $\left(\frac{a}{b}\right)^n = \frac{a^n}{b^n}$

EXAMPLE 7 Using the Rules of Exponents

Use the rules of exponents to write each expression without negative exponents.

a. $(-4x^2y^3)(7x^3y)$ b. $\left(\frac{x^5}{2y^{-3}}\right)^{-3}$

10 Chapter P Basic Concepts of Algebra

SIDE NOTE

There are many different ways to simplify exponential expressions correctly. The order in which you apply the rules for exponents is a matter of personal preference.

Solution

a. $(-4x^2y^3)(7x^3y) = (-4)(7)x^2x^3y^3y$
$= -28x^{2+3}y^{3+1}$
$= -28x^5y^4$

b. $\left(\dfrac{x^5}{2y^{-3}}\right)^{-3} = \dfrac{(x^5)^{-3}}{(2y^{-3})^{-3}}$ $\left(\dfrac{a}{b}\right)^n = \dfrac{a^n}{b^n}$

$= \dfrac{(x^5)^{-3}}{2^{-3}(y^{-3})^{-3}}$ $(ab)^n = a^n b^n$

$= \dfrac{x^{-15}}{2^{-3}y^{(-3)(-3)}}$ $(a^m)^n = a^{mn}$

$= \dfrac{x^{-15}}{2^{-3}y^9}$ Simplify.

$= \dfrac{x^{15}x^{-15}2^3}{x^{15}2^3 2^{-3}y^9}$ $\dfrac{a}{b} = \dfrac{ca}{cb}, c \neq 0$

$= \dfrac{2^3}{x^{15}y^9}$ $a^n a^{-n} = a^{n-n} = a^0 = 1$

$= \dfrac{8}{x^{15}y^9}$

Practice Problem 7 Simplify each expression.

a. $(2x^4)^{-2}$ b. $\dfrac{x^2(-y)^3}{(xy^2)^3}$

SECTION P.1 Exercises

Basic Concepts and Skills

In Exercises 1–4, write each of the following rational numbers as a decimal and state whether the decimal is repeating or terminating.

1. $-\dfrac{4}{5}$
2. $-\dfrac{3}{12}$
3. $\dfrac{3}{11}$
4. $\dfrac{41}{15}$

In Exercises 5–10, classify each of the following numbers as rational or irrational.

5. -207
6. $-\sqrt{25}$
7. $\sqrt{32}$
8. $5 + \sqrt{18}$
9. 0.321
10. $5.8\overline{2}$

In Exercises 11–16, convert each decimal to a quotient of two integers in lowest terms.

11. 3.75
12. -2.35
13. $2.\overline{13}$
14. $3.\overline{23}$
15. $4.2\overline{32}$
16. $1.42\overline{35}$

In Exercises 17–22, find the union and the intersection of the given intervals.

17. $I_1 = (-2, 3]; I_2 = [1, 5)$
18. $I_1 = [1, 7]; I_2 = (3, 5)$
19. $I_1 = (-6, 2); I_2 = [2, 10)$
20. $I_1 = (-\infty, -3]; I_2 = (-3, \infty)$
21. $I_1 = (-\infty, 7); I_2 = (-\infty, 3)$
22. $I_1 = (-2, \infty); I_2 = (0, \infty)$

In Exercises 23–30, rewrite each expression without absolute value bars.

23. $-|-4|$
24. $\left|\dfrac{5}{-7}\right|$
25. $|\sqrt{2} - 5|$
26. $\dfrac{8}{|-8|}$
27. $|5 + |-7||$
28. $|5 - |-7||$
29. $||7| - |4||$
30. $||4| - |7||$

In Exercises 31–34, use the absolute value to express the distance between the points with coordinates a and b on the number line. Then determine this distance by evaluating the absolute value expression.

31. $a = 3$ and $b = 8$
32. $a = -12$ and $b = 3$
33. $a = -20$ and $b = -6$
34. $a = \dfrac{22}{7}$ and $b = -\dfrac{4}{7}$

In Exercises 35–44, evaluate each expression for $x = 3$ and $y = -5$.

35. $2(x + y) - 3y$
36. $-2(x + y) + 5y$
37. $3|x| - 2|y|$
38. $7|x - y|$
39. $\dfrac{x - 3y}{2} + xy$
40. $\dfrac{y + 3}{x} - xy$
41. $\dfrac{2(1 - 2x)}{y} - (-x)y$
42. $\dfrac{3(2 - x)}{y} - (1 - xy)$
43. $\dfrac{\frac{14}{x} + \frac{1}{2}}{\frac{-y}{4}}$
44. $\dfrac{\frac{4}{-y} + \frac{8}{x}}{\frac{y}{2}}$

In Exercises 45–60, simplify each expression. Write your answers without negative exponents. Whenever an exponent is negative or zero, assume that the base is not zero.

45. $x^4 y^0$
46. $x^{-1} y$
47. $-8x^{-1}$
48. $x^{-1}(3y^0)$
49. $x^{-1} y^{-2}$
50. $(x^{-3})^4$
51. $(x^{-11})^{-3}$
52. $-3(xy)^5$
53. $4(xy^{-1})^2$
54. $3(x^{-1}y)^{-5}$
55. $\dfrac{(x^3)^2}{(x^2)^5}$
56. $\left(\dfrac{2xy}{x}\right)^3$
57. $\left(\dfrac{-3x^2 y}{x}\right)^5$
58. $\left(\dfrac{-3x}{5}\right)^{-2}$
59. $\dfrac{x^3 y^{-3}}{x^{-2} y}$
60. $\dfrac{27 x^{-3} y^5}{9 x^{-4} y^7}$

Applying the Concepts

61. Blood pressure. A group of college students had systolic blood pressure readings that ranged from 119.5 to 134.5 inclusive. Let x represent the value of the systolic blood pressure readings. Use inequalities to describe this range of values and graph the corresponding interval on a number line.

62. Population projections. Population projections suggest that by the year 2050, the number of people 60 years old and older in the United States will be about 107 million. In 1950, the number of people 60 years old and older in the United States was 30 million. Let x represent the number (in millions) of people in the United States who are 60 years old and older. Use inequalities to describe this population range from 1950 to 2050 and graph the corresponding interval on a number line. *Source:* U.S. Census Bureau

63. Heart rate. For exercise to be most beneficial, the optimum heart rate for a 20-year-old person is 120 beats per minute. Use absolute value notation to write an expression that describes the difference between the heart rate achieved by each of the following 20-year-old people and the ideal exercise heart rate. Then evaluate that expression.
  a. Latasha: 124 beats per minute
  b. Frances: 137 beats per minute
  c. Ignacio: 114 beats per minute

64. Downloading music. To download a 4 MB song with a 56 Kbs modem takes an average of 15 minutes. Use absolute value notation to write an expression that describes the difference between this average time and the actual time it took to download the following songs. Then evaluate that expression.
  a. *Believe* (Cher): 14 minutes
  b. *Caged Bird* (Alicia Keys): 17.5 minutes
  c. *Somewhere* (Barbra Streisand): 15 minutes

65. Square area. The area A of a square with side of length x is given by $A = x^2$. Use this relationship to
  a. verify that doubling the length of the side of a square floor increases the area of the floor by a factor of 2^2.
  b. verify that tripling the length of the side of a square floor increases the area of the floor by a factor of 3^2.

66. Circle area. The area A of a circular disc with diameter d is given by $A = \pi\left(\dfrac{d}{2}\right)^2$. Use this relationship to
  a. verify that doubling the length of the diameter of a circular skating rink increases the area of the rink by a factor of 2^2.
  b. verify that tripling the length of the diameter of a circular skating rink increases the area of the rink by a factor of 3^2.

Maintaining Skills

67. Write:
  a. $\dfrac{2}{3}$ with a denominator of 21
  b. $\dfrac{3}{5}$ with a denominator of 20

68. Perform each operation and write the answer in lowest terms.
  a. $\dfrac{5}{7} - \dfrac{2}{7}$
  b. $\dfrac{3}{8} + \dfrac{5}{12}$
  c. $\dfrac{2}{3} + \dfrac{5}{6} - \dfrac{4}{9}$
  d. $\dfrac{3}{8} \cdot \dfrac{4}{9}$
  e. $\dfrac{3}{4} \div \dfrac{12}{5}$
  f. $\dfrac{6}{7} \cdot \dfrac{21}{4} \div \dfrac{65}{4}$

69. Simplify: $2[3 + 4(7 - 1)]$

70. Simplify: $\dfrac{9 + 3 \cdot 2}{2^3 - 3}$

SECTION P.2

Radicals and Rational Exponents

BEFORE STARTING THIS SECTION, REVIEW

1. Absolute value (Section P.1, page 7)

OBJECTIVES

1. Define and evaluate *n*th roots.
2. Simplify expressions involving radicals.
3. Add, subtract, and multiply radical expressions.
4. Rationalize denominators or numerators.
5. Define and evaluate rational exponents.

1 Define and evaluate *n*th roots.

Square Roots

When we raise the number 5 to the power 2, we write $5^2 = 25$. The reverse of this squaring process is called finding a **square root**. In the case of 25, we say that 5 is a square root of 25 and write $\sqrt{25} = 5$. In general, we say that b is a square root of a if $b^2 = a$. We can't call 5 *the* square root of 25 because it is also true that $(-5)^2 = 25$, so -5 is *another* square root of 25.

The symbol $\sqrt{}$, called a **radical sign**, is used to distinguish between 5 and -5, the two square roots of 25. We write $5 = \sqrt{25}$ and $-5 = -\sqrt{25}$ and call $\sqrt{25}$ the *principal square root* of 25.

Principal Square Root

$$\sqrt{a} = b \quad \text{means} \quad (1)\ b^2 = a \quad \text{and} \quad (2)\ b \geq 0.$$

If $\sqrt{a} = b$, then $a = b^2 \geq 0$. Consequently, the symbol $\sqrt{a}$ denotes a real number only when $a \geq 0$. We say that the domain of the expression $\sqrt{x}$ is $\{x \mid x \geq 0\}$, or in interval notation, $[0, \infty)$.

TECHNOLOGY CONNECTION

The square root symbol on a graphing calculator does not have a horizontal bar that goes completely over the number or expression whose square root is to be evaluated. You must enclose a rational number in parentheses before taking its square root and then use the "Frac" feature to have the answer appear as a fraction.

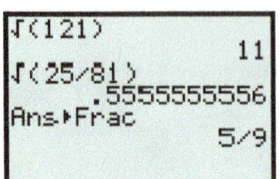

For any real number x,
$$\sqrt{x^2} = |x|$$
and $(\sqrt{x})^2 = x$ if $x \geq 0$.

For example, $\sqrt{121} = \sqrt{11^2} = 11$, $\sqrt{\dfrac{25}{81}} = \sqrt{\dfrac{5^2}{9^2}} = \sqrt{\left(\dfrac{5}{9}\right)^2} = \dfrac{5}{9}$, and $(\sqrt{7})^2 = 7$.

Other Roots

If n is a positive integer, we say that b is an *n*th root of a if $b^n = a$. We define the **principal *n*th root** of a real number a, denoted by the symbol $\sqrt[n]{a}$, as follows.

12 Chapter P Basic Concepts of Algebra

THE PRINCIPAL nth ROOT OF A REAL NUMBER

1. If a is positive ($a > 0$), then $\sqrt[n]{a} = b$ provided that $b^n = a$ and $b > 0$.
2. If a is negative ($a < 0$) and n is odd, then $\sqrt[n]{a} = b$ provided that $b^n = a$.
3. If a is negative ($a < 0$) and n is even, then $\sqrt[n]{a}$ is not a real number.
4. If $a = 0$, then $\sqrt[n]{a} = 0$.

We use the same vocabulary for nth roots that we used for square roots. That is, $\sqrt[n]{a}$ is called a **radical expression**, $\sqrt{}$ is the **radical sign**, and a is the **radicand**. The positive integer n is called the **index**. The index 2 is not written; we write $\sqrt{a}$ rather than $\sqrt[2]{a}$ for the principal square root of a. It is also common practice to call $\sqrt[3]{a}$ the **cube root** of a.

EXAMPLE 1 Finding Principal nth Roots

Find each root.

a. $\sqrt[3]{27}$ b. $\sqrt[3]{-64}$ c. $\sqrt[4]{16}$ d. $\sqrt[4]{(-3)^4}$ e. $\sqrt[8]{-46}$

Solution

a. The radicand, 27, is *positive*; so $\sqrt[3]{27} = 3$ because $3^3 = 27$ and $3 > 0$.
b. The radicand, -64, is *negative*; so $\sqrt[3]{-64} = -4$ because $(-4)^3 = -64$.
c. The radicand, 16, is *positive*; so $\sqrt[4]{16} = 2$ because $2^4 = 16$ and $2 > 0$.
d. The radicand, $(-3)^4$, is *positive*; so $\sqrt[4]{(-3)^4} = 3$ because $3^4 = (-3)^4$ and $3 > 0$.
e. The radicand, -46, is *negative*, and the index, 8, is even; so $\sqrt[8]{-46}$ is not a real number.

Practice Problem 1 Find each root.

a. $\sqrt[3]{-8}$ b. $\sqrt[5]{32}$ c. $\sqrt[4]{81}$ d. $\sqrt[6]{-4}$

Simplifying Radical Expressions

2 Simplify expressions involving radicals.

The following properties are often used to simplify expressions involving radicals.

PROPERTIES OF RADICALS

Let m and n be natural numbers with $n > 1$ and a and b be real numbers. Assume that all roots are defined and denominators are nonzero.

1. $(\sqrt[n]{a})^n = a$
2. $\sqrt[n]{a^n} = \begin{cases} |a|, & \text{if } n \text{ is even} \\ a, & \text{if } n \text{ is odd} \end{cases}$
3. $\sqrt[n]{ab} = \sqrt[n]{a} \cdot \sqrt[n]{b}$
4. $\sqrt[n]{\dfrac{a}{b}} = \dfrac{\sqrt[n]{a}}{\sqrt[n]{b}}$
5. $\sqrt[n]{a^m} = (\sqrt[n]{a})^m$
6. $\sqrt[m]{\sqrt[n]{a}} = \sqrt[mn]{a}$

EXAMPLE 2 Simplifying nth Roots Using the Properties of Radicals

Simplify.

a. $\sqrt[3]{135}$ b. $\sqrt[4]{162a^4}$ c. $\sqrt[5]{(32)^3}$

Solution

a. Because $135 = 27 \cdot 5$ and 27 is a perfect cube ($3^3 = 27$), we have
$$\sqrt[3]{135} = \sqrt[3]{27 \cdot 5} = \sqrt[3]{27} \cdot \sqrt[3]{5} = 3\sqrt[3]{5}.$$

b. Because $162 = 81 \cdot 2$ and 81 is a perfect fourth power ($3^4 = 81$), we have
$$\sqrt[4]{162a^4} = \sqrt[4]{162}\sqrt[4]{a^4} = \sqrt[4]{81 \cdot 2}|a| = \sqrt[4]{81}\sqrt[4]{2}|a| = 3\sqrt[4]{2}|a|.$$

c. Because $2^5 = 32$, $\sqrt[5]{32} = 2$. We have
$$\sqrt[5]{(32)^3} = (\sqrt[5]{32})^3 = (2)^3 = 8.$$

Practice Problem 2 Simplify.

a. $\sqrt[3]{72}$ b. $\sqrt[4]{48a^2}$ c. $\sqrt[4]{a^3}\sqrt[4]{a^5}, a \geq 0$

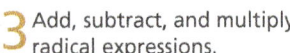

3 Add, subtract, and multiply radical expressions.

Like Radicals

Radicals that have the same index and the same radicand are called **like radicals**. Like radicals can be combined with the use of the distributive property.

EXAMPLE 3   Adding and Subtracting Radical Expressions

Simplify.

a. $\sqrt{45} + 7\sqrt{20}$ b. $5\sqrt[3]{80x} - 3\sqrt[3]{270x}$

Solution

a. We find perfect-square factors for both 45 and 20.
$$\begin{aligned}\sqrt{45} + 7\sqrt{20} &= \sqrt{9 \cdot 5} + 7\sqrt{4 \cdot 5} \\ &= \sqrt{9}\sqrt{5} + 7\sqrt{4}\sqrt{5} &&\text{Product rule} \\ &= 3\sqrt{5} + 7 \cdot 2\sqrt{5} \\ &= 3\sqrt{5} + 14\sqrt{5} &&\text{Like radicals} \\ &= (3 + 14)\sqrt{5} &&\text{Distributive property} \\ &= 17\sqrt{5}\end{aligned}$$

b. This time we find perfect-cube factors for both 80 and 270.
$$\begin{aligned}5\sqrt[3]{80x} - 3\sqrt[3]{270x} &= 5\sqrt[3]{8 \cdot 10x} - 3\sqrt[3]{27 \cdot 10x} \\ &= 5\sqrt[3]{8} \cdot \sqrt[3]{10x} - 3\sqrt[3]{27} \cdot \sqrt[3]{10x} &&\text{Product rule} \\ &= 5 \cdot 2 \cdot \sqrt[3]{10x} - 3 \cdot 3\sqrt[3]{10x} \\ &= 10\sqrt[3]{10x} - 9\sqrt[3]{10x} &&\text{Like radicals} \\ &= (10 - 9)\sqrt[3]{10x} &&\text{Distributive property} \\ &= \sqrt[3]{10x}\end{aligned}$$

Practice Problem 3 Simplify.

a. $3\sqrt{12} + 7\sqrt{3}$ b. $2\sqrt[3]{135x} - 3\sqrt[3]{40x}$

Radicals with Different Indexes

Suppose a is a positive real number and m, n, and r are positive integers. Then the property
$$\sqrt[n]{a^m} = \sqrt[nr]{a^{mr}}$$
is used to multiply radicals with different indexes.

Section P.2 ■ Radicals and Rational Exponents   15

EXAMPLE 4 **Multiplying Radicals with Different Indexes**

Write $\sqrt{2}\sqrt[3]{5}$ as a single radical.

Solution

Because $\sqrt{2}$ can also be written as $\sqrt[2]{2}$, we have
$$\sqrt{2} = \sqrt[2]{2} = \sqrt[2\cdot3]{2^3} = \sqrt[6]{2^3}.$$

Similarly,
$$\sqrt[3]{5} = \sqrt[2\cdot3]{5^2} = \sqrt[6]{5^2}.$$
So
$$\sqrt{2}\sqrt[3]{5} = \sqrt[6]{2^3}\sqrt[6]{5^2} = \sqrt[6]{2^3 5^2} = \sqrt[6]{200}.$$

Practice Problem 4 Write $\sqrt{3}\sqrt[5]{2}$ as a single radical.

4 Rationalize denominators or numerators.

Rationalizing

Removing radicals in the denominator or the numerator of a fraction is called **rationalizing the denominator** or **rationalizing the numerator**, respectively. The procedure for rationalizing involves multiplying the fraction by 1 in a special way so as to obtain a perfect nth power.

EXAMPLE 5 **Rationalizing the Denominator**

Rationalize each denominator.

a. $\sqrt{\dfrac{5}{2}}$ **b.** $\sqrt[3]{\dfrac{2x^4}{9y^5}}$

Solution

a. We multiply the numerator and the denominator of the radicand so that the denominator becomes a perfect square. So
$$\sqrt{\dfrac{5}{2}} = \sqrt{\dfrac{5\cdot 2}{2\cdot 2}} = \sqrt{\dfrac{10}{2^2}} = \dfrac{\sqrt{10}}{\sqrt{2^2}} = \dfrac{\sqrt{10}}{2}.$$

b. We multiply the numerator and the denominator of the radicand so that the denominator becomes a perfect cube.

$$\sqrt[3]{\dfrac{2x^4}{9y^5}} = \sqrt[3]{\dfrac{2x^4 \cdot 3y}{9y^5 \cdot 3y}} \qquad 9\cdot 3 = 27 = (3)^3 \text{ and } y^5 \cdot y = y^6 = (y^2)^3$$

$$= \dfrac{\sqrt[3]{6x^4 y}}{\sqrt[3]{27y^6}} \qquad \sqrt[n]{\dfrac{a}{b}} = \dfrac{\sqrt[n]{a}}{\sqrt[n]{b}}$$

$$= \dfrac{\sqrt[3]{x^3(6xy)}}{\sqrt[3]{(3y^2)^3}} \qquad x^4 = x^3 \cdot x;\ 27y^6 = (3y^2)^3$$

$$= \dfrac{\sqrt[3]{x^3}\sqrt[3]{6xy}}{\sqrt[3]{(3y^2)^3}} \qquad \sqrt[3]{ab} = \sqrt[3]{a}\sqrt[3]{b}$$

$$= \dfrac{x\sqrt[3]{6xy}}{3y^2} \qquad \sqrt[n]{a^n} = a,\ n\ \text{odd}$$

Practice Problem 5 Rationalize each denominator.

a. $\dfrac{7}{\sqrt{8}} = 7 \cdot \sqrt{\dfrac{1}{8}}$ b. $\sqrt[3]{\dfrac{a}{4b^4}}$

Conjugates

Pairs of expressions of the form $\sqrt{x} + \sqrt{y}$ and $\sqrt{x} - \sqrt{y}$ are called **conjugates**. We can form the product of these two conjugates as follows:

$$\begin{aligned}(\sqrt{x} + \sqrt{y})(\sqrt{x} - \sqrt{y}) &= \sqrt{x}(\sqrt{x} - \sqrt{y}) + \sqrt{y}(\sqrt{x} - \sqrt{y}) \\ &= (\sqrt{x})^2 - \sqrt{x}\sqrt{y} + \sqrt{y}\sqrt{x} - (\sqrt{y})^2 \\ &= x - y\end{aligned}$$

We see that the product contains no radicals. So if a fraction contains an expression of the form $\sqrt{x} + \sqrt{y}$ or $\sqrt{x} - \sqrt{y}$, conjugates can be used to rationalize the denominator or the numerator.

EXAMPLE 6   Rationalizing the Numerator

Rationalize the numerator and simplify:

$$\dfrac{\sqrt{x + h + 2} - \sqrt{x + 2}}{h}, h \neq 0$$

Solution

$\dfrac{\sqrt{x + h + 2} - \sqrt{x + 2}}{h}$

$= \dfrac{(\sqrt{x + h + 2} - \sqrt{x + 2})(\sqrt{x + h + 2} + \sqrt{x + 2})}{h(\sqrt{x + h + 2} + \sqrt{x + 2})}$ Multiply and divide by the conjugate.

$= \dfrac{(x + h + 2) - (x + 2)}{h(\sqrt{x + h + 2} + \sqrt{x + 2})}$ $(\sqrt{a} - \sqrt{b})(\sqrt{a} + \sqrt{b}) = a - b$

$= \dfrac{x + h + 2 - x - 2}{h(\sqrt{x + h + 2} + \sqrt{x + 2})}$ Distribute.

$= \dfrac{\cancel{h}}{\cancel{h}(\sqrt{x + h + 2} + \sqrt{x + 2})}$ Remove h.

$= \dfrac{1}{\sqrt{x + h + 2} + \sqrt{x + 2}}$

Practice Problem 6 Rationalize the numerator and simplify:

$$\dfrac{\sqrt{x} - \sqrt{3}}{x^2 - 9}$$

Section P.2 ■ Radicals and Rational Exponents 17

5 Define and evaluate rational exponents.

Rational Exponents

We know what 2^3 means, but what about $2^{1/3}$? The exponent $1/3$ is known as a **rational exponent**.

The Exponent $\dfrac{1}{n}$

For any real number a and any integer $n > 1$,
$$a^{1/n} = \sqrt[n]{a}.$$
When n is even and $a < 0$, $\sqrt[n]{a}$ and $a^{1/n}$ are not real numbers.

We use this definition to evaluate some expressions with rational exponents.
$$16^{1/2} = \sqrt{16} = 4$$
$$(-27)^{1/3} = \sqrt[3]{-27} = \sqrt[3]{(-3)^3} = -3$$
$$\left(\frac{1}{16}\right)^{1/4} = \sqrt[4]{\frac{1}{16}} = \sqrt[4]{\left(\frac{1}{2}\right)^4} = \frac{1}{2}$$
$$32^{1/5} = \sqrt[5]{32} = \sqrt[5]{2^5} = 2$$

$(-5)^{1/4}$ is not a real number because the base is negative and 4 is even.
We next define $a^{m/n}$, where m and n are integers, when $n > 1$ and $\sqrt[n]{a}$ is a real number.

Rational Exponents

$$a^{m/n} = (\sqrt[n]{a})^m = \sqrt[n]{a^m}$$

provided that m and n are integers with no common factors, $n > 1$, and $\sqrt[n]{a}$ is a real number.

Note that the numerator m of the exponent $\dfrac{m}{n}$ is the *exponent of the radical expression* and the denominator n is the *index of the radical*. When n is even and $a < 0$, $a^{m/n}$ is not a real number.

We change rational exponents with negative denominators such as $\dfrac{3}{-5}$ and $\dfrac{-2}{-7}$ to $-\dfrac{3}{5}$ and $\dfrac{2}{7}$, respectively, before applying the definition. Note that $(-8)^{2/6} \neq [(-8)^{1/6}]^2$ because $(-8)^{1/6} = \sqrt[6]{-8}$ is not a real number but $(-8)^{2/6}$ is a real number. Writing $\dfrac{2}{6} = \dfrac{1}{3}$, we have

$$(-8)^{2/6} = (-8)^{1/3} = \sqrt[3]{-8} = -2.$$

We can now evaluate expressions having rational exponents.
$$8^{2/3} = (\sqrt[3]{8})^2 = 2^2 = 4$$
$$-16^{5/2} = -(\sqrt{16})^5 = -4^5 = -1024$$
$$100^{-3/2} = (\sqrt{100})^{-3} = \frac{1}{(\sqrt{100})^3} = \frac{1}{10^3} = \frac{1}{1000}$$
$$(-25)^{7/2} = [(-25)^{1/2}]^7 = (\sqrt{-25})^7, \text{ which is not a real number.}$$

EXAMPLE 7 Simplifying Expressions Having Rational Exponents

Express your answer with only positive exponents. Assume that x represents a positive real number. Simplify.

a. $2x^{1/3} \cdot 5x^{1/4}$ **b.** $\dfrac{21x^{-2/3}}{7x^{1/5}}$ **c.** $(x^{3/5})^{-1/6}$

Solution

a. $2x^{1/3} \cdot 5x^{1/4} = 2 \cdot 5 x^{1/3} \cdot x^{1/4} = 10x^{1/3+1/4} = 10x^{4/12+3/12} = 10x^{7/12}$

b. $\dfrac{21x^{-2/3}}{7x^{1/5}} = \left(\dfrac{21}{7}\right)\left(\dfrac{x^{-2/3}}{x^{1/5}}\right) = 3x^{-2/3-1/5} = 3x^{-10/15-3/15} = 3x^{-13/15} = 3 \cdot \left(\dfrac{1}{x^{13/15}}\right)$

$= \dfrac{3}{x^{13/15}}$

c. $(x^{3/5})^{-1/6} = x^{(3/5)(-1/6)} = x^{-1/10} = \dfrac{1}{x^{1/10}}$

Practice Problem 7 Simplify. Express the answer with only positive exponents.

a. $4x^{1/2} \cdot 3x^{1/5}$ **b.** $\dfrac{25x^{-1/4}}{5x^{1/3}}, x > 0$ **c.** $(x^{2/3})^{-1/5}, x \ne 0$

SIDE NOTE

In calculus, you will learn that the expression in Example 8 is the "derivative" of $2x\sqrt{x+1}$.

EXAMPLE 8 Simplifying an Expression Involving Negative Rational Exponents

Simplify $x(x + 1)^{-1/2} + 2(x + 1)^{1/2}$. Express your answer with only positive exponents and then rationalize the denominator. Assume that $(x + 1)^{-1/2}$ is defined.

Solution

$x(x + 1)^{-1/2} + 2(x + 1)^{1/2} = \dfrac{x}{(x+1)^{1/2}} + \dfrac{2(x+1)^{1/2}}{1}$ $(x+1)^{-1/2} = \dfrac{1}{(x+1)^{1/2}}$

$= \dfrac{x}{(x+1)^{1/2}} + \dfrac{2(x+1)^{1/2}(x+1)^{1/2}}{(x+1)^{1/2}}$

$= \dfrac{x + 2(x+1)}{(x+1)^{1/2}} = \dfrac{3x+2}{(x+1)^{1/2}}$ $(x+1)^{1/2}(x+1)^{1/2} = x+1$

To rationalize the denominator $(x + 1)^{1/2} = \sqrt{x+1}$, multiply both the numerator and denominator by $(x + 1)^{1/2}$.

$\dfrac{3x+2}{(x+1)^{1/2}} = \dfrac{(3x+2)(x+1)^{1/2}}{(x+1)^{1/2}(x+1)^{1/2}} = \dfrac{(3x+2)(x+1)^{1/2}}{x+1}$

Practice Problem 8 Simplify $\dfrac{1}{2}x(x + 3)^{-1/2} + (x + 3)^{1/2}$.

EXAMPLE 9 Factoring an Expression Involving Rational Exponents

Factor $\dfrac{4}{3}x^{1/3}(3x - 1) + 3x^{4/3}$.

Solution

Begin by recognizing that $x^{4/3} = x \cdot x^{1/3}$; then factor out $\frac{1}{3}x^{1/3}$.

$$\frac{4}{3}x^{1/3}(3x - 1) + 3x^{4/3} = \frac{4}{3}x^{1/3}(3x - 1) + \frac{9}{3}x \cdot x^{1/3}$$

$$= \frac{1}{3}x^{1/3}[4(3x - 1) + 9x] = \frac{1}{3}x^{1/3}(21x - 4)$$

Practice Problem 9 Factor $\frac{4}{3}x^{1/3}(x + 5) + x^{4/3}$.

SECTION P.2 Exercises

Basic Concepts and Skills

In Exercises 1–16, evaluate each root or state that the root is not a real number.

1. $\sqrt{64}$
2. $\sqrt{100}$
3. $\sqrt[3]{64}$
4. $\sqrt[3]{125}$
5. $\sqrt[3]{-27}$
6. $\sqrt[3]{-216}$
7. $\sqrt[3]{-\frac{1}{8}}$
8. $\sqrt[3]{\frac{27}{64}}$
9. $\sqrt{(-3)^2}$
10. $-\sqrt{4(-3)^2}$
11. $\sqrt[4]{-16}$
12. $\sqrt[6]{-64}$
13. $-\sqrt[5]{-1}$
14. $\sqrt[7]{-1}$
15. $\sqrt[5]{(-7)^5}$
16. $-\sqrt[5]{(-4)^5}$

In Exercises 17–30, simplify each expression. Assume that all variables represent nonnegative real numbers.

17. $2\sqrt{3} + 5\sqrt{3}$
18. $7\sqrt{2} - \sqrt{2}$
19. $6\sqrt{5} - \sqrt{5} + 4\sqrt{5}$
20. $5\sqrt{7} + 3\sqrt{7} - 2\sqrt{7}$
21. $\sqrt{98x} - \sqrt{32x}$
22. $\sqrt{45x} + \sqrt{20x}$
23. $\sqrt[3]{24} - \sqrt[3]{81}$
24. $\sqrt[3]{54} + \sqrt[3]{16}$
25. $\sqrt[3]{3x} - 2\sqrt[3]{24x} + \sqrt[3]{375x}$
26. $\sqrt[3]{16x} + 3\sqrt[3]{54x} - \sqrt[3]{2x}$
27. $\sqrt{2x^5} - 5\sqrt{32x} + \sqrt{18x^3}$
28. $2\sqrt{50x^5} + 7\sqrt{2x^3} - 3\sqrt{72x}$
29. $\sqrt{48x^5y} - 4y\sqrt{3x^3y} + y\sqrt{3xy^3}$
30. $x\sqrt{8xy} + 4\sqrt{2xy^3} - \sqrt{18x^5y}$

In Exercises 31–40, rationalize the denominator of each expression.

31. $\dfrac{2}{\sqrt{3}}$
32. $\dfrac{10}{\sqrt{5}}$
33. $\dfrac{1}{\sqrt{2} + x}$
34. $\dfrac{1}{\sqrt{5} + 2x}$
35. $\dfrac{1}{\sqrt{3} + \sqrt{2}}$
36. $\dfrac{6}{\sqrt{5} - \sqrt{3}}$
37. $\dfrac{\sqrt{5} - \sqrt{2}}{\sqrt{5} + \sqrt{2}}$
38. $\dfrac{\sqrt{7} + \sqrt{3}}{\sqrt{7} - \sqrt{3}}$
39. $\dfrac{\sqrt{x + h} - \sqrt{x}}{\sqrt{x + h} + \sqrt{x}}$
40. $\dfrac{a\sqrt{x} + 3}{a\sqrt{x} - 3}$

In Exercises 41–48, rationalize the numerator of each expression.

41. $\dfrac{\sqrt{4 + h} - 2}{h}$
42. $\dfrac{\sqrt{y + 9} - 3}{y}$
43. $\dfrac{2 - \sqrt{4 - x}}{x}$
44. $\dfrac{\sqrt{x + 2} - \sqrt{2}}{x}$
45. $\dfrac{\sqrt{x} - 2}{x - 4}$
46. $\dfrac{\sqrt{x} - \sqrt{5}}{x^2 - 25}$
47. $\dfrac{\sqrt{x^2 + 4x} - x}{2}$
48. $\dfrac{\sqrt{x^2 + 2x + 3} - \sqrt{x^2 + 3}}{2}$

In Exercises 49–58, evaluate each expression without using a calculator.

49. $25^{1/2}$
50. $144^{1/2}$
51. $(-8)^{1/3}$
52. $(-27)^{1/3}$
53. $8^{2/3}$
54. $16^{3/2}$
55. $-25^{-3/2}$
56. $-9^{-3/2}$
57. $\left(\dfrac{9}{25}\right)^{-3/2}$
58. $\left(\dfrac{1}{27}\right)^{-2/3}$

In Exercises 59–70, simplify each expression, leaving your answer with only positive exponents. Assume that all variables represent positive numbers.

59. $x^{1/2} \cdot x^{2/5}$
60. $x^{3/5} \cdot 5x^{2/3}$
61. $x^{3/5} \cdot x^{-1/2}$
62. $x^{5/3} \cdot x^{-3/4}$
63. $(8x^6)^{2/3}$
64. $(16x^3)^{1/2}$
65. $(27x^6y^3)^{-2/3}$
66. $(16x^4y^6)^{-3/2}$
67. $\dfrac{15x^{3/2}}{3x^{1/4}}$
68. $\dfrac{20x^{5/2}}{4x^{2/3}}$
69. $\left(\dfrac{x^{-1/4}}{y^{-2/3}}\right)^{-12}$
70. $\left(\dfrac{27x^{-5/2}}{y^{-3}}\right)^{-1/3}$

In Exercises 71–78, convert each radical expression to its rational exponent form and then simplify. Assume that all variables represent positive numbers.

71. $\sqrt[4]{3^2}$
72. $\sqrt[4]{5^2}$
73. $\sqrt[3]{x^9}$
74. $\sqrt[3]{x^{12}}$
75. $\sqrt[3]{x^6 y^9}$
76. $\sqrt[3]{8x^3 y^{12}}$
77. $\sqrt[4]{9}\sqrt{3}$
78. $\sqrt[4]{49}\sqrt{7}$

In Exercises 79–84, factor each expression. Use only positive exponents in your answer.

79. $\dfrac{4}{3}x^{1/3}(2x - 3) + 2x^{4/3}$
80. $\dfrac{4}{3}x^{1/3}(7x + 1) + 7x^{4/3}$
81. $4(3x + 1)^{1/3}(2x - 1) + 2(3x + 1)^{4/3}$
82. $4(3x - 1)^{1/3}(x + 2) + (3x - 1)^{4/3}$
83. $3x(x^2 + 1)^{1/2}(2x^2 - x) + 2(x^2 + 1)^{3/2}(4x - 1)$
84. $3x(x^2 + 2)^{1/2}(x^2 - 2x) + (x^2 + 2)^{3/2}(2x - 2)$

Applying the Concepts

85. **Sign dimensions.** A sign in the shape of an equilateral triangle of area A has sides of length $\sqrt{\dfrac{4A}{\sqrt{3}}}$ centimeters. Find the length of a side if the sign has an area of 692 square centimeters. (Use the approximation $\sqrt{3} \approx 1.73$.)

86. **Return on investment.** For an initial investment of P dollars to mature to S dollars after two years when the interest is compounded annually, an annual interest rate of $r = \sqrt{\dfrac{S}{P}} - 1$ is required. Find r if $P = \$1{,}210{,}000$ and $S = \$1{,}411{,}344$.

87. **Terminal speed.** The terminal speed V of a steel ball that weighs w grams and is falling in a cylinder of oil is given by the equation $V = \sqrt{\dfrac{w}{1.5}}$ centimeters per second. Find the terminal speed of a steel ball weighing 6 grams.

88. **Electronic game current.** The current I (in amperes) in a circuit for an electronic game using W watts and having a resistance of R ohms is given by $I = \sqrt{\dfrac{W}{R}}$. Find the current in a game circuit using 1058 watts and having a resistance of 2 ohms.

Maintaining Skills

In Exercises 89–96, perform indicated operations and simplify.

89. $5^2 + 4 - 31$
90. $10^2 - 7^2$
91. $\dfrac{3(-2)^2 - 10}{-7 + 5}$
92. $\dfrac{|2 - 5| - |7 - 1|}{|3(-4)|}$
93. $(x - 3)^2$
94. $(x + 4)(x + 3)$
95. $(2x + 3)(2x - 3)$
96. $\left(3x - \dfrac{1}{2}\right)\left(3x + \dfrac{1}{2}\right)$

SECTION P.3 Solving Equations

OBJECTIVES

1. Solve equations in one variable.
2. Solve linear equations.
3. Solve quadratic equations.
4. Solve polynomial equations by factoring.
5. Solve rational equations.
6. Solve radical equations.
7. Solve equations that are quadratic in form.
8. Solve equations involving absolute value.

1 Solve equations in one variable.

RECALL

When finding the domain of a variable, remember that

1. Division by 0 is undefined.
2. The square root (or any even root) of a negative number is not a real number.

Definitions

An **equation in one variable** is a statement that two expressions, with at least one containing the variable, are equal. For example, $2x - 3 = 7$ is an equation in the variable x. The expressions $2x - 3$ and 7 are the *sides* of the equation. The **domain** of the variable in an equation is the set of all real numbers for which both sides of the equation are defined.

When the variable in an equation is replaced by a specific value from its domain, the resulting statement may be true or false. For example, in the equation $2x - 3 = 7$, if we let $x = 1$, the equation is false. However, if we replace x with 5, the equation is true. Those values (if any) of the variable that result in a true statement are called **solutions** or **roots** of the equation. So, 5 is a solution (or root) of the equation $2x - 3 = 7$. We also say that 5 **satisfies** the equation $2x - 3 = 7$. To **solve** an equation means to find all solutions of the equation; the set of all solutions of an equation is called its **solution set**.

An equation that is satisfied by every real number in the domain of the variable is called an **identity**. The equations

$$2(x + 3) = 2x + 6 \quad \text{Distributive property}$$
$$x^2 - 9 = (x + 3)(x - 3) \quad \text{Difference of squares}$$
$$\text{and} \quad \frac{1}{x^2 - 3x + 2} = \frac{1}{(x - 1)(x - 2)} \quad \text{Factor}$$

are examples of identities. Some equations, such as $x = x + 5$, have no solution. Such equations are called **inconsistent equations** and their solution set is designated by the empty set symbol, $\emptyset$.

Equivalent Equations

Equations that have the same solution set are called **equivalent equations**. For example, equations $x = 4$ and $3x = 12$ are equivalent equations with solution set $\{4\}$.

In general, to solve an equation in one variable, we replace the given equation with a sequence of equivalent equations until we obtain an equation whose solution is obvious, such as $x = 4$. The following operations yield equivalent equations.

Generating Equivalent Equations

Operations	Given Equation	Equivalent Equation
1. Simplify expressions on either side by eliminating parentheses, combining like terms, etc.	$(2x - 1) - (x + 1) = 4$	$2x - 1 - x - 1 = 4$ or $x - 2 = 4$
2. Add (or subtract) the same expression on *both* sides of the equation.	$x - 2 = 4$	$x - 2 + 2 = 4 + 2$ or $x = 6$
3. Multiply (or divide) *both* sides of the equation by the same *nonzero* expression.	$2x = 8$	$\frac{1}{2} \cdot 2x = \frac{1}{2} \cdot 8$ or $x = 4$
4. Interchange the two sides of the equation.	$-3 = x$	$x = -3$

2 Solve linear equations.

Solving Linear Equations in One Variable

Now we look at a special type of equation in one variable called a **linear equation**.

Linear Equations

A **conditional linear equation in one variable**, such as x, is an equation that can be written in the *standard form*

$$ax + b = 0,$$

where a and b are real numbers with $a \neq 0$.

EXAMPLE 1 Solving a Linear Equation in One Variable

Solve: $6x - [3x - 2(x - 2)] = 11$

Solution

$6x - [3x - 2(x - 2)] = 11$	Original equation
$6x - [3x - 2x + 4] = 11$	Remove innermost parentheses by distributing $-2(x - 2) = -2x + 4$.
$6x - 3x + 2x - 4 = 11$	Remove square brackets and change the sign of each enclosed term.
$5x - 4 = 11$	Combine like terms.
$5x - 4 + 4 = 11 + 4$	Add 4 to both sides.
$5x = 15$	Simplify both sides.
$\dfrac{5x}{5} = \dfrac{15}{5}$	Divide both sides by 5.
$x = 3$	Simplify.

The apparent solution is 3.

Check: Substitute $x = 3$ into the original equation. You should obtain $11 = 11$. Thus, 3 is the only solution of the original equation; so $\{3\}$ is its solution set.

Practice Problem 1 Solve $3x - [2x - 6(x + 1)] = -1$.

EXAMPLE 2 Solving a Linear Equation in One Variable

Solve: $1 - 5y + 2(y + 7) = 2y + 5(3 - y)$

Solution

$1 - 5y + 2(y + 7) = 2y + 5(3 - y)$	Original equation
$1 - 5y + 2y + 14 = 2y + 15 - 5y$	Distributive property
$15 - 3y = 15 - 3y$	Collect like terms and combine constants on each side.
$15 - 3y + 3y = 15 - 3y + 3y$	Add $3y$ to both sides.
$15 = 15$	Simplify.
$0 = 0$	Subtract 15 from both sides.

We have shown that $0 = 0$ is equivalent to the original equation. The equation $0 = 0$ is always true; its solution set is the set of all real numbers. Therefore, the solution set of the original equation is also the set of all real numbers. Thus, the original equation is an identity.

Practice Problem 2 Solve the equation $2(3x - 6) + 5 = 12 - (x + 5) + 7x$.

TYPES OF EQUATIONS

There are three types of equations.

1. An equation that is satisfied by all values in the domain of the variable is an *identity*. For example, $2(x - 1) = 2x - 2$. When you try to solve an identity, you get a true statement such as $3 = 3$ or $0 = 0$.

2. An equation that is not an identity but is satisfied by at least one number in the domain of the variable is a *conditional* equation. For example, $2x = 6$ is a linear conditional equation.

3. An equation that is not satisfied by any value of the variable is an *inconsistent* equation. For example, $x = x + 2$ is an inconsistent equation. Because no number is 2 more than itself, the solution set of the equation $x = x + 2$ is $\varnothing$. When you try to solve an inconsistent equation, you obtain a false statement such as $0 = 2$.

---

3 Solve quadratic equations.

Solving Quadratic Equations

In the variable x,

1. Linear equations are first-degree equations because the exponent on x is 1.
2. Quadratic equations are second-degree equations because the exponent on x is 2.

DO YOU KNOW?

The word *quadratic* comes from the Latin *quadratus*, meaning square.

Quadratic Equation

A quadratic equation in the variable x is an equation equivalent to the equation

$$ax^2 + bx + c = 0,$$

where a, b, and c are real numbers and $a \neq 0$.

A quadratic equation written in the form $ax^2 + bx + c = 0$ is said to be in **standard form**. Here are some examples of quadratic equations that are not in standard form.

$$2x^2 - 5x + 7 = x^2 + 3x - 1$$
$$3y^2 + 4y + 1 = 6y - 5$$
$$x^2 - 2x = 3$$

We discuss three methods for solving quadratic equations: (1) by factoring, (2) by completing the square, and (3) by using the quadratic formula.

We begin by studying real number solutions of quadratic equations.

Factoring Method

Some quadratic equations in the standard form $ax^2 + bx + c = 0$ can be solved by factoring and using the **zero-product property**.

ZERO-PRODUCT PROPERTY

Let A and B be two algebraic expressions. Then $AB = 0$ if and only if $A = 0$ or $B = 0$.

EXAMPLE 3 Solving a Quadratic Equation by Factoring

Solve by factoring: $2x^2 + 5x = 3$

Solution

$$2x^2 + 5x = 3 \quad \text{Original equation}$$
$$2x^2 + 5x - 3 = 0 \quad \text{Subtract 3 from both sides.}$$
$$(2x - 1)(x + 3) = 0 \quad \text{Factor the left side.}$$

We now set each factor equal to 0 and solve the resulting linear equations. The vertical line separates the computations leading to the two solutions.

$2x - 1 = 0$	$x + 3 = 0$	Zero-product property
$2x = 1$	$x = -3$	Isolate the x term on one side.
$x = \dfrac{1}{2}$	$x = -3$	Solve for x.

Check: You should check the solutions $x = \dfrac{1}{2}$ and $x = -3$ in the original equation.

The solution set is $\left\{\dfrac{1}{2}, -3\right\}$.

Practice Problem 3 Solve by factoring: $x^2 + 25x = -84$

Completing the Square

A method called **completing the square** can be used to solve quadratic equations that cannot be solved by factoring. The method requires two steps. First, we write a given quadratic equation in the form $(x + k)^2 = d$. Then we solve the equation $(x + k)^2 = d$ by taking the square root of both sides to get $x + k = \pm\sqrt{d}$. When $d \geq 0$ we have real number solutions.

Recall that

$$(x + k)^2 = x^2 + 2kx + k^2.$$

Notice that the coefficient of x on the right side is $2k$, and half of this coefficient is k; that is, $\frac{1}{2}(2k) = k$. We see that the constant term, k^2, in the trinomial $x^2 + 2kx + k^2$ is the *square of one-half the coefficient of x.*

Perfect-Square Trinomial

A quadratic trinomial in x with coefficient of x^2 equal to 1 is a **perfect-square trinomial** if the constant term is the square of one-half the coefficient of x.

When the constant term of a quadratic equation is *not* the square of half the coefficient of x, we subtract the constant term from both sides and add a new constant term that will result in a perfect-square trinomial.

Let's look at some examples to learn what number should be added to $x^2 + bx$ to create a perfect-square trinomial.

$x^2 + bx$	b	$\frac{b}{2}$	Add $\left(\frac{b}{2}\right)^2$	Result is $\left(x + \frac{b}{2}\right)^2$
$x^2 + 6x$	6	3	$3^2 = 9$	$x^2 + 6x + 9 = (x + 3)^2$
$x^2 - 4x$	-4	-2	$(-2)^2 = 4$	$x^2 - 4x + 4 = (x - 2)^2$
$x^2 + 3x$	3	$\frac{3}{2}$	$\left(\frac{3}{2}\right)^2 = \frac{9}{4}$	$x^2 + 3x + \frac{9}{4} = \left(x + \frac{3}{2}\right)^2$
$x^2 - x$	-1	$-\frac{1}{2}$	$\left(-\frac{1}{2}\right)^2 = \frac{1}{4}$	$x^2 - x + \frac{1}{4} = \left(x - \frac{1}{2}\right)^2$

Thus, to make $x^2 + bx$ into a perfect square, we need to add

$$\left[\frac{1}{2}(\text{coefficient of } x)\right]^2 = \left[\frac{1}{2}(b)\right]^2 = \frac{b^2}{4}$$

so that

$$x^2 + bx + \frac{b^2}{4} = \left(x + \frac{b}{2}\right)^2.$$

We say that $\frac{b^2}{4}$ was added to $x^2 + bx$ to *complete the square*. See Figure P.14.

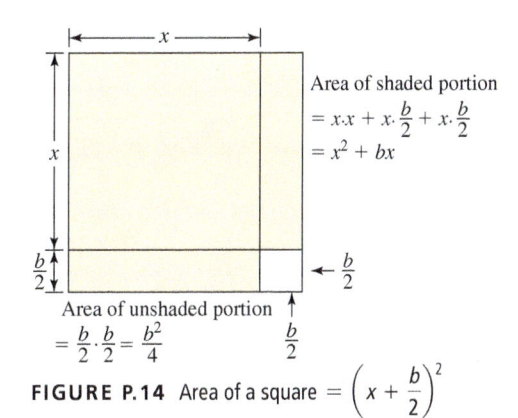

Area of shaded portion
$= x \cdot x + x \cdot \frac{b}{2} + x \cdot \frac{b}{2}$
$= x^2 + bx$

Area of unshaded portion
$= \frac{b}{2} \cdot \frac{b}{2} = \frac{b^2}{4}$

FIGURE P.14 Area of a square $= \left(x + \frac{b}{2}\right)^2$

EXAMPLE 4 Solving a Quadratic Equation by Completing the Square

Solve by completing the square: $x^2 + 10x + 8 = 0$

Solution

First, isolate the constant on the right side.

$$x^2 + 10x + 8 = 0 \quad \text{Original equation}$$
$$x^2 + 10x = -8 \quad \text{Subtract 8 from both sides.}$$

Next, add $\left[\frac{1}{2}(10)\right]^2 = 5^2 = 25$ to both sides to complete the square on the left side.

$$x^2 + 10x + 25 = -8 + 25 \quad \text{Add 25 to each side.}$$
$$(x + 5)^2 = 17 \quad x^2 + 10x + 25 = (x+5)^2$$
$$u^2 = 17 \quad \text{Let } u = x + 5.$$
$$u = \pm\sqrt{17} \quad \text{Solve for } u.$$
$$x + 5 = \pm\sqrt{17} \quad \text{Replace } u \text{ with } x + 5.$$
$$x = -5 \pm \sqrt{17} \quad \text{Subtract 5 from each side.}$$

Thus, $x = -5 + \sqrt{17}$ or $x = -5 - \sqrt{17}$, and the solution set is $\{-5 + \sqrt{17}, -5 - \sqrt{17}\}$.

Practice Problem 4 Solve by completing the square: $x^2 - 6x + 7 = 0$

The Quadratic Formula

We can generalize the method of completing the square to derive a formula that gives a solution of *any* quadratic equation.

We solve the standard form of the quadratic equation by *completing the square*.

$$ax^2 + bx + c = 0, a \neq 0$$

$$ax^2 + bx = -c \qquad \text{Isolate the constant on the right side.}$$

$$x^2 + \frac{b}{a}x = -\frac{c}{a} \qquad \text{Divide both sides by } a.$$

$$x^2 + \frac{b}{a}x + \left(\frac{b}{2a}\right)^2 = \left(\frac{b}{2a}\right)^2 - \frac{c}{a} \qquad \text{Add the square of one-half the coefficient of } x \text{ to both sides}$$

$$\left(x + \frac{b}{2a}\right)^2 = \frac{b^2}{4a^2} - \frac{c}{a} \qquad \text{The left side is a perfect square.}$$

$$\left(x + \frac{b}{2a}\right)^2 = \frac{b^2 - 4ac}{4a^2} \qquad \text{Combine fractions on the right side.}$$

$$x + \frac{b}{2a} = \pm\sqrt{\frac{b^2 - 4ac}{4a^2}} \qquad \text{Take the square roots of both sides.}$$

$$x = -\frac{b}{2a} \pm \frac{\sqrt{b^2 - 4ac}}{2a} \qquad \text{Add } -\frac{b}{2a} \text{ to both sides;} \quad \sqrt{4a^2} = \pm 2|a| = \pm 2a.$$

$$x = \frac{-b \pm \sqrt{b^2 - 4ac}}{2a} \qquad \text{Combine fractions.}$$

This formula is called the **quadratic formula**.

Section P.3 ■ Solving Equations   27

> **THE QUADRATIC FORMULA**
>
> The solutions of the quadratic equation in the standard form $ax^2 + bx + c = 0$ with $a \neq 0$ are given by the formula
> $$x = \frac{-b \pm \sqrt{b^2 - 4ac}}{2a}.$$

⚠ WARNING To use the quadratic formula, write the given quadratic equation in standard form. Then determine the values of *a* (coefficient of x^2), *b* (coefficient of *x*), and *c* (constant term).

EXAMPLE 5 Solving a Quadratic Equation by Using the Quadratic Formula

Solve $3x^2 = 5x + 2$ by using the quadratic formula.

Solution

We first rewrite the equation in standard form.

$3x^2 - 5x - 2 = 0$ Subtract $5x + 2$ from both sides.

$3x^2 + (-5)x + (-2) = 0$ Identify values of *a*, *b*, and *c* to be used in the quadratic formula.
   ↑   ↑   ↑
   a   *b*   *c*

$x = \dfrac{-b \pm \sqrt{b^2 - 4ac}}{2a}$ Quadratic formula

$x = \dfrac{-(-5) \pm \sqrt{(-5)^2 - 4(3)(-2)}}{2(3)}$ Substitute 3 for *a*, −5 for *b*, and −2 for *c*.

$= \dfrac{5 \pm \sqrt{25 + 24}}{6}$ Simplify.

$= \dfrac{5 \pm \sqrt{49}}{6} = \dfrac{5 \pm 7}{6}$ Simplify.

Then,

$x = \dfrac{5 + 7}{6} = \dfrac{12}{6} = 2$ or $x = \dfrac{5 - 7}{6} = \dfrac{-2}{6} = -\dfrac{1}{3}$; the solution set is $\left\{-\dfrac{1}{3}, 2\right\}$.

Practice Problem 5 Solve by using the quadratic formula: $6x^2 - x - 2 = 0$

4 Solve polynomial equations by factoring.

Solving Equations by Factoring

The method of factoring used in solving quadratic equations can also be used to solve certain polynomial equations of higher degree. Specifically, we use the zero-product property for solving those equations that can be expressed in factored form with 0 *on one side*.

EXAMPLE 6 Solving an Equation by Factoring

Solve by factoring:

a. $x^4 = 9x^2$ **b.** $x^3 - 2x^2 = x - 2$

Solution

a. We move all terms to the left side, factor, and set each factor equal to zero.

$$x^4 = 9x^2 \quad \text{Original equation}$$
$$x^4 - 9x^2 = 0 \quad \text{Subtract } 9x^2 \text{ from both sides.}$$
$$x^2(x^2 - 9) = 0 \quad x^4 - 9x^2 = x^2(x^2 - 9)$$
$$x^2(x + 3)(x - 3) = 0 \quad x^2 - 9 = (x + 3)(x - 3)$$
$$x^2 = 0 \quad \text{or} \quad x + 3 = 0 \quad \text{or} \quad x - 3 = 0 \quad \text{Zero-product property}$$
$$x = 0 \quad \text{or} \quad x = -3 \quad \text{or} \quad x = 3 \quad \text{Solve each equation for } x.$$

> **RECALL**
>
> $A^2 - B^2 = (A + B)(A - B)$
>
> Difference of two squares

Check each possible solution.

Let $x = 0$	Let $x = -3$	Let $x = 3$
$(0)^4 \stackrel{?}{=} 9(0)^2$	$(-3)^4 \stackrel{?}{=} 9(-3)^2$	$(3)^4 \stackrel{?}{=} 9(3)^2$
$0 = 0$ ✓	$81 = 81$ ✓	$81 = 81$ ✓

b. Move all terms to the left side and factor by grouping.

$$x^3 - 2x^2 = x - 2 \quad \text{Original equation}$$
$$x^3 - 2x^2 - x + 2 = 0 \quad \text{Add } x - 2 \text{ to both sides and simplify.}$$
$$(x^3 - 2x^2) - (x - 2) = 0 \quad \text{Group terms.}$$
$$x^2(x - 2) - 1 \cdot (x - 2) = 0 \quad \text{Factor } x^2 \text{ from the first two terms and rewrite } x - 2 \text{ as } 1 \cdot (x - 2).$$
$$(x - 2)(x^2 - 1) = 0 \quad \text{Factor out the common factor } (x - 2).$$
$$(x - 2)(x + 1)(x - 1) = 0 \quad \text{Factor } x^2 - 1.$$
$$x - 2 = 0 \quad | \quad x + 1 = 0 \quad | \quad x - 1 = 0 \quad \text{Set each factor}$$
$$x = 2 \quad | \quad x = -1 \quad | \quad x = 1 \quad \text{equal to zero.}$$

We ask you to check each possible solution.
The solution set of the equation is $\{2, 1, -1\}$.

Practice Problem 6 Solve by factoring.

a. $x^4 = 4x^2$ **b.** $x^3 - 5x^2 = 4x - 20$

Rational Equations

5 Solve rational equations.

Rational equations have at least one algebraic expression with the variable in the denominator. When we multiply a rational equation by an expression containing the variable, we may introduce a solution that satisfies the new equation but does not satisfy the original equation. Any such solution is called an **extraneous solution** or **extraneous root** (see Exercise 11).

EXAMPLE 7   Solving a Rational Equation

Solve: $\dfrac{1}{6} + \dfrac{1}{x + 1} = \dfrac{1}{x}$

Solution

The LCD from the denominators 6, $x + 1$, and x is $6x(x + 1)$. Multiply both sides of the equation by the LCD and make the right side 0.

$$6x(x+1)\left[\frac{1}{6} + \frac{1}{1+x}\right] = 6x(x+1)\left[\frac{1}{x}\right] \quad \text{Multiply both sides by the LCD.}$$

$$\frac{6x(x+1)}{6} + \frac{6x(x+1)}{x+1} = \frac{6x(x+1)}{x} \quad \text{Distributive property}$$

$$x(x+1) + 6x = 6(x+1) \quad \text{Simplify.}$$

$$x^2 + x + 6x = 6x + 6 \quad \text{Distributive property}$$

$$x^2 + x - 6 = 0 \quad \text{Subtract } 6x + 6 \text{ from both sides.}$$

$$(x+3)(x-2) = 0 \quad \text{Factor.}$$

$$x + 3 = 0 \quad \text{or} \quad x - 2 = 0 \quad \text{Set each factor equal to zero.}$$

$$x = -3 \quad \text{or} \quad x = 2 \quad \text{Solve for } x.$$

Check the solutions, -3 and 2, in the original equation. The solution set is $\{-3, 2\}$.

Practice Problem 7 Solve: $\dfrac{1}{x} - \dfrac{12}{5x+10} = \dfrac{1}{5}$

Radical Equations

6 Solve radical equations.

If an equation involves radicals or rational exponents, the method of raising both sides to a positive integer power is often used to remove them. When this is done, the solutions of the new equation always contain the solutions of the original equation. However, in some cases, the new equation has *more* solutions than the original equation. For instance, consider the equation $x = 3$. If we square both sides, we get $x^2 = 9$. Notice that the given equation, $x = 3$, has only one solution, namely, 3, whereas the new equation, $x^2 = 9$, has two solutions: 3 and -3. Extraneous solutions may be introduced whenever we raise both sides of an equation to an *even* power. So *we must check all* solutions after doing so.

EXAMPLE 8 Solving a Radical Equation

Solve $\sqrt{2x+1} + 1 = x$.

Solution

First, we isolate the radical on one side of the equation.

$$\sqrt{2x+1} = x - 1. \quad \text{Subtract 1 from both sides.}$$

Next, we square both sides and simplify.

$$(\sqrt{2x+1})^2 = (x-1)^2$$

$$2x + 1 = x^2 - 2x + 1 \quad (\sqrt{a})^2 = a$$

$$2x + 1 - 2x - 1 = x^2 - 2x + 1 - 2x - 1 \quad \text{Subtract } 2x + 1 \text{ from both sides.}$$

$$0 = x^2 - 4x \quad \text{Simplify.}$$

$$0 = x(x - 4) \quad \text{Factor.}$$

$$x = 0 \quad \text{or} \quad x - 4 = 0 \quad \text{Set each factor equal to zero.}$$

$$x = 0 \quad \text{or} \quad x = 4 \quad \text{Solve for } x.$$

Check:

$$\sqrt{2(0)+1} + 1 \stackrel{?}{=} 0 \quad \bigg| \quad \sqrt{2(4)+1} + 1 \stackrel{?}{=} 4$$

$$\sqrt{1} + 1 \stackrel{?}{=} 0 \quad \bigg| \quad \sqrt{9} + 1 \stackrel{?}{=} 4$$

$$2 \neq 0 \quad \bigg| \quad 4 = 4 \checkmark$$

Because 0 is an extraneous root, the only solution is 4; the solution set is $\{4\}$.

Practice Problem 8 Solve $\sqrt{6x+4}+2 = x$.

EXAMPLE 9 Solving a Radical Equation

Solve: $\sqrt[3]{2x+1}+4 = 3$

Solution

$$\sqrt[3]{2x+1} = -1 \quad \text{Subtract 4 from both sides.}$$
$$(\sqrt[3]{2x+1})^3 = (-1)^3 \quad \text{Cube both sides.}$$
$$2x+1 = -1 \quad (\sqrt[3]{a})^3 = a$$
$$2x = -2 \quad \text{Isolate the } x \text{ term.}$$
$$x = -1 \quad \text{Solve for } x.$$

You should always check your answers to avoid errors. The solution set is $\{-1\}$.

Practice Problem 9 Solve: $\sqrt[5]{3x-7} = 2$

Equations That Are Quadratic in Form

7 Solve equations that are quadratic in form.

An equation in a variable x is **quadratic in form** if it can be written as

$$au^2 + bu + c = 0 \quad (a \neq 0),$$

where u is an expression in the variable x. We solve the equation $au^2 + bu + c = 0$ for u. Then the solutions of the original equation can be obtained by replacing u with the expression in x that u represents and solving for x.

EXAMPLE 10 Solving by Substitution an Equation That is Quadratic in Form

Solve $x^{2/3} - 7x^{1/3} + 6 = 0$.

Solution

The equation becomes quadratic if we replace $x^{1/3}$ with u.

$$x^{2/3} - 7x^{1/3} + 6 = 0 \quad \text{Original equation}$$
$$u^2 - 7u + 6 = 0 \quad \text{Replace } x^{1/3} \text{ with } u.$$
$$(u-1)(u-6) = 0 \quad \text{Factor.}$$
$$u - 1 = 0 \quad \text{or} \quad u - 6 = 0 \quad \text{Zero-product property}$$
$$u = 1 \quad \quad u = 6 \quad \text{Solve for } u.$$
$$x^{1/3} = 1 \quad \quad x^{1/3} = 6 \quad \text{Replace } u \text{ with } x^{1/3}.$$
$$x = 1^3 = 1 \quad \quad x = 6^3 = 216 \quad \text{Raise both sides to power 3; } (x^{1/3})^3 = x.$$

You should check that both $x = 1$ and $x = 216$ are solutions of the original equation. The solution set is $\{1, 216\}$.

Practice Problem 10 Solve $x - 6\sqrt{x} + 8 = 0$.

Equations Involving Absolute Value

8 Solve equations involving absolute value.

Recall that geometrically, $|a|$ is the distance between the origin and a.

Absolute Value

$$|a| = a \text{ if } a \geq 0 \quad \text{and} \quad |a| = -a \text{ if } a < 0$$

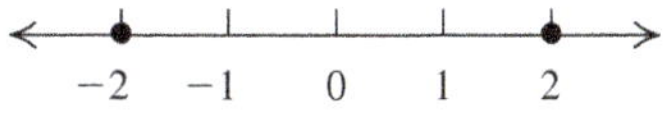

Because the only two numbers on the number line that are exactly two units from the origin are 2 and −2, they are the only solutions of the equation $|x| = 2$. See Figure P.15.

FIGURE P.15

EXAMPLE 11 Solving an Equation Involving Absolute Value

Solve the equation $|2x - 3| - 5 = 8$.

Solution

First, isolate the absolute value expression to one side of the equation.

$$|2x - 3| = 13 \quad \text{Add 5 to both sides.}$$
$$2x - 3 = 13 \quad \text{or} \quad 2x - 3 = -13. \quad \text{Definition of absolute value}$$
$$2x = 16 \quad\quad\quad 2x = -10 \quad \text{Add 3 to both sides of each equation.}$$
$$x = 8 \quad\quad\quad\quad x = -5 \quad \text{Divide both sides by 2.}$$

We leave it to you to check these solutions. The solution set is $\{-5, 8\}$.

Practice Problem 11 Solve the equation $|6x - 3| - 8 = 1$.

SECTION P.3 Exercises

Basic Concepts and Skills

In Exercises 1–50, find the solution set of each equation. If the equation has no solution, write $\varnothing$. If the equation is an identity, state so.

1. $5 - (6y + 9) + 2y = 2(y + 1)$
2. $3(4y - 3) = 4[y - (4y - 3)]$
3. $2(3x + 4) = 6(x + 2) - 4$
4. $2(x - 2) + 3x = 6x + 7 - (x - 3)$
5. $\dfrac{2x + 1}{9} - \dfrac{x + 4}{6} = 1$
6. $\dfrac{2 - 3x}{7} + \dfrac{x - 1}{3} = \dfrac{3x}{7}$
7. $\dfrac{1}{3x} + \dfrac{1}{2x} = \dfrac{1}{6} - \dfrac{1}{x}$
8. $\dfrac{2}{x - 1} = \dfrac{3}{x + 1}$
9. $\dfrac{t}{t - 2} = -\dfrac{2}{3}$
10. $\dfrac{2}{x + 1} + 3 = \dfrac{8}{x + 1}$
11. $\dfrac{5x}{x - 1} = \dfrac{5}{x - 1} + 3$
12. $\dfrac{3x}{x + 1} - 3 + \dfrac{3}{2(x + 1)} = 0$
13. $\dfrac{1 + x}{x} = \dfrac{1}{x} + 1$
14. $\dfrac{x - 1}{x - 2} - 1 = \dfrac{1}{x - 2}$
15. $x^2 - 5x = 0$
16. $x^2 - 5x + 4 = 0$
17. $x^2 = 5x + 6$
18. $x = x^2 - 12$
19. $x^2 + 2x - 5 = 0$
20. $x^2 - x - 3 = 0$
21. $3(t^2 - 1) = 2t^2 + 4t + 1$
22. $8(y^2 - y) = y^2 + 3$
23. $x^3 = 2x^2$
24. $3x^4 - 27x^2 = 0$
25. $x^4 - x^3 = x^2 - x$
26. $x^3 - 36x = 16(x - 6)$
27. $\dfrac{x + 3}{x - 1} + \dfrac{x + 4}{x + 1} = \dfrac{8x + 5}{x^2 - 1}$
28. $\dfrac{6}{2x - 2} - \dfrac{1}{x + 1} = \dfrac{2}{2x + 2} + 1$
29. $\dfrac{x}{x - 4} - \dfrac{4}{x + 4} = \dfrac{2x}{x^2 - 16}$
30. $\dfrac{x}{x - 3} + \dfrac{3}{x + 3} = \dfrac{6x}{x^2 - 9}$
31. $\dfrac{1}{x - 1} + \dfrac{x}{x + 3} = \dfrac{4}{x^2 + 2x - 3}$
32. $\dfrac{2x}{x + 3} - \dfrac{x}{x - 1} = \dfrac{14}{x^2 + 2x - 3}$
33. $t - \sqrt{3t + 6} = -2$
34. $\sqrt{5y^2 - 10y + 9} = 2y - 1$
35. $\sqrt{6t - 11} = 2t - 7$
36. $\sqrt{3x + 1} = x - 1$
37. $\sqrt{x - 3} = \sqrt{2x - 5} - 1$
38. $\sqrt{7x + 1} - \sqrt{5x + 4} = 1$
39. $x^{2/3} - 6x^{1/3} + 8 = 0$
40. $x^{2/5} + x^{1/5} - 2 = 0$
41. $2x^{1/2} - x^{1/4} - 1 = 0$
42. $81y^4 + 1 = 18y^2$
43. $(x^2 - 4)^2 - 3(x^2 - 4) - 4 = 0$
44. $(x^2 - 4x)^2 + 7(x^2 - 4x) + 12 = 0$

45. $\left(\dfrac{x}{x+1}\right)^2 - \dfrac{2x}{x+1} - 8 = 0$

46. $8\sqrt{\dfrac{x}{x+3}} - \sqrt{\dfrac{x+3}{x}} = 2$

47. $|x + 3| = 2$

48. $|2x - 3| - 1 = 0$

49. $|2x - 1| + 1 = 0$

50. $|x^2 - 4| = 0$

51. Solve for x: $\dfrac{x+b}{x-b} = \dfrac{x-5b}{2x-5b}$

52. Solve for x: $3a - \sqrt{ax} = \sqrt{a(3a+x)}$

53. Solve for x: $x = \sqrt{1 + \sqrt{1 + \sqrt{1 + \cdots}}}$
    (Hint: $x = \sqrt{1 + x}$.)

54. Solve for x: $x = 2 + \dfrac{1}{2 + \dfrac{1}{2 + \cdots}}$

In Exercises 55 and 56, find the values of k for which the given equation has equal roots.

55. $2x^2 + kx + k = 0$

56. $x^2 + k^2 = 2(k+1)x$

57. If r and s are the roots of the quadratic equation $ax^2 + bx + c = 0$, show that
$$r + s = -\dfrac{b}{a} \quad \text{and} \quad r \cdot s = \dfrac{c}{a}.$$

58. Find the sum and the product of the roots of the following equations without solving the equations.
    a. $3x^2 + 5x - 5 = 0$
    b. $3x^2 - 7x = 1$
    c. $\sqrt{3}x^2 = 3x + 4$
    d. $(1 + \sqrt{2})x^2 - \sqrt{2}x + 5 = 0$

In Exercises 59 and 60, determine k so that the sum and the product of the roots are equal.

59. $2x^2 + (k-3)x + 3k - 5 = 0$

60. $5x^2 + (2k-3)x - 2k + 3 = 0$

61. Suppose r and s are the roots of the quadratic equation $ax^2 + bx + c = 0$. Show that you can write $ax^2 + bx + c = a(x - r)(x - s)$. Thus, the quadratic trinomial can be written in factored form.
    [*Hint:* From Exercise 57, $b = -a(r + s)$ and $c = ars$. Substitute $ax^2 + bx + c$ and factor.]

62. Use Exercise 61 to factor each trinomial.
    a. $4x^2 + 4x - 5$
    b. $25x^2 + 40x + 11$
    c. $72x^2 + 95x - 1000$

Applying the Concepts

63. **Manufacturing.** The surface area A of a cylinder with height h and radius r is given by the equation $A = 2\pi rh + 2\pi r^2$. A company makes soup cans by using 32π square inches of aluminum sheet for each can. If the height of the can is 6 inches, find the radius of the can.

64. **Making a box.** A 2-inch square is cut from each corner of a rectangular piece of cardboard whose length exceeds the width by 4 inches. The sides are then turned up to form an open box. If the volume of the box is 64 cubic inches, find the dimensions of the box.

65. **Travel time.** Aya rode her bicycle from her house to a friend's house, averaging 16 kilometers per hour. It became dark before Aya was ready to return home, so her friend drove her home at a rate of 80 kilometers per hour. If Aya's total traveling time was 3 hours, how far away did her friend live?

66. **Interest rates.** Mr. Kaplan invests one sum of money at a certain rate of interest and half that sum at twice the first rate of interest. This arrangement turns out to yield 8% on the total investment. What are the two rates of interest?

67. **Mixture.** A mixture contains alcohol and water in a 5:1 ratio. Upon the addition of 5 liters of water, the ratio of alcohol to water becomes 5:2. Find the quantity of alcohol in the original mixture.

68. **Age problem.** A man and a woman have the same birthday. When he was as old as she is now, the man was twice as old as the woman. When she becomes as old as he is now, the sum of their ages will be 119. How old is the man now?

Maintaining Skills

In Exercises 69–73, simplify each expression.

69. $3(x + 2) - (2x - 4)$

70. $2(3x - 4) + 3(5 - 4x)$

71. $-2(5x - 3) + 7x - 5$

72. $0.5(4x - 3) - 0.2(2x - 1)$

73. $\dfrac{1}{3}(6 - 9x) - (2x - 1)$

74. If $x = -2$ and $y = 3$, find the value of $yx - y^2 + |2xy|$.

SECTION P.4 Inequalities

OBJECTIVES

1. Learn the vocabulary for discussing inequalities.
2. Solve and graph linear inequalities.
3. Solve and graph compound inequalities.
4. Solve and graph inequalities using test points.
5. Solve inequalities involving absolute value.

1 Learn the vocabulary for discussing inequalities.

Inequalities

An **inequality** in one variable is a statement that one algebraic expression is less than or is less than or equal to another algebraic expression. Some examples of inequalities are

$$4 < 5, \quad 3x + 2 \leq 14, \quad 5x + 7 > 3x + 23, \quad \text{and} \quad x^2 \geq 0.$$

The real numbers that result in a true statement when those numbers are substituted for the variable in the inequality are called **solutions** of the inequality. Because replacing x with 1 in the inequality $3x + 2 \leq 14$ results in the true statement $3(1) + 2 \leq 14$, the number 1 is a solution of the inequality $3x + 2 \leq 14$. We also say that 1 *satisfies* the inequality $3x + 2 \leq 14$. To **solve** an inequality means to find all solutions of the inequality—that is, its solution set.

The following properties of inequalities will be helpful in solving inequalities.

RECALL

The inequality symbols are:
$<, \leq, >, \geq$

INEQUALITY PROPERTIES

For real numbers, $a, b, c,$ and $d,$

1. **Trichotomy Property:** Exactly one of the following is true:

   $$a < b, a = b, \text{ or } a > b.$$

2. **Transitive Property:** If $a < b$ and $b < c,$ then $a < c.$
3. **Addition Property:** (i) If $a < b,$ then $a + c < b + c.$
   (ii) If $a < b$ and $c < d,$ then $a + c < b + d.$
4. **Multiplication Property:** Suppose $a < b.$
   a. If $c > 0,$ then $ac < bc.$
   b. If $c < 0,$ then $ac > bc.$

Similar results hold when $<$ is replaced throughout by any of the symbols $\leq, >,$ or $\geq.$

The inequality properties can be verified by considering the relative positions of the graphs of the real numbers $a, b, c,$ and d on a number line.

WARNING

Note that in Property 4, the inequality symbol must be reversed ($<$ to $>$ or $>$ to $<$) when we multiply both sides of the inequality by a negative number. For example, if $2 < 5$, then $(-3)(2) > (-3)(5)$, or $-6 > -15$. The statement $-6 > -15$ can be seen to be true geometrically because -15 is to the left of -6 on a number line.

The inequality properties can be used to establish inequalities involving opposites and inverses of a and b.

INEQUALITY	EXAMPLE
1. If $a < b$, then $-a > -b$.	$2 < 5$, so $-2 > -5$
2. If $a > b$, then $-a < -b$.	$6 > 3$, so $-6 < -3$
3. Suppose a and b are both positive or both negative. If $a < b$, then $\frac{1}{a} > \frac{1}{b}$.	$2 < 5$, so $\frac{1}{2} > \frac{1}{5}$ $-3 < -2$, so $-\frac{1}{3} > -\frac{1}{2}$
4. If a is negative and b is positive (that is, $a < 0 < b$), then $\frac{1}{a} < \frac{1}{b}$.	$-2 < 3$, so $-\frac{1}{2} < \frac{1}{3}$

Two inequalities that have exactly the same solution set are called **equivalent inequalities**. The basic method of solving inequalities is similar to the method for solving equations: We replace a given inequality by a series of equivalent inequalities until we arrive at an equivalent inequality, such as $x < 5$, whose solution set we already know.

The following operations produce equivalent inequalities.

1. Simplifying one or both sides of an inequality by combining like terms and eliminating parentheses.
2. Using the addition or multiplication property on both sides of the inequality.

2 Solve and graph linear inequalities.

Linear Inequalities

A **linear inequality in one variable** is an inequality that is equivalent to one of the forms

$$ax + b < 0 \quad \text{or} \quad ax + b \leq 0,$$

where a and b represent real numbers and $a \neq 0$.

Inequalities such as $2x - 1 > 0$ and $2x - 1 \geq 0$ are linear inequalities because they are equivalent to $-2x + 1 < 0$ and $-2x + 1 \leq 0$, respectively.

EXAMPLE 1 Solving and Graphing Linear Inequalities

Solve each inequality and graph its solution set.

a. $7x - 11 < 2(x - 3)$ **b.** $8 - 3x \leq 2$

Solution

a. $7x - 11 < 2(x - 3)$

$7x - 11 < 2x - 6$	Distributive property
$7x - 11 + 11 < 2x - 6 + 11$	Add 11 to both sides.
$7x < 2x + 5$	Simplify.
$7x - 2x < 2x + 5 - 2x$	Subtract $2x$ from both sides.
$5x < 5$	Simplify.
$\dfrac{5x}{5} < \dfrac{5}{5}$	Divide both sides by 5.
$x < 1$	Simplify.

The solution set is $\{x \mid x < 1\}$, or in interval notation, $(-\infty, 1)$. The graph is shown in Figure P.16.

$x < 1$, or $(-\infty, 1)$
FIGURE P.16

b. $8 - 3x \leq 2$

$8 - 3x - 8 \leq 2 - 8$	Subtract 8 from both sides.
$-3x \leq -6$	Simplify.
$\dfrac{-3x}{-3} \geq \dfrac{-6}{-3}$	Divide both sides by -3. (Reverse the direction of the inequality symbol.)
$x \geq 2$	Simplify.

The solution set is $\{x \mid x \geq 2\}$, or in interval notation, $[2, \infty)$. The graph is shown in Figure P.17.

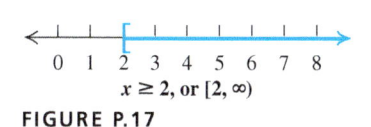

$x \geq 2$, or $[2, \infty)$
FIGURE P.17

Practice Problem 1 Solve each inequality and graph its solution set.

a. $4x + 9 > 2(x + 6) + 1$ **b.** $7 - 2x \geq -3$

If an inequality is true for all real numbers, its solution set is $(-\infty, \infty)$. If an inequality has no solution, its solution set is $\varnothing$.

EXAMPLE 2   Solving a Linear Inequality in One Variable

Write the solution set of each inequality.

a. $7(x + 2) - 20 - 4x < 3(x - 1)$ **b.** $2(x + 5) + 3x < 5(x - 1) + 3$

Solution

a.
$7(x + 2) - 20 - 4x < 3(x - 1)$	Original inequality
$7x + 14 - 20 - 4x < 3x - 3$	Distributive property
$3x - 6 < 3x - 3$	Combine like terms.
$-6 < -3$	Add $-3x$ to both sides and simplify.

The last inequality is equivalent to the original inequality and is true. So the solution set of the original inequality is $(-\infty, \infty)$.

b.
$2(x + 5) + 3x < 5(x - 1) + 3$	Original inequality
$2x + 10 + 3x < 5x - 5 + 3$	Distributive property
$5x + 10 < 5x - 2$	Combine like terms.
$10 < -2$	Add $-5x$ and simplify.

The resulting inequality is equivalent to the original inequality and is false. So the solution set of the original inequality is $\varnothing$.

Practice Problem 2 Write the solution set of each inequality.

a. $2(4 - x) + 6x < 4(x + 1) + 7$
b. $3(x - 2) + 5 \geq 7(x - 1) - 4(x - 2)$

Combining Two Inequalities

3 Solve and graph compound inequalities.

Sometimes we are interested in the solution set of two or more inequalities. The combination of two or more inequalities is called a **compound inequality**.

Suppose E_1 is an inequality with solution set (interval) I_1 and E_2 is another inequality with solution set I_2. Then the solution set of the compound inequality "E_1 and E_2" is $I_1 \cap I_2$, and the solution set of the compound inequality "E_1 or E_2" is $I_1 \cup I_2$.

EXAMPLE 3   Solving and Graphing a Compound Inequality

Graph and write the solution set of the compound inequality.

$$2x + 7 \leq 1 \quad \text{or} \quad 3x - 2 < 4(x - 1)$$

Solution

$2x + 7 \leq 1$	or	$3x - 2 < 4(x - 1)$	Original inequalities
$2x + 7 \leq 1$	or	$3x - 2 < 4x - 4$	Distributive property
$2x \leq -6$	or	$-x < -2$	Isolate variable term.
$x \leq -3$	or	$x > 2$	Solve for x: reverse second inequality symbol.

Graph each inequality and select the *union* of the two intervals.

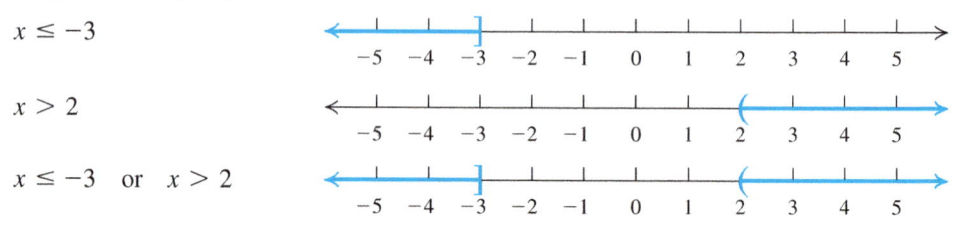

$x \leq -3$

$x > 2$

$x \leq -3$ or $x > 2$

The solution set of the compound inequality is $(-\infty, -3] \cup (2, \infty)$.

Practice Problem 3 Write the solution set of the compound inequality.

$$3x - 5 \geq 7 \quad \text{or} \quad 5 - 2x \geq 1$$

Sometimes we are interested in a joint inequality, such as $-5 < 2x + 3 \leq 9$. This use of two inequality symbols in a single expression is shorthand for $-5 < 2x + 3$ *and* $2x + 3 \leq 9$. Fortunately, solving such inequalities requires no new principles, as we see in the next example.

EXAMPLE 4   Solving and Graphing a Compound Inequality

Solve the inequality $-5 < 2x + 3 \leq 9$ and graph its solution set.

Solution

We must find all real numbers that are solutions of *both* inequalities

$$-5 < 2x + 3 \quad \text{and} \quad 2x + 3 \leq 9.$$

Let's see what happens when we solve these inequalities separately.

$-5 < 2x + 3$	$2x + 3 \leq 9$	Original inequalities
$-5 - 3 < 2x + 3 - 3$	$2x + 3 - 3 \leq 9 - 3$	Subtract 3 from both sides.
$-8 < 2x$	$2x \leq 6$	Simplify.
$\dfrac{-8}{2} < \dfrac{2x}{2}$	$\dfrac{2x}{2} \leq \dfrac{6}{2}$	Divide both sides by 2.
$-4 < x$	$x \leq 3$	Simplify.

The solution of the original pair of inequalities consists of all real numbers x such that $-4 < x$ and $x \leq 3$. This solution may be written more compactly as $\{x \mid -4 < x \leq 3\}$. In interval notation, we write $(-4, 3]$. The graph is shown in Figure P.18.

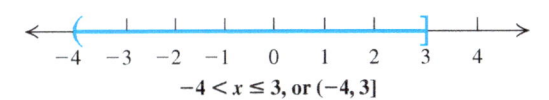

$-4 < x \leq 3$, or $(-4, 3]$

FIGURE P.18

Notice that we did the same thing to both inequalities in each step of the solution process. We can accomplish this simultaneous solution more efficiently by working on both inequalities at the same time, as follows.

$-5 < 2x + 3 \leq 9$	Original inequality
$-5 - 3 < 2x + 3 - 3 \leq 9 - 3$	Subtract 3 from each part.
$-8 < 2x \leq 6$	Simplify each part.
$\dfrac{-8}{2} < \dfrac{2x}{2} \leq \dfrac{6}{2}$	Divide each part by 2.
$-4 < x \leq 3$	Simplify each part.

The solution set is $(-4, 3]$, the same solution set we found previously.

Practice Problem 4 Solve and graph $-6 \leq 4x - 2 < 4$.

If you know that a particular quantity, represented by x, can vary between two values, then the procedures for working with inequalities can be used to find the values between which a linear expression $mx + k$ will vary. We show how to do this in Example 5.

EXAMPLE 5 Finding a Fahrenheit Temperature from a Celsius Range

The weather in London is predicted to range between 10° and 20° Celsius during the three-week period you will be working there. To decide what kind of clothes to take, you want to convert the temperature range to Fahrenheit temperatures. The formula for converting Celsius temperature C to Fahrenheit temperature F is $F = \dfrac{9}{5}C + 32$. What range of Fahrenheit temperatures might you find in London during your stay?

Solution

First, we express the Celsius temperature range as an inequality. Let C = temperature in Celsius degrees.

For the three weeks under consideration, $10 \leq C \leq 20$.

We want to know the range of $F = \dfrac{9}{5}C + 32$ when $10 \leq C \leq 20$.

$$10 \leq C \leq 20$$

$$\left(\dfrac{9}{5}\right)(10) \leq \dfrac{9}{5}C \leq \left(\dfrac{9}{5}\right)(20) \quad \text{Multiply each part by } \dfrac{9}{5}.$$

$$18 \leq \dfrac{9}{5}C \leq 36 \quad \text{Simplify.}$$

$$18 + 32 \leq \dfrac{9}{5}C + 32 \leq 36 + 32 \quad \text{Add 32 to each part.}$$

$$50 \leq \dfrac{9}{5}C + 32 \leq 68 \quad \text{Simplify.}$$

$$50 \leq F \leq 68 \quad F = \dfrac{9}{5}C + 32$$

So the temperature range from 10° to 20° Celsius corresponds to a range from 50° to 68° Fahrenheit.

Practice Problem 5 If $-2 < x < 5$, find real numbers a and b so that $a < 3x - 5 < b$.

Using Test Points to Solve Inequalities

4 Solve and graph inequalities using test points.

The **test-point** method, also known as the **sign-chart** method, involves writing an inequality (by rearranging if necessary) so that the expression on the left side of the inequality symbol is in factored form and the right side is 0. In Example 6, we illustrate this method to solve a **quadratic inequality**.

PROCEDURE IN ACTION

EXAMPLE 6 Using the Test-Point Method to Solve an Inequality

OBJECTIVE
Solve an inequality by using the test-point method.

EXAMPLE
Solve: $x^2 + 4x \geq 5x + 6$

Step 1 Rewrite (if necessary) the inequality so that the right side of the inequality is 0. Simplify the left side.

1. $x^2 + 4x - 5x - 6 \geq 0$ Subtract $5x + 6$ from both sides.
$x^2 - x - 6 \geq 0$ Simplify.

Step 2 Factor the left side.

2. $(x + 2)(x - 3) \geq 0$ Factor.

Step 3 Draw a number line. Find and plot the points where each factor is 0.

The n points so obtained divide the number line into $(n + 1)$ intervals.

3. $x + 2 = 0$ for $x = -2$, and $x - 3 = 0$ for $x = 3$.

[number line from -3 to 4 with points marked at -2 and 3]

The three intervals determined by the two points -2 and 3 are $(-\infty, -2)$, $(-2, 3)$, and $(3, \infty)$.

(continued)

Step 4 The final expression on the left side of the inequality in Step 1 is always positive or always negative on each of the $(n + 1)$ intervals of Step 3. Select convenient "test points" in each interval to determine the sign of the inequality.

4.

Test Interval	Test Point	Value of $x^2 - x - 6$	Result
$(-\infty, -2)$	-3	$(-3)^2 - (-3) - 6 = 6$	Positive
$(-2, 3)$	0	$0^2 - 0 - 6 = -6$	Negative
$(3, \infty)$	4	$4^2 - 4 - 6 = 6$	Positive

Step 5 Draw a sign chart. Show the information from Steps 3 and 4 on a number line.

5.
$x^2 - x - 6$ + + + + + + 0 − − − − − 0 + + + + +

Step 6 From the sign chart in Step 5, write the solution set of the inequality. Graph the solution set.

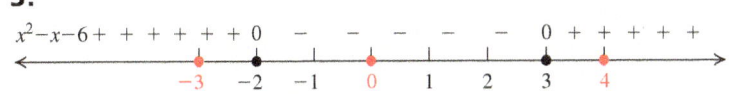

6. To find where $x^2 - x - 6 \geq 0$ means to find where $x^2 - x - 6 > 0$ (indicated by + signs) or $x^2 - x - 6 = 0$ (indicated by 0) on the sign chart. So the solution set of the inequality is $(-\infty, -2] \cup [3, \infty)$. This set is shown below.

Practice Problem 6 Solve $x^2 + 2 < 3x + 6$.

A **rational inequality** is an inequality that includes one or more rational expressions.

EXAMPLE 7 Solving a Rational Inequality

Solve $\dfrac{x^2 + 2x - 15}{x - 1} \geq 3$

Step 1
$\dfrac{x^2 + 2x - 15}{x - 1} - 3 \geq 0$ Rearrange so that the right side is 0.

$\dfrac{x^2 + 2x - 15}{x - 1} - \dfrac{3(x - 1)}{x - 1} \geq 0$ Common denominator

$\dfrac{(x^2 + 2x - 15) - 3(x - 1)}{x - 1} \geq 0$ $\dfrac{a}{c} \pm \dfrac{b}{c} = \dfrac{a \pm b}{c}$

$\dfrac{x^2 + 2x - 15 - 3x + 3}{x - 1} \geq 0$ Distribute.

$\dfrac{x^2 - x - 12}{x - 1} \geq 0$ Simplify.

Step 2 $\dfrac{(x + 3)(x - 4)}{x - 1} \geq 0$ Factor the numerator.

Step 3 $x + 3 = 0$ for $x = -3$
$x - 4 = 0$ for $x = 4$
$x - 1 = 0$ for $x = 1$ The rational expression is undefined at $x = 1$.

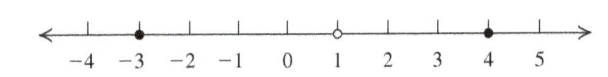

The four intervals determined by the three points -3, 1, and 4 are

$(-\infty, -3), (-3, 1), (1, 4),$ and $(4, \infty)$.

Step 4

Test Interval	Test Point	Sign of $\dfrac{(x+3)(x-4)}{(x-1)}$	Result
$(-\infty, -3)$	-4	$\dfrac{(-)(-)}{(-)}$	Negative
$(-3, 1)$	0	$\dfrac{(+)(-)}{(-)}$	Positive
$(1, 4)$	2	$\dfrac{(+)(-)}{(+)}$	Negative
$(4, \infty)$	5	$\dfrac{(+)(+)}{(+)}$	Positive

Step 5

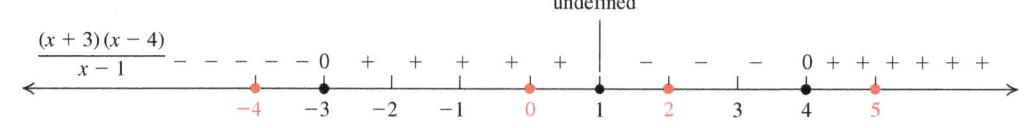

Step 6 $\dfrac{(x+3)(x-4)}{(x-1)} \geq 0$ on the set $[-3, 1) \cup [4, \infty)$, with the following graph:

Practice Problem 7 Solve: $\dfrac{2x+5}{x-1} \leq 1$

5 Solve inequalities involving absolute value.

Inequalities Involving Absolute Value

If $|x| < 2$, then x is closer than two units to the origin. See Figure P.19(a). If $|x| > 2$, then x is farther than two units from the origin and is in the interval $(-\infty, -2)$ or the interval $(2, \infty)$, that is, either $x < -2$ or $x > 2$. See Figure P.19(b).

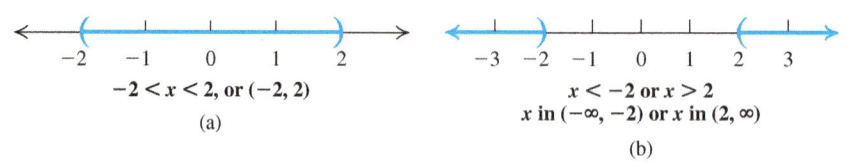

$-2 < x < 2$, or $(-2, 2)$
(a)

$x < -2$ or $x > 2$
x in $(-\infty, -2)$ or x in $(2, \infty)$
(b)

FIGURE P.19

This discussion suggests the following rules.

RULES FOR SOLVING ABSOLUTE VALUE INEQUALITIES

If $a > 0$ and u is an algebraic expression, then

1. $|u| < a$ is equivalent to $-a < u < a$.
2. $|u| \leq a$ is equivalent to $-a \leq u \leq a$.
3. $|u| > a$ is equivalent to $u < -a$ or $u > a$.
4. $|u| \geq a$ is equivalent to $u \leq -a$ or $u \geq a$.

Section P.4 ■ Inequalities 41

EXAMPLE 8 Solving an Inequality Involving Absolute Value

Solve the inequality $|4x - 1| \leq 9$ and graph the solution set.

Solution
Rule 2 applies here, with $u = 4x - 1$ and $a = 9$.

$|4x - 1| \leq 9$ is equivalent to

$$-9 \leq 4x - 1 \leq 9 \quad \text{Rule 2, } -a \leq u \leq a$$
$$1 - 9 \leq 4x - 1 + 1 \leq 9 + 1 \quad \text{Add 1 to each part.}$$
$$-8 \leq 4x \leq 10 \quad \text{Simplify.}$$
$$-\frac{8}{4} \leq \frac{4x}{4} \leq \frac{10}{4} \quad \text{Divide by 4: the sense of the inequality is unchanged.}$$
$$-2 \leq x \leq \frac{5}{2} \quad \text{Simplify.}$$

The solution set is $\left\{x \mid -2 \leq x \leq \frac{5}{2}\right\}$; that is, the solution set is the closed interval $\left[-2, \frac{5}{2}\right]$. See Figure P.20.

$-2 \leq x \leq \frac{5}{2}$, or $\left[-2, \frac{5}{2}\right]$

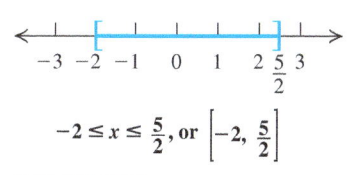

FIGURE P.20

Practice Problem 8 Solve $|3x + 3| \leq 6$ and graph the solution set.

EXAMPLE 9 Solving an Inequality Involving Absolute Value

Solve the inequality $|2x - 8| \geq 4$ and graph the solution set.

Solution

$|2x - 8| \geq 4$ is equivalent to

$2x - 8 \leq -4$	or	$2x - 8 \geq 4.$	Rule 4, with $u = 2x - 8$ and $a = 4.$
$2x - 8 + 8 \leq -4 + 8$		$2x - 8 + 8 \geq 4 + 8$	Add 8 to each part.
$2x \leq 4$		$2x \geq 12$	Simplify.
$\frac{2x}{2} \leq \frac{4}{2}$		$\frac{2x}{2} \geq \frac{12}{2}$	Divide both sides of each part by 2.
$x \leq 2$		$x \geq 6$	Simplify.

The solution set is $\{x \mid x \leq 2 \text{ or } x \geq 6\}$; that is, the solution set is the set of all real numbers x in either of the intervals $(-\infty, 2]$ or $[6, \infty)$. This set can also be written as $(-\infty, 2] \cup [6, \infty)$. See Figure P.21.

$x \leq 2$ or $x \geq 6$
$(-\infty, 2] \cup [6, \infty)$

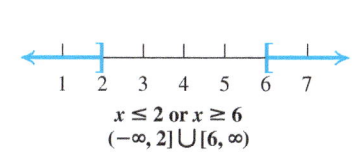

FIGURE P.21

Practice Problem 9 Solve $|2x + 3| \geq 6$ and graph the solution set.

EXAMPLE 10 Solving Special Cases of Absolute Value Inequalities

Solve each inequality.

a. $|3x - 2| > -5$ **b.** $|5x + 3| \leq -2$

Solution

a. Because the absolute value is always nonnegative, $|3x - 2| > -5$ is true for all real numbers x. The solution set is the set of all real numbers; in interval notation, we write $(-\infty, \infty)$.

b. There is no real number with absolute value ≤ -2 because the absolute value of any number is nonnegative. The solution set for $|5x + 3| \leq -2$ is the empty set, $\emptyset$.

Practice Problem 10 Solve.

a. $|5 - 9x| > -3$ **b.** $|7x - 4| \leq -1$

The next example shows that the test-point method can also be used to solve certain absolute value inequalities.

EXAMPLE 11   Solving an Inequality of the Form $|u| < |v|$

Solve: $|x + 1| < 3|x - 1|$

Solution

$$|x + 1| < 3|x - 1| \quad \text{Original inequality}$$

$$\frac{|x + 1|}{|x - 1|} < 3 \quad \text{For } x \neq 1, \text{ divide both sides by } |x - 1|.$$

$$\left|\frac{x + 1}{x - 1}\right| < 3 \quad \frac{|a|}{|b|} = \left|\frac{a}{b}\right|$$

$$-3 < \frac{x + 1}{x - 1} < 3 \quad |u| < a \text{ implies } -a < u < a.$$

We obtain two inequalities:

$$-3 < \frac{x + 1}{x - 1} \quad \text{and} \quad \frac{x + 1}{x - 1} < 3$$

$$0 < \frac{x + 1}{x - 1} + 3 \qquad\qquad \frac{x + 1}{x - 1} - 3 < 0$$

$$0 < \frac{(x + 1) + 3(x - 1)}{x - 1} \qquad \frac{(x + 1) - 3(x - 1)}{x - 1} < 0 \quad \text{Combine.}$$

$$0 < \frac{2(2x - 1)}{x - 1} \qquad\qquad \frac{-2(x - 2)}{x - 1} < 0 \quad \text{Simplify.}$$

$$0 < \frac{2x - 1}{x - 1} \qquad\qquad \frac{-(x - 2)}{x - 1} < 0 \quad \text{Divide by 2.}$$

$$\frac{2x - 1}{x - 1} > 0 \qquad\qquad \frac{x - 2}{x - 1} > 0 \quad \text{Rewrite.}$$

$S_1 = \left(-\infty, \frac{1}{2}\right) \cup (1, \infty)$

$S_2 = (-\infty, 1) \cup (2, \infty)$

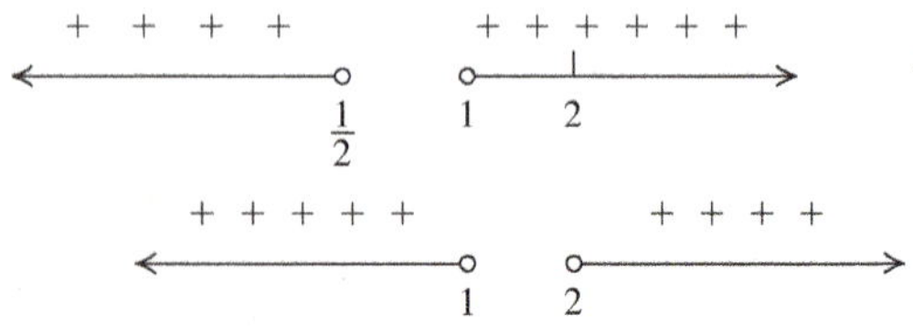

FIGURE P.22

From Figure P.22, we see that both inequalities are true on $S_1 \cap S_2 = \left(-\infty, \frac{1}{2}\right) \cup (2, \infty)$.

So the solution set of the given inequality is $\left(-\infty, \frac{1}{2}\right) \cup (2, \infty)$.

Practice Problem 11 Solve: $|x - 2| < 4|x + 4|$

EXAMPLE 12 Inequalities Useful in Calculus

Find a positive number δ (Greek letter "delta") such that if $|x - 1| < \delta$, then $|(3x + 4) - 7| < 0.12$.

Solution

We usually solve such problems as follows:

Step 1 Start with the conclusion. $\qquad |(3x + 4) - 7| < 0.12$

Step 2 Simplify until we obtain $\qquad |3x + 4 - 7| < 0.12$

$|x - 1|$ less than some number. $\qquad |3x - 3| < 0.12$

Let δ equal that number. $\qquad 3|x - 1| < 0.12$

$$|x - 1| < \frac{0.12}{3} = 0.04$$

Let $\delta = 0.04$.

Step 3 Verify that the δ in Step 2 is the desired number. This can be done by retracing your steps in Step 2 to obtain $|(3x + 4) - 7| < 0.12$.

Practice Problem 12 Find a positive number δ such that if $|x - 2| < \delta$, then $|(2x + 5) - 9| < 0.1$.

SECTION P.4 Exercises

Basic Concepts and Skills

In Exercises 1–6, solve each inequality. Write the solution in interval notation and graph the solution set.

1. $3(x + 2) < 2x + 5$
2. $4(x - 4) > 3(x - 5)$
3. $3(x + 2) \geq 5x + 18$
4. $5(x + 2) \leq 3(x + 1) + 10$
5. $3(x - 1) + (7 - x) < 2(x + 11) - 10$
6. $2(3x + 5) - 2(x + 1) > 4(x + 5) - 2$

In Exercises 7–12, solve each compound *or* inequality.

7. $2x + 5 < 1$ or $2 + x > 4$
8. $3x - 2 > 7$ or $2(1 - x) > 1$
9. $\frac{2x - 3}{4} \leq 2$ or $\frac{4 - 3x}{2} \geq 2$
10. $\frac{5 - 3x}{5} \geq \frac{1}{6}$ or $\frac{x - 1}{3} \leq 1$
11. $\frac{2x + 1}{3} \geq x + 1$ or $\frac{x}{2} - 1 > \frac{x}{3}$
12. $\frac{x - 1}{2} < \frac{x}{3} - 1$ or $\frac{2x + 5}{3} \leq \frac{x + 1}{6}$

In Exercises 13–18, solve each compound *and* inequality.

13. $3x - 5 < 10$ and $3 - 2x \leq 1$
14. $6 - x \leq 3x + 10$ and $7x - 14 \leq 3x + 14$
15. $2(x + 1) - 3 > 7$ and $3(2x + 1) + 1 < 10$
16. $5(x + 2) + 7 < 2$ and $2(5 - 3x) + 3 < 17$
17. $5 + 3(x - 1) < 3 + 3(x + 1)$ and $3x - 7 \leq 8$
18. $3(x + 3) + 7 > 2(x - 1) + 5$ and $2x + 1 \geq 3(5 - x) + 11$

In Exercises 19–24, solve each combined inequality.

19. $3 < x + 5 < 4$
20. $-4 \leq x - 2 < 2$
21. $0 \leq 1 - \frac{x}{3} < 2$
22. $0 < 5 - \frac{x}{2} \leq 3$
23. $5x \leq 3x + 1 < 4x + 2$
24. $3x + 2 < 2x + 3 < 4x - 1$

In Exercises 25–30, find *a* and *b*.

25. If $-2 < x < -1$, then $a < x + 7 < b$.
26. If $1 < x < 5$, then $a < 2x + 3 < b$.

27. If $-1 < x < 1$, then $a < 2 - x < b$.
28. If $3 < x < 7$, then $a < 1 - 3x < b$.
29. If $0 < x < 4$, then $a < 5x - 1 < b$.
30. If $-4 < x < 0$, then $a < 3x + 4 < b$.

In Exercises 31–42, solve each inequality. Write the solution set in interval notation.

31. $x^2 + 4x - 12 \leq 0$
32. $x^2 - 8x + 7 \geq 0$
33. $x^3 - x^2 - 4x + 4 > 0$
34. $x^3 + x^2 - 9x - 9 < 0$
35. $x^4 < 2x^2$
36. $x^3 < -8$
37. $\dfrac{(x-3)(x+1)}{(x+2)(x-4)} \leq 0$
38. $\dfrac{x(x-4)}{(x-2)(x+3)} \leq 0$
39. $\dfrac{2x-3}{x+3} \leq 1$
40. $\dfrac{2x+6}{2x+1} \geq 3$
41. $\dfrac{x-1}{x+1} \leq \dfrac{x+2}{x-3}$
42. $\dfrac{x+3}{x-1} \geq \dfrac{x-1}{x-2}$

In Exercises 43–52, solve each inequality.

43. $|x+1| < 3$
44. $|x-4| < 1$
45. $|2x-3| < 4$
46. $|4x-6| \leq 6$
47. $|2x-5| > 3$
48. $|3x-3| \geq 15$
49. $|2x-15| < 0$
50. $|x+5| \leq -3$
51. $|2x-3| > -4$
52. $|2x-6| \leq 0$

In Exercises 53–60, solve each absolute value inequality.

53. $\left|\dfrac{x-2}{x+3}\right| < 1$
54. $\left|\dfrac{x+3}{x-1}\right| < 2$
55. $\left|\dfrac{2x-3}{x+1}\right| \leq 3$
56. $\left|\dfrac{2x-1}{3x+2}\right| \leq 1$
57. $\left|\dfrac{x-1}{x+2}\right| \geq 2$
58. $\left|\dfrac{x+3}{x-2}\right| \geq 3$
59. $\left|\dfrac{2x+1}{x-1}\right| > 4$
60. $\left|\dfrac{2x-1}{3x+2}\right| > 5$

In Exercises 61–64, find a positive number δ.

61. If $|x-2| < \delta$, then $|(2x+3) - 7| < 0.1$.
62. If $|x-3| < \delta$, then $|(3x-4) - 5| < 0.01$.
63. If $|x+1| < \delta$, then $|(5x+3) + 2| < 0.15$.
64. If $|x+2| < \delta$, then $|(3x+7) - 1| < 0.02$.

Applying the Concepts

65. **Appliance markup.** The markup over the dealer's cost on a new refrigerator ranges from 15% to 20%. If the dealer's cost is $1750, over what range will the selling price vary?
66. **Return on investment.** An investor has $5000 to invest for a period of 1 year. Find the range of per annum simple interest rates required to generate interest between $200 and $275 inclusive.
67. **Average grade.** Sean has taken three exams and earned scores of 85, 72, and 77 out of a possible 100 points. He needs an average of at least 80 to earn a B in the course. What range of scores on the fourth (and last) 100-point test will guarantee a B in the course?
68. **Temperature conversion.** The formula for converting Fahrenheit temperature F to Celsius temperature C is $C = \dfrac{5}{9}(F - 32)$. What range in Celsius degrees corresponds to a range of 68° to 86° Fahrenheit?
69. **Parking expense.** The parking cost at the local airport (in dollars) is $C = 2 + 1.75(h - 1)$, where h is the number of hours the car is parked. For what range of hours is the parking cost between $37 and $51?
70. **Plumbing charges.** A plumber charges $42 per hour. A repair estimate includes a fixed cost of $147 for parts and a labor cost that is determined by how long it takes to complete the job. If the total estimate is for at least $210 and at most $294, what interval was estimated for the job?

Maintaining Skills

In Exercises 71–76, simplify each expression. Write your answers using positive exponents only.

71. $\dfrac{2}{x^{-3}}$
72. $\dfrac{x^{-3}}{y^{-5}}$
73. $\left(\dfrac{2}{3}\right)^{-3}$
74. $(2y^2)^3 y^4 - 5y^3 y^7$
75. $(3x^{3z})^2$
76. $a^{3m-1} \cdot (a^{m+2})^2$

SECTION P.5 Complex Numbers

OBJECTIVES

1. Define a complex number.
2. Add and subtract complex numbers.
3. Multiply and divide complex numbers.

1 Define a complex number.

Complex Numbers

Because the squares of real numbers are nonnegative (that is, $x^2 \geq 0$ for any real number x), the quadratic equation $x^2 = -1$ has no solution in the set of real numbers. To solve this equation, we extend the real number system to a larger system called the complex number system. We define a new number, i, with the following properties:

$$i = \sqrt{-1}, \quad i^2 = -1$$

Complex Numbers

A **complex number** is a number of the form

$$a + bi,$$

where a and b are real numbers and $i^2 = -1$.

The **real part** of the complex number $a + bi$ is a.

The **imaginary part** of the complex number $a + bi$ is b.

Two complex numbers are **equal** if and only if their real parts are equal and their imaginary parts are equal.

TECHNOLOGY CONNECTION

Most graphing calculators allow you to work with complex numbers by changing from "real" to $a + bi$ mode. Once the $a + bi$ mode is set, the $\sqrt{-1}$ is recognized as i. The i key is used to enter complex numbers such as $2 + 3i$.

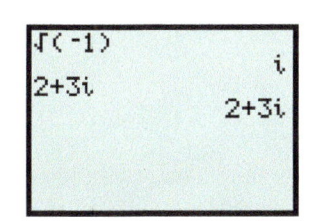

```
√(-1)
              i
2+3i
           2+3i
```

The real and imaginary parts of a complex number can then be found.

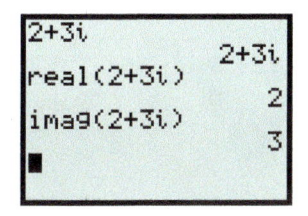

```
2+3i
           2+3i
real(2+3i)
              2
imag(2+3i)
              3
```

A complex number written in the form $a + bi$, where a and b are real numbers, is said to be in **standard form**. A complex number with real part 0, written as just bi, is called a **pure imaginary number**. Real numbers form a subset of complex numbers with imaginary part 0. For example, $-3 = -3 + 0i$.

We can express the **principal square root** of any negative number as the product of a real number and i.

If $a > 0$ is any real number, then $\sqrt{-a} = (\sqrt{a})i$.

In the following chart, we identify the real and the imaginary parts of each complex number.

	Real Part	Imaginary Part
$2 + 5i$	2	5
$7 - 8i$	7	-8
$3i$	0	3
-9	-9	0
$3 + \sqrt{-25}$	3	5 (because $\sqrt{-25} = i\sqrt{25} = 5i$)

2 Add and subtract complex numbers.

Addition and Subtraction

We find the sum or difference of two complex numbers by treating them as binomials in which i is the variable.

> **ADDITION AND SUBTRACTION OF COMPLEX NUMBERS**
>
> For all real numbers $a, b, c,$ and $d,$
> $$(a + bi) + (c + di) = (a + c) + (b + d)i \text{ and}$$
> $$(a + bi) - (c + di) = (a - c) + (b - d)i.$$

EXAMPLE 1 **Adding and Subtracting Complex Numbers**

Write the sum or difference of two complex numbers in standard form.

a. $(3 + 7i) + (2 - 4i)$ **b.** $(5 + 9i) - (6 - 8i)$

Solution

a. $(3 + 7i) + (2 - 4i) = (3 + 2) + [7 + (-4)]i = 5 + 3i$
b. $(5 + 9i) - (6 - 8i) = (5 - 6) + [9 - (-8)]i = -1 + 17i$

Practice Problem 1 Write the following complex numbers in standard form.

a. $(1 - 4i) + (3 + 2i)$ **b.** $(4 + 3i) - (5 - i)$

WARNING

Recall that if a and b are nonnegative real numbers, then
$$\sqrt{a}\sqrt{b} = \sqrt{ab}.$$
However, this property is not true for nonreal numbers. For example,
$$\sqrt{-9}\sqrt{-9} = (3i)(3i) = 9i^2 = 9(-1) = -9,$$
but
$$\sqrt{(-9)(-9)} = \sqrt{81} = 9.$$
Thus,
$$\sqrt{-9}\sqrt{-9} \neq \sqrt{(-9)(-9)}.$$

---

3 Multiply and divide complex numbers.

Multiplying and Dividing Complex Numbers

We multiply complex numbers as we do with real numbers and then replacing i^2 with -1.

> **MULTIPLYING COMPLEX NUMBERS**
>
> For all real numbers $a, b, c,$ and $d,$
> $$(a + bi)(c + di) = (ac - bd) + (ad + bc)i.$$

EXAMPLE 2 **Multiplying Complex Numbers**

Write the following products in standard form.

a. $(3 - 5i)(2 + 7i)$ **b.** $-2i(5 - 9i)$

Solution

a. $(3 - 5i)(2 + 7i) = \overset{F}{6} + \overset{O}{21i} - \overset{I}{10i} - \overset{L}{35i^2}$

$\qquad\qquad\qquad = 6 + 11i + 35 \qquad$ Because $i^2 = -1, -35i^2 = 35.$

$\qquad\qquad\qquad = 41 + 11i \qquad\qquad$ Combine terms.

b. $-2i(5 - 9i) = -10i + 18i^2 \qquad$ Distributive property

$\qquad\qquad\quad = -10i - 18 \qquad\qquad$ Because $i^2 = -1, 18i^2 = -18.$

$\qquad\qquad\quad = -18 - 10i \qquad\qquad$ Standard form

Practice Problem 2 Write the following products in standard form.

a. $(2 - 6i)(1 + 4i)$ b. $-3i(7 - 5i)$

The Conjugate of a Complex Number

If $z = a + bi$, then the conjugate (or *complex* conjugate) of z is denoted by $\bar{z}$ and defined by $\bar{z} = \overline{a + bi} = a - bi.$

For example, $\overline{2 + 7i} = 2 - 7i$ and $\overline{5 - 3i} = \overline{5 + (-3)i} = 5 - (-3)i = 5 + 3i.$

EXAMPLE 3   Multiplying a Complex Number by Its Conjugate

Find the product $z\bar{z}$ for the complex number $z = 2 + 5i.$

Solution

If $z = 2 + 5i$, then $\bar{z} = 2 - 5i.$

$z\bar{z} = (2 + 5i)(2 - 5i) = 2^2 - (5i)^2 \qquad$ Difference of squares

$\qquad\qquad\qquad\qquad = 4 - 25i^2 \qquad\qquad (5i)^2 = 5^2 i^2 = 25i^2$

$\qquad\qquad\qquad\qquad = 4 - (-25) \qquad\quad$ Because $i^2 = -1, 25i^2 = -25.$

$\qquad\qquad\qquad\qquad = 29 \qquad\qquad\qquad\;$ Simplify.

Practice Problem 3 Find the product $z\bar{z}$ for the complex number $z = 1 + 6i.$

Notice that the product $z\bar{z}$ in Example 3 is a real number. To write the reciprocal of a nonzero complex number (or the quotient of two complex numbers) in standard form, multiply the numerator and denominator by the conjugate of the denominator. The resulting denominator is a real number, as the next theorem states.

COMPLEX CONJUGATE PRODUCT THEOREM

If $z = a + bi$, then

$$z\bar{z} = a^2 + b^2.$$

The complex conjugate product theorem provides a procedure for writing the quotient of two complex numbers in standard form. You do not need to memorize the final result.

DIVIDING COMPLEX NUMBERS

Let $z = a + bi$ ($z \neq 0$) and $w = c + di$ be two complex numbers in standard form. Then

$$\frac{w}{z} = \frac{w\bar{z}}{z\bar{z}} \quad \text{Multiply numerator and denominator by } \bar{z}.$$

$$= \frac{(c + di)(a - bi)}{a^2 + b^2} = \frac{ac + bd + (ad - bc)i}{a^2 + b^2}$$

$$= \left(\frac{ac + bd}{a^2 + b^2}\right) + \left(\frac{ad - bc}{a^2 + b^2}\right)i$$

EXAMPLE 4   **Dividing Complex Numbers**

Write the following quotients in standard form.

a. $\dfrac{1}{2 + i}$ **b.** $\dfrac{4 + \sqrt{-25}}{2 - \sqrt{-9}}$

Solution

a. The denominator is $2 + i$, so its conjugate is $2 - i$.

$$\frac{1}{2 + i} = \frac{1(2 - i)}{(2 + i)(2 - i)} \qquad \text{Multiply numerator and denominator by } 2 - i.$$

$$= \frac{2 - i}{2^2 + 1^2} = \frac{2 - i}{5} = \frac{2}{5} - \frac{1}{5}i$$

b. Write $\dfrac{4 + \sqrt{-25}}{2 - \sqrt{-9}}$ as $\dfrac{4 + 5i}{2 - 3i}$.

$$\frac{4 + 5i}{2 - 3i} = \frac{(4 + 5i)(2 + 3i)}{(2 - 3i)(2 + 3i)} \qquad \text{Multiply numerator and denominator by } 2 + 3i.$$

$$= \frac{8 + 12i + 10i + 15i^2}{2^2 + 3^2} \qquad \text{Use FOIL in the numerator: } (2 - 3i)(2 + 3i) = 2^2 + 3^2.$$

$$= \frac{-7 + 22i}{13} = -\frac{7}{13} + \frac{22}{13}i \qquad i^2 = -1, 8 + 15i^2 = 8 - 15 = -7$$

Practice Problem 4 Write the following quotients in standard form.

a. $\dfrac{2}{1 - i}$ **b.** $\dfrac{-3i}{4 + \sqrt{-25}}$

Powers of *i*

We already know the first two powers of i: $i^1 = i$ and $i^2 = -1$. For the following powers of i, notice the pattern.

$$i^1 = i \qquad\qquad\qquad i^5 = i^4 \cdot i = (1)i = i$$
$$i^2 = -1 \qquad\qquad\quad i^6 = i^4 \cdot i^2 = (1) \cdot i^2 = i^2 = -1$$
$$i^3 = i^2 \cdot i = (-1)i = -i \qquad i^7 = i^4 \cdot i^3 = (1) \cdot i^3 = i^3 = -i$$
$$i^4 = i^2 \cdot i^2 = (-1)(-1) = 1 \qquad i^8 = i^4 \cdot i^4 = (1)(1) = 1$$

Section P.5 ■ Complex Numbers   49

After i^4, the powers of i cycle indefinitely through the values i, -1, $-i$, and 1. If n is a positive integer, a quick way to evaluate i^n is to divide n by 4; then $i^n = i^r$, where r is the remainder: 0, 1, 2, or 3.

To see how this works, we evaluate i^{1003}. Dividing 1003 by 4, we get $1003 = 4(250) + 3$. Then $i^{1003} = i^{4(250)+3} = i^{4(250)} \cdot i^3 = (i^4)^{250} \cdot i^3 = (1)^{250} \cdot i^3 = 1 \cdot i^3 = i^3 = -i$. Similarly, dividing 26 by 4 gives the remainder 2; so $i^{26} = i^2 = -1$.

SECTION P.5 Exercises

Basic Concepts and Skills

In Exercises 1–4, use the definition of equality of complex numbers to find the real numbers x and y such that the equation is true.

1. $2 + xi = y + 3i$
2. $x - 2i = 7 + yi$
3. $x - \sqrt{-16} = 2 + yi$
4. $3 + yi = x - \sqrt{-25}$

In Exercises 5–22, perform each operation and write the result in the standard form $a + bi$.

5. $(5 + 2i) + (3 + i)$
6. $(4 - 3i) - (5 + 3i)$
7. $(3 - 5i) - (3 + 2i)$
8. $(-2 - 3i) + (-3 - 2i)$
9. $3(5 + 2i)$
10. $-4(2 - 3i)$
11. $3i(5 + i)$
12. $-3i(5 - 2i)$
13. $(3 + i)(2 + 3i)$
14. $(4 + 3i)(2 + 5i)$
15. $(2 - 3i)(2 + 3i)$
16. $(4 - 3i)(4 + 3i)$
17. $(3 + 4i)(4 - 3i)$
18. $(-2 + 3i)(-3 + 10i)$
19. $(\sqrt{3} - 12i)^2$
20. $(-\sqrt{5} - 13i)^2$
21. $(1 + 3i)^3$
22. $(1 - 2i)^3$

In Exercises 23–28, write the conjugate $\bar{z}$ of each complex number z. Then find $z\bar{z}$.

23. $z = 2 - 3i$
24. $z = \dfrac{1}{2} - 2i$
25. $z = 4 + 5i$
26. $z = \dfrac{2}{3} + \dfrac{1}{2}i$
27. $z = \sqrt{2} - 3i$
28. $z = \sqrt{5} + \sqrt{3}i$

In Exercises 29–34, write each quotient in the standard form $a + bi$.

29. $\dfrac{5}{-i}$
30. $\dfrac{-1}{1+i}$
31. $\dfrac{5i}{2+i}$
32. $\dfrac{2+3i}{1+i}$
33. $\dfrac{2-5i}{4-7i}$
34. $\dfrac{2+\sqrt{-4}}{1+i}$

In Exercises 35–38, find each power of i and simplify the expression.

35. i^{17}
36. i^{125}
37. i^{-7}
38. i^{-24}

In Exercises 39–42, let $z_1 = a + bi$ and $z_2 = c + di$.

39. Prove that $\overline{\overline{z_1}} = z_1$.
40. Prove that $\overline{z_1 + z_2} = \overline{z_1} + \overline{z_2}$.
41. Prove that $\overline{z_1 z_2} = \overline{z_1}\, \overline{z_2}$. Use this fact to prove that $\overline{z^2} = (\bar{z})^2$.
42. Prove that $z_1 + \overline{z_1} = 2a$ and that $z_1 - \overline{z_1} = 2bi$.

In Exercises 43–46, let $z = 2 + 4i$ and $w = 3 - 2i$. Write the value of each expression in standard form.

43. $\overline{wz}$
44. $\overline{w}z$
45. $\overline{z}w$
46. $\dfrac{w}{z}$

Applying the Concepts

47. **Series circuits.** If the impedance of a resistor in a circuit is $Z_1 = 4 + 3i$ ohms and the impedance of a second resistor is $Z_2 = 5 - 2i$ ohms, find the total impedance of the two resistors when they are placed in series (the sum of the two impedances).

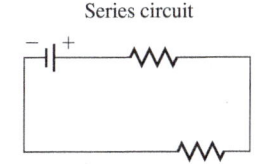

Series circuit

48. Parallel circuits. If the two resistors in Exercise 47 are connected in parallel, the total impedance is given by

$$\frac{Z_1 Z_2}{Z_1 + Z_2}.$$

Find the total impedance if the resistors in Exercise 47 are connected in parallel.

Parallel circuit

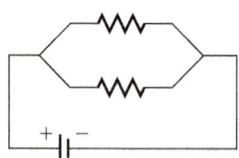

As with impedance, the current I and voltage V in a circuit can be represented by complex numbers. The three quantities (voltage V, impedance Z, and current I) are related by the equation $Z = \dfrac{V}{I}$. Thus, if two of these values are given, the value of the third can be found from this equation. In Exercises 49–54, find the value of I, V, or Z, that is not specified.

49. Finding impedance. $I = 7 + 5i$ $V = 35 + 70i$
50. Finding impedance. $I = 7 + 4i$ $V = 45 + 88i$
51. Finding voltage. $Z = 5 - 7i$ $I = 2 + 5i$
52. Finding voltage. $Z = 7 - 8i$ $I = \dfrac{1}{3} + \dfrac{1}{6}i$
53. Finding current. $V = 12 + 10i$ $Z = 12 + 6i$
54. Finding current. $V = 29 + 18i$ $Z = 25 + 6i$

Maintaining Skills

55. Solve $V = \dfrac{1}{3}\pi r^2 h$ for h.

56. Graph: $3x - 5 < 10$ and $2x + 1 \geq -5$

57. Rationalize the denominator and simplify: $\dfrac{\sqrt{5} + \sqrt{3}}{\sqrt{5} - \sqrt{3}}$

58. The velocity of an object is given by $v = t^2 - 10t + 12$. Find the time(s) when $v = 0$.

CHAPTER 1

Graphs and Functions

TOPICS

1.1 Graphs of Equations
1.2 Lines
1.3 Functions
1.4 A Library of Functions
1.5 Transformations of Functions
1.6 Combining Functions; Composite Functions
1.7 Inverse Functions

Hurricanes are tracked with the use of coordinate grids, and data from fields as diverse as medicine and sports are related and analyzed by means of functions. The material in this chapter will introduce you to the versatile concepts that are the everyday tools of all dynamic industries.

SECTION 1.1 Graphs of Equations

BEFORE STARTING THIS SECTION, REVIEW

1. The real number line (Section P.1, page 5)
2. Equivalent equations (Section P.3, page 21)
3. Completing squares (Section P.3, page 24)
4. Interval Notation (Section P.1, page 6)

OBJECTIVES

1. Plot points in the Cartesian coordinate plane.
2. Find the distance between two points.
3. Find the midpoint of a line segment.
4. Sketch a graph by plotting points.
5. Find the intercepts of a graph.
6. Find the symmetries in a graph.
7. Find the equation of a circle.

◆ A Fly on the Ceiling

One day the French mathematician René Descartes noticed a fly buzzing around on a ceiling made of square tiles. He watched the fly and wondered how he could mathematically describe its location. Finally, he realized that he could describe the fly's position by its distance from the walls of the room. Descartes had just discovered the coordinate plane. In fact, the coordinate plane is sometimes called the Cartesian plane in his honor. The discovery led to the development of analytic geometry, the first blending of algebra and geometry.

Although the basic idea of graphing with coordinate axes dates all the way back to Apollonius in the second century B.C., Descartes, who lived in the 1600s, gets the credit for coming up with the two-axis system we use today.

1 Plot points in the Cartesian coordinate plane.

The Coordinate Plane

A visually powerful device for exploring relationships between numbers is the Cartesian plane. A pair of real numbers in which the order is specified is called an **ordered pair** of real numbers. The ordered pair (a, b) has *first component a* and *second component b*. Two ordered pairs (x, y) and (a, b) are **equal**, and we write $(x, y) = (a, b)$ if and only if $x = a$ and $y = b$.

Just as the real numbers are identified with points on a line, called the number line or the coordinate line, the sets of ordered pairs of real numbers are identified with points on a plane called the **coordinate plane** or the **Cartesian plane**.

We begin with two coordinate lines, one horizontal and one vertical, that intersect at their zero points. The horizontal line (with positive numbers to the right) is usually called the **x-axis**, and the vertical line (with positive numbers up) is usually called the **y-axis**. Their point of intersection is called the **origin**. The x-axis and y-axis are called **coordinate axes**, and the plane they form is sometimes called the **xy-plane**. The axes divide the plane into four regions called **quadrants**, which are numbered as shown in Figure 1.1. The points on the axes themselves do not belong to any of the quadrants.

The notation $P(a, b)$, or $P = (a, b)$, designates the point P whose first component is a and whose second component is b. In an xy-plane, the first component, a, is called the **x-coordinate** of $P(a, b)$ and the second component, b, is called the **y-coordinate**

52 Chapter 1 Graphs and Functions

HISTORICAL NOTE

René Descartes (1596–1650)

Descartes was born at La Haye, near Tours in southern France. He is often called the father of modern science. Descartes established a new, clear way of thinking about philosophy and science by accepting only those ideas that could be proved by or deduced from first principles. He took as his philosophical starting point the statement **Cogito ergo sum**: "I think; therefore, I am." Descartes made major contributions to modern mathematics, including the Cartesian coordinate system and the theory of equations.

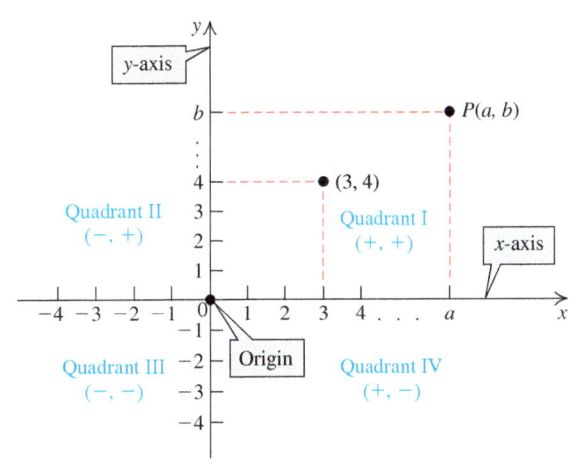

FIGURE 1.1 Quadrants in a plane

of $P(a, b)$. The signs of the x- and y-coordinates for each quadrant are shown in Figure 1.1. The point corresponding to the ordered pair (a, b) is called the **graph of the ordered pair (a, b)**. However, we frequently ignore the distinction between an ordered pair and its graph.

EXAMPLE 1 Graphing Points

Graph the following points in the xy-plane:

$A(3, 1), B(-2, 4), C(-3, -4), D(2, -3)$, and $E(-3, 0)$

Solution

Figure 1.2 shows a coordinate plane along with the graph of the given points. These points are located by moving left, right, up, or down starting from the origin $(0, 0)$.

$A(3, 1)$	3 units right, 1 unit up	$D(2, -3)$	2 units right, 3 units down
$B(-2, 4)$	2 units left, 4 units up	$E(-3, 0)$	3 units left
$C(-3, -4)$	3 units left, 4 units down		

Practice Problem 1 Graph the following points in the xy-plane:

$$P(-2, 2), Q(4, 0), R(5, -3), S(0, -3), \text{ and } T\left(-2, \frac{1}{2}\right)$$

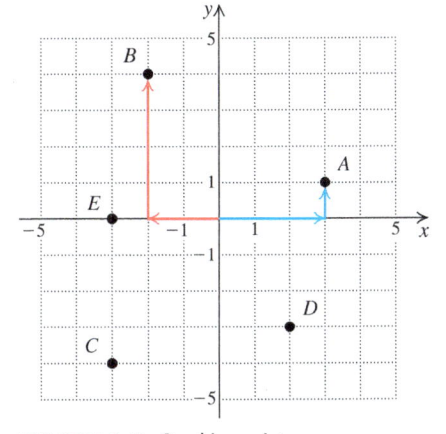

FIGURE 1.2 Graphing points

The Distance Formula

2 Find the distance between two points.

If a Cartesian coordinate system has the same unit of measurement, such as inches or centimeters, on both axes, we can calculate the distance between any two points in that plane.

Recall that the Pythagorean Theorem states that in a right triangle with hypotenuse of length c and the legs of lengths a and b,

$$a^2 + b^2 = c^2, \quad \text{Pythagorean Theorem}$$

as shown in Figure 1.3.

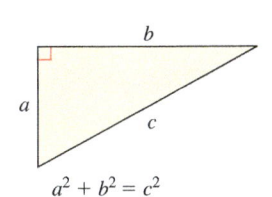

FIGURE 1.3 Pythagorean theorem

54 Chapter 1 Graphs and Functions

Suppose we want to compute the distance between the two points $P(x_1, y_1)$ and $Q(x_2, y_2)$. We draw a horizontal line through the point Q and a vertical line through the point P to form the right triangle PQS, as shown in Figure 1.4.

The length of the horizontal side of the triangle is $|x_2 - x_1|$, and the length of the vertical side is $|y_2 - y_1|$. The distance between P and Q, denoted $d(P, Q)$, is the length of the hypotenuse.

$$[d(P, Q)]^2 = |x_2 - x_1|^2 + |y_2 - y_1|^2 \qquad \text{Pythagorean Theorem}$$
$$d(P, Q) = \sqrt{|x_2 - x_1|^2 + |y_2 - y_1|^2} \qquad \text{Take the square root of both sides.}$$
$$d(P, Q) = \sqrt{(x_2 - x_1)^2 + (y_2 - y_1)^2} \qquad |a|^2 = a^2$$

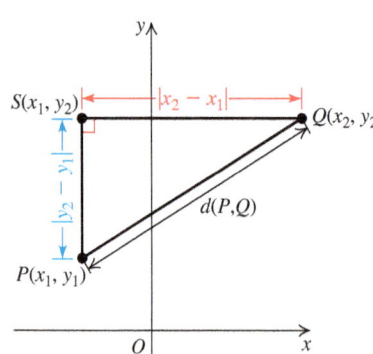

FIGURE 1.4 Visualizing the distance formula

THE DISTANCE FORMULA IN THE COORDINATE PLANE

Let $P = (x_1, y_1)$ and $Q = (x_2, y_2)$ be any two points in the coordinate plane. Then the distance between P and Q, denoted $d(P, Q)$, is given by the **distance formula**

$$d(P, Q) = \sqrt{(x_2 - x_1)^2 + (y_2 - y_1)^2}.$$

EXAMPLE 2   Finding the Distance Between Two Points

Find the distance between the points $P = (-2, 5)$ and $Q = (3, -4)$.

Solution

Let $(x_1, y_1) = (-2, 5)$ and $(x_2, y_2) = (3, -4)$. Then

$$x_1 = -2, y_1 = 5, x_2 = 3, \text{ and } y_2 = -4.$$

$$\begin{aligned} d(P, Q) &= \sqrt{(x_2 - x_1)^2 + (y_2 - y_1)^2} && \text{Distance formula} \\ &= \sqrt{[3 - (-2)]^2 + (-4 - 5)^2} && \text{Substitute the values for } x_1, x_2, y_1, y_2. \\ &= \sqrt{5^2 + (-9)^2} \approx 10.3 && \text{Use a calculator.} \end{aligned}$$

Practice Problem 2 Find the distance between the points $(-5, 2)$ and $(-4, 1)$.

---

3 Find the midpoint of a line segment.

The Midpoint Formula

Recall that M is the **midpoint** of a line segment $\overline{PQ}$ if M is a point on $\overline{PQ}$ and $d(P, M) = d(M, Q)$. We provide the **midpoint formula** in the xy-plane.

THE MIDPOINT FORMULA

The coordinates of the midpoint $M = (x, y)$ on the line segment joining $P = (x_1, y_1)$ and $Q = (x_2, y_2)$ are given by (see Figure 1.5)

$$M = (x, y) = \left(\frac{x_1 + x_2}{2}, \frac{y_1 + y_2}{2} \right).$$

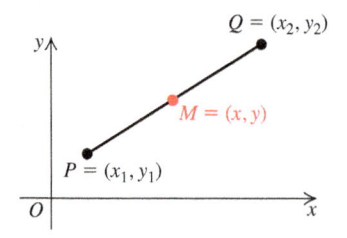

FIGURE 1.5 Midpoint of a segment

SIDE NOTE

The coordinates of the midpoint of a line segment are found by taking the average of the x-coordinates and the average of the y-coordinates of the endpoints.

The midpoint formula can be proved by showing that $d(P, M) = d(Q, M)$.

EXAMPLE 3   Finding the Midpoint of a Line Segment

Find the midpoint of the line segment joining the points $P(-3, 6)$ and $Q(1, 4)$.

Solution

Let $(x_1, y_1) = (-3, 6)$ and $(x_2, y_2) = (1, 4)$. Then

$x_1 = -3, y_1 = 6, x_2 = 1,$ and $y_2 = 4$.

$$\text{Midpoint} = \left(\frac{x_1 + x_2}{2}, \frac{y_1 + y_2}{2} \right) = \left(\frac{-3 + 1}{2}, \frac{6 + 4}{2} \right) \quad \text{Substitute values for } x_1, x_2, y_1, y_2.$$

$$= (-1, 5) \quad \text{Simplify.}$$

The midpoint is $M = (-1, 5)$.

Practice Problem 3 Find the midpoint of the line segment whose endpoints are $(5, -2)$ and $(6, -1)$.

4 Sketch a graph by plotting points.

Graph of an Equation

An ordered pair (a, b) is said to **satisfy** an equation with variables x and y if, when a is substituted for x and b is substituted for y in the equation, the resulting statement is true. For example, the ordered pair $(2, 5)$ satisfies the equation $y = 2x + 1$ because replacing x with 2 and y with 5 yields $5 = 2(2) + 1$, which is a true statement. The ordered pair $(5, -2)$ does not satisfy this equation because replacing x with 5 and y with -2 yields $-2 = 2(5) + 1$, which is false. An ordered pair that satisfies an equation is called a **solution** of the equation.

In an equation involving x and y, if the value of y can be found given the value of x, then we say that y is the **dependent variable** and x is the **independent variable**. In the equation $y = 2x + 1$, for any real number x, there is a corresponding value of y. Hence, we have infinitely many solutions of the equation $y = 2x + 1$. When these solutions are graphed or plotted as points in the coordinate plane, they constitute the *graph of the equation*. Therefore, the graph of an equation is a geometric picture of its solution set.

Graph of an Equation

The **graph of an equation** in two variables, such as x and y, is the graph of all ordered pairs (a, b) in the coordinate plane that satisfy the equation.

EXAMPLE 4 Sketching a Graph by Plotting Points

Sketch the graph of $y = x^2 - 3$.

Solution

The equation has infinitely many solutions. To find a few, we choose some values of x between -3 and 3. Then we find the corresponding values of y as shown in Table 1.1.

TABLE 1.1

x	$y = x^2 - 3$	(x, y)
-3	$y = (-3)^2 - 3 = 9 - 3 = 6$	$(-3, 6)$
-2	$y = (-2)^2 - 3 = 4 - 3 = 1$	$(-2, 1)$
-1	$y = (-1)^2 - 3 = 1 - 3 = -2$	$(-1, -2)$
0	$y = 0^2 - 3 = 0 - 3 = -3$	$(0, -3)$
1	$y = 1^2 - 3 = 1 - 3 = -2$	$(1, -2)$
2	$y = 2^2 - 3 = 4 - 3 = 1$	$(2, 1)$
3	$y = 3^2 - 3 = 9 - 3 = 6$	$(3, 6)$

56 Chapter 1 Graphs and Functions

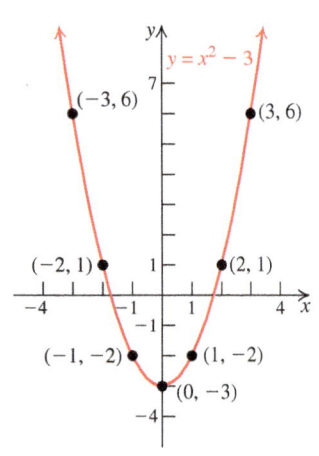

FIGURE 1.6 Graph of an equation

TECHNOLOGY CONNECTION

The calculator graph of $y = x^2 - 3$ reinforces the result of Example 4.

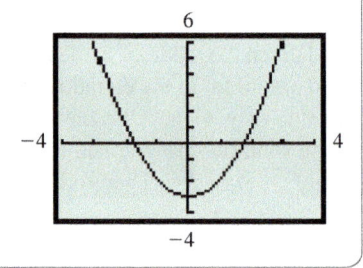

We plot the solutions (x, y) and join them with a smooth curve to sketch the graph of $y = x^2 - 3$ shown in Figure 1.6.

Practice Problem 4 Sketch the graph of $y = -x^2 + 1$.

The bowl-shaped curve sketched in Figure 1.6 is called a *parabola*. You can find parabolic shapes in everyday settings, such as the path of a thrown ball or in the reflector behind a car's headlight.

Example 4 suggests how to sketch the graph of any equation by plotting points. We summarize the steps for this technique.

SKETCHING A GRAPH BY PLOTTING POINTS

Step 1 Make a representative table of solutions of the equation.

Step 2 Plot the solutions as ordered pairs in the Cartesian coordinate plane.

Step 3 Connect the solutions in Step 2 with a smooth curve.

Comment This technique has obvious pitfalls. For instance, many different curves pass through the same four points in Figure 1.7. Assume that these points are solutions of a given equation. We cannot guarantee that any curve through these points is the actual graph of the equation. However, in general, the more solutions plotted, the more accurate the graph.

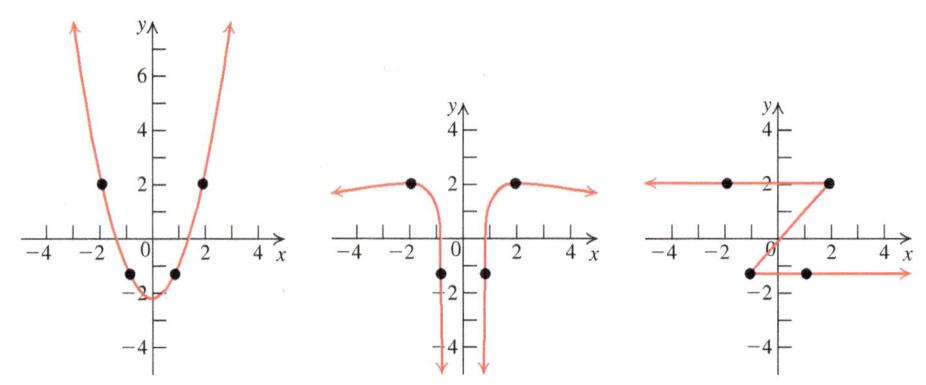

FIGURE 1.7 Several graphs through the same four points

5 Find the intercepts of a graph.

Intercepts

Let's examine the points where a graph intersects (crosses or touches) the coordinate axes. Because all points on the x-axis have a y-coordinate of 0, any point where a graph intersects the x-axis has the form $(a, 0)$. See Figure 1.8. The number a is called an **x-intercept** of the graph. Similarly, any point where a graph intersects the y-axis has the form $(0, b)$, and the number b is called a **y-intercept** of the graph. Together, the x-intercepts and the y-intercepts are referred to as the **intercepts** of the graph.

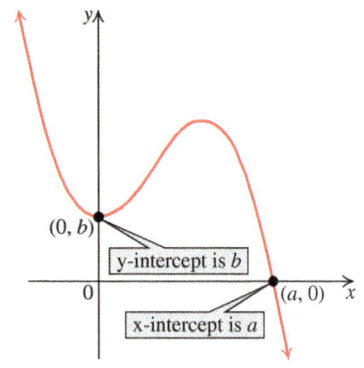

FIGURE 1.8 Intercepts of a graph

FINDING THE INTERCEPTS OF THE GRAPH OF AN EQUATION

Step 1 To find the x-intercepts of the graph of an equation, set $y = 0$ in the equation and solve for x.

Step 2 To find the y-intercepts of the graph of an equation, set $x = 0$ in the equation and solve for y.

Section 1.1 ■ Graphs of Equations   57

WARNING Do not try to calculate the *x*-intercept by setting $x = 0$. An *x*-intercept is the *x*-coordinate of a point where the graph touches or crosses the *x*-axis, so the *y*-coordinate must be 0.

EXAMPLE 5 Finding Intercepts

Find the *x*- and *y*-intercepts of the graph of the equation $y = x^2 - x - 2$.

Solution

Step 1 Set $y = 0$ in the equation and solve for *x*.

$$0 = x^2 - x - 2 \qquad \text{Set } y = 0.$$
$$0 = (x + 1)(x - 2) \qquad \text{Factor.}$$
$$x + 1 = 0 \quad \text{or} \quad x - 2 = 0 \qquad \text{Zero-product property}$$
$$x = -1 \quad \text{or} \quad x = 2 \qquad \text{Solve each equation for } x.$$

The *x*-intercepts are -1 and 2.

Step 2 Set $x = 0$ in the equation and solve for *y*.

$$y = 0^2 - 0 - 2 \qquad \text{Set } x = 0.$$
$$y = -2 \qquad \text{Solve for } y.$$

The *y*-intercept is -2.

The graph of the equation $y = x^2 - x - 2$ is shown in Figure 1.9.

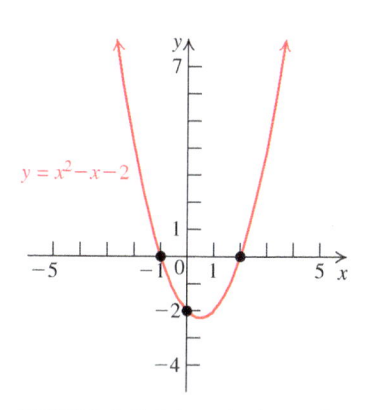

FIGURE 1.9 Intercepts of a graph

Practice Problem 5 Find the intercepts of the graph of $y = 2x^2 + 3x - 2$.

6 Find the symmetries in a graph.

Symmetry

The concept of **symmetry** helps us sketch graphs of equations. A graph has symmetry if one portion of the graph is a *mirror image* of another portion. As shown in Figure 1.10(b), if a line ℓ is an **axis of symmetry**, or **line of symmetry**, we can construct the mirror image of any point *P* not on ℓ by first drawing the perpendicular line segment from *P* to ℓ. Then we extend this segment an equal distance on the other side to a point P' so that the line ℓ perpendicularly bisects the line segment $\overline{PP'}$. In Figure 1.10(b), we say that the point P' is the *symmetric image* of the point *P* with respect to the line ℓ.

Two points *M* and M' are **symmetric about a point** *Q* if *Q* is the midpoint of the line segment $\overline{MM'}$. Figure 1.10(a) illustrates the symmetry about the origin *O*. Symmetry lets us use information about part of the graph to draw the remainder of the graph.

(a) Symmetric points about the origin

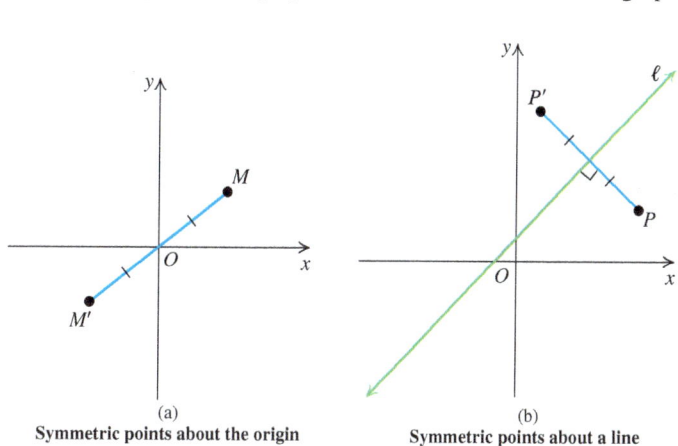

(b) Symmetric points about a line

FIGURE 1.10

58 Chapter 1 Graphs and Functions

The following three types of symmetries are frequently used in graphing an equation.

SYMMETRIES

1. A graph is **symmetric with respect to (or about) the y-axis** if for every point (x, y) on the graph, the point $(-x, y)$ is also on the graph. See Figure 1.11(a).
2. A graph is **symmetric with respect to (or about) the x-axis** if for every point (x, y) on the graph, the point $(x, -y)$ is also on the graph. See Figure 1.11(b).
3. A graph is **symmetric with respect to (or about) the origin** if for every point (x, y) on the graph, the point $(-x, -y)$ is also on the graph. See Figure 1.11(c).

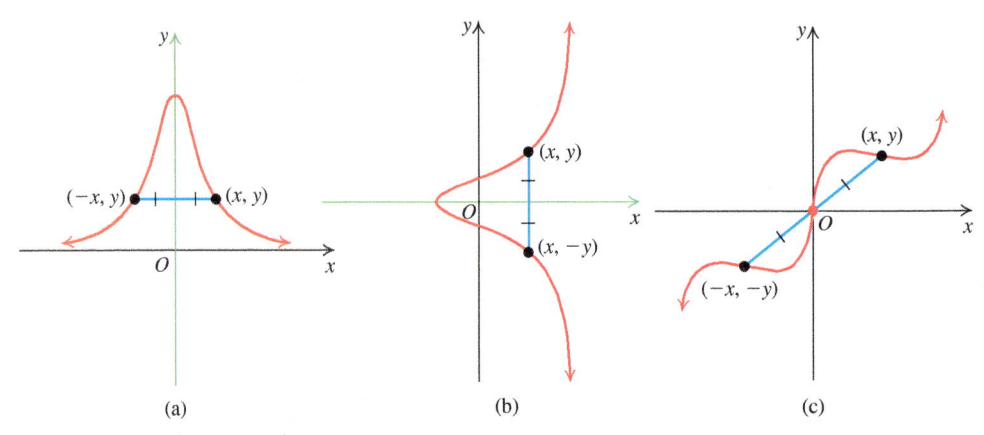

FIGURE 1.11 Three types of symmetry

TESTS FOR SYMMETRY

1. The graph of an equation is symmetric about the y-axis if replacing x with $-x$ results in an equivalent equation.
2. The graph of an equation is symmetric about the x-axis if replacing y with $-y$ results in an equivalent equation.
3. The graph of an equation is symmetric about the origin if replacing x with $-x$ and y with $-y$ results in an equivalent equation.

EXAMPLE 6 **Checking for Symmetry**

Determine whether the graph of the equation $y = \dfrac{1}{x^2 + 5}$ is symmetric with respect to the y-axis.

Solution

Replace x with $-x$ to see if $(-x, y)$ also satisfies the equation.

$$y = \frac{1}{x^2 + 5} \qquad \text{Original equation}$$

$$y = \frac{1}{(-x)^2 + 5} \qquad \text{Replace } x \text{ with } -x.$$

$$y = \frac{1}{x^2 + 5} \qquad \text{Simplify: } (-x)^2 = x^2.$$

Section 1.1 ■ Graphs of Equations 59

SIDE NOTE

Note that if *only* even powers of x appear in an equation, then the graph is symmetric with respect to the y-axis because for any integer n, $(-x)^{2n} = x^{2n}$.

Because replacing x with $-x$ gives us the original equation, the graph of $y = \dfrac{1}{x^2 + 5}$ is symmetric with respect to the y-axis.

Practice Problem 6 Check the graph of $x^2 - y^2 = 1$ for symmetry with respect to the y-axis.

PROCEDURE IN ACTION

EXAMPLE 7 Sketching a Graph Using Symmetry

OBJECTIVE

Use symmetry to sketch the graph of an equation.

EXAMPLE

Use symmetry to sketch the graph of $y = 4x - x^3$.

Step 1 Test for all three symmetries.
About the x-axis: Replace y with $-y$.
About the y-axis: Replace x with $-x$.
About the origin: Replace x with $-x$ and y with $-y$.

1.

x-axis	y-axis	origin
Replace y with $-y$	Replace x with $-x$	Replace x with $-x$ and y with $-y$
$-y = 4x - x^3$ $y = -4x + x^3$	$y = 4(-x) - (-x)^3$ $y = -4x + x^3$	$-y = 4(-x) - (-x)^3$ $-y = -4x + x^3$ $y = 4x - x^3$
No	No	Yes

Step 2 Make a table of values using any symmetries found in Step 1.

2. Origin symmetry: If (x, y) is on the graph, so is $(-x, -y)$. Use only positive x values in the table.

x	0	0.5	1	1.5	2	2.5
$y = 4x - x^3$	0	1.875	3	2.625	0	−5.625

Step 3 Plot the points from the table and draw a smooth curve through them.

3.

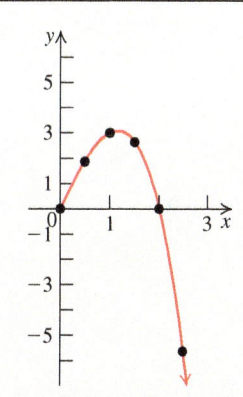

Graph of $y = 4x - x^3, x \geq 0$

Step 4 Extend the portion of the graph found in Step 3 using symmetries.

4.

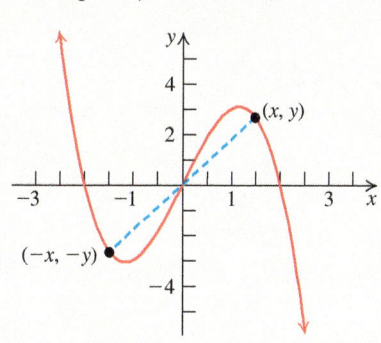

Graph of $y = 4x - x^3$

Practice Problem 7 Use symmetry to sketch the graph of $y = x^4 - 4x^2$.

Circles

Sometimes a curve that is described geometrically can also be described by an algebraic equation. We illustrate this situation in the case of a circle.

7 Find the equation of a circle.

Circle

A **circle** is a set of points in a Cartesian coordinate plane that are at a fixed distance r from a specified point (h, k). The fixed distance r is called the **radius** of the circle, and the specified point (h, k) is called the **center** of the circle.

Standard Form A point $P(x, y)$ is on the circle if and only if its distance from the center $C(h, k)$ is r. Using the notation for the distance between the points P and C, we have

$$d(P, C) = r$$
$$\sqrt{(x - h)^2 + (y - k)^2} = r \quad \text{Distance formula}$$
$$(x - h)^2 + (y - k)^2 = r^2 \quad \text{Square both sides.}$$

The equation $(x - h)^2 + (y - k)^2 = r^2$ is an equation of a circle with radius r and center (h, k). A point (x, y) is on the circle of radius r and center $C(h, k)$ if and only if it satisfies this equation. Figure 1.12 is the graph of a circle with center $C(h, k)$ and radius r.

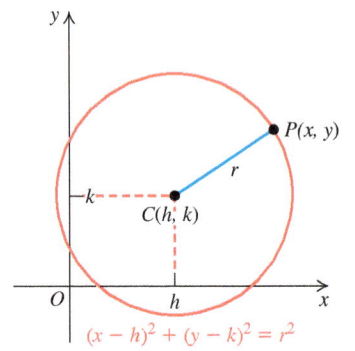

FIGURE 1.12

The Standard Form for the Equation of a Circle

The equation of a circle with center (h, k) and radius r is

$$(1) \qquad (x - h)^2 + (y - k)^2 = r^2$$

and is called the **standard form** of an equation of a circle.

RECALL

The **diameter of a circle** is a line segment that passes through the center of the circle and whose endpoints are on the circle.

EXAMPLE 8 Finding the Standard Form of the Equation of a Circle

Find the standard form of the equation of a circle.

a. Center $(-3, 4)$ and radius 7
b. Diameter with endpoints $(-1, -6)$ and $(7, 2)$

Solution

a.
$$(x - h)^2 + (y - k)^2 = r^2 \quad \text{Standard form}$$
$$[x - (-3)]^2 + (y - 4)^2 = 7^2 \quad \text{Replace } h \text{ with } -3, k \text{ with } 4, \text{ and } r \text{ with } 7.$$
$$(x + 3)^2 + (y - 4)^2 = 49 \quad -(-3) = 3$$

b. Find the length d of the diameter:

$$d = \sqrt{(7 - (-1))^2 + (2 - (-6))^2} \quad \text{Distance formula}$$
$$= \sqrt{64 + 64} = 8\sqrt{2} \quad \text{Simplify.}$$

$$\text{Radius } r = \frac{1}{2}d = \frac{1}{2}(8\sqrt{2}) = 4\sqrt{2}$$

$$\text{Center } C = \left(\frac{-1 + 7}{2}, \frac{-6 + 2}{2}\right) = (3, -2) \quad \text{Midpoint formula}$$

Equation of the circle in standard form with center $(3, -2)$ and radius $4\sqrt{2}$ is

$$(x - 3)^2 + (y + 2)^2 = 32.$$

Section 1.1 ■ Graphs of Equations 61

Practice Problem 8 Find the standard form of the equation of the circle having diameter with endpoints $(-3, -14)$ and $(9, 2)$.

If an equation in two variables can be written in standard form given in equation (1), then its graph is a circle with center (h, k) and radius r.

EXAMPLE 9   Graphing a Circle

Specify the center and radius and graph each circle.

a. $x^2 + y^2 = 1$
b. $(x + 2)^2 + (y - 3)^2 = 25$

Solution

a. The equation $x^2 + y^2 = 1$ can be rewritten as

$$(x - 0)^2 + (y - 0)^2 = 1^2.$$

Comparing this equation with equation (1), we conclude that the given equation is an equation of a circle with center $(0, 0)$ and radius 1. The graph is shown in Figure 1.13. This circle is called the **unit circle**.

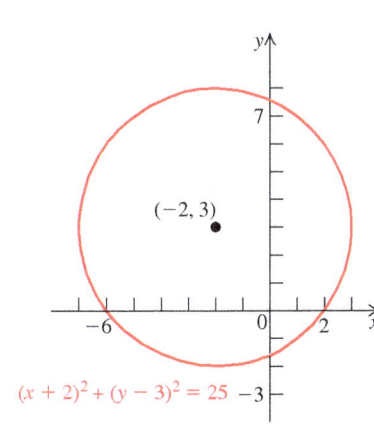

FIGURE 1.13 The unit circle

b. Rewriting the equation $(x + 2)^2 + (y - 3)^2 = 25$ as

$$[x - (-2)]^2 + (y - 3)^2 = 5^2, \quad x + 2 = x - (-2)$$

we see that the graph of this equation is a circle with center $(-2, 3)$ and radius 5. The graph is shown in Figure 1.14.

Practice Problem 9 Graph the equation $(x - 2)^2 + (y + 1)^2 = 36$.

FIGURE 1.14 Circle with radius 5 and center $(-2, 3)$

General Form If we expand the squared expressions in the standard equation of a circle,

(1) $$(x - h)^2 + (y - k)^2 = r^2,$$

and then simplify, we obtain an equation of the form

(2) $$x^2 + y^2 + ax + by + c = 0.$$

Equation (2) is called the *general form* of the equation of a circle. The graph of the equation $Ax^2 + By^2 + Cx + Dy + E = 0$ is also a circle if $A \neq 0$ and $A = B$. You can convert this equation to the general form by dividing both sides by A.

General Form of the Equation of a Circle

The **general form** of the equation of a circle is

$$x^2 + y^2 + ax + by + c = 0.$$

On the other hand, if we are given an equation in general form, we can convert it to standard form by completing the squares on the x- and y-terms. This gives

(3) $$(x - h)^2 + (y - k)^2 = d.$$

If $d > 0$, the graph of equation (3) is a circle with center (h, k) and radius $\sqrt{d}$. If $d = 0$, the graph of equation (3) is the point (h, k). If $d < 0$, there is no graph.

EXAMPLE 10 Converting the General Form to Standard Form

Find the center and radius of the circle with equation
$$x^2 + y^2 - 6x + 8y + 10 = 0.$$

Solution

Complete the squares on both the *x*-terms and *y*-terms to get standard form.

$x^2 + y^2 - 6x + 8y + 10 = 0$	Original equation
$(x^2 - 6x) + (y^2 + 8y) = -10$	Group the *x*-terms and *y*-terms.
$(x^2 - 6x + 9) + (y^2 + 8y + 16) = -10 + 9 + 16$	Complete the squares by adding 9 and 16 to both sides.
$(x - 3)^2 + (y + 4)^2 = 15$	Factor and simplify.
$(x - 3)^2 + [y - (-4)]^2 = (\sqrt{15})^2$	Rewrite.

The last equation tells us that we have $h = 3$, $k = -4$, and $r = \sqrt{15}$. Therefore, the circle has center $(3, -4)$ and radius $\sqrt{15} \approx 3.9$.

Practice Problem 10 Find the center and radius of the circle with equation $x^2 + y^2 + 4x - 6y - 12 = 0.$

RECALL

To make $x^2 + bx$ a perfect square, add $\left(\dfrac{1}{2}b\right)^2$ to get

$x^2 + bx + \left(\dfrac{1}{2}b\right)^2 = \left(x + \dfrac{b}{2}\right)^2.$

This completes the square.

SECTION 1.1 Exercises

Basic Concepts and Skills

1. Plot and label each of the given points in a Cartesian coordinate plane and state the quadrant, if any, in which each point is located. $(2, 2), (3, -1), (-1, 0), (-2, -5), (0, 0), (-7, 4), (0, 3), (-4, 2)$

2. **a.** Write the coordinates of any five points on the *x*-axis. What do these points have in common?
   b. Graph the points $(-2, 1), (0, 1), (0.5, 1), (1, 1)$, and $(2, 1)$. Describe the set of all points of the form $(x, 1)$, where x is a real number.

3. **a.** If the *x*-coordinate of a point is 0, where does that point lie?
   b. Graph the points $(-1, 1), (-1, 1.5), (-1, 2), (-1, 3)$, and $(-1, 4)$. Describe the set of all points of the form $(-1, y)$, where y is a real number.

4. Let $P(x, y)$ be a point in a coordinate plane. In which quadrant does P lie
   a. if both x and y are negative?
   b. if both x and y are positive?
   c. if x is positive and y is negative?
   d. if x is negative and y is positive?

In Exercises 5–8, find (a) the distance between P and Q and (b) the coordinates of the midpoint of the line segment PQ.

5. $P(3, 5), Q(-2, 5)$
6. $P(-1, -5), Q(2, -3)$
7. $P(x, y), Q(-2, 3)$
8. $P(t, k), Q(k, t)$

9. Find the point that is $\dfrac{1}{4}$ th the distance from $(-3, 6)$ to $(7, 10)$.

10. Find the point that is $\dfrac{1}{8}$ th the distance from $(9, 12)$ to $(-5, 4)$.

In Exercises 11–14, determine whether the given points are collinear. Points are *collinear* if they can be labeled P, Q, and R so that $d(P, Q) + d(Q, R) = d(P, R)$. If the points are noncollinear, find the perimeter of the triangle PQR.

11. $(0, 0), (1, 2), (-1, -2)$
12. $(3, 4), (0, 0), (-3, -4)$
13. $(9, 6), (0, -3), (3, 1)$
14. $(-2, 3), (3, 1), (2, -1)$

In Exercises 15–20, identify the triangle PQR as an *isosceles* (two sides of equal length), *equilateral* (three sides of equal length), or *scalene* (three sides of different lengths) triangle.

15. $P(-5, 5), Q(-1, 4), R(-4, 1)$
16. $P(3, 2), Q(6, 6), R(-1, 5)$
17. $P(-4, 8), Q(0, 7), R(-3, 5)$
18. $P(6, 6), Q(-1, -1), R(-5, 3)$
19. $P(1, -1), Q(-1, 1), R(-\sqrt{3}, -\sqrt{3})$
20. $P(-.5, -1), Q(-1.5, 1), R(\sqrt{3} - 1, \sqrt{3}/2)$

In Exercises 21–30, graph each equation by plotting points. Let $x = -3, -2, -1, 0, 1, 2,$ and 3 where applicable.

21. $y = x + 1$
22. $y = 2x$
23. $y = |x|$
24. $y = |x + 1|$
25. $y = 4 - x^2$
26. $y = x^2 - 4$
27. $y = \sqrt{9 - x^2}$
28. $y = -\sqrt{9 - x^2}$
29. $y = x^3$
30. $y = -x^3$

31. Write the x- and y-intercepts of the graph.

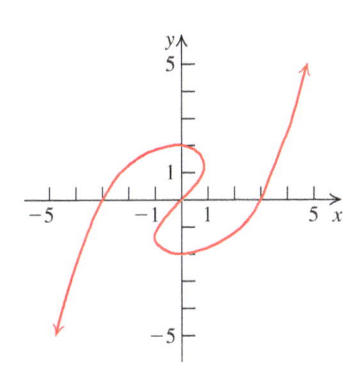

32. Write the x- and y-intercepts of the graph.

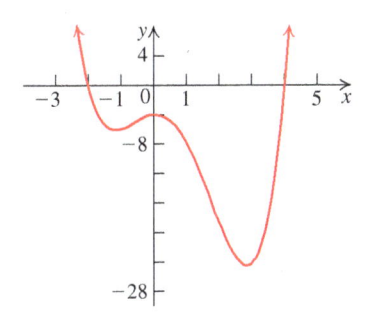

In Exercises 33–40, find the x- and y-intercepts of the graph of each equation.

33. $3x + 4y = 12$
34. $\dfrac{x}{2} - \dfrac{y}{3} = 1$
35. $y = x^2 - 6x + 8$
36. $x = y^2 - 5y + 6$
37. $x^2 + y^2 = 4$
38. $y = \sqrt{9 - x^2}$
39. $y = \sqrt{x^2 - 1}$
40. $xy = 1$

In Exercises 41–48, test each equation for symmetry with respect to the x-axis, the y-axis, and the origin.

41. $y = x^2 + 1$
42. $x = y^2 + 1$
43. $y = x^3 + x$
44. $y = 2x^3 - x$
45. $y = 5x^4 + 2x^2$
46. $y = -3x^6 + 2x^4 + x^2$
47. $y = -3x^5 + 2x^3$
48. $y = 2x^2 - |x|$

In Exercises 49 and 50, specify the center and the radius of each circle.

49. $(x + 1)^2 + (y - 3)^2 = 16$
50. $(x + 2)^2 + (y + 3)^2 = 11$

In Exercises 51–56, find the standard form of the equation of a circle that satisfies the given conditions.

51. Center $(3, 1)$; radius 3
52. Center $(-1, 2)$; radius 2
53. Center $(-3, -2)$ and touches the y-axis
54. Center $(3, -4)$ and passes through the point $(-1, 5)$
55. Diameter with endpoints $(2, -3)$ and $(8, 5)$
56. Diameter with endpoints $(7, 4)$ and $(-3, 6)$

In Exercises 57–60, a. find the center and radius of each circle, and b. find the x- and y-intercepts of the graph of each circle.

57. $x^2 + y^2 - 2x - 2y - 4 = 0$
58. $x^2 + y^2 - 4x - 2y - 15 = 0$
59. $2x^2 + 2y^2 + 4y = 0$
60. $3x^2 + 3y^2 + 6x = 0$
61. $x^2 + y^2 - 2x + 4y + 5 = 0$
62. $x^2 + y^2 + 4x - 6y + 17 = 0$

Applying the Concepts

In Exercises 63–66, a graph is described *geometrically* as the path of a point $P(x, y)$ on the graph. Find an equation for the graph described.

63. **Geometry.** $P(x, y)$ is on the graph if and only if the distance from $P(x, y)$ to the x-axis is equal to its distance from the y-axis.

64. **Geometry.** $P(x, y)$ is the same distance from the two points $(1, 2)$ and $(3, -4)$.

65. **Geometry.** $P(x, y)$ is the same distance from the point $(2, 0)$ and the y-axis.

66. **Geometry.** $P(x, y)$ is the same distance from the point $(0, 4)$ and the x-axis.

67. **Corporate profits.** The equation $P = -0.5t^2 - 3t + 8$ describes the monthly profits (in millions of dollars) of ABCD Corp. for 2012, with $t = 0$ representing July 2012.
    a. How much profit did the corporation make in March 2012?
    b. How much profit did the corporation make in October 2012?
    c. Sketch the graph of the equation.
    d. Find the t-intercepts. What do they represent?
    e. Find the P-intercept. What does it represent?

68. **Female students in colleges.** The equation
    $$P = -0.002t^2 + 0.093t + 8.18$$
    models the approximate number (in millions) of female college students in the United States for the academic years 1995–2001, with $t = 0$ representing 1995.
    a. Sketch the graph of the equation.
    b. Find the positive t-intercept. What does it represent?
    c. Find the P-intercept. What does it represent?
    (*Source: Statistical Abstracts of the United States.*)

69. **Motion.** An object is thrown up from the top of a building that is 320 feet high. The equation $y = -16t^2 + 128t + 320$ gives the object's height (in feet) above the ground at any time t (in seconds) after the object is thrown.
   a. What is the height of the object after 0, 1, 2, 3, 4, 5, and 6 seconds?
   b. Sketch the graph of the equation $y = -16t^2 + 128t + 320$.
   c. What part of the graph represents the physical aspects of the problem?
   d. What are the intercepts of this graph? What do they mean?

70. **Diving for treasure.** A treasure-hunting team of divers is placed in a computer-controlled diving cage. The equation $d = \frac{40}{3}t - \frac{2}{9}t^2$ describes the depth d (in feet) that the cage will descend in t minutes.
   a. Sketch the graph of the equation $d = \frac{40}{3}t - \frac{2}{9}t^2$.
   b. What part of the graph represents the physical aspects of the problem?
   c. What is the total time of the entire diving experiment?

Critical Thinking / Discussion / Writing

71. Sketch the graph of $y^2 = 2x$ and explain how this graph is related to the graphs of $y = \sqrt{2x}$ and $y = -\sqrt{2x}$.

72. Show that a graph that is symmetric with respect to the x-axis and y-axis must be symmetric with respect to the origin. Give an example to show that the converse is not true.

73. **True or False.** To find the x-intercepts of the graph of an equation, set $x = 0$ and solve for y. Explain your answer.

74. Write an equation whose graph has x-intercepts -2 and 3 and y-intercept 6. (Many answers are possible.)

75. a. Show that a circle with diameter having endpoints $A(0, 1)$ and $B(6, 8)$ intersects the x-axis at the roots of the equation $x^2 - 6x + 8 = 0$.
    b. Show that a circle with diameter having endpoints $A(0, 1)$ and $B(a, b)$ intersects the x-axis at the roots of the equation $x^2 - ax + b = 0$.
    c. Use graph paper, ruler, and compass to approximate the roots of the equation $x^2 - 3x + 1 = 0$.

Maintaining Skills

In Exercises 76–82, perform the indicated operations.

76. $\dfrac{3}{8} + \dfrac{5}{12}$

77. $\dfrac{1}{4} - \dfrac{1}{6}$

78. $\dfrac{9}{7} \cdot \dfrac{14}{27}$

79. $\dfrac{\frac{7}{8}}{\frac{21}{16}}$

80. $(x + 1)(x + 2)$

81. $(x + 2)(2x + 3)$

82. $(2x + 3)(2x - 3)$

SECTION 1.2

Lines

BEFORE STARTING THIS SECTION, REVIEW

1. Graph of an equation (Section 1.1, page 55)

OBJECTIVES

1. Find the slope of a line.
2. Write the point–slope form of the equation of a line.
3. Write the slope–intercept form of the equation of a line.
4. Recognize the equations of horizontal and vertical lines.
5. Recognize the general form of the equation of a line.
6. Find equations of parallel and perpendicular lines.

◆ A Texan's Tall Tale

Gunslinger Wild Bill Longley's *first* burial took place on October 11, 1878, after he was hanged before a crowd of thousands in Giddings, Texas.

Rumors persisted that Longley's hanging had been a hoax and that he had somehow faked his death and escaped execution. In 2001, Longley's descendants had his grave opened to determine whether the remains matched Wild Bill's description: a tall white male, age 27. Both the skeleton and some personal effects suggested that this was indeed Wild Bill. Modern science lent a hand too: the DNA of Wild Bill's sister's descendant Helen Chapman was a perfect match.

Now the notorious gunman could be buried back in the Giddings cemetery—for the *second* time. How did the scientists conclude from the skeletal remains that Wild Bill was approximately 6 feet tall? See Example 8.

1 Find the slope of a line.

Slope of a Line

In this section, we study various forms of first-degree equations in two variables. Because the graphs of these equations are straight lines, they are called **linear equations**. Just as we measure weight or temperature by a number, we measure the "steepness" of a line by a number called its **slope**.

Consider two points $P(x_1, y_1)$ and $Q(x_2, y_2)$ on a line, as shown in Figure 1.15. We say that the *rise* is the change in y-coordinates between the points (x_1, y_1) and (x_2, y_2) and that the *run* is the corresponding change in the x-coordinates. A positive run indicates change to the right, as shown in Figure 1.15; a negative run indicates change to the left. Similarly, a positive rise indicates upward change; a negative rise indicates downward change.

66   Chapter 1   Graphs and Functions

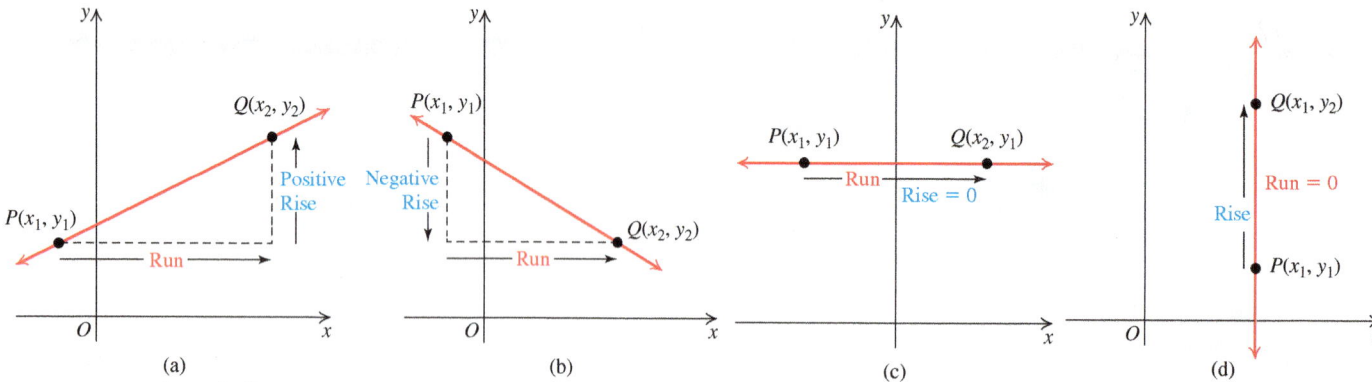

(a) (b) (c) (d)

FIGURE 1.15 Slope of a line

Slope of a Line

The **slope** of a nonvertical line that passes through the points $P(x_1, y_1)$ and $Q(x_2, y_2)$ is denoted by m and is defined by

$$m = \frac{\text{rise}}{\text{run}} = \frac{y_2 - y_1}{x_2 - x_1}, \quad x_1 \neq x_2.$$

The slope of a vertical line is undefined.

Because $m = \dfrac{\text{rise}}{\text{run}} = \dfrac{\text{change in } y\text{-coordinates}}{\text{change in } x\text{-coordinates}}$, if the change in x is one unit to the right, or positive 1, then the change in y equals the slope of the line. The slope of a line is therefore the **change in y per unit change in x**. In other words, the slope of a line measures the rate of change of y with respect to x. In calculus, the ratio $\dfrac{\text{change in } y}{\text{change in } x}$ is written as $\dfrac{\Delta y}{\Delta x}$. The symbol Δ is the Greek letter *delta* that represents a change in the variable. So the slope $m = \dfrac{\Delta y}{\Delta x}$ (read m equals "change in y over change in x"). Roofs, staircases, graded landscapes, and mountainous roads all have slopes. For example, if the pitch (slope) of a section of a roof is 0.4, then for every horizontal distance of 10 feet in that section, the roof ascends 4 feet.

EXAMPLE 1   Finding and Interpreting the Slope of a Line

Sketch the graph of the line that passes through the points $P(1, -1)$ and $Q(3, 3)$. Find and interpret the slope of the line.

Solution

The graph of the line is sketched in Figure 1.16. The slope m of this line is given by

$$m = \frac{\text{rise}}{\text{run}} = \frac{\text{change in } y\text{-coordinates}}{\text{change in } x\text{-coordinates}}$$

$$= \frac{(y\text{-coordinate of } Q) - (y\text{-coordinate of } P)}{(x\text{-coordinate of } Q) - (x\text{-coordinate of } P)}$$

$$= \frac{(3) - (-1)}{(3) - (1)} = \frac{3 + 1}{3 - 1} = \frac{4}{2} = 2.$$

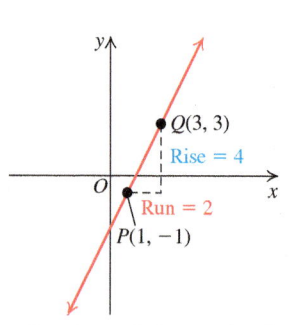

FIGURE 1.16 Interpreting slope

Interpretation

A slope of 2 means that the value of y increases two units for every one unit increase in the value of x.

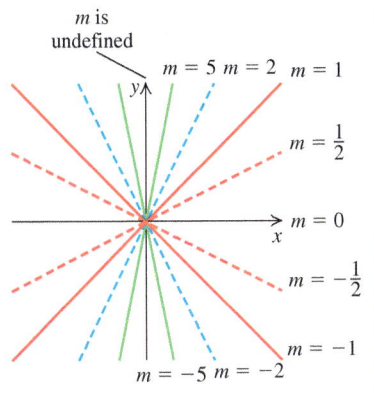

FIGURE 1.17 Slopes of lines

Practice Problem 1 Find and interpret the slope of the line containing the points $(-7, 5)$ and $(6, -3)$.

Figure 1.17 shows several lines passing through the origin. The slope of each line is labeled.

MAIN FACTS ABOUT SLOPES OF LINES

1. Scanning graphs from left to right, lines with positive slopes rise and lines with negative slopes fall.
2. The greater the absolute value of the slope, the steeper the line.
3. The slope of a vertical line is undefined.
4. The slope of a horizontal line is zero.

WARNING Be careful when using the formula for finding the slope of a line joining the points (x_1, y_1) and (x_2, y_2). Be sure to subtract coordinates in the same order.
You can use either

$$m = \frac{y_2 - y_1}{x_2 - x_1} \quad \text{or} \quad m = \frac{y_1 - y_2}{x_1 - x_2},$$

but it is incorrect to use

$$m = \frac{y_1 - y_2}{x_2 - x_1} \quad \text{or} \quad m = \frac{y_2 - y_1}{x_1 - x_2}.$$

2 Write the point–slope form of the equation of a line.

Point–Slope Form

We now find the equation of a line ℓ that passes through the point $A(x_1, y_1)$ and has slope m. Let $P(x, y)$, with $x \neq x_1$, be any point in the plane. Then $P(x, y)$ is on the line ℓ if and only if the slope of the line passing through $P(x, y)$ and $A(x_1, y_1)$ is m. This is true if and only if

$$\frac{y - y_1}{x - x_1} = m.$$

See Figure 1.18.
Multiplying both sides by $x - x_1$ gives $y - y_1 = m(x - x_1)$, which is also satisfied when $x = x_1$ and $y = y_1$.

The Point–Slope Form of the Equation of a Line

If a line has slope m and passes through the point (x_1, y_1), then the **point-slope form** of an equation of the line is

$$y - y_1 = m(x - x_1).$$

FIGURE 1.18 Point–slope form

EXAMPLE 2 Finding an Equation of a Line with Given Point and Slope

Find the point–slope form of the equation of the line passing through the point $(1, -2)$ with slope $m = 3$. Then solve for y.

Solution

$$y - y_1 = m(x - x_1) \quad \text{Point–slope form}$$
$$y - (-2) = 3(x - 1) \quad \text{Substitute } x_1 = 1, y_1 = -2, \text{ and } m = 3.$$
$$y + 2 = 3x - 3 \quad \text{Simplify.}$$
$$y = 3x - 5 \quad \text{Solve for } y.$$

Practice Problem 2 Find the point–slope form of the equation of the line passing through the point $(-2, -3)$ with slope $-\frac{2}{3}$. Then solve for y.

EXAMPLE 3 Finding an Equation of a Line Passing Through Two Given Points

Find the point–slope form of the equation of the line ℓ passing through the points $(-2, 1)$ and $(3, 7)$. Then solve for y to find the slope-intercept form.

Solution

We first find the slope m of the line ℓ.

$$m = \frac{7 - 1}{3 - (-2)} = \frac{6}{3 + 2} = \frac{6}{5} \qquad m = \frac{y_2 - y_1}{x_2 - x_1}$$

We use $m = \frac{6}{5}$ and either of the two given points when substituting into the point-slope form:

$$y - y_1 = m(x - x_1) \qquad \text{Point–slope form}$$

With $(x_1, y_1) = (3, 7)$ \qquad With $(x_1, y_1) = (-2, 1)$

$$y - 7 = \frac{6}{5}(x - 3) \qquad\qquad y - 1 = \frac{6}{5}(x - (-2))$$

$$y = \frac{6}{5}x + \frac{17}{5} \quad \text{Simplify.} \qquad y = \frac{6}{5}x + \frac{17}{5} \quad \text{Simplify.}$$

——————— Same result ———————

Practice Problem 3 Find the point–slope form of the equation of a line passing through the points $(-3, -4)$ and $(-1, 6)$. Then solve for y to find the slope-intercept form.

3 Write the slope–intercept form of the equation of a line.

Slope–Intercept Form

If we know the slope of a line and a given point that is the y-intercept, then we can also use the point–slope form to create another form of the equation of the line.

EXAMPLE 4 Finding an Equation of a Line with a Given Slope and y-intercept

Find the point–slope form of the equation of the line with slope m and y-intercept b. Then solve for y.

Solution

Because the line has y-intercept b, the line passes through the point $(0, b)$. See Figure 1.19.

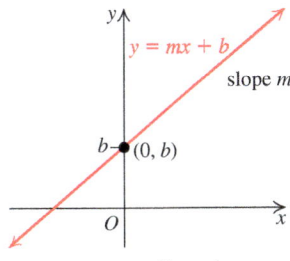

$$y - y_1 = m(x - x_1) \quad \text{Point–slope form}$$
$$y - b = m(x - 0) \quad \text{Substitute } x_1 = 0 \text{ and } y_1 = b.$$
$$y - b = mx \quad \text{Simplify.}$$
$$y = mx + b \quad \text{Solve for } y.$$

FIGURE 1.19 Slope–intercept form

Practice Problem 4 Find the point–slope form of the equation of the line with slope 2 and y-intercept -3. Then solve for y.

HISTORICAL NOTE

No one is sure why the letter m is used for slope. Many people suggest that m comes from the French **monter** (to climb), but Descartes, who was French, did not use m. Professor John Conway of Princeton University has suggested that m could stand for "modulus of slope."

Slope–Intercept Form of the Equation of a Line

The **slope–intercept form** of the equation of the line with slope m and y-intercept b is

$$y = mx + b.$$

The linear equation in the form $y = mx + b$ displays the slope m (the coefficient of x) and the y-intercept b (the constant term). The number m tells which way and how much the line is tilted; the number b tells where the line intersects the y-axis.

EXAMPLE 5 Graph Using the Slope and y-intercept

Graph the line whose equation is $y = \frac{2}{3}x + 2$.

Solution

The equation

$$y = \frac{2}{3}x + 2$$

is in the slope–intercept form with slope $\frac{2}{3}$ and y-intercept 2. To sketch the graph, find two points on the line and draw a line through the two points. Use the y-intercept as one of the points; then use the slope to locate a second point. Because $m = \frac{2}{3}$, let 2 be the rise and 3 be the run. From the point $(0, 2)$, move three units to the right (run) and two units up (rise). This gives $(0 + \text{run}, 2 + \text{rise}) = (0 + 3, 2 + 2) = (3, 4)$ as the second point.

The line we want joins the points $(0, 2)$ and $(3, 4)$ and is shown in Figure 1.20.

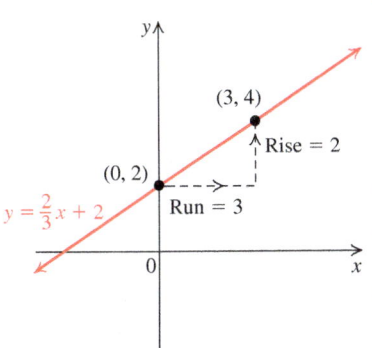

FIGURE 1.20 Locating a second point on a line

Practice Problem 5 Graph the line with slope $-\frac{2}{3}$ that contains the point $(0, 4)$.

4 Recognize the equations of horizontal and vertical lines.

Equations of Horizontal and Vertical Lines

Consider the horizontal line through the point (h, k). The y-coordinate of every point on this line is k. So we can write its equation as $y = k$. This line has slope $m = 0$ and y-intercept k. Similarly, an equation of the vertical line through the point (h, k) is $x = h$. This line has undefined slope and x-intercept h. See Figure 1.21.

70 Chapter 1 Graphs and Functions

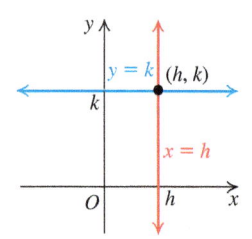

FIGURE 1.21 Horizontal and vertical lines

MAIN FACTS

Horizontal and Vertical Lines

An equation of a horizontal line through the point (h, k) is $y = k$.
An equation of a vertical line through the point (h, k) is $x = h$.

EXAMPLE 6 Recognizing Horizontal and Vertical Lines

Discuss the graph of each equation in the xy-plane.

a. $y = 2$ b. $x = 4$

Solution

a. The equation $y = 2$ may be considered as an equation in two variables x and y by writing

$$0 \cdot x + y = 2.$$

Any ordered pair of the form $(x, 2)$ is a solution of this equation. Some solutions are $(-1, 2), (0, 2), (2, 2),$ and $(7, 2)$. It follows that the graph is a line parallel to the x-axis and two units above it, as shown in Figure 1.22. Its slope is 0.

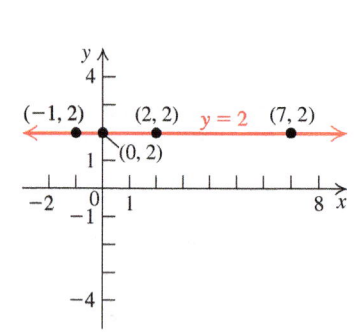

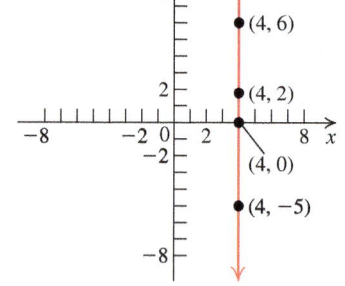

FIGURE 1.22 Horizontal line

FIGURE 1.23 Vertical line

b. The equation $x = 4$ may be written as

$$x + 0 \cdot y = 4.$$

Any ordered pair of the form $(4, y)$ is a solution of this equation. Some solutions are $(4, -5), (4, 0), (4, 2),$ and $(4, 6)$. The graph is a line parallel to the y-axis and four units to the right of it, as shown in Figure 1.23. Its slope is undefined.

Practice Problem 6 Sketch the graphs of the lines $x = -3$ and $y = 7$.

TECHNOLOGY CONNECTION

Calculator graph of $y = 2$

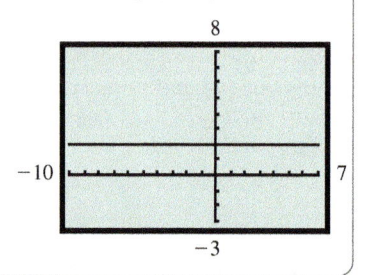

SIDE NOTE

Students sometimes find it easy to remember that if a linear equation

(1) does not have y in it, the graph is parallel to the y-axis or
(2) does not have x in it, the graph is parallel to the x-axis.

TECHNOLOGY CONNECTION

You cannot graph $x = 4$ by using the $\boxed{Y=}$ key on your calculator.

5 Recognize the general form of the equation of a line.

General Form of the Equation of a Line

An equation of the form

$$ax + by + c = 0,$$

where a, b, and c are constants and a and b are not both zero, is called the *general form* of a linear equation. Consider two possible cases: $b \neq 0$ and $b = 0$.

Suppose $b \neq 0$: we can isolate y on one side and rewrite the equation in slope–intercept form.

$ax + by + c = 0$ Original equation
$by = -ax - c$ Subtract $ax + c$ from both sides.
$y = -\dfrac{a}{b}x - \dfrac{c}{b}.$ Divide by b.

The result is an equation of a line in slope–intercept form with slope $-\dfrac{a}{b}$ and y-intercept $-\dfrac{c}{b}$.

Suppose $b = 0$. We know that $a \neq 0$ because a and b are not both zero, and we can solve for x.

$$ax + 0 \cdot y + c = 0 \qquad \text{Replace } b \text{ with } 0.$$
$$ax + c = 0 \qquad \text{Simplify.}$$
$$x = -\dfrac{c}{a} \qquad \text{Solve for } x.$$

The graph of the equation $x = -\dfrac{c}{a}$ is a vertical line.

General Form of the Equation of a Line

The graph of every linear equation
$$ax + by + c = 0,$$
where a, b, and c are constants and a and b are not both zero, is a line. The equation $ax + by + c = 0$ is called the **general form** of the equation of a line.

TECHNOLOGY CONNECTION

Calculator graph of $y = \dfrac{3}{4}x + 3$

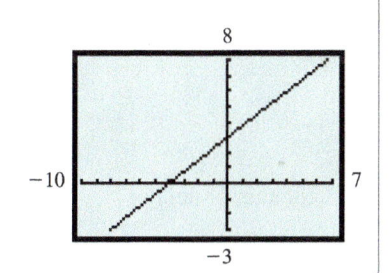

EXAMPLE 7   Graphing a Linear Equation Using Intercepts

Find the slope, y-intercept, and x-intercept of the line with equation
$$3x - 4y + 12 = 0.$$
Then sketch the graph.

Solution

First, solve for y to write the equation in slope–intercept form.

$$3x - 4y + 12 = 0 \qquad \text{Original equation}$$
$$4y = 3x + 12 \qquad \text{Subtract } 3x + 12; \text{ multiply by } -1.$$
$$y = \dfrac{3}{4}x + 3 \qquad \text{Divide by } 4.$$

This equation tells us that the slope m is $\dfrac{3}{4}$ and the y-intercept is 3. To find the x-intercept, set $y = 0$ in the original equation and obtain $3x + 12 = 0$, or $x = -4$. So the x-intercept is -4.

We can sketch the graph of the equation by finding two points on the graph. We use the intercepts and sketch the line joining the points $(-4, 0)$ and $(0, 3)$. The graph is shown in Figure 1.24.

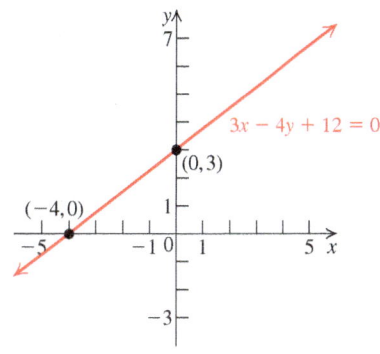

FIGURE 1.24 Graphing a line using intercepts

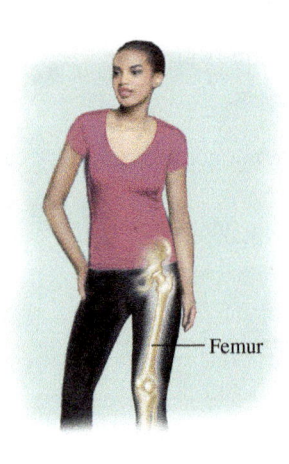

Practice Problem 7 Sketch the graph of the equation $3x + 4y = 24$.

Forensic scientists use various clues to determine the identity of deceased individuals. Scientists know that certain bones serve as excellent sources of information about a person's height. The length of the *femur,* the long bone that stretches from the hip (pelvis) socket to the kneecap (patella), can be used to estimate a person's height. Knowing both the gender and race of the individual increases the accuracy of this relationship between the length of the bone and the height of the person.

EXAMPLE 8 Inferring Height from the Femur

The height H of a human male is related to the length x of his femur by the formula

$$H = 2.6x + 65,$$

where all measurements are in centimeters. The femur of Wild Bill Longley measured between 45 and 46 centimeters. Estimate the height of Wild Bill.

Solution
Substituting $x = 45$ and $x = 46$ into the preceding formula, we have possible heights

$$H_1 = (2.6)(45) + 65 = 182$$
$$\text{and} \quad H_2 = (2.6)(46) + 65 = 184.6.$$

Therefore, Wild Bill's height was between 182 and 184.6 centimeters.

Converting to inches, we conclude that his height was between $\dfrac{182}{2.54} \approx 71.7$ inches and $\dfrac{184.6}{2.54} \approx 72.7$ inches. He was approximately 6 feet tall.

RECALL
1 inch = 2.54 centimeters

Practice Problem 8 Suppose the femur of a man measures between 43 and 44 centimeters. Estimate the height of the man in centimeters.

6 Find equations of parallel and perpendicular lines.

Parallel and Perpendicular Lines

We can use slope to decide whether two lines are parallel, perpendicular, or neither.

Parallel and Perpendicular Lines

Let ℓ_1 and ℓ_2 be two distinct lines with slopes m_1 and m_2, respectively. Then

ℓ_1 is parallel to ℓ_2 if and only if $m_1 = m_2$ and

ℓ_1 is perpendicular to ℓ_2 if and only if $m_1 \cdot m_2 = -1$.

Any two vertical lines are parallel, as are any two horizontal lines. Any horizontal line is perpendicular to any vertical line.

Figure 1.25 shows examples of the slopes of a pair of parallel lines and a pair of perpendicular lines.

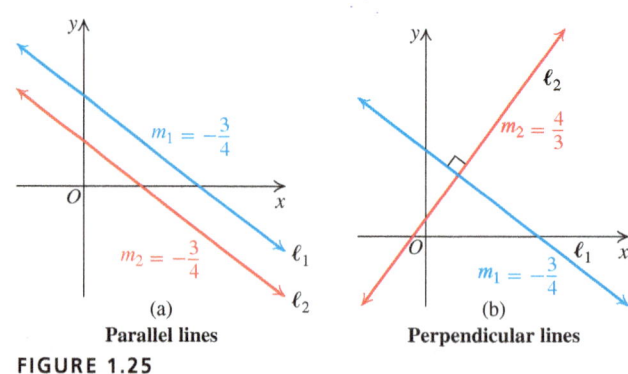

(a) Parallel lines: $m_1 = -\dfrac{3}{4}$, $m_2 = -\dfrac{3}{4}$

(b) Perpendicular lines: $m_2 = \dfrac{4}{3}$, $m_1 = -\dfrac{3}{4}$

FIGURE 1.25

PROCEDURE IN ACTION

EXAMPLE 9 — Finding Equations of Parallel and Perpendicular Lines

OBJECTIVE
Let $\ell : ax + by + c = 0$. Find the equation of each line through the point (x_1, y_1):

(a) ℓ_1 parallel to ℓ
(b) ℓ_2 perpendicular to ℓ

EXAMPLE
Let $\ell : 2x - 3y + 6 = 0$. Find the general form of the equation of each line through the point $(2, 8)$:

(a) ℓ_1 parallel to ℓ
(b) ℓ_2 perpendicular to ℓ

Step 1 Find the slope m of ℓ.

Write ℓ in the form $y = mx + b$.

1. $\ell : 2x - 3y + 6 = 0$ Original equation
   $-3y = -2x - 6$ Subtract $2x + 6$.
   $y = \dfrac{2}{3}x + 2$ Divide by -3.

   The slope of ℓ is $m = \dfrac{2}{3}$.

Step 2 Write slope of ℓ_1 and ℓ_2.

The slope m_1 of ℓ_1 is m.
The slope m_2 of ℓ_2 is $-\dfrac{1}{m}$.

2. $m_1 = \dfrac{2}{3}$ ℓ_1 parallel to ℓ
   $m_2 = -\dfrac{3}{2}$ ℓ_2 perpendicular to ℓ

Step 3 Write equations of ℓ_1 and ℓ_2.

Use point–slope form to write equations of ℓ_1 and ℓ_2.
Simplify to write equations in the requested form.

3. $\ell_1 : y - 8 = \dfrac{2}{3}(x - 2)$ Point–slope form
   $2x - 3y + 20 = 0$ Simplify.
   $\ell_2 : y - 8 = -\dfrac{3}{2}(x - 2)$ Point–slope form
   $3x + 2y - 22 = 0$ Simplify.

Practice Problem 9 Find the general form of the equation of the line

a. through $(-2, 5)$ and parallel to the line containing $(2, 3)$ and $(5, 7)$.
b. through $(3, -4)$ and perpendicular to the line $4x + 5y + 1 = 0$.

SUMMARY OF MAIN FACTS

1. We summarize different forms of equations whose graphs are lines.

Form	Equation	Slope	Other Information
General	$ax + by + c = 0$ $a \neq 0$, or $b \neq 0$	$-\dfrac{a}{b}$ if $b \neq 0$ Undefined if $b = 0$	x-intercept is $-\dfrac{c}{a}$, $a \neq 0$. y-intercept is $-\dfrac{c}{b}$, $b \neq 0$.
Point–slope	$y - y_1 = m(x - x_1)$	m	Line passes through (x_1, y_1).
Slope–intercept	$y = mx + b$	m	x-intercept is $-\dfrac{b}{m}$ if $m \neq 0$; y-intercept is b.
Vertical line through (h, k)	$x = h$	undefined	When $h \neq 0$, no y-intercept; x-intercept is h. When $h = 0$, the graph is the y-axis.
Horizontal line through (h, k)	$y = k$	0	When $k \neq 0$, no x-intercept; y-intercept is k. When $k = 0$, the graph is the x-axis.

2. **Parallel and perpendicular lines:** Two lines with slopes m_1 and m_2 are
   (a) parallel if $m_1 = m_2$ and (b) perpendicular if $m_1 \cdot m_2 = -1$.

SECTION 1.2 Exercises

Basic Concepts and Skills

In Exercises 1–4, find the slope of the line through the given pair of points. Without plotting any points, state whether the line is rising, falling, horizontal, or vertical.

1. $(1, 3), (4, 7)$
2. $(0, 4), (2, 0)$
3. $(3, -2), (6, -2)$
4. $(-3, 7), (-3, -4)$

5. In the figure, identify the line with the given slope m.
   a. $m = 1$
   b. $m = -1$
   c. $m = 0$
   d. m is undefined.

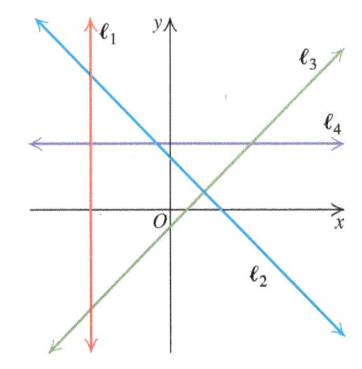

6. Find the slope of each line. (The scale is the same on both axes.)

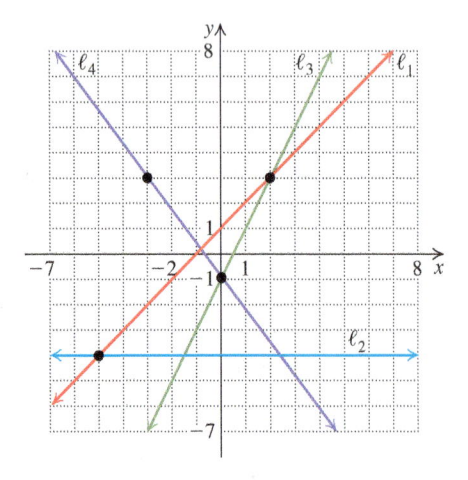

In Exercises 7 and 8, write an equation whose graph is described.

7. a. Graph is the x-axis.
   b. Graph is the y-axis.
8. a. Graph consists of all points with a y-coordinate of -4.
   b. Graph consists of all points with an x-coordinate of 5.

In Exercises 9–14, find an equation in slope–intercept form of the line that passes through the given point and has slope m. Also sketch the graph of the line by locating the second point with the rise-and-run method.

9. $(0, 4); m = \dfrac{1}{2}$
10. $(0, 4); m = -\dfrac{1}{2}$
11. $(2, 1); m = -\dfrac{3}{2}$
12. $(-1, 0); m = \dfrac{2}{5}$
13. $(5, -4); m = 0$
14. $(5, -4); m$ is undefined.

In Exercises 15–30, use the given conditions to find an equation in slope–intercept form of each nonvertical line. Write vertical lines in the form $x = h$.

15. Passing through $(0, 1)$ and $(1, 3)$
16. Passing through $(-1, 3)$ and $(3, 3)$
17. Passing through $(-1, -3)$ and $(6, -9)$
18. Passing through $\left(\dfrac{1}{2}, \dfrac{1}{4}\right)$ and $(0, 2)$
19. A vertical line through $(5, 1.7)$
20. A horizontal line through $(1.4, 1.5)$
21. A horizontal line through $(0, 0)$
22. A vertical line through $(0, 0)$
23. $m = 2$; y-intercept $= 5$
24. $m = -\dfrac{2}{3}$; y-intercept $= -4$
25. x-intercept $= -3$; y-intercept $= 4$
26. x-intercept $= -5$; y-intercept $= -2$
27. Parallel to $y = 5$; passing through $(4, 7)$
28. Parallel to $x = 5$; passing through $(4, 7)$
29. Perpendicular to $x = -4$; passing through $(-3, -5)$
30. Perpendicular to $y = -4$; passing through $(-3, -5)$
31. Let ℓ_1 be a line with slope $m = -2$. Determine whether the given line ℓ_2 is parallel to ℓ_1, perpendicular to ℓ_1, or neither.
    a. ℓ_2 is the line through the points $(1, 5)$ and $(3, 1)$.
    b. ℓ_2 is the line through the points $(7, 3)$ and $(5, 4)$.
    c. ℓ_2 is the line through the points $(2, 3)$ and $(4, 4)$.
32. A particle initially at $(-1, 3)$ moves along a line of slope $m = 4$ to a new position (x, y).
    a. Find y if $x = 2$.
    b. Find x if $y = 9$.

In Exercises 33–40, find the slope and intercepts from the equation of the line. Sketch the graph of each equation.

33. $x + 2y - 4 = 0$
34. $3x - 2y + 6 = 0$
35. $x = 3y - 9$
36. $2x = -4y + 15$
37. $x - 5 = 0$
38. $2y + 5 = 0$
39. $x = 0$
40. $y = 0$

41. Show that an equation of a line through the points $(a, 0)$ and $(0, b)$, with a and b nonzero, can be written in the form

$$\frac{x}{a} + \frac{y}{b} = 1.$$

This form is called the **two-intercept form** of the equation of a line.

42. Find an equation of the line whose x-intercept is 4 and y-intercept is 3.

In Exercises 43–46, use the two-intercept form of the equation of a line given in Exercise 41.

43. Find the x- and y-intercepts of the graph of the equation $2x + 3y = 6$.

44. Repeat Exercise 43 for the equation $3x - 4y + 12 = 0$.

45. Find an equation of the line that has equal intercepts and passes through the point $(3, -5)$.

46. Find an equation of the line making intercepts on the axes equal in magnitude but opposite in sign and passing through the point $(-5, -8)$.

47. Find an equation of the line passing through the points $(2, 4)$ and $(7, 9)$. Use the equation to show that the three points $(2, 4)$, $(7, 9)$, and $(-1, 1)$ are on the same line.

48. Use the technique of Exercise 47 to check whether the points $(7, 2)$, $(2, -3)$, and $(5, 1)$ are on the same line.

In Exercises 49–56, determine whether each pair of lines are parallel, perpendicular, or neither.

49. $x = -1$ and $2x + 7 = 9$
50. $x = 0$ and $y = 0$
51. $2x + 3y = 7$ and $y = 2$
52. $10x + 2y = 3$ and $y + 1 = -5x$
53. $y = 3x + 1$ and $6y + 2x = 0$
54. $4x + 3y = 1$ and $3 + y = 2x$
55. $3x + 8y = 7$ and $5x - 7y = 0$
56. $x = 4y + 8$ and $y = -4x + 1$

In Exercises 57–62, find the equation of the line in slope–intercept form satisfying the given conditions.

57. Parallel to $x + y = 1$; passing through $(1, 1)$
58. Parallel to $y = 6x + 5$; y-intercept of -2
59. Perpendicular to $3x - 9y = 18$; passing through $(-2, 4)$
60. Perpendicular to $-2x + y = 14$; passing through $(0, 0)$
61. Perpendicular to $y = 6x + 5$; y-intercept of 4
62. Parallel to $-2x + 3y - 7 = 0$; passing through $(1, 0)$

Applying the Concepts

In Exercises 63–64, write an equation of a line in slope–intercept form. To describe this situation, interpret (a) the variables x and y and (b) the meaning of the slope and the y-intercept.

63. Christmas savings account. You currently have $130 in a Christmas savings account. You deposit $7 per week into this account. (Ignore interest.)

64. Manufacturing. A manufacturer produces 50 TVs at a cost of $17,500 and 75 TVs at a cost of $21,250.

65. Fahrenheit and Celsius. In the Fahrenheit temperature scale, water boils at 212°F and freezes at 32°F. In the Celsius scale, water boils at 100°C and freezes at 0°C. Assume that the Fahrenheit temperature F and Celsius temperature C are linearly related.
  a. Find the equation in the slope–intercept form relating F and C, with C as the independent variable.
  b. What is the meaning of slope in part (a)?
  c. Find the Fahrenheit temperatures, to the nearest degree, corresponding to 40°C, 25°C, −5°C, and −10°C.
  d. Find the Celsius temperatures, to the nearest degree, corresponding to 100°F, 90°F, 75°F, −10°F, and −20°F.
  e. The normal body temperature in humans ranges from 97.6°F to 99.6°F. Convert this temperature range to degrees Celsius.
  f. When is the Celsius temperature the same numerical value as the Fahrenheit temperature?

66. Demand equation. A demand equation expresses a relationship between the demand q (the number of items demanded) and the price p per unit. A linear demand equation has the form $q = mp + b$. The slope m (usually negative) measures the change in demand per unit change in price. The y-intercept b gives the demand if the items were given away. A merchant can sell 480 T-shirts per week at a price of $4 each, but can sell only 400 per week if the price per T-shirt is increased to $4.50.
  a. Find a linear demand equation for T-shirts.
  b. Predict the demand for T-shirts if the price per T-shirt is $5.

67. Cost. The cost of producing four modems is $210.20, and the cost of producing ten modems is $348.80.
  a. Write a linear equation in slope–intercept form relating cost to the number of modems produced. Sketch a graph of the equation.
  b. What is the practical meaning of the slope and the intercepts of the equation in part (a)?
  c. Predict the cost of producing 12 modems.

68. Viewers of a TV show. After five months, the number of viewers of *The Weekly Show* on TV was 5.73 million. After eight months, the number of viewers of the show rose to 6.27 million. Assume that the linear model applies.
  a. Write an equation expressing the number of viewers V after x months.
  b. Interpret the slope and the intercepts of the line in part (a).
  c. Predict the number of viewers of the show after 11 months.

Critical Thinking / Discussion / Writing

69. Geometry. Show that an equation of a circle with (x_1, y_1) and (x_2, y_2) as endpoints of a diameter can be written in the form

$$(x - x_1)(x - x_2) + (y - y_1)(y - y_2) = 0.$$

[*Hint:* Use the fact that when the two endpoints of a diameter and any other point on the circle are used as vertices for a triangle, the angle opposite the diameter is a right angle.]

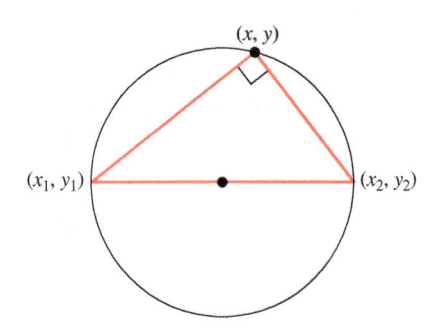

70. Geometry. Find an equation of the tangent line to the circle $x^2 + y^2 = 169$ at the point $(-5, 12)$. (See the accompanying figure.)

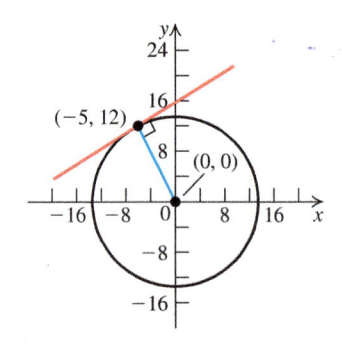

71. By considering slopes, show that the points $(1, -1), (-2, 5),$ and $(3, -5)$ lie on the same line.

72. a. The equation $y + 2x + k = 0$ describes a family of lines for each value of k. Sketch the graphs of the members of this family for $k = -3, 0,$ and 2. What is the common characteristic of the family?
  b. Repeat part (a) for the family of lines $y - kx + 4 = 0$.

Maintaining Skills

In Exercises 73–78, perform the indicated operation.

73. $\dfrac{3}{2-x} + \dfrac{x}{x-2}$

74. $\dfrac{2}{x+3} - \dfrac{5}{(x+3)^2}$

75. $\dfrac{2}{x-1} - \dfrac{x}{x^2-1}$

76. $\dfrac{2}{x-2} + \dfrac{3x}{x^2-6x+8}$

77. $\dfrac{6x+4}{2x-8} \cdot \dfrac{x-4}{9x+6}$

78. $\dfrac{\dfrac{1}{x} - x}{1 - \dfrac{1}{x^2}}$

SECTION 1.3

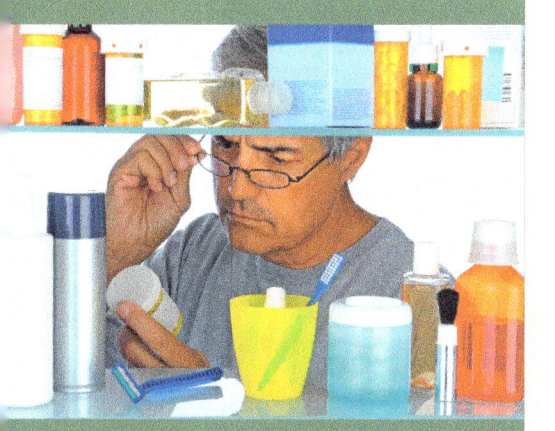

Functions

BEFORE STARTING THIS SECTION, REVIEW

1. Graph of an equation (Section 1.1, page 55)
2. Equation of a circle (Section 1.1, page 60)

OBJECTIVES

1. Use functional notation and find function values.
2. Find the domain and range of a function.
3. Identify the graph of a function.
4. Get information about a function from its graph.
5. Solve applied problems by using functions.

◆ The Ups and Downs of Drug Levels

Medicines in the form of a pill have a much harder time getting to work than you do. First, they are swallowed and dumped into a pool of acid in your stomach. Then they dissolve; leave the stomach; and begin to be absorbed, mostly through the lining of the small intestine. Next, the blood around the intestine carries the medicines to the liver, an organ designed to metabolize (break down) and remove foreign material from the blood. Because of this property of the liver, most drugs have a tough time getting past it. The amount of drug that makes it into your bloodstream, compared with the amount that you put in your mouth, is called the drug's *bioavailability*. If a drug has low bioavailability, then much of the drug is destroyed before it reaches the bloodstream. The amount of drug you take in a pill is calculated to correct for this deficiency so that the amount you need actually ends up in your blood.

Once the drugs are past the liver, the bloodstream carries them throughout the rest of the body in about one minute. The study of the ways drugs are absorbed and move through the body is called *pharmacokinetics*, or PK for short. PK measures the ups and downs of drug levels in your body. In Example 10, we consider an application of *functions* to the levels of drugs in the bloodstream.

(Adapted from *The Ups and Downs of Drug Levels*, by Bob Munk, www.thebody.com/content/treat/art1062.html).

1 Use functional notation and find function values.

Functions

There are many ways of expressing a relationship between two quantities. For example, if you are paid $10 per hour, then the relation between the number of hours, x, that you work and the amount of money, y, that you earn may be expressed by the equation $y = 10x$. Replacing x with 40 yields $y = 400$ and indicates that 40 hours of work corresponds to a $400 paycheck. A special relationship such as $y = 10x$ in which to each element x in one set there corresponds a unique element y in another set is called a *function*. We say that your pay is a function of the number of hours you work. Because the value of y depends on the value of x, y is called the **dependent variable** and x is called the **independent variable**. Any symbol may be used for the dependent and independent variable.

Function

A **function** from a set X to a set Y is a rule that assigns to each element of X exactly one element of Y. The set X is the **domain** of the function. The set of those elements of Y that correspond (are assigned) to the elements of X is the **range** of the function.

A rule of assignment may be described by a **correspondence diagram**, in which an arrow points from each domain element to the corresponding element or elements in the range.

To decide whether a correspondence diagram defines a function, we must check whether *any* domain element is paired with more than one range element. If this happens, that domain element does not have a *unique* corresponding element of Y and the correspondence is not a function. See the correspondence diagrams in Figure 1.26.

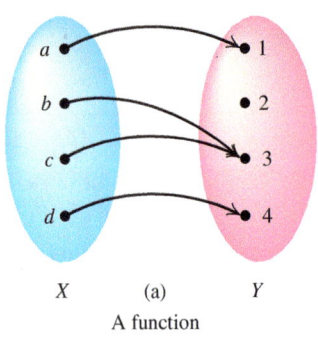

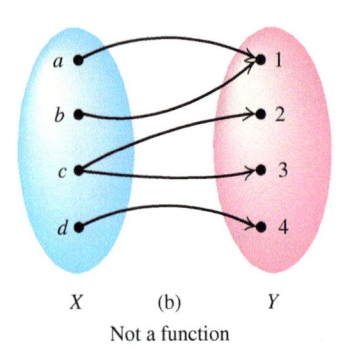

(a) A function
Domain: $\{a, b, c, d\}$
Range: $\{1, 3, 4\}$

(b) Not a function

FIGURE 1.26 Correspondence diagrams

MAIN FACTS FROM THE DEFINITION OF A FUNCTION

Let f be a function from a set X to a set Y.

1. The domain of f is the entire set X. That is, *each* element of X must be assigned to some element of Y.
2. The range of f is the set of those elements in Y that are assigned to some element of X. So the range of f may not be the entire set Y.
3. It *is permissible* to assign the same element of Y to two or more elements of X. See Figure 1.26(a).
4. It *is not permissible* to assign an element of X to two or more elements of Y. See Figure 1.26(b).

DO YOU KNOW?

The functional notation $y = f(x)$ was first used by the great Swiss mathematician Leonhard Euler in the *Commentarii Academia Petropolitanae*, published in 1735.

Function Notation

We usually use single letters such as $f, F, g, G, h,$ and H as the name of a function. Some special function names use more than one letter, such as *sin* (the sine function) and *log* (the logarithm function). If we choose f as the name of a function, then for each x in the domain of f, there corresponds a unique y in its range. The number y is denoted by $f(x)$, read as "f of x" or as "f at x." We call $f(x)$ the **value of f at the number x** and say that f *assigns* the value $f(x)$ to x. If the function shown in Figure 1.26(a) is named f, then the correspondence diagram may be described by:

$$f(a) = 1, f(b) = 3, f(c) = 3, \text{ and } f(d) = 4.$$

FIGURE 1.27 The function f, as a machine

Yet another way to illustrate the concept of a function is by a "machine," as shown in Figure 1.27. If x is in the domain of a function, then the machine accepts it as an "input" and produces the value $f(x)$ as the "output." In other words, output is a function of the input. A calculator is an example of a function machine. If an input number (say, 9) is entered and some key (such as $\sqrt{\ }$) is pressed, the display (3) outputs the function value.

WARNING

In writing $y = f(x)$, do not confuse f, which is the name of the function, with $f(x)$, which is a number in the range of f that the function assigns to x.
The symbol $f(x)$ does not mean "f times x."

Representations of Functions

We have seen that the rule used in describing a function may be expressed in the form of a correspondence diagram. Here are some additional forms.

Functions Defined by Ordered Pairs Any set of ordered pairs is called a **relation**. The set of first components is the **domain of the relation**, and the set of second components is the **range of the relation**. In the ordered pair (x, y), we say that y corresponds to x. Then a *function* is a relation in which each element of the domain corresponds to exactly one element of the range. In other words, a function is a relation in which no two distinct ordered pairs have the same first component.

EXAMPLE 1 Functions Defined by Ordered Pairs

Determine whether each relation defines a function.
a. $r = \{(-1, 2), (1, 3), (5, 2), (-1, -3)\}$
b. $s = \{(-2, 1), (0, 2), (2, 1), (-1, -3)\}$

Solution

a. The domain of the relation r is $\{-1, 1, 5\}$, and its range is $\{2, 3, -3\}$. It is not a function because the ordered pairs $(-1, 2)$ and $(-1, -3)$ have the same first component. See the correspondence diagram for the relation r in Figure 1.28.
b. The domain of the relation s is $\{-2, -1, 0, 2\}$, and its range is $\{-3, 1, 2\}$. The relation s is a function because no two ordered pairs of s have the same first component. Figure 1.29 shows the correspondence diagram for the function s.

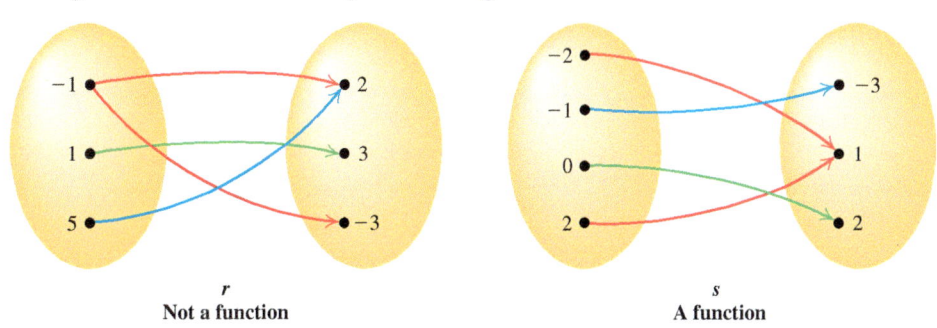

r
Not a function
FIGURE 1.28

s
A function
FIGURE 1.29

Practice Problem 1 Repeat Example 1 for each set of ordered pairs.
a. $R = \{(2, 1), (-2, 1), (3, 2)\}$
b. $S = \{(2, 5), (3, -2), (3, 5)\}$

80 Chapter 1 Graphs and Functions

TABLE 1.2

x	y
−1	2
1	3
5	2
−1	−3

Relation r

TABLE 1.3

x	y
−2	1
−1	−3
0	2
2	1

Relation s

Functions Defined by Tables and Graphs Tables and graphs can be used to describe relations and functions. We can display in tabular form the list of ordered pairs (x, y) of a relation, where x is in the domain and the corresponding y is in the range. For example, the relations r and s of Example 1 described in tabular form are shown in Tables 1.2 and 1.3, respectively.

Alternatively, we can display geometrically the list of ordered pairs (x, y) in a relation by plotting each pair (x, y) as a point in the coordinate plane. The geometric display is the graph of the relation. Graphs of the relations r and s are given in Figure 1.30.

Relation r
(a)

Relation s
(b)

FIGURE 1.30 Graphs of relations

The graph of the discrete data points of a relation or function is called a **scatterplot.**

SIDE NOTE

The definition of a function is independent of the letters used to denote the function and the variables. For example,

$s(x) = x^2$, $g(y) = y^2$, and $h(t) = t^2$

represent the same function.

Functions Defined by Equations When the relation defined by an equation in two variables is a function, we can often solve the equation for the dependent variable in terms of the independent variable. For example, the equation $y - x^2 = 0$ can be solved for y in terms of x:

$$y = x^2.$$

Now we can replace the dependent variable, in this case, y, with functional notation $f(x)$ and express the function as

$$f(x) = x^2 \quad (\text{read ``}f\text{ of } x \text{ equals } x^2\text{''}).$$

Here $f(x)$ plays the role of y and is the value of the function f at x. For example, if $x = 3$ is an element (input value) in the domain of f, the corresponding element (output value) in the range is found by replacing x with 3 in the equation

$$f(x) = x^2$$

so that

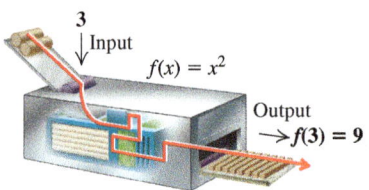

$$f(3) = 3^2 = 9. \quad \text{Replace } x \text{ with 3.}$$

FIGURE 1.31 The function f, as a machine

We say that the value of the function f at 3 is 9, or 9 is the evaluated function value $f(3)$. See Figure 1.31. In other words, the number 9 in the range corresponds to the number 3 in the domain and the ordered pair $(3, 9)$ is an ordered pair of the function f.

If a function g is defined by an equation such as $y = x^2 - 6x + 8$, the notations

$$y = x^2 - 6x + 8 \quad \text{and} \quad g(x) = x^2 - 6x + 8$$

define the same function.

Section 1.3 ■ Functions 81

EXAMPLE 2 Determining Whether an Equation Defines a Function

Determine whether y is a function of x for each equation.

a. $6x^2 - 3y = 12$ **b.** $y^2 - x^2 = 4$

Solution

Solve each equation for y in terms of x. If more than one value of y corresponds to the same value of x, then y is not a function of x.

a.
$$6x^2 - 3y = 12 \quad \text{Original equation}$$
$$6x^2 - 3y + 3y - 12 = 12 + 3y - 12 \quad \text{Add } 3y - 12 \text{ to both sides.}$$
$$6x^2 - 12 = 3y \quad \text{Simplify.}$$
$$2x^2 - 4 = y \quad \text{Divide by 3.}$$

The last equation shows that only one value of y corresponds to each value of x. For example, if $x = 0$, then $y = 0 - 4 = -4$. So y is a function of x.

b.
$$y^2 - x^2 = 4 \quad \text{Original equation}$$
$$y^2 - x^2 + x^2 = 4 + x^2 \quad \text{Add } x^2 \text{ to both sides.}$$
$$y^2 = x^2 + 4 \quad \text{Simplify.}$$
$$y = \pm\sqrt{x^2 + 4} \quad \text{Square root property}$$

The last equation shows that two values of y correspond to each value of x. For example, if $x = 0$, then $y = \pm\sqrt{0^2 + 4} = \pm\sqrt{4} = \pm 2$. Both $y = 2$ and $y = -2$ correspond to $x = 0$. Therefore, y is not a function of x. See Figure 1.32. However, the equation $y^2 - x^2 = 4$ defines two functions:

$$f_1(x) = \sqrt{x^2 + 4}; \text{ domain: } (-\infty, \infty); \text{ range: } [2, \infty), \text{ and}$$
$$f_2(x) = -\sqrt{x^2 + 4}; \text{ domain: } (-\infty, \infty); \text{ range } (-\infty, -2].$$

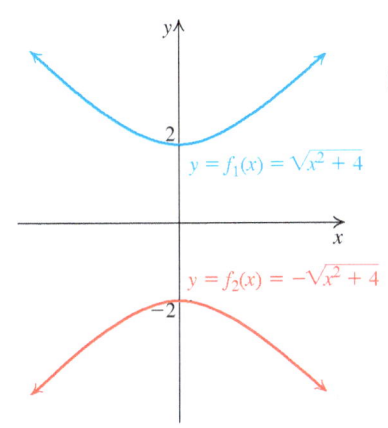

FIGURE 1.32

Practice Problem 2 Determine whether y is a function of x for each equation.

a. $2x^2 - y^2 = 1$ **b.** $x - 2y = 5$

EXAMPLE 3 Evaluating a Function

Let g be the function defined by the equation

$$y = x^2 - 6x + 8.$$

Final each function value.

a. $g(3)$ **b.** $g(-2)$ **c.** $g\left(\dfrac{1}{2}\right)$ **d.** $g(a + 2)$ **e.** $g(x + h)$

Solution

Because the function is named g, we replace y with $g(x)$ and write

$$g(x) = x^2 - 6x + 8 \quad \text{Replace } y \text{ with } g(x).$$

In this notation, the independent variable x is a placeholder. We can write $g(x) = x^2 - 6x + 8$ as

$$g(\ ) = (\ )^2 - 6(\ ) + 8.$$

a. $g(x) = x^2 - 6x + 8$ The given equation
$g(3) = 3^2 - 6(3) + 8$ Replace x with 3 at each occurrence of x.
$= 9 - 18 + 8 = -1$ Simplify.

SIDE NOTE

Problems like Examples **3d** and **3e** will arise frequently. We emphasize that you must take whatever appears in the parentheses and substitute for the independent variable.

The statement $g(3) = -1$ means that the value of the function g at 3 is -1. Just as in part **a**, we can evaluate g at any value x in its domain.

$g(x) = x^2 - 6x + 8$ Original equation

b. $g(-2) = (-2)^2 - 6(-2) + 8 = 24$ Replace x with -2 and simplify.

c. $g\left(\dfrac{1}{2}\right) = \left(\dfrac{1}{2}\right)^2 - 6\left(\dfrac{1}{2}\right) + 8 = \dfrac{21}{4}$ Replace x with $\dfrac{1}{2}$ and simplify.

d. $g(a + 2) = (a + 2)^2 - 6(a + 2) + 8$ Replace x with $(a + 2)$ in $g(x)$.
$= a^2 + 4a + 4 - 6a - 12 + 8$ Recall: $(x + y)^2 = x^2 + 2xy + y^2$.
$= a^2 - 2a$ Simplify.

e. $g(x + h) = (x + h)^2 - 6(x + h) + 8$ Replace x with $(x + h)$ in $g(x)$.
$= x^2 + 2xh + h^2 - 6x - 6h + 8$ Simplify.

Practice Problem 3 Let g be the function defined by the equation $y = -2x^2 + 5x$. Find each function value.

a. $g(0)$ **b.** $g(-1)$ **c.** $g(x + h)$

EXAMPLE 4   **Finding the Area of a Rectangle**

In Figure 1.33, find the area of the rectangle *PLMN*.

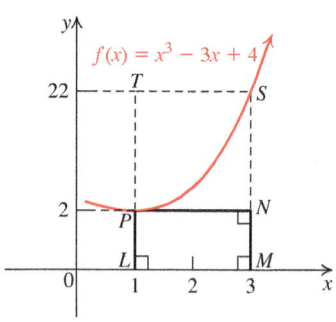

FIGURE 1.33

Solution

Area of the rectangle $PLMN$ = (length)(height)
$= (|3 - 1|)(f(1))$ $P = (1, f(1))$
$= (2)(2)$ $f(1) = 1^3 - 3(1) + 4 = 2$
$= 4$ sq. units

Practice Problem 4 In Figure 1.33, find the area of the rectangle *TLMS*.

2 Find the domain and range of a function.

The Domain of a Function

Sometimes a function does not have a specified domain.

AGREEMENT ON DOMAIN

If the domain of a function that is defined by an equation is not specified, then we agree that the domain of the function is the largest set of real numbers that results in real numbers as outputs.

DO YOU KNOW?

The word *domain* comes from the Latin word *dominus*, or *master*, which in turn is related to *domus*, meaning "home." *Domus* is also the root for the words *domestic* and *domicile*.

When we use our agreement to find the domain of a function, we usually find the values of the variable that do not result in real number outputs. Then we exclude those numbers from the domain. Remember that

1. Division by zero is undefined.
2. The square root (or any even root) of a negative number is not a real number.

EXAMPLE 5 Finding the Domain of a Function

Find the domain of each function.

a. $f(x) = \dfrac{1}{1 - x^2}$ **b.** $g(x) = \sqrt{x}$

c. $h(x) = \dfrac{1}{\sqrt{x - 1}}$ **d.** $P(x) = \sqrt{x^2 - x - 6}$

Solution

a. The function f is not defined when the denominator $1 - x^2$ is 0. Because $1 - x^2 = 0$ if $x = 1$ or $x = -1$, the domain of f is the set $\{x | x \neq -1 \text{ and } x \neq 1\}$, which in interval notation is written $(-\infty, -1) \cup (-1, 1) \cup (1, \infty)$.

b. Because the square root of a negative number is not a real number, negative numbers are excluded from the domain of g. The domain of $g(x) = \sqrt{x}$ is $\{x | x \geq 0\}$, or $[0, \infty)$ in interval notation.

c. The function $h(x) = \dfrac{1}{\sqrt{x - 1}}$ has *two* restrictions. The square root of a negative number is not a real number, so $\sqrt{x - 1}$ is a real number only if $x - 1 \geq 0$. However, we cannot allow $x - 1 = 0$ because $\sqrt{x - 1}$ is in the denominator. Therefore, we must have $x - 1 > 0$, or $x > 1$. The domain of h must be $\{x | x > 1\}$, or $(1, \infty)$ in interval notation.

d. The function $P(x)$ is defined when the expression under the radical sign is nonnegative. The sign graph on page 39 shows that $x^2 - x - 6 = (x + 2)(x - 3) \geq 0$ on the two intervals $(-\infty, -2]$ and $[3, \infty)$. So, the domain of the function $P(x)$ in interval notation is $(-\infty, -2] \cup [3, \infty)$.

Practice Problem 5 Find the domain of each function.

a. $f(x) = \dfrac{1}{\sqrt{1 - x}}$. **b.** $g(x) = \sqrt{\dfrac{x + 2}{x - 3}}$

The Range of a Function

Suppose a function f is defined by an equation. To say that a number y is in the range of f means to find at least one solution x of the equation $f(x) = y$ in the domain of f.

EXAMPLE 6 Finding the Range of a Function

Let $f(x) = x^2$ with domain $X = [3, 5]$.

a. Is 10 in the range of f?
b. Is 4 in the range of f?
c. Find the range of f.

84 Chapter 1 Graphs and Functions

Solution

a. We find possible solutions of the equation

$$f(x) = 10.$$
$$x^2 = 10 \quad \text{Replace } f(x) \text{ with } x^2.$$
$$x = \pm\sqrt{10} \quad \text{Square root property}$$

Because $3 < \sqrt{10} < 5$, the number $x = \sqrt{10}$ is in the interval $X = [3, 5]$. So the equation $f(x) = 10$ has at least one solution in the domain of f. Therefore, 10 is in the range of f.

b. The solutions of the equation $x^2 = 4$ are $x = \pm 2$. Neither of these numbers is in the domain $X = [3, 5]$. So 4 is not in the range of f.

c. The range of f is the interval $[9, 25]$ because for each number y in the interval $[9, 25]$, the number $x = \sqrt{y}$ is in the interval $[3, 5]$ so that $f(x) = f(\sqrt{y}) = (\sqrt{y})^2 = y$.

Practice Problem 6 Repeat Example 6 for $f(x) = x^2$ with domain $X = [-3, 3]$.

Finding the range of a function such as $f(x) = \dfrac{x+1}{x-2}$ is a little more complicated. We discuss this problem in Section 1.7.

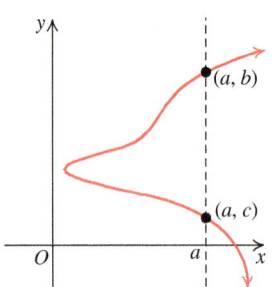

3 Identify the graph of a function.

Graphs of Functions

The **graph of a function** f is the set of ordered pairs $(x, f(x))$ such that x is in the domain of f. That is, the graph of f is the graph of the equation $y = f(x)$. We sketch the graph of $y = f(x)$ by plotting points and joining them with a smooth curve. The graph of a function provides valuable visual information about the function.

Figure 1.34 shows that not every curve in the plane is the graph of a function. In Figure 1.34, a vertical line intersects the curve at two distinct points (a, b) and (a, c). This curve cannot be the graph of $y = f(x)$ for any function f because having $f(a) = b$ and $f(a) = c$ means that f assigns two different range values to the same domain element a. We can express this statement, called the **vertical-line test**, as follows:

FIGURE 1.34 Graph does not represent a function

VERTICAL-LINE TEST

If no vertical line intersects the graph of a relation at more than one point, then the graph of the relation is the graph of a function.

EXAMPLE 7 Identifying the Graph of a Function

Determine which graphs in Figure 1.35 are graphs of functions.

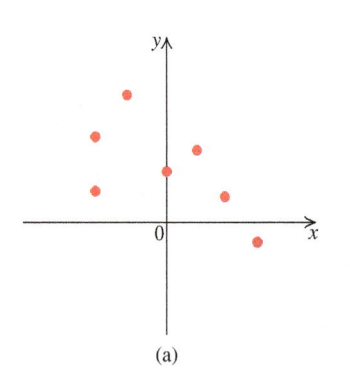

(a) (b) (c) (d)

FIGURE 1.35 The vertical-line test

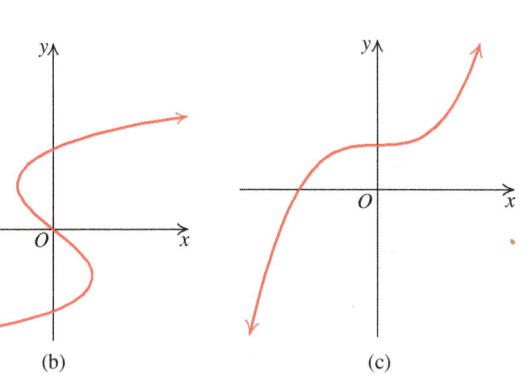

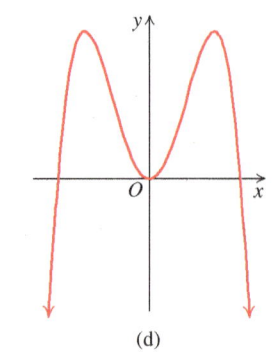

Section 1.3 ■ Functions 85

Solution

The graphs in Figures 1.35(a) and 1.35(b) are not graphs of functions because a vertical line can be drawn through the two points farthest to the left in Figure 1.35(a) and the y-axis is one of many vertical lines that contain more than one point on the graph in Figure 1.35(b). The graphs in Figures 1.35(c) and 1.35(d) are the graphs of functions because no vertical line intersects either graph at more than one point.

Practice Problem 7 Decide whether the graph in Figure 1.36 is the graph of a function.

FIGURE 1.36

EXAMPLE 8 Examining the Graph of a Function

Let $f(x) = x^2 - 2x - 3$.

a. Is the point $(1, -3)$ on the graph of f?
b. Find all values of x so that $(x, 5)$ is on the graph of f.
c. Find all y-intercepts of the graph of f.
d. Find all x-intercepts of the graph of f.

Solution

a. We check whether $(1, -3)$ satisfies the equation $y = x^2 - 2x - 3$.

$$-3 \stackrel{?}{=} (1)^2 - 2(1) - 3 \quad \text{Replace } x \text{ with 1 and } y \text{ with } -3.$$
$$-3 \stackrel{?}{=} -4 \quad \text{No!}$$

So $(1, -3)$ is *not* on the graph of f. See Figure 1.37.

b. Substitute 5 for y in $y = x^2 - 2x - 3$ and solve for x.

$$5 = x^2 - 2x - 3$$
$$0 = x^2 - 2x - 8 \quad \text{Subtract 5 from both sides.}$$
$$0 = (x - 4)(x + 2) \quad \text{Factor.}$$
$$x - 4 = 0 \quad \text{or} \quad x + 2 = 0 \quad \text{Zero-product property}$$
$$x = 4 \quad \text{or} \quad x = -2 \quad \text{Solve for } x.$$

The point $(x, 5)$ is on the graph of f only when $x = -2$ or $x = 4$. Both points $(-2, 5)$ and $(4, 5)$ are on the graph of f. See Figure 1.37.

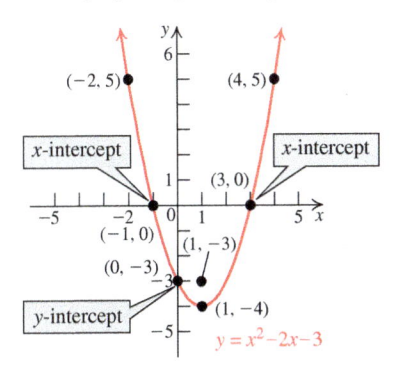

FIGURE 1.37

c. Find all points (x, y) with $x = 0$ in $y = x^2 - 2x - 3$.

$$y = 0^2 - 2(0) - 3 \quad \text{Replace } x \text{ with 0 to find the y-intercept.}$$
$$y = -3 \quad \text{Simplify.}$$

The only y-intercept is -3. (See Figure 1.37.)

d. Find all points (x, y) with $y = 0$ in $y = x^2 - 2x - 3$.

$$0 = x^2 - 2x - 3$$
$$0 = (x + 1)(x - 3) \quad \text{Factor.}$$
$$x + 1 = 0 \quad \text{or} \quad x - 3 = 0 \quad \text{Zero-product property}$$
$$x = -1 \quad \text{or} \quad x = 3 \quad \text{Solve for } x.$$

The x-intercepts of the graph of f are -1 and 3. See Figure 1.37.

Practice Problem 8 Let $y = f(x) = x^2 + 4x - 5$.

a. Is the point $(2, 7)$ on the graph of f?

b. Find all values of x so that $(x, -8)$ is on the graph of f.

c. Find the y-intercept of the graph of f.

d. Find any x-intercepts of the graph of f.

4 Get information about a function from its graph.

Function Information from Its Graph

Given a graph in the xy-plane, we use the vertical line test to determine whether the graph is that of a function. We can obtain the following information from the graph of a function.

1. **Point on a graph**

   A point (a, b) is on the graph of f means that a is in the domain of f and the value of f at a is b; that is, $f(a) = b$. We can visually determine whether a given point is on the graph of a function. In Figure 1.38, the point (a, b) is on the graph of f and the point (c, d) is not on the graph of f.

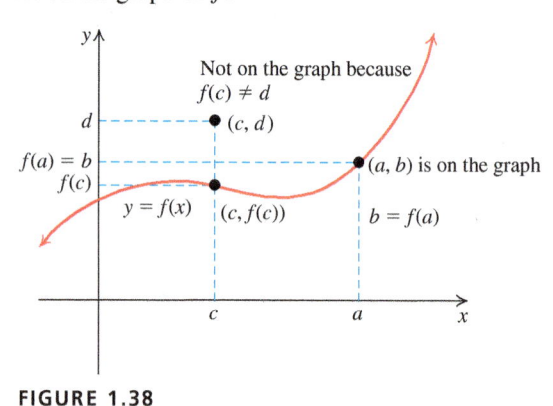

FIGURE 1.38

2. **Domain and range from a graph**

   To determine the *domain* of a function, we look for the portion on the x-axis that is used in graphing f. We can find this portion by projecting (collapsing) the graph on the x-axis. This projection is the domain of f.

   The range of f is the projection of its graph on the y-axis. See Figure 1.39.

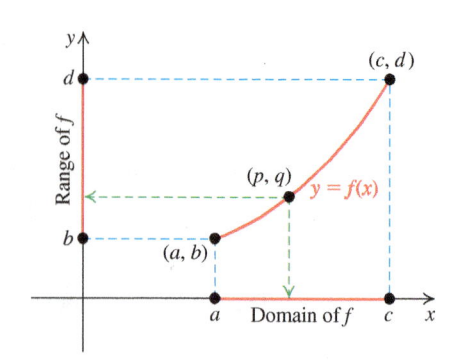

FIGURE 1.39 Domain and range of f

Section 1.3 ■ Functions 87

EXAMPLE 9 Finding the Domain and Range from a Graph

Use the graphs in Figure 1.40 to find the domain and the range of each function.

(a)

(b)

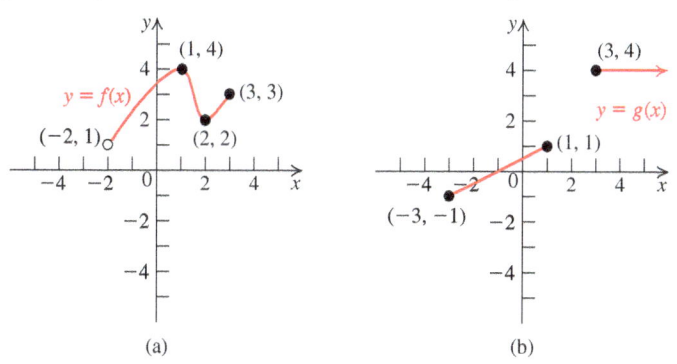

FIGURE 1.40 Domain and range from a graph

Solution

a. In Figure 1.40(a), the open circle at $(-2, 1)$ indicates that the point $(-2, 1)$ does not belong to the graph of f, and the full circle at the point $(3, 3)$ indicates that the point $(3, 3)$ is part of the graph. When we project the graph of $y = f(x)$ onto the x-axis, we obtain the interval $(-2, 3]$. The domain of f in interval notation is $(-2, 3]$. Similarly, the projection of the graph of f onto the y-axis gives the interval $(1, 4]$, which is the range of f.

b. The projection of the graph of g in Figure 1.40(b) onto the x-axis is made up of the two intervals $[-3, 1]$ and $[3, \infty)$. So the domain of g is $[-3, 1] \cup [3, \infty)$.

To find the range of g, we project its graph onto the y-axis. The projection of the line segment joining $(-3, -1)$ and $(1, 1)$ onto the y-axis is the interval $[-1, 1]$. The projection of the horizontal ray starting at the point $(3, 4)$ onto the y-axis is just a single point at $y = 4$. Therefore, the range of g is $[-1, 1] \cup \{4\}$.

Practice Problem 9 Find the domain and the range of $y = h(x)$ in Figure 1.41.

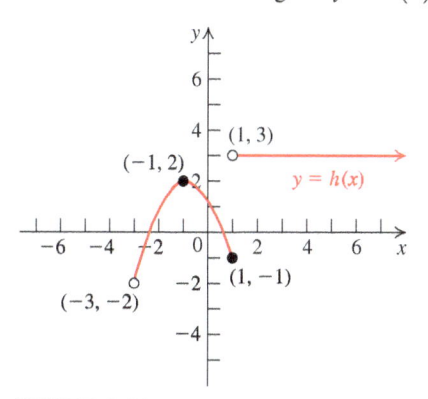

FIGURE 1.41

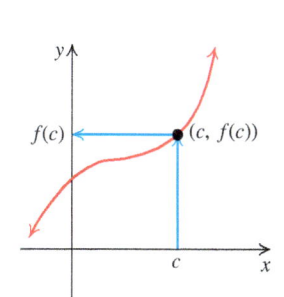

FIGURE 1.42 Locating $f(c)$ graphically

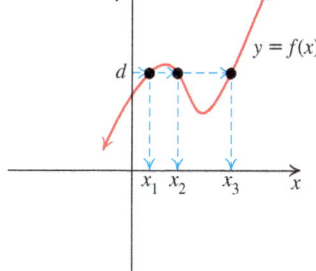

FIGURE 1.43 Graphically solving $f(x) = d$

3. **Evaluations**

    a. **Finding $f(c)$** Given a number c in the domain of f, we find $f(c)$ from the graph of f.

    First, locate the number c on the x-axis. Draw a vertical line through c. It intersects the graph at the point $(c, f(c))$. Through this point, draw a horizontal line to intersect the y-axis. Read the value $f(c)$ on the y-axis. See Figure 1.42.

    b. **Solving $f(x) = d$** Given d in the range of f, find values of x for which $f(x) = d$. Locate the number d on the y-axis. Draw a horizontal line through d. This line intersects the graph at point(s) (x, d). Through each of these points, draw vertical lines to intersect the x-axis. Read the values on the x-axis. These are the values of x for which $f(x) = d$. Figure 1.43 shows three solutions—x_1, x_2, and x_3—of the equation $f(x) = d$.

SUMMARY OF MAIN FACTS

A function is usually described in one or more of the following six ways:

- A correspondence diagram
- A set of ordered pairs
- A table of values
- An equation or a formula
- A scatterplot or a graph
- In the words "*y* is a function of *x*"

Input	Output
x	*y* or *f*(*x*)
First coordinate	Second coordinate
Independent variable	Dependent variable
Domain is the set of all inputs.	Range is the set of all outputs.

5 Solve applied problems by using functions.

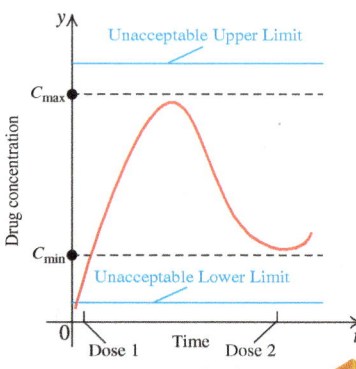

FIGURE 1.44 Drug levels

Building Functions

Level of Drugs in Bloodstream When you take a dose of medication, the drug level in the blood goes up quickly and soon reaches its *peak*, called C_{max} (the maximum concentration). As your liver or kidneys remove the drug, your blood's drug levels drop until the next dose enters your bloodstream. The lowest drug level is called the *trough*, or C_{min} (the minimum concentration). The ideal dose should be strong enough to be effective without causing too many side effects. We start by setting upper and lower limits on the drug's blood levels, shown by the two horizontal lines on a pharmacokinetics (PK) graph. See Figure 1.44. The upper line represents the drug level at which people start to develop serious side effects. The lower line represents the minimum drug level that provides the desired effect.

EXAMPLE 10   Cholesterol-Reducing Drugs

Many drugs used to lower high blood cholesterol levels are called *statins*. These drugs, along with proper diet and exercise, help prevent heart attacks and strokes. Recall from the introduction to this section that bioavailability is the amount of a drug you have ingested that makes it into your bloodstream. A statin with a bioavailability of 30% has been prescribed for Boris to treat his cholesterol levels. He is to take 20 milligrams daily. Every day his body filters out half the statin. Find the maximum concentration of the statin in the bloodstream on each of the first ten days of using the drug and graph the result.

Solution

Because the statin has 30% bioavailability and Boris takes 20 milligrams per day, the maximum concentration in the bloodstream is 30% of 20 milligrams, or 20(0.3) = 6 milligrams from each day's prescription. Because half the statin is filtered out of the body each day, the daily maximum concentration is

$$\frac{1}{2}(\text{previous days maximum concentration}) + 6.$$

So if $C(n)$ denotes the maximum concentration on the *n*th day, then

(1) $$C(n) = \frac{1}{2}C(n-1) + 6, \quad C(0) = 0.$$

Equation (1) is an example of a **recursive definition** of a function.

Table 1.4 shows the maximum concentration of the drug for each of the first ten days. After the first day, each number in the second column is computed (to three decimal places) by adding 6 to one-half the number above it. The graph is shown in Figure 1.45.

TABLE 1.4

Day	Maximum Concentration
1	6.000
2	9.000
3	10.500
4	11.250
5	11.625
6	11.813
7	11.906
8	11.953
9	11.977
10	11.988

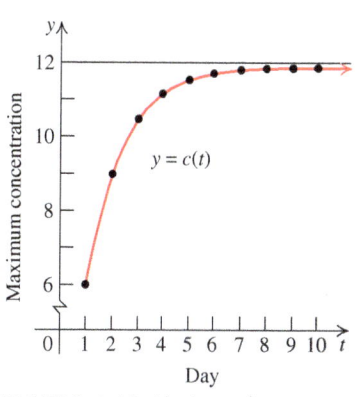

FIGURE 1.45 Maximum drug concentration

The graph shows that the maximum concentration of the statin in the bloodstream approaches 12 milligrams if Boris continues to take one pill every day.

Practice Problem 10 In Example 10, use Table 1.4 to find the domain and the range of the function. Also compute the function value at 11.

EXAMPLE 11 Cost of a Fiber-Optic Cable

Two points A and B are opposite each other on the banks of a straight river that is 500 feet wide. The point D is on the same side as B but is 1200 feet up the river from B. The local Internet cable company wants to lay a fiber-optic cable from A to D. The cost per foot of cable is \$5 per foot under water and \$3 per foot on land. To save money, the company lays the cable under water from A to P and then on land from P to D. In Figure 1.46, write the total cost C as a function of x.

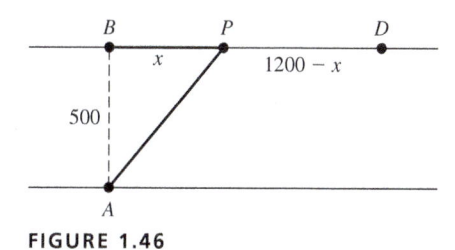

FIGURE 1.46

Solution

In Figure 1.46, we have

$$d(A, P) = AP = \sqrt{(500)^2 + x^2} \text{ feet} \quad \text{Pythagorean Theorem}$$
$$d(P, D) = PD = 1200 - x \text{ feet}$$

So the total cost C is given by

$$C = 5(AP) + 3(PD)$$
$$= 5\sqrt{(500)^2 + x^2} + 3(1200 - x).$$

Practice Problem 11 Repeat Example 11 assuming that the cost is 30% more under water than it is on land.

SECTION 1.3 Exercises

Basic Concepts and Skills

In Exercises 1–6, determine the domain and range of each relation. Explain why each non-function is not a function.

1. 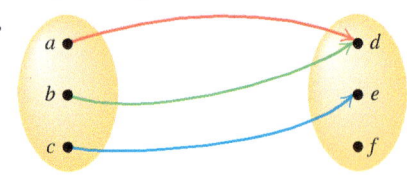
2. 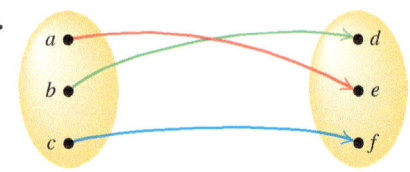
3. 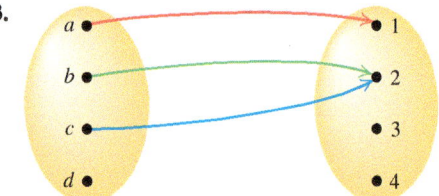
4.

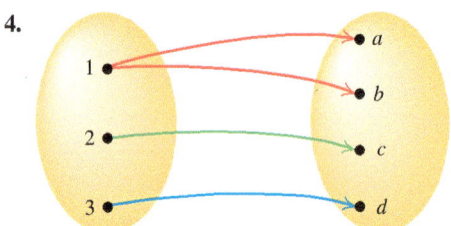

5.

x	0	3	8	0	3	8
y	−1	−2	−3	1	2	2

6.

x	−3	−1	0	1	2	3
y	−8	0	1	0	−3	−8

In Exercises 7–18, determine whether each equation defines y as a function of x.

7. $x + y = 2$
8. $x = y - 1$
9. $y = \dfrac{1}{x}$
10. $xy = -1$
11. $y^2 = x^2$
12. $x = |y|$
13. $y = \dfrac{1}{\sqrt{2x - 5}}$
14. $y = \dfrac{1}{\sqrt{x^2 - 1}}$
15. $x^2 + y^2 = 8$
16. $x = y^2$
17. $x^2 + y^3 = 5$
18. $x + y^3 = 8$

19. Let $f(x) = \dfrac{2x}{\sqrt{4 - x^2}}$. Find each function value.
    a. $f(0)$
    b. $f(1)$
    c. $f(2)$
    d. $f(-2)$
    e. $f(-x)$

20. Let $g(x) = 2x + \sqrt{x^2 - 4}$. Find each function value.
    a. $g(0)$
    b. $g(1)$
    c. $g(2)$
    d. $g(-3)$
    e. $g(-x)$

21. In the figure for Exercise 21, find the sum of the areas of the shaded rectangles.

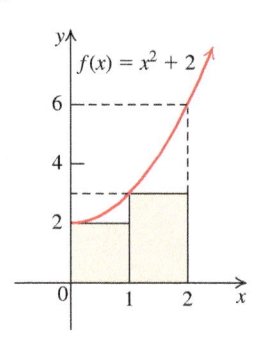

22. In the figure for Exercise 22, find the sum of the areas of the shaded rectangles.

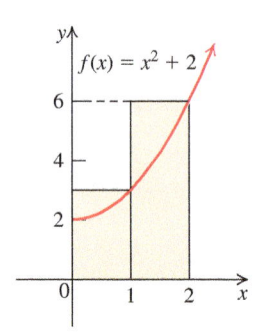

In Exercises 23–30, find the domain of each function.

23. $f(x) = \dfrac{1}{x - 9}$
24. $f(x) = \dfrac{1}{x + 9}$
25. $h(x) = \dfrac{2x}{x^2 - 1}$
26. $h(x) = \dfrac{x - 3}{x^2 - 4}$
27. $G(x) = \dfrac{\sqrt{x - 3}}{x + 2}$
28. $f(x) = \dfrac{x}{\sqrt{2 - x}}$
29. $F(x) = \dfrac{x + 4}{x^2 + 3x + 2}$
30. $F(x) = \dfrac{1 - x}{x^2 + 5x + 6}$

In Exercises 31–34, find the range of each function $f(x)$ with the given domain X.

31. $f(x) = 2x + 1; X = [1, 5]$
32. $f(x) = -2x + 3; \quad X = [-2, 4]$
33. $f(x) = -x^2 + 2; \quad X = [-2, 3]$
34. $f(x) = x^3 - 3; X = [-1, 2]$
35. Let $f(x) = |x - 1|$ with domain $X = [6, 12]$.
    a. Is 10 in the range of f?
    b. Is 12 in the range of f?
    c. Find the range of f.
36. Let $g(x) = \sqrt{x + 3} - 2$ with domain $X = [-3, 13]$.
    a. Is 3 in the range of g?
    b. Is -1 in the range of g?
    c. Find the range of g.

In Exercises 37–42, use the vertical-line test to determine whether the given graph represents a function.

37. 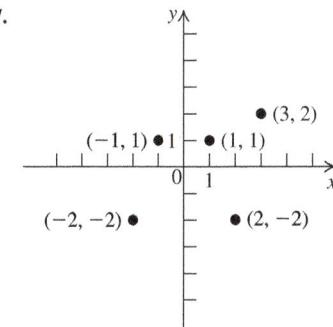
38. 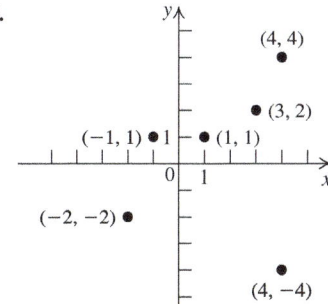
39. 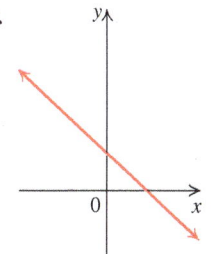
40. 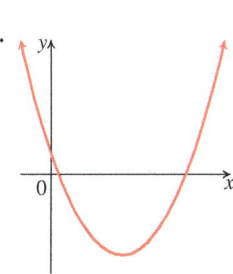
41. 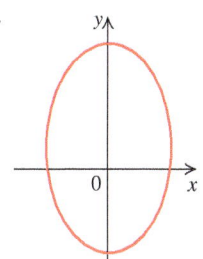
42.

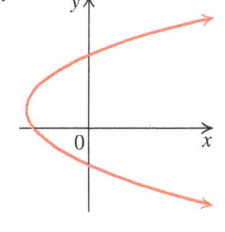

43. Let $h(x) = x^2 - x + 1$. Find x such that $(x, 7)$ is on the graph of h.
44. Let $H(x) = x^2 + x + 8$. Find x such that $(x, 7)$ is on the graph of H.
45. Let $f(x) = -2(x + 1)^2 + 7$.
    a. Is $(1, 1)$ a point of the graph of f?
    b. Find x such that $(x, 1)$ is on the graph of f.
    c. Find all y-intercepts of the graph of f.
    d. Find all x-intercepts of the graph of f.
46. Let $f(x) = -3x^2 - 12x$.
    a. Is $(-2, 10)$ a point of the graph of f?
    b. Find x such that $(x, 12)$ is on the graph of g.
    c. Find all y-intercepts of the graph of f.
    d. Find all x-intercepts of the graph of f.

For Exercises 47–54, refer to the graph of $y = f(x)$ in the figure. The axes are marked off in one-unit intervals.

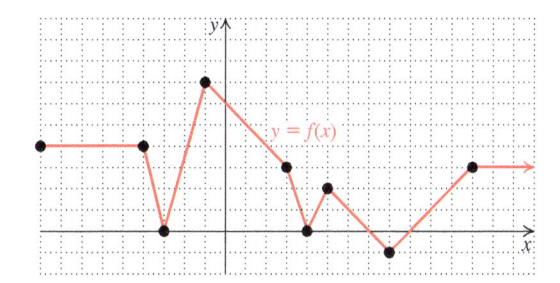

47. Find the domain of f.
48. Find the range of f.
49. Find the x-intercepts of f.
50. Find the y-intercepts of f.
51. Find $f(-7), f(1)$, and $f(5)$.
52. Find $f(-4), f(-1)$, and $f(3)$.
53. Solve the equation $f(x) = 3$.
54. Solve the equation $f(x) = -2$.

In Exercises 55–60, refer to the graph of $y = g(x)$ in the figure. The axes are marked off in one-unit intervals.

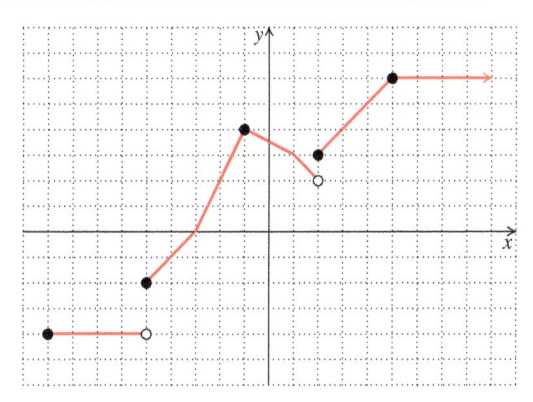

55. Find the domain of g.
56. Find the range of g.
57. Find $g(-4), g(1)$, and $g(3)$.
58. Find $|g(-5) - g(5)|$.
59. Solve $g(x) = -4$.
60. Solve $g(x) = 6$.

Applying the Concepts

61. **Square tiles.** The area $A(x)$ of a square tile is a function of the length x of the square's side. Write a function rule for the area of a square tile. Find and interpret $A(4)$.

62. **Surface area.** Is the total surface area S of a cube a function of the edge length x of the cube? If it is not a function, explain why not. If it is a function, write the function rule $S(x)$ and evaluate $S(3)$.

63. **Breaking even.** Peerless Publishing Company intends to publish Harriet Henrita's next novel. The estimated cost is $75,000 plus $5.50 for each copy printed. The selling price per copy is $15. The bookstores retain 40% of the selling price as commission. Let x represent the number of copies printed and sold.
    a. Find the cost function $C(x)$.
    b. Find the revenue function $R(x)$.
    c. Find the profit function $P(x)$.
    d. How many copies of the novel must be sold for Peerless to break even?
    e. What is the company profit if 46,000 copies are sold?

64. **Motion of a projectile.** A stone thrown upward with an initial velocity of 128 feet per second will attain a height of h feet in t seconds, where
    $$h(t) = 128t - 16t^2, \quad 0 \le t \le 8.$$
    a. What is the domain of h?
    b. Find $h(2)$, $h(4)$, and $h(6)$.
    c. How long will it take the stone to hit the ground?
    d. Sketch a graph of $y = h(t)$.

65. **Drug prescription.** A certain drug has been prescribed to treat an infection. The patient receives an injection of 16 milliliters of the drug every four hours. During this same four-hour period, the kidneys filter out one-fourth of the drug. Find the concentration of the drug after 4 hours, 8 hours, 12 hours, 16 hours, and 20 hours. Sketch the graph of the concentration of the drug in the bloodstream as a function of time.

66. **Drug prescription.** Every day Jane takes one 500-milligram aspirin tablet that has 80% bioavailability. During the same day, three-fourths of the aspirin is metabolized. Find the maximum concentration of the aspirin in the bloodstream on each of the first ten days of using the drug and graph the result.

67. **Numbers.** The sum of two numbers x and y is 28. Write the product P of these numbers as a function of x.

68. **Area of a rectangle.** The dimensions of a rectangle are x and y, and its perimeter is 60 meters. Write the area A of the rectangle as a function of x.

69. **Volume of a box.** A closed box with a square base of side x inches is to hold 64 in^3. Write the surface area S of the box as a function of x.

70. **Inscribed rectangle.** In the figure, a rectangle is inscribed in a semicircle of diameter $2r$.
    a. Write the perimeter P of the rectangle as a function of x.
    b. Write the area A of the rectangle as a function of x.

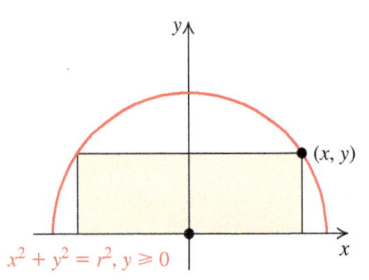

71. **Piece of wire.** A piece of wire 20 inches long is to be cut into two pieces, one to form a circle and the other to form a square. Write A, the sum of the areas of the two figures, as a function of x.

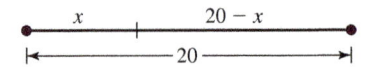

72. **Area of sheet metal.** An open cylindrical tank with circular base of radius r is to be constructed of metal to contain a volume of 64 in^3. Write the area A of the metal as a function of r.

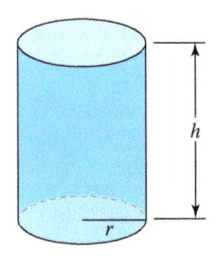

73. **Cost of a pool.** A 288-ft^3 pool is to be built with a square bottom of side length x. The sides are built with tiles; the bottom, with cement. The cost per unit of tiles is $6/ft^2, and the cost of the bottom is $2/ft^2. Write the total cost C as a function of x.

74. **Distance between cars.** Two cars are traveling along two roads that cross each other at right angles at point A. Both cars are traveling toward A for t seconds at 30 ft/sec. Initially, their distances from A are 1500 feet and 2100 feet, respectively. Write the distance d between the cars as a function of t.

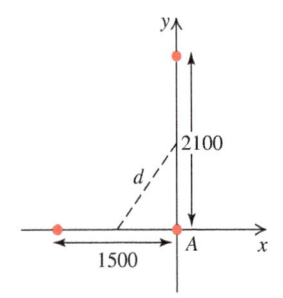

75. Distance. Write the distance d from the point $(2, 1)$ to the point (x, y) on the graph of $y = f(x) = x^3 - 3x + 6$ as a function of x.

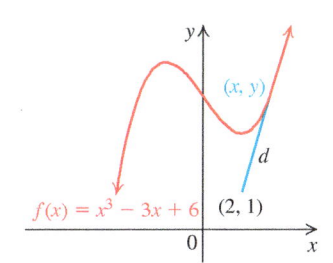

76. Time of travel. An island is at point A, 5 miles from the nearest point B on a straight beach. A store is at point C, 8 miles up the beach from point B. Julio can row 4 mi/hr and walk 5 mi/hr. He rows to point P, x miles up the beach from point B, and walks from P to C. Write the total time T of travel from A to C as a function of x.

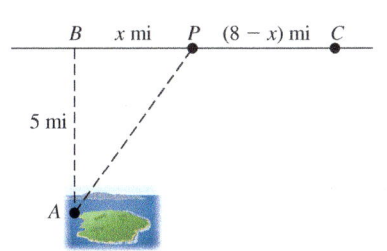

Critical Thinking / Discussion / Writing

77. Assuming that $f(x) = \dfrac{x-1}{x+1}$, show that
$$f\left(\dfrac{x-1}{x+1}\right) = -\dfrac{1}{x}.$$

78. Write an equation of a function with each domain. Answers will vary.
   a. $[2, \infty)$ **b.** $(2, \infty)$
   c. $(-\infty, 2]$ **d.** $(-\infty, 2)$

79. Consider the graph of the function
$$y = f(x) = ax^2 + bx + c, a \neq 0.$$
   a. Write an equation whose solution yields the x-intercepts.
   b. Write an equation whose solution is the y-intercept.
   c. Write (if possible) a condition under which the graph of $y = f(x)$ has no x-intercepts.
   d. Write (if possible) a condition under which the graph of $y = f(x)$ has no y-intercepts.

80. Let $X = \{a, b\}$ and $Y = \{1, 2, 3\}$.
   a. How many functions are there from X to Y?
   b. How many functions are there from Y to X?

81. If a set X has m elements and a set Y has n elements, how many functions can be defined from X to Y? Justify.

Maintaining Skills

In Exercises 82–87, factor each polynomial.

82. $3x^2 + 12x + 2x + 8$

83. $x^2 - 6x + 8$

84. $x^3 + 7x^2 + 12x$

85. $25x^2 - 16$

86. $x^3 + 5x^2 - x - 5$

87. $6x^2 + x - 12$

SECTION 1.4 A Library of Functions

BEFORE STARTING THIS SECTION, REVIEW

1. Equation of a line (Section 1.2, page 73)
2. Absolute value (Section P.1, page 7)
3. Graph of an equation (Section 1.1, page 55)

OBJECTIVES

1. Define linear functions.
2. Discuss important properties of functions.
3. Graph square root and cube root functions.
4. Evaluate and graph piecewise functions.
5. Graph basic functions.
6. Find the average rate of change of a function.

◆ The Algebra in Coughing

Coughing is the body's way of removing foreign material or mucus from the lungs and upper airway passages or reacting to an irritated airway. When you cough, the windpipe contracts to increase the velocity of the outgoing air.

The average flow velocity, v, can be modeled by the equation

$$v = c(r_0 - r)r^2 \text{ mm/sec,}$$

where r_0 is the rest radius of the windpipe in millimeters, r is its contracted radius, and c is a positive constant. In Example 3, we find the value of r that maximizes the velocity of the air going out. The diameter of the windpipe of a normal adult male varies between 25 mm and 27 mm; that of a normal adult female varies between 21 mm and 23 mm.

1 Define linear functions.

Linear Functions

We know from Section 1.1 that the graph of a linear equation $y = mx + b$ is a straight line with slope m and y-intercept b. A *linear function* has a similar definition.

Linear Functions

Let m and b be real numbers. The function $f(x) = mx + b$ is called a **linear function**. If $m = 0$, the function $f(x) = b$ is called a **constant function**. If $m = 1$ and $b = 0$, the resulting function $f(x) = x$ is called the **identity function**. See Figure 1.47.

The domain of a linear function is the interval $(-\infty, \infty)$ because the expression $mx + b$ is defined for any real number x. Figure 1.47 shows that the graph extends indefinitely to the left and right. The range of a nonconstant linear function also is the interval $(-\infty, \infty)$ because the graph extends indefinitely upward and downward. The range of a constant function $f(x) = b$ is the single real number b. See Figure 1.47(c).

Section 1.4 ■ A Library of Functions 95

GRAPH OF $f(x) = mx + b$

(a) $m > 0$

(b) $m < 0$

(c) $m = 0$, $f(x) = b$

(d) $m = 1, b = 0$, $f(x) = x$

FIGURE 1.47 The graph of a linear function is a nonvertical line with slope m and y-intercept b.

EXAMPLE 1 Writing a Linear Function

Write a linear function g for which $g(1) = 4$ and $g(-3) = -2$.

Solution

First, we need to find the equation of the line passing through the two points:

$$(x_1, y_1) = (1, 4) \text{ and } (x_2, y_2) = (-3, -2)$$

The slope of the line is given by

$$m = \frac{y_2 - y_1}{x_2 - x_1} = \frac{-2 - 4}{-3 - 1} = \frac{-6}{-4} = \frac{3}{2}.$$

Next, we use the point–slope form of a line.

$y - y_1 = m(x - x_1)$ Point–slope form of a line

$y - 4 = \frac{3}{2}(x - 1)$ Substitute $x_1 = 1$, $y_1 = 4$, and $m = \frac{3}{2}$.

$y - 4 = \frac{3}{2}x - \frac{3}{2}$ Distributive property

$y = \frac{3}{2}x + \frac{5}{2}$ Add $4 = \frac{8}{2}$ to both sides and simplify.

$g(x) = \frac{3}{2}x + \frac{5}{2}$ Function notation

The graph of the function $y = g(x)$ is shown in Figure 1.48.

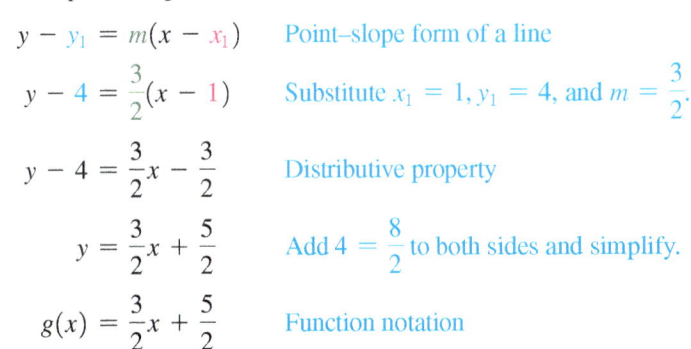

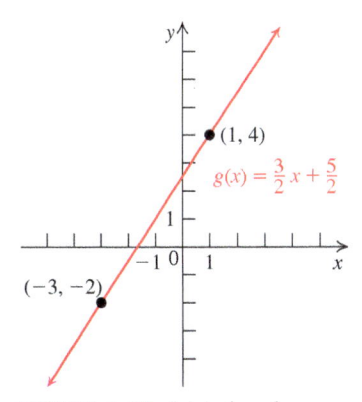

FIGURE 1.48 Point–slope form

Practice Problem 1 Write a linear function g for which $g(-2) = 2$ and $g(1) = 8$.

2 Discuss important properties of functions.

Increasing and Decreasing Functions

Classifying functions according to how one variable changes with respect to the other variable can be very useful. Imagine a particle moving from left to right along the graph of a function. This means that the x-coordinate of the point on the graph is getting larger. If the corresponding y-coordinate of the point is getting larger, getting smaller, or staying the same, then the function is called *increasing, decreasing,* or *constant,* respectively.

96　Chapter 1　Graphs and Functions

Increasing, Decreasing, and Constant Functions

Let f be a function and let x_1 and x_2 be any two numbers in an open interval (a, b) contained in the domain of f. See Figure 1.49. The symbols a and b may represent real numbers, $-\infty$, or ∞. Then

(i)　f is called an **increasing function** on (a, b) if $x_1 < x_2$ implies $f(x_1) < f(x_2)$.

(ii)　f is called a **decreasing function** on (a, b) if $x_1 < x_2$ implies $f(x_1) > f(x_2)$.

(iii)　f is **constant** on (a, b) if $x_1 < x_2$ implies $f(x_1) = f(x_2)$.

Increasing function on (a, b)
(a)

Decreasing function on (a, b)
(b)

Constant function on (a, b)
(c)

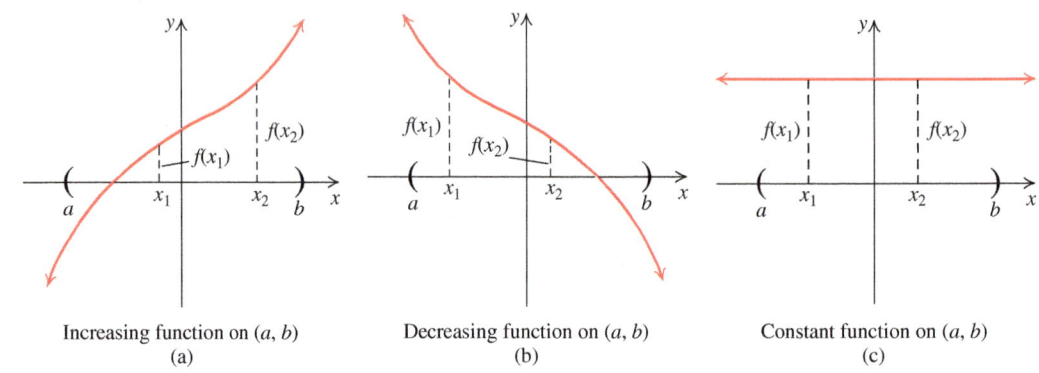

FIGURE 1.49 Increasing, decreasing, and constant functions

FIGURE 1.50 The function $f(t)$ increases on $(0, 3)$ and decreases on $(3, 6)$.

Geometrically, the graph of an increasing function on an open interval (a, b) rises as x-values increase on (a, b). This is because, by definition, as x-values increase from x_1 to x_2, the y-values also increase, from $f(x_1)$ to $f(x_2)$. Similarly, the graph of a decreasing function on (a, b) falls as x-values increase. The graph of a constant function is horizontal, or flat, over the interval (a, b).

Not every function always increases or decreases. The function that assigns the height of a ball tossed in the air to the length of time it is in the air is an example of a function that increases, but also decreases, over different intervals of its domain. See Figure 1.50.

EXAMPLE 2　Tracking the Behavior of a Function

From the graph of the function g in Figure 1.51(a), find the intervals over which g is increasing, is decreasing, or is constant.

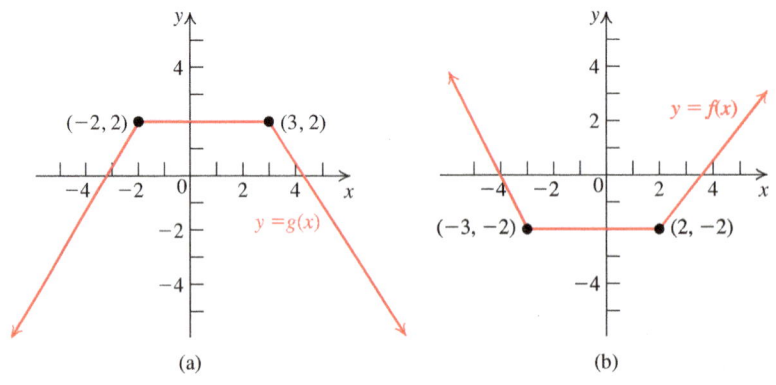

FIGURE 1.51

Solution

The function g is increasing on the interval $(-\infty, -2)$. It is constant on the interval $(-2, 3)$. It is decreasing on the interval $(3, \infty)$.

Practice Problem 2　Using the graph in Figure 1.51(b), find the intervals over which f is increasing, is decreasing, or is constant.

SIDE NOTE

When specifying the intervals over which a function f is increasing, decreasing, or constant, you need to use the intervals in the *domain* of f, *not* in the range of f.

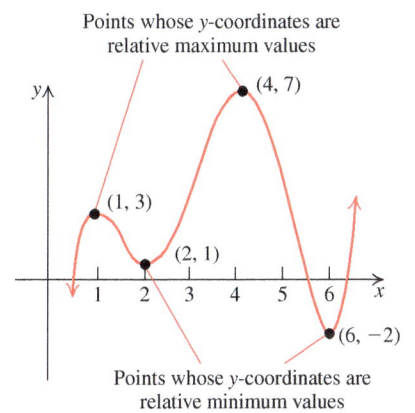

Relative Maximum and Minimum Values

The y-coordinate of a point that is not lower (not higher) than any nearby points on a graph is called a *relative maximum* (*relative minimum*) value of the function. These points are called the *relative maximum* (*relative minimum*) points of the graph. Relative maximum and minimum values are the y-coordinates of points corresponding to the peaks and troughs, respectively, of the graph.

Relative Maximum and Relative Minimum

If a is in the domain of a function f, we say that the value $f(a)$ is a **relative maximum of f** if there is an interval (x_1, x_2) containing a such that

$$f(a) \geq f(x) \text{ for every } x \text{ in the interval } (x_1, x_2).$$

We say that the value $f(a)$ is a **relative minimum of f** if there is an interval (x_1, x_2) containing a such that

$$f(a) \leq f(x) \text{ for every } x \text{ in the interval } (x_1, x_2).$$

The value $f(a)$ is called an **extreme value** of f if it is either a relative maximum value or a relative minimum value.

The graph of a function can help you find or approximate maximum or minimum values (if any) of the function. For example, the graph of a function f in Figure 1.52 shows that

- f has two relative maxima:
  (i) at $x = 1$ with relative maximum value $f(1) = 3$
  (ii) at $x = 4$ with relative maximum value $f(4) = 7$
- f has two relative minima:
  (i) at $x = 2$ with relative minimum value $f(2) = 1$,
  (ii) at $x = 6$ with relative minimum value $f(6) = -2$

In Figure 1.52, each of the points $(1, 3)$, $(2, 1)$, $(4, 7)$, and $(6, -2)$ is called a **turning point**. At a turning point, a graph changes direction from increasing to decreasing or from decreasing to increasing. In calculus, you study techniques for finding the exact points at which a function has a relative maximum or a relative minimum value (if any). For the present, however, we may settle for approximations of these points by using a graphing utility.

FIGURE 1.52 Relative maximum and minimum values

EXAMPLE 3   Algebra in Coughing

The average flow velocity, v, of outgoing air through the windpipe is modeled by

$$v = c(r_0 - r)r^2 \text{ mm/sec} \qquad \frac{r_0}{2} \leq r \leq r_0,$$

where r_0 is the rest radius of the windpipe, r is its contracted radius, and c is a positive constant. For Mr. Osborn, assume that $c = 1$ and $r_0 = 13$ mm. Use a graphing utility to estimate the value of r that will maximize the airflow v when Mr. Osborn coughs.

Solution

The graph of v for $0 \leq r \leq 13$ is shown in Figure 1.53. By using the TRACE and ZOOM features, we see that the maximum point on the graph is estimated at $(8.67, 325)$. So Mr. Osborn's windpipe contracts to a radius of 8.67 mm to maximize the airflow velocity.

FIGURE 1.53 Flow velocity

Practice Problem 3 Repeat Example 3 for Mrs. Osborn. Again, let $c = 1$ but $r_0 = 11$.

Even–Odd Functions and Symmetry

This topic, *even–odd functions*, uses ideas of symmetry discussed in Section 1.1.

Even–Odd Functions

A function f is called an **even function** if for each x in the domain of f, $-x$ is also in the domain of f and

$$f(-x) = f(x).$$

The graph of an even function is symmetric about the y-axis.

A function f is an **odd function** if for each x in the domain of f, $-x$ is also in the domain of f and

$$f(-x) = -f(x).$$

The graph of an odd function is symmetric about the origin.

EXAMPLE 4 Testing for Evenness and Oddness

Determine whether each function is even, odd, or neither.

a. $g(x) = x^3 - 4x$ **b.** $f(x) = x^2 + 5$ **c.** $h(x) = 2x^3 + x^2$

Solution

a. $g(x)$ is an odd function because

$g(-x) = (-x)^3 - 4(-x)$ Replace x with $-x$.
$= -x^3 + 4x$ Simplify.
$= -(x^3 - 4x)$ Distributive property
$= -g(x)$

b. $f(x)$ is an even function because

$f(-x) = (-x)^2 + 5$ Replace x with $-x$.
$= x^2 + 5$ Simplify.
$= f(x)$

c. $h(-x) = 2(-x)^3 + (-x)^2$ Replace x with $-x$.
$= -2x^3 + x^2$ Simplify.

Comparing $h(x) = 2x^3 + x^2$, $h(-x) = -2x^3 + x^2$, and $-h(x) = -2x^3 - x^2$, we note that $h(-x) \neq h(x)$ and $h(-x) \neq -h(x)$. Hence, $h(x)$ is neither even nor odd.

Practice Problem 4 Repeat Example 4 for each function.

a. $g(x) = 3x^4 - 5x^2$ **b.** $f(x) = 4x^5 + 2x^3$ **c.** $h(x) = 2x + 1$

3 Graph square root and cube root functions.

Square Root and Cube Root Functions

So far, we have learned to graph linear functions: $y = mx + b$. We now add some common functions to this list.

Square Root Function Recall that for every nonnegative real number x, there is only one principal square root denoted by $\sqrt{x}$. Therefore, the equation $y = \sqrt{x}$ defines a function, called the **square root function**.

EXAMPLE 5 Graphing the Square Root Function

Graph $f(x) = \sqrt{x}$.

Solution

To sketch the graph of $f(x) = \sqrt{x}$, we make a table of values. For convenience, we have selected the values of x that are perfect squares. See Table 1.5. We then plot the ordered pairs (x, y) and draw a smooth curve through the plotted points. See Figure 1.54.

TABLE 1.5

x	$y = f(x)$	(x, y)
0	$\sqrt{0} = 0$	$(0, 0)$
1	$\sqrt{1} = 1$	$(1, 1)$
4	$\sqrt{4} = 2$	$(4, 2)$
9	$\sqrt{9} = 3$	$(9, 3)$
16	$\sqrt{16} = 4$	$(16, 4)$

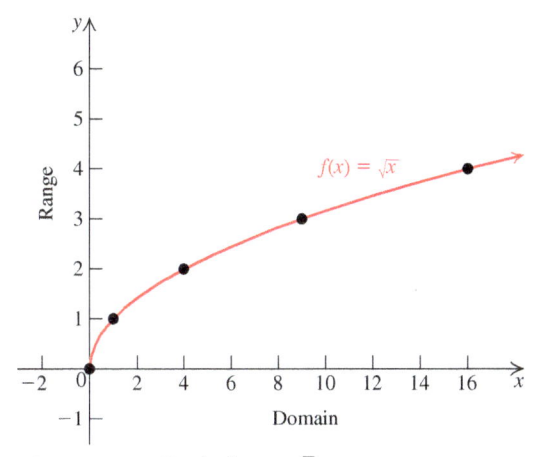

FIGURE 1.54 Graph of $y = \sqrt{x}$

The domain of $f(x) = \sqrt{x}$ is $[0, \infty)$, and its range also is $[0, \infty)$.

Practice Problem 5 Graph $g(x) = \sqrt{-x}$ and find its domain and range.

Cube Root Function There is only one cube root for each real number x. Therefore, the equation $y = \sqrt[3]{x}$ defines a function, called the **cube root function**.

EXAMPLE 6 Graphing the Cube Root Function

Graph $f(x) = \sqrt[3]{x}$.

Solution

We use the point plotting method to graph $f(x) = \sqrt[3]{x}$. For convenience, we select values of x that are perfect cubes. See Table 1.6. The graph of $f(x) = \sqrt[3]{x}$ is shown in Figure 1.55.

TABLE 1.6

x	$y = f(x)$	(x, y)
-8	$\sqrt[3]{-8} = -2$	$(-8, -2)$
-1	$\sqrt[3]{-1} = -1$	$(-1, -1)$
0	$\sqrt[3]{0} = 0$	$(0, 0)$
1	$\sqrt[3]{1} = 1$	$(1, 1)$
8	$\sqrt[3]{8} = 2$	$(8, 2)$

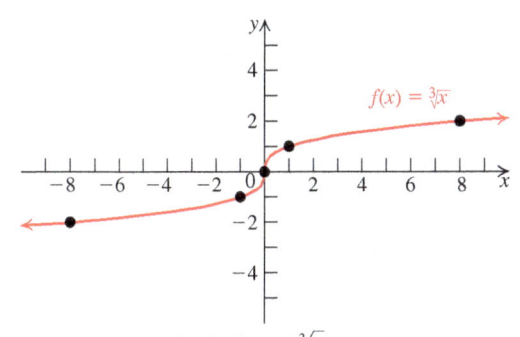

FIGURE 1.55 Graph of $y = \sqrt[3]{x}$

The domain of $f(x) = \sqrt[3]{x}$ is $(-\infty, \infty)$, and its range also is $(-\infty, \infty)$.

Practice Problem 6 Graph $g(x) = \sqrt[3]{-x}$ and find its domain and range.

4 Evaluate and graph piecewise functions.

Piecewise Functions

In the definition of some functions, different rules for assigning output values are used on different parts of the domain. Such functions are called **piecewise functions**. For example, in Peach County, Georgia, a section of the interstate highway has a speed limit of 55 miles per hour (mph). If you are caught speeding between 56 and 74 mph, your fine is $50 plus $3 for every mile per hour over 55 mph. For 75 mph and higher, your fine is $150 plus $5 for every mile per hour over 75 mph.

Let $f(x)$ be the piecewise function that represents your fine for speeding at x miles per hour. We express $f(x)$ as a piecewise function:

$$f(x) = \begin{cases} 50 + 3(x - 55), & 56 \leq x < 75 \\ 150 + 5(x - 75), & x \geq 75 \end{cases}$$

The first line of the function means that if $56 \leq x < 75$, your fine is $f(x) = 50 + 3(x - 55)$. Suppose you are caught driving 60 mph. Then we have

$f(x) = 50 + 3(x - 55)$ Expression used for $56 \leq x < 75$
$f(60) = 50 + 3(60 - 55)$ Substitute 60 for x.
$= 65$ Simplify.

Your fine for speeding at 60 mph is $65. The second line of the function means that if $x \geq 75$, your fine is $f(x) = 150 + 5(x - 75)$. Suppose you are caught driving 90 mph. Then we have

$f(x) = 150 + 5(x - 75)$ Expression used for $x \geq 75$
$f(90) = 150 + 5(90 - 75)$ Substitute 90 for x.
$= 225$ Simplify.

Your fine for speeding at 90 mph is $225.

PROCEDURE IN ACTION

EXAMPLE 7 Evaluating a Piecewise Function

OBJECTIVE
Evaluate $F(a)$ for a piecewise function F.

EXAMPLE
Let

$$F(x) = \begin{cases} x^2 & \text{if } x < 1 \\ 2x + 1 & \text{if } x \geq 1. \end{cases}$$

Find $F(0)$ and $F(2)$.

Step 1 Determine which line of the function applies to the number a.

1. Let $a = 0$. Because $0 < 1$, use the first line, $F(x) = x^2$.
Let $a = 2$. Because $2 > 1$, use the second line, $F(x) = 2x + 1$.

Step 2 Evaluate $F(a)$ using the line chosen in Step 1.

2. $F(0) = (0)^2 = 0$
$F(2) = 2(2) + 1 = 5$

Practice Problem 7

Let $f(x) = \begin{cases} x^2 & \text{if } x \leq -1 \\ 2x & \text{if } x > -1 \end{cases}$. Find $f(-2)$ and $f(3)$.

Graphing Piecewise Functions

Let's consider how to graph a piecewise function. The **absolute value function**, $f(x) = |x|$, can be expressed as a piecewise function by using the definition of absolute value:

$$f(x) = |x| = \begin{cases} -x & \text{if } x < 0 \\ x & \text{if } x \geq 0 \end{cases}$$

The first line in the function means that if $x < 0$, we use the equation $y = f(x) = -x$. So if $x = -3$, then

$$y = f(-3) = -(-3) \quad \text{Substitute } -3 \text{ for } x \text{ in } f(x) = -x.$$
$$= 3 \quad \text{Simplify.}$$

Thus, $(-3, 3)$ is a point on the graph of $y = |x|$. However, if $x \geq 0$, we use the second line in the function, which is $y = f(x) = x$. So if $x = 2$, then

$$y = f(2) = 2 \quad \text{Substitute 2 for } x \text{ in } f(x) = x.$$

So $(2, 2)$ is a point on the graph of $y = |x|$. The two pieces $y = -x$ and $y = x$ are linear functions. We graph the appropriate parts of these lines ($y = -x$ for $x < 0$ and $y = x$ for $x \geq 0$) to form the graph of $y = |x|$. See Figure 1.56.

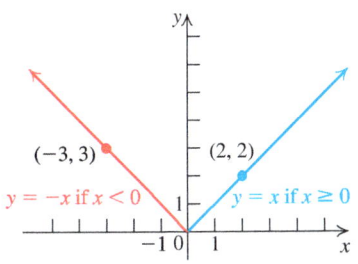

FIGURE 1.56 Graph of $y = |x|$

EXAMPLE 8 Graphing a Piecewise Function

Let

$$F(x) = \begin{cases} -2x + 1 & \text{if } x < 1 \\ 3x + 1 & \text{if } x \geq 1 \end{cases}$$

Sketch the graph of $y = F(x)$.

Solution

In the definition of F, the formula changes at $x = 1$. We call such numbers the **transition points** of the formula. For the function F, the only transition point is 1. Generally, to graph a piecewise function, we graph the function separately over the open intervals determined by the transition points and then graph the function at the transition points themselves. For the function $y = F(x)$, we graph the equation $y = -2x + 1$ on the interval $(-\infty, 1)$. See Figure 1.57(a). Next, we graph the equation $y = 3x + 1$ on the interval $(1, \infty)$ and at the transition point 1, where $y = F(1) = 3(1) + 1 = 4$. See Figure 1.57(b).

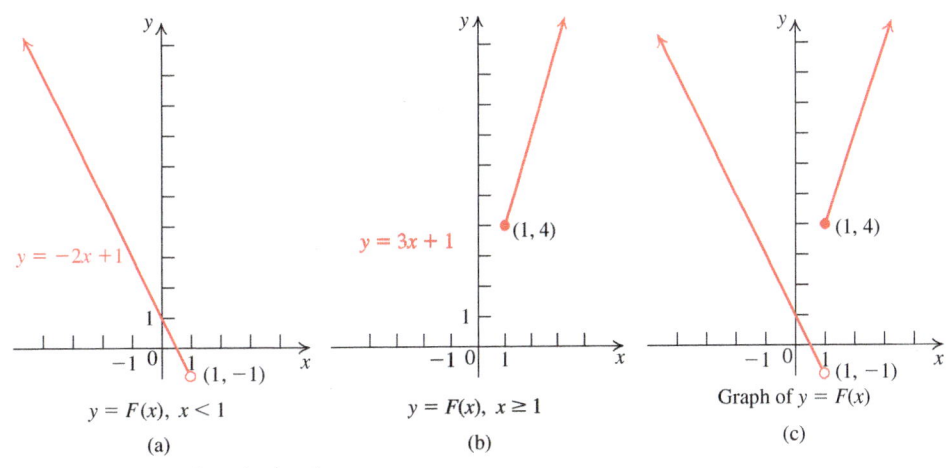

(a) $y = F(x)$, $x < 1$

(b) $y = F(x)$, $x \geq 1$

(c) Graph of $y = F(x)$

FIGURE 1.57 A piecewise function

Combining these portions, we obtain the graph of $y = F(x)$, shown in Figure 1.57(c). When we come to the edge of the part of the graph we are working with, we draw

(i) A closed circle if that point is included.
(ii) An open circle if the point is excluded.

You may find it helpful to think of this procedure as following the graph of $y = -2x + 1$ when x is less than 1 and then jumping to the graph of $y = 3x + 1$ when x is equal to or greater than 1.

Practice Problem 8 Let $F(x) = \begin{cases} -3x & \text{if } x \leq -1 \\ 2x & \text{if } x > -1 \end{cases}$. Sketch the graph of $y = F(x)$.

Some piecewise functions are called **step functions**. Their graphs look like the steps of a staircase.

The **greatest integer function** is denoted by $[\![x]\!]$, or $\text{int}(x)$, where $[\![x]\!]$ = the greatest integer less than or equal to x. For example,

$$[\![2]\!] = 2, \quad [\![2.3]\!] = 2, \quad [\![2.7]\!] = 2, \quad [\![2.99]\!] = 2$$

because 2 is the greatest integer less than or equal to 2, 2.3, 2.7, and 2.99. Similarly,

$$[\![-2]\!] = -2, \quad [\![-1.9]\!] = -2, \quad [\![-1.1]\!] = -2, \quad [\![-1.001]\!] = -2$$

because -2 is the greatest integer less than or equal to $-2, -1.9, -1.1,$ and -1.001.

In general, if m is an integer such that $m \leq x < m + 1$, then $[\![x]\!] = m$. In other words, if x is between two consecutive integers m and $m + 1$, then $[\![x]\!]$ is assigned the smaller integer m.

EXAMPLE 9 Graphing a Step Function

Graph the greatest integer function $f(x) = [\![x]\!]$.

Solution

Choose a typical closed interval between two consecutive integers—say, the interval $[2, 3]$. We know that between 2 and 3, the greatest integer function's value is 2. In symbols, if $2 \leq x < 3$, then $[\![x]\!] = 2$. Similarly, if $1 \leq x < 2$, then $[\![x]\!] = 1$. Therefore, the values of $[\![x]\!]$ are constant between each pair of consecutive integers and jump by one unit at each integer. The graph of $f(x) = [\![x]\!]$ is shown in Figure 1.58.

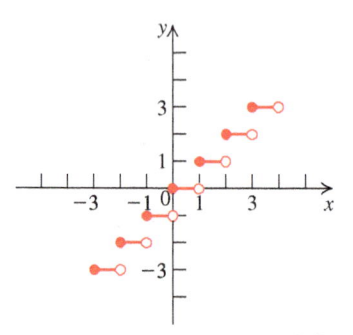

FIGURE 1.58 Graph of $y = [\![x]\!]$

Practice Problem 9 Find the values of $f(x) = [\![x]\!]$ for $x = -3.4$ and $x = 4.7$.

The greatest integer function f can be interpreted as a piecewise function:

$$f(x) = [\![x]\!] = \begin{cases} \vdots \\ -2 & \text{if } -2 \leq x < -1 \\ -1 & \text{if } -1 \leq x < 0 \\ 0 & \text{if } 0 \leq x < 1 \\ 1 & \text{if } 1 \leq x < 2 \\ \vdots \end{cases}$$

5 Graph basic functions.

Basic Functions

As you progress through this course and future mathematics courses, you will repeatedly come across a small list of basic functions. The following box lists some of these common algebraic functions, along with their properties. You should try to produce these graphs by plotting points and using symmetries. The unit length in all of the graphs shown is the same on both axes.

SUMMARY OF MAIN FACTS

A Library of Basic Functions

Constant Function
$f(x) = c$

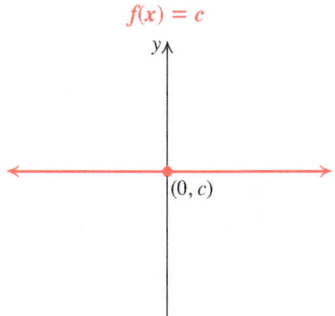

Domain: $(-\infty, \infty)$
Range: $\{c\}$
Constant on $(-\infty, \infty)$
Even function (y-axis symmetry)

Identity Function
$f(x) = x$

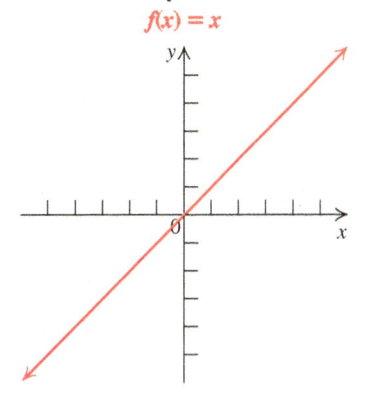

Domain: $(-\infty, \infty)$
Range: $(-\infty, \infty)$
Increasing on $(-\infty, \infty)$
Odd function (origin symmetry)

Squaring Function
$f(x) = x^2$

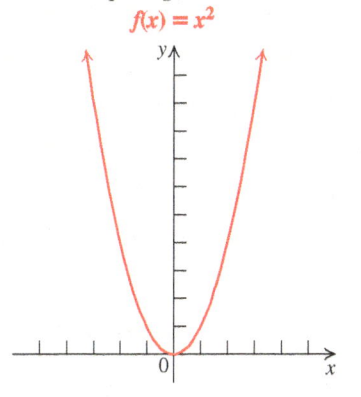

Domain: $(-\infty, \infty)$
Range: $[0, \infty)$
Decreasing on $(-\infty, 0)$
Increasing on $(0, \infty)$
Even function (y-axis symmetry)

Cubing Function
$f(x) = x^3$

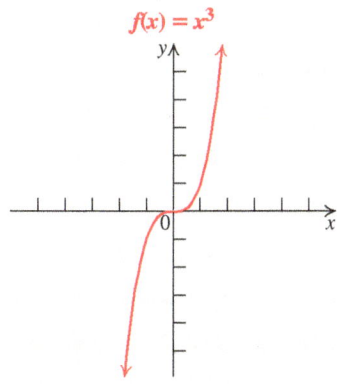

Domain: $(-\infty, \infty)$
Range: $(-\infty, \infty)$
Increasing on $(-\infty, \infty)$
Odd function (origin symmetry)

Absolute Value Function
$f(x) = |x|$

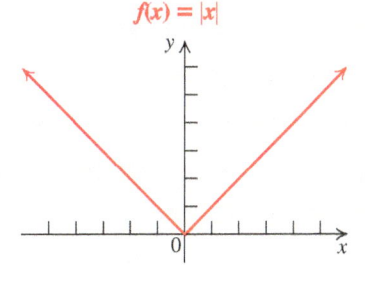

Domain: $(-\infty, \infty)$
Range: $[0, \infty)$
Decreasing on $(-\infty, 0)$
Increasing on $(0, \infty)$
Even function (y-axis symmetry)

Square Root Function
$f(x) = \sqrt{x} = x^{1/2}$

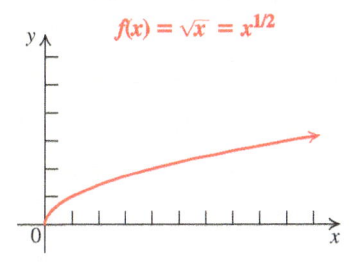

Domain: $[0, \infty)$
Range: $[0, \infty)$
Increasing on $(0, \infty)$
Neither even nor odd (no symmetry)

(continued)

Cube Root Function

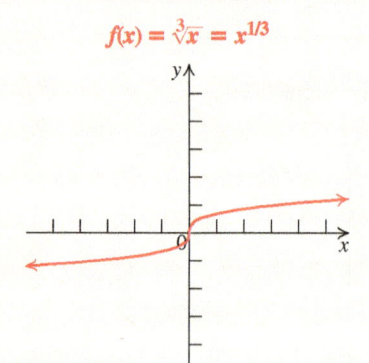

$f(x) = \sqrt[3]{x} = x^{1/3}$

Domain: $(-\infty, \infty)$
Range: $(-\infty, \infty)$
Increasing on $(-\infty, \infty)$
Odd function (origin symmetry)

Reciprocal Function

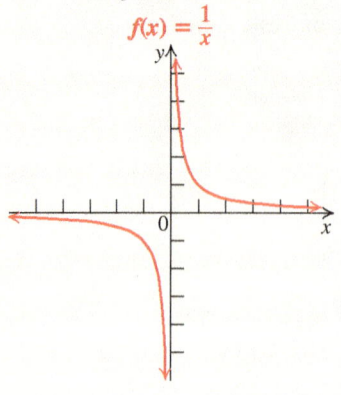

$f(x) = \dfrac{1}{x}$

Domain: $(-\infty, 0) \cup (0, \infty)$
Range: $(-\infty, 0) \cup (0, \infty)$
Decreasing on $(-\infty, 0) \cup (0, \infty)$
Odd function (origin symmetry)

Reciprocal Square Function

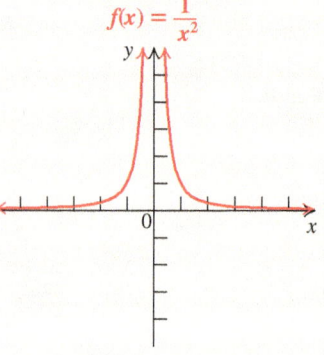

$f(x) = \dfrac{1}{x^2}$

Domain: $(-\infty, 0) \cup (0, \infty)$
Range: $(0, \infty)$
Increasing on $(-\infty, 0)$
Decreasing on $(0, \infty)$
Even function (y-axis symmetry)

Rational Power Functions

$f(x) = x^{\frac{3}{2}} = \left(x^{\frac{1}{2}}\right)^3$

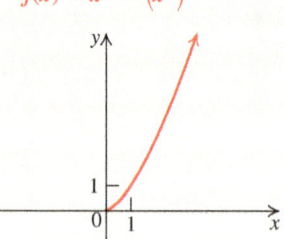

Domain: $[0, \infty)$
Range: $[0, \infty)$
Increasing on $(0, \infty)$
Neither even nor odd
(no symmetry)

$f(x) = x^{\frac{2}{3}} = \left(x^{\frac{1}{3}}\right)^2$

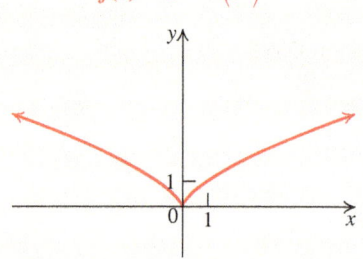

Domain: $(-\infty, \infty)$
Range: $[0, \infty)$
Decreasing on $(-\infty, 0)$
Increasing on $(0, \infty)$
Even function (y-axis symmetry)

Greatest Integer Function

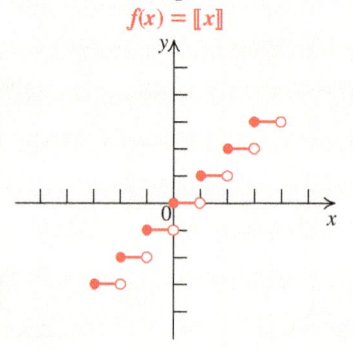

$f(x) = [\![x]\!]$

Domain: $(-\infty, \infty)$
Range: $\{\ldots -3, -2, -1, 0, 1, 2, 3, \ldots\}$
Neither even nor odd (no symmetry)

6 Find the average rate of change of a function.

Average Rate of Change

Suppose your salary increases from $25,000 a year to $40,000 a year over a five-year period. You then have the following:

Change in salary: $40,000 − $25,000 = $15,000

Average rate of change in salary:

$$\dfrac{\$40{,}000 - \$25{,}000}{5} = \dfrac{\$15{,}000}{5} = \$3000 \text{ per year}$$

Regardless of when you actually received the raises during the five-year period, the final salary is the same as if you received your average annual increase of $3000 *each* year.

The average rate of change can be defined in a more general setting.

Section 1.4 ■ A Library of Functions   105

Average Rate of Change of a Function

Let $(a, f(a))$ and $(b, f(b))$ be two points on the graph of a function f. Then the **average rate of change** of $f(x)$ as x changes from a to b is defined by

$$\frac{f(b) - f(a)}{b - a}, \quad a \neq b.$$

Note that the average rate of change of f as x changes from a to b is the slope of the **secant line** passing through the two points $(a, f(a))$ and $(b, f(b))$ on the graph of f. See Figure 1.59.

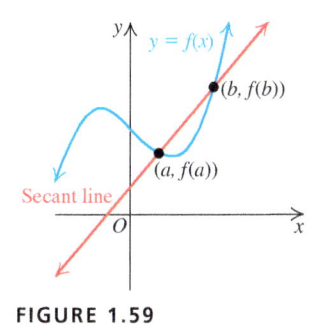

FIGURE 1.59

PROCEDURE
IN ACTION

EXAMPLE 10   Finding the Average Rate of Change

OBJECTIVE

Find the average rate of change of a function f as x changes from a to b.

Step 1 Find $f(a)$ and $f(b)$.

Step 2 Use the values from Step 1 in the definition of average rate of change.

EXAMPLE

Find the average rate of change of $f(x) = 2 - 3x^2$ as x changes from $x = 1$ to $x = 3$.

1. $f(1) = 2 - 3(1)^2 = -1$ $a = 1$
    $f(3) = 2 - 3(3)^2 = -25$ $b = 3$

2. $\dfrac{f(b) - f(a)}{b - a} = \dfrac{-25 - (-1)}{3 - 1} = -\dfrac{24}{2}$
    $= -12$

Practice Problem 10 Find the average rate of change of $f(x) = 1 - x^2$ as x changes from 2 to 4.

EXAMPLE 11   Finding the Average Rate of Change

Find the average rate of change of $f(t) = 2t^2 - 3$ as t changes from $t = 5$ to $t = x, x \neq 5$.

Solution

$$\text{Average rate of change} = \frac{f(b) - f(a)}{b - a} \qquad \text{Definition}$$

$$= \frac{f(x) - f(5)}{x - 5} \qquad b = x, a = 5$$

$$\text{Average rate of change} = \frac{(2x^2 - 3) - (2 \cdot 5^2 - 3)}{x - 5} \qquad \text{Substitute for } f(x) \text{ and } f(5).$$

$$= \frac{2x^2 - 3 - 2 \cdot 5^2 + 3}{x - 5}$$

$$= \frac{2(x^2 - 5^2)}{x - 5} \qquad \text{Simplify.}$$

$$= \frac{2(x + 5)(x - 5)}{x - 5} \qquad \text{Difference of squares}$$

$$= 2x + 10 \qquad \text{Simplify: } x - 5 \neq 0.$$

Practice Problem 11 Find the average rate of change of $f(t) = 1 - t$ as t changes from $t = 2$ to $t = x, x \neq 2$.

The average rate of change calculated in Example 11 is a *difference quotient* at 5. The difference quotient is an important concept in calculus.

Difference Quotient

For a function f, the **difference quotient** is

$$\frac{f(x+h) - f(x)}{h}, \quad h \neq 0.$$

EXAMPLE 12   Evaluating and Simplifying a Difference Quotient

Let $f(x) = 2x^2 - 3x + 5$. Find and simplify $\dfrac{f(x+h) - f(x)}{h}, h \neq 0$.

Solution

First, we find $f(x + h)$.

$$\begin{aligned}
f(x) &= 2x^2 - 3x + 5 & &\text{Original equation} \\
f(x+h) &= 2(x+h)^2 - 3(x+h) + 5 & &\text{Replace } x \text{ with } (x+h). \\
&= 2(x^2 + 2xh + h^2) - 3(x+h) + 5 & &(x+h)^2 = x^2 + 2xh + h^2 \\
&= 2x^2 + 4xh + 2h^2 - 3x - 3h + 5 & &\text{Use the distributive property.}
\end{aligned}$$

Then we substitute into the difference quotient.

$$\begin{aligned}
\frac{f(x+h) - f(x)}{h} &= \frac{\overbrace{(2x^2 + 4xh + 2h^2 - 3x - 3h + 5)}^{f(x+h)} - \overbrace{(2x^2 - 3x + 5)}^{f(x)}}{h} \\
&= \frac{2x^2 + 4xh + 2h^2 - 3x - 3h + 5 - 2x^2 + 3x - 5}{h} & &\text{Subtract.} \\
&= \frac{4xh - 3h + 2h^2}{h} & &\text{Simplify.} \\
&= \frac{h(4x - 3 + 2h)}{h} & &\text{Factor out } h. \\
&= 4x - 3 + 2h & &\text{Remove the common factor } h.
\end{aligned}$$

Practice Problem 12   Repeat Example 12 for $f(x) = -x^2 + x - 3$.

SECTION 1.4   Exercises

Basic Concepts and Skills

In Exercises 1–10, write a linear function f that has the indicated values. Sketch the graph of f.

1. $f(0) = 1, f(-1) = 0$
2. $f(1) = 0, f(2) = 1$
3. $f(-1) = 1, f(2) = 7$
4. $f(-1) = -5, f(2) = 4$
5. $f(1) = 1, f(2) = -2$
6. $f(1) = -1, f(3) = 5$
7. $f(-2) = 2, f(2) = 4$
8. $f(2) = 2, f(4) = 5$
9. $f(0) = -1, f(3) = -3$
10. $f(1) = \dfrac{1}{4}, f(4) = -2$

In Exercises 11–20, sketch the graph and find the intervals over which the given function is increasing, is decreasing, or is constant.

11. $f(x) = 2$
12. $g(x) = -3$
13. $h(x) = 3x + 4$
14. $g(x) = -2x + 5$
15. $f(x) = 5x^2$
16. $h(x) = -x^3$
17. $g(x) = 2|x|$
18. $f(x) = -|x|$
19. $f(x) = -\sqrt[3]{x}$
20. $g(x) = -\sqrt{x}$

In Exercises 21–34, determine whether the given function is even, odd, or neither. The domain for each of the functions in Exercises 21–24 is $\{-3, -2, -1, 0, 1, 2, 3\}$.

	x	-3	-2	-1	0	1	2	3
21.	$f(x)$	5	-2	1	0	1	-2	5
22.	$g(x)$	4	-3	0	2	0	3	-4
23.	$h(x)$	3	2	1	0	5	2	-3
24.	$p(x)$	2	3	4	0	-4	-3	-2

25. $f(x) = 2x^4 + 4$
26. $g(x) = 3x^4 - 5$
27. $f(x) = 5x^3 - 3x$
28. $g(x) = 2x^3 + 4x$
29. $f(x) = \dfrac{1}{x^2 + 4}$
30. $g(x) = \dfrac{x}{x^2 + 1}$
31. $f(x) = \dfrac{x^3}{x^2 + 1}$
32. $g(x) = \dfrac{x^4 + 3}{2x^3 - 3x}$
33. $f(x) = \dfrac{x^2 - 2x}{5x^4 + 4x^2 + 7}$
34. $g(x) = \dfrac{x^2 + 7}{3x^4 + 16x^2 + 9}$

In Exercises 35–42, the graph of a function is given. Use the graph to find each of the following:
a. The domain and the range of the function
b. The intercepts, if any
c. The intervals on which the function is increasing, is decreasing, or is constant
d. Whether the function is even, odd, or neither

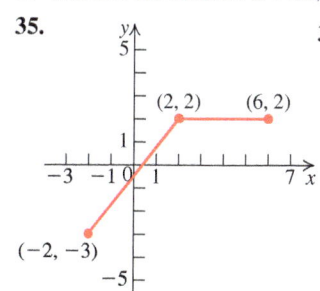

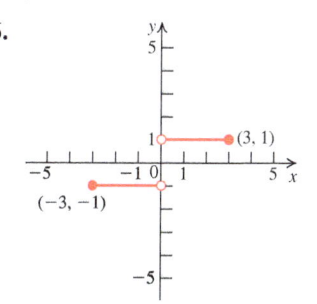

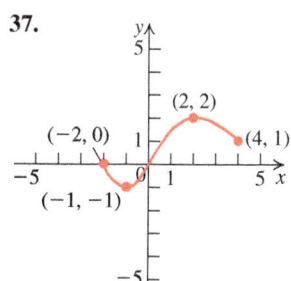

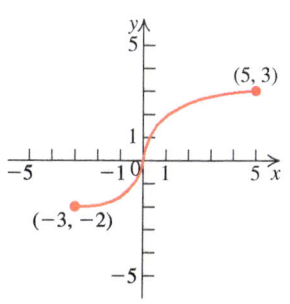

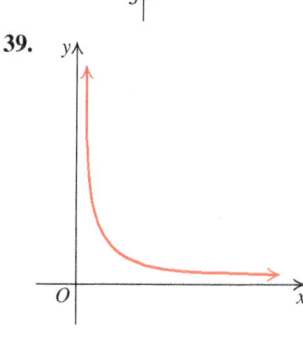

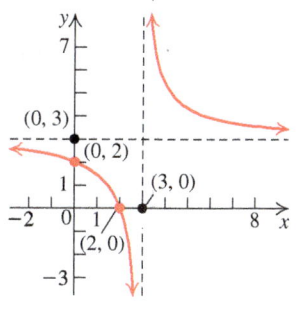

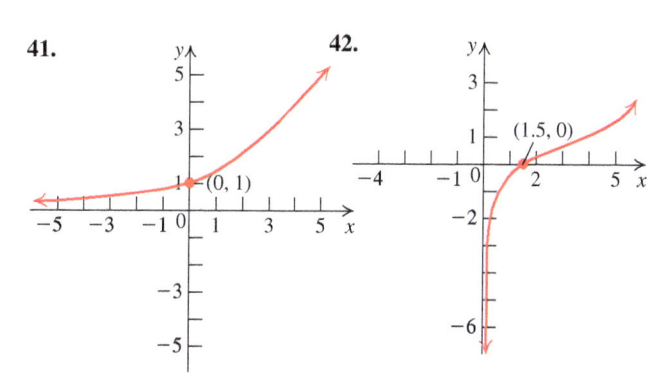

35. [graph with points $(-2, -3)$, $(2, 2)$, $(6, 2)$]
36. [graph with points $(-3, -1)$, $(3, 1)$]
37. [graph with points $(-2, 0)$, $(-1, -1)$, $(2, 2)$, $(4, 1)$]
38. [graph with points $(-3, -2)$, $(5, 3)$]
39. [graph]
40. [graph with points $(0, 3)$, $(0, 2)$, $(3, 0)$, $(2, 0)$]
41. [graph with point $(0, 1)$]
42. [graph with point $(1.5, 0)$]

43. Let
$$f(x) = \begin{cases} x \text{ if } x \geq 2 \\ 2 \text{ if } x < 2 \end{cases}$$
a. Find $f(1), f(2)$, and $f(3)$.
b. Sketch the graph of $y = f(x)$.

44. Let
$$g(x) = \begin{cases} 2x \text{ if } x < 0 \\ x \text{ if } x \geq 0 \end{cases}$$
a. Find $g(-1), g(0)$, and $g(1)$.
b. Sketch the graph of $y = g(x)$.

45. Let
$$f(x) = \begin{cases} 1 \text{ if } x > 0 \\ -1 \text{ if } x < 0 \end{cases}$$
a. Find $f(-15)$ and $f(12)$.
b. Sketch the graph of $y = f(x)$.
c. Find the domain and the range of f.

46. Let
$$g(x) = \begin{cases} 2x + 4 \text{ if } x > 1 \\ x + 2 \text{ if } x \leq 1 \end{cases}$$
a. Find $g(-3), g(1)$, and $g(3)$.
b. Sketch the graph of $y = g(x)$.
c. Find the domain and the range of g.

In Exercises 47–52, sketch the graph of each piecewise function. From the graphs find the range of each function.

47. $f(x) = \begin{cases} x^2 & \text{if } x \geq 2 \\ 3x - 2 & \text{if } x < 2 \end{cases}$
48. $g(x) = \begin{cases} 2x^2 & \text{if } x \geq 1 \\ |x| & \text{if } x < 1 \end{cases}$
49. $g(x) = \begin{cases} \sqrt[3]{x} & \text{if } x \geq 8 \\ -|x| & \text{if } x < 8 \end{cases}$
50. $h(x) = \begin{cases} \dfrac{1}{x} & \text{if } x \geq 2 \\ x & \text{if } x < 2 \end{cases}$
51. $f(x) = \begin{cases} 2 & \text{if } x \geq 3 \\ [x] & \text{if } 1 \leq x < 3 \\ -|x| & \text{if } x < 1 \end{cases}$
52. $f(x) = \begin{cases} -x^2 & \text{if } x \geq 1 \\ |x| & \text{if } -2 \leq x < 1 \\ 2x & \text{if } x < -2 \end{cases}$

In Exercises 53–56, find the average rate of change of the function as x changes from a to b.

53. $f(x) = -2x + 7; a = -1, b = 3$
54. $f(x) = 4x - 9; a = -2, b = 2$

55. $g(x) = 2x^2; a = 0, b = 5$
56. $h(x) = 2 - x^2; a = 3, b = 4$

In Exercises 57–60, find and simplify the difference quotient of the form $\dfrac{f(x) - f(a)}{x - a}, x \neq a$.

57. $f(x) = 3x^2 + x, a = 2$
58. $f(x) = -2x^2 + x, a = 3$
59. $f(x) = \dfrac{4}{x}, a = 1$
60. $f(x) = -\dfrac{4}{x}, a = 1$

In Exercises 61–64, find and simplify the difference quotient of the form $\dfrac{f(x + h) - f(x)}{h}, h \neq 0$.

61. $f(x) = 2x^2 + 3x$
62. $f(x) = 3x^2 - 2x + 5$
63. $f(x) = 4$
64. $f(x) = \dfrac{1}{x}$

Applying the Concepts

Motion from a graph. For Exercises 65–74, use the accompanying graph of a particle moving on a coordinate line with velocity $v = f(t)$ in ft/sec at time t seconds. The axes are marked off at one-unit intervals. Use these terms to describe the motion state: moving forward/backward, increasing/decreasing speed, and resting. Recall that speed $= |\text{velocity}|$.

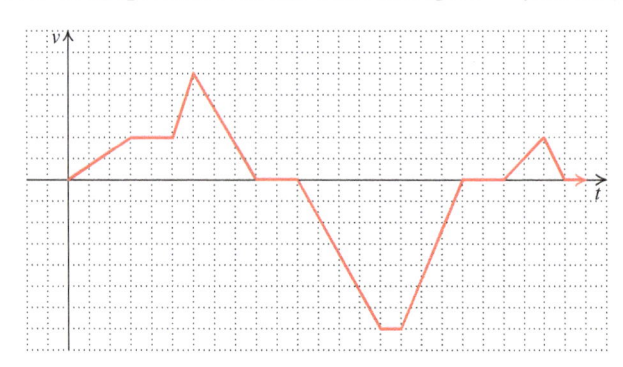

65. Find the domain of $v = f(t)$. Interpret.
66. Find the range of $v = f(t)$. Interpret.
67. Find the intervals over which the graph is above the t-axis. Interpret.
68. Find the intervals over which the graph is below the t-axis. Interpret.
69. Find the intervals over which $|v| = |f(t)|$ is increasing. Interpret for $v > 0$ and for $v < 0$.
70. Find the intervals over which $|v| = |f(t)|$ is decreasing. Interpret for $v > 0$ and for $v < 0$.
71. When does the particle have maximum speed?
72. When does the particle have minimum speed?
73. Describe the motion of the particle between $t = 5$ and $t = 6$.
74. Describe the motion of the particle between $t = 16$ and $t = 19$.
75. **State income tax.** Suppose a state's income tax code states that the tax liability T on x dollars of taxable income is as follows:

$$T(x) = \begin{cases} 0.04x & \text{if } 0 \leq x < 20{,}000 \\ 800 + 0.06(x - 20{,}000) & \text{if } x \geq 20{,}000 \end{cases}$$

   a. Graph the function $y = T(x)$.
   b. Find the tax liability on each taxable income.
      (i) $12,000
      (ii) $20,000
      (iii) $50,000
   c. Find your taxable income if you had each tax liability.
      (i) $600
      (ii) $1400
      (iii) $2300

76. **Parking cost.** The cost C of parking a car at the metropolitan airport is $4 for the first hour and $2 for each additional hour or fraction thereof. Write the cost function $C(x)$, where x is the number of parking hours, in terms of the greatest integer function.

Critical Thinking / Discussion / Writing

77. Give an example (if possible) of a function matching each description. There are many different correct answers.
    a. Its graph is symmetric with respect to the y-axis.
    b. Its graph is symmetric with respect to the x-axis.
    c. Its graph is symmetric with respect to the origin.
    d. Its graph consists of a single point.
    e. Its graph is a horizontal line.
    f. Its graph is a vertical line.
78. What function is both odd and even?

Let f be a function with domain $[a, c]$ or $[c, b]$. Then $a, b,$ and c are endpoints of this domain.

79. Define relative maximum at c.
80. Define relative minimum at c.
81. Give an example (if possible) of a function f matching each description.
    a. f has endpoint maximum and endpoint minimum.
    b. f has endpoint minimum but no endpoint maximum.
    c. f has endpoint maximum but no endpoint minimum.
    d. f has neither endpoint maximum nor endpoint minimum.

Maintaining Skills

In Exercises 82–87, simplify each expression.

82. $3\sqrt{2} + 2\sqrt{2}$
83. $\sqrt[3]{54} + \sqrt[3]{16}$
84. $\dfrac{2 \pm \sqrt{8}}{2}$
85. $\dfrac{4 \pm \sqrt{12}}{6}$
86. $\sqrt[3]{-x^6}$
87. $\sqrt{\sqrt{x^8}}$

SECTION 1.5

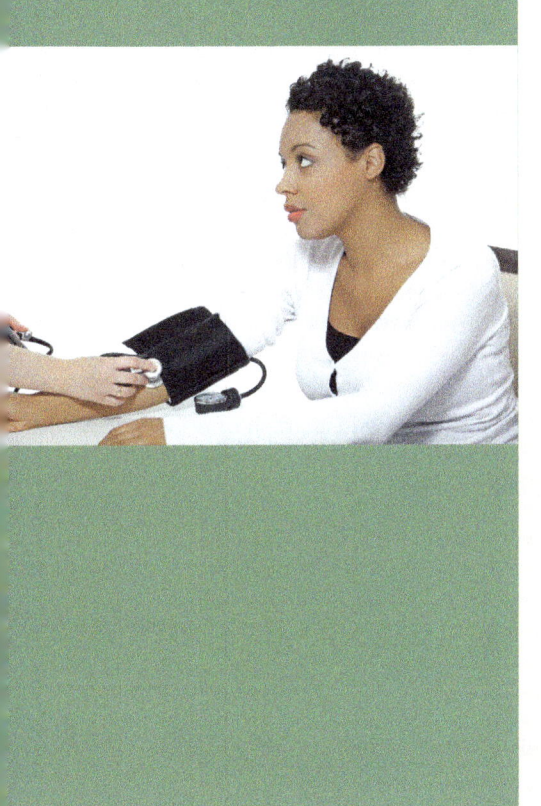

Transformations of Functions

BEFORE STARTING THIS SECTION, REVIEW

1. Graphs of basic functions (Section 1.4, pages 103 and 104)
2. Symmetry (Section 1.1, page 57)
3. Completing the square (Section P.3, page 24)

OBJECTIVES

1. Define transformations of functions.
2. Use vertical or horizontal shifts to graph functions.
3. Use reflections to graph functions.
4. Use stretching or compressing to graph functions.

◆ Measuring Blood Pressure

Blood pressure is the force of blood per unit area against the walls of the arteries. Blood pressure is recorded as two numbers: the systolic pressure (as the heart beats) and the diastolic pressure (as the heart relaxes between beats). If your blood pressure is "120 over 80," it is rising to a maximum of 120 millimeters of mercury as the heart beats and is falling to a minimum of 80 millimeters of mercury as the heart relaxes.

The most precise way to measure blood pressure is to place a small glass tube in an artery and let the blood flow out and up as high as the heart can pump it. That was the way Stephen Hales (1677–1761), the first investigator of blood pressure, learned that a horse's heart could pump blood 8 feet, 3 inches, up a tall tube. Such a test, however, consumed a great deal of blood. Jean Louis Marie Poiseuille (1797–1869) greatly improved the process by using a mercury-filled manometer, which allowed a smaller, narrower tube.

In Example 9, we investigate Poiseuille's Law for arterial blood flow.

1 Define transformations of functions.

Transformations

If a new function is formed by performing certain operations on a given function f, then the graph of the new function is called a **transformation** of the graph of f. For example, the graphs of $y = |x| + 2$ and $y = |x| - 3$ are transformations of the graph of $y = |x|$; that is, the graph of each is a special modification of the graph of $y = |x|$.

2 Use vertical or horizontal shifts to graph functions.

Vertical and Horizontal Shifts

A transformation that changes only the position of a graph but not its shape is called a **rigid transformation**. We first consider rigid transformations involving **vertical shifts** and **horizontal shifts** of a graph of a function.

EXAMPLE 1 Graphing Vertical Shifts

Let $f(x) = |x|$, $g(x) = |x| + 2$, and $h(x) = |x| - 3$. Sketch the graphs of these functions on the same coordinate plane. Describe how the graphs of g and h relate to the graph of f.

110 Chapter 1 Graphs and Functions

Solution

Make a table of values and graph the equations $y = f(x)$, $y = g(x)$, and $y = h(x)$.

TABLE 1.7

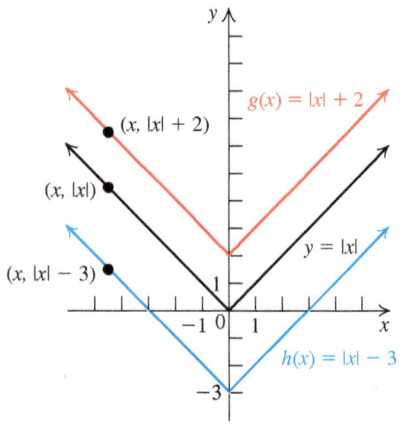

x	$y = |x|$	$g(x) = |x| + 2$	$h(x) = |x| - 3$
-5	5	7	2
-3	3	5	0
-1	1	3	-2
0	0	2	-3
1	1	3	-2
3	3	5	0
5	5	7	2

FIGURE 1.60 Vertical shifts of $y = |x|$

Notice that in Table 1.7 and Figure 1.60, each point $(x, |x|)$ on the graph of f has a corresponding point $(x, |x| + 2)$ on the graph of g, and $(x, |x| - 3)$ on the graph of h. So the graph of $g(x) = |x| + 2$ is the graph of $y = |x|$ shifted two units up, and the graph of $h(x) = |x| - 3$ is the graph of $y = |x|$ shifted three units down.

Practice Problem 1 Let

$$f(x) = x^3, g(x) = x^3 + 1, \text{ and } h(x) = x^3 - 2.$$

Sketch the graphs of all three functions on the same coordinate plane. Describe how the graphs of g and h relate to the graph of f.

Example 1 illustrates the concept of the vertical (up or down) shift of a graph.

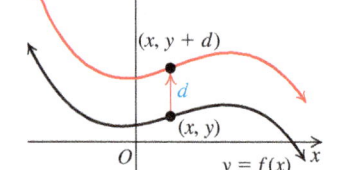

VERTICAL SHIFT

Let $d > 0$. The graph of $y = f(x) + d$ is the graph of $y = f(x)$ shifted d units *up*, and the graph of $y = f(x) - d$ is the graph of $y = f(x)$ shifted d units *down*. See Figure 1.61.

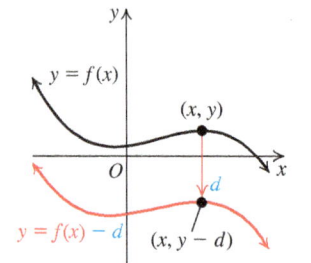

So if (x, y) is a point on the graph of $y = f(x)$ and the graph of f is shifted by $d > 0$ upward, then the point $(x, y + d)$ is on the graph of $y = f(x) + d$. That is, the x-coordinate of a point is unchanged when we apply a vertical shift.

Next, we consider the operation that shifts a graph horizontally.

EXAMPLE 2 Writing Functions for Horizontal Shifts

Let $f(x) = x^2$, $g(x) = (x - 2)^2$, and $h(x) = (x + 3)^2$. A table of values for f, g, and h is given in Table 1.8. The three functions f, g, and h are graphed on the same coordinate plane in Figure 1.62. Describe how the graphs of g and h relate to the graph of f.

FIGURE 1.61 Vertical shifts

Section 1.5 ■ Transformations of Functions 111

TABLE 1.8

x	$y = x^2$	$y = (x - 2)^2$	x	$y = x^2$	$y = (x + 3)^2$
−4	16	36	−4	16	1
−3	9	25	−3	9	0
−2	4	16	−2	4	1
−1	1	9	−1	1	4
0	0	4	0	0	9
1	1	1	1	1	16
2	4	0	2	4	25
3	9	1	3	9	36
4	16	4	4	16	49

(a)　　　　　　　　　　　　(b)

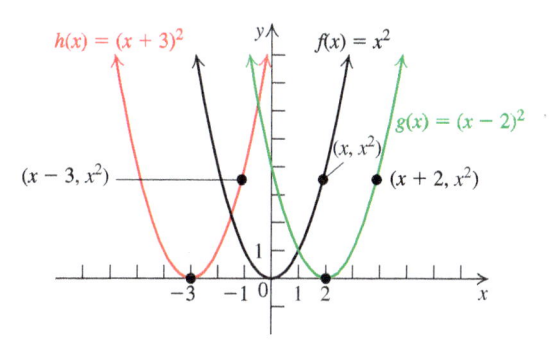

FIGURE 1.62 Horizontal shifts

Solution

First, notice that all three functions are squaring functions.

a. Each point (x, x^2) on the graph of f has a corresponding point $(x + 2, x^2)$ on the graph of g because $g(x + 2) = (x + 2 - 2)^2 = x^2$. So the graph of $g(x) = (x - 2)^2$ is just the graph of $f(x) = x^2$ shifted two units to the *right*. Noticing that $f(0) = 0$ and $g(2) = 0$ will help you remember which way to shift the graph. Table 1.8(a) and Figure 1.62 illustrate these ideas.

b. Each point (x, x^2) on the graph of f has a corresponding point $(x - 3, x^2)$ on the graph of h because $h(x - 3) = (x - 3 + 3)^2 = x^2$. So the graph of $h(x) = (x + 3)^2$ is just the graph of $f(x) = x^2$ shifted three units to the *left*. Noticing that $f(0) = 0$ and $h(-3) = 0$ will help you remember.
Table 1.8(b) and Figure 1.62 confirm these considerations.

Practice Problem 2 Let

$$f(x) = x^3, g(x) = (x - 1)^3, \text{ and } h(x) = (x + 2)^3.$$

Sketch the graphs of all three functions on the same coordinate plane. Describe how the graphs of g and h relate to the graph of f.

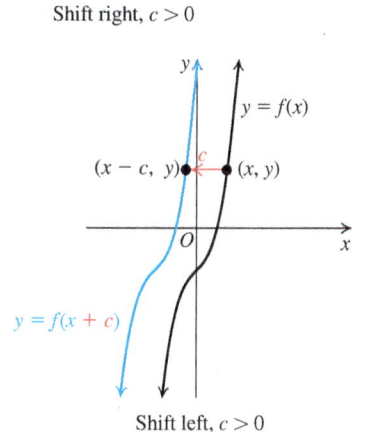

Shift right, $c > 0$

Shift left, $c > 0$
FIGURE 1.63 Horizontal shifts

HORIZONTAL SHIFTS

Let $c > 0$. The graph of $y = f(x - c)$ is the graph of $y = f(x)$ shifted c units to the right. The graph of $y = f(x + c)$ is the graph of $y = f(x)$ shifted c units to the left. See Figure 1.63. If (x, y) is a point on the graph of f, then the corresponding point $(x + c, y)$ is on the graph of $y = f(x - c)$, and the point $(x - c, y)$ is on the graph of $y = f(x + c)$.

112 Chapter 1 Graphs and Functions

Replacing x with $x - c$ in the equation $y = f(x)$ results in a function whose graph is the graph of f shifted c units to the right ($c > 0$). Similarly, replacing x with $x + c$ in the equation $y = f(x)$ results in a function whose graph is the graph of f shifted c units to the left ($c > 0$).

PROCEDURE IN ACTION

EXAMPLE 3   Graphing Combined Vertical and Horizontal Shifts

OBJECTIVE
Sketch the graph of $g(x) = f(x - c) + d$, where f is a function whose graph is known.

Step 1 Identify and graph the known function f.

Step 2 Identify the constants d and c.

Step 3 Let $c > 0$,
(i) graph $y = f(x - c)$ by shifting the graph of f horizontally c units to the right
(ii) graph $y = f(x + c)$ by shifting the graph of f horizontally c units to the left.

Step 4 Let $d > 0$,
(i) graph $y = f(x \pm c) + d$ by shifting the graph of $y = f(x \pm c)$ vertically up d units.
(ii) graph $y = f(x \pm c) - d$ by shifting the graph of $y = f(x \pm c)$ vertically down d units.

EXAMPLE
Sketch the graph of $g(x) = \sqrt{x + 2} - 3$.

1. Choose $f(x) = \sqrt{x}$.
   The graph of $y = \sqrt{x}$ is shown in Step 4.

2. $g(x) = \sqrt{x + 2} - 3$, so $c = 2$ and $d = 3$.

3. Because $c = 2 > 0$, the graph of $y = \sqrt{x + 2}$ is the graph of f shifted horizontally two units to the left. See the blue graph in Step 4.

4. Graph $y = \sqrt{x + 2} - 3$ by shifting the graph of $y = \sqrt{x + 2}$ three units down. Domain: $[-2, \infty)$; range: $[-3, \infty)$.

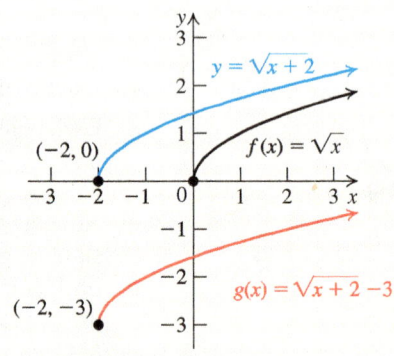

Horizontal and vertical shifts

Practice Problem 3 Sketch the graph of
$$f(x) = \sqrt{x - 2} + 3.$$

3 Use reflections to graph functions.

Reflections

We now consider rigid transformations that reflect a graph about a coordinate axis.

Comparing the Graphs of $y = f(x)$ and $y = -f(x)$ Consider the graph of $f(x) = x^2$, shown in Figure 1.64. Table 1.9 gives some values for $f(x)$ and $g(x) = -f(x)$. Note that the y-coordinate of each point in the graph of $g(x) = -f(x)$ is the opposite of the y-coordinate of the corresponding point on the graph of $f(x)$. So the graph of $y = -x^2$ is the reflection of the graph of $y = x^2$ about the x-axis. This means that the

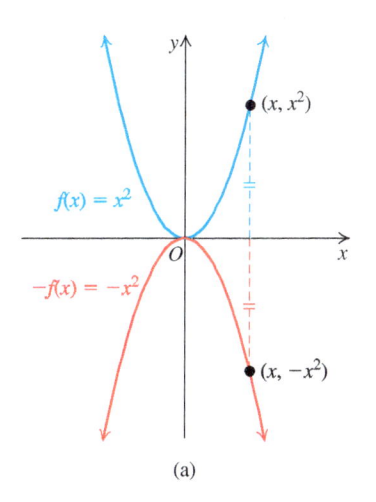

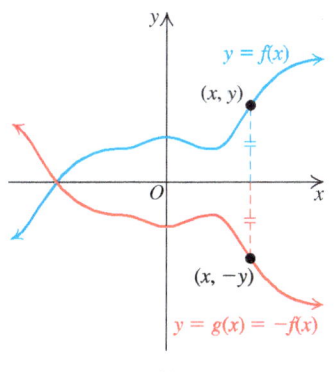

Section 1.5 ■ Transformations of Functions 113

points (x, x^2) and $(x, -x^2)$ are the same distance from but on opposite sides of the x-axis. See Figure 1.64(a).

TABLE 1.9

x	$f(x) = x^2$	$g(x) = -f(x) = -x^2$
-3	9	-9
-2	4	-4
-1	1	-1
0	0	0
1	1	-1
2	4	-4
3	9	-9

FIGURE 1.64 Reflection about the x-axis

> **REFLECTION ABOUT THE x-AXIS**
>
> The graph of $g(x) = -f(x)$ is a reflection of the graph of $y = f(x)$ about the x-axis. If a point (x, y) is on the graph of f, then the point $(x, -y)$ is on the graph of g. See Figure 1.64(b).

Comparing the Graphs of $y = f(x)$ and $y = f(-x)$ To compare the graphs of $f(x)$ and $g(x) = f(-x)$, consider the graph of $f(x) = \sqrt{x}$. Then $f(-x) = \sqrt{-x}$. See Figure 1.65. Table 1.10 gives some values for $f(x)$ and $g(x) = f(-x)$. The domain of f is $[0, \infty)$, and the domain of g is $(-\infty, 0]$. Each point (x, y) on the graph of f has a corresponding point $(-x, y)$ on the graph of g. So the graph of $y = f(-x)$ is the reflection of the graph of $y = f(x)$ about the y-axis. This means that the points $(x, \sqrt{x})$ and $(-x, \sqrt{x})$ are the same distance from but on opposite sides of the y-axis. See Figure 1.65.

TABLE 1.10

x	$f(x) = \sqrt{x}$	$-x$	$g(x) = \sqrt{-x}$
-4	Undefined	4	2
-1	Undefined	1	1
0	0	0	0
1	1	-1	Undefined
4	2	-4	Undefined

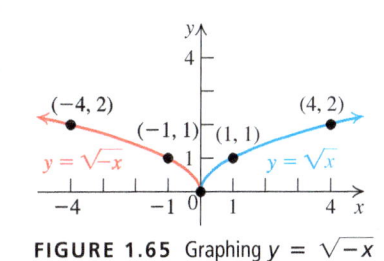

FIGURE 1.65 Graphing $y = \sqrt{-x}$

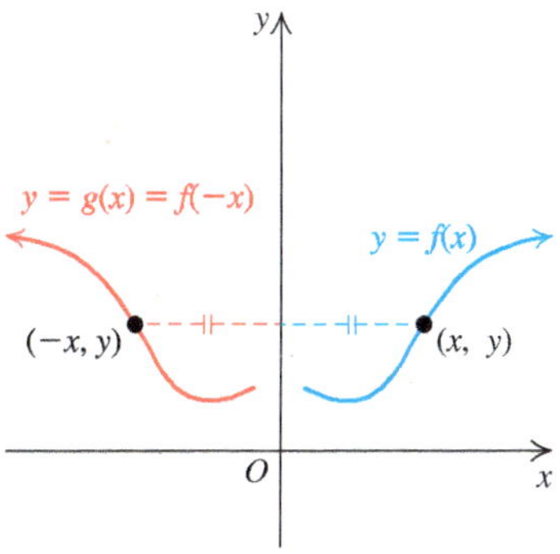

FIGURE 1.66 Reflection about the y-axis

> **REFLECTION ABOUT THE y-AXIS**
>
> The graph of $g(x) = f(-x)$ is a reflection of the graph of $y = f(x)$ about the y-axis. If a point (x, y) is on the graph of f, then the point $(-x, y)$ is on the graph of g. See Figure 1.66.

114 Chapter 1 Graphs and Functions

EXAMPLE 4 Combining Transformations

Explain how the graph of $y = -|x - 2| + 3$ can be obtained from the graph of $y = |x|$.

Solution

Start with the graph of $y = |x|$. Follow the point $(0, 0)$ on $y = |x|$. See Figure 1.67(a).

Step 1 Shift the graph of $y = |x|$ two units to the right to obtain the graph of $y = |x - 2|$. The point $(0, 0)$ moves to $(2, 0)$. See Figure 1.67(b).

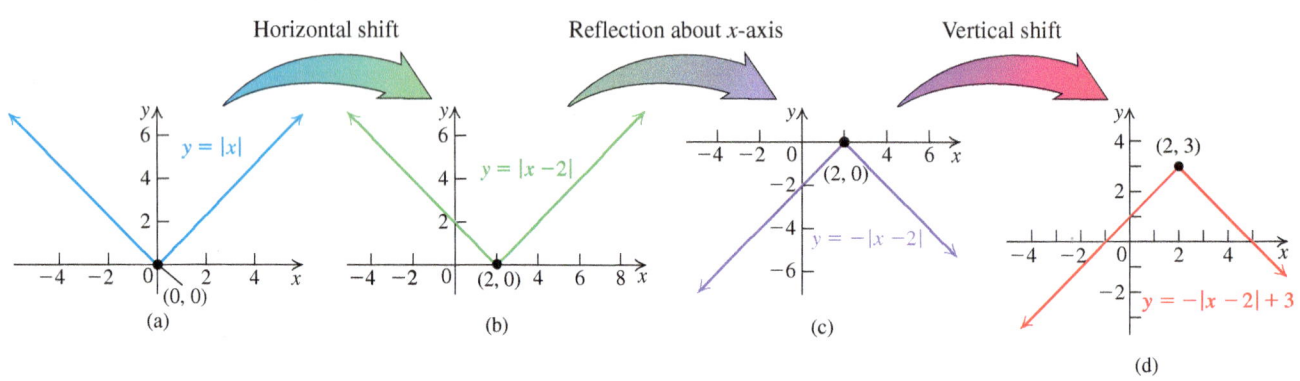

FIGURE 1.67 Transformations of $y = |x|$

Step 2 Reflect the graph of $y = |x - 2|$ about the x-axis to obtain the graph of $y = -|x - 2|$. The point $(2, 0)$ remains $(2, 0)$. See Figure 1.67(c).

Step 3 Finally, shift the graph of $y = -|x - 2|$ three units up to obtain the graph of $y = -|x - 2| + 3$. The point $(2, 0)$ moves to $(2, 3)$. See Figure 1.67(d).

Practice Problem 4 Explain how the graph of $y = -(x - 1)^2 + 2$ can be obtained from the graph of $y = x^2$.

EXAMPLE 5 Graphing $y = |f(x)|$

Use the graph of $f(x) = (x + 1)^2 - 4$ to sketch the graph of $y = |f(x)|$.

Solution

The graph of $y = (x + 1)^2 - 4$ is obtained by shifting the graph of $y = x^2$ left one unit and then shifting the resulting graph down four units. See Figure 1.68(a). The function f has domain $(-\infty, \infty)$, and range $[-4, \infty)$.

We know that

$$|y| = \begin{cases} y & \text{if } y \geq 0 \\ -y & \text{if } y < 0. \end{cases}$$

This means that the portion of the graph on or above the x-axis ($y \geq 0$) is unchanged and that the portion of the graph below the x-axis ($y < 0$) is reflected above the x-axis. See Figure 1.68(b). The function $|f|$ has domain $(-\infty, \infty)$, and range $[0, \infty)$.

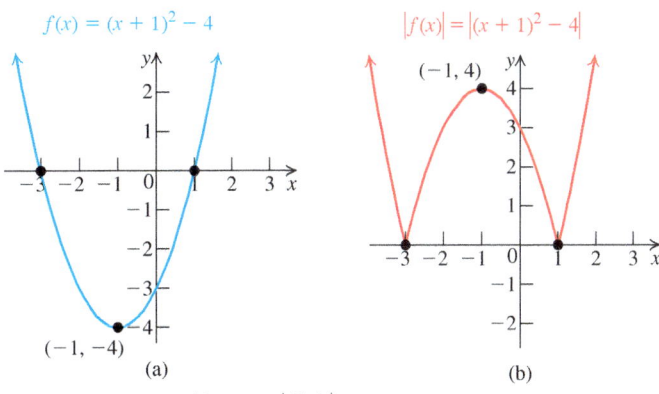

FIGURE 1.68 Graphing $y = |f(x)|$

Practice Problem 5 Use the graph of $f(x) = 2x - 4$ to sketch the graph of $y = |f(x)|$.

Stretching or Compressing

4 Use stretching or compressing to graph functions.

We now look at transformations that distort the shape of a graph, called **nonrigid transformations**. We consider the relationship of the graphs of $y = af(x)$ and $y = f(bx)$ to the graph of $y = f(x)$.

Comparing the Graphs of $y = f(x)$ and $y = af(x)$

EXAMPLE 6 Stretching or Compressing a Function Vertically

Let $f(x) = |x|$, $g(x) = 2|x|$, and $h(x) = \frac{1}{2}|x|$. Sketch the graphs of f, g, and h on the same coordinate plane and describe how the graphs of g and h are related to the graph of f.

Solution

The graphs of $y = |x|$, $y = 2|x|$, and $y = \frac{1}{2}|x|$ are sketched in Figure 1.69. Table 1.11 gives some typical function values.

The graph of $y = 2|x|$ is the graph of $y = |x|$ vertically stretched (expanded) by multiplying each of its y-coordinates by 2. It is twice as high as the graph of $|x|$ at every real number x. The result is a taller V-shaped curve. See Figure 1.69.

The graph $y = \frac{1}{2}|x|$ is the graph of $y = |x|$ vertically compressed (shrunk) by multiplying each of its y-coordinates by $\frac{1}{2}$. It is half as high as the graph of $|x|$ at every real number x. The result is a flatter V-shaped curve. See Figure 1.69.

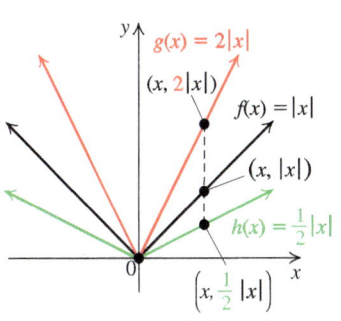

FIGURE 1.69 Vertical stretch and compression

TABLE 1.11

x	$f(x) = |x|$	$g(x) = 2|x|$	$h(x) = \frac{1}{2}|x|$
-2	2	4	1
-1	1	2	$\frac{1}{2}$
0	0	0	0
1	1	2	$\frac{1}{2}$
2	2	4	1

116 Chapter 1 Graphs and Functions

Practice Problem 6 Let $f(x) = \sqrt{x}$ and $g(x) = 2\sqrt{x}$. Sketch the graphs of f and g on the same coordinate plane and describe how the graph of g is related to the graph of f.

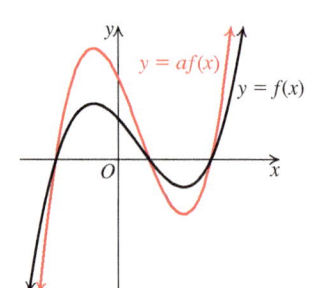

> ### VERTICAL STRETCHING OR COMPRESSING
>
> The graph of $g(x) = af(x)$ is obtained from the graph of $y = f(x)$ by multiplying the y-coordinate of each point on the graph of $y = f(x)$ by a and leaving the x-coordinate unchanged. The result (see Figure 1.70) is as follows:
>
> 1. A **vertical stretch** away from the x-axis if $a > 1$
> 2. A **vertical compression** toward the x-axis if $0 < a < 1$
>
> If $a < 0$, graph $y = |a|f(x)$ by stretching or compressing the graph of $y = f(x)$ vertically. Then reflect the resulting graph about the x-axis.
>
> So, if (x, y) is a point on the graph of f, then the point (x, ay) is on the graph of g.

$a > 1$: stretch vertically

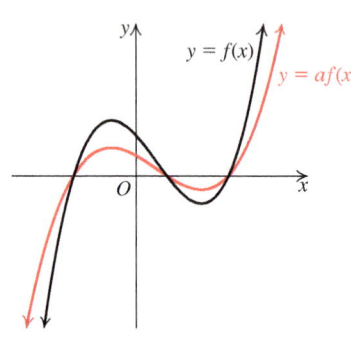

$0 < a < 1$: compress vertically

FIGURE 1.70 Vertical stretch or compression

Comparing the Graphs of $y = f(x)$ and $y = f(bx)$ Given a function $f(x)$, let's explore the effect of the constant b in graphing the function $y = f(bx)$. Consider the graphs of $y = f(x)$ and $y = f(2x)$ in Figure 1.71(a). Multiplying the *independent* variable x by 2 compresses the graph $y = f(x)$ horizontally toward the y-axis. Because the value $2x$ is twice the value of x, a point on the x-axis will be only half as far from the origin when $y = f(2x)$ has the same y value as $y = f(x)$. See Figure 1.71(a).

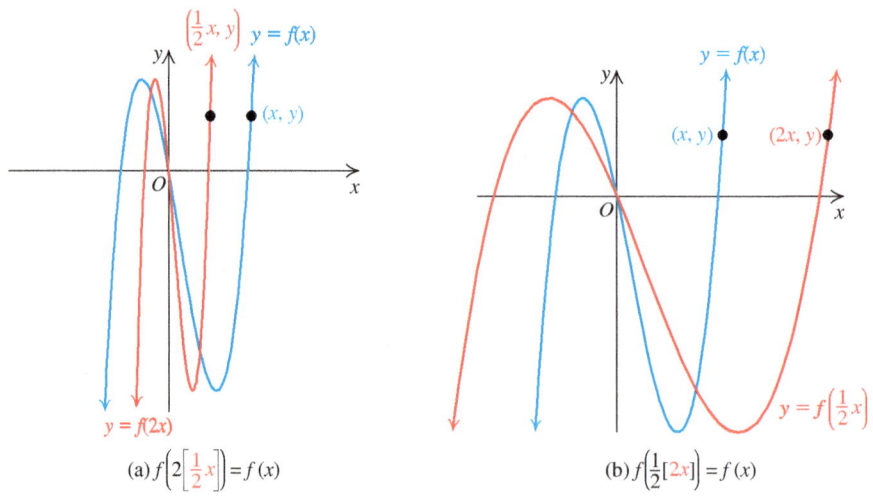

(a) $f\left[2\left[\frac{1}{2}x\right]\right] = f(x)$

(b) $f\left[\frac{1}{2}[2x]\right] = f(x)$

FIGURE 1.71 Horizontal stretch or compression

> **SIDE NOTE**
>
> Notice that replacing the variable x with $x \pm c$ or bx in $y = f(x)$ results in a horizontal change in the graphs. However, replacing the y value $f(x)$ by $f(x) \pm d$, or $af(x)$, results in a vertical change in the graph of $y = f(x)$.

Now consider the graphs of $y = f(x)$ and $y = f\left(\frac{1}{2}x\right)$ in Figure 1.71(b). Multiplying the *independent* variable x by $\frac{1}{2}$ stretches the graph of $y = f(x)$ horizontally away from the y-axis. The value $\frac{1}{2}x$ is half the value of x so that a point on the x-axis will be twice as far from the origin for $y = f\left(\frac{1}{2}x\right)$ to have the same y value as $y = f(x)$. See Figure 1.71(b).

HORIZONTAL STRETCHING OR COMPRESSING

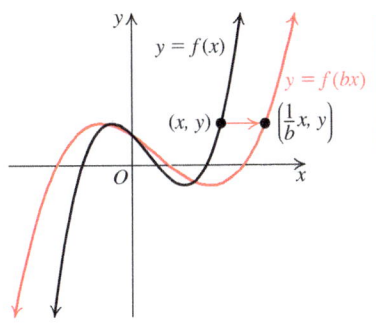

The graph of $g(x) = f(bx)$ is obtained from the graph of $y = f(x)$ by multiplying the x-coordinate of each point on the graph of $y = f(x)$ by $\frac{1}{b}$ and leaving the y-coordinate unchanged. The result (see Figure 1.72) is as follows:

1. A **horizontal stretch** away from the y-axis if $0 < b < 1$.
2. A **horizontal compression** toward the y-axis if $b > 1$.

If $b < 0$, graph $f(|b|x)$ by stretching or compressing the graph of $y = f(x)$ horizontally. Then reflect the graph of $y = f(|b|x)$ about the y-axis.

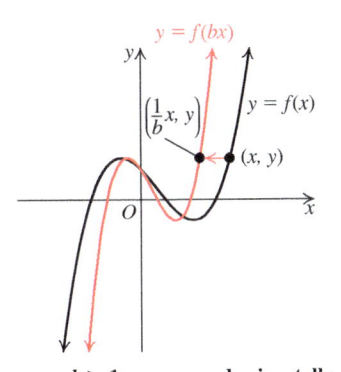

If (x, y) is a point on the graph of $y = f(x)$, then the point $\left(\frac{1}{b}x, y\right)$ is on the graph of g.

FIGURE 1.72 Horizontal stretch and compression

EXAMPLE 7 Stretching or Compressing a Function Horizontally

The graph of the function $y = f(x)$ whose equation is not given is shown in Figure 1.73. Sketch the following graphs.

a. $f\left(\frac{1}{2}x\right)$ **b.** $f(2x)$ **c.** $f(-2x)$

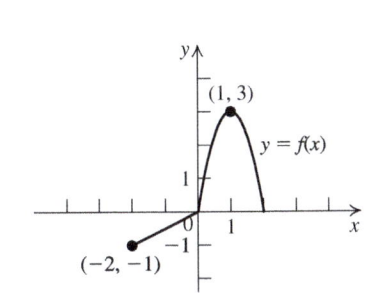

FIGURE 1.73

Solution

Note that the domain of f is $[-2, 2]$ and its range is $[-1, 3]$.

a. To graph $y = f\left(\frac{1}{2}x\right)$, we stretch the graph of $y = f(x)$ horizontally by a factor of 2. In other words, we transform each point (x, y) in Figure 1.73 to the point $(2x, y)$ in Figure 1.74(a).

b. To graph $y = f(2x)$, we compress the graph of $y = f(x)$ horizontally by a factor of $\frac{1}{2}$. Therefore, we transform each point (x, y) in Figure 1.73 to $\left(\frac{1}{2}x, y\right)$ in Figure 1.74(b).

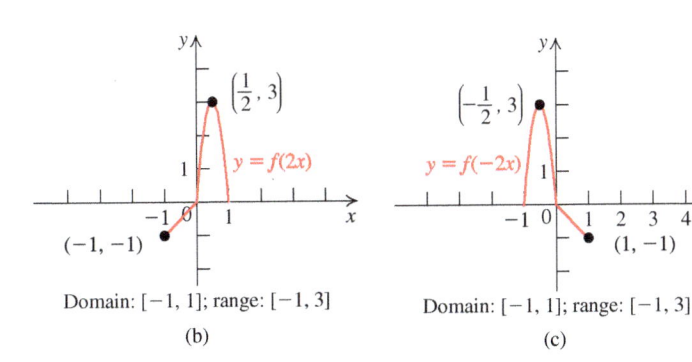

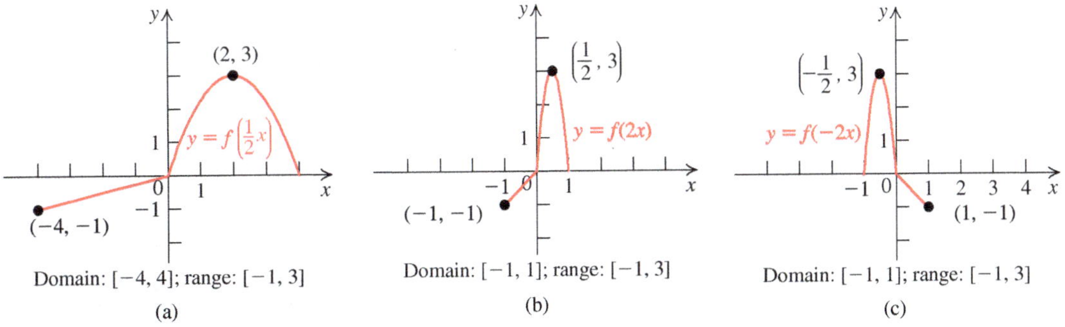

Domain: $[-4, 4]$; range: $[-1, 3]$
(a)

Domain: $[-1, 1]$; range: $[-1, 3]$
(b)

Domain: $[-1, 1]$; range: $[-1, 3]$
(c)

FIGURE 1.74

c. To graph $y = f(-2x)$, we reflect the graph of $y = f(2x)$ in Figure 1.74(b) about the y-axis. So we transform each point (x, y) in Figure 1.74(b) to the point $(-x, y)$ in Figure 1.74(c).

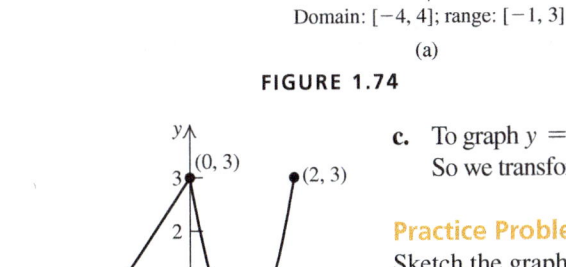

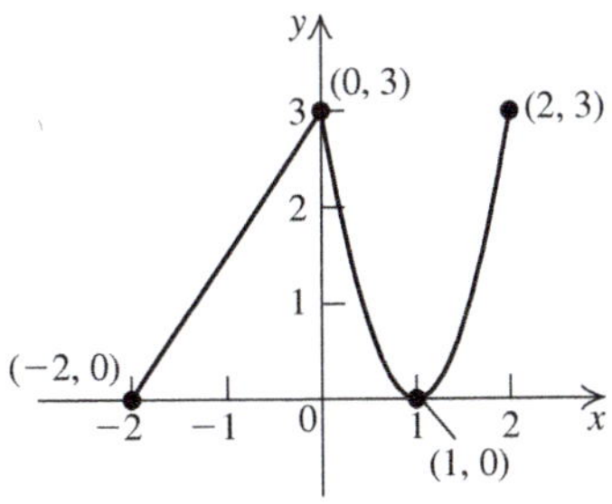

Practice Problem 7 The graph of a function $y = f(x)$ is given in the margin. Sketch the graphs of the following functions.

a. $f\left(\frac{1}{2}x\right)$ **b.** $f(2x)$

118 Chapter 1 Graphs and Functions

Multiple Transformations in Sequence

When graphing requires more than one transformation of a basic function, it is helpful to perform transformations in the following order:

1. Horizontal shift   2. Stretch or compression   3. Reflection   4. Vertical shift

EXAMPLE 8   Combining Transformations

Sketch the graph of the function $f(x) = 3 - 2(x-1)^2$.

Solution

Begin with the basic function $y = x^2$. Then apply the necessary transformations in a sequence of steps. The result of each step is shown in Figure 1.75.

Step 1 $y = x^2$ — Identify a related function whose graph is familiar. In this case, use $y = x^2$. See Figure 1.75(a).

Step 2 $y = (x-1)^2$ — Replace x with $x - 1$; shift the graph of $y = x^2$ one unit to the right. See Figure 1.75(b).

Step 3 $y = 2(x-1)^2$ — Multiply by 2; stretch the graph of $y = (x-1)^2$ vertically by a factor of 2. See Figure 1.75(c).

Step 4 $y = -2(x-1)^2$ — Multiply by -1. Reflect the graph of $y = 2(x-1)^2$ about the x-axis. See Figure 1.75(d).

Step 5 $y = 3 - 2(x-1)^2$ — Add 3. Shift the graph of $y = -2(x-1)^2$ three units up. See Figure 1.75(e).

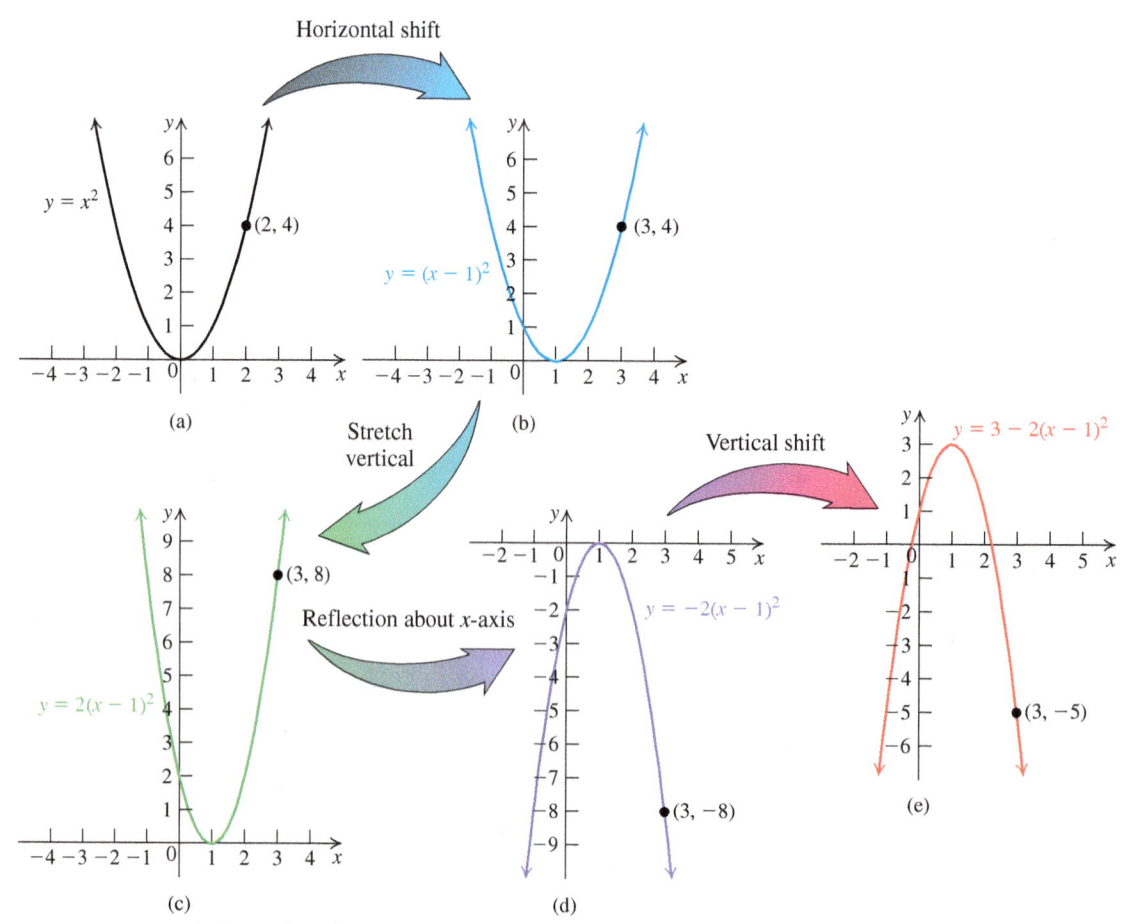

FIGURE 1.75 Multiple transformations

Practice Problem 8 Sketch the graph of the function
$$f(x) = 3\sqrt{x+1} - 2.$$

EXAMPLE 9 Using Poiseuille's Law for Arterial Blood Flow

For an artery with radius R, the velocity v of the blood flow at a distance r from the center of the artery is given by
$$v = c(R^2 - r^2), 0 \leq r \leq R,$$
where c is a constant that is determined for a particular artery. See Figure 1.76.

To emphasize that the velocity v depends on the distance r from the center of the artery, we write $v = v(r)$, or $v(r) = c(R^2 - r^2)$. Starting with the graph of $y = r^2$, sketch the graph of $y = v(r)$, where the artery has radius $R = 3$ and $c = 10^4$.

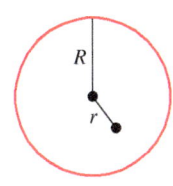

FIGURE 1.76

Solution

$$v(r) = c(R^2 - r^2) \quad \text{Original equation}$$
$$v(r) = 10^4(9 - r^2) \quad \text{Substitute } c = 10^4 \text{ and } R = 3.$$
$$v(r) = 9 \cdot 10^4 - 10^4 r^2 \quad \text{Distributive property}$$

Sketch the graph of $y = v(r)$ in Figure 1.77 through a sequence of transformations: stretch the graph of $y = r^2$ vertically by a factor of 10^4, reflect the resulting graph about the x-axis, and shift this graph $9 \cdot 10^4$ units up. Figure 1.77 shows the graph of $y = v(r) = 10^4(9 - r^2)$. Only the section of the graph with $0 \leq r \leq 3$ has a physical meaning.

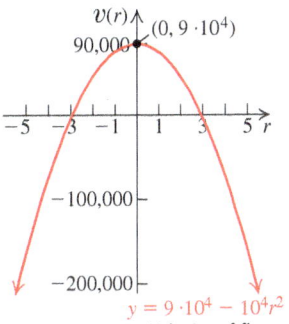

FIGURE 1.77 Velocity of flow, r units from the artery's center

The velocity of the blood decreases from the center of the artery ($r = 0$) to the artery wall ($r = 3$), where it ceases to flow. The portions of the graph in Figure 1.77 corresponding to negative values of r, or values of r greater than 3, do not have any physical significance.

Practice Problem 9 Suppose in Example 9 the artery has radius $R = 3$ and that $c = 10^3$. Sketch the graph of $y = v(r)$.

SUMMARY OF MAIN FACTS

Summary of Transformations of $y = f(x)$

To graph	Draw the graph of f and	Make these changes to the equation $y = f(x)$	Change the graph point (x, y) to
Vertical shift			
For $d > 0$,	Shift the graph of f		
$y = f(x) + d$	up d units.	Add d to y.	$(x, y + d)$
$y = f(x) - d$	down d units.	Subtract d from y.	$(x, y - d)$
Horizontal shift			
For $c > 0$,	Shift the graph of f		
$y = f(x - c)$	to the right c units	Replace x with $x - c$.	$(x + c, y)$
$y = f(x + c)$	to the left c units	Replace x with $x + c$.	$(x - c, y)$
Reflection about the x-axis			
$y = -f(x)$	Reflect the graph of f about the x-axis.	Multiply y by -1.	$(x, -y)$
Reflection about the y-axis			
$y = f(-x)$	Reflect the graph of f about the y-axis.	Replace x with $-x$.	$(-x, y)$
Vertical stretching or compressing:			
$y = af(x)$	Multiply each y-coordinate of $y = f(x)$ by a. The graph of $y = f(x)$ is stretched vertically away from the x-axis if $a > 1$ and is compressed vertically toward the x-axis if $0 < a < 1$. If $a < 0$, sketch the graph of $y = \lvert a \rvert f(x)$ and then reflect it about the x-axis.	Multiply y by a.	(x, ay)
Horizontal stretching or compressing:			
$y = f(bx)$	Multiply each x-coordinate of $y = f(x)$ by $\frac{1}{b}$. The graph of $y = f(x)$ is stretched away from the y-axis if $0 < b < 1$ and is compressed toward the y-axis if $b > 1$. If $b < 0$, sketch the graph of $y = f(\lvert b \rvert x)$ and then reflect it about the y-axis.	Replace x with $\frac{x}{b}$.	$\left(\frac{x}{b}, y\right)$

(continued)

Summary of Combining Transformations in Sequence

To graph $y = af(bx + c) + d$, with $b > 0$

- Starting with the graph of $y = f(x)$, perform the indicated sequence of transformations. Here is one possible sequence:

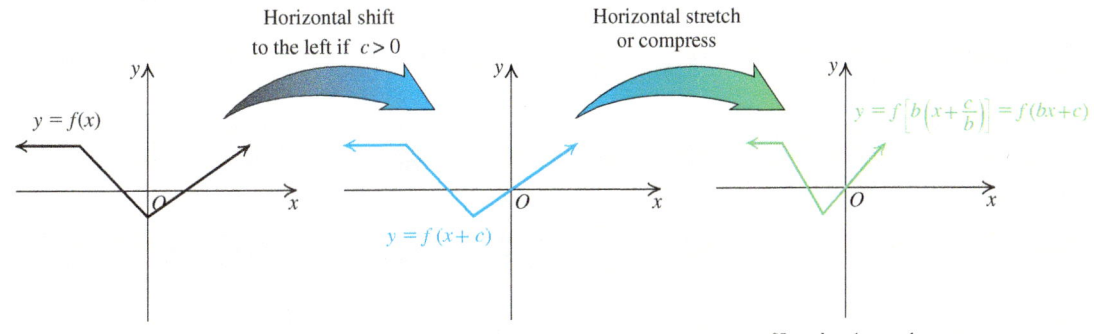

Horizontal shift to the left if $c > 0$ → $y = f(x + c)$ (Here $c < 0$; graph shifts left.)

Horizontal stretch or compress → $y = f\left[b\left(x + \frac{c}{b}\right)\right] = f(bx + c)$ (Here $b > 1$; graph compresses horizontally.)

Vertical stretch or compress → $y = |a|f\left[b\left(x + \frac{c}{b}\right)\right]$ (Here $|a| > 1$; graph stretches vertically.)

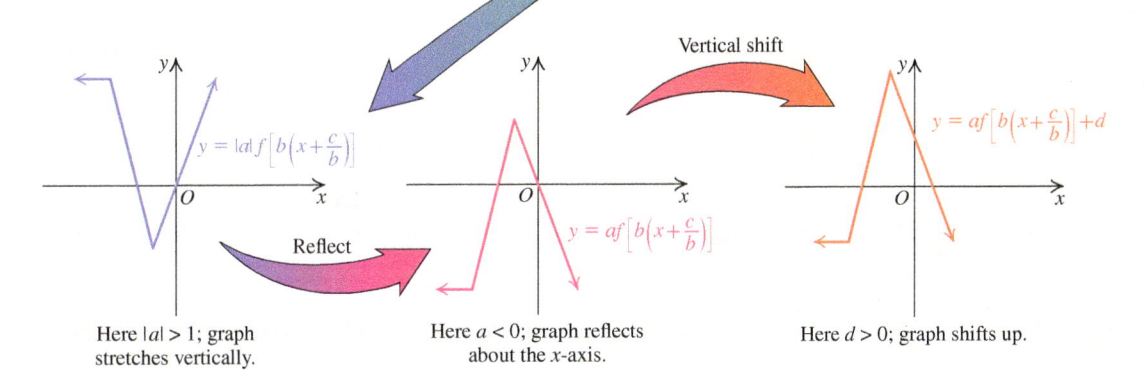

Reflect → $y = af\left[b\left(x + \frac{c}{b}\right)\right]$ (Here $a < 0$; graph reflects about the x-axis.)

Vertical shift → $y = af\left[b\left(x + \frac{c}{b}\right)\right] + d$ (Here $d > 0$; graph shifts up.)

Note: If $b < 0$ and $c > 0$, graph $y = af(|b|x + c) + d$, but reflect this graph about the y-axis before completing the last step, the vertical shift, d.

SECTION 1.5 Exercises

Basic Concepts and Skills

In Exercises 1–14, describe the transformations that produce the graphs of g and h from the graph of f.

1. $f(x) = \sqrt{x}$
   a. $g(x) = \sqrt{x} + 2$
   b. $h(x) = \sqrt{x} - 1$

2. $f(x) = |x|$
   a. $g(x) = |x| + 1$
   b. $h(x) = |x| - 2$

3. $f(x) = x^2$
   a. $g(x) = (x + 1)^2$
   b. $h(x) = (x - 2)^2$

4. $f(x) = \dfrac{1}{x}$
   a. $g(x) = \dfrac{1}{x + 2}$
   b. $h(x) = \dfrac{1}{x - 3}$

5. $f(x) = \sqrt{x}$
   a. $g(x) = \sqrt{x + 1} - 2$
   b. $h(x) = \sqrt{x - 1} + 3$

6. $f(x) = x^2$
   a. $g(x) = -x^2$
   b. $h(x) = (-x)^2$

7. $f(x) = |x|$
   a. $g(x) = -|x|$
   b. $h(x) = |-x|$
8. $f(x) = \sqrt{x}$
   a. $g(x) = 2\sqrt{x}$
   b. $h(x) = \sqrt{2x}$
9. $f(x) = \dfrac{1}{x}$
   a. $g(x) = \dfrac{2}{x}$
   b. $h(x) = \dfrac{1}{2x}$
10. $f(x) = x^3$
    a. $g(x) = (x-2)^3 + 1$
    b. $h(x) = -(x+1)^3 + 2$
11. $f(x) = \sqrt{x}$
    a. $g(x) = -\sqrt{x} + 1$
    b. $h(x) = \sqrt{-x} + 1$
12. $f(x) = [\![x]\!]$
    a. $g(x) = [\![x-1]\!] + 2$
    b. $h(x) = 3[\![x]\!] - 1$
13. $f(x) = \sqrt[3]{x}$
    a. $g(x) = \sqrt[3]{x} + 1$
    b. $h(x) = \sqrt[3]{x+1}$
14. $f(x) = \sqrt[3]{x}$
    a. $g(x) = 2\sqrt[3]{1-x} + 4$
    b. $h(x) = -\sqrt[3]{x-1} + 3$

In Exercises 15–44, graph each function by starting with a function from the library of functions and then using the techniques of shifting, compressing, stretching, and/or reflecting.

15. $f(x) = x^2 - 2$
16. $f(x) = x^2 + 3$
17. $g(x) = \sqrt{x} + 1$
18. $g(x) = \sqrt{x} - 4$
19. $h(x) = |x+1|$
20. $h(x) = |x-2|$
21. $f(x) = (x-3)^3$
22. $f(x) = (x+2)^3$
23. $g(x) = (x-2)^2 + 1$
24. $g(x) = (x+3)^2 - 5$
25. $h(x) = -\sqrt{x}$
26. $h(x) = \sqrt{-x}$
27. $f(x) = -\dfrac{1}{x}$
28. $f(x) = -\dfrac{1}{2x}$
29. $g(x) = \dfrac{1}{2}|x|$
30. $g(x) = 4|x|$
31. $h(x) = -x^3 + 1$
32. $h(x) = -(x+1)^3$
33. $f(x) = 2(x+1)^2 - 1$
34. $f(x) = -(x-1)^2$
35. $g(x) = 5 - x^2$
36. $g(x) = 2 - (x+3)^2$
37. $h(x) = |1-x|$
38. $h(x) = -2\sqrt{x-1}$
39. $f(x) = -|x+3| + 1$
40. $f(x) = 2 - \sqrt{x}$
41. $g(x) = -\sqrt{-x} + 2$
42. $g(x) = 3\sqrt{2-x}$
43. $h(x) = 2[\![x+1]\!]$
44. $h(x) = [\![-x]\!] + 1$

In Exercises 45–52, write an equation for a function whose graph fits the given description.

45. The graph of $f(x) = x^3$ is shifted two units up.
46. The graph of $f(x) = \sqrt{x}$ is shifted three units left.
47. The graph of $f(x) = |x|$ is reflected about the x-axis.
48. $f(x) = \sqrt{x}$ is reflected about the y-axis.
49. The graph of $f(x) = x^2$ is shifted three units right and two units up.
50. The graph of $f(x) = \sqrt{x}$ is shifted three units left, reflected about the x-axis, and shifted two units down.
51. The graph of $f(x) = x^3$ is shifted four units left, stretched vertically by a factor of 3, reflected about the y-axis, and shifted two units up.
52. The graph of $f(x) = |x|$ is shifted four units right, stretched vertically by a factor of 2, reflected about the x-axis, and shifted three units down.

In Exercises 53–60, graph (a) $y = g(x)$ and (b) $y = |g(x)|$, given the following graph of $y = f(x)$.

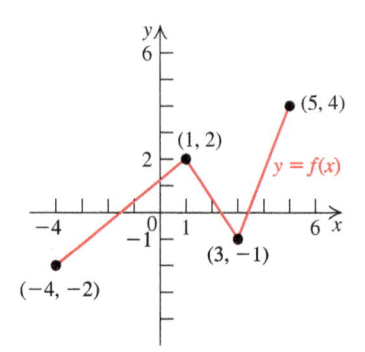

53. $g(x) = f(x) + 1$
54. $g(x) = -2f(x)$
55. $g(x) = f\left(\dfrac{1}{2}x\right)$
56. $g(x) = f(-2x)$
57. $g(x) = f(x-1)$
58. $g(x) = f(2-x)$
59. $g(x) = -2f(x+1) + 3$
60. $g(x) = -f(-x+1) - 2$

Applying the Concepts

In Exercises 61–64, let f be the function that associates the employee number x of each employee of the ABC Corporation with his or her annual salary $f(x)$ in dollars.

61. **Across-the-board raise.** Each employee was awarded an across-the-board raise of $800 per year. Write a function $g(x)$ to describe the new salary.
62. **Percentage raise.** Suppose each employee was awarded a 5% raise. Write a function $h(x)$ to describe the new salary.
63. **Across-the-board and percentage raises.** Suppose each employee was awarded a $500 across-the-board raise and an additional 2% of his or her increased salary. Write a function $p(x)$ to describe these new salaries.
64. **Across-the-board percentage raise.** Suppose the employees making $30,000 or more received a 2% raise, and those making less than $30,000 received a 10% raise. Write a piecewise function to describe these new salaries.

65. Daylight. At 60° north latitude, the graph of $y = f(t)$ gives the number of hours of daylight. (On the t-axis, note that 1 = January and 12 = December.)

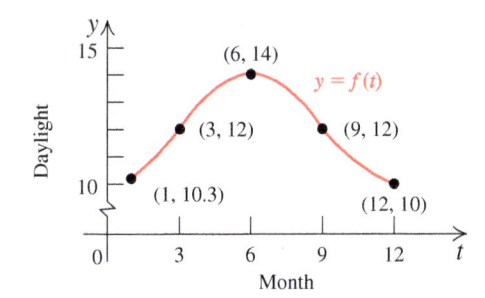

Sketch the graph of $y = f(t) - 12$.

66. Use the graph of $y = f(t)$ of Exercise 65 to sketch the graph of $y = 24 - f(t)$. Interpret the result.

Critical Thinking / Discussion / Writing

67. If f is a function with x-intercept -2 and y-intercept 3, find the corresponding x-intercept and y-intercept, if possible, for
 a. $y = f(x - 3)$ **b.** $y = 5f(x)$
 c. $y = -f(x)$ **d.** $y = f(-x)$

68. If f is a function with x-intercept 4 and y-intercept -1, find the corresponding x-intercept and y-intercept, if possible, for
 a. $y = f(x + 2)$ **b.** $y = 2f(x)$
 c. $y = -f(x)$ **d.** $y = f(-x)$

69. Suppose the graph of $y = f(x)$ is given. Using transformations, explain how the graph of $y = g(x)$ differs from the graph of $y = h(x)$.
 a. $g(x) = f(x) + 3$, $h(x) = f(x + 3)$
 b. $g(x) = f(x) - 1$, $h(x) = f(x - 1)$
 c. $g(x) = 2f(x)$, $h(x) = f(2x)$
 d. $g(x) = -3f(x)$, $h(x) = f(-3x)$

70. Suppose the graph of $y = f(-4x)$ is given. Explain how to obtain the graph of $y = f(x)$.

Maintaining Skills

In Exercises 71–74, rationalize the denominator.

71. $\dfrac{1}{3 + \sqrt{5}}$

72. $\dfrac{1}{\sqrt{5} - \sqrt{3}}$

73. $\dfrac{\sqrt{2} + \sqrt{3}}{\sqrt{2} - \sqrt{3}}$

74. $\dfrac{\sqrt{x + h} - \sqrt{x}}{\sqrt{x + h} + \sqrt{x}}$

In Exercises 75 and 76, rationalize the numerator.

75. $\dfrac{\sqrt{x + h} - \sqrt{x}}{h}$

76. $\left(\dfrac{\sqrt{x^2 + x + 1} - \sqrt{x^2 + 1}}{2} \right)$

SECTION 1.6 Combining Functions; Composite Functions

BEFORE STARTING THIS SECTION, REVIEW

1. Domain of function (Section 1.3, page 82)
2. Rational inequalities (Section P.4, page 39)

OBJECTIVES

1. Define basic operations on functions.
2. Form composite functions.
3. Find the domain of a composite function.
4. Decompose a function.
5. Apply composition to practical problems.

◆ EXXON VALDEZ Oil Spill Disaster

The oil tanker *Exxon Valdez* departed from the Trans-Alaska Pipeline terminal at 9:12 P.M., March 23, 1989. After passing through the Valdez Narrows, the tanker encountered icebergs in the shipping lanes, and Captain Hazelwood ordered the vessel to be taken out of the shipping lanes to go around the icebergs. For reasons that remain unclear, the ship failed to make the turn back into the shipping lanes, and the tanker ran aground on Bligh Reef at 12:04 A.M., March 24, 1989.

At the time of the accident, the *Exxon Valdez* was carrying 53 million gallons of oil, and approximately 11 million gallons were spilled, an amount equivalent to 125 Olympic-sized swimming pools. This spill covered a large area, stretching for 460 miles from Bligh Reef to the tiny village of Chignik on the Alaska Peninsula. In Example 7, we calculate the area covered by an oil spill.

1 Define basic operations on functions.

Combining Functions

Just as numbers can be added, subtracted, multiplied, and divided to produce new numbers, functions can be added, subtracted, multiplied, and divided to produce other functions. To add, subtract, multiply, or divide functions, we simply add, subtract, multiply, or divide their range, or output values.

For example, if $f(x) = x^2 - 2x + 4$ and $g(x) = x + 2$, then

$$f(x) + g(x) = (x^2 - 2x + 4) + (x + 2) = x^2 - x + 6.$$

This new function $y = x^2 - x + 6$ is called the sum function $f + g$.

Sum, Difference, Product, and Quotient of Functions

Let f and g be two functions. The **sum** $f + g$, the **difference** $f - g$, the **product** fg, and the **quotient** $\dfrac{f}{g}$ are functions defined as follows:

- Sum $\quad (f + g)(x) = f(x) + g(x)$
- Difference $\quad (f - g)(x) = f(x) - g(x)$
- Product $\quad fg(x) = f(x) \cdot g(x)$; also written as $(fg)(x)$
- Quotient $\quad \dfrac{f}{g}(x) = \dfrac{f(x)}{g(x)}$, provided that $g(x) \neq 0$; also written as $\left(\dfrac{f}{g}\right)(x)$

The **domain** of each of the new functions consists of those values of x that are common to the domains of f and g, except that for $\dfrac{f}{g}$, all x for which $g(x) = 0$ must also be excluded.

EXAMPLE 1 Combining Functions

Let $f(x) = x^2 - 6x + 8$ and $g(x) = x - 2$. Find each of the following functions or function values.

a. $(f - g)(4)$
b. $\left(\dfrac{f}{g}\right)(3)$
c. $(f + g)(x)$
d. $(f - g)(x)$
e. $(fg)(x)$
f. $\left(\dfrac{f}{g}\right)(x)$

Solution

a. $f(4) = 4^2 - 6(4) + 8 = 0$
$g(4) = 4 - 2 = 2$
$(f - g)(4) = f(4) - g(4)$
$= 0 - 2 = -2$

b. $f(3) = 3^2 - 6(3) + 8 = -1$
$g(3) = 3 - 2 = 1$
$\left(\dfrac{f}{g}\right)(3) = \dfrac{f(3)}{g(3)} = \dfrac{-1}{1} = -1$

c. $(f + g)(x) = f(x) + g(x)$ — Definition of sum
$= (x^2 - 6x + 8) + (x - 2)$ — Add $f(x)$ and $g(x)$.
$= x^2 - 5x + 6$ — Combine like terms.

d. $(f - g)(x) = f(x) - g(x)$ — Definition of difference
$= (x^2 - 6x + 8) - (x - 2)$ — Subtract $g(x)$ from $f(x)$.
$= x^2 - 7x + 10$ — Combine like terms.

e. $(fg)(x) = f(x) \cdot g(x)$ — Definition of product
$= (x^2 - 6x + 8)(x - 2)$ — Multiply $f(x)$ and $g(x)$.
$= x^2(x - 2) - 6x(x - 2) + 8(x - 2)$ — Distributive property
$= x^3 - 2x^2 - 6x^2 + 12x + 8x - 16$ — Distributive property
$= x^3 - 8x^2 + 20x - 16$ — Combine like terms.

f. $\left(\dfrac{f}{g}\right)(x) = \dfrac{f(x)}{g(x)}, \quad g(x) \neq 0$ Definition of quotient

    $= \dfrac{x^2 - 6x + 8}{x - 2}, \quad x - 2 \neq 0$ Divide $f(x)$ by $g(x)$.

    $= \dfrac{(x - 2)(x - 4)}{x - 2}, \quad x \neq 2$ Factor the numerator.

    $= x - 4, \quad x \neq 2$ Factor out $(x - 2)$.

Domain Because f and g are polynomials, the domain of f and g is the set of all real numbers, or in interval notation, $(-\infty, \infty)$. So the domain for $f + g$, $f - g$, and fg is $(-\infty, \infty)$. However, for $\dfrac{f}{g}$, we must exclude 2 because $g(2) = 0$. The domain for $\dfrac{f}{g}$ is the set of all real numbers x, $x \neq 2$, or in interval notation, $(-\infty, 2) \cup (2, \infty)$.

Practice Problem 1 Let $f(x) = 3x - 1$ and $g(x) = x^2 + 2$. Find $(f + g)(x)$, $(f - g)(x)$, $(fg)(x)$, and $\left(\dfrac{f}{g}\right)(x)$.

WARNING

When rewriting a function such as

$$\left(\dfrac{f}{g}\right)(x) = \dfrac{(x - 2)(x - 4)}{x - 2}, \quad x \neq 2,$$

by removing the common factor, remember to specify "$x \neq 2$."
We identify the domain before we simplify and write

$$\left(\dfrac{f}{g}\right)(x) = x - 4, \quad x \neq 2.$$

---

2 Form composite functions.

Composition of Functions

There is yet another way to construct a new function from two functions. Figure 1.78 shows what happens when you apply one function $g(x) = x^2 - 1$ to an input (domain) value x and then apply a second function $f(x) = \sqrt{x}$ to the output value, $g(x)$, from the first function. In this case, the output from the first function becomes the input for the second function.

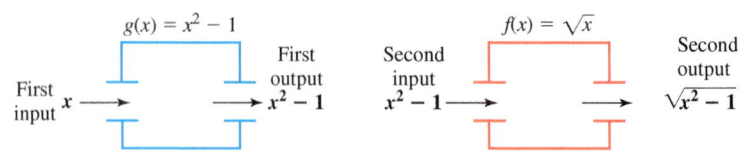

First input x → $g(x) = x^2 - 1$ → First output $x^2 - 1$ → Second input $x^2 - 1$ → $f(x) = \sqrt{x}$ → Second output $\sqrt{x^2 - 1}$

FIGURE 1.78 Composition of functions

Composition of Functions

If f and g are two functions, the composition of the function f with the function g is written as $f \circ g$ and is defined by the equation

$$(f \circ g)(x) = f(g(x)) \quad \text{(read "}f\text{ of }g\text{ of }x\text{")},$$

where the domain of $f \circ g$ consists of those values x in the domain of g for which $g(x)$ is in the domain of f.

Figure 1.79 may help clarify the definition of $f \circ g$.

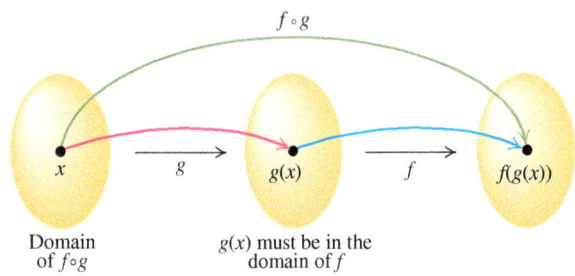

FIGURE 1.79 Composition of function f with g

Evaluating $(f \circ g)(x)$ By definition,

$$(f \circ g)(x) = f(g(x)).$$

with $g(x)$ as the inner function and f as the outer function.

Thus, to evaluate $(f \circ g)(x)$, we must

(i) Evaluate the inner function g at x.
(ii) Use the result $g(x)$ as the input for the outer function f.

EXAMPLE 2   Evaluating a Composite Function

Let $f(x) = x^3$ and $g(x) = x + 1$. Find each composite function value.

a. $(f \circ g)(1)$ **b.** $(g \circ f)(1)$ **c.** $(f \circ f)(-1)$

> **SIDE NOTE**
> When computing $(f \circ g)(x) = f(g(x))$, you must replace every x in $f(x)$ with $g(x)$.

Solution

a. $(f \circ g)(1) = f(g(1))$ Definition of $f \circ g$
$\qquad \qquad \quad = f(2)$ $g(1) = 1 + 1 = 2$
$\qquad \qquad \quad = 2^3 = 8$ Evaluate $f(2)$ and simplify.

b. $(g \circ f)(1) = g(f(1))$ Definition of $g \circ f$
$\qquad \qquad \quad = g(1)$ $f(1) = 1^3 = 1$
$\qquad \qquad \quad = 1 + 1 = 2$ Evaluate $g(1)$ and simplify.

c. $(f \circ f)(-1) = f(f(-1))$ Definition of $f \circ f$
$\qquad \qquad \quad\; = f(-1)$ $f(-1) = (-1)^3 = -1$
$\qquad \qquad \quad\; = (-1)^3 = -1$ Evaluate $f(-1)$ and simplify.

Practice Problem 2 Let $f(x) = -5x$ and $g(x) = x^2 + 1$. Find each composite function value.

a. $(f \circ g)(0)$ **b.** $(g \circ f)(0)$

EXAMPLE 3   Finding Composite Functions

Let $f(x) = 2x + 1$ and $g(x) = x^2 - 3$. Find each composite function.

a. $(f \circ g)(x)$ **b.** $(g \circ f)(x)$ **c.** $(f \circ f)(x)$

Solution

a. $(f \circ g)(x) = f(g(x))$ Definition of $f \circ g$
$\qquad \qquad \quad = f(x^2 - 3)$ Replace $g(x)$ with $x^2 - 3$.
$\qquad \qquad \quad = 2(x^2 - 3) + 1$ Replace x with $x^2 - 3$ in $f(x) = 2x + 1$.
$\qquad \qquad \quad = 2x^2 - 5$ Simplify.

b. $(g \circ f)(x) = g(f(x))$ Definition of $g \circ f$
$ = g(2x + 1)$ Replace $f(x)$ with $2x + 1$.
$ = (2x + 1)^2 - 3$ Replace x with $2x + 1$ in $g(x) = x^2 - 3$.
$ = 4x^2 + 4x + 1 - 3$ $(A + B)^2 = A^2 + 2AB + B^2$
$ = 4x^2 + 4x - 2$ Simplify.

c. $(f \circ f)(x) = f(f(x))$ Definition of $f \circ f$
$ = f(2x + 1)$ Replace $f(x)$ with $2x + 1$.
$ = 2(2x + 1) + 1$ Replace x with $2x + 1$ in $f(x) = 2x + 1$.
$ = 4x + 3$ Simplify.

Practice Problem 3 Let $f(x) = 2 - x$ and $g(x) = 2x^2 + 1$. Find each of the following.

a. $(g \circ f)(x)$ **b.** $(f \circ g)(x)$ **c.** $(g \circ g)(x)$

Parts **a** and **b** of Example 3 illustrate that in general, $f \circ g \neq g \circ f$. In other words, the composition of functions is not commutative.

WARNING   Note that $f(g(x))$ is not $f(x) \cdot g(x)$. In $f(g(x))$, the output of g is *used as an input* for f; in contrast, in $f(x) \cdot g(x)$, the output $f(x)$ is *multiplied* by the output $g(x)$.

3 Find the domain of a composite function.

Domain of Composite Functions

Let f and g be two functions. The domain of the composite function $f \circ g$ consists of those values of x in the domain of g for which $g(x)$ is in the domain of f. So to find the domain of $f \circ g$, we first find the domain of g. Then from the domain of g, we exclude those values of x such that $g(x)$ is not in the domain of f.

SIDE NOTE

If the function $f \circ g$ can be simplified, determine the domain *before* simplifying. For example, if $f(x) = x^2$ and $g(x) = \sqrt{x}$,

$(f \circ g)(x) = f(g(x)) = f(\sqrt{x})$.

Before simplifying:

$(f \circ g)(x) = (\sqrt{x})^2, x \geq 0$

The domain of f is $(-\infty, \infty)$, and $(f \circ g)(x)$ is defined on the interval $[0, \infty)$. So the domain of $f \circ g$ is $[0, \infty)$. After simplifying:

$(f \circ g)(x) = x$

Here it appears that the domain of $f \circ g$ is $(-\infty, \infty)$. However, this is not correct.

EXAMPLE 4   **Finding the Domain of a Composite Function**

Let $f(x) = x + 1$ and $g(x) = \dfrac{1}{x}$.

a. Find $(f \circ g)(x)$ and its domain.   **b.** Find $(g \circ f)(x)$ and its domain.

Solution

a. $(f \circ g)(x) = f(g(x))$ Definition of $f \circ g$
$ = f\left(\dfrac{1}{x}\right)$ Replace $g(x)$ with $\dfrac{1}{x}$.
$ = \dfrac{1}{x} + 1$ Replace x with $\dfrac{1}{x}$ in $f(x) = x + 1$.

Because 0 is not in the domain of $g(x) = \dfrac{1}{x}$, we must exclude 0 from the domain of $f \circ g$. This is the only restriction on the domain of $f \circ g$ because the domain of f is $(-\infty, \infty)$. So the domain of $f \circ g$ is all real $x, x \neq 0$. In interval notation, the domain of $f \circ g$ is $(-\infty, 0) \cup (0, \infty)$.

b. $(g \circ f)(x) = g(f(x))$ Definition of $g \circ f$
$= g(x + 1)$ Replace $f(x)$ with $x + 1$.
$= \dfrac{1}{x + 1}$ Replace x with $x + 1$ in $g(x) = \dfrac{1}{x}$.

The domain of f is $(-\infty, \infty)$, so we need to exclude from the domain of $g \circ f$ only values of x for which $f(x)$ is not in the domain of g, that is, values of x such that $f(x) = 0$. If $f(x) = 0$, then $x + 1 = 0$ and $x = -1$. Therefore, the domain of $g \circ f$ is all real numbers $x, x \ne -1$, or $(-\infty, -1) \cup (-1, \infty)$.

Practice Problem 4 Let $f(x) = \sqrt{x}$ and $g(x) = 2x + 5$. Find $(g \circ f)(x)$ and its domain.

EXAMPLE 5   Finding the Domain of Composite Functions

Let $f(x) = \sqrt{x - 2}$, $g(x) = \sqrt{4 - x}$, and $h(x) = x^2$. Find the following functions and their domains.

a. $f \circ g$ **b.** $g \circ f$ **c.** $f \circ h$ **d.** $g \circ h$ **e.** $h \circ g$

Solution

a. $(f \circ g)(x) = f(g(x))$
$= f(\sqrt{4 - x})$ Replace $g(x)$ with $\sqrt{4 - x}$.
$= \sqrt{\sqrt{4 - x} - 2}$ Evaluate $f(\sqrt{4 - x})$.

The function $g(x) = \sqrt{4 - x}$ is defined for $4 - x \geq 0$ or $x \leq 4$. And $(f \circ g)(x)$ is defined for

$\sqrt{4 - x} - 2 \geq 0$
$\sqrt{4 - x} \geq 2$ Add 2 to both sides.
$4 - x \geq 4$ Square both sides.
or $x \leq 0$ Solve for x.

So combining $x \leq 4$ and $x \leq 0$, we conclude that the domain of $f \circ g$ is the interval $(-\infty, 0]$.

b. $(g \circ f)(x) = g(f(x))$
$= g(\sqrt{x - 2})$ Replace $f(x)$ with $\sqrt{x - 2}$.
$= \sqrt{4 - \sqrt{x - 2}}$ Evaluate $g(\sqrt{x - 2})$.

The function $f(x) = \sqrt{x - 2}$ is defined for $x \geq 2$. And $(g \circ f)(x)$ is defined for $4 - \sqrt{x - 2} \geq 0$ or $\sqrt{x - 2} \leq 4$, or $x - 2 \leq 16$, or $x \leq 18$. So we must have $2 \leq x \leq 18$. The domain of $g \circ f$ is the interval $[2, 18]$.

c. $(f \circ h)(x) = f(h(x))$
$= f(x^2)$
$= \sqrt{x^2 - 2}$

The function $h(x) = x^2$ is defined on $(-\infty, \infty)$. The function $(f \circ h)(x) = \sqrt{x^2 - 2}$ is defined for $x^2 - 2 \geq 0$ or $x^2 \geq 2$. This implies that either $x \geq \sqrt{2}$ or $x \leq -\sqrt{2}$. So the domain of $f \circ h$ is

$(-\infty, -\sqrt{2}] \cup [\sqrt{2}, \infty)$.

d. $(g \circ h)(x) = g(h(x))$
$= g(x^2)$
$= \sqrt{4 - x^2}$

The expression $\sqrt{4 - x^2}$ is defined for $4 - x^2 \geq 0$ or $x^2 \leq 4$. This implies that $-2 \leq x \leq 2$. So the domain of $g \circ h$ is $[-2, 2]$.

e. $(h \circ g)(x) = h(g(x))$
$= h(\sqrt{4-x})$
$= (\sqrt{4-x})^2$
$= 4 - x$

Although the last expression is defined for all real numbers, the interval $(-\infty, \infty)$ is not the domain of $h \circ g$. To find the domain of $h \circ g$, we first consider the domain of g, which is given by $4 - x \geq 0$ or $x \leq 4$. The domain of $h \circ g$ is $(-\infty, 4]$.

Practice Problem 5 For the functions f and h of Example 5, find $h \circ f$ and its domain.

Decomposition of a Function

4 Decompose a function.

In the composition of two functions, we combine two functions and create a new function. However, sometimes it is more useful to use the concept of composition to **decompose** a function into simpler functions. For example, consider the function $H(x) = \sqrt{x^2 - 2}$. A natural way to write $H(x)$ as the composition of two functions is to let $f(x) = \sqrt{x}$ and $g(x) = x^2 - 2$ so that $f(g(x)) = f(x^2 - 2) = \sqrt{x^2 - 2} = H(x)$.

A function may be decomposed into simpler functions in several different ways by choosing various "inner" functions.

PROCEDURE IN ACTION

EXAMPLE 6   Decomposing a Function

OBJECTIVE
Write a function H as a composite of simpler functions f and g so that $H = f \circ g$.

EXAMPLE
Write
$$H(x) = \frac{1}{\sqrt{2x^2 + 1}}$$
as $f \circ g$.

Step 1 Define $g(x)$ as any expression in the defining equation for H.

1. Let $g(x) = 2x^2 + 1$.

Step 2 To get $f(x)$ from the defining equation for H,

(1) Replace the letter H with f.
(2) Replace the expression chosen for $g(x)$ with x.

2. $H(x) = \dfrac{1}{\sqrt{2x^2 + 1}}$

becomes

$f(x) = \dfrac{1}{\sqrt{x}}$ Replace the expression $2x^2 + 1$ from Step 1 with x and replace H with f.

Step 3 Now we have
$H(x) = f(g(x))$
$= f \circ g(x).$

3. $f(g(x)) = f(2x^2 + 1)$ $f(x) = \dfrac{1}{\sqrt{x}}, g(x) = 2x^2 + 1$
$= \dfrac{1}{\sqrt{2x^2 + 1}}$
$= H(x)$

Note: Other choices include $g(x) = 2x^2$ and $f(x) = \dfrac{1}{\sqrt{x + 1}}$.

Practice Problem 6 Repeat Example 6 letting $g(x) = \sqrt{2x^2 + 1}$. Find $f(x)$ and write $H(x) = (f \circ g)(x)$.

Applications of Composite Functions

5 Apply composition to practical problems.

EXAMPLE 7 Calculating the Area of an Oil Spill from a Tanker

Suppose oil spills from a tanker into the Pacific Ocean and the area of the spill is a perfect circle. The radius of this oil slick increases at the rate of 2 miles per hour.

a. Express the area of the oil slick as a function of time.
b. Calculate the area covered by the oil slick in six hours.

Solution

The area A of the oil slick is a function of its radius r:

$$A = f(r) = \pi r^2$$

The radius is also a function of time. We know that r increases at the rate of 2 miles per hour. So in t hours, the radius r of the oil slick is given by the function

$$r = g(t) = 2t \qquad \text{Distance} = \text{Rate} \times \text{Time}$$

a. Therefore, the area A of the oil slick is a function of the radius, which is itself a function of time. It is a composite function given by

$$A = f(g(t)) = f(2t) = \pi(2t)^2$$
$$= 4\pi t^2.$$

b. Substitute $t = 6$ into A to find the area covered by the oil in six hours:

$$A = 4\pi(6)^2 = 4\pi(36)$$
$$= 144\pi \approx 452 \text{ square miles}$$

Practice Problem 7 What would the result of Example 7 be if the radius of the oil slick increased at a rate of 3 miles per hour?

EXAMPLE 8 Applying Composition to Sales

A car dealer offers an 8% discount off the manufacturer's suggested retail price (MSRP) of x dollars for any new car on his lot. At the same time, the manufacturer offers a $4000 rebate for each purchase of a car.

a. Write a function $r(x)$ that represents the price after only the rebate.
b. Write a function $d(x)$ that represents the price after only the dealer's discount.
c. Write the functions $(r \circ d)(x)$ and $(d \circ r)(x)$. What do they represent?
d. Calculate $(d \circ r)(x) - (r \circ d)(x)$. Interpret this expression.

Solution

a. The MSRP is x dollars, and the manufacturer's rebate is $4000 for each car; so

$$r(x) = x - 4000$$

represents the price of a car after the rebate.

b. The dealer's discount is 8% of x, or $0.08x$; therefore,

$$d(x) = x - 0.08x = 0.92x$$

represents the price of the car after the dealer's discount.

c. (i) $(r \circ d)(x) = r(d(x))$ Apply the dealer's discount first.
$= r(0.92x)$ $d(x) = 0.92x$
$= 0.92x - 4000$ Evaluate $r(0.92x)$ using $r(x) = x - 4000$.

Then $(r \circ d)(x) = 0.92x - 4000$ represents the price when the dealer's discount is applied first.

(ii) $(d \circ r)(x) = d(r(x))$ Apply the manufacturer's rebate first.
$= d(x - 4000)$ $r(x) = x - 4000$
$= 0.92(x - 4000)$ Evaluate $d(x - 4000)$ using $d(x) = 0.92x$.
$= 0.92x - 3680$

Thus, $(d \circ r)(x) = 0.92x - 3680$ represents the price when the manufacturer's rebate is applied first.

d. $(d \circ r)(x) - (r \circ d)(x) = (0.92x - 3680) - (0.92x - 4000)$ Subtract function values.
$= \$320$

This equation shows that any car, regardless of its price, will cost \$320 more if you apply the rebate first and then the discount.

Practice Problem 8 How will the result of Example 8 change if the dealer offers a 6% discount and the manufacturer offers a \$4500 rebate?

SECTION 1.6 Exercises

Basic Concepts and Skills

In Exercises 1–4, functions f and g are given. Find each of the given values.
a. $(f + g)(-1)$ **b.** $(f - g)(0)$
c. $(f \cdot g)(2)$ **d.** $\left(\dfrac{f}{g}\right)(1)$

1. $f(x) = 2x; g(x) = -x$
2. $f(x) = 1 - x^2; g(x) = x + 1$
3. $f(x) = \dfrac{1}{\sqrt{x + 2}}; g(x) = 2x + 1$
4. $f(x) = \dfrac{x}{x^2 - 6x + 8}; g(x) = 3 - x$

In Exercises 5–12, functions f and g are given. Find each of the following functions and state its domain.
a. $f + g$ **b.** $f - g$ **c.** $f \cdot g$
d. $\dfrac{f}{g}$ **e.** $\dfrac{g}{f}$

5. $f(x) = x - 3; g(x) = x^2$
6. $f(x) = 2x - 1; g(x) = x^2$
7. $f(x) = x^3 - 1; g(x) = 2x^2 + 5$
8. $f(x) = x^2 - 4; g(x) = x^2 - 6x + 8$
9. $f(x) = 2x - 1; g(x) = \sqrt{x}$
10. $f(x) = 1 - \dfrac{1}{x}; g(x) = \dfrac{1}{x}$
11. $f(x) = \dfrac{2}{x + 1}; g(x) = \dfrac{x}{x + 1}$
12. $f(x) = \dfrac{1}{\sqrt{x}}$
    $g(x) = \dfrac{\sqrt{x}}{x - 1}$

In Exercises 13 and 14, use each diagram to evaluate $(g \circ f)(x)$. Then evaluate $(g \circ f)(2)$ and $(g \circ f)(-3)$.

13. $x \rightarrow f(x) = x^2 - 1 \rightarrow \rightarrow g(x) = 2x + 3 \rightarrow$

14. $x \rightarrow f(x) = |x + 1| \rightarrow \rightarrow g(x) = 3x^2 - 1 \rightarrow$

In Exercises 15–26, let $f(x) = 2x + 1$ and $g(x) = 2x^2 - 3$. Evaluate each expression.

15. $(f \circ g)(2)$ 16. $(g \circ f)(2)$
17. $(f \circ g)(-3)$ 18. $(g \circ f)(-5)$
19. $(f \circ g)(0)$ 20. $(g \circ f)\left(\dfrac{1}{2}\right)$
21. $(f \circ g)(-c)$ 22. $(f \circ g)(c)$

23. $(g \circ f)(a)$ **24.** $(g \circ f)(-a)$
25. $(f \circ f)(1)$ **26.** $(g \circ g)(-1)$

In Exercises 27–32, the functions f and g are given. Find $f \circ g$ and its domain.

27. $f(x) = \dfrac{2}{x+1}$; $g(x) = \dfrac{1}{x}$

28. $f(x) = \dfrac{1}{x-1}$; $g(x) = \dfrac{2}{x+3}$

29. $f(x) = \sqrt{x-3}$; $g(x) = 2 - 3x$

30. $f(x) = \dfrac{x}{x-1}$; $g(x) = 2 + 5x$

31. $f(x) = \sqrt{x-1}$; $g(x) = x^2$

32. $f(x) = \sqrt{4-x}$; $g(x) = x^2$

In Exercises 33–40, the functions f and g are given. Find each composite function and describe its domain.
  a. $f \circ g$ b. $g \circ f$
  c. $f \circ f$ d. $g \circ g$

33. $f(x) = 2x^2 + 3x$; $g(x) = 2x - 1$
34. $f(x) = x^2 + 3x$; $g(x) = 2x$
35. $f(x) = x^2$; $g(x) = \sqrt{x}$
36. $f(x) = x^2 + 2x$; $g(x) = \sqrt{x+2}$
37. $f(x) = \sqrt{x-1}$; $g(x) = \sqrt{4-x}$
38. $f(x) = x^2 - 4$; $g(x) = \sqrt{4-x^2}$
39. $f(x) = \dfrac{3}{|x|}$; $g(x) = 2x + 3$
40. $f(x) = 1 + \dfrac{1}{x}$; $g(x) = \dfrac{1+x}{1-x}$

In Exercises 41–50, express the given function H as a composition of two functions f and g such that $H(x) = (f \circ g)(x)$.

41. $H(x) = \sqrt{x+2}$
42. $H(x) = |3x + 2|$
43. $H(x) = (x^2 - 3)^{10}$
44. $H(x) = \sqrt{3x^2 + 5}$
45. $H(x) = \dfrac{1}{3x-5}$
46. $H(x) = \dfrac{5}{2x+3}$
47. $H(x) = \sqrt[3]{x^2 - 7}$
48. $H(x) = \sqrt[4]{x^2 + x + 1}$
49. $H(x) = \dfrac{1}{|x^3 - 1|}$
50. $H(x) = \sqrt[3]{1 + \sqrt{x}}$

Applying the Concepts

51. Cost, revenue, and profit. A retailer purchases x shirts from a wholesaler at a price of $12 per shirt. Her selling price for each shirt is $22. The state has 7% sales tax. Interpret each of the following functions.
  a. $f(x) = 12x$
  b. $g(x) = 22x$
  c. $h(x) = g(x) + 0.07g(x)$
  d. $P(x) = g(x) - f(x)$

52. Consumer issues. You work in a department store in which employees are entitled to a 30% discount on their purchases. You also have a coupon worth $5 off any item.
  a. Write a function $f(x)$ that models discounting an item by 30%.
  b. Write a function $g(x)$ that models applying the coupon.
  c. Use a composition of your two functions from parts (a) and (b) to model your cost for an item if the clerk applies the discount first and then the coupon.
  d. Use a composition of your two functions from parts (a) and (b) to model your cost for an item if the clerk applies the coupon first and then the discount.
  e. Use the composite functions from parts (c) and (d) to find how much more an item costs if the clerk applies the coupon first.

53. Test grades Professor Harsh gave a test to his college algebra class, and nobody got more than 80 points (out of 100) on the test. One problem worth 8 points had insufficient data, so nobody could solve that problem. The professor adjusted the grades for the class by (1) increasing everyone's score by 10% and (2) giving everyone 8 bonus points. Let x represent the original score of a student.
  a. Write statements (1) and (2) as functions $f(x)$ and $g(x)$, respectively.
  b. Find $(f \circ g)(x)$ and explain what it means.
  c. Find $(g \circ f)(x)$ and explain what it means.
  d. Evaluate $(f \circ g)(70)$ and $(g \circ f)(70)$.
  e. Does $(f \circ g)(x) = (g \circ f)(x)$?
  f. Suppose a score of 90 or better results in an A. What is the lowest original score that will result in an A if the professor uses
   (i) $(f \circ g)(x)$?
   (ii) $(g \circ f)(x)$?

54. Sales commissions. Henrita works as a salesperson in a department store. Her weekly salary is $200 plus a 3% bonus on weekly sales over $8000. Suppose her sales in a week are x dollars.
  a. Let $f(x) = 0.03x$. What does this mean?
  b. Let $g(x) = x - 8000$. What does this mean?
  c. Which composite function, $(f \circ g)(x)$ or $(g \circ f)(x)$, represents Henrita's bonus?
  d. What was Henrita's salary for the week in which her sales were $17,500?
  e. What were Henrita's sales for the week in which her salary was $521?

55. Enhancing a fountain. A circular fountain has a radius of x feet. A circular fence is installed around the fountain at a distance of 30 feet from its edge.
   a. Write the function $f(x)$ that represents the area of the fountain.
   b. Write the function $g(x)$ that represents the entire area enclosed by the fence.
   c. What area does the function $g(x) - f(x)$ represent?
   d. The cost of the fence was $4200, installed at the rate of $10.50 per running foot (per perimeter foot). You are to prepare an estimate for paving the area between the fence and the fountain at $1.75 per square foot. To the nearest dollar, what should your estimate be?

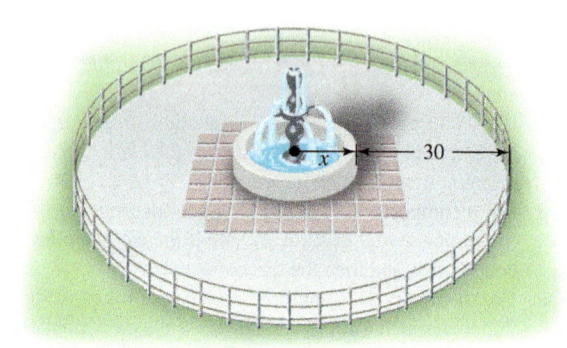

56. Running track design. An outdoor running track with semicircular ends and parallel sides is constructed. The length of each straight portion of the sides is 180 meters. The track has a uniform width of 4 meters throughout. The inner radius of each semicircular end is x meters.
   a. Write the function $f(x)$ that represents the area enclosed within the *outer* edge of the running track.
   b. Write the function $g(x)$ that represents the area of the field enclosed within the *inner* edge of the running track.
   c. What does the function $f(x) - g(x)$ represent?
   d. Suppose the inner perimeter of the track is 900 meters.
      (i) Find the area of the track.
      (ii) Find the outer perimeter of the track.

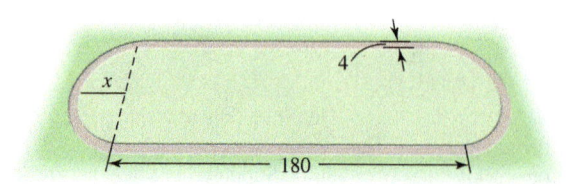

57. Area of a disk. The area A of a circular disk of radius r units is given by $A = f(r) = \pi r^2$. Suppose a metal disk is being heated and its radius r is increasing according to the equation $r = g(t) = 2t + 1$, where t is time in hours.
   a. Find $(f \circ g)(t)$.
   b. Determine A as a function of time.
   c. Compare parts (a) and (b).

58. Volume of a balloon. The volume V of a spherical balloon of radius r inches is given by the formula $V = f(r) = \frac{4}{3}\pi r^3$.
Suppose the balloon is being inflated and its radius is increasing at the rate of 2 inches per second. That is, $r = g(t) = 2t$, where t is time in seconds.
   a. Find $(f \circ g)(t)$.
   b. Determine V as a function of time.
   c. Compare parts (a) and (b).

59. Let $f = \{(-2, 3), (1, 2), (2, 1), (3, 0)\}$; and $g = \{(-2, 0), (0, 2), (1, -2), (3, 2)\}$.
Find each function and its domain.
   a. $f + g$ **b.** fg **c.** $\dfrac{f}{g}$ **d.** $f \circ g$

60. Let
$$f(x) = \begin{cases} 2x & \text{if } -2 \leq x \leq 1 \\ x + 1 & \text{if } 1 < x \leq 3 \end{cases}; \text{ and}$$
$$g(x) = \begin{cases} x + 1 & \text{if } -3 \leq x < 2 \\ 2x - 1 & \text{if } 2 \leq x \leq 3 \end{cases}$$

Find each function and its domain.
   a. $(f + g)(x)$ **b.** $(f \circ g)(x)$ **c.** $(f \circ f)(x)$

Critical Thinking / Discussion / Writing

61. Let $h(x)$ be any function whose domain contains $-x$ whenever it contains x. Show that
   a. $f(x) = h(x) + h(-x)$ is an even function.
   b. $g(x) = h(x) - h(-x)$ is an odd function.
   c. $h(x)$ can always be expressed as the sum of an even function and an odd function.

62. Use Exercise 61 to write each of the following functions as the sum of an even function and an odd function.
   a. $h(x) = x^2 - 2x + 3$
   b. $h(x) = [x] + x$

63. Let $f(x) = \dfrac{1}{[x]}$ and $g(x) = \sqrt{x(2-x)}$. Find the domain of each of the following functions.
   a. $f(x)$ **b.** $g(x)$
   c. $f(x) + g(x)$ **d.** $\dfrac{f(x)}{g(x)}$

64. For each of the following functions, find the domain of $f \circ f$.
  a. $f(x) = \dfrac{1}{\sqrt{-x}}$
  b. $f(x) = \dfrac{1}{\sqrt{1-x}}$

65. Even and odd functions. State whether each of the following functions is an odd function, an even function, or neither. Justify your statement.
  a. The sum of two even functions
  b. The sum of two odd functions
  c. The sum of an even function and an odd function
  d. The product of two even functions
  e. The product of two odd functions
  f. The product of an even function and an odd function

66. Even and odd functions. State whether the composite function $f \circ g$ is an odd function, an even function, or neither in the following situations.
  a. f and g are odd functions.
  b. f and g are even functions.
  c. f is odd and g is even.
  d. f is even and g is odd.

Maintaining Skills

In Exercises 67–72, solve each equation.

67. $3(2 - 3x) - 4x = 3x - 10$

68. $\dfrac{x}{x-2} - \dfrac{1}{x+2} = 1$

69. $x^2 + 1 = -3$

70. $x^2 + 2x - 5 = 0$

71. $4x^2 + 12x + 9 = 0$

72. $\sqrt{2x+1} + 1 = x$

SECTION 1.7

Inverse Functions

BEFORE STARTING THIS SECTION, REVIEW

1. Vertical-line test (Section 1.3, page 84)
2. Domain and range of a function (Section 1.3, pages 82 and 83)
3. Composition of functions (Section 1.6, page 128)
4. Line of symmetry (Section 1.1, page 57)
5. Solving linear and quadratic equations (Section P.3, page 21)

OBJECTIVES

1. Define an inverse function.
2. Find the inverse function.
3. Use inverse functions to find the range of a function.
4. Apply inverse functions in the real world.

◆ Water Pressure on Underwater Devices

The pressure at any depth in the ocean is the pressure at sea level (1 atm) plus the pressure due to the weight of the overlying water: the deeper you go, the greater the pressure. Pressure is a vital consideration in the design of all tools and vehicles that work beneath the water's surface, and pressure gauges are built to measure the pressure due to the overlying water. Every 11 feet of depth increases the pressure by approximately 5 pounds per square inch. Computing the pressure precisely requires a somewhat complicated equation that takes into account the water's salinity, temperature, and slight compressibility. However, a basic formula is given by "pressure equals depth times 5, divided by 11 plus 15." So we have pressure (psi) = depth (ft) × 5 psi/11(ft) + 15, where *psi* means "pounds per square inch."

For example, at 99 feet below the surface, the pressure is 45 psi. At 1 mile (5280 feet) below the surface, the pressure is (5280 × 5)/11 + 15 = 2415 psi. Suppose the pressure gauge on a diving bell breaks and shows a reading of 1800 psi. We can determine the bell's depth when the gauge failed by using an inverse function. We will return to this problem in Example 9.

1 Define an inverse function.

Inverses

Before we define inverse functions, we need to look at a special type of function called a one-to-one function.

One-to-One Function

A function is a **one-to-one function** if each *y*-value in its range corresponds to only one *x*-value in its domain.

Let f be a one-to-one function. Then the preceding definition says that for any two numbers x_1 and x_2 in the domain of f, if $x_1 \neq x_2$, then $f(x_1) \neq f(x_2)$, or equivalently, if $f(x_1) = f(x_2)$, then $x_1 = x_2$. That is, *f is a one-to-one function if different x-values correspond to different y-values.*

136 Chapter 1 Graphs and Functions

Section 1.7 ■ Inverse Functions   137

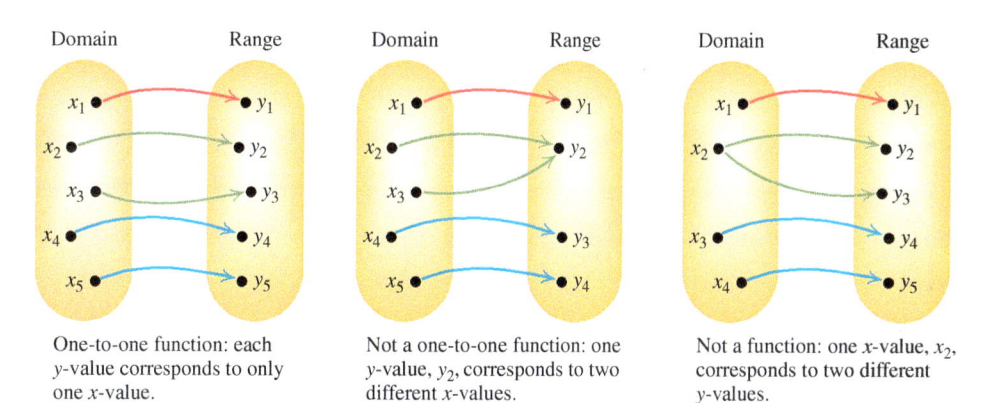

Domain   Range   Domain   Range   Domain   Range

One-to-one function: each   Not a one-to-one function: one   Not a function: one x-value, x_2,
y-value corresponds to only   y-value, y_2, corresponds to two   corresponds to two different
one x-value.   different x-values.   y-values.
(a)   (b)   (c)

FIGURE 1.80 Deciding whether a diagram represents a function, a one-to-one function, or neither

Figure 1.80(a) represents a one-to-one function, and Figure 1.80(b) represents a function that is not one-to-one. The relation shown in Figure 1.80(c) is not a function.

Recall that with a one-to-one function, different x-values correspond to different y-values. Consequently, if a function f is *not* one-to-one, then there are at least two numbers x_1 and x_2 in the domain of f such that $x_1 \neq x_2$ and $f(x_1) = f(x_2)$. Geometrically, this means that the horizontal line passing through the two points $(x_1, f(x_1))$ and $(x_2, f(x_2))$ contains these *two* points on the graph of f. See Figure 1.81. The geometric interpretation of a one-to-one function is called the **horizontal-line test**.

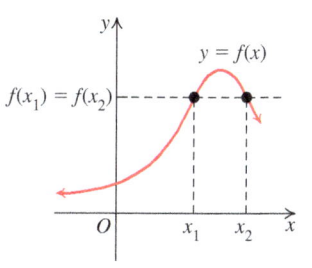

FIGURE 1.81 The function f is *not* one-to-one.

Horizontal-Line Test

A function f is one-to-one if no horizontal line intersects the graph of f in more than one point.

EXAMPLE 1   Using the Horizontal-Line Test

Determine which of the following functions are one-to-one.

a. $f(x) = 2x + 5$ **b.** $g(x) = x^2 - 1$ **c.** $h(x) = 2\sqrt{x}$

Solution

Algebraic Approach   Let x_1 and x_2 be two numbers in the domain of each function.

a. Suppose $f(x_1) = f(x_2)$. Then $2x_1 + 5 = 2x_2 + 5$

$\quad\quad\quad\quad\quad\quad\quad\quad 2x_1 = 2x_2$ Subtract 5 from each side.

$\quad\quad\quad\quad\quad\quad\quad\quad x_1 = x_2$ Divide both sides by 2.

So $f(x_1) = f(x_2)$ implies that $x_1 = x_2$; the function f is one-to-one.

b. Suppose $g(x_1) = g(x_2)$. Then $x_1^2 - 1 = x_2^2 - 1$

$\quad\quad\quad\quad\quad\quad\quad\quad x_1^2 = x_2^2$ Add 1 to both sides.

$\quad\quad\quad\quad\quad\quad\quad\quad x_1 = \pm x_2$ Square root property

This means that for two distinct numbers x_1 and $x_2 = -x_1$, $g(x_1) = g(x_2)$; so g is not a one-to-one function.

c. Suppose $h(x_1) = h(x_2)$. Then $2\sqrt{x_1} = 2\sqrt{x_2}$

$\quad\quad\quad\quad\quad\quad\quad\quad \sqrt{x_1} = \sqrt{x_2}$ Divide both sides by 2.

$\quad\quad\quad\quad\quad\quad\quad\quad x_1 = x_2$ Square both sides.

So $h(x_1) = h(x_2)$ implies that $x_1 = x_2$; the function h is one-to-one.

Geometric Approach (horizontal-line test). Figure 1.82 shows the graphs of functions f, g, and h. We see that no horizontal line intersects the graphs of f [Figure 1.82(a)] or h [Figure 1.82(c)] in more than one point; therefore, the functions f and h are one-to-one. The function g is not one-to-one because the horizontal line $y = 3$ (among others) in Figure 1.82(b) intersects the graph of g at more than one point.

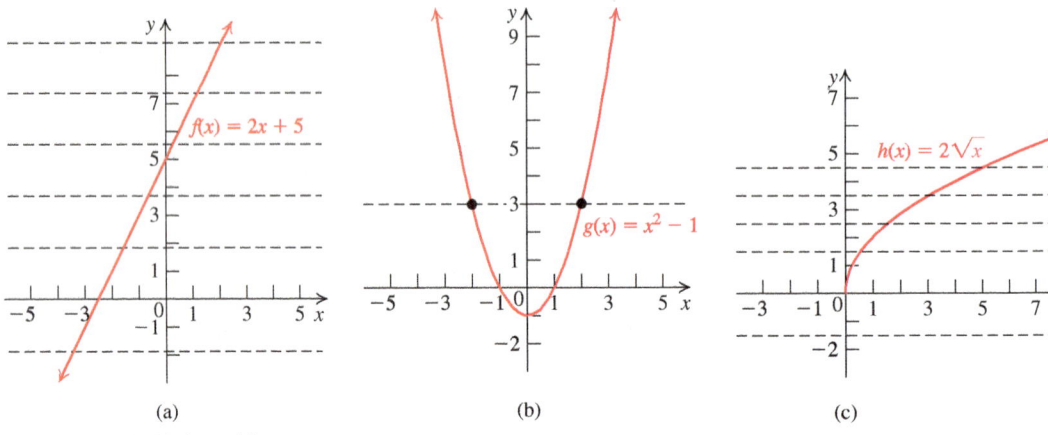

FIGURE 1.82 Horizontal-line test

Practice Problem 1 Determine whether $f(x) = (x-1)^2$ is a one-to-one function.

Inverse Function

SIDE NOTE

The reason only a one-to-one function can have an inverse is that if $x_1 \neq x_2$ but $f(x_1) = y$ and $f(x_2) = y$, then $f^{-1}(y)$ must be x_1 as well as x_2. This is not possible because $x_1 \neq x_2$.

Let f represent a one-to-one function. Then if y is in the range of f, there is only one value of x in the domain of f such that $f(x) = y$. We define the inverse of f, called the **inverse function of f**, denoted f^{-1}, by $f^{-1}(y) = x$ if and only if $y = f(x)$.

From this definition we have the following:

$$\text{Domain of } f = \text{Range of } f^{-1} \quad \text{and} \quad \text{Range of } f = \text{Domain of } f^{-1}$$

WARNING The notation $f^{-1}(x)$ does not mean $\dfrac{1}{f(x)}$. The expression $\dfrac{1}{f(x)}$ represents the reciprocal of $f(x)$ and is sometimes written as $[f(x)]^{-1}$.

EXAMPLE 2 Relating the Values of a Function and Its Inverse

Assume that f is a one-to-one function.

a. If $f(3) = 5$, find $f^{-1}(5)$. **b.** If $f^{-1}(-1) = 7$, find $f(7)$.

Solution

By definition, $f^{-1}(y) = x$ if and only if $y = f(x)$.

a. Let $x = 3$ and $y = 5$. Now reading the definition from right to left, $5 = f(3)$ if and only if $f^{-1}(5) = 3$. Thus, $f^{-1}(5) = 3$.

b. Let $y = -1$ and $x = 7$. Now $f^{-1}(-1) = 7$ if and only if $f(7) = -1$. Thus, $f(7) = -1$.

Practice Problem 2 Assume that f is a one-to-one function.

a. If $f(-3) = 12$, find $f^{-1}(12)$. **b.** If $f^{-1}(4) = 9$, find $f(9)$.

Consider the following input–output diagram and Figure 1.83 for $f^{-1} \circ f$.

INPUT–OUTPUT DIAGRAM

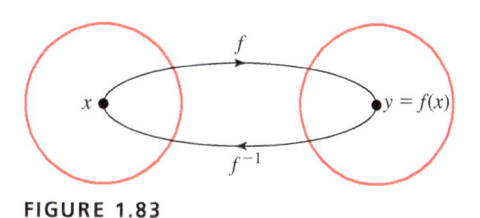

$$y = f(x) \qquad\qquad f^{-1}(y) = x$$

input x $\xrightarrow{\;f\;}$ output y $\quad$ input y $\xrightarrow{\;f^{-1}\;}$ output x

input x $\xrightarrow{\qquad f^{-1} \circ f \qquad}$ output x

FIGURE 1.83

Figure 1.83 suggests the following:

INVERSE FUNCTION PROPERTY

Let f denote a one-to-one function. Then

1. $f^{-1}(f(x)) = x$ for every x in the domain of f.
2. $f(f^{-1}(x)) = x$ for every x in the domain of f^{-1}.

Further, if g is any function such that (for all values of x in the domain of the inner function)

$$f(g(x)) = x \text{ and } g(f(x)) = x, \text{ then } g = f^{-1}.$$

One interpretation of the equation $(f^{-1} \circ f)(x) = x$ is that f^{-1} undoes anything that f does to x. For example, let

$$f(x) = x + 2 \qquad f \text{ adds 2 to any input } x.$$

To undo what f does to x, we should subtract 2 from x. That is, the inverse of f should be

$$g(x) = x - 2 \qquad g \text{ subtracts 2 from } x.$$

Let's verify that $g(x) = x - 2$ is indeed the inverse function of x.

$$\begin{aligned}(f \circ g)(x) &= f(g(x)) & &\text{Definition of } f \circ g \\ &= f(x - 2) & &\text{Replace } g(x) \text{ with } x - 2. \\ &= (x - 2) + 2 & &\text{Replace } x \text{ with } x - 2 \text{ in } f(x) = x + 2. \\ &= x & &\text{Simplify.}\end{aligned}$$

We leave it for you to check that $(g \circ f)(x) = x$.

EXAMPLE 3 **Verifying Inverse Functions**

Verify that the following pairs of functions are inverses of each other:

$$f(x) = 2x + 3 \quad \text{and} \quad g(x) = \frac{x - 3}{2}$$

Solution

$$(f \circ g)(x) = f(g(x)) \quad \text{Definition of } f \circ g$$

$$= f\left(\frac{x-3}{2}\right) \quad \text{Replace } g(x) \text{ with } \frac{x-3}{2}.$$

$$= 2\left(\frac{x-3}{2}\right) + 3 \quad \text{Replace } x \text{ with } \frac{x-3}{2} \text{ in } f(x) = 2x + 3.$$

$$= x \quad \text{Simplify.}$$

Thus, $f(g(x)) = x$.

$$g(x) = \frac{x-3}{2} \quad \text{Given function } g$$

$$(g \circ f)(x) = g(f(x)) \quad \text{Definition of } g \text{ of } f$$

$$= g(2x + 3) \quad \text{Replace } f(x) \text{ with } 2x + 3.$$

$$= \frac{(2x+3) - 3}{2} \quad \text{Replace } x \text{ with } 2x + 3 \text{ in } g(x) = \frac{x-3}{2}.$$

$$= x \quad \text{Simplify.}$$

Thus, $g(f(x)) = x$. Because $f(g(x)) = g(f(x)) = x$, f and g are inverses of each other.

Practice Problem 3 Verify that $f(x) = 3x - 1$ and $g(x) = \frac{x+1}{3}$ are inverses of each other.

2 Find the inverse function.

Finding the Inverse Function

Let $y = f(x)$ be a one-to-one function; then f has an inverse function. Suppose (a, b) is a point on the graph of f. Then $b = f(a)$. This means that $a = f^{-1}(b)$, so (b, a) is a point on the graph of f^{-1}. The points (a, b) and (b, a) are symmetric with respect to the line $y = x$, as shown in Figure 1.84. That is, if the graph paper is folded along the line $y = x$, the points (a, b) and (b, a) will coincide. Therefore, we have the following property:

> **SYMMETRY PROPERTY OF THE GRAPHS OF f AND f^{-1}**
>
> The graph of a one-to-one function f and the graph of f^{-1} are symmetric with respect to the line $y = x$.

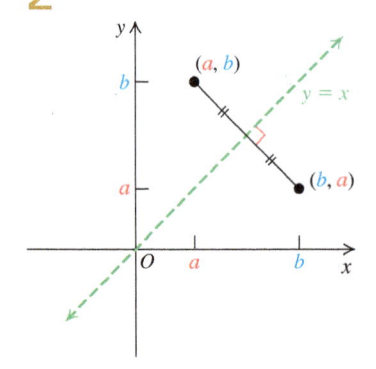

FIGURE 1.84 Relationship between points (a, b) and (b, a)

EXAMPLE 4 Finding the Graph of f^{-1} from the Graph of f

The graph of a function f is shown in Figure 1.85. Sketch the graph of f^{-1}.

Solution

By the horizontal-line test, f is a one-to-one function; so its inverse will also be a function.

The graph of f consists of two line segments: one joining the points $(-3, -5)$ and $(-1, 2)$ and the other joining the points $(-1, 2)$ and $(4, 3)$.

The graph of f^{-1} is the reflection of the graph of f about the line $y = x$. The reflections of the points $(-3, -5), (-1, 2),$ and $(4, 3)$ about the line $y = x$ are $(-5, -3)$, $(2, -1)$, and $(3, 4)$, respectively.

Section 1.7 ■ Inverse Functions 141

SIDE NOTE

It is helpful to notice that points on the x-axis, (a, 0), reflect to points on the y-axis, (0, a), and conversely. Also notice that points on the line y = x are unaffected by reflection about the line y = x.

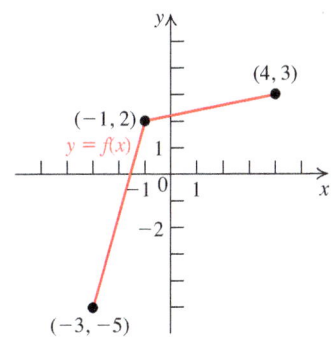

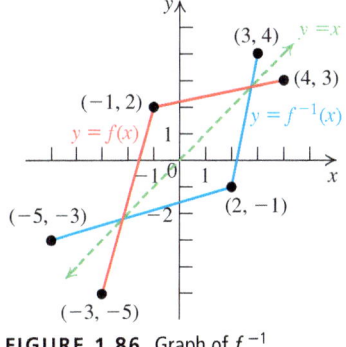

FIGURE 1.85 Graph of f

FIGURE 1.86 Graph of f^{-1}

The graph of f^{-1} consists of two line segments: one joining the points $(-5, -3)$ and $(2, -1)$ and the other joining the points $(2, -1)$ and $(3, 4)$. See Figure 1.86.

Practice Problem 4 Use the graph of a function f in Figure 1.87 to sketch the graph of f^{-1}.

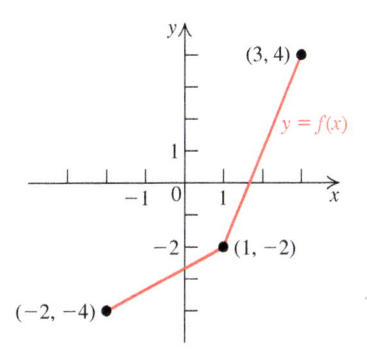

FIGURE 1.87 Graphing f^{-1} from the graph of f

The symmetry between the graphs of f and f^{-1} about the line $y = x$ tells us that we can find an equation for the inverse function $y = f^{-1}(x)$ from the equation of a one-to-one function $y = f(x)$ by interchanging the roles of x and y in the equation $y = f(x)$. This results in the equation $x = f(y)$. Then we solve the equation $x = f(y)$ for y in terms of x to get $y = f^{-1}(x)$.

PROCEDURE IN ACTION

EXAMPLE 5 Finding an Equation for f^{-1}

OBJECTIVE
Find the inverse of a function.

EXAMPLE
Find the inverse of $f(x) = 3x - 4$.

Step 1 Replace $f(x)$ with y in the equation defining $f(x)$.

1. $y = 3x - 4$ Replace $f(x)$ with y.

Step 2 Interchange x and y.

2. $x = 3y - 4$ Interchange x and y.

Step 3 Solve the equation in Step 2 for y.

3. $x + 4 = 3y$ Add 4 to both sides.

   $\dfrac{x + 4}{3} = y$ Divide both sides by 3.

Step 4 Write $f^{-1}(x)$ for y.

4. $f^{-1}(x) = \dfrac{x + 4}{3}$ We usually end with $f^{-1}(x)$ on the left.

Practice Problem 5 Find the inverse of $f(x) = -2x + 3$.

142 Chapter 1 Graphs and Functions

> **SIDE NOTE**
> Interchanging x and y in Step 2 is done so that we can write f^{-1} in the usual format with x as the independent variable and y as the dependent variable.

EXAMPLE 6 Finding the Inverse Function

Find the inverse of the one-to-one function

$$f(x) = \frac{x+1}{x-2}, \quad x \neq 2.$$

Solution

Step 1 $\quad y = \dfrac{x+1}{x-2}$ $\qquad$ Replace $f(x)$ with y.

Step 2 $\quad x = \dfrac{y+1}{y-2}$ $\qquad$ Interchange x and y.

Step 3 Solve $x = \dfrac{y+1}{y-2}$ for y. This is the most challenging step.

$\quad x(y-2) = y+1$ $\qquad$ Multiply both sides by $y-2$.
$\quad xy - 2x = y+1$ $\qquad$ Distributive property
$\quad xy - 2x + 2x - y = y + 1 + 2x - y$ $\qquad$ Add $2x - y$ to both sides.
$\quad xy - y = 2x + 1$ $\qquad$ Simplify.
$\quad y(x-1) = 2x+1$ $\qquad$ Factor out y.
$\quad y = \dfrac{2x+1}{x-1}$ $\qquad$ Divide both sides by $x-1$, assuming that $x \neq 1$.

Step 4 $\quad f^{-1}(x) = \dfrac{2x+1}{x-1}, x \neq 1$ $\qquad$ Replace y with $f^{-1}(x)$.

To see if our calculations are accurate, we compute $f(f^{-1}(x))$ and $f^{-1}(f(x))$.

$$f^{-1}(f(x)) = f^{-1}\left(\frac{x+1}{x-2}\right) = \frac{2\left(\dfrac{x+1}{x-2}\right)+1}{\dfrac{x+1}{x-2}-1} = \frac{2x+2+x-2}{x+1-x+2} = \frac{3x}{3} = x$$

You should also check that $f(f^{-1}(x)) = x$.

Practice Problem 6 Find the inverse of the one-to-one function $f(x) = \dfrac{x}{x+3}$, $x \neq -3$.

3 Use inverse functions to find the range of a function.

Finding the Range of a One-to-One Function

It is not always easy to determine the range of a function that is defined by an equation. However, for a one-to-one function f, we can find the range of f by finding the domain of f^{-1}.

EXAMPLE 7 Finding the Domain and Range

> **RECALL**
> Remember that the domain of a function f given by a formula is the largest set of real numbers for which the values $f(x)$ are real numbers.

Find the domain and the range of the function $f(x) = \dfrac{x+1}{x-2}$ of Example 6.

Solution

The domain of $f(x) = \dfrac{x+1}{x-2}$ is the set of all real numbers x such that $x \neq 2$. In interval notation, the domain of f is $(-\infty, 2) \cup (2, \infty)$.

From Example 6, $f^{-1}(x) = \dfrac{2x+1}{x-1}, x \neq 1$; therefore,

$$\text{Range of } f = \text{Domain of } f^{-1} = \{x \mid x \neq 1\}.$$

In interval notation, the range of f is $(-\infty, 1) \cup (1, \infty)$.

Section 1.7 ■ Inverse Functions 143

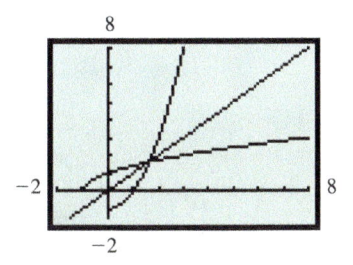

Practice Problem 7 Find the domain and the range of the function $f(x) = \dfrac{x}{x+3}$.

If a function f is not one-to-one, then it does not have an inverse function. Sometimes by changing its domain, we can produce an interesting function that does have an inverse. (This technique is frequently used in trigonometry.) We saw in Example 1(b) that $g(x) = x^2 - 1$ is not a one-to-one function; so g does not have an inverse function. However, the horizontal-line test shows that the function

$$G(x) = x^2 - 1, \quad x \geq 0$$

with domain $[0, \infty)$ is one-to-one. See Figure 1.88. Therefore, G has an inverse function G^{-1}.

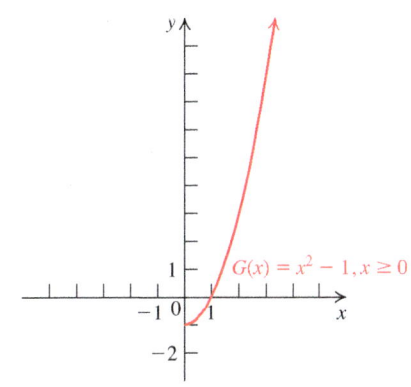

TECHNOLOGY CONNECTION

With a graphing calculator, you can also see that the graphs of $G(x) = x^2 - 1$, $x \geq 0$ and $G^{-1}(x) = \sqrt{x+1}$ are symmetric about the line $y = x$.

Enter Y_1, Y_2, and Y_3 as $G(x)$, $G^{-1}(x)$, and $f(x) = x$, respectively.

FIGURE 1.88 The function G has an inverse

EXAMPLE 8 Finding an Inverse Function

Find the inverse of $G(x) = x^2 - 1, x \geq 0$.

Solution

Step 1	$y = x^2 - 1, \quad x \geq 0$	Replace $G(x)$ with y.
Step 2	$x = y^2 - 1, \quad y \geq 0$	Interchange x and y.
Step 3	$x + 1 = y^2, \quad y \geq 0$	Add 1 to both sides.
	$y = \sqrt{x+1}$ because $y \geq 0$	Solve for y.
Step 4	$G^{-1}(x) = \sqrt{x+1}$	Replace y with $G^{-1}(x)$.

The graphs of G and G^{-1} are shown in Figure 1.89.

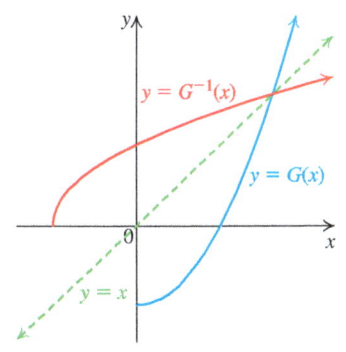

FIGURE 1.89 Graphs of G and G^{-1}

Practice Problem 8 Find the inverse of $G(x) = x^2 - 1, x \leq 0$.

4 Apply inverse functions in the real world.

Applications

Functions that model real-life situations are frequently expressed as formulas with letters that remind you of the variable they represent. In finding the inverse of a function expressed by a formula, interchanging the letters could be very confusing. Accordingly, we omit Step 2, keep the same letters, and just solve the formula for the other variable.

EXAMPLE 9 Water Pressure on Underwater Devices

At the beginning of this section, we wrote the formula for finding the water pressure p (in pounds per square inch) at a depth d (in feet) below the surface. This formula can be written as $p = \dfrac{5d}{11} + 15$. Suppose the pressure gauge on a diving bell breaks and shows a reading of 1800 psi. How far below the surface was the bell when the gauge failed?

Solution

We want to find the unknown depth in terms of the known pressure. This depth is given by the inverse of the function $p = \dfrac{5d}{11} + 15$. To find the inverse, we solve the given equation for d.

$$p = \dfrac{5d}{11} + 15 \quad \text{Original equation}$$

$$11p = 5d + 165 \quad \text{Multiply both sides by 11.}$$

$$d = \dfrac{11p}{5} - 33 \quad \text{Solve for } d.$$

Now we can use this formula to find the depth when the gauge read 1800 psi. We let $p = 1800$.

$$d = \dfrac{11p}{5} - 33 \quad \text{Depth from pressure equation}$$

$$d = \dfrac{11(1800)}{5} - 33 \quad \text{Replace } p \text{ with 1800.}$$

$$d = 3927$$

The device was 3927 feet below the surface when the gauge failed.

Practice Problem 9 In Example 9, suppose the pressure gauge showed a reading of 1650 psi. Determine the depth of the bell when the gauge failed.

SECTION 1.7 Exercises

Basic Concepts and Skills

In Exercises 1–8, the graph of a function is given. Use the horizontal-line test to determine whether the function is one-to-one.

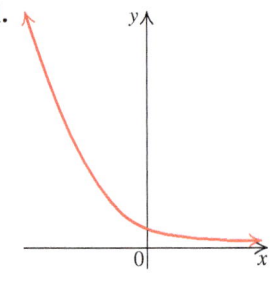

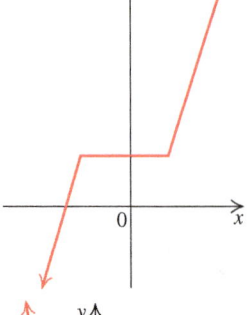

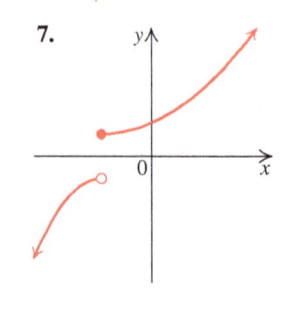

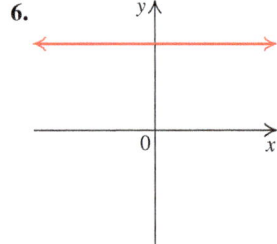

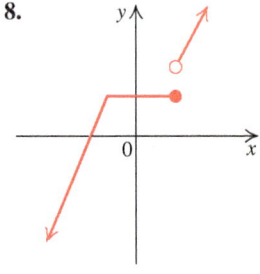

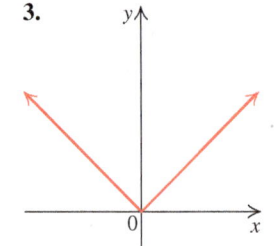

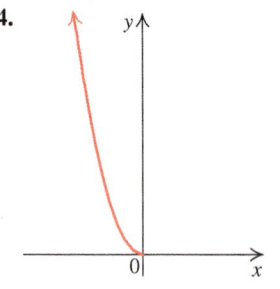

In Exercises 9–18, assume that the function f is one-to-one.

9. If $f(2) = 7$, find $f^{-1}(7)$.
10. If $f^{-1}(4) = -7$, find $f(-7)$.
11. If $f(-1) = 2$, find $f^{-1}(2)$.
12. If $f^{-1}(-3) = 5$, find $f(5)$.
13. If $f(a) = b$, find $f^{-1}(b)$.
14. If $f^{-1}(c) = d$, find $f(d)$.
15. Find $(f^{-1} \circ f)(337)$.
16. Find $(f \circ f^{-1})(25\pi)$.
17. Find $(f \circ f^{-1})(-1580)$.
18. Find $(f^{-1} \circ f)(9728)$.
19. For $f(x) = 2x - 3$, find each of the following.
    a. $f(3)$ b. $f^{-1}(3)$
    c. $(f \circ f^{-1})(19)$ d. $(f \circ f^{-1})(5)$
20. For $f(x) = x^3$, find each of the following.
    a. $f(2)$ b. $f^{-1}(8)$
    c. $(f \circ f^{-1})(15)$ d. $(f^{-1} \circ f)(27)$
21. For $f(x) = x^3 + 1$, find each of the following.
    a. $f(1)$
    b. $f^{-1}(2)$
    c. $(f \circ f^{-1})(269)$
22. For $g(x) = \sqrt[3]{2x^3 - 1}$, find each of the following.
    a. $g(1)$
    b. $g^{-1}(1)$
    c. $(g^{-1} \circ g)(135)$

In Exercises 23–28, show that f and g are inverses of each other by verifying that $f(g(x)) = x = g(f(x))$.

23. $f(x) = 3x + 1$; $g(x) = \dfrac{x-1}{3}$
24. $f(x) = 2 - 3x$; $g(x) = \dfrac{2-x}{3}$
25. $f(x) = x^3$; $g(x) = \sqrt[3]{x}$
26. $f(x) = \dfrac{1}{x}$; $g(x) = \dfrac{1}{x}$
27. $f(x) = \dfrac{x-1}{x+2}$; $g(x) = \dfrac{1+2x}{1-x}$
28. $f(x) = \dfrac{3x+2}{x-1}$; $g(x) = \dfrac{x+2}{x-3}$

In Exercises 29–34, the graph of a function f is given. Sketch the graph of f^{-1}.

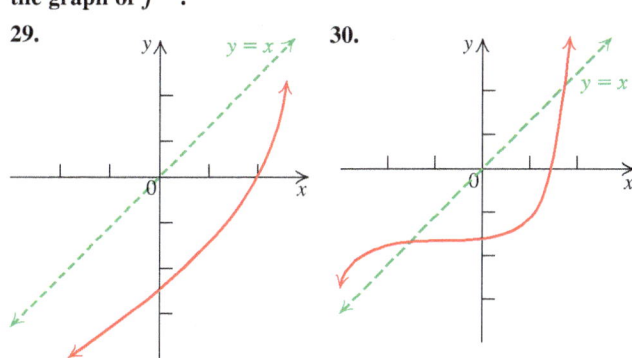

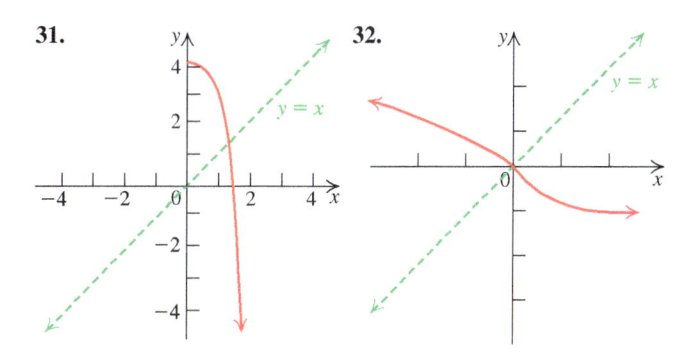

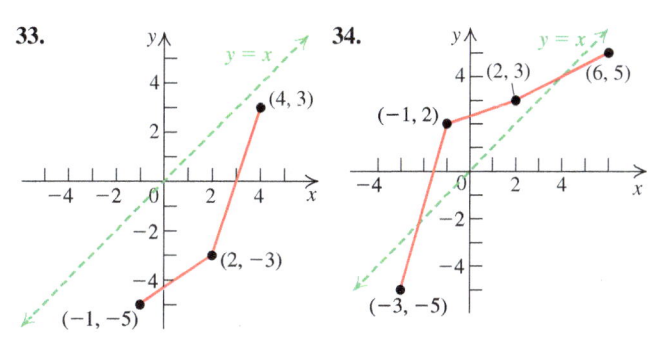

29.

30.

31.

32.

33.

34.

In Exercises 35–46,

a. Determine algebraically whether the given function is a one-to-one function.
b. If the function is one-to-one, find its inverse.
c. Sketch the graph of the function and its inverse on the same coordinate axes.
d. Give the domain and intercepts of each one-to-one function and its inverse function.

35. $f(x) = 15 - 3x$ 36. $g(x) = 2x + 5$
37. $f(x) = \sqrt{4 - x^2}$ 38. $f(x) = -\sqrt{9 - x^2}$
39. $f(x) = \sqrt{x} + 3$ 40. $f(x) = 4 - \sqrt{x}$
41. $g(x) = \sqrt[3]{x+1}$ 42. $h(x) = \sqrt[3]{1-x}$
43. $f(x) = \dfrac{1}{x-1}, x \neq 1$ 44. $g(x) = 1 - \dfrac{1}{x}, x \neq 0$
45. $f(x) = 2 + \sqrt{x+1}$ 46. $f(x) = -1 + \sqrt{x+2}$

In Exercises 47–50, assume that the given function is one-to-one. Find the inverse of the function. Also find the domain and the range of the given function.

47. $f(x) = \dfrac{x+1}{x-2}, x \neq 2$ 48. $g(x) = \dfrac{x+2}{x+1}, x \neq -1$
49. $f(x) = \dfrac{1-2x}{1+x}, x \neq -1$ 50. $h(x) = \dfrac{x-1}{x-3}, x \neq 3$

In Exercises 51–58, find the inverse of each function and sketch the graph of the function and its inverse on the same coordinate axes.

51. $f(x) = -x^2, x \geq 0$ 52. $g(x) = -x^2, x \leq 0$
53. $f(x) = |x|, x \geq 0$ 54. $g(x) = |x|, x \leq 0$
55. $f(x) = x^2 + 1, x \leq 0$ 56. $g(x) = x^2 + 5, x \geq 0$
57. $f(x) = -x^2 + 2, x \leq 0$ 58. $g(x) = -x^2 - 1, x \geq 0$

Applying the Concepts

59. Currency exchange. Alisha went to Europe last summer. She discovered that when she exchanged her U.S. dollars for euros, she received 25% fewer euros than the number of dollars she exchanged. (She got 75 euros for every 100 U.S. dollars.) When she returned to the United States, she got 25% more dollars than the number of euros she exchanged.
   a. Write each conversion function.
   b. Show that in part (a), the two functions are not inverse functions.
   c. Does Alisha gain or lose money after converting both ways?

60. Hourly wages. Anwar is a short-order cook in a diner. He is paid $4 per hour plus 5% of all food sales per hour. His average hourly wage w in terms of the food sales of x dollars is $w = 4 + 0.05x$.
   a. Write the inverse function. What does it mean?
   b. Use the inverse function to estimate the hourly sales at the diner if Anwar averages $12 per hour.

61. Hourly wages. In Exercise 60, suppose in addition that Anwar is guaranteed a minimum wage of $7 per hour.
   a. Write a function expressing his hourly wage w in terms of food sales per hour. [*Hint:* Use a piecewise function.]
   b. Does the function in part (a) have an inverse? Explain.
   c. If the answer in part (b) is yes, find the inverse function. If the answer is no, restrict the domain so that the new function has an inverse.

62. Simple pendulum. If a pendulum is released at a certain point, the **period** is the time it takes for the pendulum to swing along its path and return to the point from which it was released. The period T (in seconds) of a simple pendulum is a function of its length l (in feet) and is given by $T = (1.11)\sqrt{l}$.
   a. Find the inverse function. What does it mean?
   b. Use the inverse function to calculate the length of the pendulum if its period is two seconds.
   c. The convention center in Portland, Oregon, has the longest pendulum in the United States. The pendulum's length is 90 feet. Find the period.

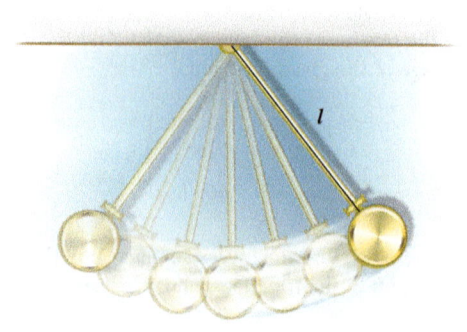

63. Water supply. Suppose x is the height of the water above the opening at the base of a water tank. The velocity V of water that flows from the opening at the base is a function of x and is given by $V(x) = 8\sqrt{x}$.
   a. Find the inverse function. What does it mean?
   b. Use the inverse function to calculate the height of the water in the tank when the flow is **(i)** 30 feet per second and **(ii)** 20 feet per second.

64. Physics. A projectile is fired from the origin over horizontal ground. Its altitude y (in feet) is a function of its horizontal distance x (in feet) and is given by $y = 64x - 2x^2$.
   a. Find the inverse function.
   b. Use the inverse function to compute the horizontal distance when the altitude of the projectile is **(i)** 32 feet, **(ii)** 256 feet, and **(iii)** 512 feet.

Critical Thinking / Discussion / Writing

65. Does every odd function have an inverse? Explain.

66. Is there an even function that has an inverse? Explain.

67. Does every increasing or decreasing function have an inverse? Explain.

68. A relation R is a set of ordered pairs (x, y). The inverse of R is the set of ordered pairs (y, x).
   a. Give an example of a function whose inverse relation is not a function.
   b. Give an example of a relation R whose inverse is a function.

Maintaining Skills

In Exercises 69–74, find the domain and the range of each function.

69. $f(x) = -2x + 5$
70. $f(x) = x^2 - 5$
71. $f(x) = \sqrt{1 - x}$
72. $f(x) = \sqrt{4 - x^2}$
73. $f(x) = \dfrac{x}{x + 2}$
74. $f(x) = \sqrt{\dfrac{x + 1}{x - 2}}$

SUMMARY Definitions, Concepts, and Formulas

1.1 Graphs of Equations

i. **Ordered Pair.** A pair of numbers in which the order is specified is called an ordered pair of numbers.

ii. **The Distance Formula.** The distance between two points $P(x_1, y_1)$ and $Q(x_2, y_2)$, denoted by $d(P, Q)$, is given by
$$d(P, Q) = \sqrt{(x_2 - x_1)^2 + (y_2 - y_1)^2}.$$

iii. **The Midpoint Formula.** The coordinates of the midpoint $M(x, y)$ of the line segment joining $P(x_1, y_1)$ and $Q(x_2, y_2)$ are given by
$$M = (x, y) = \left(\frac{x_1 + x_2}{2}, \frac{y_1 + y_2}{2} \right).$$

iv. The graph of an equation in two variables, say, x and y, is the set of all ordered pairs (a, b) in the coordinate plane that satisfy the given equation. A graph of an equation, then, is a picture of its solution set.

v. **Sketching the graph of an equation by plotting points**
   Step 1 Make a representative table of solutions of the equation.
   Step 2 Plot the solutions in Step 1 as ordered pairs in the coordinate plane.
   Step 3 Connect the representative solutions in Step 2 by a smooth curve.

vi. **Intercepts**
   1. The x-intercept is the x-coordinate of a point on the graph where the graph touches or crosses the x-axis. To find the x-intercepts, set $y = 0$ in the equation and solve for x.
   2. The y-intercept is the y-coordinate of a point on the graph where the graph touches or crosses the y-axis. To find the y-intercepts, set $x = 0$ in the equation and solve for y.

vii. **Symmetry**
   A graph is symmetric with respect to
   a. the y-axis if for every point (x, y) on the graph the point $(-x, y)$ is also on the graph. That is, replacing x with $-x$ in the equation produces an equivalent equation.
   b. the x-axis if for every point (x, y) on the graph the point $(x, -y)$ is also on the graph. That is, replacing y with $-y$ in the equation produces an equivalent equation.
   c. the *origin* if for every point (x, y) on the graph the point $(-x, -y)$ is also on the graph. That is, replacing x with $-x$ and y with $-y$ in the equation produces an equivalent equation.

viii. **Circle**
A circle is a set of points in the coordinate plane that is at a fixed distance r from a fixed point (h, k). The fixed distance r is called the radius of the circle, and the fixed point (h, k) is called the center of the circle. The equation
$$(x - h)^2 + (y - k)^2 = r^2$$
is called the *standard equation* of a circle.

1.2 Lines

i. The slope m of a nonvertical line through the points $P(x_1, y_1)$ and $Q(x_2, y_2)$ is
$$m = \frac{\text{rise}}{\text{run}} = \frac{(y\text{-coordinate of } Q) - (y\text{-coordinate of } P)}{(x\text{-coordinate of } Q) - (x\text{-coordinate of } P)}$$
$$= \frac{y_2 - y_1}{x_2 - x_1}.$$

The slope of a vertical line is undefined. The slope of a horizontal line is 0.

ii. **Equation of a line**

$y - y_1 = m(x - x_1)$	Point–slope form
$y = mx + b$	Slope–intercept form
$y = k$	Horizontal line
$x = k$	Vertical line
$ax + by + c = 0$	General form

iii. **Parallel and Perpendicular Lines**
Two distinct lines with respective slopes m_1 and m_2 are
1. parallel if $m_1 = m_2$.
2. perpendicular if $m_1 m_2 = -1$.

1.3 Functions

i. A **function** from a set X to a set Y is a rule that assigns to each element of X one and only one corresponding element of Y. The set X is the **domain** of the function. The set of those elements of Y that correspond (are assigned) to the elements of X is the **range** of the function.

ii. Any set of ordered pairs is called a **relation**. A **function** is a relation in which no two ordered pairs have the same first component.

iii. If a function f is defined by an equation, then its domain is the largest set of real numbers for which $f(x)$ is a real number.

iv. **Vertical-line test.** If each vertical line intersects a graph at no more than one point, the graph is the graph of a function.

1.4 A Library of Functions

i. A function f is **increasing, decreasing,** or **constant** on an interval depending on whether its graph is respectively rising, falling, or staying the same as you move from left to right on the graph.

ii. A function f is **even** if $f(-x) = f(x)$. The graph of an even function is symmetric with respect to the y-axis. A function f is **odd** if $f(-x) = -f(x)$. The graph of an odd function is symmetric with respect to the origin.

iii. For **piecewise functions**, different rules are used in different parts of the domain.

148 Chapter 1 Graphs and Functions

iv. **Basic Functions**

Identity function	$f(x) = x$
Constant function	$f(x) = c$
Squaring function	$f(x) = x^2$
Cubing function	$f(x) = x^3$
Absolute value function	$f(x) = |x|$
Square root function	$f(x) = \sqrt{x}$
Cube root function	$f(x) = \sqrt[3]{x}$
Reciprocal function	$f(x) = \dfrac{1}{x}$
Reciprocal square function	$f(x) = \dfrac{1}{x^2}$
Rational power function	$f(x) = x^{m/n}$
Greatest integer function	$f(x) = [\![x]\!]$

v. The quantity $\dfrac{f(x+h) - f(x)}{h}, h \neq 0$, is called the **difference quotient**.

1.5 Transformations of Functions

i. **Vertical and horizontal shifts:** Let f be a function and c be a positive number.

To Graph	Shift the Graph of f by c Units
$f(x) + c$	up
$f(x) - c$	down
$f(x - c)$	right
$f(x + c)$	left

ii. **Reflections:**

To Graph	Reflect the Graph of f About the
$-f(x)$	x-axis
$f(-x)$	y-axis

iii. **Stretching and compressing:** The graph of $g(x) = af(x)$ for $a > 0$ has the same shape as the graph of $f(x)$ and is taller if $a > 1$ and flatter if $0 < a < 1$. If a is negative, the graph of $y = |a|f(x)$ is reflected about the x-axis.

The graph of $g(x) = f(bx)$ for $b > 0$ is obtained from the graph of f by stretching away from the y-axis if $0 < b < 1$ and compressing horizontally toward the y-axis if $b > 1$. If b is negative, the graph of $y = f(|b|x)$ is reflected about the y-axis.

1.6 Combining Functions; Composite Functions

i. Given two functions f and g, for all values of x for which both $f(x)$ and $g(x)$ are defined, the functions can be combined to form **sum**, **difference**, **product**, and **quotient** functions. See page 125 for the domains of these functions.

ii. The **composition** of f and g is defined by
$(f \circ g)(x) = f(g(x))$.

The input of f is the output of g. The domain of $f \circ g$ is the set of all x's in the domain of g such that $g(x)$ is in the domain of f.

In general $f \circ g \neq g \circ f$.

1.7 Inverse Functions

i. A function f is **one-to-one** if for any x_1 and x_2 in the domain of $f, f(x_1) = f(x_2)$ implies $x_1 = x_2$.

ii. **Horizontal-line test.** If each horizontal line intersects the graph of a function f in, at most, one point, then f is a one-to-one function.

iii. **Inverse function.** Let f be a one-to-one function. Then g is the inverse of f, and we write $g = f^{-1}$ if $(f \circ g)(x) = x$ for every x in the domain of g and $(g \circ f)(x) = x$ for every x in the domain of f. The graph of f^{-1} is the reflection of the graph of f about the line $y = x$.

REVIEW EXERCISES

Basic Concepts and Skills

In Exercises 1–8, state whether the given statement is true or false.

1. $(0, 5)$ is the midpoint of the line segment joining $(-3, 1)$ and $(3, 11)$.
2. The equation $(x + 2)^2 + (y + 3)^2 = 5$ is the equation of a circle with center $(2, 3)$ and radius 5.
3. If a graph is symmetric with respect to the x-axis and the y-axis, then it must be symmetric with respect to the origin.
4. If a graph is symmetric with respect to the origin, then it must be symmetric with respect to the x-axis and the y-axis.
5. In the graph of the equation of the line $3y = 4x + 9$, the slope is 4 and the y-intercept is 9.
6. If the slope of a line is 2, then the slope of any line perpendicular to it is $\dfrac{1}{2}$.
7. The slope of a vertical line is undefined.
8. The equation $(x - 2)^2 + (y + 3)^2 = -25$ is the equation of a circle with center $(2, -3)$ and radius 5.

In Exercises 9–14, find
a. the distance between the point P and Q.
b. the coordinates of the midpoint of the line segment $\overline{PQ}$.
c. the slope of the line containing the points P and Q.

9. $P(3, 5), Q(-1, 3)$
10. $P(-3, 5), Q(3, -1)$
11. $P(4, -3), Q(9, -8)$
12. $P(2, 3), Q(-7, -8)$
13. $P(2, -7), Q(5, -2)$
14. $P(-5, 4), Q(10, -3)$
15. Show that the points $A(0, 5), B(-2, -3)$, and $C(3, 0)$ are the vertices of a right triangle.
16. Show that the points $A(1, 2), B(4, 8), C(7, -1)$, and $D(10, 5)$ are the vertices of a rhombus.

17. Which of the points $(-6, 3)$ and $(4, 5)$ is closer to the origin?

18. Which of the points $(-6, 4)$ and $(5, 10)$ is closer to the point $(2, 3)$?

19. Find a point on the *x*-axis that is equidistant from the points $(-5, 3)$ and $(4, 7)$.

20. Find a point on the *y*-axis that is equidistant from the points $(-3, -2)$ and $(2, -1)$.

In Exercises 21–24, specify whether the given graph has axis symmetry or origin symmetry (or neither).

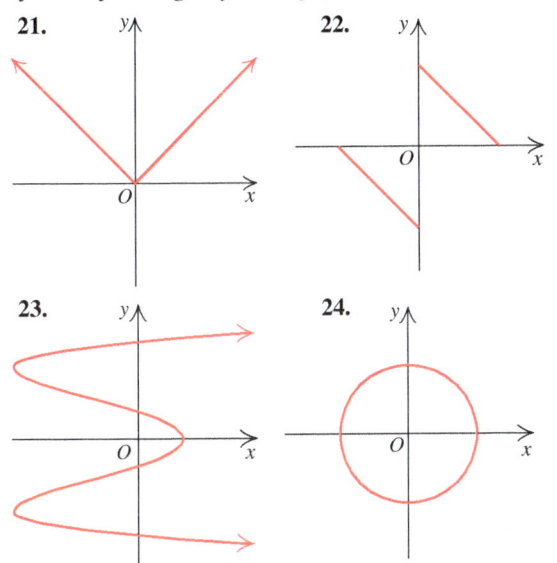

21.

22.

23.

24.

In Exercises 25–34, sketch the graph of the given equation. List all intercepts and describe any symmetry of the graph.

25. $x + 2y = 4$
26. $3x - 4y = 12$
27. $y = -2x^2$
28. $x = -y^2$
29. $y = x^3$
30. $x = -y^3$
31. $y = x^2 + 2$
32. $y = 1 - x^2$
33. $x^2 + y^2 = 16$
34. $y = x^4 - 4$

In Exercises 35–37, find the standard form of the equation of the circle that satisfies the given conditions.

35. Center $(2, -3)$, radius 5

36. Diameter with endpoints $(5, 2)$ and $(-5, 4)$

37. Center $(-2, -5)$, touching the *y*-axis

In Exercises 38–42, describe and sketch the graph of the given equation and give the *x*- and *y*-intercepts.

38. $2x - 5y = 10$
39. $\dfrac{x}{2} - \dfrac{y}{5} = 1$
40. $(x + 1)^2 + (y - 3)^2 = 16$
41. $x^2 + y^2 - 2x + 4y - 4 = 0$
42. $3x^2 + 3y^2 - 6x - 6 = 0$

In Exercises 43–47, find the slope–intercept form of the equation of the line that satisfies the given conditions.

43. Passing through $(1, 2)$ with slope -2

44. *x*-intercept 2, *y*-intercept 5

45. Passing through $(1, 3)$ and $(-1, 7)$

46. Passing through $(1, 2)$, parallel to $2x + 3y = 7$

47. Passing through $(1, 3)$, perpendicular to $3x - 4y = 12$

48. Determine whether the lines in each pair are parallel, perpendicular, or neither.
  a. $y = 3x - 2$ and $y = 3x + 2$
  b. $3x - 5y + 7 = 0$ and $5x - 3y + 2 = 0$
  c. $ax + by + c = 0$ and $bx - ay + d = 0$
  d. $y + 2 = \dfrac{1}{3}(x - 3)$ and $y - 5 = 3(x - 3)$

In Exercises 49–58, graph each equation. State which equations determine *y* as a function of *x*.

49. $x = y^2$
50. $y = x^2$
51. $x - y = 1$
52. $y = \sqrt{x - 2}$
53. $x^2 + y^2 = 0.04$
54. $x = -\sqrt{y}$
55. $x = 1$
56. $y = 2$
57. $y = |x + 1|$
58. $x = y^2 + 1$

In Exercises 59–76, let $f(x) = 3x + 1$ and $g(x) = x^2 - 2$. Find each of the following.

59. $f(-2)$
60. $g(-2)$
61. x if $f(x) = 4$
62. x if $g(x) = 2$
63. $(f + g)(1)$
64. $(f - g)(-1)$
65. $(f \cdot g)(-2)$
66. $(g \cdot f)(0)$
67. $(f \circ g)(3)$
68. $(g \circ f)(-2)$
69. $(f \circ g)(x)$
70. $(g \circ f)(x)$
71. $(f \circ f)(x)$
72. $(g \circ g)(x)$
73. $\dfrac{f(x + h) - f(x)}{h}$
74. $\dfrac{g(x + h) - g(x)}{h}$

75. Let $f = \{(0, 1), (1, 3), (2, 5)\}$; and $g = \{(-1, 0), (1, 2), (2, 3)\}$. Find $f \circ g$ and its domain.

76. Let $f(x) = \sqrt{x + 1}$; and $g(x) = \dfrac{1}{x}$. Find $f \circ g$ and its domain.

In Exercises 77–82, graph each function and state its domain and range. Determine the intervals over which the function is increasing, decreasing, or constant.

77. $f(x) = -3$
78. $f(x) = x^2 - 2$
79. $g(x) = \sqrt{3x - 2}$
80. $h(x) = \sqrt{36 - x^2}$
81. $f(x) = \begin{cases} x + 1 & \text{if } x \geq 0 \\ -x + 1 & \text{if } x < 0 \end{cases}$
82. $g(x) = \begin{cases} x & \text{if } x \geq 0 \\ x^2 & \text{if } x < 0 \end{cases}$

In Exercises 83–86, use transformations to graph each pair of functions on the same coordinate axes.

83. $f(x) = \sqrt{x}, g(x) = \sqrt{x + 1}$
84. $f(x) = |x|, g(x) = 2|x - 1| + 3$
85. $f(x) = \dfrac{1}{x}, g(x) = -\dfrac{1}{x - 2}$
86. $f(x) = x^2, g(x) = (x + 1)^2 - 2$

In Exercises 87–92, state whether each function is odd, even, or neither. Discuss the symmetry of each graph.

87. $f(x) = x^2 - x^4$
88. $f(x) = x^3 + x$
89. $f(x) = |x| + 3$
90. $f(x) = 3x + 5$
91. $f(x) = \sqrt{x}$
92. $f(x) = \dfrac{2}{x}$

In Exercises 93–96, express each function as a composition of two functions.

93. $f(x) = \sqrt{x^2 - 4}$
94. $g(x) = (x^2 - x + 2)^{50}$
95. $h(x) = \sqrt{\dfrac{x-3}{2x+5}}$
96. $H(x) = (2x - 1)^3 + 5$

In Exercises 97–100, determine whether the given function is one-to-one. If the function is one-to-one, find its inverse and sketch the graph of the function and its inverse on the same coordinate axes.

97. $f(x) = x + 2$
98. $f(x) = -2x + 3$
99. $f(x) = \sqrt[3]{x - 2}$
100. $f(x) = 8x^3 - 1$

In Exercises 101 and 102, assume that f is a one-to-one function. Find f^{-1} and find the domain and range of f.

101. $f(x) = 4 + \sqrt{x - 1}$
102. $f(x) = -3 + \sqrt{x + 2}$

103. For the following graph of a function f
   a. write a formula for f as a piecewise function.
   b. find the domain and range of f.
   c. find the intercepts of the graph of f.
   d. draw the graph of $y = f(-x)$.
   e. draw the graph of $y = -f(x)$.
   f. draw the graph of $y = f(x) + 1$.
   g. draw the graph of $y = f(x + 1)$.
   h. draw the graph of $y = 2f(x)$.
   i. draw the graph of $y = f(2x)$.
   j. draw the graph of $y = f\left(\dfrac{1}{2}x\right)$.
   k. explain why f is one-to-one.
   l. draw the graph of $y = f^{-1}(x)$.

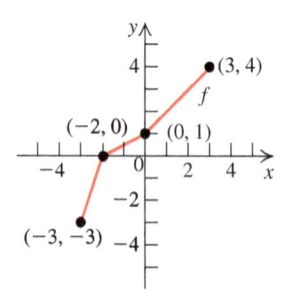

Applying the Concepts

104. **Scuba diving.** The pressure P on the body of a scuba diver increases linearly as she descends to greater depths d in seawater. The pressure at a depth of 10 feet is 19.2 pounds per square inch, and the pressure at a depth of 25 feet is 25.95 pounds per square inch.
   a. Write the equation relating P and d (with d as independent variable) in slope–intercept form.
   b. What is the meaning of the slope and the y-intercept in part (a)?
   c. Determine the pressure on a scuba diver who is at a depth of 160 feet.
   d. Find the depth at which the pressure on the body of the scuba diver is 104.7 pounds per square inch.

105. **Waste disposal.** The Sioux city council considered the following data regarding the cost of disposing of waste material:

Year	2007	2011
Waste (in pounds)	87,000	223,000
Cost	$54,000	$173,000

Assume that the cost C (in dollars) is linearly related to the waste w (in pounds).
   a. Write the linear equation relating C and w in slope–intercept form.
   b. Explain the meaning of the slope and the intercepts of the equation in part (a).
   c. Suppose the city has projected 609,000 pounds of waste for 2012. Calculate the projected cost of disposing of the waste for 2012.
   d. Suppose the city's projected budget for waste collection in 2012 is $1 million. How many pounds of waste can the city handle?

106. **Checking your speedometer.** Zoe checks the accuracy of her speedometer by driving a few measured (not odometer) miles at a constant 60 miles per hour while her friend checks his watch at the beginning and end of the measured distance.
   a. How does a watch tell Zoe's friend whether the speedometer is accurate? What mathematics is involved?
   b. Why shouldn't Zoe use her odometer to measure the number of miles?

107. **Playing blackjack.** Chloe, a bright mathematician, read a book about counting cards in blackjack and went to Las Vegas to try her luck. She started with $100 and discovered that the amount of money she had at time t hours after the start of the game could be expressed by the function $f(t) = 100 + 55t - 3t^2$.
   a. What amount of money had Chloe won or lost in the first two hours?
   b. What was the average rate at which Chloe was winning or losing money during the first two hours?
   c. Did she lose all of her money? If so, when?
   d. If she played until she lost all of her money, what was the average rate at which she was losing her money?

108. **Volume discounting.** Major cola distributors sell a case (containing 24 cans) of cola to the retailer at a price of $4. They offer a discount of 20% for purchases over 100 cases and a discount of 25% for purchases over 500 cases. Write a piecewise function that describes this pricing scheme. Use x as the number of cases purchased and $f(x)$ as the price paid.

109. Air pollution. The daily level L of carbon monoxide in a city is a function of the number of automobiles in the city. Suppose $L(x) = 0.5\sqrt{x^2 + 4}$, where x is the number of automobiles (in hundred thousands). Suppose further that the number of automobiles in a given city is growing according to the formula $x(t) = 1 + 0.002t^2$, where t is time in years measured from now.
  a. Form a composite function describing the daily pollution level as a function of time.
  b. What pollution level is expected in five years?

110. Toy manufacturing. After being in business for t years, a toy manufacturer is making $x = 5000 + 50t + 10t^2$ GI Jimmy toys per year. The sale price p in dollars per toy has risen according to the formula $p = 10 + 0.5t$. Write a formula for the manufacturer's yearly revenue as
  a. a function of time t.
  b. a function of price p.

EXERCISES FOR CALCULUS

1. An open cardboard box is to be made by cutting small square corners out of a 40×24-inch piece of rectangular cardboard and folding up the sides. Write the volume V of the box as a function of the length x of the side of the cut-out portion. What is the domain of V?

2. A salesperson's annual income I is based on a fixed salary of $25{,}000 plus commission. She receives 6% commission for the first $100{,}000 in sales and 8% of any amount above it. Write I as a function of sales. What is the domain of I?

3. Rationalize the numerator and simplify.
   a. $\dfrac{\sqrt{x} - 2}{x - 4}, \quad x \neq 4$
   b. $\dfrac{\sqrt{x^2 + 8} - 3}{x^2 - 1}, \quad |x| \neq 1$
   c. $\dfrac{\sqrt{x + h} - \sqrt{x}}{h}, \quad h \neq 0$
   d. $\dfrac{x - \sqrt{x}}{x - 1}, \quad x \neq 1$

4. Compute the average rate of change for the function $f(x) = 2x^2 - 3x + 4$ between $x = 1$ and $x = 3$.

5. Compute the average rate of change for the linear function $f(x) = mx + b$ between $x = x_1$ and $x = x_2$.

6. Simplify: $\dfrac{1}{h}\left[\dfrac{1}{x + h} - \dfrac{1}{x}\right], \, h \neq 0$

7. A function of the form $f(x) = ax^b$, where a and b are real numbers, is called a *power function*. Convert the following to power functions.
   a. $\dfrac{2}{3x^4}$
   b. $\dfrac{6}{5\sqrt{x}}$
   c. $\dfrac{2}{\sqrt[4]{x^3}}$
   d. $\dfrac{\sqrt[3]{x}}{\sqrt[4]{x}}$

8. Write $g(x) = 2\sqrt{x} + \dfrac{1}{x^2} - \dfrac{5}{x^3}$ as a sum of the power functions.

9. Find the intervals over which $f(x)$ is positive and negative.
   a. $f(x) = x^3 - x^2 - 2x$
   b. $f(x) = \dfrac{(x + 2)(x - 3)}{(x - 1)^2}$

10. Sketch the graph of each function.
    a. $f(x) = \dfrac{|x|}{x}$
    b. $f(x) = \dfrac{x + 1}{|x + 1|}$

PRACTICE TEST A

1. Determine which symmetries the graph of the equation $3x + 2xy^2 = 1$ has.

2. Find the x- and y-intercepts of the graph of $y = x^2(x - 3)(x + 1)$.

3. Graph the equation $x^2 + y^2 - 2x - 2y = 2$ and give the x- and y-intercepts.

4. Write the slope–intercept form of the equation of the line with slope -1 and passing through the point $(2, 7)$.

5. Write an equation of the line parallel to the line $8x - 2y = 7$ and passing through $(2, -1)$.

6. Use $f(x) = -2x + 1$ and $g(x) = x^2 + 3x + 2$ to find $(fg)(2)$.

7. Use $f(x) = 2x - 3$ and $g(x) = 1 - 2x^2$ to evaluate $g(f(2))$.

8. Use $f(x) = x^2 - 2x$ to find $(f \circ f)(x)$.

152 Chapter 1 Graphs and Functions

9. If $f(x) = \begin{cases} x^3 - 2 & \text{if } x \leq 0 \\ 1 - 2x^2 & \text{if } x > 0 \end{cases}$,
   find (a) $f(-1)$, (b) $f(0)$, and (c) $f(1)$.

10. Find the domain of the function $f(x) = \dfrac{\sqrt{x}}{\sqrt{1-x}}$.

11. Find the domain of the function $f(x) = \sqrt{x^2 + x - 6}$.

12. Determine the average rate of change of the function $f(x) = 2x + 7$ between $x = 1$ and $x = 4$.

13. Determine whether the function $f(x) = 2x^4 - \dfrac{3}{x^2}$ is even, odd, or neither.

14. Find the intervals where the function shown is increasing or decreasing.

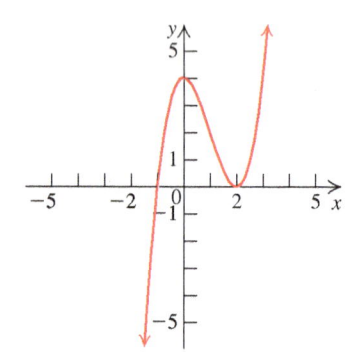

15. Suppose the graph of f is given. Describe how the graph of $y = f(x - 3)$ can be obtained from the graph of f.

16. A ball is thrown upward from the ground. After t seconds, the height h (in feet) above the ground is given by $h = 25 - (2t - 5)^2$. How many seconds does it take the ball to reach a height of 25 feet?

17. If f is a one-to-one function and $f(2) = 7$, find $f^{-1}(7)$.

18. Find the inverse function $f^{-1}(x)$ of the one-to-one function $f(x) = 1 + \dfrac{1}{x}$.

19. Now that Jo has saved $1000 for a car, her dad takes over and deposits $100 into her account each month. Write an equation that relates the total amount of money, A, in Jo's account to the number of months, x, that Jo's dad had been putting money into her account.

20. The cost C in dollars for renting a car for one day is a function of the number of miles traveled, m. For a car renting for $30.00 per day and $0.25 per mile, this function is given by
$$C(m) = 0.25m + 30.$$
    a. Find the cost of renting the car for one day and driving 230 miles.
    b. If the charge for renting the car for one day is $57.50, how many miles were driven?

PRACTICE TEST B

1. The graph of the equation $|x| + 2|y| = 2$ is symmetric with respect to which of the following?
   I. the origin   II. the x-axis   III. the y-axis
   a. I only   b. II only
   c. III only   d. I, II, and III

2. Which are the x- and y-intercepts of the graph of $y = x^2 - 9$?
   a. x-intercept 3, y-intercept -9
   b. x-intercepts ± 3, y-intercept -9
   c. x-intercept 3, y-intercept 9
   d. x-intercepts ± 3, y-intercepts ± 9

3. The slope of a line is undefined if the line is parallel to
   a. the x-axis   b. the line $x - y = 0$
   c. the line $x + y = 0$ d. the y-axis

4. Which is an equation of the circle with center $(0, -5)$ and radius 7?
   a. $x^2 + y^2 - 10y = 0$ b. $x^2 + (y - 5)^2 = 7$
   c. $x^2 + (y - 5)^2 = 49$ d. $x^2 + y^2 + 10y = 24$

5. Which is an equation of the line passing through $(2, -3)$ and having slope -1?
   a. $y + 3 = -(x + 2)$ b. $y - 3 = -(x - 2)$
   c. $y + 3 = -(x - 2)$ d. $y - 3 = -(x + 2)$

6. What is a second point on the line through $(3, 2)$ and having slope $-\dfrac{1}{2}$?
   a. $(2, 4)$ b. $(5, 1)$
   c. $(7, 2)$ d. $(4, 4)$

7. Which is an equation of the line parallel to the line $6x - 3y = 5$ and passing through $(-1, 2)$?
   a. $y - 2 = -2(x + 1)$ b. $y - 2 = 6(x + 1)$
   c. $y - 2 = -6(x + 1)$ d. $y - 2 = 2(x + 1)$

8. Use $f(x) = 3x - 5$ and $g(x) = 2 - x^2$ to find $(f \circ g)(x)$.
   a. $-9x^2 + 30x - 23$ b. $-3x^2 + 1$
   c. $3x^2 + 1$ d. $9x^2 - 30x + 23$

9. Use $f(x) = 2x^2 - x$ to find $(f \circ f)(x)$.
   a. $8x^4 - 8x^3 + x$ b. $8x^4 + x$
   c. $4x^4 - 4x^3 + x^2 - x$ d. $4x^4 - x^3 + x^2$
   e. $-x^4 + 6x^2 - x$

10. If $g(t) = \dfrac{1-t}{1+t}$, find $g(a - 1)$.
    a. $\dfrac{2-a}{2+a}$ b. $\dfrac{1-a}{1+a}$ c. $\dfrac{2-a}{a}$ d. -1

11. Which of the following is the domain of the function $f(x) = \sqrt{x} + \sqrt{1-x}$?
    a. $[0, 1]$ b. $(-\infty, -1] \cup [0, +\infty)$
    c. $(-\infty, 1]$ d. $[0, +\infty)$

12. What is the domain of the function $f(x) = \sqrt{x^2 + 6x - 7}$?
    a. $(-\infty, -7) \cup (1, \infty)$ b. $(-\infty, -7] \cup [1, \infty)$
    c. $[-7, 1]$ d. $(1, \infty)$

13. Which description of the behavior of the graph shown here is correct?
    a. increasing on $(0, 3)$; decreasing on $(-3, 0) \cup (3, 4)$
    b. increasing on $(-1, 3)$; decreasing on $(2, -1) \cup (3, 1)$
    c. increasing on $(-3, 3)$; decreasing on $(2, -1) \cup (3, 1)$
    d. increasing on $(1.2, 4)$; decreasing on $(-3, -1.4)$

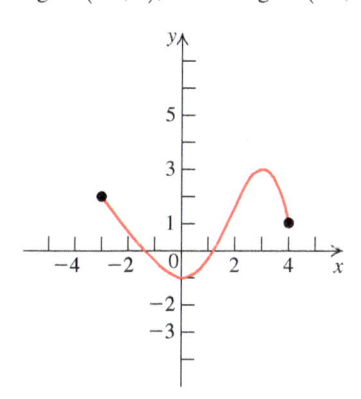

14. Suppose the graph of f is given. Describe how to obtain the graph of $y = f(x + 4)$ from the graph of f.
    a. Shift the graph of f four units to the left.
    b. Shift the graph of f four units down.
    c. Shift the graph of f four units up.
    d. Shift the graph of f four units to the right.
    e. Reflect the graph of f about the x-axis.

15. Suppose the graph of f is given. Describe how to obtain the graph of $y = -2f(x - 4)$ from the graph of f.
    a. Shift the graph of f four units to the left, shrink it vertically by a factor of $\frac{1}{2}$, and reflect it about the x-axis.
    b. Shift the graph of f four units to the right, stretch it vertically by a factor of 2, and reflect it about the x-axis.
    c. Shift the graph of f four units to the left, shrink it vertically by a factor of $\frac{1}{2}$, and reflect it about the x-axis.
    d. Shift the graph of f four units to the right; then stretch it vertically by a factor of 2 and reflect it about the y-axis.

16. Which of the following is a one-to-one function?
    a. $f(x) = |x + 3|$
    b. $f(x) = x^2$
    c. $f(x) = \sqrt{x^2 + 9}$
    d. $f(x) = x^3 + 2$

17. If f is a one-to-one function and $f(3) = 5$, then $f^{-1}(5) =$
    a. $\frac{1}{5}$
    b. $\frac{1}{3}$
    c. 3
    d. -5

18. Find the inverse function $f^{-1}(x)$ of $f(x) = \dfrac{1 - 3x}{5 + 2x}$.
    a. $\dfrac{1 + 3x}{5 - 2x}$
    b. $\dfrac{5 + 2x}{1 - 3x}$
    c. $\dfrac{1 - 5x}{3 + 2x}$
    d. $\dfrac{3 + 2x}{1 - 5x}$

19. The ideal weight w (in pounds) for a man of height x inches is found by subtracting 190 from 5 times his height. Write a linear equation that gives the ideal weight of a man of height x inches. Then find the ideal weight of a man whose height is 70 inches.
    a. $w = \dfrac{x + 190}{5}$, 160 lb
    b. $w = 5x - 190$, 160 lb
    c. $w = 190 - \dfrac{x}{5}$, 176 lb
    d. $w = \dfrac{x + 190}{5}$, 52 lb

20. Cassie rented an intermediate sedan at $25.00 per day plus $0.20 per mile. How many miles can Cassie travel for $50?
    a. 125
    b. 1025
    c. 250
    d. 175

CHAPTER 2

Polynomial and Rational Functions

TOPICS

- **2.1** Quadratic Functions
- **2.2** Polynomial Functions
- **2.3** Dividing Polynomials and the Rational Zeros Test
- **2.4** Zeros of a Polynomial Function
- **2.5** Rational Functions

Polynomials and rational functions are used in navigation, in computer-aided geometric design, in the study of supply and demand in business, in the design of suspension bridges, in computations determining the force of gravity on planets, and in virtually every area of inquiry requiring numerical approximations. Their applicability is so widespread that it is hard to imagine an area in which they have not made an impact.

SECTION 2.1

Quadratic Functions

BEFORE STARTING THIS SECTION, REVIEW

1. Linear functions (Section 1.4, page 94)
2. Completing the square (Section P.3, page 24)
3. Quadratic formula (Section P.3, page 27)
4. Transformations (Section 1.5, page 109)

OBJECTIVES

1. Graph a quadratic function in standard form.
2. Graph any quadratic function.
3. Solve problems modeled by quadratic functions.

Leaning Tower of Pisa

◆ Galileo and Free-Falling Objects

Legend has it that Galileo Galilei (1564–1642) disproved the theory that heavy objects fall faster than lighter objects by dropping two balls of equal size, one made of lead and the other of balsa wood, from the top of Italy's Leaning Tower of Pisa. Both balls are said to have reached the ground at the same instant after they were released simultaneously from the top of the Tower.

In fact, such experiments "work" properly only in a vacuum, where objects of different mass do fall at the same rate. In the absence of a vacuum, air resistance may greatly affect the rate at which different objects fall. It turns out that on Earth (ignoring air resistance), regardless of the weight of the object we hold at rest and then drop, the distance in feet that the object falls in t seconds is $16t^2$. If we throw the object down with an initial velocity of v_0 feet per second, the distance the object travels in t seconds is $16t^2 + v_0 t$. If an object is propelled upward at an initial velocity of v_0 feet per second, the distance it travels upward in t seconds is $-16t^2 + v_0 t$. (See Exercises 59 and 60.)

1 Graph a quadratic function in standard form.

Quadratic Functions

A function whose defining equation is a polynomial is called a *polynomial function*. In this section, you will study **quadratic functions**, which are polynomial functions of degree 2.

Quadratic Function

A function of the form
$$f(x) = ax^2 + bx + c,$$
where a, b, and c, are real numbers with $a \neq 0$, is called a **quadratic function**.

Note that the squaring function $s(x) = x^2$ (whose graph is a parabola) is the quadratic function $f(x) = ax^2 + bx + c$ with $a = 1$, $b = 0$, and $c = 0$. We will show that we can graph any quadratic function by transforming the graph of $s(x) = x^2$. Therefore, because the graph of $y = x^2$ is a parabola, the graph of any quadratic function is also a parabola.

First, we compare the graphs of $f(x) = ax^2$ for $a > 0$ and $g(x) = ax^2$ for $a < 0$. These are quadratic functions with $b = 0$ and $c = 0$.

156 Chapter 2 Polynomial and Rational Functions

$f(x) = ax^2, a > 0$	$g(x) = ax^2, a < 0$
The graph of $f(x) = ax^2$ is obtained from the graph of the squaring function $s(x) = x^2$ by vertically stretching or compressing it. Here are three such graphs on the same axes for $a = 2, a = 1,$ and $a = \frac{1}{2}$.	The graph of $g(x) = ax^2$ is the reflection of the graph of $f(x) = \|a\|x^2$ about the x-axis. Here are three graphs for $a = -2, a = -1,$ and $a = -\frac{1}{2}$.

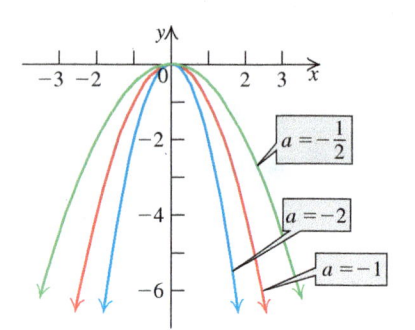

Standard Form of a Quadratic Function

Any quadratic function $f(x) = ax^2 + bx + c$ with $a \neq 0$ can be rewritten in the form $f(x) = a(x - h)^2 + k$, called the **standard form** of the quadratic function f. By using transformations (see Section 1.5), we can obtain the graph of $f(x) = a(x - h)^2 + k$ from the graph of the squaring function $s(x) = x^2$ as follows:

$s(x) = x^2$ The squaring function
$\downarrow$
$g(x) = (x - h)^2$ Horizontal translation to the right if $h > 0$ and to the left if $h < 0$
$\downarrow$
$G(x) = a(x - h)^2$ Vertical stretching or compressing. If $a < 0$, the graph is also reflected about the x-axis.
$\downarrow$
$f(x) = a(x - h)^2 + k$ Vertical translation, up if $k > 0$ and down if $k < 0$

These transformations show that the graph of a quadratic function $f(x) = a(x - h)^2 + k$ is a parabola that opens up if $a > 0$ and down if $a < 0$. Its domain is $(-\infty, \infty)$. The parabola is symmetric with respect to the vertical line $x = h$. The line of symmetry, $x = h$, is called the *axis* (or *axis of symmetry*) of the parabola. The point (h, k) where the axis meets the parabola is called the *vertex* of the parabola. The vertex of any parabola that opens up is the *lowest point* of the graph, and the vertex of any parabola that opens down is the *highest point* of the graph. Therefore, k is the *minimum value* for any parabola that opens up and k is the *maximum value* for any parabola that opens down. The standard form is particularly useful because it immediately identifies the vertex (h, k).

The Standard Form of a Quadratic Function

The quadratic function

$$f(x) = a(x - h)^2 + k, a \neq 0$$

is in **standard form**. The graph of f is a parabola with **vertex** (h, k). See Figure 2.1. The parabola is symmetric with respect to the line $x = h$, called the **axis** of the parabola. If $a > 0$, the parabola opens up, and if $a < 0$, the parabola opens down. If $a > 0$, k is the **minimum value** of f, and if $a < 0$, k is the **maximum value** of f.

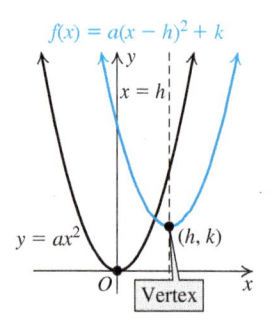

FIGURE 2.1 Quadratic functions

EXAMPLE 1 Writing the Equation of a Quadratic Function

Find the standard form of the quadratic function f whose graph has vertex $(-3, 4)$ and passes through the point $(-4, 7)$. Does f have a maximum or a minimum value?

Solution

$f(x) = a(x - h)^2 + k$	Standard form of quadratic function
$f(x) = a[x - (-3)]^2 + 4$	Replace h with -3 and k with 4.
$y = a(x + 3)^2 + 4$	Replace $f(x)$ with y and simplify.
$7 = a(-4 + 3)^2 + 4$	Replace x with -4 and y with 7 because $(-4, 7)$ lies on the graph of $y = f(x)$.
$7 = a + 4$	Simplify.
$a = 3$.	Solve for a.

The standard form is

$$f(x) = 3(x + 3)^2 + 4. \quad a = 3, h = -3, k = 4$$

Because $a = 3 > 0$, f has a minimum value of 4 at $x = -3$.

Practice Problem 1 Find the standard form of the quadratic function f whose graph has vertex $(1, -5)$ and passes through the point $(3, 7)$. Does f have a maximum or a minimum value?

PROCEDURE IN ACTION

EXAMPLE 2 Graphing a Quadratic Function in Standard Form

OBJECTIVE

Sketch the graph of $f(x) = a(x - h)^2 + k$.

Step 1 The graph is a parabola because the function has the form $f(x) = a(x - h)^2 + k$. Identify a, h, and k.

Step 2 Determine how the parabola opens. If $a > 0$, the parabola opens *up*. If $a < 0$, the parabola opens *down*.

Step 3 Find the vertex (h, k). If $a > 0$ (or $a < 0$), the function f has a minimum (or a maximum) value k at $x = h$.

Step 4 Find the x-intercepts (if any). Set $f(x) = 0$ and solve the equation $a(x - h)^2 + k = 0$ for x. If the solutions are real numbers, they are the x-intercepts. If not, the parabola lies above the x-axis (when $a > 0$) or below the x-axis (when $a < 0$).

EXAMPLE

Sketch the graph of $f(x) = -3(x + 2)^2 + 12$.

1. The graph of $f(x) = -3(x + 2)^2 + 12$
$$= (-3)[x - (-2)]^2 + 12$$
$$\uparrow\uparrow\uparrow$$
$$ahk$$

is a parabola; $a = -3$, $h = -2$, and $k = 12$.

2. Because $a = -3 < 0$, the parabola opens down.

3. The vertex is $(h, k) = (-2, 12)$. Because the parabola opens down, the function f has a maximum value of 12 at $x = -2$.

4.
$0 = -3(x + 2)^2 + 12$	Set $f(x) = 0$.
$3(x + 2)^2 = 12$	Add $3(x + 2)^2$ to both sides.
$(x + 2)^2 = 4$	Divide both sides by 3.
$x + 2 = \pm 2$	Square root property
$x = -2 \pm 2$	Subtract 2 from both sides.
$x = 0$ or $x = -4$	Solve for x.

The x-intercepts are 0 and -4. The parabola passes through the points $(0, 0)$ and $(-4, 0)$.

(continued)

Step 5 Find the y-intercept. Replace x with 0. Then $f(0) = ah^2 + k$ is the y-intercept.

Step 6 Sketch the graph. Plot the points found in Steps 3–5 and join them to form a parabola. Show the axis $x = h$ of the parabola by drawing a dashed vertical line.

If there are no x-intercepts, draw the half of the parabola that passes through the vertex and a second point, such as the y-intercept. Then use the axis of symmetry to draw the other half.

5. $f(0) = -3(0 + 2)^2 + 12 = -3(4) + 12 = 0$. The y-intercept is 0. As already discussed, the parabola passes through $(0, 0)$.

6. The axis of the parabola is the vertical line $x = -2$. The graph of the parabola is shown in the figure.

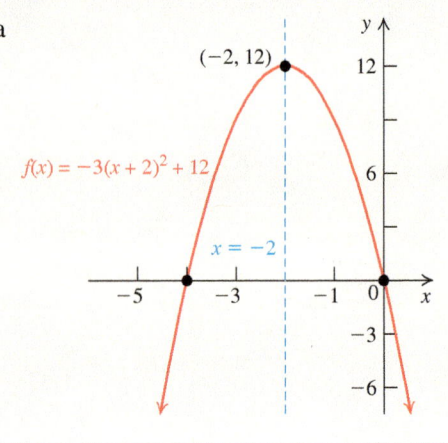

Practice Problem 2 Graph the quadratic function $f(x) = -2(x + 1)^2 + 3$. Find the vertex, maximum or minimum value, and x- and y-intercepts.

2 Graph any quadratic function.

Graphing a Quadratic Function $f(x) = ax^2 + bx + c$

A quadratic function $f(x) = ax^2 + bx + c$ can be changed to the standard form $f(x) = a(x - h)^2 + k$ by *completing the square*.

Converting $f(x) = ax^2 + bx + c$ to Standard Form

$f(x) = ax^2 + bx + c$ Original function

$= a\left(x^2 + \dfrac{b}{a}x\right) + c$ Factor out a from $ax^2 + bx$.

$= a\left(x^2 + \dfrac{b}{a}x + \dfrac{b^2}{4a^2} - \dfrac{b^2}{4a^2}\right) + c$ To complete the square, add and subtract $\left(\dfrac{1}{2} \cdot \dfrac{b}{a}\right)^2 = \dfrac{b^2}{4a^2}$ within the parentheses.

$= a\left(x^2 + \dfrac{b}{a}x + \dfrac{b^2}{4a^2}\right) - a \cdot \dfrac{b^2}{4a^2} + c$ Distribute a to $-\dfrac{b^2}{4a^2}$; regroup terms.

$= a\left(x + \dfrac{b}{2a}\right)^2 + c - \dfrac{b^2}{4a}$ $\left(x + \dfrac{b}{2a}\right)^2 = x^2 + \dfrac{b}{a}x + \dfrac{b^2}{4a^2}$

Comparing this form with the standard form $f(x) = a(x - h)^2 + k$, we have:

$$f(x) = a\left(x - \left(-\dfrac{b}{2a}\right)\right)^2 + c - \dfrac{b^2}{4a}$$

$$h = -\dfrac{b}{2a} \qquad k = c - \dfrac{b^2}{4a}$$

Notice that $f(h) = f\left(-\dfrac{b}{2a}\right) = a\left(-\dfrac{b}{2a} + \dfrac{b}{2a}\right)^2 + c - \dfrac{b^2}{4a}$

$= c - \dfrac{b^2}{4a} = k.$

Section 2.1 ■ Quadratic Functions 159

> **FINDING THE VERTEX OF $f(x) = ax^2 + bx + c$, $a \neq 0$**
>
> To find the vertex (h, k) of $f(x) = ax^2 + bx + c, a \neq 0$, use the equations
>
> $$h = -\frac{b}{2a} \quad \text{and} \quad k = f\left(-\frac{b}{2a}\right).$$

PROCEDURE IN ACTION

EXAMPLE 3   Graphing a Quadratic Function

OBJECTIVE
Graph $f(x) = ax^2 + bx + c, a \neq 0$.

EXAMPLE
Sketch the graph of $f(x) = 2x^2 + 8x - 10$.

Step 1 Identify $a, b,$ and c.

1. In the equation $y = f(x) = 2x^2 + 8x - 10$,
   $a = 2, b = 8,$ and $c = -10$.

Step 2 Determine how the parabola opens.
If $a > 0$, the parabola opens *up*;
if $a < 0$, the parabola opens *down*.

2. Because $a = 2 > 0$, the parabola opens up.

Step 3 Find the vertex (h, k). Use the formulas
$$h = -\frac{b}{2a},$$
$$k = f\left(-\frac{b}{2a}\right)$$

3. $h = -\dfrac{b}{2a} = -\dfrac{8}{2(2)} = -2$ $a = 2, b = 8$

   $k = f(h) = f(-2)$
   $ = 2(-2)^2 + 8(-2) - 10$ Replace h with -2.
   $ = -18$ Simplify.

   The vertex is $(-2, -18)$.

   The function f has a minimum value of -18 at $x = -2$.

Step 4 Find the *x*-intercepts (if any). Let $f(x) = 0$ and solve $ax^2 + bx + c = 0$. If the solutions are real numbers, they are the *x*-intercepts. If not, the parabola lies entirely above the *x*-axis (when $a > 0$) or entirely below the *x*-axis (when $a < 0$).

4. $2x^2 + 8x - 10 = 0$ Set $f(x) = 0$.
   $2(x^2 + 4x - 5) = 0$ Factor out 2.
   $2(x + 5)(x - 1) = 0$ Factor.
   $x + 5 = 0$ or $x - 1 = 0$ Zero-product property
   $x = -5$ or $x = 1$ Solve for x.

   The *x*-intercepts are -5 and 1, the parabola passes through the points $(-5, 0)$ and $(1, 0)$.

Step 5 Find the *y*-intercept. Let $x = 0$. The result, $f(0) = c$, is the *y*-intercept.

5. Set $x = 0$ to obtain $f(0) = 2(0)^2 + 8(0) - 10 = -10$. The *y*-intercept is -10.

Step 6 The parabola is symmetric with respect to its axis, $x = -\dfrac{b}{2a}$. Use this symmetry to find additional points.

6. The axis of symmetry is $x = -2$. The symmetric image of $(0, -10)$ with respect to the axis $x = -2$ is $(-4, -10)$.

Step 7 Draw a parabola through the points found in Steps 3–6.

7. The parabola passing through the points in Steps 3–6 is sketched in the figure.

If there are no *x*-intercepts, draw the half of the parabola that passes through the vertex and a second point, such as the *y*-intercept. Then use the axis of symmetry to draw the other half.

Practice Problem 3   Graph the quadratic function $f(x) = 3x^2 - 3x - 6$.

160 Chapter 2 Polynomial and Rational Functions

As the graph in Example 3 suggests, when $a > 0$, every value of y, with $y \geq k$, is the second coordinate for some point on the graph of $f(x) = ax^2 + bx + c$. So when $a > 0$, the interval $[k, \infty)$ is the range of f. Similarly, when $a < 0$, the range of f is the interval $(-\infty, k]$.

EXAMPLE 4 Identifying the Characteristics of a Quadratic Function from Its Graph

The graph of the function $f(x) = -2x^2 + 8x - 5$ is shown in Figure 2.2.

a. Find the domain and range of f.
b. Solve the inequality $-2x^2 + 8x - 5 > 0$.

Solution

The graph in Figure 2.2 is a parabola that opens down and whose vertex is $(2, 3)$.

a. The domain of f is $(-\infty, \infty)$. Because the parabola opens down, the maximum value of f is the y-coordinate, 3, of its vertex $(2, 3)$. So the range of f is $(-\infty, 3]$.

b. The x-intercepts of f are $\dfrac{4 - \sqrt{6}}{2} \approx 0.775$ and $\dfrac{4 + \sqrt{6}}{2} \approx 3.225$. The graph of f is above the x-axis between these intercepts. So $y = -2x^2 + 8x - 5 > 0$ for the x-values in the interval $\left(\dfrac{4 - \sqrt{6}}{2}, \dfrac{4 + \sqrt{6}}{2} \right)$. The solution set of the inequality $-2x^2 + 8x - 5 > 0$ is the interval $\left(\dfrac{4 - \sqrt{6}}{2}, \dfrac{4 + \sqrt{6}}{2} \right)$.

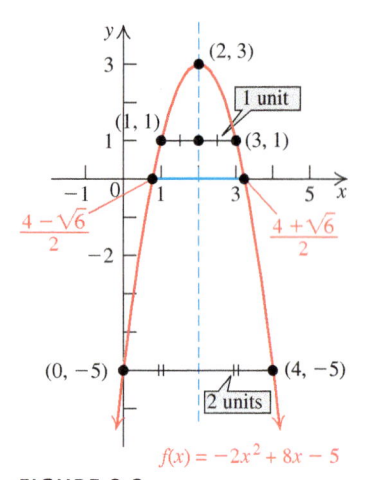

FIGURE 2.2

Practice Problem 4 Graph the function $f(x) = 3x^2 - 6x - 1$ and solve the inequality $3x^2 - 6x - 1 \leq 0$.

3 Solve problems modeled by quadratic functions.

Applications

Many applications of quadratic functions involve finding the maximum or minimum value of the function.

EXAMPLE 5 Finding the Maximum Area of a Rectangle

Find the dimensions of a rectangular garden of maximum area if the perimeter of the garden is 80 feet. What is the maximum area of such a rectangle?

Solution

Let x = width of the rectangle and y = length of the rectangle.

Then $\quad p = 2x + 2y$ = perimeter of the rectangle.

So $\quad 80 = 2x + 2y \quad$ Given $p = 80$

$\quad 40 = x + y \quad$ Divide both sides by 2.

$\quad y = 40 - x \quad$ Subtract x from both sides and interchange sides.

The area, A, of the rectangle is given by

$$A = A(x) = xy$$
$$A(x) = x(40 - x) \quad \text{Replace } y \text{ with } 40 - x.$$
$$= -x^2 + 40x. \quad \text{Simplify.}$$

RECALL

For a quadratic function $f(x) = ax^2 + bx + c$, the vertex is $(h, k) = \left(-\dfrac{b}{2a}, f\left(-\dfrac{b}{2a} \right) \right)$ and k is the function's minimum value if $a > 0$ or its maximum value if $a < 0$.

The graph of $A(x) = -x^2 + 40x$ is a parabola with $a = -1, b = 40, c = 0$ and vertex (h, k) where

$$h = -\frac{b}{2a} = (-40)/2(-1) = 20 \text{ and}$$
$$k = A(h) = A(20) = -(20)^2 + 40(20) = 400.$$

The maximum area, k, of this rectangle is 400 square feet, and its dimensions are

$x =$ width $= h = 20$ feet and
$y =$ length $= 20$ feet.   Substitute $x = 20$ in $y = 40 - x$.

So, the rectangle of maximum area with a given perimeter is a square.

Practice Problem 5 A rancher with 120 meters of fence intends to enclose a rectangular region along a river (which serves as a natural boundary requiring no fence). Find the maximum area that can be enclosed.

SECTION 2.1 Exercises

Basic Concepts and Skills

In Exercises 1–8, match each quadratic function with its graph.

1. $y = -\frac{1}{3}x^2$
2. $y = -3x^2$
3. $y = -3(x + 1)^2$
4. $y = 2(x + 1)^2$
5. $y = (x - 1)^2 + 2$
6. $y = (x - 1)^2 - 3$
7. $y = 2(x + 1)^2 - 3$
8. $y = -3(x + 1)^2 + 2$

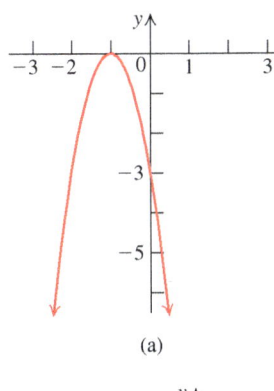

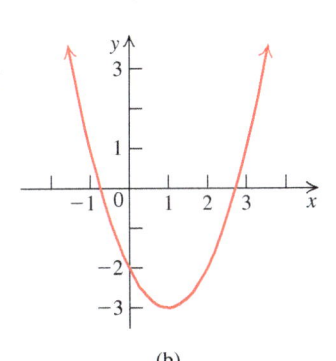

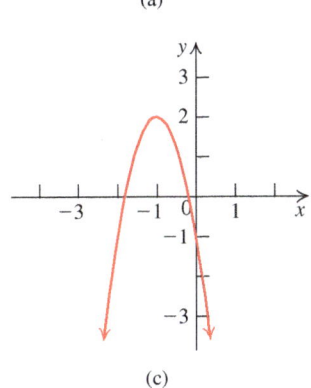

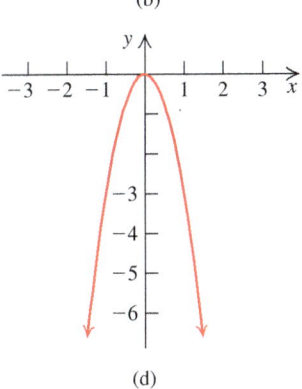

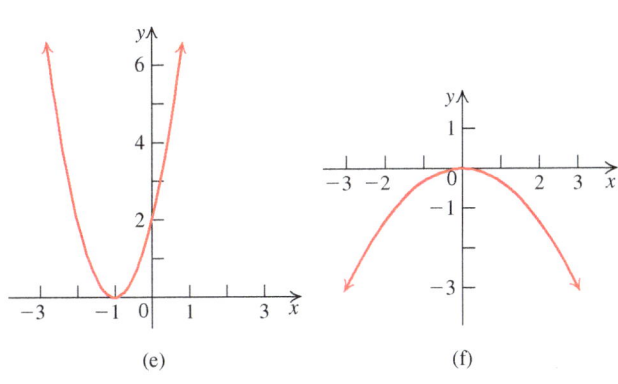

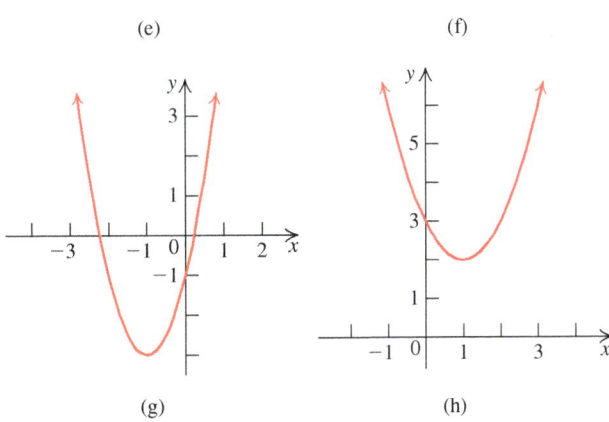

(a)   (b)   (c)   (d)   (e)   (f)   (g)   (h)

In Exercises 9–12, find a quadratic function of the form $y = ax^2$ that passes through the given point.

9. $(2, -8)$
10. $(-3, 3)$
11. $(2, 20)$
12. $(-3, -6)$

In Exercises 13–22, find the quadratic function $y = f(x)$ that has the given vertex and whose graph passes through the given point. Write the function in standard form.

13. Vertex $(0, 0)$; passing through $(-2, 8)$
14. Vertex $(2, 0)$; passing through $(1, 3)$

15. Vertex $(-3, 0)$; passing through $(-5, -4)$
16. Vertex $(0, 1)$; passing through $(-1, 0)$
17. Vertex $(2, 5)$; passing through $(3, 7)$
18. Vertex $(-3, 4)$; passing through $(0, 0)$
19. Vertex $(2, -3)$; passing through $(-5, 8)$
20. Vertex $(-3, -2)$; passing through $(0, -8)$
21. Vertex $\left(\dfrac{1}{2}, \dfrac{1}{2}\right)$; passing through $\left(\dfrac{3}{4}, -\dfrac{1}{4}\right)$
22. Vertex $\left(-\dfrac{3}{2}, -\dfrac{5}{2}\right)$; passing through $\left(1, \dfrac{55}{8}\right)$

In Exercises 23–26, the graph of a quadratic function $y = f(x)$ is given. Find the standard form of the function.

23.

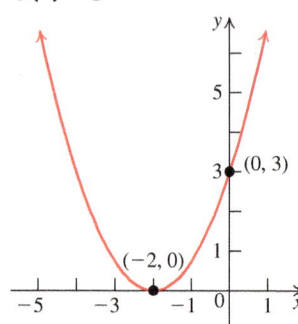

24.

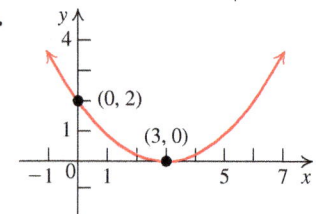

25.

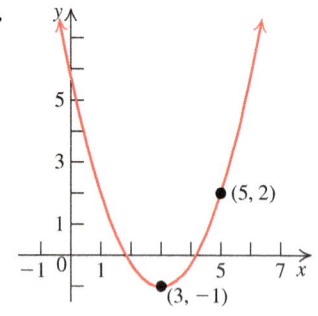

26.

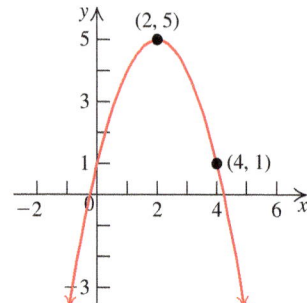

In Exercises 27–34, graph the given function by writing it in the standard form $y = a(x - h)^2 + k$ and then using transformations on $y = x^2$. Find the vertex, the axis of symmetry, and the x- and y-intercepts.

27. $y = x^2 + 4x$
28. $y = x^2 - 2x + 2$
29. $y = 6x - 10 - x^2$
30. $y = 8 + 3x - x^2$
31. $y = 2x^2 - 8x + 9$
32. $y = 3x^2 + 12x - 7$
33. $y = -3x^2 + 18x - 11$
34. $y = -5x^2 - 20x + 13$

In Exercises 35–42, (a) determine whether the graph of the given quadratic function opens up or down, (b) find the vertex, (c) find the axis of symmetry, (d) find the x- and y-intercepts, and (e) sketch the graph of the function.

35. $y = x^2 - 8x + 15$
36. $y = x^2 + 8x + 13$
37. $y = x^2 - x - 6$
38. $y = x^2 + x - 2$
39. $y = x^2 - 2x + 4$
40. $y = x^2 - 4x + 5$
41. $y = 6 - 2x - x^2$
42. $y = 2 + 5x - 3x^2$

In Exercises 43–50, a quadratic function f is given. (a) Determine whether the given quadratic function has a maximum value or a minimum value. Then find this value. (b) Find the range of f.

43. $f(x) = x^2 - 4x + 3$
44. $f(x) = -x^2 + 6x - 8$
45. $f(x) = -4 + 4x - x^2$
46. $f(x) = x^2 - 6x + 9$
47. $f(x) = 2x^2 - 8x + 3$
48. $f(x) = 3x^2 + 12x - 5$
49. $f(x) = -4x^2 + 12x + 7$
50. $f(x) = 8x - 5 - 2x^2$

In Exercises 51–56, solve the given quadratic inequality by sketching the graph of the corresponding quadratic function.

51. $x^2 - 4 \leq 0$
52. $x^2 + 4x + 5 < 0$
53. $x^2 - 4x + 3 > 0$
54. $x^2 + x - 2 > 0$
55. $-6x^2 + x + 7 > 0$
56. $-5x^2 + 9x - 4 \leq 0$

Applying the Concepts

Note: The domain of a function that models a real-life problem may consist of only integer values. However, in solving these problems, we frequently find it convenient to extend the domain to a suitable interval of real numbers containing the relevant integers.

57. **Enclosing playing fields.** You have 600 meters of fencing, and you plan to lay out two identical playing fields. The fields can be side-by-side and separated by a fence, as shown in the figure. Find the dimensions and the maximum area of each field.

58. Enclosing area on a budget. You budget $2400 for constructing a rectangular enclosure that consists of a high surrounding fence and a lower inside fence that divides the enclosure in half. The high fence costs $8 per foot, and the low fence costs $4 per foot. Find the dimensions and the maximum area of each half of the enclosure.

59. Motion of a projectile. A projectile is fired straight up with a velocity of 64 ft/sec. Its altitude (height) h after t seconds is given by

$$h(t) = -16t^2 + 64t.$$

a. What is the maximum height of the projectile?
b. When does the projectile hit the ground?

60. Motion of a projectile. From the top of a 400-foot-high building, a projectile is shot up with a velocity of 64 ft/sec. Its altitude (height above the ground) h after t seconds is given by

$$h(t) = -16t^2 + 64t + 400.$$

a. What is the maximum height of the projectile?
b. When does the projectile hit the ground?

61. Find two numbers whose sum equals 6 and whose product equals 7.

62. Find two numbers whose sum equals 10 and whose product equals 22.

Critical Thinking / Discussion / Writing

63. Symmetry about the line $x = h$. The graph of a polynomial function $y = f(x)$ is symmetric about the line $x = h$ if $f(h + p) = f(h - p)$ for every number p. For $f(x) = ax^2 + bx + c, a \neq 0$, set $f(h + p) = f(h - p)$ and show that $h = -\dfrac{b}{2a}$. This proves that the graph of a quadratic function is symmetric about its axis: $x = -\dfrac{b}{2a}$.

64. In the graph of $y = 2x^2 - 8x + 9$, find the coordinates of the point symmetric to the point $(-1, 19)$ about the axis of symmetry.

Maintaining Skills

In Exercises 65–72, simplify each expression. Write your answers without negative exponents.

65. -5^0

66. $(-4)^{-2}$

67. -4^{-2}

68. $(-2x^2)(-3x^4)$

69. $x^7 \cdot x^{-7}$

70. $(2x^2)(3x)(-x^3)$

71. $x^2\left(3 - \dfrac{3}{4}\right)$

72. $x^4\left(1 + \dfrac{3}{x^2} - \dfrac{5}{x^3}\right)$

SECTION 2.2

Polynomial Functions

BEFORE STARTING THIS SECTION, REVIEW

1. Graphs of equations (Section 1.1, page 55)
2. Solving linear equations (Section P.3, page 22)
3. Solving quadratic equations (Section P.3, page 23)
4. Transformations (Section 1.5, page 109)
5. Relative maximum and minimum values (Section 1.4, page 97)

OBJECTIVES

1. Learn properties of the graphs of polynomial functions.
2. Determine the end behavior of polynomial functions.
3. Find the zeros of a polynomial function by factoring.
4. Identify the relationship between degrees, real zeros, and turning points.
5. Graph polynomial functions.

Johannes Kepler (1571–1630)
Kepler was born in Weil in southern Germany and studied at the University of Tübingen. He studied with Michael Mastlin (professor of mathematics), who privately taught Kepler the Copernican theory of the sun-centered universe while hardly daring to recognize it openly in his lectures. Kepler knew that to work out a detailed version of Copernicus's theory, he had to have access to the observations of the Danish astronomer Tycho Brahe (1546–1601). Kepler's correspondence with Brahe resulted in his appointment as Brahe's assistant. Brahe died about 18 months after Kepler's arrival. During those 18 months, Kepler learned enough about Brahe's work to create his own major project. He produced the famous three laws of planetary motion published in 1609 in his book *Nova Astronomia*.

◆ Kepler's Wedding Reception

Kepler is best known as an astronomer who discovered the three laws of planetary motion. However, Kepler's primary field was not astronomy, but mathematics. Legend has it that at his own wedding reception, Kepler observed that the Austrian vintners could quickly and mysteriously compute the volumes of a variety of wine barrels. Each barrel had a hole, called a *bunghole*, in the middle of its side. The vintner would insert a rod, called the *bungrod*, into the hole until it hit the far corner. Then he would announce the volume. During the reception, Kepler occupied his mind with the problem of finding the volume of a wine barrel with a bungrod.

In Example 10, we show you how he solved the problem for barrels that are perfect cylinders. But barrels are not perfect cylinders; they are wider in the middle and narrower at the ends, and in between, the sides make a graceful curve that appears to be an arc of a circle. What could be the formula for the volume of such a shape? Kepler tackled this problem in his important book *Nova Stereometria Doliorum Vinariorum* (1615) and developed a related mathematical theory.

1 Learn properties of the graphs of polynomial functions.

Polynomial Functions

We begin with some terminology. A **polynomial function of degree** n is a function that can be written in the form

$$f(x) = a_n x^n + a_{n-1} x^{n-1} + \cdots + a_2 x^2 + a_1 x + a_0,$$

where n is a nonnegative integer and the *coefficients* $a_n, a_{n-1}, \ldots, a_2, a_1, a_0$ are real numbers with $a_n \neq 0$. The term $a_n x^n$ is called the **leading term**, the number a_n (the coefficient of x^n) is called the **leading coefficient**, and a_0 is the **constant term**. A constant function $f(x) = a \, (a \neq 0)$, which may be written as $f(x) = ax^0$, is a polynomial of degree 0. Because the zero function $f(x) = 0$ can be written in several ways, such as $0 = 0x^6 + 0x = 0x^5 + 0 = 0x^4 + 0x^3 + 0$, no degree is assigned to it.

164 Chapter 2 Polynomial and Rational Functions

Section 2.2 ■ Polynomial Functions   **165**

In Section 1.4, we discussed linear functions, which are polynomial functions of degree 0 and 1. In Section 2.1, we studied quadratic functions, which are polynomial functions of degree 2. Polynomials of degree 3, 4, and 5 are also called **cubic, quartic**, and **quintic polynomials**, respectively. In this section, we concentrate on graphing polynomial functions of degree 3 or more.

Here are some common properties shared by all polynomial functions.

COMMON PROPERTIES OF POLYNOMIAL FUNCTIONS

1. The domain of a polynomial function is the set of all real numbers.
2. The graph of a polynomial function is a **continuous curve**. This means that the graph has no holes or gaps and can be drawn on a sheet of paper without lifting the pencil. See Figure 2.3.

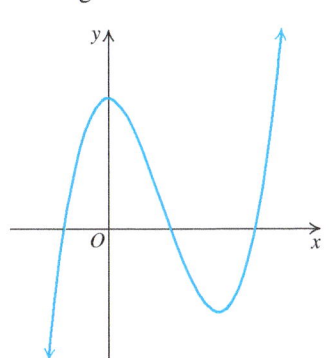

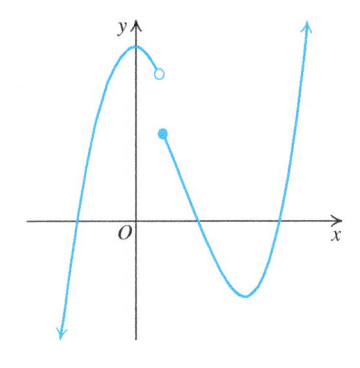

A continuous curve
(a)

A discontinuous curve; cannot be the graph of a polynomial function
(b)

FIGURE 2.3

3. The graph of a polynomial function is a **smooth curve**. This means that the graph does not contain any sharp corners. See Figure 2.4.

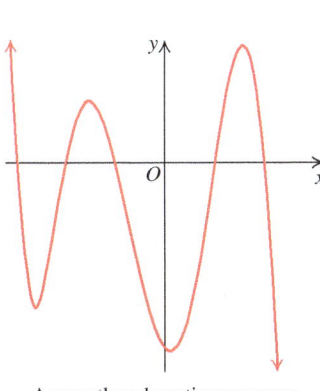

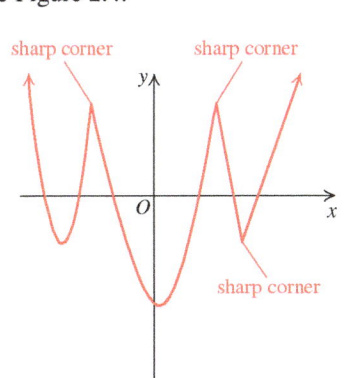

A smooth and continuous curve
(a)

A continuous but not a smooth curve; cannot be the graph of a polynomial function
(b)

FIGURE 2.4

EXAMPLE 1   Polynomial Functions

State which functions are polynomial functions. For each polynomial function, find its degree, the leading term, and the leading coefficient.

a. $f(x) = 5x^4 - 2x + 7$ **b.** $g(x) = 7x^2 - x + 1, \ 1 \leq x \leq 5$

166 Chapter 2 Polynomial and Rational Functions

Solution

a. $f(x) = 5x^4 - 2x + 7$ is a polynomial function. Its degree is 4, the leading term is $5x^4$, and the leading coefficient is 5.

b. $g(x) = 7x^2 - x + 1, 1 \leq x \leq 5$, is not a polynomial function because its domain is not $(-\infty, \infty)$.

Practice Problem 1 Repeat Example 1 for each function.

a. $f(x) = \dfrac{x^2 + 1}{x - 1}$ b. $g(x) = 2x^7 + 5x^2 - 17$

Power Functions

A function of the form $f(x) = ax^n$ is the simplest nth-degree polynomial function, and it is called a power function.

Power Function

A function of the form

$$f(x) = ax^n$$

is called a **power function of degree n**, where a is a nonzero real number and n is a positive integer.

The graph of a power function can be obtained by stretching or shrinking the graph of $y = x^n$ (and reflecting it about the x-axis if $a < 0$). So the basic shape of the graph of a power function $f(x) = ax^n$ is similar to the shape of the graphs of the equations $y = \pm x^n$.

Power Functions of Even Degree Let $f(x) = ax^n$. If n is even, then $(-x)^n = x^n$. So in this case, $f(-x) = a(-x)^n = ax^n = f(x)$. **Therefore, a power function is an even function when n is even; so its graph is symmetric with respect to the y-axis.**

The graph of $y = x^n$ (when n is even) is similar to the graph of $y = x^2$; the graphs of $y = x^2, y = x^4$, and $y = x^6$ are shown in Figure 2.5.

Notice that each of these graphs passes through the points $(-1, 1), (0, 0)$, and $(1, 1)$. On the interval $-1 < x < 1$, the larger the exponent, the flatter the graph. However, on the intervals $(-\infty, -1)$ and $(1, \infty)$, the larger the exponent is, the more rapidly the graph rises.

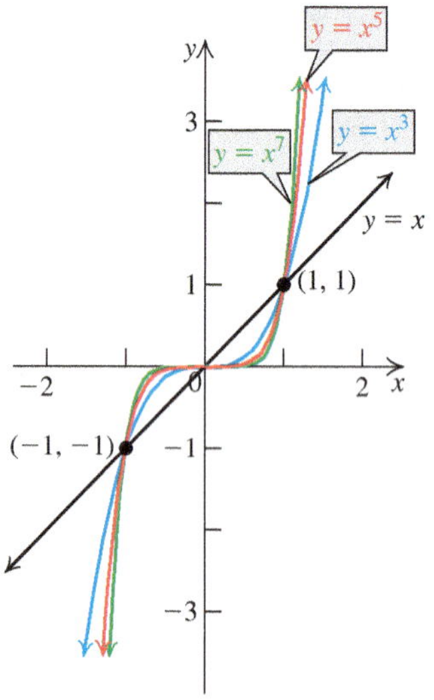

Power Functions of Odd Degree Again, let $f(x) = ax^n$. If n is odd, then $(-x)^n = -x^n$. In this case, we have $f(-x) = a(-x)^n = -ax^n = -f(x)$. **So the power function $f(x) = ax^n$ is an odd function when n is odd; its graph is symmetric with respect to the origin.** When n is odd, and $n \geq 3$, the graph of $f(x) = x^n$ is similar to the graph of $y = x^3$. The graphs of $y = x, y = x^3, y = x^5$, and $y = x^7$ are shown in Figure 2.6. If n is an odd positive integer, then the graph of $y = x^n$ passes through the points $(-1, -1), (0, 0)$, and $(1, 1)$. The larger the value of n is, the flatter the graph is near the origin (in the interval $-1 < x < 1$) and the more rapidly it rises (or falls) on the intervals $(-\infty, -1)$ and $(1, \infty)$.

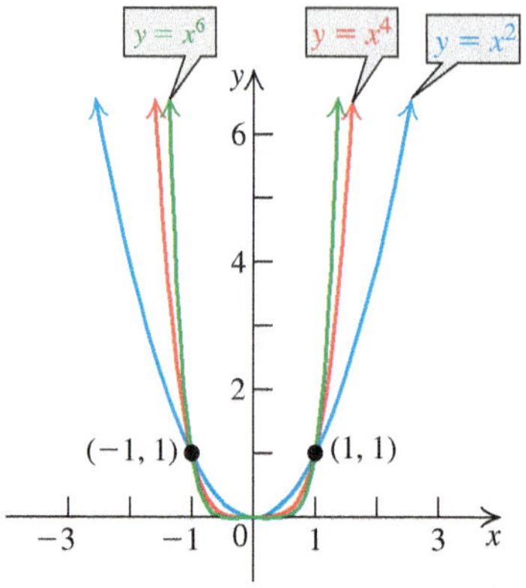

FIGURE 2.5 Power functions of even degree

FIGURE 2.6 Power functions of odd degree

2 Determine the end behavior of polynomial functions.

End Behavior of Polynomial Functions

Let's study the behavior of the function $y = f(x)$ when the independent variable x is large in absolute value. The notation $x \to \infty$ is read "x approaches infinity" and means that x gets larger and larger without bound: it can assume values greater than 1, 10, 100, 1000, . . . ,

SIDE NOTE

Here is a scheme for remembering the axis directions associated with the symbols $-\infty$ and ∞.

$x \to -\infty$	Left
$x \to \infty$	Right
$y \to -\infty$	Down
$y \to \infty$	Up

without end. Similarly, $x \to -\infty$ (read "x approaches negative infinity") means that x can assume values less than $-1, -10, -100, -1000, \ldots$, without end. Of course, x may be replaced by any other independent variable.

The behavior of a function $y = f(x)$ as $x \to \infty$ or $x \to -\infty$ is called the **end behavior** of the function. In the accompanying box, we describe the behavior of the power function $y = f(x) = ax^n$ as $x \to \infty$ or as $x \to -\infty$. **Every polynomial function exhibits end behavior similar to one of the four functions** $y = x^2, y = -x^2, y = x^3,$ or $y = -x^3$.

END BEHAVIOR FOR THE GRAPH OF $y = f(x) = ax^n$

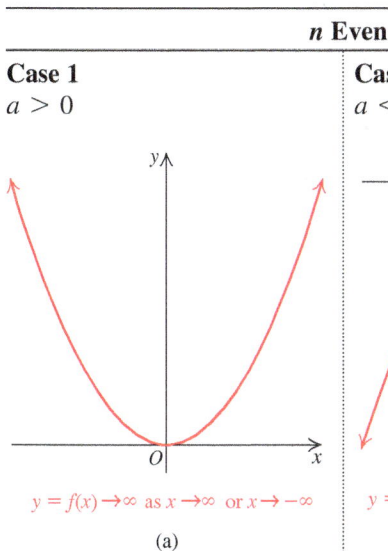

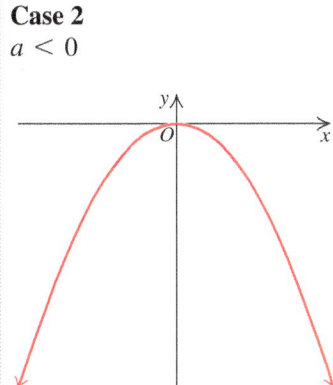

n Even		n Odd	
Case 1 $a > 0$	**Case 2** $a < 0$	**Case 3** $a > 0$	**Case 4** $a < 0$
$y = f(x) \to \infty$ as $x \to \infty$ or $x \to -\infty$ (a)	$y = f(x) \to -\infty$ as $x \to \infty$ or $x \to -\infty$ (b)	$y = f(x) \to \infty$ as $x \to \infty$; $y \to -\infty$ as $x \to -\infty$ (c)	$y = f(x) \to -\infty$ as $x \to \infty$; $y \to \infty$ as $x \to -\infty$ (d)
The graph rises to the left and right, similar to $y = x^2$.	The graph falls to the left and right, similar to $y = -x^2$.	The graph rises to the right and falls to the left, similar to $y = x^3$.	The graph rises to the left and falls to the right, similar to $y = -x^3$.

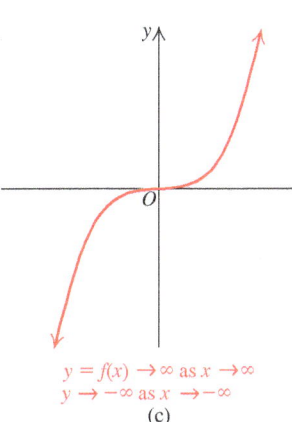

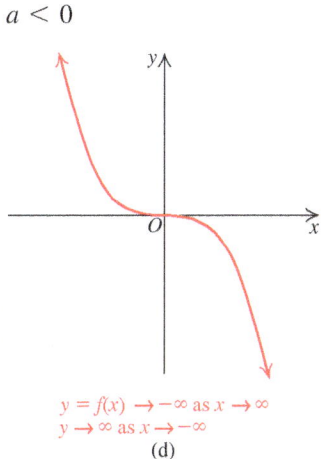

TECHNOLOGY CONNECTION

Calculator graph of $P(x) = 2x^3 + 5x^2 - 7x + 11$

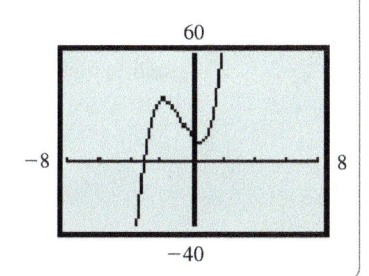

TABLE 2.1

x	$\dfrac{5}{x}$
10	0.5
10^2	0.05
10^3	0.005
10^4	0.0005

In the next example, we show that the end behavior of the given polynomial function is determined by its leading term. If a polynomial $P(x)$ has approximately the same values as $f(x) = ax^n$ when $|x|$ is very large, we write $P(x) \approx ax^n$ when $|x|$ is very large.

EXAMPLE 2 Understanding the End Behavior of a Polynomial Function

Let $P(x) = 2x^3 + 5x^2 - 7x + 11$ be a polynomial function of degree 3. Show that $P(x) \approx 2x^3$ when $|x|$ is very large.

Solution

$$P(x) = 2x^3 + 5x^2 - 7x + 11 \quad \text{Given polynomial}$$
$$= x^3\left(2 + \frac{5}{x} - \frac{7}{x^2} + \frac{11}{x^3}\right) \quad \text{Distributive property}$$

When $|x|$ is very large, the terms $\dfrac{5}{x}, -\dfrac{7}{x^2},$ and $\dfrac{11}{x^3}$ are close to 0. Table 2.1 shows some values of $\dfrac{5}{x}$ as x increases. (You should compute the values of $-\dfrac{7}{x^2}$ and $\dfrac{11}{x^3}$ for $x = 10, 10^2, 10^3,$ and 10^4.) When $|x|$ is very large, we have

$$P(x) = x^3\left(2 + \frac{5}{x} - \frac{7}{x^2} + \frac{11}{x^3}\right) \approx x^3(2 + 0 - 0 + 0)$$
$$= 2x^3.$$

So the polynomial $P(x) \approx 2x^3$ when $|x|$ is very large.

168 Chapter 2 Polynomial and Rational Functions

Practice Problem 2 Let $P(x) = 4x^3 + 2x^2 + 5x - 17$. Show that $P(x) \approx 4x^3$ when $|x|$ is very large.

It is possible to use the technique of Example 2 to show that the end behavior of any polynomial function is determined by its leading term.

The end behaviors of polynomial functions are shown in the following graphs.

THE LEADING-TERM TEST

Let $f(x) = a_n x^n + a_{n-1} x^{n-1} + \cdots + a_1 x + a_0 \, (a_n \neq 0)$ be a polynomial function. Its leading term is $a_n x^n$. The behavior of the graph of f as $x \to \infty$ or $x \to -\infty$ is similar to one of the following four graphs and is described as shown in each case.

n Even		*n* Odd	
Case 1 $a_n > 0$	**Case 2** $a_n < 0$	**Case 3** $a_n > 0$	**Case 4** $a_n < 0$
$y \to \infty$ as $x \to -\infty$ $y \to \infty$ as $x \to \infty$ (a)	$y \to -\infty$ as $x \to -\infty$ $y \to -\infty$ as $x \to \infty$ (b)	$y \to -\infty$ as $x \to -\infty$ $y \to \infty$ as $x \to \infty$ (c)	$y \to \infty$ as $x \to -\infty$ $y \to -\infty$ as $x \to \infty$ (d)

This test does not describe the middle portion of each graph, shown by the dashed lines.

EXAMPLE 3 Using the Leading-Term Test

Use the leading-term test to determine the end behavior of the graph of

$$y = f(x) = -2x^3 + 3x + 4.$$

Solution

Here we have degree $n = 3$ (odd) and leading coefficient $a_n = -2 < 0$. Case 4 applies. The graph of f rises to the left and falls to the right. See Figure 2.7. This end behavior is described as $y \to \infty$ as $x \to -\infty$ and $y \to -\infty$ as $x \to \infty$.

Practice Problem 3 What is the end behavior of $f(x) = -2x^4 + 5x^2 + 3$?

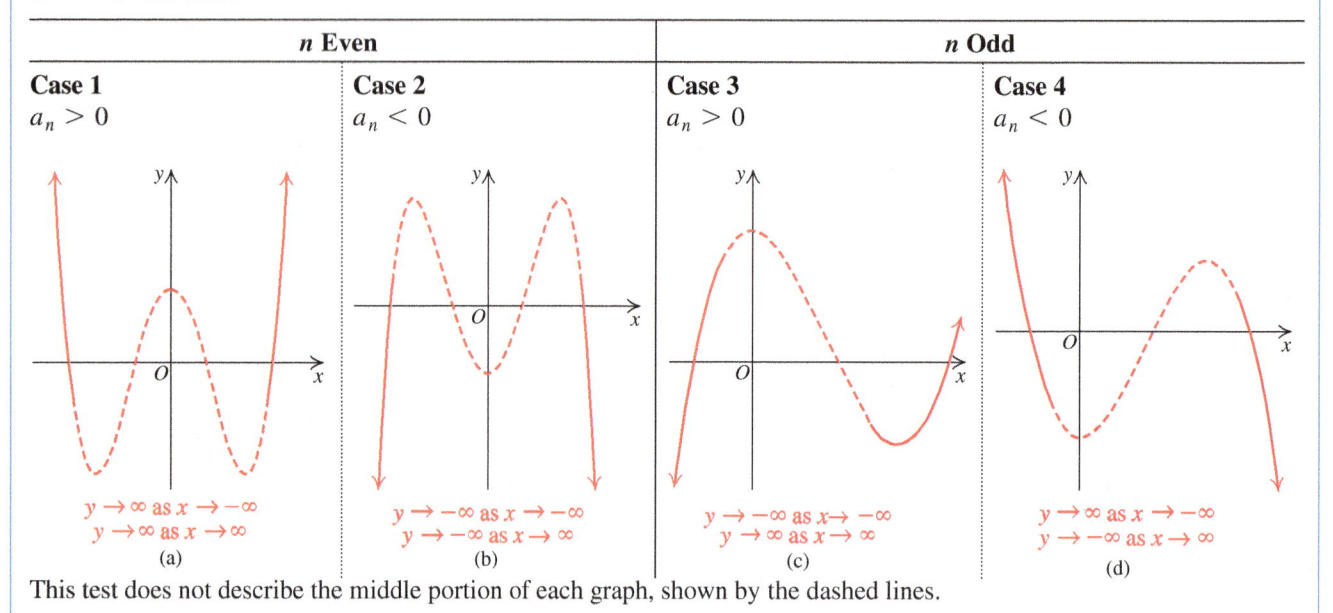

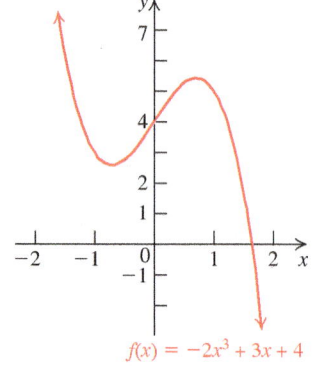

FIGURE 2.7 End behavior of $f(x) = -2x^3 + 3x + 4$

3 Find the zeros of a polynomial function by factoring.

Zeros of a Function

Let f be a function. An input c in the domain of f that produces output 0 is called a **zero** of the function f. For example, if $f(x) = (x - 3)^2$, then 3 is a zero of f because $f(3) = (3 - 3)^2 = 0$. In other words, a number c is a **zero** of f if $f(c) = 0$. The zeros of a function f can be obtained by solving the equation $f(x) = 0$. One way to solve a polynomial equation $f(x) = 0$ is to factor $f(x)$ and then use the zero-product property.

Section 2.2 ■ Polynomial Functions   169

If c is a real number and $f(c) = 0$, then c is called a **real zero** of f. Geometrically, this means that the graph of f has an x-intercept at $x = c$.

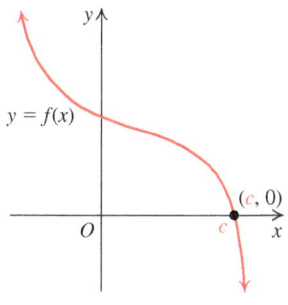

> ### REAL ZEROS OF POLYNOMIAL FUNCTIONS
>
> If f is a polynomial function and c is a real number, then the following statements are equivalent:
>
> 1. c is a **zero** of f.
> 2. c is a **solution** (or **root**) of the equation $f(x) = 0$.
> 3. c is an **x-intercept** of the graph of f. The point $(c, 0)$ is on the graph of f. See Figure 2.8.

FIGURE 2.8 A zero, c, of f

TECHNOLOGY CONNECTION

$f(x) = x^4 - 2x^3 - 3x^2$

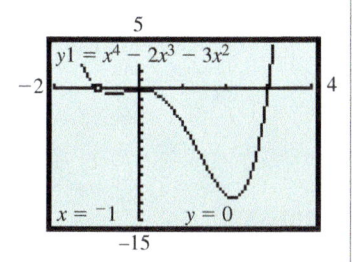

We can verify that the zeros of f are 0, -1, and 3 using the TRACE function.

$g(x) = x^3 - 2x^2 + x - 2$

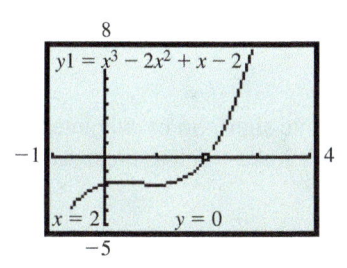

EXAMPLE 4   Finding the Zeros of a Polynomial Function

Find all real zeros of each polynomial function.

a. $f(x) = x^4 - 2x^3 - 3x^2$ **b.** $g(x) = x^3 - 2x^2 + x - 2$

Solution

a. We factor $f(x)$ and then solve the equation $f(x) = 0$.

$$\begin{aligned} f(x) &= x^4 - 2x^3 - 3x^2 & &\text{Given function} \\ &= x^2(x^2 - 2x - 3) & &\text{Factor out } x^2. \\ &= x^2(x + 1)(x - 3) & &\text{Factor } x^2 - 2x - 3. \end{aligned}$$

$$\begin{aligned} x^2(x + 1)(x - 3) &= 0 & &\text{Set } f(x) = 0. \\ x^2 = 0, \quad x + 1 &= 0, \quad \text{or} \quad x - 3 = 0 & &\text{Zero-product property} \\ x = 0, \quad x &= -1, \quad \text{or} \quad x = 3 & &\text{Solve each equation.} \end{aligned}$$

The zeros of $f(x)$ are 0, -1, and 3. (See the calculator graph of f in the margin.)

b. By grouping terms, we factor $g(x)$ to obtain

$$\begin{aligned} g(x) &= x^3 - 2x^2 + x - 2 \\ &= x^2(x - 2) + 1 \cdot (x - 2) & &\text{Group terms.} \\ &= (x - 2)(x^2 + 1) & &\text{Factor out } x - 2. \\ 0 &= (x - 2)(x^2 + 1) & &\text{Set } g(x) = 0. \\ x - 2 &= 0, \quad \text{or} \quad x^2 + 1 = 0 & &\text{Zero-product property} \end{aligned}$$

Solving for x, we obtain 2 as the only real zero of $g(x)$ because $x^2 + 1 > 0$ for all real numbers x. The calculator graph of g in the margin supports our conclusion.

Practice Problem 4 Find all real zeros of $f(x) = 2x^3 - 3x^2 + 4x - 6$.

Finding the zeros of a polynomial of degree 3 or more is one of the most important problems of mathematics. You will learn more about this topic in Sections 2.3 and 2.4. For now, we will approximate the zeros by using an *Intermediate Value Theorem*. Recall that the graph of a polynomial function $f(x)$ is a continuous curve. Suppose you locate two numbers a and b such that the function values $f(a)$ and $f(b)$ have opposite signs (one positive and the other negative). Then the continuous graph of f connecting the points $(a, f(a))$ and $(b, f(b))$ must cross the x-axis (at least once) somewhere between a and b. In other words, the function f has a zero between a and b. See Figure 2.9 on page 170.

AN INTERMEDIATE VALUE THEOREM

Let a and b be two numbers with $a < b$. If f is a polynomial function such that $f(a)$ and $f(b)$ have opposite signs, then there is at least one number c, with $a < c < b$, for which $f(c) = 0$.

Notice that Figure 2.9(c) shows more than one zero of f. That is why this Intermediate Value Theorem says that f has *at least one* zero between a and b.

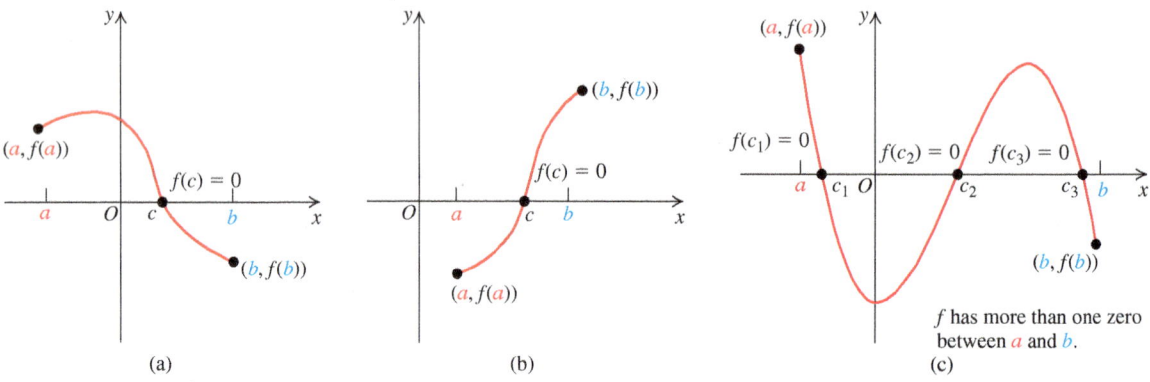

FIGURE 2.9 Each function has at least one zero between a and b.

EXAMPLE 5 Using an Intermediate Value Theorem

Show that the function $f(x) = -2x^3 + 4x + 5$ has a real zero between 1 and 2.

Solution

$f(1) = -2(1)^3 + 4(1) + 5$ Replace x with 1 in $f(x)$.
$ = -2 + 4 + 5 = 7$ Simplify.
$f(2) = -2(2)^3 + 4(2) + 5$ Replace x with 2.
$ = -2(8) + 8 + 5 = -3$ Simplify.

Because $f(1)$ is positive and $f(2)$ is negative, they have opposite signs. So by this Intermediate Value Theorem, the polynomial function $f(x) = -2x^3 + 4x + 5$ has a real zero between 1 and 2. See Figure 2.10.

Practice Problem 5 Show that $f(x) = 2x^3 - 3x - 6$ has a real zero between 1 and 2.

In Example 5, if you want to find the zero between 1 and 2 more accurately, you can divide the interval $[1, 2]$ into tenths and find the value of the function at each of the points $1, 1.1, 1.2, \ldots, 1.9$, and 2. You will find that

$$f(1.8) = 0.536 \quad \text{and} \quad f(1.9) = -1.118.$$

So by this Intermediate Value Theorem, f has a zero between 1.8 and 1.9. If you divide the interval $[1.8, 1.9]$ into tenths, you will find that $f(1.83) > 0$ and $f(1.84) < 0$; so f has a zero between 1.83 and 1.84. By continuing this process, you can find the zero of f between 1.8 and 1.9 to any desired degree of accuracy.

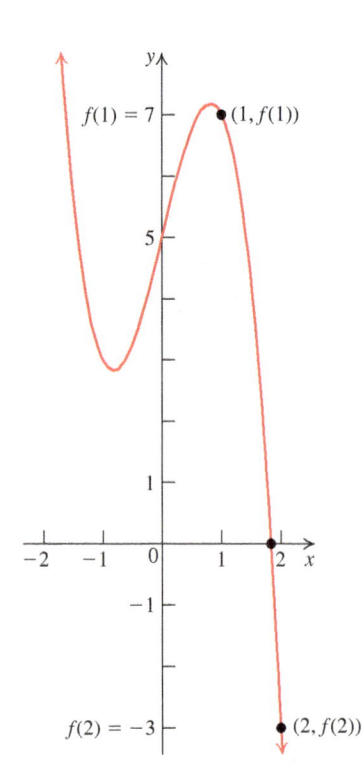

FIGURE 2.10 A zero lies between 1 and 2.

4 Identify the relationship between degrees, real zeros, and turning points.

Zeros and Turning Points

In Section 2.4, you will study the Fundamental Theorem of Algebra. One of its consequences is the following fact:

REAL ZEROS OF A POLYNOMIAL FUNCTION

A polynomial function of degree n with real coefficients has, *at most*, n real zeros.

SIDE NOTE

A polynomial of degree n can have no more than n real zeros. In Examples 6b and 6c, we see two polynomials of degree 3 each having *fewer* than three zeros.

EXAMPLE 6 Finding the Number of Real Zeros

Find the number of distinct real zeros of the following polynomial functions of degree 3.

a. $f(x) = (x - 1)(x + 2)(x - 3)$ **b.** $g(x) = (x + 1)(x^2 + 1)$
c. $h(x) = (x - 3)^2(x + 1)$

Solution

a. $f(x) = (x - 1)(x + 2)(x - 3) = 0$ Set $f(x) = 0$.
   $x - 1 = 0$ or $x + 2 = 0$ or $x - 3 = 0$ Zero-product property
   $x = 1$ or $x = -2$ or $x = 3$ Solve for x.

So $f(x)$ has three real zeros: $1, -2,$ and 3.

b. $g(x) = (x + 1)(x^2 + 1) = 0$ Set $g(x) = 0$.
   $x + 1 = 0$ or $x^2 + 1 = 0$ Zero-product property
   $x = -1$ Solve for x.

Because $x^2 + 1 > 0$ for all real x, the equation $x^2 + 1 = 0$ has no real solution. So -1 is the only real zero of $g(x)$.

c. $h(x) = (x - 3)^2(x + 1) = 0$ Set $h(x) = 0$.
   $(x - 3)^2 = 0$ or $x + 1 = 0$ Zero-product property
   $x = 3$ or $x = -1$ Solve for x.

The real zeros of $h(x)$ are 3 and -1. Thus, $h(x)$ has two distinct real zeros.

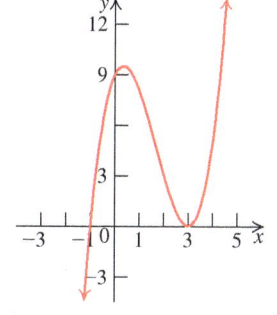

Practice Problem 6 Find the distinct real zeros of

$$f(x) = (x + 1)^2(x - 3)(x + 5).$$

In Example 6c, the factor $(x - 3)$ appears twice in $h(x)$. We say that 3 is a zero repeated *twice*, or that 3 is a *zero of multiplicity* 2. Notice that the graph of $y = h(x)$ in Figure 2.11 *touches* the x-axis at $x = 3$, where h has a zero of multiplicity 2 (an even number). In contrast, the graph of h *crosses* the x-axis at $x = -1$, where it has a zero of multiplicity 1 (an odd number). We next consider the geometric significance of the multiplicity of a zero.

FIGURE 2.11 Graph of $h(x) = (x - 3)^2(x + 1)$

Multiplicity of a Zero

If c is a zero of a polynomial function $f(x)$ and the corresponding factor $(x - c)$ occurs exactly m times when $f(x)$ is factored as $f(x) = (x - c)^m q(x)$, with $q(c) \neq 0$, then c is called a **zero of multiplicity** m.

1. If m is odd, the graph of f crosses the x-axis at $x = c$.
2. If m is even, the graph of f touches but does not cross the x-axis at $x = c$.

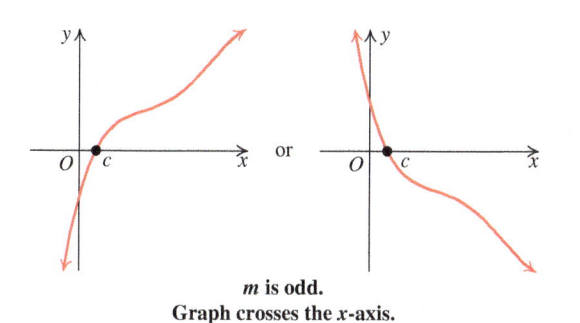

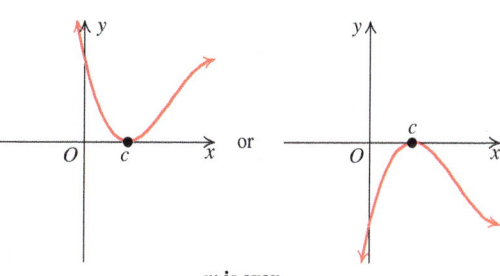

m is odd.
Graph crosses the x-axis.

m is even.
Graph touches but does not cross the x-axis.

172 Chapter 2 Polynomial and Rational Functions

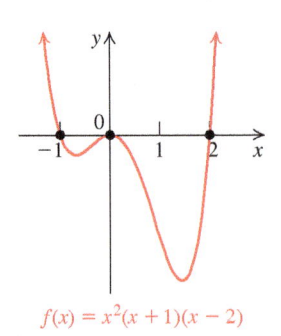

$f(x) = x^2(x+1)(x-2)$
FIGURE 2.12

The graph of $f(x) = 2x^3 - 3x^2 - 12x + 5$ has turning points at $(-1, 12)$ and $(2, -15)$.
FIGURE 2.13 Turning points

EXAMPLE 7 Finding the Zeros and Their Multiplicities

Find the zeros of the polynomial function $f(x) = x^2(x+1)(x-2)$ and give the multiplicity of each zero.

Solution

The polynomial $f(x)$ is already in factored form.

$$f(x) = x^2(x+1)(x-2) = 0 \quad \text{Set } f(x) = 0.$$
$$x^2 = 0 \quad \text{or} \quad x+1 = 0 \quad \text{or} \quad x-2 = 0 \quad \text{Zero-product property}$$
$$x = 0 \quad \text{or} \quad x = -1 \quad \text{or} \quad x = 2 \quad \text{Solve for } x.$$

The function has three distinct zeros: $0, -1,$ and 2. The zero $x = 0$ has multiplicity 2, and each of the zeros -1 and 2 has multiplicity 1. The graph of f is given in Figure 2.12. Notice that the graph of f in Figure 2.12 crosses the x-axis at -1 and 2 (the zeros of odd multiplicity), and it touches but does not cross the x-axis at $x = 0$ (the zero of even multiplicity).

Practice Problem 7 Write the zeros of $f(x) = (x-1)^2(x+3)(x+5)$ and give the multiplicity of each.

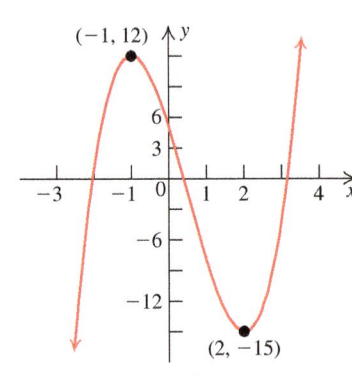

Turning Points The graph of $f(x) = 2x^3 - 3x^2 - 12x + 5$ is shown in Figure 2.13. The value $f(-1) = 12$ is a **local** (or **relative**) **maximum** value of f. Similarly, the value $f(2) = -15$ is a **local** (or **relative**) **minimum** value of f. (See Section 1.4 for definitions.) The graph points corresponding to the local maximum and local minimum values of a polynomial function are the **turning points**. At each turning point, the graph changes direction from increasing to decreasing or from decreasing to increasing. One of the successes of calculus is finding the maximum and minimum values for many functions.

NUMBER OF TURNING POINTS

If $f(x)$ is a polynomial of degree n, then the graph of f has, *at most*, $(n-1)$ turning points.

EXAMPLE 8 Finding the Number of Turning Points

Use a graphing calculator and the window $-10 \leq x \leq 10; -30 \leq y \leq 30$ to find the number of turning points of the graph of each polynomial function.

a. $f(x) = x^4 - 7x^2 - 18$
b. $g(x) = x^3 + x^2 - 12x$
c. $h(x) = x^3 - 3x^2 + 3x - 1$

Solution

The graphs of $f, g,$ and h are shown in Figure 2.14.

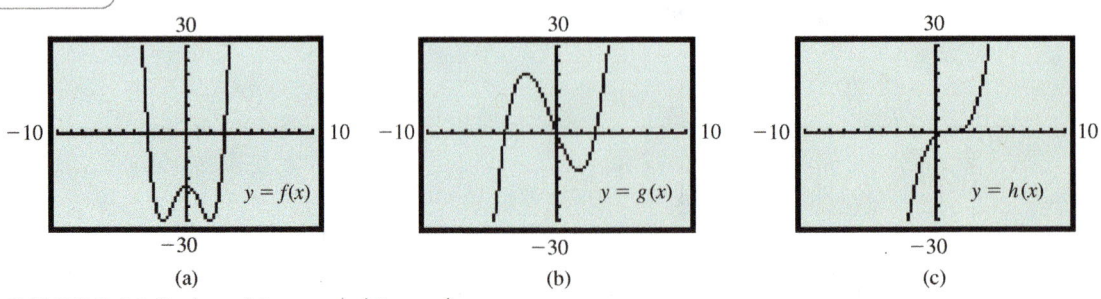

SIDE NOTE
Notice that in Figure 2.14(c), the graph seems to "turn" but the graph is always increasing. There is no change from increasing to decreasing, so it has no turning points.

(a) $y = f(x)$ (b) $y = g(x)$ (c) $y = h(x)$

FIGURE 2.14 Turning points on a calculator graph

Section 2.2 ■ Polynomial Functions 173

a. Function f has three turning points: two local minima and one local maximum.

b. Function g has two turning points: one local maximum and one local minimum.

c. Function h has no turning points: it is increasing on the interval $(-\infty, \infty)$.

Practice Problem 8 Find the number of turning points of the graph of $f(x) = -x^4 + 3x^2 - 2$.

5 Graph polynomial functions.

Graphing a Polynomial Function

You can graph a polynomial function by constructing a table of values for a large number of points, plotting the points, and connecting the points with a smooth curve. (In fact, a graphing calculator uses this method.) Generally, however, this process is a poor way to graph a function by hand because it is tedious, is error-prone, and may give misleading results. Similarly, a graphing calculator may give misleading results if the viewing window is not chosen appropriately. Next, we discuss a better strategy for graphing a polynomial function.

PROCEDURE IN ACTION

EXAMPLE 9 Graphing a Polynomial Function

OBJECTIVE
Sketch the graph of a polynomial function.

EXAMPLE
Sketch the graph of $f(x) = -x^3 - 4x^2 + 4x + 16$.

Step 1 Determine the end behavior. Apply the leading-term test.

1. Because the degree, 3, of f is odd and the leading coefficient, -1, is negative, the end behavior (Case 4, page 168) is similar to that of $y = -x^3$ as shown in this sketch.

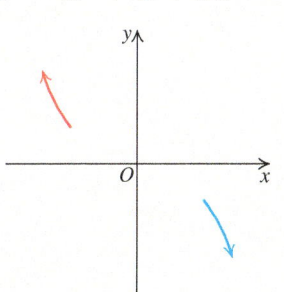

Step 2 Find the real zeros of the polynomial function. Set $f(x) = 0$. Solve the equation $f(x) = 0$. The real zeros (with their multiplicities) will give you the x-intercepts of the graph and determine whether or not the graph crosses the x-axis there.

2. $-x^3 - 4x^2 + 4x + 16 = 0$
$-(x + 4)(x + 2)(x - 2) = 0$
$x = -4, x = -2, \text{ or } x = 2$

There are three zeros, each of multiplicity 1; so the graph crosses the x-axis at each zero.

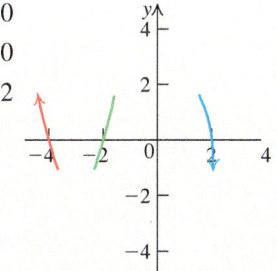

Step 3 Find the y-intercept by computing $f(0)$.

3. The y-intercept is $f(0) = 16$. The graph passes through the point $(0, 16)$.

Step 4 Use symmetry (if any). Check whether the function is odd, even, or neither.

4. $f(-x) = -(-x)^3 - 4(-x)^2 + 4(-x) + 16$
$= x^3 - 4x^2 - 4x + 16$

Because $f(-x) \neq f(x)$, there is no symmetry about the y-axis. Also, because $f(-x) \neq -f(x)$, there is no symmetry with respect to the origin.

(continued)

Step 5 Determine the sign of $f(x)$ by using arbitrarily chosen "test numbers" in the intervals defined by the x-intercepts. Find on which of these intervals the graph lies above or below the x-axis.

5. The three zeros $-4, -2,$ and 2 divide the x-axis into four intervals: $(-\infty, -4), (-4, -2), (-2, 2),$ and $(2, \infty)$. We choose $-5, -3, 0,$ and 3 as test numbers.

Test number	-5	-3	0	3
Value of f	21	-5	16	-35
Sign graph	$+++$	$---$	$+++$	$---$
Graph of f	Above x-axis	Below x-axis	Above x-axis	Below x-axis
Graph point	$(-5, 21)$	$(-3, -5)$	$(0, 16)$	$(3, -35)$

Step 6 Draw the graph using points from steps 1–5. To check whether the graph is drawn correctly, use the fact that the number of turning points is less than the degree of the polynomial.

6. The graph connects the points we have found with a smooth curve.

The number of turning points is 2, which is less than 3, the degree of f.

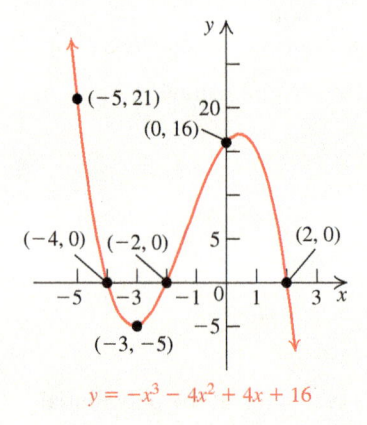

$y = -x^3 - 4x^2 + 4x + 16$

Practice Problem 9 Graph $f(x) = -x^4 + 5x^2 - 4$.

◆ **Volume of a Wine Barrel** Kepler first analyzed the problem of finding the volume of a cylindrical wine barrel. (Review the introduction to this section and see Figure 2.15.) The volume V of a cylinder with radius r and height h is given by the formula $V = \pi r^2 h$. In Figure 2.15, $h = 2z$; so $V = 2\pi r^2 z$. Let x represent the value measured by the rod. Then by the Pythagorean Theorem,

$$x^2 = 4r^2 + z^2. \quad (2r)^2 = 4r^2$$

Kepler's challenge was to calculate V, given only x—in other words, to express V as a function of x. To solve this problem, Kepler observed the key fact that Austrian wine barrels were all made with the same height-to-diameter ratio. Kepler assumed that the winemakers would have made a smart choice for this ratio, one that would maximize the volume for a fixed value of r. He calculated the height-to-diameter ratio to be $\sqrt{2}$.

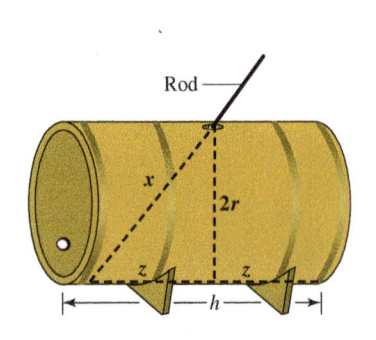

FIGURE 2.15 A wine barrel

EXAMPLE 10 Volume of a Wine Barrel

Express the volume V of the wine barrel in Figure 2.15 as a function of x. Assume that $\dfrac{\text{height}}{\text{diameter}} = \dfrac{2z}{2r} = \dfrac{z}{r} = \sqrt{2}$.

Solution

We have the following relationships:

$V = 2\pi r^2 z$ Volume with radius r and height $2z$

$\dfrac{z}{r} = \sqrt{2}$ Given ratio

$x^2 = 4r^2 + z^2$ Pythagorean Theorem (Figure 2.15)

$\dfrac{x^2}{r^2} = 4 + \dfrac{z^2}{r^2}$ Divide both sides by r^2.

$\dfrac{x^2}{r^2} = 4 + 2 = 6$ Because $\dfrac{z}{r} = \sqrt{2}, \dfrac{z^2}{r^2} = \left(\dfrac{z}{r}\right)^2 = (\sqrt{2})^2 = 2.$

From $\dfrac{x^2}{r^2} = 6$; so $r = \dfrac{x}{\sqrt{6}}$ $(r > 0)$. Using $\dfrac{z}{r} = \sqrt{2}$, we obtain

$z = \sqrt{2}\, r = \sqrt{2}\, \dfrac{x}{\sqrt{6}}$ Replace r with $\dfrac{x}{\sqrt{6}}$.

$= \dfrac{x}{\sqrt{3}}$ $\dfrac{\sqrt{2}}{\sqrt{6}} = \dfrac{\sqrt{2}}{\sqrt{2}\cdot\sqrt{3}} = \dfrac{1}{\sqrt{3}}$

Now, $V = 2\pi r^2 z$

$= 2\pi \left(\dfrac{x^2}{6}\right)\left(\dfrac{x}{\sqrt{3}}\right)$ Replace z with $\dfrac{x}{\sqrt{3}}$ and r^2 with $\dfrac{x^2}{6}$.

$= \dfrac{\pi}{3\sqrt{3}} x^3$ Simplify.

$\approx 0.6046 x^3$ Use a calculator.

Consequently, Austrian vintners could easily calculate the volume of a wine barrel by using the formula $V = 0.6046 x^3$ where x was the length measured by the rod. For a quick estimate, use $V \approx 0.6 x^3$.

Practice Problem 10 Use Kepler's formula to compute the volume of wine (in liters) in a barrel with $x = 70$ cm. Note: 1 decimeter (dm) $= 10$ cm and 1 liter $= (1\text{ dm})^3$.

SECTION 2.2 Exercises

Basic Concepts and Skills

In Exercises 1–6, for each polynomial function, find the degree, the leading term, and the leading coefficient.

1. $f(x) = 2x^5 - 5x^2$
2. $f(x) = 3 - 5x - 7x^4$
3. $f(x) = \dfrac{2x^3 + 7x}{3}$
4. $f(x) = -7x + 11 + \sqrt{2}x^3$
5. $f(x) = \pi x^4 + 1 - x^2$
6. $f(x) = 5$

In Exercises 7–14, explain why the given function is not a polynomial function.

7. $f(x) = x^2 + 3|x| - 7$
8. $f(x) = 2x^3 - 5x, \ 0 \le x \le 2$
9. $f(x) = \dfrac{x^2 - 1}{x + 5}$

176 Chapter 2 • Polynomial and Rational Functions

10. $f(x) = \begin{cases} 2x + 3, & x \neq 0 \\ 1, & x = 0 \end{cases}$

11. $f(x) = 5x^2 - 7\sqrt{x}$

12. $f(x) = x^{4.5} - 3x^2 + 7$

13. $f(x) = \dfrac{x^2 - 1}{x - 1}, x \neq 1$

14. $f(x) = 3x^2 - 4x + 7x^{-2}$

In Exercises 15–20, explain why the given graph cannot be the graph of a polynomial function.

15.

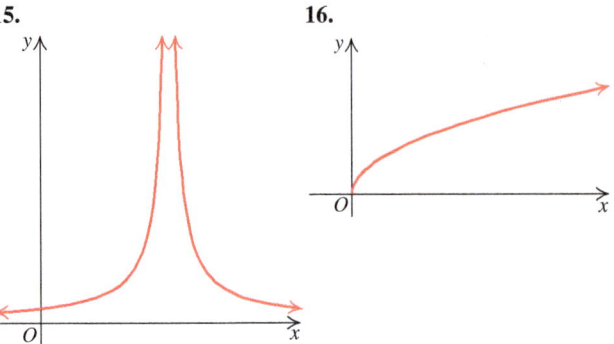

16.

17.

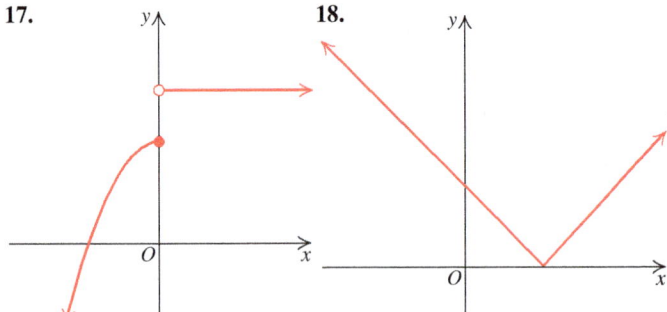

18.

19.

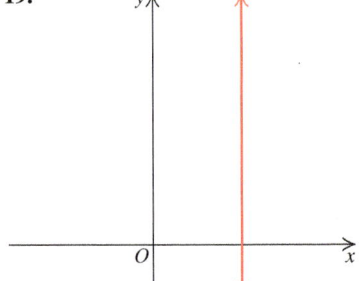

20.

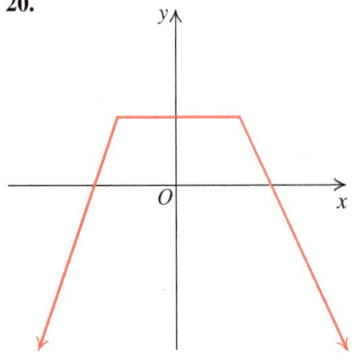

In Exercises 21–26, match the polynomial function with its graph. Use the leading-term test and the y-intercept.

21. $f(x) = -x^4 + 3x^2 + 4$

22. $f(x) = x^6 - 7x^4 + 7x^2 + 15$

23. $f(x) = x^4 - 10x^2 + 9$

24. $f(x) = x^3 + x^2 - 17x + 15$

25. $f(x) = x^3 + 6x^2 + 12x + 8$

26. $f(x) = -x^3 - 2x^2 + 11x + 12$

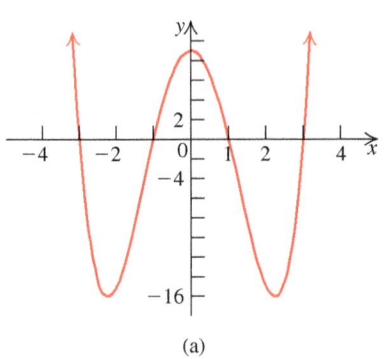

(a)

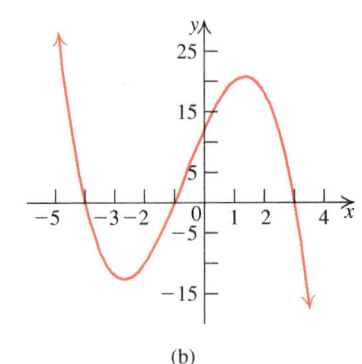

(b)

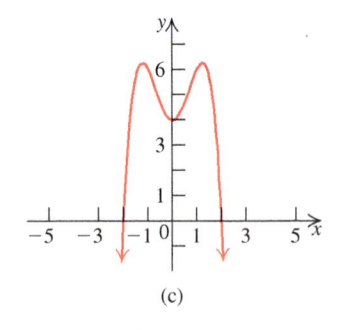

(c)

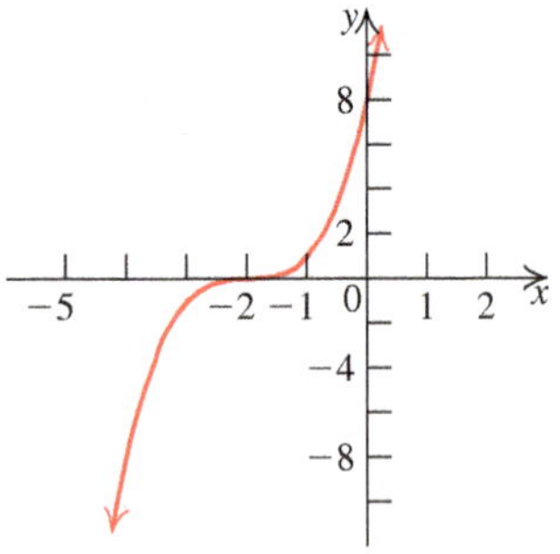

(d)

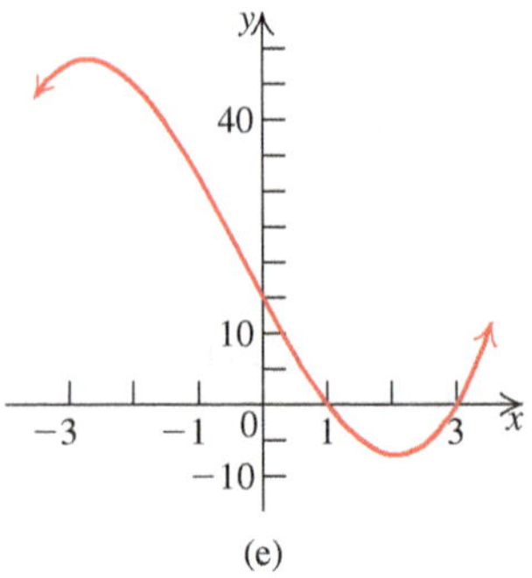

(e)

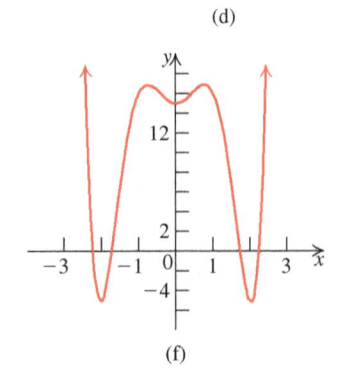

(f)

In Exercises 27–34, (a) find the real zeros of each polynomial function and state the multiplicity for each zero and (b) state whether the graph crosses or touches but does not cross the x-axis at each x-intercept. (c) What is the maximum possible number of turning points?

27. $f(x) = 5(x - 1)(x + 5)$
28. $f(x) = 3(x + 2)^2(x - 3)$
29. $f(x) = (x + 1)^2(x - 1)^3$
30. $f(x) = 2(x + 3)^2(x + 1)(x - 6)$
31. $f(x) = -5\left(x + \dfrac{2}{3}\right)\left(x - \dfrac{1}{2}\right)^2$
32. $f(x) = x^3 - 6x^2 + 8x$
33. $f(x) = -x^4 + 6x^3 - 9x^2$
34. $f(x) = x^4 - 5x^2 - 36$

In Exercises 35–38, use the Intermediate Value Theorem to show that each polynomial $P(x)$ has a real zero in the specified interval. Approximate this zero to two decimal places.

35. $P(x) = x^4 - x^3 - 10;\ [2, 3]$
36. $P(x) = x^4 - x^2 - 2x - 5;\ [1, 2]$
37. $P(x) = x^5 - 9x^2 - 15;\ [2, 3]$
38. $P(x) = x^5 + 5x^4 + 8x^3 + 4x^2 - x - 5;\ [0, 1]$

In Exercises 39–52, for each polynomial function f,
a. describe the end behavior of f.
b. find the real zeros of f. Determine whether the graph of f crosses or touches but does not cross the x-axis at each x-intercept.
c. use the zeros of f and test numbers to find the intervals over which the graph of f is above or below the x-axis.
d. determine the y-intercept.
e. determine whether the graph has symmetry about the y-axis or about the origin.
f. determine the maximum possible number of turning points.
g. sketch the graph of f.

39. $f(x) = x^2 + 3$
40. $f(x) = x^2 - 4x + 4$
41. $f(x) = x^2 + 4x - 21$
42. $f(x) = -x^2 + 4x + 12$
43. $f(x) = -2x^2(x + 1)$
44. $f(x) = x - x^3$
45. $f(x) = x^2(x - 1)^2$
46. $f(x) = x^2(x + 1)(x - 2)$
47. $f(x) = (x - 1)^2(x + 3)(x - 4)$
48. $f(x) = (x + 1)^2(x^2 + 1)$
49. $f(x) = -x^2(x^2 - 1)(x + 1)$
50. $f(x) = -x^2(x^2 - 4)(x + 2)$
51. $f(x) = x(x + 1)(x - 1)(x + 2)$
52. $f(x) = x^2(x^2 + 1)(x - 2)$

Applying the Concepts

53. **Drug reaction.** Let $f(x) = 3x^2(4 - x)$.
    a. Find the zeros of f and their multiplicities.
    b. Sketch the graph of $y = f(x)$.
    c. How many turning points are in the graph of f?
    d. A patient's reaction $R(x)\,(R(x) \geq 0)$ to a certain drug, the size of whose dose is x, is observed to be $R(x) = 3x^2(4 - x)$. What is the domain of the function $R(x)$? What portion of the graph of $f(x)$ constitutes the graph of $R(x)$?

54. In Exercise 53, use a graphing calculator to estimate the value of x for which $R(x)$ is a maximum.

55. **Cough and air velocity.** The normal radius of the windpipe of a human adult at rest is approximately 1 cm. When you cough, your windpipe contracts to, say, a radius r. The velocity v at which the air is expelled depends on the radius r and is given by the formula $v = a(1 - r)r^2$, where a is a positive constant.
    a. What is the domain of v?
    b. Sketch the graph of $v(r)$.

56. In Exercise 55, use a graphing calculator to estimate the value of r for which v is a maximum.

Critical Thinking / Discussion / Writing

In Exercises 57–60, find the smallest possible degree of a polynomial with the given graph. Explain your reasoning.

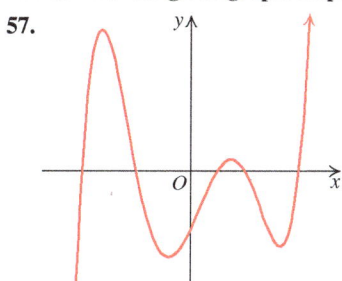

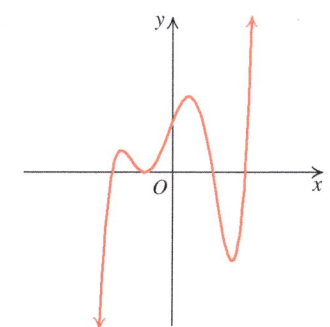

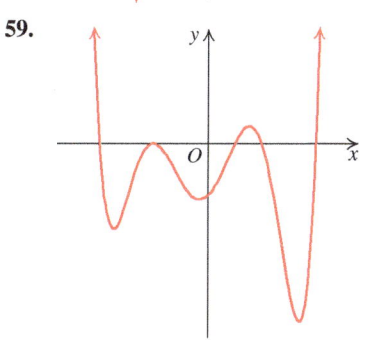

57.

58.

59.

60.

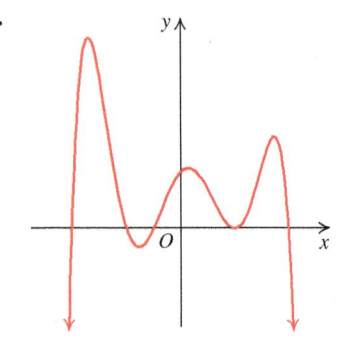

61. Is it possible for the graph of a polynomial function to have no y-intercepts? Explain.

62. Is it possible for the graph of a polynomial function to have no x-intercepts? Explain.

63. Is it possible for the graph of a polynomial function of degree 3 to have exactly one local maximum and no local minimum? Explain.

64. Is it possible for the graph of a polynomial function of degree 4 to have exactly one local maximum and exactly one local minimum? Explain.

Maintaining Skills

In Exercises 65–69, simplify each expression. (Do not use negative exponents.)

65. $\dfrac{12x^5}{-3x^3}$

66. $\dfrac{x^5}{x^7}$

67. $\dfrac{3x^4}{2x^2}$

68. $\dfrac{x^5 - 3x^2}{x^2}$

69. $\dfrac{5x^3 + 2x^2 - x}{x}$

In Exercises 70–72, recall that an integer a is a factor of an integer b is there is an integer c so that $b = ac$.

70. **a.** Show that 1 is a factor of 7.
   b. Show that 7 is a factor of 7.
   c. Show that -1 is a factor of 7.
   d. Show that -7 is a factor of 7.

71. Find all integer factors of 7.

72. Find all integer factors of 120.

SECTION 2.3 Dividing Polynomials and the Rational Zeros Test

BEFORE STARTING THIS SECTION, REVIEW

1. Evaluating a function (Section 1.3, page 81)

OBJECTIVES

1. Learn the Division Algorithm.
2. Use the Remainder and Factor Theorems.
3. Use the Rational Zeros Test.

◆ Greenhouse Effect and Global Warming

The **greenhouse effect** refers to the fact that Earth is surrounded by an atmosphere that acts like the transparent cover of a greenhouse, allowing sunlight to filter through while trapping heat. It is becoming increasingly clear that we are experiencing *global warming*. This means that Earth's atmosphere is becoming warmer near its surface. Scientists who study climate agree that global warming is due largely to the emission of "greenhouse gases," such as carbon dioxide (CO_2), chlorofluorocarbons (CFCs), methane (CH_4), ozone (O_3), and nitrous oxide (N_2O). The temperature of Earth's surface increased by 1°F over the twentieth century. The late 1990s and early 2000s were some of the hottest years ever recorded. Projections of future warming suggest a global temperature increase of between 2.5°F and 10.4°F by 2100.

Global warming will result in long-term changes in climate, including a rise in average temperatures, unusual rainfalls, and more storms and floods. The sea level may rise so much that people will have to move away from coastal areas. Some regions of the world may become too dry for farming.

The production and consumption of energy is one of the major causes of greenhouse gas emissions. In Example 5, we look at the pattern for the consumption of petroleum in the United States.

---

1 Learn the Division Algorithm.

The Division Algorithm

Dividing polynomials is similar to dividing integers. The fact that 5 divides 10 evenly is expressed as $\dfrac{10}{5} = 2$ or as $10 = 5 \cdot 2$. By comparison, in equation

RECALL

A polynomial $P(x)$ is an expression of the form

$a_n x^n + \cdots + a_1 x + a_0$.

(1) $x^3 - 1 = (x - 1)(x^2 + x + 1)$ Factor, using difference of cubes.

Dividing both sides by $x - 1$, we obtain

$$\dfrac{x^3 - 1}{x - 1} = x^2 + x + 1, \quad x \neq 1.$$

SIDE NOTE

It is helpful to compare the division process for integers with that of polynomials. Notice that

$$\begin{array}{r} 2 \\ 3\overline{)7} \\ \underline{6} \\ 1 \end{array}$$

can also be written as

$$7 = 3 \cdot 2 + 1$$

or

$$\frac{7}{3} = 2 + \frac{1}{3}.$$

DO YOU KNOW?

An *algorithm* is a finite sequence of precise steps that lead to a solution of a problem.

Polynomial Factor

A polynomial $D(x)$ is a **factor** of a polynomial $F(x)$ if there is a polynomial $Q(x)$ such that $F(x) = D(x) \cdot Q(x)$.

Next, consider the equation

(2) $x^3 - 2 = (x - 1)(x^2 + x + 1) - 1$ Subtract 1 from both sides of equation (1).

Dividing both sides by $(x - 1)$, we can rewrite equation (2) in the form

$$\frac{x^3 - 2}{x - 1} = x^2 + x + 1 - \frac{1}{x - 1}, \quad x \neq 1.$$

Here we say that when the **dividend** $x^3 - 2$ is divided by the **divisor** $x - 1$, the **quotient** is $x^2 + x + 1$ and the **remainder** is -1. This result is an example of the *division algorithm*.

THE DIVISION ALGORITHM

If a polynomial $F(x)$ is divided by a polynomial $D(x)$ and $D(x)$ is not the zero polynomial, then there are unique polynomials $Q(x)$ and $R(x)$ such that

$$\frac{F(x)}{D(x)} = Q(x) + \frac{R(x)}{D(x)}$$

or

$$\underbrace{F(x)}_{\text{Dividend}} = \underbrace{D(x)}_{\text{Divisor}} \cdot \underbrace{Q(x)}_{\text{Quotient}} + \underbrace{R(x)}_{\text{Remainder}}$$

Either $R(x)$ is the *zero polynomial* or the degree of $R(x)$ is less than the degree of $D(x)$.

The division algorithm says that the dividend is equal to the divisor times the quotient plus the remainder. The expression $\dfrac{F(x)}{D(x)}$ is **improper** if

degree of $F(x) \geq$ degree of $D(x)$.

However, the expression $\dfrac{R(x)}{D(x)}$ is always **proper** because by the division algorithm, degree of $R(x) <$ degree of $D(x)$.

For an improper expression $\dfrac{F(x)}{D(x)}$, we use the **long division** or **synthetic division** processes to find the quotient and remainder.

EXAMPLE 1 Using Long Division and Synthetic Division

Use both long division and synthetic division to find the quotient and remainder when $2x^4 + x^3 - 16x^2 + 18$ is divided by $x + 2$.

Solution

Long Division

$$\begin{array}{r} 2x^3 - 3x^2 - 10x + 20 \\ x + 2 \overline{\smash{\big)}\, 2x^4 + x^3 - 16x^2 + 0x + 18} \\ \underline{2x^4 + 4x^3 } \\ -3x^3 - 16x^2 + 0x + 18 \\ \underline{-3x^3 - 6x^2 } \\ -10x^2 + 0x + 18 \\ \underline{-10x^2 - 20x } \\ 20x + 18 \\ \underline{20x + 40} \\ -22 \end{array}$$

Synthetic Division

$$\begin{array}{r|rrrrr} -2 & 2 & 1 & -16 & 0 & 18 \\ & & -4 & 6 & 20 & -40 \\ \hline & 2 & -3 & -10 & 20 & \mid -22 \end{array}$$

The quotient is $2x^3 - 3x^2 - 10x + 20$, and the remainder is -22. So the result is

$$\frac{2x^4 + x^3 - 16x^2 + 18}{x + 2} = 2x^3 - 3x^2 - 10x + 20 + \frac{-22}{x + 2}$$

$$= 2x^3 - 3x^2 - 10x + 20 - \frac{22}{x + 2}.$$

Practice Problem 1 Use synthetic division to find the quotient and remainder when $2x^3 + x^2 - 18x - 7$ is divided by $x - 3$.

2 Use the Remainder and Factor Theorems.

The Remainder and Factor Theorems

In Example 1, we discovered that when the polynomial

$$F(x) = 2x^4 + x^3 - 16x^2 + 18 \text{ is divided by } x + 2,$$

the remainder is -22, a constant.

Now evaluate F at -2. We have

$F(-2) = 2(-2)^4 + (-2)^3 - 16(-2)^2 + 18$ Replace x with -2 in $F(x)$.
$= 2(16) - 8 - 16(4) + 18 = -22$ Simplify.

We see that $F(-2) = -22$, the remainder we found when we divided $F(x)$ by $x + 2$. The following theorem shows that this result is not a coincidence.

THE REMAINDER THEOREM

If a polynomial $F(x)$ is divided by $x - a$, then the remainder R is given by

$$R = F(a).$$

Note that in the statement of the Remainder Theorem, $F(x)$ plays the dual role of representing a polynomial and the function defined by the same polynomial.

The Remainder Theorem has a very simple proof: the division algorithm says that if a polynomial $F(x)$ is divided by a polynomial $D(x) \neq 0$, then

$$F(x) = D(x) \cdot Q(x) + R(x),$$

where $Q(x)$ is the quotient and $R(x)$ is the remainder. Suppose $D(x)$ is a linear polynomial of the form $x - a$. Because if $R(x) \neq 0$, its degree is less than the degree of $D(x)$, $R(x)$ is a constant. Call this constant R. In this case, by the division algorithm, we have

$$F(x) = (x - a)Q(x) + R.$$

Replacing x with a in $F(x)$, we obtain

$$F(a) = (a - a)Q(a) + R = 0 + R = R.$$

This proves the Remainder Theorem.

EXAMPLE 2 Using the Remainder Theorem

Find the remainder when the polynomial

$$F(x) = 2x^5 - 4x^3 + 5x^2 - 7x + 2$$

is divided by $x - 1$.

Solution

We could find the remainder by using long division or synthetic division, but a quicker way is to evaluate $F(x)$ when $x = 1$. By the Remainder Theorem, $F(1)$ is the remainder.

$$F(1) = 2(1)^5 - 4(1)^3 + 5(1)^2 - 7(1) + 2 \quad \text{Replace } x \text{ with 1 in } F(x).$$
$$= 2 - 4 + 5 - 7 + 2 = -2$$

The remainder is -2.

Practice Problem 2 Use the Remainder Theorem to find the remainder when

$$F(x) = x^{110} - 2x^{57} + 5 \text{ is divided by } x - 1.$$

Sometimes the Remainder Theorem is used in the other direction: to evaluate a polynomial at a specific value of x. This use is illustrated in the next example.

EXAMPLE 3 Using the Remainder Theorem

Let $f(x) = x^4 + 3x^3 - 5x^2 + 8x + 75$. Find $f(-3)$.

Solution

One way of solving this problem is to evaluate $f(x)$ when $x = -3$.

$$f(-3) = (-3)^4 + 3(-3)^3 - 5(-3)^2 + 8(-3) + 75 \quad \text{Replace } x \text{ with } -3.$$
$$= 6 \quad \text{Simplify.}$$

Another way is to use synthetic division to find the remainder when the polynomial $f(x)$ is divided by $(x + 3)$.

```
-3 | 1    3   -5    8    75
   |     -3    0   15   -69
   ------------------------
     1    0   -5   23   |6
```

The remainder is 6. By the Remainder Theorem, $f(-3) = 6$.

Practice Problem 3 Use synthetic division to find $f(-2)$ where

$$f(x) = x^4 + 10x^2 + 2x - 20.$$

Note that if we divide any polynomial $F(x)$ by $x - a$, we have

$$F(x) = (x - a)Q(x) + R \qquad \text{Division algorithm}$$
$$F(x) = (x - a)Q(x) + F(a) \qquad R = F(a) \text{ by the Remainder Theorem}$$

The last equation says that if $F(a) = 0$, then

$$F(x) = (x - a)Q(x).$$

so, $(x - a)$ is a factor of $F(x)$. Conversely, if $(x - a)$ is a factor of $F(x)$, then dividing $F(x)$ by $(x - a)$ gives a remainder of 0. Therefore, by the Remainder Theorem, $F(a) = 0$. This argument proves the Factor Theorem.

> **THE FACTOR THEOREM**
>
> A polynomial $F(x)$ has $(x - a)$ as a factor if and only if $F(a) = 0$.

The Factor Theorem shows that if $F(x)$ is a polynomial, then the following problems are equivalent:

1. Factoring $F(x)$ and then applying the Zero-Product property
2. Finding the zeros of the function $F(x)$ defined by the polynomial expression
3. Solving (or finding the roots of) the polynomial equation $F(x) = 0$

Note that if a is a zero of the polynomial function $F(x)$, then $F(x) = (x - a)Q(x)$. Any solution of the equation $Q(x) = 0$ is also a zero of $F(x)$. Because the degree of $Q(x)$ is less than the degree of $F(x)$, it is simpler to solve the equation $Q(x) = 0$ to find other possible zeros of $F(x)$. The equation $Q(x) = 0$ is called the **depressed equation** of $F(x)$.

The next example illustrates how to use the Factor Theorem and the depressed equation to solve a polynomial equation.

EXAMPLE 4   **Using the Factor Theorem**

Given that 2 is a zero of the function $f(x) = 3x^3 + 2x^2 - 19x + 6$, solve the polynomial equation $3x^3 + 2x^2 - 19x + 6 = 0$.

Solution

Because 2 is a zero of $f(x)$, we have $f(2) = 0$. The Factor Theorem tells us that $(x - 2)$ is a factor of $f(x)$. Next, we use synthetic division to divide $f(x)$ by $(x - 2)$.

$$\begin{array}{r|rrrr} 2 & 3 & 2 & -19 & 6 \\ & & 6 & 16 & -6 \\ \hline & 3 & 8 & -3 & 0 \end{array} \longleftarrow \text{Remainder}$$

$\underbrace{}_{\text{coefficients of the quotient}}$

The coefficients of the quotient give the depressed equation, $3x^2 + 8x - 3 = 0$. Because the remainder is 0, we have:

$$f(x) = 3x^3 + 2x^2 - 19x + 6 = (x - 2)\underbrace{(3x^2 + 8x - 3)}_{\text{quotient}}.$$

Any solution of the depressed equation $3x^2 + 8x - 3 = 0$ is a zero of f. Because this equation is of degree 2, any method of solving a quadratic equation may be used to solve it.

$$\begin{aligned} 3x^2 + 8x - 3 &= 0 && \text{Depressed equation} \\ (3x - 1)(x + 3) &= 0 && \text{Factor.} \\ 3x - 1 = 0 \text{ or } x + 3 &= 0 && \text{Zero-product property} \\ x = \frac{1}{3} \text{ or } x &= -3 && \text{Solve each equation for } x. \end{aligned}$$

The solution set is $\left\{-3, \frac{1}{3}, 2\right\}$.

Practice Problem 4 Solve the equation $3x^3 - x^2 - 20x - 12 = 0$ given that one solution is -2.

◆ **Energy Units** Energy is measured in British thermal units (BTU). One BTU is the amount of energy required to raise the temperature of 1 pound of water 1°F at or near 39.2°F; 1 million BTU is equivalent to 8 gallons of gasoline. Fuels such as coal, petroleum, and natural gas are measured in quadrillion BTU (1 quadrillion = 10^{15}), abbreviated "quads."

EXAMPLE 5   Petroleum Consumption

The combined petroleum consumption C (in quads) by a group of the most economically developed countries in the world during 1996–2011 can be modeled by the function

$$C(x) = 0.021x^3 - 0.973x^2 + 11.347x + 202.965,$$

where $x = 0$ represents 1996, $x = 1$ represents 1997, and so on.

The model indicates that $C(5) = 238.0$ quads of petroleum were consumed in 2001. Find another year between 2001 and 2011 when the model indicates that 238.0 quads of petroleum were consumed.

Solution

We are given that $C(5) = 238.0$. Therefore, $x = 5$ is a zero of

$$F(x) = C(x) - 238.0 = 0.021x^3 - 0.973x^2 + 11.347x - 35.035$$

so that $F(x) = (x - 5)Q(x)$.

Finding another year between 2001 and 2011 when the petroleum consumption was 238.0 quads requires us to find another zero of $F(x)$ that is between $x = 5$ and $x = 15$. Dividing $F(x)$ by $(x - 5)$ using, say, synthetic division yields the quotient

$$Q(x) = 0.021x^2 - 0.868x + 7.007.$$

Solving the depressed equation $Q(x) = 0$ by the quadratic formula, we obtain

$$x = \frac{0.868 \pm \sqrt{(-0.868)^2 - 4(0.021)(7.007)}}{2(0.021)} = 11.0 \quad \text{or} \quad 30.\overline{3}.$$

Because x needs to be between 5 and 15, this model tells us that in 2007 ($x = 11$), the combined petroleum consumption by the most economically developed group of countries was also 238.0 quads.

Practice Problem 5 The value of $C(x) = 0.23x^3 - 4.255x^2 + 0.345x + 41.05$ is 10 when $x = 3$. Find another positive number x for which $C(x) = 10$.

3 Use the Rational Zeros Test.

The Rational Zeros Test

We use the Factor Theorem to determine whether a number a is a zero of a polynomial function $F(x)$. We use the following test (see Exercise 62 for a proof) to find *possible* rational zeros of a polynomial function with integer coefficients.

THE RATIONAL ZEROS TEST

If $F(x) = a_n x^n + a_{n-1} x^{n-1} + \cdots + a_2 x^2 + a_1 x + a_0$ is a polynomial function with integer coefficients ($a_n \neq 0, a_0 \neq 0$) and $\dfrac{p}{q}$ is a rational number in lowest terms that is a zero of $F(x)$, then

1. p is a factor of the constant term a_0.
2. q is a factor of the leading coefficient a_n.

EXAMPLE 6 Using the Rational Zeros Test

Find all rational zeros of $F(x) = 2x^3 + 5x^2 - 4x - 3$.

Solution

First, we list all *possible* rational zeros of $F(x)$.

$$\text{Possible rational zeros} = \frac{p}{q} = \frac{\text{Factors of the constant term, } -3}{\text{Factors of the leading coefficient, } 2}$$

Factors of -3: $\pm 1, \pm 3$

Factors of 2: $\pm 1, \pm 2$

All combinations can be found by using just the *positive* factors of the leading coefficient. The possible rational zeros are

$$\frac{\pm 1}{1}, \frac{\pm 1}{2}, \frac{\pm 3}{1}, \frac{\pm 3}{2}$$

or

$$\pm 1, \pm \frac{1}{2}, \pm 3, \pm \frac{3}{2}.$$

We use synthetic division to determine whether a rational zero exists among these eight candidates. We start by testing 1. If 1 is not a rational zero, we will test other possibilities.

$$\begin{array}{r|rrrr} 1 & 2 & 5 & -4 & -3 \\ & & 2 & 7 & 3 \\ \hline & 2 & 7 & 3 & 0 \end{array}$$

coefficients of the depressed equation

The zero remainder tells us that $x = 1$ is a zero of $F(x)$ and $(x - 1)$ is a factor of $F(x)$, with the other factor being $2x^2 + 7x + 3$. Solving the depressed equation

$$2x^2 + 7x + 3 = 0,$$

we have

$(2x + 1)(x + 3) = 0$ Factor.

$2x + 1 = 0$ or $x + 3 = 0$ Zero-product property

$x = -\dfrac{1}{2}$ or $x = -3$ Solve each equation for x.

The solution set is $\left\{1, -\dfrac{1}{2}, -3\right\}$. The rational zeros of F are $-3, -\dfrac{1}{2}$, and 1.

Practice Problem 6 Find all rational zeros of $F(x) = 2x^3 + 3x^2 - 6x - 8$.

SECTION 2.3 Exercises

Basic Concepts and Skills

In Exercises 1–10, use synthetic division to find the quotient and the remainder when the first polynomial is divided by the second polynomial.

1. $x^3 - x^2 - 7x + 2$; $x - 1$
2. $2x^3 - 3x^2 - x + 2$; $x + 2$
3. $x^3 + 4x^2 - 7x - 10$; $x + 2$
4. $x^3 + x^2 - 13x + 2$; $x - 3$
5. $x^4 - 3x^3 + 2x^2 + 4x + 5$; $x - 2$
6. $x^4 - 5x^3 - 3x^2 + 10$; $x - 1$

7. $2x^3 + 4x^2 - 3x + 1$; $x - \dfrac{1}{2}$

8. $3x^3 + 8x^2 + x + 1$; $x - \dfrac{1}{3}$

9. $2x^3 - 5x^2 + 3x + 2$; $x + \dfrac{1}{2}$

10. $3x^3 - 2x^2 + 8x + 2$; $x + \dfrac{1}{3}$

In Exercises 11–14, use synthetic division and the Remainder Theorem to find each function value. Check your answer by evaluating the function at the given x-value.

11. $f(x) = x^3 + 3x^2 + 1$
    a. $f(1)$ b. $f(-1)$
    c. $f\left(\dfrac{1}{2}\right)$ d. $f(10)$

12. $g(x) = 2x^3 - 3x^2 + 1$
    a. $g(-2)$ b. $g(-1)$
    c. $g\left(-\dfrac{1}{2}\right)$ d. $g(7)$

13. $h(x) = x^4 + 5x^3 - 3x^2 - 20$
    a. $h(1)$ b. $h(-1)$
    c. $h(-2)$ d. $h(2)$

14. $f(x) = x^4 + 0.5x^3 - 0.3x^2 - 20$
    a. $f(0.1)$ b. $f(0.5)$
    c. $f(1.7)$ d. $f(-2.3)$

In Exercises 15–22, use the Factor Theorem to show that the first polynomial is a factor of the second polynomial. Check your answer by synthetic division.

15. $x - 1$; $2x^3 + 3x^2 - 6x + 1$
16. $x - 3$; $3x^3 - 9x^2 - 4x + 12$
17. $x + 1$; $5x^4 + 8x^3 + x^2 + 2x + 4$
18. $x + 3$; $3x^4 + 9x^3 - 4x^2 - 9x + 9$
19. $x - 2$; $x^4 + x^3 - x^2 - x - 18$
20. $x + 3$; $x^5 + 3x^4 + x^2 + 8x + 15$
21. $x + 2$; $x^6 - x^5 - 7x^4 + x^3 + 8x^2 + 5x + 2$
22. $x - 2$; $2x^6 - 5x^5 + 4x^4 + x^3 - 7x^2 - 7x + 2$

In Exercises 23–26, find the value of k for which the first polynomial is a factor of the second polynomial.

23. $x + 1$; $x^3 + 3x^2 + x + k$ [*Hint:* Let $f(x) = x^3 + 3x^2 + x + k$. Then $f(-1) = 0$.]
24. $x - 1$; $-x^3 + 4x^2 + kx - 2$
25. $x - 2$; $2x^3 + kx^2 - kx - 2$
26. $x - 1$; $k^2 - 3kx^2 - 2kx + 6$

In Exercises 27–30, use the Factor Theorem to show that the first polynomial is not a factor of the second polynomial. [*Hint:* Show that the remainder is not zero.]

27. $x - 2$; $-2x^3 + 4x^2 - 4x + 9$
28. $x + 3$; $-3x^3 - 9x^2 + 5x + 12$
29. $x + 2$; $4x^4 + 9x^3 + 3x^2 + x + 4$
30. $x - 3$; $3x^4 - 8x^3 + 5x^2 + 7x - 3$

In Exercises 31–34, the graph of a polynomial function f is given. Assume that when f is completely factored, each zero, c, corresponds to a factor of the form $(x - c)^m$. Find the equation of least degree for f. [*Hint:* Use the y-intercept to find the leading coefficient.]

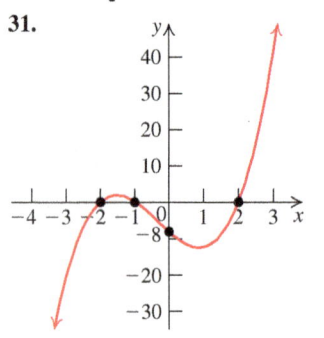

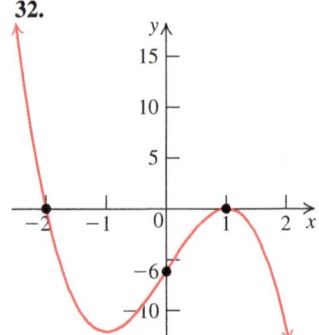

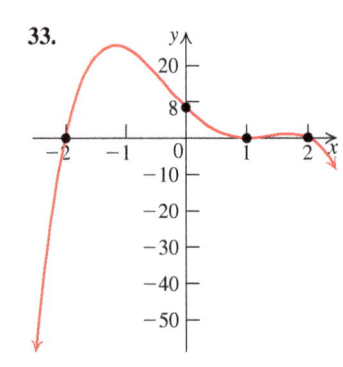

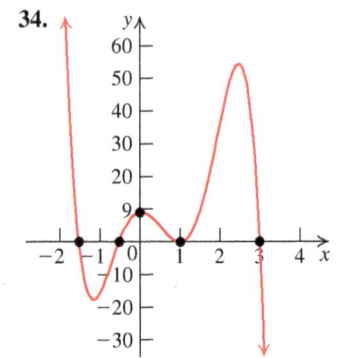

31.

32.

33.

34.

In Exercises 35–38, find the set of possible rational zeros of the given function.

35. $f(x) = 3x^3 - 4x^2 + 5$
36. $g(x) = 2x^4 - 5x^2 - 2x + 1$
37. $h(x) = 4x^4 - 9x^2 + x + 6$
38. $F(x) = 6x^6 + 5x^5 + x - 35$

In Exercises 39–50, find all rational zeros of the given polynomial function.

39. $f(x) = x^3 - x^2 - 4x + 4$
40. $f(x) = x^3 - 4x^2 + x + 6$
41. $g(x) = 3x^3 - 2x^2 + 3x - 2$
42. $g(x) = 6x^3 + 13x^2 + x - 2$
43. $g(x) = 2x^3 + 3x^2 + 4x + 6$
44. $h(x) = 3x^3 + 7x^2 + 8x + 2$
45. $h(x) = x^4 - x^3 - x^2 - x - 2$
46. $h(x) = 2x^4 + 3x^3 + 8x^2 + 3x - 4$
47. $F(x) = x^4 - x^3 - 13x^2 + x + 12$
48. $G(x) = 3x^4 + 5x^3 + x^2 + 5x - 2$
49. $F(x) = x^4 - 2x^3 + 10x^2 - x + 1$
50. $G(x) = x^6 + 2x^4 + x^2 + 2$

In Exercises 51–54, find all rational zeros of f; then use the depressed equation (see Example 4) to find all roots of the equation $f(x) = 0$.

51. $f(x) = x^3 + 5x^2 - 8x + 2$
52. $f(x) = x^3 - 7x^2 - 5x + 3$
53. $f(x) = x^4 - 3x^3 + 3x - 1$
54. $f(x) = x^4 - 6x^3 - 7x^2 + 54x - 18$

Applying the Concepts

55. **Geometry.** The area of a rectangle is $(2x^4 - 2x^3 + 5x^2 - x + 2)$ square centimeters. Its length is $(x^2 - x + 2)$ centimeters. Find its width.

56. **Geometry.** The volume of a rectangular solid is $(x^4 + 3x^3 + x + 3)$ cubic inches. Its length and width are $(x + 3)$ and $(x + 1)$ inches, respectively. Find its height.

57. **Geometry.** Find a point on the parabola $y = x^2$ whose distance from $(18, 0)$ is $4\sqrt{17}$.

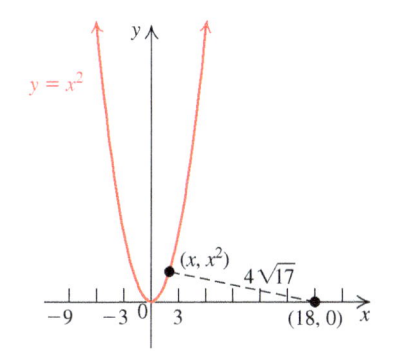

58. **Making a box.** A square piece of tin 18 inches on each side is to be made into a box, without a top, by cutting a square from each corner and folding up the flaps to form the sides. What size corners should be cut so that the volume of the box is 432 cubic inches?

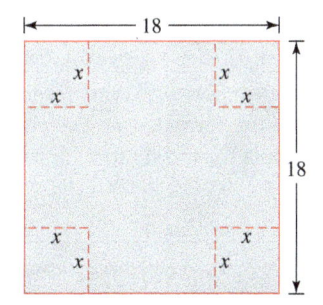

Critical Thinking / Discussion / Writing

59. Identify each statement as true or false. Explain your answer.
    a. $\dfrac{1}{3}$ is a possible zero of $P(x) = x^5 - 3x^2 + x + 3$.
    b. $-\dfrac{2}{5}$ is a possible zero of $P(x) = 2x^7 - 5x^4 - 17x + 25$.

60. If n is a positive integer, find the values of n for which each of the following is true.
    a. $x + a$ is a factor of $x^n + a^n$.
    b. $x + a$ is a factor of $x^n - a^n$.
    c. $x - a$ is a factor of $x^n + a^n$.
    d. $x - a$ is a factor of $x^n - a^n$.

61. Use Exercise 60 to prove the following:
    a. $7^{11} - 2^{11}$ is exactly divisible by 5.
    b. $(19)^{20} - (10)^{20}$ is exactly divisible by 261.

62. To prove the Rational Zeros Test, we assume the **Fundamental Theorem of Arithmetic**, which states that "Every integer has a unique prime factorization." Supply reasons for each step in the following proof:

    (i) $F\left(\dfrac{p}{q}\right) = 0$ and $\dfrac{p}{q}$ is in lowest terms.
    (ii) $a_n\left(\dfrac{p}{q}\right)^n + a_{n-1}\left(\dfrac{p}{q}\right)^{n-1} + \cdots + a_1\left(\dfrac{p}{q}\right) + a_0 = 0$
    (iii) $a_n p^n + a_{n-1}p^{n-1}q + \cdots + a_1 pq^{n-1} + a_0 q^n = 0$
    (iv) $a_n p^n + a_{n-1}p^{n-1}q + \cdots + a_1 pq^{n-1} = -a_0 q^n$
    (v) p is a factor of the left side of (iv).
    (vi) p is a factor of $-a_0 q^n$.
    (vii) p is a factor of a_0.
    (viii) $a_{n-1}p^{n-1}q + \cdots + a_1 pq^{n-1} + a_0 q^n = -a_n p^n$
    (ix) q is a factor of the left side of the equation in (viii).
    (x) q is a factor of $-a_n p^n$.
    (xi) q is a factor of a_n.

63. Find all rational roots of the equation
$$2x^4 + 2x^3 + \dfrac{1}{2}x^2 + 2x - \dfrac{3}{2} = 0.$$

[*Hint:* Multiply both sides of the equation by the lowest common denominator.]

Maintaining Skills

In Exercises 64–67, use synthetic division to find each quotient and remainder.

64. $x^4 - x^3 + 2x^2 + x - 3$ is divided by $x - 1$.
65. $x^3 + x^2 - x + 2$ is divided by $x + 2$.
66. $-x^6 + 5x^5 + 2x^2 - 10x + 3$ is divided by $x - 5$.
67. $x^4 + 3x^3 + 3x^2 + 2x + 7$ is divided by $x + 1$.

In Exercises 68–71, find all of the factors of each integer.

68. 11
69. 18
70. 51
71. 72

SECTION 2.4

Zeros of a Polynomial Function

BEFORE STARTING THIS SECTION, REVIEW

1. Rational zeros test (Section 2.3, page 184)
2. Complex numbers (Section P.5, page 45)
3. Conjugate of a complex number (Section P.5, page 47)

OBJECTIVES

1. Use Descartes's Rule of Signs to find the possible number of positive and negative zeros of polynomials.
2. Find the bounds on the real zeros of polynomials.
3. Learn basic facts about the complex zeros of polynomials.
4. Use the Conjugate Pairs Theorem to find zeros of polynomials.

Gerolamo Cardano (1501–1576)

Cardano was trained as a physician, but his range of interests included philosophy, music, gambling, mathematics, astrology, and the occult sciences. Cardano was a colorful man, full of contradictions. His lifestyle was deplorable, and his personal life was filled with tragedies. His wife died in 1546. He saw one son executed for wife poisoning. He cropped the ears of a second son who attempted the same crime. In 1570, he was imprisoned for six months for heresy when he published the horoscope of Jesus. Cardano spent the last few years in Rome, where he wrote his autobiography, *De Propria Vita*, in which he did not spare the most shameful revelations.

◆ Cardano Accused of Stealing Formula

In sixteenth-century Italy, getting and keeping a teaching job at a university was difficult. University appointments were mostly temporary. One way a professor could convince the administration that he was worthy of continuing in his position was by winning public challenges. Two contenders for a position would present each other with a list of problems, and later in a public forum, each person would present his solutions to the other's problems. As a result, scholars often kept secret their new methods of solving problems.

In 1535, in a public contest primarily about solving the cubic equation, Nicolo Tartaglia (1499–1557) won over his opponent Antonio Maria Fiore. Gerolamo Cardano (1501–1576), in the hope of learning the secret of solving a cubic equation, invited Tartaglia to visit him. After many promises and much flattery, Tartaglia revealed his method on the promise, probably given under oath, that Cardano would keep it confidential. When Cardano's book *Ars Magna* (The Great Art) appeared in 1545, the formula and the method of proof were fully disclosed. Angered at this apparent breach of a solemn oath and feeling cheated out of the rewards of his monumental work, Tartaglia accused Cardano of being a liar and a thief for stealing his formula. Thus began the bitterest feud in the history of mathematics, carried on with name-calling and mudslinging of the lowest order. As Tartaglia feared, the formula (see Exercises 66 and 67) has forever since been known as Cardano's formula for the solution of the cubic equation.

---

1 Use Descartes's Rule of Signs to find the possible number of positive and negative zeros of polynomials.

Descartes's Rule of Signs

In Section 2.3, we used the Rational Zeros Test to find the possible rational zeros of a polynomial function with integer coefficients. We now investigate briefly other useful tests for the real zeros of polynomial functions. One such test is called *Descartes's Rule of Signs*.

If a polynomial (with nonzero coefficients) is written in descending power order, we say that a **variation of sign** occurs when the signs of two consecutive terms differ. For example, in the polynomial $2x^5 - 5x^3 - 6x^2 + 7x + 3$, the signs of the terms are $+\ -\ -\ +\ +$. Therefore, there are two variations of sign, as follows:

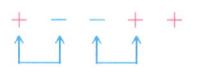

188 Chapter 2 Polynomial and Rational Functions

There are three variations of sign in
$$x^5 - 3x^3 + 2x^2 + 5x - 7.$$

DESCARTES'S RULE OF SIGNS

Let $F(x)$ be a polynomial function with real (nonzero) coefficients and with terms written in descending power order.

1. The number of positive zeros of F is equal to the number of variations of sign of $F(x)$ or is less than that number by an even integer.
2. The number of negative zeros of F is equal to the number of variations of sign of $F(-x)$ or is less than that number by an even integer.

When using Descartes's Rule, a zero of multiplicity m should be counted as m zeros.

EXAMPLE 1 Using Descartes's Rule of Signs

Find the possible number of positive and negative zeros of
$$f(x) = x^5 - 3x^3 - 5x^2 + 9x - 7.$$

Solution
There are three variations of sign in $f(x)$:
$$f(x) = x^5 - 3x^3 - 5x^2 + 9x - 7$$

The number of positive zeros of $f(x)$ must be either 3 or $3 - 2 = 1$. Next,
$$f(-x) = (-x)^5 - 3(-x)^3 - 5(-x)^2 + 9(-x) - 7$$
$$= -x^5 + 3x^3 - 5x^2 - 9x - 7.$$

The polynomial $f(-x)$ has two variations of sign; therefore, the number of negative zeros of $f(x)$ is either 2 or $2 - 2 = 0$.

Practice Problem 1 Find the possible number of positive and negative zeros of
$$f(x) = 2x^5 + 3x^2 + 5x - 1.$$

Bounds on the Real Zeros

2 Find the bounds on the real zeros of polynomials.

If a polynomial function $F(x)$ has no zeros greater than a number k, then k is called an **upper bound** on the zeros of $F(x)$. Similarly, if $F(x)$ has no zeros less than a number k, then k is called a **lower bound** on the zeros of $F(x)$. These bounds have the following rules.

RULES FOR BOUNDS

Let $F(x)$ be a polynomial function with real coefficients and a positive leading coefficient. Suppose $F(x)$ is synthetically divided by $x - k$.

1. If $k > 0$ and each number in the last row is zero or positive, then k is an upper bound on the zeros of $F(x)$.
2. If $k < 0$ and numbers in the last row alternate in sign (zeros in the last row can be regarded as positive or negative), then k is a lower bound on the zeros of $F(x)$.

EXAMPLE 2 Finding the Bounds on the Zeros

Find upper and lower bounds on the zeros of
$$F(x) = x^4 - x^3 - 5x^2 - x - 6.$$

Solution

By the Rational Zeros Test, the possible rational zeros of $F(x)$ are $\pm 1, \pm 2, \pm 3,$ and ± 6. There is only one variation of sign in $F(x) = x^4 - x^3 - 5x^2 - x - 6$, so $F(x)$ has only one positive zero. First, we try synthetic division by $x - k$ with $k = 1, 2, 3, \ldots$. The first integer that makes all numbers in the last row 0 or positive is an upper bound on the zeros of $F(x)$.

```
1 | 1  -1  -5  -1  -6      With k = 1        2 | 1  -1  -5  -1  -6      With k = 2
  |      1   0  -5  -6                         |      2   2  -6 -14
  | 1   0  -5  -6 -12                          | 1   1  -3  -7 -20
```

```
       3 | 1  -1  -5  -1  -6      Synthetic division with k = 3
         |      3   6   3   6
         | 1   2   1   2   0  ←── All numbers are 0 or positive.
```

By the rule for bounds, 3 is an upper bound on the zeros of $F(x)$.

We now try synthetic division by $x - k$ with $k = -1, -2, -3, \ldots$. The first negative integer for which the numbers in the last row alternate in sign is a lower bound on the zeros of $F(x)$.

```
-1 | 1  -1  -5  -1  -6      With k = -1
   |     -1   2   3  -2
   | 1  -2  -3   2  -8
```

```
-2 | 1  -1  -5  -1  -6      With k = -2
   |     -2   6  -2   6
   | 1  -3   1  -3   0  ←── These numbers alternate in
                            sign if we count 0 as positive.
```

By the rule of bounds, -2 is a lower bound on the zeros of $F(x)$.

Practice Problem 2 Find upper and lower bounds on the zeros of
$$f(x) = x^3 - x^2 - 4x + 4.$$

SIDE NOTE

A graphing calculator may be used to obtain the approximate location of the real zeros. For a rational zero, verify by synthetic division.

STEPS FOR GATHERING INFORMATION AND LOCATING THE REAL ZEROS OF A POLYNOMIAL FUNCTION

Step 1 Find the maximum number of real zeros by using the degree of the polynomial function.

Step 2 Find the possible number of positive and negative zeros by using Descartes's Rule of Signs.

Step 3 Write the set of possible rational zeros.

Step 4 Test the smallest positive integer in the set in Step 3, then the next larger, and so on, until an integer zero or an upper bound of the zeros is found.
   a. If a zero is found, use the function represented by the depressed equation in further calculations.
   b. If an upper bound is found, discard all larger numbers in the set of possible rational zeros.

Step 5 Test the positive fractions that remain in the set in Step 3 after considering any bound that has been found.

Step 6 Use modified Steps 4 and 5 for negative numbers in the set in Step 3.

Step 7 If a depressed equation is quadratic, use any method (including the quadratic formula) to solve this equation.

3 Learn basic facts about the complex zeros of polynomials.

Complex Zeros of Polynomials

The Factor Theorem connects the concept of *factors* and *zeros* of a polynomial, and we continue to investigate this connection.

Because the equation $x^2 + 1 = 0$ has no real roots, the polynomial function $P(x) = x^2 + 1$ has no real zeros. However, if we replace x with the complex number i, we have

$$P(i) = i^2 + 1 = (-1) + 1 = 0.$$

Consequently, i is a complex number for which $P(i) = 0$; that is, i is a *complex zero* of $P(x)$. In Section 2.2, we stated that a polynomial of degree n can have, at most, n real zeros. We now extend our number system to allow the coefficients of polynomials and variables to represent complex numbers. When we want to emphasize that complex numbers may be used in this way, we call the polynomial $P(x)$ a **complex polynomial**. If $P(z) = 0$ for a complex number z, we say that z is a **zero** or a **complex zero** of $P(x)$. In the complex number system, every nth-degree polynomial equation has *exactly* n roots and every nth-degree polynomial can be factored into exactly n linear factors. This fact follows from the Fundamental Theorem of Algebra.

> **RECALL**
>
> A number z of the form $z = a + bi$ is a **complex number**, where a and b are real numbers and $i = \sqrt{-1}$.

> **FUNDAMENTAL THEOREM OF ALGEBRA**
>
> Every polynomial
>
> $$P(x) = a_n x^n + a_{n-1} x^{n-1} + \cdots + a_1 x + a_0 \quad (n \geq 1, a_n \neq 0)$$
>
> with complex coefficients $a_n, a_{n-1}, \ldots, a_1, a_0$ has at least one complex zero.

This theorem was proved by the mathematician Karl Friedrich Gauss at age 20. The proof is beyond the scope of this book, but if we are allowed to use complex numbers, we can use the theorem to prove that every polynomial has a complete factorization.

> **RECALL**
>
> If $b = 0$, the complex number $a + bi$ is a real number. So the set of real numbers is a subset of the set of complex numbers.

> **FACTORIZATION THEOREM FOR POLYNOMIALS**
>
> If $P(x)$ is a complex polynomial of degree $n \geq 1$, it can be factored into n (not necessarily distinct) linear factors of the form
>
> $$P(x) = a(x - r_1)(x - r_2) \cdots (x - r_n),$$
>
> where $a, r_1, r_2, \ldots, r_n$ are complex numbers.

To prove the Factorization Theorem for Polynomials, we notice that if $P(x)$ is a polynomial of degree 1 or higher with leading coefficient a, then the Fundamental Theorem of Algebra states that there is a zero r_1 for which $P(r_1) = 0$. By the Factor Theorem, which also holds for complex polynomials, $(x - r_1)$ is a factor of $P(x)$ and

$$P(x) = (x - r_1) q_1(x),$$

where the degree of $q_1(x)$ is $n - 1$ and its leading coefficient is a. If $n - 1$ is positive, then we apply the Fundamental Theorem to $q_1(x)$; this gives us a zero, say, r_2, of $q_1(x)$. By the Factor Theorem, $(x - r_2)$ is a factor of $q_1(x)$ and

$$q_1(x) = (x - r_2) q_2(x),$$

where the degree of $q_2(x)$ is $n - 2$ and its leading coefficient is a. Then

$$P(x) = (x - r_1)(x - r_2) q_2(x).$$

This process can be continued until $P(x)$ is completely factored as

$$P(x) = a(x - r_1)(x - r_2) \cdots (x - r_n),$$

where a is the leading coefficient and $r_1, r_2, \ldots, r_n$ are zeros of $P(x)$. The proof is now complete.

In general, the numbers $r_1, r_2, \ldots, r_n$ may not be distinct. As with the polynomials we encountered previously, if the factor $(x - r)$ appears m times in the complete factorization of $P(x)$, then we say that r is a zero of *multiplicity m*. The polynomial $P(x) = (x - 2)^3(x - i)^2$ has zeros:

$$2 \text{ (of multiplicity 3) and } i \text{ (of multiplicity 2)}.$$

The polynomial $P(x) = (x - 2)^3(x - i)^2$ has only 2 and i as its zeros, although it is of degree 5. We have the following theorem:

NUMBER OF ZEROS THEOREM

Any polynomial of degree n has exactly n zeros, provided a zero of multiplicity k is counted k times.

EXAMPLE 3 Constructing a Polynomial Whose Zeros Are Given

Find a polynomial $P(x)$ of degree 4 with a leading coefficient of 2 and zeros $-1, 3, i$, and $-i$. Write $P(x)$

a. in completely factored form. **b.** by expanding the product found in part **a.**

Solution

a. Because $P(x)$ has degree 4, we write

$$P(x) = a(x - r_1)(x - r_2)(x - r_3)(x - r_4) \quad \text{Completely factored form}$$

Now replace the leading coefficient a with 2 and the zeros $r_1, r_2, r_3,$ and r_4 with $-1, 3, i,$ and $-i$ (in any order). Then

$$P(x) = 2[x - (-1)](x - 3)(x - i)[x - (-i)]$$
$$= 2(x + 1)(x - 3)(x - i)(x + i) \quad \text{Simplify.}$$

b. $P(x) = 2(x + 1)(x - 3)(x - i)(x + i)$
$ = 2(x + 1)(x - 3)(x^2 + 1) \quad$ Expand $(x - i)(x + i); i^2 = -1.$
$ = 2(x + 1)(x^3 - 3x^2 + x - 3) \quad$ Expand $(x - 3)(x^2 + 1).$
$ = 2(x^4 - 2x^3 - 2x^2 - 2x - 3) \quad$ Expand $(x + 1)(x^3 - 3x^2 + x - 3).$
$ = 2x^4 - 4x^3 - 4x^2 - 4x - 6 \quad$ Distributive property

Practice Problem 3 Find a polynomial $P(x)$ of degree 4 with a leading coefficient of 3 and zeros $-2, 1, 1 + i,$ and $1 - i$. Write $P(x)$

a. in completely factored form. **b.** by expanding the product found in part **a.**

4 Use the Conjugate Pairs Theorem to find zeros of polynomials.

RECALL

The conjugate of a complex number $z = a + bi$ is $\bar{z} = a - bi$. For two complex numbers z_1 and z_2, we have (Section P.5)

$$\overline{z_1 + z_2} = \bar{z_1} + \bar{z_2}$$
$$\overline{z_1 z_2} = \bar{z_1} \bar{z_2}.$$

Conjugate Pairs Theorem

In Example 3, we constructed a polynomial that had both i and its conjugate, $-i$, among its zeros. Our next result says that for all polynomials with real coefficients, nonreal zeros occur in complex conjugate pairs.

CONJUGATE PAIRS THEOREM

If $P(x)$ is a polynomial function whose coefficients are real numbers and if $z = a + bi$ is a zero of P, then its conjugate, $\bar{z} = a - bi$, is also a zero of P.

To prove the Conjugate Pairs Theorem, let

$$P(x) = a_n x^n + a_{n-1} x^{n-1} + \cdots + a_1 x + a_0,$$

where $a_n, a_{n-1}, \ldots, a_1, a_0$ are real numbers. Now suppose $P(z) = 0$. Then we must show that $P(\bar{z}) = 0$ also. We will use the fact that the conjugate of a real number $a = a + 0i$ is identical to the real number a (because $\bar{a} = a - 0i = a$) and then use the conjugate sum and product properties $\overline{z_1 + z_2} = \bar{z_1} + \bar{z_2}$, and $\overline{z_1 z_2} = \bar{z_1}\, \bar{z_2}$.

$$\begin{aligned}
P(z) &= a_n z^n + a_{n-1} z^{n-1} + \cdots + a_1 z + a_0 && \text{Replace } x \text{ with } z. \\
P(\bar{z}) &= a_n (\bar{z})^n + a_{n-1}(\bar{z})^{n-1} + \cdots + a_1 \bar{z} + a_0 && \text{Replace } z \text{ with } \bar{z}. \\
&= \overline{a_n}\, \overline{z^n} + \overline{a_{n-1}}\, \overline{z^{n-1}} + \cdots + \overline{a_1}\, \bar{z} + \overline{a_0} && (\bar{z})^k = \overline{z^k},\ \overline{a_k} = a_k \\
&= \overline{a_n z^n} + \overline{a_{n-1} z^{n-1}} + \cdots + \overline{a_1 z} + \overline{a_0} && \overline{z_1}\, \overline{z_2} = \overline{z_1 z_2} \\
&= \overline{a_n z^n + a_{n-1} z^{n-1} + \cdots + a_1 z + a_0} && \overline{z_1} + \overline{z_2} = \overline{z_1 + z_2} \\
&= \overline{P(z)} = \bar{0} = 0 && P(z) = 0
\end{aligned}$$

We see that if $P(z) = 0$, then $P(\bar{z}) = 0$, and the theorem is proved.

This theorem has several uses. If, for example, we know that $2 - 5i$ is a zero of a polynomial with real coefficients, then we know that $2 + 5i$ is also a zero. Because nonreal zeros occur in conjugate pairs, we know that there will always be an *even* number of nonreal zeros. Therefore, any polynomial of odd degree with real coefficients has at least one zero that is a real number.

> **ODD-DEGREE POLYNOMIALS WITH REAL ZEROS**
>
> Any polynomial $P(x)$ of odd degree with real coefficients must have at least one real zero.

For example, the polynomial $P(x) = 5x^7 + 2x^4 + 9x - 12$ has at least one zero that is a real number because $P(x)$ has degree 7 (odd degree) and has real numbers as coefficients.

EXAMPLE 4   **Using the Conjugate Pairs Theorem**

A polynomial $P(x)$ of degree 9 with real coefficients has the following zeros: 2, of multiplicity 3; $4 + 5i$, of multiplicity 2; and $3 - 7i$. First, determine the number of zeros; then write the zeros of $P(x)$.

Solution

Because complex zeros occur in conjugate pairs, the conjugate $4 - 5i$ of $4 + 5i$ is a zero of multiplicity 2 and the conjugate $3 + 7i$ of $3 - 7i$ is a zero of $P(x)$. Counting multiplicities, this gives nine zeros of $P(x)$:

$$2, 2, 2, 4 + 5i, 4 + 5i, 4 - 5i, 4 - 5i, 3 + 7i, \text{ and } 3 - 7i.$$

SIDE NOTE

In problems such as Example 4, once you determine all of the zeros, you can use the Factorization Theorem to write a polynomial with those zeros.

Practice Problem 4 A polynomial $P(x)$ of degree 8 with real coefficients has the following zeros: -3 and $2 - 3i$, each of multiplicity 2, and i. First, determine the number of zeros; then write the zeros of $P(x)$.

Every polynomial of degree n has exactly n zeros and can be factored into a product of n linear factors. If the polynomial has real coefficients, then, by the Conjugate Pairs Theorem, its nonreal zeros occur as conjugate pairs. Consequently, if $r = a + bi$, with $b \neq 0$, is a zero, then so is $\bar{r} = a - bi$, and both $x - r$ and $x - \bar{r}$ are linear factors of the polynomial. Multiplying these factors, we have

$$\begin{aligned}
(x - r)(x - \bar{r}) &= x^2 - (r + \bar{r})x + r\bar{r} \\
&= x^2 - 2ax + (a^2 + b^2) \quad r + \bar{r} = 2a;\ r\bar{r} = a^2 + b^2
\end{aligned}$$

194 Chapter 2 Polynomial and Rational Functions

The quadratic polynomial $x^2 - 2ax + (a^2 + b^2)$ has real coefficients and is *irreducible* (cannot be factored any further) *over the real numbers* because its discriminant $(-2a)^2 - 4(a^2 + b^2) = -4b^2 < 0$. This result shows that each pair of nonreal conjugate zeros are the zeros of one *irreducible* quadratic factor.

> **FACTORIZATION THEOREM FOR A POLYNOMIAL WITH REAL COEFFICIENTS**
>
> Every polynomial with real coefficients can be uniquely factored over the real numbers as a product of linear factors and/or irreducible quadratic factors.

EXAMPLE 5 Finding the Zeros of a Polynomial from a Given Complex Zero

Given that $2 - i$ is a zero of $P(x) = x^4 - 6x^3 + 14x^2 - 14x + 5$, find the remaining zeros.

Solution

Because $P(x)$ has real coefficients, the conjugate $\overline{2 - i} = 2 + i$ is also a zero. By the Factorization Theorem, the linear factors $[x - (2 - i)]$ and $[x - (2 + i)]$ appear in the factorization of $P(x)$. Consequently, their product

$$
\begin{aligned}
[x - (2 - i)][x - (2 + i)] &= (x - 2 + i)(x - 2 - i) \\
&= [(x - 2) + i][(x - 2) - i] && \text{Regroup.} \\
&= (x - 2)^2 - i^2 && (A + B)(A - B) = A^2 - B^2 \\
&= x^2 - 4x + 4 + 1 && \text{Expand } (x - 2)^2, i^2 = -1. \\
&= x^2 - 4x + 5 && \text{Simplify.}
\end{aligned}
$$

is also a factor of $P(x)$. We divide $P(x)$ by $x^2 - 4x + 5$ to find the other factor.

> **RECALL**
>
> Dividend =
> Quotient · Divisor + Remainder

$$
\begin{array}{r}
x^2 - 2x + 1 \quad \leftarrow \text{Quotient} \\
\text{Divisor} \rightarrow x^2 - 4x + 5 \overline{) x^4 - 6x^3 + 14x^2 - 14x + 5} \quad \leftarrow \text{Dividend} \\
\underline{x^4 - 4x^3 + 5x^2} \\
-2x^3 + 9x^2 - 14x \\
\underline{-2x^3 + 8x^2 - 10x} \\
x^2 - 4x + 5 \\
\underline{x^2 - 4x + 5} \\
0 \quad \leftarrow \text{Remainder}
\end{array}
$$

$$
\begin{aligned}
P(x) &= (x^2 - 2x + 1)(x^2 - 4x + 5) && P(x) = \text{Quotient} \cdot \text{Divisor} \\
&= (x - 1)(x - 1)(x^2 - 4x + 5) && \text{Factor } x^2 - 2x + 1. \\
&= (x - 1)(x - 1)[x - (2 - i)][x - (2 + i)] && \text{Factor } x^2 - 4x + 5.
\end{aligned}
$$

The zeros of $P(x)$ are 1 (of multiplicity 2), $2 - i$, and $2 + i$.

Practice Problem 5 Given that $2i$ is a zero of $P(x) = x^4 - 3x^3 + 6x^2 - 12x + 8$, find the remaining zeros.

EXAMPLE 6 Finding the Zeros of a Polynomial

Find all zeros of the polynomial $P(x) = x^4 - x^3 + 7x^2 - 9x - 18$.

Solution

Because the degree of $P(x)$ is 4, $P(x)$ has four zeros. The Rational Zeros Test tells us that the possible rational zeros are

$$\pm 1, \pm 2, \pm 3, \pm 6, \pm 9, \text{ and } \pm 18.$$

Testing these possible zeros by synthetic division, we find that 2 is a zero.

$$\begin{array}{r|rrrr} 2 & 1 & -1 & 7 & -9 & -18 \\ & & 2 & 2 & 18 & 18 \\ \hline & 1 & 1 & 9 & 9 & 0 \end{array}$$

$P(2) = 0$, so $(x - 2)$ is a factor of $P(x)$ and the quotient is $x^3 + x^2 + 9x + 9$.
We can solve the depressed equation $x^3 + x^2 + 9x + 9 = 0$ using factoring by grouping.

$$\begin{aligned} x^2(x + 1) + 9(x + 1) &= 0 && \text{Group terms, distributive property} \\ (x^2 + 9)(x + 1) &= 0 && \text{Distributive property} \\ x^2 + 9 = 0 \text{ or } x + 1 &= 0 && \text{Zero-product property} \\ x^2 = -9 \text{ or } x &= -1 && \text{Solve each equation.} \\ x = \pm\sqrt{-9} &= \pm 3i && \text{Solve } x^2 = -9 \text{ for } x. \end{aligned}$$

The four zeros of $P(x)$ are $-1, 2, 3i,$ and $-3i$. The complete factorization of $P(x)$ is

$$P(x) = x^4 - x^3 + 7x^2 - 9x - 18 = (x + 1)(x - 2)(x - 3i)(x + 3i).$$

Practice Problem 6 Find all zeros of the polynomial

$$P(x) = x^4 - 8x^3 + 22x^2 - 28x + 16.$$

SECTION 2.4 Exercises

Basic Concepts and Skills

In Exercises 1–8, determine the possible number of positive and negative zeros of the given function by using Descartes's Rule of Signs.

1. $f(x) = 5x^3 - 2x^2 - 3x + 4$
2. $g(x) = 3x^3 + x^2 - 9x - 3$
3. $f(x) = 2x^3 + 5x^2 - x + 2$
4. $g(x) = 3x^4 + 8x^3 - 5x^2 + 2x - 3$
5. $h(x) = 2x^5 - 5x^3 + 3x^2 + 2x - 1$
6. $F(x) = 5x^6 - 7x^4 + 2x^3 - 1$
7. $G(x) = -3x^4 - 4x^3 + 5x^2 - 3x + 7$
8. $H(x) = -5x^5 + 3x^3 - 2x^2 - 7x + 4$

In Exercises 9–16, determine upper and lower bounds on the zeros of the given function.

9. $f(x) = 3x^3 - x^2 + 9x - 3$
10. $g(x) = 2x^3 - 3x^2 - 14x + 21$
11. $F(x) = 3x^3 + 2x^2 + 5x + 7$
12. $G(x) = x^3 + 3x^2 + x - 4$
13. $h(x) = x^4 + 3x^3 - 15x^2 - 9x + 31$
14. $H(x) = 3x^4 - 20x^3 + 28x^2 + 19x - 13$
15. $f(x) = 6x^4 + x^3 - 43x^2 - 7x + 7$
16. $g(x) = 6x^4 + 23x^3 + 25x^2 - 9x - 5$

In Exercises 17–26, find all solutions of the equation in the complex number system.

17. $x^2 + 25 = 0$
18. $9x^2 + 16 = 0$
19. $(x - 2)^2 + 9 = 0$
20. $(x - 1)^2 + 27 = 0$
21. $x^2 + 4x + 4 = -9$
22. $x^2 + 2x + 1 = -16$
23. $x^3 - 8 = 0$
24. $x^4 - 1 = 0$
25. $(x - 2)(x - 3i)(x + 3i) = 0$
26. $(x - 1)(x - 2i)(2x - 6i)(2x + 6i) = 0$

In Exercises 27–32, find the remaining zeros of a polynomial $P(x)$ with real coefficients and with the specified degree and zeros.

27. Degree 3; zeros: $2, 3 + i$
28. Degree 3; zeros: $2, 2 - i$
29. Degree 4; zeros: $0, 1, 2 - i$
30. Degree 4; zeros: $-i, 1 - i$
31. Degree 6; zeros: $0, 5, i, 3i$
32. Degree 6; zeros: $2i, 4 + i, i - 1$

In Exercises 33–36, find the polynomial $P(x)$ with real coefficients having the specified degree, leading coefficient, and zeros.

33. Degree 4; leading coefficient 2; zeros: $5 - i, 3i$
34. Degree 4; leading coefficient -3; zeros: $2 + 3i, 1 - 4i$
35. Degree 5; leading coefficient 7; zeros: 5 (multiplicity 2), 1, $3 - i$
36. Degree 6; leading coefficient 4; zeros: 3, 0 (multiplicity 3), $2 - 3i$

In Exercises 37–40, use the given zero to find all the zeros of each polynomial function.

37. $P(x) = x^4 + x^3 + 9x^2 + 9x$, zero: $3i$
38. $P(x) = x^4 - 2x^3 + x^2 + 2x - 2$, zero: $1 - i$
39. $P(x) = x^5 - 5x^4 + 2x^3 + 22x^2 - 20x$, zero: $3 - i$
40. $P(x) = 2x^5 - 11x^4 + 19x^3 - 17x^2 + 17x - 6$, zero: i

In Exercises 41–50, find all zeros of each polynomial function.

41. $P(x) = x^3 - 9x^2 + 25x - 17$
42. $P(x) = x^3 - 5x^2 + 7x + 13$
43. $P(x) = 3x^3 - 2x^2 + 22x + 40$
44. $P(x) = 3x^3 - x^2 + 12x - 4$
45. $P(x) = 2x^4 - 10x^3 + 23x^2 - 24x + 9$
46. $P(x) = 9x^4 + 30x^3 + 14x^2 - 16x + 8$
47. $P(x) = x^4 - 4x^3 - 5x^2 + 38x - 30$
48. $P(x) = x^4 + x^3 + 7x^2 + 9x - 18$
49. $P(x) = 2x^5 - 11x^4 + 19x^3 - 17x^2 + 17x - 6$
50. $P(x) = x^5 - 2x^4 - x^3 + 8x^2 - 10x + 4$

In Exercises 51–54, find an equation of a polynomial function of least degree having the given complex zeros, intercepts, and graph.

51. f has complex zero $3i$

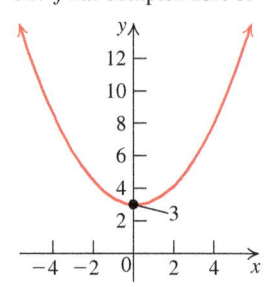

52. f has complex zero $-i$

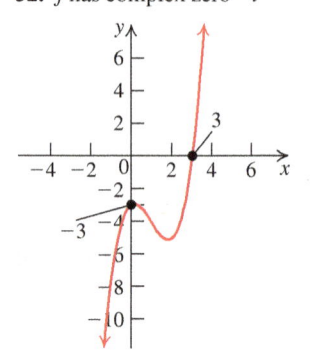

53. f has complex zeros i and $2i$

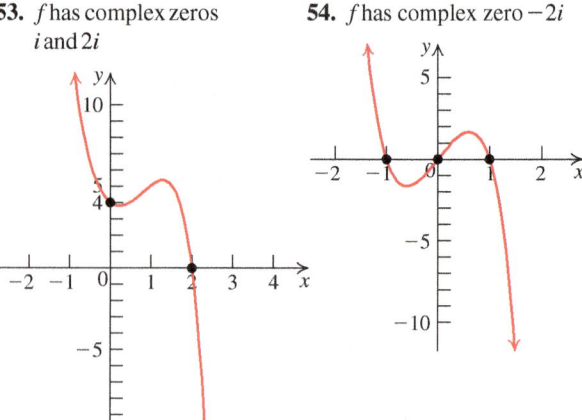

54. f has complex zero $-2i$

55. The solutions -1 and 1 of the equation $x^2 = 1$ are called the square roots of 1. The solutions of the equation $x^3 = 1$ are called the cube roots of 1. Find the cube roots of 1. How many are there?

56. The solutions of the equation $x^n = 1$, where n is a positive integer, are called the nth roots of 1 or "nth roots of unity." Explain the relationship between the solutions of the equation $x^n = 1$ and the zeros of the complex polynomial $P(x) = x^n - 1$. How many nth roots of 1 are there?

57. Show that the polynomial function $P(x) = x^6 + 2x^4 + 3x^2 + 4$ has six nonreal zeros.

58. Show that the polynomial function $F(x) = x^5 + x^3 + 2x + 1$ has four nonreal zeros.

59. Show that the polynomial function $f(x) = x^7 + 5x + 3$ has six complex zeros.

60. Show that the polynomial function $g(x) = x^3 - x^2 - 1$ has two complex zeros.

In Exercises 61–64, find an equation with real coefficients of a polynomial function f that has the given characteristics. Then write the end behavior of the graph of $y = f(x)$.

61. Degree: 3; zeros: $2, 1 + 2i$; y-intercept: 40
62. Degree: 3; zeros: $1, 2 - 3i$; y-intercept: -26
63. Degree: 4; zeros: $1, -1, 3 + i$; y-intercept: 20
64. Degree: 4; zeros: $1 - 2i, 3 - 2i$; y-intercept: 130

Critical Thinking / Discussion / Writing

65. Show that if $r_1, r_2, \ldots, r_n$ are the roots of the equation

$$a_n x^n + a_{n-1} x^{n-1} + \cdots + a_1 x + a_0 = 0 \quad (a_n \neq 0),$$

then the sum of the roots satisfies

$$r_1 + r_2 + \cdots + r_n = -\frac{a_{n-1}}{a_n}$$

and the product of the roots satisfies

$$r_1 \cdot r_2 \cdot \cdots \cdot r_n = (-1)^n \frac{a_0}{a_n}.$$

[*Hint:* Factor the polynomial; then multiply it out, using $r_1, r_2, \ldots, r_n$, and compare coefficients with those of the original polynomial.]

66. In his book *Ars Magna*, Cardano explained how to solve cubic equations. He considered the following example:

$$x^3 + 6x = 20$$

  a. Explain why this equation has exactly one real solution.
  b. Cardano explained the method as follows: "I take two cubes v^3 and u^3 whose difference is 20 and whose product is 2, that is, a third of the coefficient of x. Then, I say that $x = v - u$ is a solution of the equation." Show that if $v^3 - u^3 = 20$ and $vu = 2$, then $x = v - u$ is indeed the solution of the equation $x^3 + 6x = 20$.
  c. Solve the system

$$v^3 - u^3 = 20$$
$$vu = 2$$

  to find u and v.
  d. Consider the equation $x^3 + px = q$, where p is a positive number. Using your work in parts (a), (b), and (c) as a guide, show that the unique solution of this equation is

$$x = \sqrt[3]{\frac{q}{2} + \sqrt{\left(\frac{q}{2}\right)^2 + \left(\frac{p}{3}\right)^3}} - \sqrt[3]{-\frac{q}{2} + \sqrt{\left(\frac{q}{2}\right)^2 + \left(\frac{p}{3}\right)^3}}.$$

  e. Consider an arbitrary cubic equation

$$x^3 + ax^2 + bx + c = 0.$$

  Show that the substitution $x = y - \frac{a}{3}$ allows you to write the cubic equation as

$$y^3 + py = q.$$

67. Use Cardano's method to solve the equation
$x^3 + 6x^2 + 10x + 8 = 0.$

Maintaining Skills

In Exercises 68–75, let $f(x) = 2x + 3$, $g(x) = 3x^2 + x - 1$, $h(x) = x^3 + x$, and $s(x) = x^3$. Find each of the following.

68. $(f + g)(x)$
69. $(g + h)(x)$
70. $(f \cdot g)(x)$
71. $(g \cdot h)(x)$
72. $(h - s)(x)$
73. $(s - g)(x)$
74. $(f \circ s)(x)$
75. $(h \circ s)(x)$

SECTION 2.5

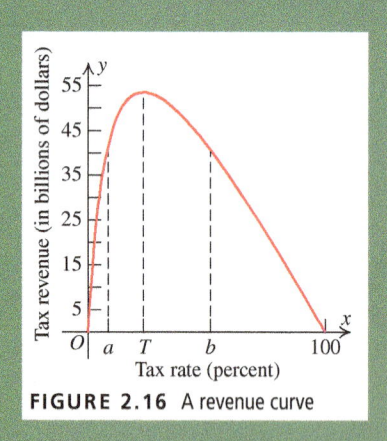

Rational Functions

BEFORE STARTING THIS SECTION, REVIEW

1. Domain of a function (Section 1.3, page 78)
2. Zeros of a function (Section 2.2, page 171)
3. Symmetry (Section 1.1, page 57)

OBJECTIVES

1. Define a rational function.
2. Define vertical and horizontal asymptotes.
3. Graph translations of $f(x) = \dfrac{1}{x}$.
4. Find vertical and horizontal asymptotes (if any).
5. Graph rational functions.
6. Graph rational functions with oblique asymptotes.
7. Graph a revenue curve.

◆ Federal Taxes and Revenues

Federal governments levy income taxes on their citizens and corporations to pay for defense, infrastructure, and social services. The relationship between tax rates and tax revenues shown in Figure 2.16 is a "revenue curve."

All economists agree that a zero tax rate produces no tax revenues and that no one would bother to work with a 100% tax rate; so tax revenue would be zero. It then follows that in a given economy, there is some optimal tax rate T between zero and 100% at which people are willing to maximize their output and still pay their taxes. This optimal rate brings in the most revenue for the government. However, because no one knows the actual shape of the revenue curve, it is impossible to find the exact value of T.

Notice in Figure 2.16 that between the two extreme rates, there are two rates (such as a and b in the figure) that will collect the same amount of revenue. In a democracy, politicians can argue that taxes are currently too high (at some point b in Figure 2.16) and should therefore be reduced to encourage incentives and harder work (this is *supply-side economics*); at the same time, the government will generate more revenue. Others can argue that we are well to the left of T (at some point a in Figure 2.16), so the tax rate should be raised (for the rich) to generate more revenue. In Example 10, we sketch the graph of a revenue curve for a hypothetical economy.

FIGURE 2.16 A revenue curve

1 Define a rational function.

Rational Functions

Recall that the sum, difference, and product of two polynomial functions is also a polynomial function. However, the quotient of two polynomial functions is generally *not* a polynomial function. In this case, we call the quotient of two polynomial functions a *rational function*.

Rational Function

A function f that can be expressed in the form

$$f(x) = \frac{N(x)}{D(x)},$$

where the numerator $N(x)$ and the denominator $D(x)$ are polynomials and $D(x)$ is not the zero polynomial, is called a **rational function**. The domain of f consists of all real numbers for which $D(x) \neq 0$.

Examples of rational functions are

$$f(x) = \frac{x-1}{2x+1}, \quad g(y) = \frac{5y}{y^2+1}, \quad h(t) = \frac{3t^2 - 4t + 1}{t-2}, \quad \text{and} \quad F(z) = 3z^4 - 5z^2 + 6z + 1.$$

($F(z)$ has the constant polynomial function $D(z) = 1$ as its denominator.) By contrast, $S(x) = \frac{\sqrt{1-x^2}}{x+1}$ and $G(x) = \frac{|x|}{x}$ are not rational functions.

EXAMPLE 1 Finding the Domain of a Rational Function

Find the domain of each rational function.

a. $f(x) = \dfrac{3x^2 - 12}{x - 1}$ **b.** $g(x) = \dfrac{x}{x^2 - 6x + 8}$ **c.** $h(x) = \dfrac{x^2 - 4}{x - 2}$

Solution

We eliminate all x for which $D(x) = 0$.

a. $D(x) = x - 1 = 0$, if $x = 1$. The domain of $f(x) = \dfrac{3x^2 - 12}{x - 1}$ is the set of all real numbers x except $x = 1$; in interval notation the domain is $(-\infty, 1) \cup (1, \infty)$.

b. The domain of g is the set of all real numbers x for which the denominator $D(x)$ is nonzero.

$$\begin{aligned}
D(x) = x^2 - 6x + 8 &= 0 && \text{Set } D(x) = 0. \\
(x - 2)(x - 4) &= 0 && \text{Factor } D(x). \\
x - 2 = 0 \text{ or } x - 4 &= 0 && \text{Zero-product property} \\
x = 2 \text{ or } x &= 4 && \text{Solve for } x.
\end{aligned}$$

The domain of g is the set of all real numbers x except 2 and 4, or in interval notation, $(-\infty, 2) \cup (2, 4) \cup (4, \infty)$.

c. The domain of $h(x) = \dfrac{x^2 - 4}{x - 2}$ is the set of all real numbers x except 2, or in interval notation, $(-\infty, 2) \cup (2, \infty)$.

Practice Problem 1 Find the domain of the rational function $f(x) = \dfrac{x - 3}{x^2 - 4x - 5}$.

In Example 1(c), note that the functions $h(x) = \dfrac{x^2 - 4}{x - 2} = \dfrac{(x-2)(x+2)}{x-2}$ and $H(x) = x + 2$ are not equal: the domain of $H(x)$ is the set of all real numbers, but the domain of $h(x)$ is the set of all real numbers x except 2. The graph of $y = H(x)$ is a line

---

RECALL

The graph of a polynomial function has no holes, gaps, or sharp corners.

with slope 1 and y-intercept 2. The graph of $y = h(x)$ is the same line as $y = H(x)$, except that the point $(2, 4)$ is missing. See Figure 2.17.

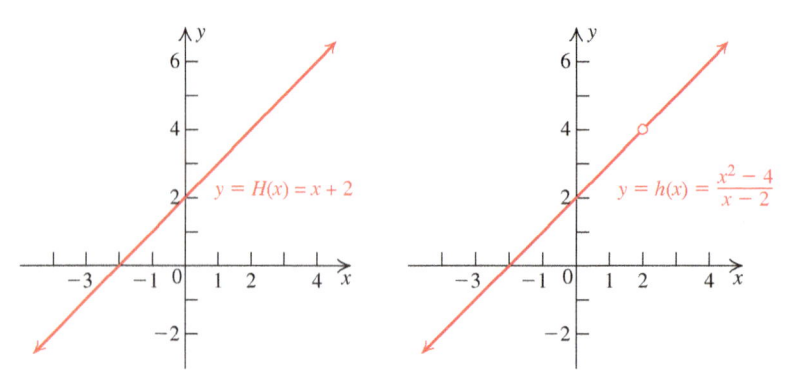

FIGURE 2.17 Functions $H(x)$ and $h(x)$ have different graphs

As with the quotient of integers, if the polynomials $N(x)$ and $D(x)$ have no common factors, then the rational function $f(x) = \dfrac{N(x)}{D(x)}$ is said to be in **lowest terms**. The zeros of $N(x)$ and $D(x)$ will play an important role in graphing the rational function f.

2 Define vertical and horizontal asymptotes.

Vertical and Horizontal Asymptotes

Consider the function $f(x) = \dfrac{1}{x}$, introduced in Section 1.4. We know that $f(0)$ is undefined, meaning that f is not assigned any value at $x = 0$. We examine more closely the behavior of $f(x)$ "near" the excluded number $x = 0$. Consider the following table:

· SIDE
NOTE

$\dfrac{1}{\text{small number}}$ = large number

x	1	0.1	0.01	0.001	0.0001	10^{-9}	10^{-50}
$f(x) = \dfrac{1}{x}$	1	10	100	1000	10,000	10^9	10^{50}

We see that as x approaches 0 (with $x > 0$), $f(x)$ increases without bound. In symbols, we write "as $x \to 0^+, f(x) \to \infty$." This statement is read "As x approaches 0 *from the right*, $f(x)$ approaches infinity."

The symbol $x \to a^+$, read "x approaches a from the right," refers only to values of x near a that are greater than a. The symbol $x \to a^-$ is read "x approaches a from the left" and refers only to values of x near a that are less than a.

For the function $f(x) = \dfrac{1}{x}$, we can make another table of values that includes negative values of x near zero, such as $x = -1, -0.1, -0.001, \ldots, -10^{-9}$, and so on. We can conclude that as $x \to 0^-$, $f(x) \to -\infty$ and say that as x approaches 0 *from the left*, $f(x)$ approaches negative infinity.

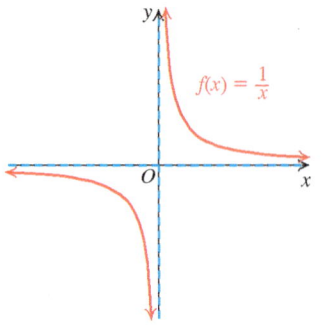

FIGURE 2.18 Reciprocal function, $f(x) = \dfrac{1}{x}$

The graph of $f(x) = \dfrac{1}{x}$ is the *reciprocal function* from Chapter 1. See Figure 2.18. The vertical line $x = 0$ (the y-axis) is a *vertical asymptote*.

Vertical Asymptote

The line $x = a$ is a **vertical asymptote** of the graph of a function f if $f(x) \to \infty$, or $f(x) \to -\infty$, as $x \to a^+$ or as $x \to a^-$.

Section 2.5 ■ Rational Functions   **201**

This definition says that if the line $x = a$ is a vertical asymptote of the graph of a function f, then the graph of f near $x = a$ behaves like any of the graphs in Figure 2.19.

As $x \to a^-, f(x) \to -\infty$ As $x \to a^+, f(x) \to \infty$ As $x \to a^-, f(x) \to \infty$ As $x \to a^+, f(x) \to -\infty$

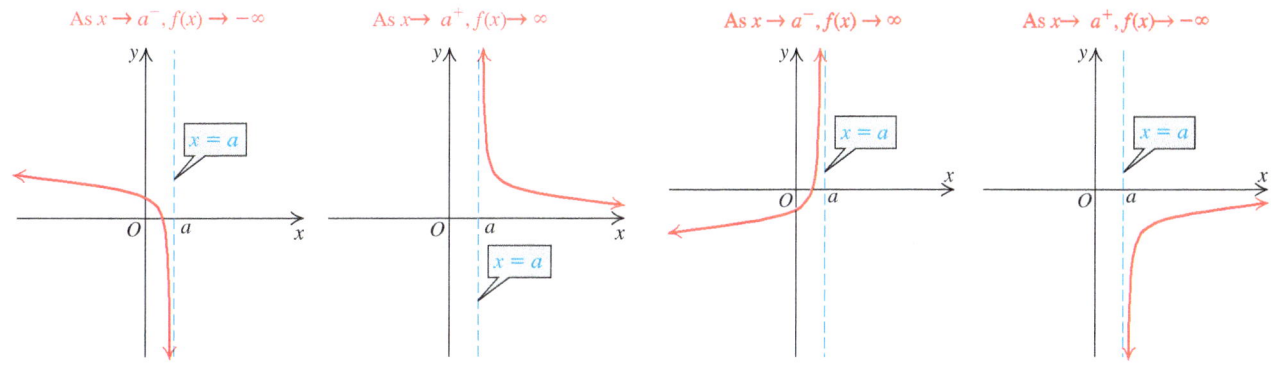

FIGURE 2.19 Behavior of a function near a vertical asymptote

If $x = a$ is a vertical asymptote for a rational function f, then a is not in the domain; so the graph of f *does not cross* the line $x = a$.

SIDE NOTE

$\dfrac{1}{\text{large number}} = \text{small number}$

Consider again the function $f(x) = \dfrac{1}{x}$. The larger we make x, the closer $\dfrac{1}{x}$ is to 0: as $x \to \infty, f(x) = \dfrac{1}{x} \to 0$. Similarly, as $x \to -\infty, f(x) = \dfrac{1}{x} \to 0$. The x-axis, with equation $y = 0$, is called a *horizontal asymptote* of the graph of $f(x) = \dfrac{1}{x}$. See Figure 2.18.

Horizontal Asymptote

The line $y = k$ is a **horizontal asymptote** of the graph of a function f if $f(x) \to k$ as $x \to \infty$ or if $f(x) \to k$ as $x \to -\infty$.

This definition states that if the line $y = k$ is a horizontal asymptote of the graph of a function f, then the end behavior of the graph of f is similar to one of the graphs in Figure 2.20.

$f(x) \to k$ as $x \to -\infty$ $f(x) \to k$ as $x \to \infty$

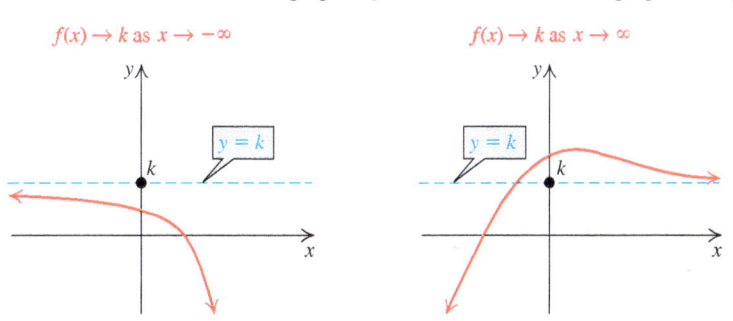

FIGURE 2.20 Behavior of a function near a horizontal asymptote

The graph of a function $y = f(x)$ can cross its horizontal asymptote (in contrast to a vertical asymptote).

3 Graph translations of $f(x) = \dfrac{1}{x}$.

Translations of $f(x) = \dfrac{1}{x}$

We can graph any function of the form

$$g(x) = \dfrac{ax + b}{cx + d}$$

starting with the graph of $f(x) = \dfrac{1}{x}$ and using the techniques of vertical stretching and compressing, shifting, and/or reflecting.

EXAMPLE 2 Graphing Rational Functions Using Translations

Graph each rational function, identify the vertical and horizontal asymptotes, and state the domain and range.

a. $g(x) = \dfrac{-2}{x+1}$ **b.** $h(x) = \dfrac{3x-2}{x-1}$

Solution

a. Let $f(x) = \dfrac{1}{x}$. We can write $g(x)$ in terms of $f(x)$.

$$g(x) = \dfrac{-2}{x+1}$$
$$= -2\left(\dfrac{1}{x+1}\right)$$
$$= -2f(x+1) \qquad \text{Replace } x \text{ with } x+1 \text{ in } f(x) = \dfrac{1}{x}.$$

The graph of $y = f(x+1)$ is the graph of $y = f(x)$ shifted one unit to the left. This moves the vertical asymptote one unit to the left. The graph of $y = -2f(x+1)$ is the graph of $y = f(x+1)$ stretched vertically two units and then reflected about the x-axis. The graph is shown in Figure 2.21. The domain of g is $(-\infty, -1) \cup (-1, \infty)$, and the range is $(-\infty, 0) \cup (0, \infty)$. The graph has vertical asymptote $x = -1$ and horizontal asymptote $y = 0$ (the x-axis).

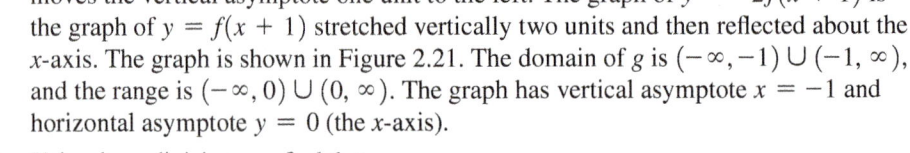

FIGURE 2.21 Graph of $g(x) = \dfrac{-2}{x+1}$

b. Using long division, we find that

$$h(x) = 3 + \dfrac{1}{x-1} \qquad \begin{array}{r} 3 \\ x-1 \overline{\smash{)}3x-2} \\ \underline{3x-3} \\ 1 \end{array} \quad \text{(long division)}$$

Then $h(x) = f(x-1) + 3 \qquad \text{Replace } x \text{ with } x-1 \text{ in } f(x) = \dfrac{1}{x}.$

We see that the graph of $y = h(x)$ is the graph of $f(x) = \dfrac{1}{x}$ shifted horizontally one unit to the right and vertically up three units. The graph is shown in Figure 2.22. The domain of h is $(-\infty, 1) \cup (1, \infty)$, and the range is $(-\infty, 3) \cup (3, \infty)$. The graph of h has vertical asymptote $x = 1$ and horizontal asymptote $y = 3$.

FIGURE 2.22 Graph of $h(x) = \dfrac{3x-2}{x-1}$

Practice Problem 2 Repeat Example 2 with

a. $g(x) = \dfrac{3}{x-2}$. **b.** $h(x) = \dfrac{2x+5}{x+1}$.

4 Find vertical and horizontal asymptotes (if any).

Vertical and Horizontal Asymptotes

> **LOCATING VERTICAL ASYMPTOTES OF RATIONAL FUNCTIONS**
>
> If $f(x) = \dfrac{N(x)}{D(x)}$ is a rational function, where $N(x)$ and $D(x)$ do not have a common factor and a is a real zero of $D(x)$, then the line with equation $x = a$ is a vertical asymptote of the graph of f.

This means that the vertical asymptotes (if any) of a rational function are found by locating the real zeros of its denominator. However, as we will see in Example 4, there will not necessarily be a vertical asymptote at every real zero of the denominator.

Section 2.5 ■ Rational Functions   203

TECHNOLOGY CONNECTION

Calculator graph for
$$h(x) = \frac{x^2 - 9}{x - 3}$$

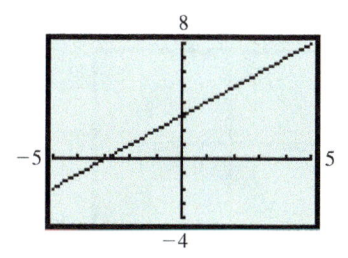

A calculator graph does not show the hole in the graph, but a calculator table explains the situation.

EXAMPLE 3   Finding Vertical Asymptotes

Find all vertical asymptotes of the graph of each rational function.

a. $f(x) = \dfrac{1}{x - 1}$ **b.** $g(x) = \dfrac{1}{x^2 - 9}$ **c.** $h(x) = \dfrac{1}{x^2 + 1}$

Solution

a. There are no common factors in the numerator and denominator of $f(x) = \dfrac{1}{x - 1}$, and the only *zero* of the denominator is 1. Therefore, $x = 1$ is a vertical asymptote of $f(x)$.

b. The rational function $g(x)$ is in lowest terms. Factoring $x^2 - 9 = (x + 3)(x - 3)$, we see that the *zeros* of the denominator are -3 and 3. Therefore, the lines $x = -3$ and $x = 3$ are the two vertical asymptotes of $g(x)$.

c. Because the denominator $x^2 + 1$ has no real zeros, the graph of $h(x)$ has *no* vertical asymptotes.

Practice Problem 3 Find the vertical asymptotes of the graph of

$$f(x) = \frac{x + 1}{x^2 + 3x - 10}.$$

The next example illustrates that the graph of a rational function may have gaps (missing points) with or without vertical asymptotes.

EXAMPLE 4   Rational Function Whose Graph Has a Hole

Find all vertical asymptotes of the graph of each rational function.

a. $h(x) = \dfrac{x^2 - 9}{x - 3}$ **b.** $g(x) = \dfrac{x + 2}{x^2 - 4}$

Solution

a. $h(x) = \dfrac{x^2 - 9}{x - 3}$ Given function

$ = \dfrac{(x + 3)(x - 3)}{x - 3}$ Factor $x^2 - 9 = (x + 3)(x - 3)$ in the numerator.

$ = x + 3, \quad \text{if } x \neq 3$ Simplify.

The graph of $h(x)$ is the line $y = x + 3$ with a gap (or hole) at $x = 3$. See Figure 2.23.

The graph of $h(x)$ has no vertical asymptote at $x = 3$, the zero of the denominator, because the numerator and the denominator have the factor $(x - 3)$ in common.

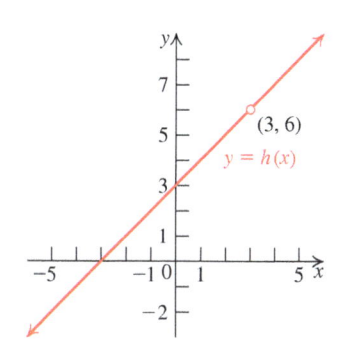

FIGURE 2.23 Graph with a hole

b. $g(x) = \dfrac{x + 2}{x^2 - 4}$ Given function

$ = \dfrac{x + 2}{(x + 2)(x - 2)}$ Factor the denominator.

$ = \dfrac{1}{x - 2}, \quad x \neq -2$ Simplify.

The graph of $g(x)$ has a hole at $x = -2$ and a vertical asymptote at $x = 2$. See Figure 2.24.

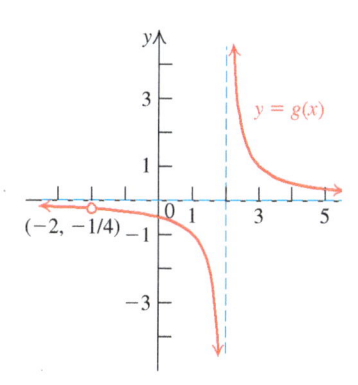

FIGURE 2.24 Graph with a hole and an asymptote

Practice Problem 4 Find all vertical asymptotes of the graph of $f(x) = \dfrac{3-x}{x^2-9}$.

TECHNOLOGY CONNECTION

Graphing calculators give very different graphs for

$$g(x) = \dfrac{1}{x^2 - 9}$$

depending on whether the mode is set to *connected* or *dot*.

In *connected* mode, the calculator connects the dots between plotted points, producing vertical lines at $x = -3$ and $x = 3$. *Dot* mode avoids graphing these vertical lines by plotting the same points but not connecting them. However, *dot* mode fails to display continuous sections of the graph.

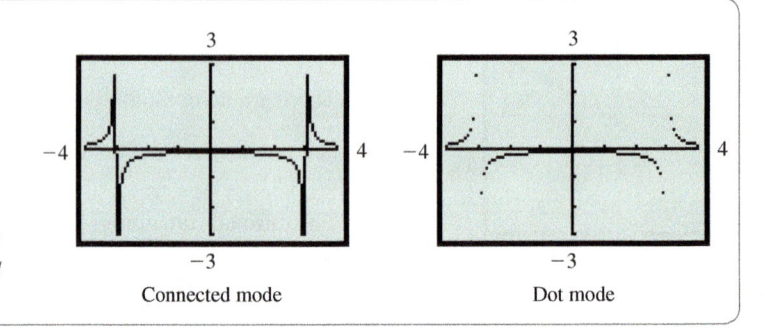

Connected mode

Dot mode

The graph of $g(x) = \dfrac{x+2}{x^2-4}$, shown in Figure 2.24, has a horizontal asymptote: $y = 0$. This can be verified algebraically. Divide the numerator and the denominator of g by x^2.

$$g(x) = \dfrac{x+2}{x^2-4} = \dfrac{\dfrac{x+2}{x^2}}{\dfrac{x^2-4}{x^2}}$$

$$= \dfrac{\dfrac{x}{x^2} + \dfrac{2}{x^2}}{\dfrac{x^2}{x^2} - \dfrac{4}{x^2}} \qquad \dfrac{a \pm b}{c} = \dfrac{a}{c} \pm \dfrac{b}{c}$$

$$= \dfrac{\dfrac{1}{x} + \dfrac{2}{x^2}}{1 - \dfrac{4}{x^2}} \qquad \text{Simplify each expression.}$$

As $|x| \to \infty$, the expressions $\dfrac{1}{x}, \dfrac{2}{x^2}$, and $\dfrac{4}{x^2}$ all approach 0; so

$$g(x) \to \dfrac{0+0}{1-0} = \dfrac{0}{1} = 0.$$

Because $g(x) \to 0$ as $|x| \to \infty$, the line $y = 0$ (the x-axis) is a horizontal asymptote.

Note that although a rational function can have more than one vertical asymptote (see Example 6), it can have, at most, one horizontal asymptote. **We can find the horizontal asymptote (if any) of a rational function by dividing the numerator and denominator by the highest power of x that appears in the denominator and investigating the resulting expression as $|x| \to \infty$.**

Alternatively, one may use the following rules.

> **RULES FOR LOCATING HORIZONTAL ASYMPTOTES**
>
> Let f be a rational function given by
>
> $$f(x) = \frac{N(x)}{D(x)} = \frac{a_n x^n + a_{n-1} x^{n-1} + \cdots + a_2 x^2 + a_1 x + a_0}{b_m x^m + b_{m-1} x^{m-1} + \cdots + b_2 x^2 + b_1 x + b_0}, a_n \neq 0, b_m \neq 0$$
>
> be in lowest terms. To find whether the graph of f has one horizontal asymptote or no horizontal asymptote, we compare the degree of the numerator, n, with that of the denominator, m:
>
> 1. If $n < m$, then the x-axis ($y = 0$) is the horizontal asymptote.
> 2. If $n = m$, then the line with equation $y = \dfrac{a_n}{b_m}$ is the horizontal asymptote.
> 3. If $n > m$, then the graph of f has *no* horizontal asymptote.

EXAMPLE 5 Finding the Horizontal Asymptote

Find the horizontal asymptote (if any) of the graph of each rational function.

a. $f(x) = \dfrac{5x + 2}{1 - 3x}$ **b.** $g(x) = \dfrac{2x}{x^2 + 1}$ **c.** $h(x) = \dfrac{3x^2 - 1}{x + 2}$

Solution

a. The numerator and denominator of $f(x) = \dfrac{5x + 2}{1 - 3x}$ are both of degree 1. The leading coefficient of the numerator is 5, and that of the denominator is -3. By Rule 2, the line $y = \dfrac{5}{-3} = -\dfrac{5}{3}$ is the horizontal asymptote of the graph of f.

b. For the function $g(x) = \dfrac{2x}{x^2 + 1}$, the degree of the numerator is 1 and that of the denominator is 2. By Rule 1, the line $y = 0$ (the x-axis) is the horizontal asymptote.

c. The degree of the numerator, 2, of $h(x)$ is greater than the degree of its denominator, 1. By Rule 3, the graph of h has no horizontal asymptote.

Practice Problem 5 Find the horizontal asymptote (if any) of the graph of each function.

a. $f(x) = \dfrac{2x - 5}{3x + 4}$ **b.** $g(x) = \dfrac{x^2 + 3}{x - 1}$

5 Graph rational functions.

Graphing Rational Functions

In the next procedure, we assume that the rational function is in lowest terms. When this is not the case, reduce the function to lowest terms, and plot the "hole" for the canceled factor (or factors).

PROCEDURE IN ACTION

EXAMPLE 6 Graphing a Rational Function

OBJECTIVE

Graph $f(x) = \dfrac{N(x)}{D(x)}$, where $f(x)$ is in lowest terms.

Step 1 Find the intercepts. Because f is in lowest terms, the x-intercepts are found by solving the equation $N(x) = 0$. The y-intercept, if there is one, is $f(0)$.

EXAMPLE

Sketch the graph of $f(x) = \dfrac{2x^2 - 2}{x^2 - 9}$.

1.
$$2x^2 - 2 = 0 \quad \text{Set } N(x) = 0.$$
$$2(x - 1)(x + 1) = 0 \quad \text{Factor.}$$
$$x - 1 = 0 \quad \text{or} \quad x + 1 = 0 \quad \text{Zero-product property}$$
$$x = 1 \quad \text{or} \quad x = -1 \quad \text{Solve for } x.$$

The x-intercepts are -1 and 1, so the graph passes through the points $(-1, 0)$ and $(1, 0)$.

$$f(0) = \dfrac{2(0)^2 - 2}{(0)^2 - 9} \quad \text{Replace } x \text{ with 0 in } f(x).$$
$$= \dfrac{2}{9} \quad \text{Simplify.}$$

The y-intercept is $\dfrac{2}{9}$, so the graph of f passes through the point $\left(0, \dfrac{2}{9}\right)$.

Step 2 Find the vertical asymptotes (if any). Solve $D(x) = 0$ to find the vertical asymptotes of the graph. Sketch the vertical asymptotes with dashed vertical lines.

2.
$$x^2 - 9 = 0 \quad \text{Set } D(x) = 0.$$
$$(x - 3)(x + 3) = 0 \quad \text{Factor.}$$
$$x - 3 = 0 \quad \text{or} \quad x + 3 = 0 \quad \text{Zero-product property}$$
$$x = 3 \quad \text{or} \quad x = -3 \quad \text{Solve for } x.$$

The vertical asymptotes of the graph of f are the lines $x = -3$ and $x = 3$.

Step 3 Find the horizontal asymptote (if any). Use the rules on page 208 for finding the horizontal asymptote of a rational function. Sketch the horizontal asymptote with a dashed horizontal line

3. Because $n = m = 2$, by Rule 2, the horizontal asymptote is
$$y = \dfrac{\text{Leading coefficient of } N(x)}{\text{Leading coefficient of } D(x)} = \dfrac{2}{1} = 2.$$

Step 4 Test for symmetry. If $f(-x) = f(x)$, then f is symmetric with respect to the y-axis. If $f(-x) = -f(x)$, then f is symmetric with respect to the origin.

4. $f(-x) = \dfrac{2(-x)^2 - 2}{(-x)^2 - 9} = \dfrac{2x^2 - 2}{x^2 - 9} = f(x)$. The graph of f is symmetric about the y-axis. This is the only symmetry.

Step 5 Locate the graph relative to the horizontal asymptote (if any). If $y = k$ is a horizontal asymptote, divide $N(x)$ by $D(x)$ and write $f(x) = \dfrac{N(x)}{D(x)} = k + \dfrac{R(x)}{D(x)}$. Use a sign graph for $f(x) - k = \dfrac{R(x)}{D(x)}$ and "test" numbers associated with the zeros of $R(x)$ and $D(x)$ to determine intervals where the graph of f is above the asymptote $y = k$ [$f(x) - k$ is *positive*] and where it is below $y = k$ [$f(x) - k$ is *negative*]. The graph of f intersects the line $y = k$ if $f(x) - k = 0$.

5. $f(x) - 2 = \dfrac{2x^2 - 2}{x^2 - 9} - 2 = \dfrac{16}{x^2 - 9}$; $R(x) = 16$ has no zeros, and $D(x)$ has zeros -3 and 3. These zeros divide the x-axis into three intervals. See the figure. We choose test points -4, 0, and 4 to test the sign of $\dfrac{16}{x^2 - 9}$.

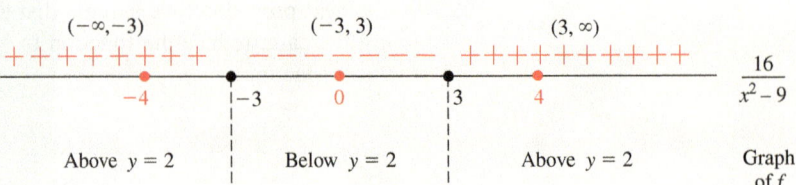

$(-\infty, -3)$	$(-3, 3)$	$(3, \infty)$	
$+++++++$	$------$	$++++++++$	$\dfrac{16}{x^2 - 9}$
$-4 \quad\; -3$	0	$3 \quad 4$	
Above $y = 2$	Below $y = 2$	Above $y = 2$	Graph of f

(continued)

Step 6 Sketch the graph. Plot the points and asymptotes found in Steps 1–5 and use symmetry to sketch the graph of f.

6.

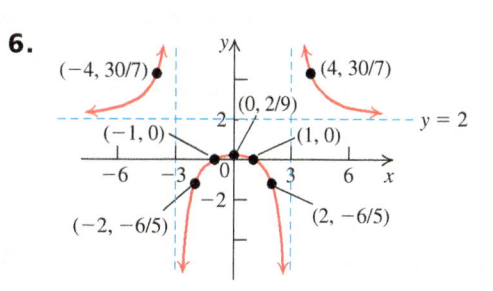

(−4, 30/7), (4, 30/7), (0, 2/9), (−1, 0), (1, 0), (−2, −6/5), (2, −6/5), $y = 2$

Practice Problem 6 Sketch the graph of $f(x) = \dfrac{2x}{x^2 - 1}$.

SIDE NOTE

It is important to reduce a rational function to lowest terms when using the given procedure for graphing a rational function.

EXAMPLE 7 Graphing a Rational Function

Sketch the graph of $f(x) = \dfrac{x^2 + 2}{(x + 2)(x - 1)}$.

Solution

Step 1 Because $x^2 + 2 > 0$, the graph has no x-intercepts.

$$f(0) = \dfrac{0^2 + 2}{(0 + 2)(0 - 1)} \qquad \text{Replace } x \text{ with } 0 \text{ in } f(x).$$

$$= -1 \qquad \text{Simplify.}$$

The y-intercept is -1.

Step 2 Set $(x + 2)(x - 1) = 0$. Solving for x, we have $x = -2$ or $x = 1$. The vertical asymptotes are the lines $x = -2$ and $x = 1$.

Step 3 $f(x) = \dfrac{x^2 + 2}{x^2 + x - 2}$. By Rule 2, page 208, the horizontal asymptote is $y = \dfrac{1}{1} = 1$ because the leading coefficient for both the numerator and the denominator is 1.

Step 4 **Symmetry.** None

Step 5 **Asymptote.** $f(x) - 1 = \dfrac{x^2 + 2}{(x + 2)(x - 1)} - 1 = \dfrac{4 - x}{(x + 2)(x - 1)}$; $R(x) = 4 - x$ has one zero, 4, and $D(x) = (x + 2)(x - 1)$ has two zeros, -2 and 1. These zeros divide the x-axis into four intervals. See the figure. We choose test points $-3, 0, 2$ and 5 to test the sign of $\dfrac{4 - x}{(x + 2)(x - 1)}$.

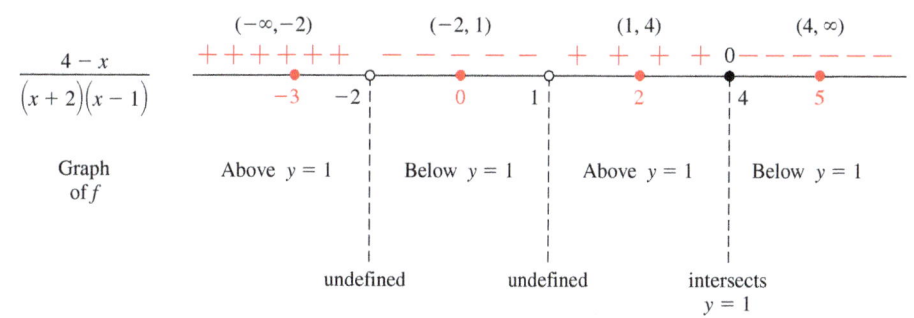

	$(-\infty, -2)$	$(-2, 1)$	$(1, 4)$	$(4, \infty)$
$\dfrac{4-x}{(x+2)(x-1)}$	$+++++$ −3	$-----$ 0	$+++ +$ 2	$0 -----$ 5
Graph of f	Above $y = 1$	Below $y = 1$	Above $y = 1$	Below $y = 1$
		undefined	undefined	intersects $y = 1$

Step 6 The graph of f is shown in Figure 2.25.

Practice Problem 7 Sketch the graph of $f(x) = \dfrac{2x^2 - 1}{2x^2 + x - 3}$.

Graphing a rational function is more complicated than graphing a polynomial. Plotting a larger number of points and checking to see if the function crosses its horizontal asymptotes (if any) gives additional detail. Notice in Figure 2.25 that the graph of f crosses the horizontal asymptote $y = 1$. To find the point of intersection, set $f(x) = 1$ and solve for x.

$$\dfrac{x^2 + 2}{x^2 + x - 2} = 1 \quad \text{Set } f(x) = 1.$$

$$x^2 + 2 = x^2 + x - 2 \quad \text{Multiply both sides by } x^2 + x - 2.$$

$$x = 4 \quad \text{Solve for } x.$$

The graph of f crosses the horizontal asymptote at the point $(4, 1)$.

We know that the graphs of polynomial functions are continuous (all in one piece). The next example shows that the graph of a rational function (other than a polynomial function) can also be continuous.

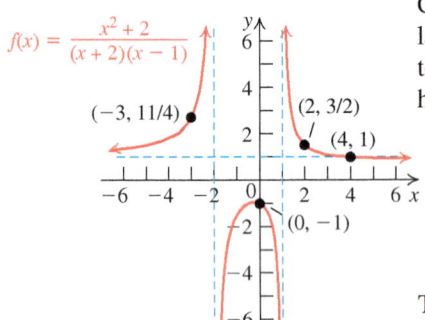

FIGURE 2.25 Graph crossing horizontal asymptote

EXAMPLE 8 Graphing a Rational Function

Sketch a graph of $f(x) = \dfrac{x^2}{x^2 + 1}$.

Solution

Step 1 Now $f(0) = 0$ and solving $f(x) = 0$, we have $x = 0$. Therefore, 0 is both the y-intercept and the x-intercept for the graph of f.

Step 2 Because $x^2 + 1 > 0$ for all x (the domain is the set of all real numbers because there are no real zeros for the denominator), there are no vertical asymptotes.

Step 3 By Rule 2 for locating horizontal asymptotes, the horizontal asymptote is $y = 1$.

Step 4 $f(-x) = \dfrac{(-x)^2}{(-x)^2 + 1} = \dfrac{x^2}{x^2 + 1} = f(x)$. The graph is symmetric with respect to the y-axis.

Step 5 $f(x) = \dfrac{x^2}{x^2 + 1} = 1 + \dfrac{-1}{x^2 + 1}$; neither $R(x) = -1$ nor $D(x) = x^2 + 1$ has zeros. Because $f(x) - 1 = \dfrac{-1}{x^2 + 1}$ is negative for all values of x, the graph of f is always below the line $y = 1$.

Step 6 The graph of $y = f(x)$ is shown in Figure 2.26.

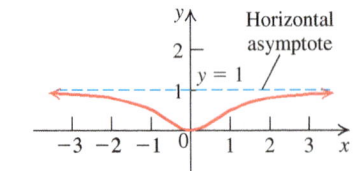

FIGURE 2.26 A continuous rational function

Practice Problem 8 Sketch the graph of $f(x) = \dfrac{x^2 + 1}{x^2 + 2}$.

6 Graph rational functions with oblique asymptotes.

Oblique Asymptotes

Suppose

$$f(x) = \dfrac{N(x)}{D(x)}$$

and the degree of $N(x)$ is greater than the degree of $D(x)$. We know from Rule 3 on page 205 that the graph of f has no horizontal asymptote. Use either long division or synthetic division (if applicable) to obtain

$$f(x) = \frac{N(x)}{D(x)} = Q(x) + \frac{R(x)}{D(x)},$$

where the degree of $R(x)$ is less than the degree of $D(x)$. Rule 1 on page 205 tells us that the x-axis is a horizontal asymptote for $\frac{R(x)}{D(x)}$; that is, as $x \to \infty$ or as $x \to -\infty$, the expression $\frac{R(x)}{D(x)} \to 0$.

Therefore, as $x \to \pm \infty$,

$$f(x) \to Q(x) + 0 = Q(x).$$

This means that as $|x|$ gets very large, the graph of $f(x)$ behaves like the graph of the polynomial $Q(x)$. If the degree of $N(x)$ is exactly one more than the degree of $D(x)$, then $Q(x)$ will have the linear form $mx + b$. In this case, $f(x)$ is said to have an *oblique* (or *slant*) asymptote. Consider, for example, the function

$$f(x) = \frac{x^2 + x}{x - 1} = x + 2 + \frac{2}{x - 1}.$$

Now as $x \to \pm \infty$, the difference $f(x) - (x + 2) = \frac{2}{x - 1} \to 0$. So that the graph of f approaches the graph of the oblique asymptote: the line $y = x + 2$, as shown in Figure 2.27.

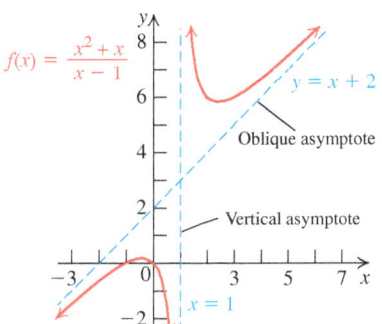

FIGURE 2.27 Graph with vertical and oblique asymptotes

Oblique Asymptote

The line $y = mx + b$ is an **oblique asymptote** of the graph of a function f if the difference $f(x) - (mx + b) \to 0$ as $x \to -\infty$ or as $x \to \infty$.

EXAMPLE 9 Graphing a Rational Function with an Oblique Asymptote

Sketch the graph of $f(x) = \frac{x^2 - 4}{x + 1}$.

Solution

Step 1 Intercepts. $f(0) = \frac{0 - 4}{0 + 1} = -4$, so the y-intercept is -4. Set $x^2 - 4 = 0$. We have $x = -2$ and $x = 2$, so the x-intercepts are -2 and 2.

Step 2 Vertical asymptotes. Set $x + 1 = 0$. We get $x = -1$, so the line $x = -1$ is a vertical asymptote.

Step 3 Asymptotes. Because the degree of the numerator is greater than the degree of the denominator, $f(x)$ has no horizontal asymptote. However, by long division,

$$f(x) = \frac{x^2 - 4}{x + 1} = x - 1 + \frac{-3}{x + 1}$$

As $x \to \pm \infty$, the expression $\frac{-3}{x + 1} \to 0$; so the line $y = x - 1$ is an oblique asymptote: the graph gets close to the line $y = x - 1$ as $x \to \infty$ and as $x \to -\infty$.

210 Chapter 2 Polynomial and Rational Functions

Step 4 **Symmetry.** You should check that there is no y-axis or origin symmetry.

Step 5 **Asymptote:** Use the intervals determined by the zeros of the numerator and the denominator of

$$f(x) - (x - 1) = \frac{x^2 - 4}{x + 1} - (x - 1) = \frac{-3}{x + 1}$$

to create a sign graph. The numerator -3 has no zero and denominator, $x + 1$ has one zero, -1. The zero, -1, divides the x-axis into two intervals (see the figure). We choose test points -3, and 0 to test the sign of $f(x) - (x - 1) = \frac{-3}{x + 1}$.

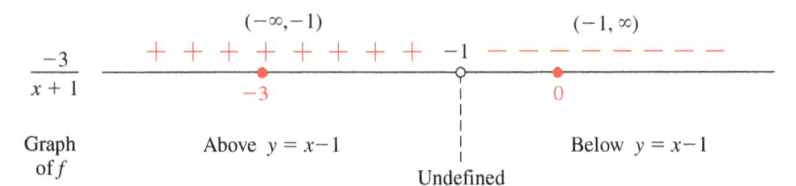

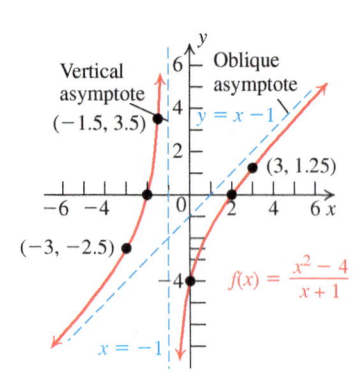

FIGURE 2.28

Step 6 The graph of f is shown in Figure 2.28.

Practice Problem 9 Sketch the graph of $f(x) = \frac{x^2 + 2}{x - 1}$.

Graph of a Revenue Curve

7 Graph a revenue curve.

TECHNOLOGY CONNECTION

TRACE
$$y = \frac{100x - x^2}{x + 10}$$
to approximate the maximum revenue.

EXAMPLE 10 Graphing a Revenue Curve

The revenue curve (see Figure 2.29) for an economy of a country is given by

$$R(x) = \frac{x(100 - x)}{x + 10},$$

where x is the tax rate in percent and $R(x)$ is the tax revenue in billions of dollars.

a. Find and interpret $R(10)$, $R(20)$, $R(30)$, $R(40)$, $R(50)$, and $R(60)$.
b. Sketch the graph of $y = R(x)$ for $0 \leq x \leq 100$.
c. Use a graphing calculator to estimate the tax rate that yields the maximum revenue.

Solution

a. $R(10) = \frac{10(100 - 10)}{10 + 10} = \$45$ billion. This means that if the income is taxed at the rate of 10%, the total revenue for the government will be \$45 billion.

Similarly, $R(20) \approx \$53.3$ billion,
$R(30) = \$52.5$ billion,
$R(40) = \$48$ billion,
$R(50) \approx \$41.7$ billion, and
$R(60) \approx \$34.3$ billion.

b. The graph of the function $y = R(x)$ for $0 \leq x \leq 100$ is shown in Figure 2.29.
c. From the calculator graph of

$$Y = \frac{100x - x^2}{x + 10},$$

by using the TRACE feature, you can see that the tax rate of about 23% produces the maximum tax revenue of about \$53.7 billion for the government.

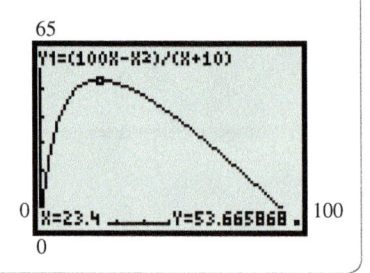

FIGURE 2.29

Practice Problem 10 Repeat Example 10 for $R(x) = \frac{x(100 - x)}{x + 20}$.

SECTION 2.5 Exercises

Basic Concepts and Skills

In Exercises 1–8, find the domain of each rational function.

1. $f(x) = \dfrac{x-3}{x+4}$

2. $f(x) = \dfrac{x+1}{x-1}$

3. $g(x) = \dfrac{x-1}{x^2+1}$

4. $g(x) = \dfrac{x+2}{x^2+4}$

5. $h(x) = \dfrac{x-3}{x^2-x-6}$

6. $h(x) = \dfrac{x-7}{x^2-6x-7}$

7. $F(x) = \dfrac{2x+3}{x^2-6x+8}$

8. $F(x) = \dfrac{3x-2}{x^2-3x+2}$

In Exercises 9–18, use the graph of the rational function $f(x)$ to complete each statement.

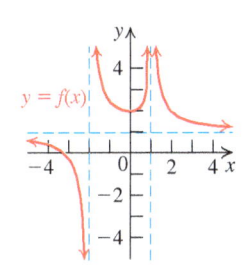

9. As $x \to 1^+$, $f(x) \to$ _____.
10. As $x \to 1^-$, $f(x) \to$ _____.
11. As $x \to -2^+$, $f(x) \to$ _____.
12. As $x \to -2^-$, $f(x) \to$ _____.
13. As $x \to \infty$, $f(x) \to$ _____.
14. As $x \to -\infty$, $f(x) \to$ _____.
15. The domain of $f(x)$ is _____.
16. There are _____ vertical asymptotes.
17. The equations of the vertical asymptotes of the graph are _____ and _____.
18. The equation of the horizontal asymptote of the graph is _____.

In Exercises 19–26, graph each rational function as a translation of $f(x) = \dfrac{1}{x}$. Identify the vertical and horizontal asymptotes and state the domain and range.

19. $f(x) = \dfrac{3}{x-4}$

20. $g(x) = \dfrac{-4}{x+3}$

21. $f(x) = \dfrac{-x}{3x+1}$

22. $g(x) = \dfrac{2x}{4x-2}$

23. $g(x) = \dfrac{-3x+2}{x+2}$

24. $f(x) = \dfrac{-x+1}{x-3}$

25. $g(x) = \dfrac{5x-3}{x-4}$

26. $f(x) = \dfrac{2x+12}{x+5}$

In Exercises 27–36, find the vertical asymptotes, if any, of the graph of each rational function.

27. $f(x) = \dfrac{x}{x-1}$

28. $f(x) = \dfrac{x+3}{x-2}$

29. $g(x) = \dfrac{(x+1)(2x-2)}{(x-3)(x+4)}$

30. $g(x) = \dfrac{(2x-1)(x+2)}{(2x+3)(3x-4)}$

31. $h(x) = \dfrac{x^2-1}{x^2+x-6}$

32. $h(x) = \dfrac{x^2-4}{3x^2+x-4}$

33. $f(x) = \dfrac{x^2-6x+8}{x^2-x-12}$

34. $f(x) = \dfrac{x^2-9}{x^3-4x}$

35. $g(x) = \dfrac{2x+1}{x^2+x+1}$

36. $g(x) = \dfrac{x^2-36}{x^2+5x+9}$

In Exercises 37–44, find the horizontal asymptote, if any, of the graph of each rational function.

37. $f(x) = \dfrac{x+1}{x^2+5}$

38. $f(x) = \dfrac{2x-1}{x^2-4}$

39. $g(x) = \dfrac{2x-3}{3x+5}$

40. $g(x) = \dfrac{3x+4}{-4x+5}$

41. $h(x) = \dfrac{x^2-49}{x+7}$

42. $h(x) = \dfrac{x+3}{x^2-9}$

43. $f(x) = \dfrac{2x^2-3x+7}{3x^3+5x+11}$

44. $f(x) = \dfrac{3x^3+2}{x^2+5x+11}$

In Exercises 45–50, match the rational function with its graph.

45. $f(x) = \dfrac{2}{x-3}$

46. $f(x) = \dfrac{x-2}{x+3}$

47. $f(x) = \dfrac{1}{x^2-2x}$

48. $f(x) = \dfrac{x}{x^2+1}$

49. $f(x) = \dfrac{x^2 + 2x}{x - 3}$

50. $f(x) = \dfrac{x^2}{x^2 - 4}$

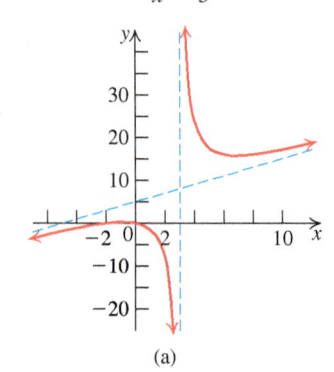

62. $h(x) = \dfrac{2x^2}{x^2 + 4}$

63. $f(x) = \dfrac{x^3 - 4x}{x^3 - 9x}$

64. $f(x) = \dfrac{x^3 + 32x}{x^3 + 8x}$

65. $g(x) = \dfrac{(x - 2)^2}{x - 2}$

66. $g(x) = \dfrac{(x - 1)^2}{x - 1}$

(a)

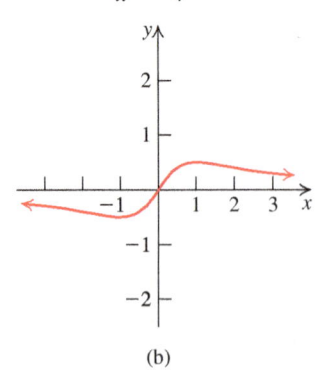

(b)

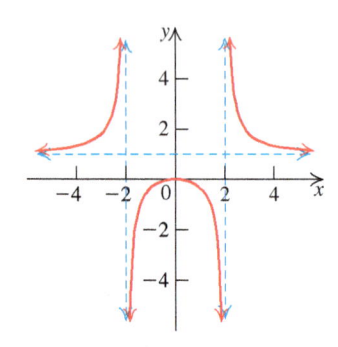

In Exercises 67–70, find an equation of a rational function having the given asymptotes, intercepts, and graph.

67.

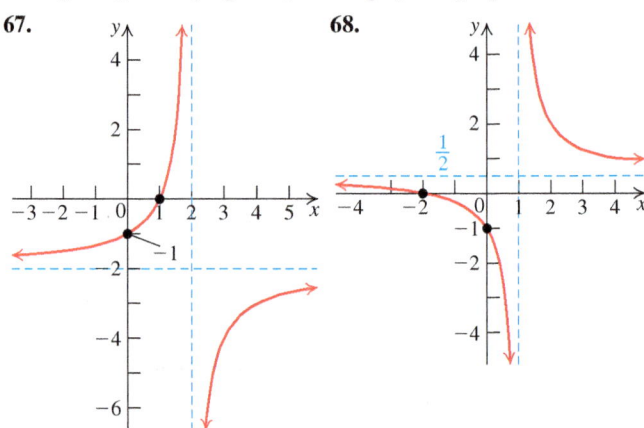

68.

(c)

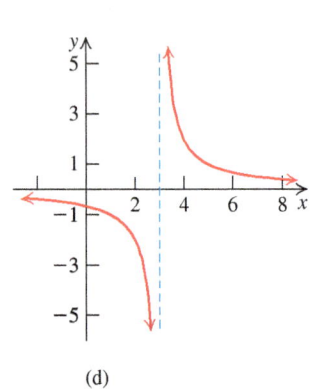

(d)

69.

70.

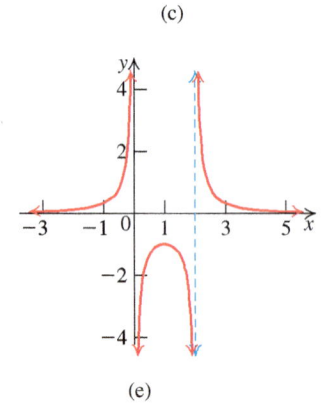

(e)

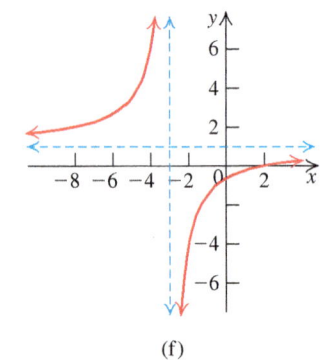

(f)

In Exercises 51–66, use the six-step procedure on pages 206–207 to graph each rational function.

51. $f(x) = \dfrac{2x}{x - 3}$

52. $f(x) = \dfrac{-x}{x - 1}$

53. $f(x) = \dfrac{x}{x^2 - 4}$

54. $f(x) = \dfrac{x}{1 - x^2}$

55. $h(x) = \dfrac{-2x^2}{x^2 - 9}$

56. $h(x) = \dfrac{4 - x^2}{x^2}$

57. $f(x) = \dfrac{2}{x^2 - 2}$

58. $f(x) = \dfrac{-2}{x^2 - 3}$

59. $g(x) = \dfrac{x + 1}{(x - 2)(x + 3)}$

60. $g(x) = \dfrac{x - 1}{(x + 1)(x - 2)}$

61. $h(x) = \dfrac{x^2}{x^2 + 1}$

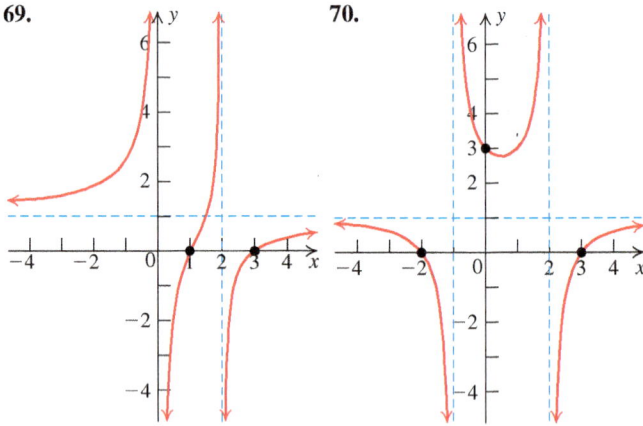

In Exercises 71–78, find the oblique asymptote and sketch the graph of each rational function.

71. $f(x) = \dfrac{2x^2 + 1}{x}$

72. $f(x) = \dfrac{x^2 - 1}{x}$

73. $g(x) = \dfrac{x^3 - 1}{x^2}$

74. $g(x) = \dfrac{2x^3 + x^2 + 1}{x^2}$

75. $h(x) = \dfrac{x^2 - x + 1}{x + 1}$

76. $h(x) = \dfrac{2x^2 - 3x + 2}{x - 1}$

77. $f(x) = \dfrac{x^3 - 2x^2 + 1}{x^2 - 1}$

78. $h(x) = \dfrac{x^3 - 1}{x^2 - 4}$

Applying the Concepts

79. Biology: birds collecting seeds. A bird is collecting seed from a field that contains 100 grams of seed. The bird collects x grams of seed in t minutes, where

$$t = f(x) = \frac{4x + 1}{100 - x}, 0 < x < 100.$$

   a. Sketch the graph of $t = f(x)$.
   b. How long does it take the bird to collect
       (i) 50 grams?
       (ii) 75 grams?
       (iii) 95 grams?
       (iv) 99 grams?
   c. Complete the following statements if applicable:
       (i) As $x \to 100^-$, $f(x) \to$ _____.
       (ii) As $x \to 100^+$, $f(x) \to$ _____.
   d. Does the bird ever collect all of the seed from the field?

80. Criminology. Suppose the city of Las Vegas decides to be crime-free. The estimated cost of catching and convicting $x\%$ of the criminals is given by

$$C(x) = \frac{1000}{100 - x} \text{ million dollars.}$$

   a. Find and interpret $C(50)$, $C(75)$, $C(90)$, and $C(99)$.
   b. Sketch a graph of $C(x)$, $0 \le x < 100$.
   c. What happens to $C(x)$ as $x \to 100^-$?
   d. What percentage of the criminals can be caught and convicted for $30 million?

81. Environment. Suppose an environmental agency decides to get rid of the impurities from the water of a polluted river. The estimated cost of removing $x\%$ of the impurities is given by

$$C(x) = \frac{3x^2 + 50}{x(100 - x)} \text{ billion dollars.}$$

   a. How much will it cost to remove 50% of the impurities?
   b. Sketch the graph of $y = C(x)$.
   c. Estimate the percentage of the impurities that can be removed at a cost of $30 billion.

82. Biology. The growth function $g(x)$ describes the growth rate of organisms as a function of some nutrient concentration x. Suppose

$$g(x) = a\frac{x}{k + x}, \quad x \ge 0,$$

where a and k are positive constants.
   a. Find the horizontal asymptote of the graph of $g(x)$. Use the asymptote to explain why a is called the *saturation level* of the nutrient.
   b. Show that k is the half-saturation constant; that is, show that if $x = k$, then $g(x) = \frac{a}{2}$.

83. Population of bacteria. The population P (in thousands) of a colony of bacteria at time t (in hours) is given by

$$P(t) = \frac{8t + 16}{2t + 1}, \quad t \ge 0.$$

   a. Find the initial population of the colony. (Find the population at $t = 0$ hours.)
   b. What is the long-term behavior of the population? (What happens when $t \to \infty$?)

Critical Thinking / Discussion / Writing

In Exercises 84–87, find an equation of a rational function f satisfying the given conditions.

84. Vertical asymptote: $x = 3$
Horizontal asymptote: $y = -1$
x-intercept: 2

85. Vertical asymptote: $x = -1, x = 1$
Horizontal asymptote: $y = 1$
x-intercept: 0

86. Vertical asymptote: $x = 0$
Oblique asymptote: $y = -x$
x-intercepts: $-1, 1$

87. Vertical asymptote: $x = 3$
Oblique asymptote: $y = x + 4$
y-intercept: 2; $f(4) = 14$

In Exercises 88–93, make up a rational function $f(x)$ that has all of the characteristics given in the exercise.

88. Has a vertical asymptote at $x = 2$, a horizontal asymptote at $y = 1$, and a y-intercept at $(0, -2)$

89. Has vertical asymptotes at $x = 2$ and $x = -1$, a horizontal asymptote at $y = 0$, a y-intercept at $(0, 2)$, and an x-intercept at 4

90. Has $f\left(\frac{1}{2}\right) = 0$; $f(x) \to 4$ as $x \to \pm\infty$,
$f(x) \to \infty$ as $x \to 1^-$, and $f(x) \to \infty$ as $x \to 1^+$

91. Has $f(0) = 0$; $f(x) \to 2$ as $x \to \pm\infty$; has no vertical asymptotes and is symmetric about the y-axis

92. Has $y = 3x + 2$ as an oblique asymptote; has a vertical asymptote at $x = 1$

93. Is it possible for the graph of a rational function to have both a horizontal and oblique asymptote? Explain.

Maintaining Skills

94. Find the equation of the line in slope–intercept form that passes through the point $(-5, 3)$ and has slope $-\frac{2}{3}$.

95. Find the equation of the line in slope–intercept form that passes through the point $(-2, -3)$ and is perpendicular to the line $4x + 5y = 6$.

96. Solve: $\dfrac{x^2 - 9}{x + 1} \ge 0$

97. Solve: $|2x - 1| = |x + 3|$

SUMMARY Definitions, Concepts, and Formulas

2.1 Quadratic Functions

i. A **quadratic function** f is a function of the form
$$f(x) = ax^2 + bx + c, a \neq 0.$$

ii. The **standard form** of a quadratic function is
$$f(x) = a(x - h)^2 + k, a \neq 0.$$

iii. The graph of a quadratic function is a transformation of the graph of $y = x^2$.

iv. The graph of a quadratic function is a parabola with vertex
$$(h, k) = \left(-\frac{b}{2a}, f\left(-\frac{b}{2a}\right)\right).$$

v. The maximum (if $a < 0$) or minimum (if $a > 0$) value of a quadratic function $f(x) = ax^2 + bx + c$ occurs at the vertex of the parabola.

2.2 Polynomial Functions

i. A function f of the form
$$f(x) = a_n x^n + a_{n-1}x^{n-1} + \cdots + a_1 x + a_0, a_n \neq 0$$
is a polynomial function of degree n.

ii. The graph of a polynomial function is smooth and continuous.

iii. The end behavior of the graph of a polynomial function depends upon the sign of the leading coefficient and the degree of the polynomial.

iv. A real number c is a **zero** of a function f if $f(c) = 0$. Geometrically, c is an x-intercept of the graph of $y = f(x)$.

v. If in the factorization of a polynomial function $f(x)$ the factor $(x - a)$ occurs exactly m times, then a is a zero of **multiplicity** m. If m is odd, the graph of $y = f(x)$ crosses the x-axis at a; if m is even, the graph touches but does not cross the x-axis at a.

vi. If the degree of a polynomial function $f(x)$ is n, then $f(x)$ has, at most, n real zeros and the graph of $f(x)$ has, at most, $(n - 1)$ turning points.

vii. An **Intermediate Value Theorem.** Let $f(x)$ be a polynomial function and a and b be two numbers such that $a < b$. If $f(a)$ and $f(b)$ have opposite signs, then there is at least one number c, with $a < c < b$, for which $f(c) = 0$.

viii. See Example 9 in Objective 5 for graphing a polynomial function.

2.3 Dividing Polynomials and the Rational Zeros Test

i. **Division Algorithm.** If a polynomial $F(x)$ is divided by a polynomial $D(x) \neq 0$, there are unique polynomials $Q(x)$ and $R(x)$ such that $F(x) = D(x)Q(x) + R(x)$, where either $R(x) = 0$ or deg $R(x) <$ deg $D(x)$. In words, "The dividend equals the product of the divisor and the quotient plus the remainder."

ii. **Remainder Theorem.** If a polynomial $F(x)$ is divided by $(x - a)$, the remainder is $F(a)$.

iii. **Factor Theorem.** A polynomial function $F(x)$ has $(x - a)$ as a factor if and only if $F(a) = 0$.

iv. **Rational Zeros Test.** If $\frac{p}{q}$ is a rational zero in lowest terms for a polynomial function with integer coefficients, then p is a factor of the constant term and q is a factor of the leading coefficient.

2.4 Zeros of a Polynomial Function

i. **Descartes's Rule of Signs.** Let $F(x)$ be a polynomial function with real coefficients.
   a. The number of positive zeros of F is equal to the number of variations of sign of $F(x)$ or is less than that number by an even integer.
   b. The number of negative zeros of F is equal to the number of variations of sign of $F(-x)$ or is less than that number by an even integer.

ii. **Rules for Bounds on the Zeros.** Suppose a polynomial $F(x)$ is synthetically divided by $x - k$.
   a. If $k > 0$ and each number in the last row is zero or positive, then k is an upper bound on the zeros of $F(x)$.
   b. If $k < 0$ and the numbers in the last row alternate in sign, then k is a lower bound on the zeros of $F(x)$.

iii. **The Fundamental Theorem of Algebra.** An nth-degree polynomial equation has at least one complex zero.

iv. **Factorization Theorem for Polynomials.** If $P(x)$ is a polynomial of degree $n \geq 1$, it can be factored into n (not necessarily distinct) linear factors of the form $P(x) = a(x - r_1)(x - r_2) \cdots (x - r_n)$, where a, $r_1, r_2, \ldots, r_n$ are complex numbers.

v. **Number of Zeros Theorem.** A polynomial of degree n has exactly n complex zeros provided a zero of multiplicity k is counted k times.

vi. **Conjugate Pairs Theorem.** If $a + bi$ is a zero of the polynomial function P (with real coefficients), then $a - bi$ is also a zero of P.

2.5 Rational Functions

i. A function $F(x) = \dfrac{N(x)}{D(x)}$, where $N(x)$ and $D(x)$ are polynomials and $D(x) \neq 0$, is called a rational function. The domain of $F(x)$ is the set of all real numbers except the real zeros of $D(x)$.

ii. The line $x = a$ is a vertical asymptote of the graph of f if, $f(x) \to \infty$, or $f(x) \to -\infty$, as $x \to a^+$ or as $x \to a^-$.

iii. If $\dfrac{N(x)}{D(x)}$ is in lowest terms, then the graph of $F(x)$ has vertical asymptotes at the real zeros of $D(x)$.

iv. The line $y = k$ is a horizontal asymptote of the graph of a function if $f(x) \to k$ as $x \to \infty$ or if $f(x) \to k$ as $x \to -\infty$.

v. The techniques of vertical stretching and compressing, shifting, and/or reflecting can be used to graph any function of the form $g(x) = \dfrac{ax + b}{cx + d}$ starting with the graph of $f(x) = \dfrac{1}{x}$.

vi. The line $y = mx + b$ is an oblique asymptote of the graph of a function f if the difference $f(x) - (mx + b) \to 0$ as $x \to \infty$ or as $x \to -\infty$.

vii. The line $y = k$ is a horizontal asymptote of the graph of f if $f(x) \to k$ as $x \to \infty$ or $f(x) \to k$ as $x \to -\infty$.

viii. A procedure for graphing rational functions is given in Example 6 in Objective 5.

REVIEW EXERCISES

Basic Concepts and Skills

In Exercises 1–10, graph each quadratic function by finding (i) whether the parabola opens up or down, (ii) its vertex, (iii) its axis, (iv) its x-intercepts, (v) its y-intercept, and (vi) the intervals over which the function is increasing and decreasing.

1. $y = (x - 1)^2 + 2$
2. $y = (x + 2)^2 - 3$
3. $y = -2(x - 3)^2 + 4$
4. $y = -\dfrac{1}{2}(x + 1)^2 + 2$
5. $y = -2x^2 + 3$
6. $y = 2x^2 + 4x - 1$
7. $y = 2x^2 - 4x + 3$
8. $y = -2x^2 - x + 3$
9. $y = 3x^2 - 2x + 1$
10. $y = 3x^2 - 5x + 4$

In Exercises 11–14, determine whether the given quadratic function has a maximum or a minimum value and then find that value.

11. $f(x) = 3 - 4x + x^2$
12. $f(x) = 8x - 4x^2 - 3$
13. $f(x) = -2x^2 - 3x + 2$
14. $f(x) = \dfrac{1}{2}x^2 - \dfrac{3}{4}x + 2$

In Exercises 15–18, graph each polynomial function by using transformations on the appropriate function $y = x^n$.

15. $f(x) = (x + 1)^3 - 2$
16. $f(x) = (x + 1)^4 + 2$
17. $f(x) = (1 - x)^3 + 1$
18. $f(x) = x^4 + 3$

In Exercises 19–24, for each polynomial function f,
i. determine the end behavior of f.
ii. determine the zeros of f. State the multiplicity of each zero. Determine whether the graph of f crosses or only touches the axis at each x-intercept.
iii. find the x- and y-intercepts of the graph of f.
iv. use test numbers to find the intervals over which the graph of f is above or below the x-axis.
v. find any symmetry.
vi. sketch the graph of $y = f(x)$.

19. $f(x) = x(x - 1)(x + 2)$
20. $f(x) = x^3 - x$
21. $f(x) = -x^2(x - 1)^2$
22. $f(x) = -x^3(x - 2)^2$
23. $f(x) = -x^2(x^2 - 1)$
24. $f(x) = -(x - 1)^2(x^2 + 1)$

In Exercises 25–28, divide by using long division.

25. $\dfrac{6x^2 + 5x - 13}{3x - 2}$
26. $\dfrac{8x^2 - 14x + 15}{2x - 3}$
27. $\dfrac{8x^4 - 4x^3 + 2x^2 - 7x + 165}{x + 1}$
28. $\dfrac{x^3 - 3x^2 + 4x + 7}{x^2 - 2x + 6}$

In Exercises 29–32, divide by using synthetic division.

29. $\dfrac{x^3 - 12x + 3}{x - 3}$
30. $\dfrac{-4x^3 + 3x^2 - 5x}{x - 6}$

31. $\dfrac{2x^4 - 3x^3 + 5x^2 - 7x + 165}{x + 1}$

32. $\dfrac{3x^5 - 2x^4 + x^2 - 16x - 132}{x + 2}$

In Exercises 33–36, a polynomial function $f(x)$ and a constant c are given. Find $f(c)$ by (i) evaluating the function and (ii) using synthetic division and the Remainder Theorem.

33. $f(x) = x^3 - 3x^2 + 11x - 29;\ c = 2$
34. $f(x) = 2x^3 + x^2 - 15x - 2;\ c = -2$
35. $f(x) = x^4 - 2x^2 - 5x + 10;\ c = -3$
36. $f(x) = x^5 + 2;\ c = 1$

In Exercises 37–40, a polynomial function $f(x)$ and a constant c are given. Use synthetic division to show that c is a zero of $f(x)$. Use the result to final all zeros of $f(x)$.

37. $f(x) = x^3 - 7x^2 + 14x - 8;\ c = 2$
38. $f(x) = 2x^3 - 3x^2 - 12x + 4;\ c = -2$
39. $f(x) = 3x^3 + 14x^2 + 13x - 6;\ c = \dfrac{1}{3}$
40. $f(x) = 4x^3 + 19x^2 - 13x + 2;\ c = \dfrac{1}{4}$

In Exercises 41 and 42, use the Rational Zeros Test to list all possible rational zeros of $f(x)$.

41. $f(x) = x^4 + 3x^3 - x^2 - 9x - 6$
42. $f(x) = 9x^3 - 36x^2 - 4x + 16$

In Exercises 43–46, use Descartes's Rule of Signs and the Rational Zeros Test to find all real zeros of each polynomial function.

43. $f(x) = 5x^3 + 11x^2 + 2x$
44. $f(x) = x^3 + 2x^2 - 5x - 6$
45. $f(x) = x^3 + 3x^2 - 4x - 12$
46. $f(x) = 2x^3 - 9x^2 + 12x - 5$

In Exercises 47–52, find all of the zeros of $f(x)$, real and nonreal, that are not given.

47. $f(x) = x^3 - 7x + 6$; one zero is 2.
48. $f(x) = x^4 + x^3 - 3x^2 - x + 2$; 1 is a zero of multiplicity 2.
49. $f(x) = x^4 - 2x^3 + 6x^2 - 18x - 27$; two zeros are -1 and 3.
50. $f(x) = 4x^3 - 19x^2 + 32x - 15$; one zero is $2 - i$.
51. $f(x) = x^4 + 2x^3 + 9x^2 + 8x + 20$; one zero is $-1 + 2i$.
52. $f(x) = x^5 - 7x^4 + 24x^3 - 32x^2 + 64$; $2 + 2i$ is a zero of multiplicity 2.

In Exercises 53–58, solve each equation in the complex number system.

53. $x^3 - x^2 - 4x + 4 = 0$
54. $2x^3 + x^2 - 12x - 6 = 0$
55. $4x^3 - 7x - 3 = 0$
56. $x^3 + x^2 - 8x - 6 = 0$
57. $x^4 - x^3 - x^2 - x - 2 = 0$
58. $x^4 - x^3 - 13x^2 + x + 12 = 0$

In Exercises 59 and 60, show that the given equation has no rational roots.

59. $x^3 + 13x^2 - 6x - 2 = 0$
60. $3x^4 - 9x^3 - 2x^2 - 15x - 5 = 0$

In Exercises 61 and 62, use an Intermediate Value Theorem to find the value of the real root between 1 and 2 of each equation to two decimal places.

61. $x^3 + 6x^2 - 28 = 0$
62. $x^3 + 3x^2 - 3x - 7 = 0$

In Exercises 63–70, graph each rational function by following the six-step procedure outlined in Section 2.5.

63. $f(x) = 1 + \dfrac{1}{x}$
64. $f(x) = \dfrac{2 - x}{x}$
65. $f(x) = \dfrac{x}{x^2 - 1}$
66. $f(x) = \dfrac{x^2 - 9}{x^2 - 4}$
67. $f(x) = \dfrac{x^3}{x^2 - 9}$
68. $f(x) = \dfrac{x + 1}{x^2 - 2x - 8}$
69. $f(x) = \dfrac{x^4}{x^2 - 4}$
70. $f(x) = \dfrac{x^2 + x - 6}{x^2 - x - 12}$

Applying the Concepts

71. **Missile path.** A missile fired from the origin of a coordinate system follows a path described by the equation $y = -\dfrac{1}{10}x^2 + 20x$, where the x-axis is at ground level. Sketch the missile's path and determine what its maximum altitude is and where it hits the ground.

72. **Minimizing area.** Suppose a wire 20 centimeters long is to be cut into two pieces, each of which will be formed into a square. Find the size of each piece that minimizes the total area.

73. **Minimizing length.** A farmer wants to fence off three identical adjoining rectangular pens, each 400 square feet in area. See the figure. What should be the width and length of each pen so that the least amount of fence is used?

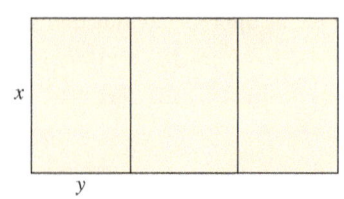

74. **Minimizing cost.** Suppose the outer boundary of the pens in Exercise 73 requires heavy fence that costs $5 per foot and two internal partitions cost $3 per foot. What dimensions x and y will minimize the cost?

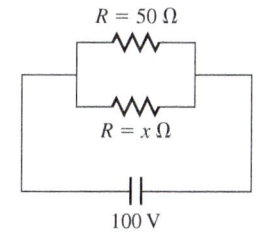

75. Electric circuit. In the circuit shown in the figure, the voltage $V = 100$ volts and the resistance $R = 50$ ohms. We want to determine the size of the remaining resistor (x ohms). The power absorbed by the circuit is given by $p(x) = \dfrac{V^2 x}{(R + x)^2}$.

$R = 50\,\Omega$

$R = x\,\Omega$

$100\,V$

  a. Graph the function $y = p(x)$.
  b. Use a graphing calculator to find the value of x that maximizes the power absorbed.

76. Maximizing area. A sheet of paper for a poster is 18 square feet in area. The margins at the top and bottom are 9 inches each, and the margin on each side is 6 inches. What should the dimensions of the paper be if the printed area is to be a maximum?

77. Meteorology. The function $p = \dfrac{69.1}{a + 2.3}$ relates the atmospheric pressure p in inches of mercury to the altitude a in miles from the surface of Earth.
  a. Find the pressure on Mount Kilimanjaro at an altitude of 19,340 feet.
  b. Is there an altitude at which the pressure is 0?

EXERCISES FOR CALCULUS

1. Simplify $2(1 - 2x + 4x^2) + (2x + 1)(8x - 2)$.
2. Simplify $\dfrac{(x - 1)(2x - 2) - (x^2 + 2x - 3)}{(x - 1)^2}$

In Exercises 3–6, find all x-coordinates for the point of intersection of the graphs of the given functions. Then for each pair of consecutive x-coordinates a and b, with $a < b$, state whether $f(x) > g(x)$ or $f(x) < g(x)$ for all x in the interval (a, b).

3. $f(x) = x^2 - 1$; $g(x) = x + 1$
4. $f(x) = x^2$; $g(x) = 8 - x^2$
5. $f(x) = x^3 - 4x^2 + 3x$; $g(x) = 0$
6. $f(x) = 3x^3 + 1$; $g(x) = 2x^3 + x^2 + 4x - 3$
7. Sketch and discuss the graphs of the rational functions of the form $y = \dfrac{a}{x^n}$, where a is a nonzero real number and n is a positive integer, in the following cases:
   a. $a > 0$ and n is odd.
   b. $a < 0$ and n is odd.
   c. $a > 0$ and n is even.
   d. $a < 0$ and n is even.
8. **Graphing the reciprocal of a polynomial function.** Let $f(x)$ be a polynomial function and $g(x) = \dfrac{1}{f(x)}$. Justify the following statements about the graphs of $f(x)$ and $g(x)$:
   a. If c is a zero of $f(x)$, then $x = c$ is a vertical asymptote of the graph of $g(x)$.
   b. If $f(x) > 0$, then $g(x) > 0$, and if $f(x) < 0$, then $g(x) < 0$.
   c. The graphs of f and g intersect for those values (and only those values) of x for which $f(x) = g(x) = \pm 1$.
   d. When f is increasing (decreasing, constant) on an interval of its domain, g is decreasing (increasing, constant) on that interval.
9. Use Exercise 8 to sketch the graphs of $f(x) = x^2 - 4$ and $g(x) = \dfrac{1}{x^2 - 4}$ on the same coordinate axes.
10. Use Exercise 8 to sketch the graphs of $f(x) = x^2 + 1$ and $g(x) = \dfrac{1}{x^2 + 1}$ on the same coordinate axes.
11. Recall that the inverse of a function $f(x)$ is written as $f^{-1}(x)$, while the reciprocal of $f(x)$ can be written as either $\dfrac{1}{f(x)}$ or $[f(x)]^{-1}$. The point of this exercise is to make clear that the reciprocal of a function has nothing to do with the inverse of a function. Let $f(x) = 2x + 3$. Find both $[f(x)]^{-1}$ and $f^{-1}(x)$. Compare the two functions. Graph all three functions on the same coordinate axes.
12. Let $f(x) = \dfrac{x - 1}{x + 2}$. Find $[f(x)]^{-1}$ and $f^{-1}(x)$. Graph all three functions on the same coordinate axes.
13. Express the surface area A, of a cylindrical can with volume $V\,\text{in}^3$ (having both top and bottom) as a function of the radius of its base, x, given in inches.

14. The tangent line to a circle at any point on the circle is the line that intersects the circle at only that point. It is perpendicular to the line containing the point and the circle's center. The graph of $x^2 + y^2 = 25$ is a circle of radius 5 with center at the origin of the coordinate plane. The graph of the half-circle with negative second coordinate is shown in the figure; the point $(3, -4)$ is on this graph.

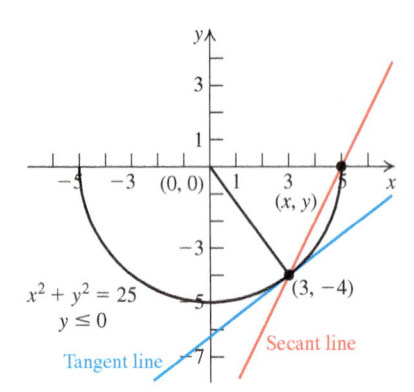

a. Use the fact that the line containing the radius from the origin to the point $(3, -4)$ is perpendicular to the tangent line to the circle through $(3, -4)$ to find the slope of this tangent line.

b. For h positive and $h \leq 1$, find an expression for the slope of the secant line containing the circle points $(3, -4)$ and $[3 + h, -\sqrt{25 - (3 + h)^2}]$ and write it in the form $\dfrac{a - b}{h}$, where a is an integer and b is an expression in h.

c. Multiply the numerator and the denominator of the expression found in part (b) by $a + b$ and simplify the result to obtain an expression for the slope of this secant line in the form $\dfrac{h(mh + k)}{h(a + b)} = \dfrac{mh + k}{a + b}, h \neq 0.$

d. Use a calculator to evaluate the expression $\dfrac{mh + k}{a + b}$ from part (c) for values of h in the table and the difference between this expression and the slope of the tangent line found in part (a) (to four decimal places). What can you say about the slopes of the secant lines compared to the slope of the tangent line at $(3, -4)$ when h is close to zero?

h	$3 + h$	Expression Value	Tangent's Slope	Difference
−0.001	2.999	0.74980	0.75	−0.00020
−0.0001	2.9999	0.74998	0.75	−0.00002
−0.00001	2.99999	0.75000	0.75	0.00000
0	3	0.75000	0.75	0.00000
0.00001	3.00001	0.75000	0.75	0.00000
0.0001	3.0001	0.75002	0.75	0.00002
0.001	3.001	0.75020	0.75	0.00020

PRACTICE TEST A

1. Find the x-intercepts of the graph of $f(x) = x^2 - 6x + 2$.
2. Graph $f(x) = 3 - (x + 2)^2$.
3. Find the vertex of the parabola described by $y = -7x^2 + 14x + 3$.
4. Find the domain of the function $f(x) = \dfrac{x^2 - 1}{x^2 + 3x - 4}.$
5. Find the quotient and remainder of $\dfrac{x^3 - 2x^2 - 5x + 6}{x + 2}.$
6. Graph the polynomial function $P(x) = x^5 - 4x^3$.
7. Find all of the zeros of $f(x) = 2x^3 - 2x^2 - 8x + 8$, given that 2 is one of the zeros.
8. Find the quotient of $\dfrac{-6x^3 + x^2 + 17x + 3}{2x + 3}.$
9. Use the Remainder Theorem to find the value $P(-2)$ of the polynomial $P(x) = x^4 + 5x^3 - 7x^2 + 9x + 17$.
10. Find all rational roots of the equation $x^3 - 5x^2 - 4x + 20 = 0$ and then find the irrational roots if there are any.
11. Find the zeros of the polynomial function $f(x) = x^4 + x^3 - 15x^2$.
12. For $P(x) = 2x^{18} - 5x^{13} + 6x^3 - 5x + 9$, list all possible rational zeros found by the Rational Zeros Test, but do not check to see which values are actually zeros.
13. Describe the end behavior of $f(x) = (x + 3)^3(x - 5)^2$.
14. Find the zeros and the multiplicity of each zero for $f(x) = (x^2 - 4)(x + 2)^2$.
15. Determine how many positive and how many negative real zeros the polynomial function $P(x) = 3x^6 + 2x^3 - 7x^2 + 8x$ can have.
16. Find the horizontal and the vertical asymptotes of the graph of
$$f(x) = \dfrac{2x^2 + 3}{x^2 - x - 20}.$$
17. Find the oblique asymptote for $f(x) = \dfrac{x^2 - 14x + 5}{x - 8}.$

18. Where does the graph of $f(x) = -3(x + 1)^4(x - 3)^5$ cross the x-axis?

19. The cost C of producing x thousand units of a product is given by

$$C = x^2 - 30x + 335 \text{ (dollars)}.$$

Find the value of x for which the cost is minimum.

20. From a rectangular 8×17 piece of cardboard, four congruent squares with sides of length x are cut out, one at each corner. The sides can then be folded to form a box. Find the volume V of the box as a function of x.

PRACTICE TEST B

1. Find the x-intercepts of the graph of $f(x) = x^2 + 5x + 3$.
  a. $\dfrac{-5 \pm i\sqrt{13}}{2}$
  b. $\dfrac{-5 \pm \sqrt{13}}{2}$
  c. $\dfrac{-5 \pm i\sqrt{37}}{2}$
  d. $\dfrac{-5 \pm \sqrt{37}}{2}$

2. Which is the graph of $f(x) = 4 - (x - 2)^2$?

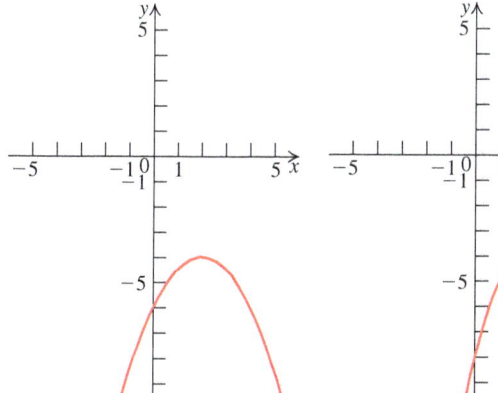

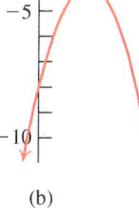

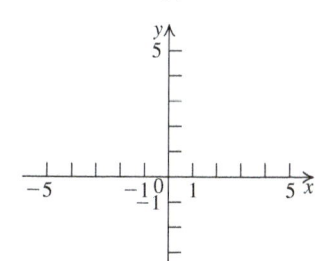

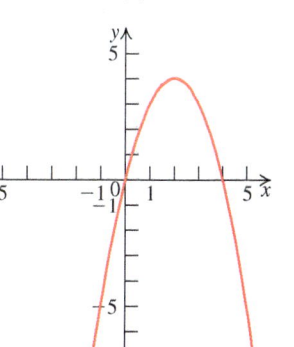

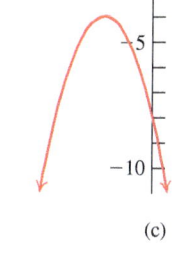

(a) (b) (c) (d)

3. Find the vertex of the parabola described by $y = 6x^2 + 12x - 5$.
  a. $(-1, -11)$
  b. $(1, -5)$
  c. $(-1, 13)$
  d. $(1, 13)$

4. Which of the following is *not* in the domain of the function $f(x) = \dfrac{x^2}{x^2 + x - 6}$?
  I. -3
  II. 0
  III. 2
  a. I and II
  b. I and III
  c. II and III
  d. I only

5. Find the quotient and remainder when $x^3 - 8x + 6$ is divided by $x + 3$.
  a. $x^2 - 8; 2$
  b. $x^2 - 8; 0$
  c. $x^2 - 3x + 1; x + 3$
  d. $x^2 - 3x + 1; 3$

6. Which is the graph of the polynomial $P(x) = x^4 + 2x^3$?

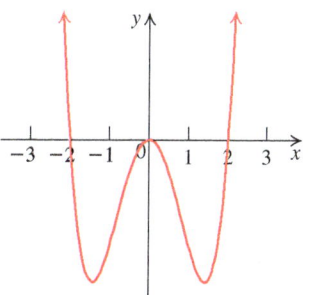

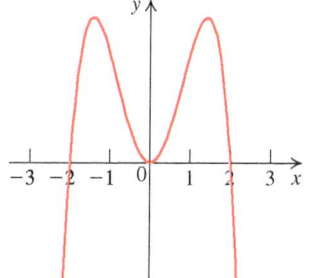

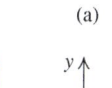

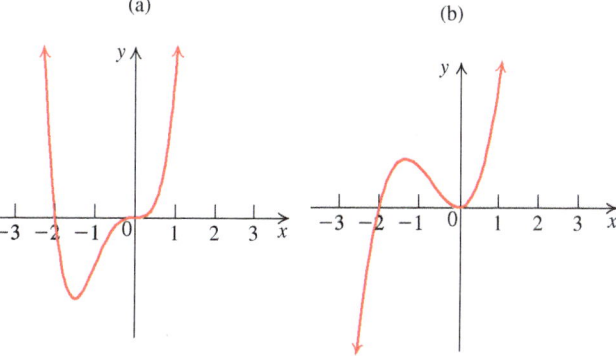

(a) (b) (c) (d)

7. Find all of the zeros of $f(x) = 3x^3 - 26x^2 + 61x - 30$ given that 3 is one of the zeros [that is, $f(3) = 0$].
  a. $3, -5, -\dfrac{2}{3}$
  b. $3, 2, \dfrac{5}{3}$
  c. $3, 5, \dfrac{2}{3}$
  d. $3, -2, \dfrac{5}{3}$

8. Find the quotient $\dfrac{-10x^3 + 21x^2 - 17x + 12}{2x - 3}$.
   a. $x^2 - 3x + 4$
   b. $-5x^2 + 3x - 4$
   c. $x^2 + 3x - 4$
   d. $-5x^2 - 4$

9. Use the Remainder Theorem to find the value $P(-3)$ of the polynomial $P(x) = x^4 + 4x^3 + 7x^2 + 10x + 15$.
   a. 13
   b. 15
   c. 21
   d. 6

10. Find all rational roots of the equation $-x^3 + x^2 + 8x - 12 = 0$ and then find the irrational roots if there are any.
    a. -3 and 2
    b. -3 and $\sqrt{2}$
    c. $-1, -2,$ and 3
    d. -3 and $\sqrt{3}$

11. Find the zeros of the polynomial function $f(x) = x^3 + x^2 - 30x$.
    a. $x = -6, x = 5, x = 0$
    b. $x = 0, x = -6$
    c. $x = 4, x = 5$
    d. $x = 0, x = 4, x = 5$

12. For $P(x) = x^{30} - 4x^{25} + 6x^2 + 60$, list all possible rational zeros found by the Rational Zeros Test (but do not check to see which values are actually zeros).
    a. $\pm 2, \pm 3, \pm 4, \pm 5, \pm 6, \pm 8, \pm 10, \pm 12, \pm 15, \pm 20, \pm 30,$ and ± 60
    b. $\pm 1, \pm 2, \pm 3, \pm 4, \pm 5, \pm 6, \pm 10, \pm 12, \pm 15, \pm 18, \pm 20, \pm 24, \pm 30,$ and ± 60
    c. $\pm 1, \pm 3, \pm 4, \pm 5, \pm 6, \pm 12, \pm 15, \pm 20, \pm 30,$ and ± 60
    d. $\pm 1, \pm 2, \pm 3, \pm 4, \pm 5, \pm 6, \pm 10, \pm 12, \pm 15, \pm 20, \pm 30,$ and ± 60

13. Which of the following correctly describes the end behavior of $f(x) = (x + 1)^2 (x - 2)^2$?
    a. $\begin{cases} y \to \infty \text{ as } x \to -\infty \\ y \to -\infty \text{ as } x \to \infty \end{cases}$
    b. $\begin{cases} y \to -\infty \text{ as } x \to -\infty \\ y \to -\infty \text{ as } x \to \infty \end{cases}$
    c. $\begin{cases} y \to \infty \text{ as } x \to -\infty \\ y \to \infty \text{ as } x \to \infty \end{cases}$
    d. $\begin{cases} y \to -\infty \text{ as } x \to -\infty \\ y \to \infty \text{ as } x \to \infty \end{cases}$

14. Find the zeros and the multiplicity of each zero for $f(x) = (x^2 - 1)(x + 1)^2$.
    a. $\begin{cases} \text{zero } 1, & \text{multiplicity 1} \\ \text{zero} -1, & \text{multiplicity 2} \end{cases}$
    b. $\begin{cases} \text{zero } 1, & \text{multiplicity 1} \\ \text{zero} -1, & \text{multiplicity 3} \end{cases}$
    c. $\begin{cases} \text{zero } 1, & \text{multiplicity 2} \\ \text{zero} -1, & \text{multiplicity 2} \end{cases}$
    d. $\begin{cases} \text{zero } i, & \text{multiplicity 2} \\ \text{zero} -1, & \text{multiplicity 2} \end{cases}$

15. Determine how many positive and how many negative real zeros the polynomial $P(x) = x^5 - 4x^3 - x^2 + 6x - 3$ can have.
    a. positive 3, negative 2
    b. positive 2 or 0, negative 2 or 0
    c. positive 3 or 1, negative 2 or 0
    d. positive 2 or 0, negative 3 or 1

16. The horizontal and vertical asymptotes of the graph of $f(x) = \dfrac{x^2 + x - 2}{x^2 + x - 12}$ are
    a. $x = -2, y = 3, y = -4$.
    b. $x = 1, y = 3, y = -4$.
    c. $y = 1, x = 3, x = -4$.
    d. $y = -2, x = 3, x = -4$.

17. Which of the following rational functions has an oblique asymptote?
    a. $r(x) = \dfrac{x^2 - 14x + 52}{x^2 - 64}$
    b. $r(x) = \dfrac{x^2 + 2x - 43}{x^2 - 9x + 8}$
    c. $r(x) = \dfrac{2x - 4}{x - 8}$
    d. $r(x) = \dfrac{x^2 - 14x + 52}{x - 8}$

18. The graph of $4(x - 1)^4(x + 2)^3$
    a. does not cross the x-axis at either -2 or 1.
    b. crosses the x-axis at both -2 and 1.
    c. touches but does not cross the x-axis at -2 and crosses the x-axis at 1.
    d. touches but does not cross the x-axis at 1 and crosses the x-axis at -2.

19. The cost C of producing x thousand units of a product is given by
    $$C = x^2 - 24x + 319 \text{ (dollars)}.$$
    Find the value of x for which the cost is minimum.
    a. 319
    b. 12
    c. 175
    d. 24

20. From a rectangular 10×12 piece of cardboard, four congruent squares with sides of length x are cut out, one at each corner. The sides can then be folded to form a box. Find the volume V of the box as a function of x.
    a. $V = 2x(10 - x)(12 - x)$
    b. $V = 2x(10 - 2x)(12 - 2x)$
    c. $V = x(10 - x)(12 - x)$
    d. $V = x(10 - 2x)(12 - 2x)$

CHAPTER 3

Exponential and Logarithmic Functions

TOPICS

3.1 Exponential Functions

3.2 Logarithmic Functions

3.3 Rules of Logarithms

3.4 Exponential and Logarithmic Equations and Inequalities

The world of finance, the growth of many populations, the molecular activity that results in an atomic explosion, and countless everyday events are modeled with exponential and logarithmic functions.

SECTION 3.1

Exponential Functions

BEFORE STARTING THIS SECTION, REVIEW

1. Integer exponents (Section P.1, page 8)
2. Rational exponents (Section P.2, page 17)
3. Graphing and transformations (Section 1.5, page 109)
4. One-to-one functions (Section 1.7, page 136)
5. Increasing and decreasing functions (Section 1.4, page 96)

OBJECTIVES

1. Define an exponential function.
2. Graph exponential functions.
3. Develop formulas for simple and compound interest.
4. Define the number *e*.
5. Graph the natural exponential function.
6. Model with exponential functions.

◆ Fooling a King

One night in northwest India, a wise man named Shashi invented a new game called "Shatranj" (chess). The next morning he took it to king Rai Bhalit, who was so impressed that he said to Shashi, "Name your reward." Shashi merely requested that 1 grain of wheat be placed on the first square of the chessboard, 2 grains on the second, 4 on the third, 8 on the fourth, and so on, for all 64 squares. The king agreed to his request, thinking the man was an eccentric fool for asking for only a few grains of wheat when he could have had gold, jewels, or even his daughter's hand in marriage.

You may know the end of this story. Much more wheat was needed to satisfy Shashi's request. The king could never fulfill his promise to Shashi, so instead he had him beheaded.

The details of this rapidly growing wheat phenomenon are given in the margin. Table 3.1 shows the number of grains of wheat that need to be stacked on each square of the chessboard.

These data can be modeled by the function

$$g(n) = 2^{n-1}$$

where $g(n)$ is the number of grains of wheat and n is the number of the square (or the *n*th square) on the chessboard. Each square has twice the number of grains as the previous square. The function *g* is an example of an *exponential function* with base 2, with *n* restricted to 1, 2, 3, ..., 64—the number of the chessboard squares. The name *exponential function* comes from the fact that the variable *n* occurs in the exponent. The base 2 is the *growth factor*.

When the growth rate of a quantity is directly proportional to the existing amount, the growth can be modeled by an exponential function. For example, exponential functions are often used to model the growth of investments and populations and the depreciation in the value of a boat or car. Example 11 discusses the growth of world populations.

TABLE 3.1 Grains of Wheat on a Chessboard

Square Number	Grains Placed on This Square
1	1 (=2^0)
2	2 (=2^1)
3	4 (=2^2)
4	8 (=2^3)
5	16 (=2^4)
6	32 (=2^5)
.	.
.	.
.	.
63	2^{62}
64	2^{63}

Exponential Functions

1 Define an exponential function.

Exponential Function

A function f of the form

$$f(x) = a^x, a > 0, \quad a \neq 1,$$

is called an **exponential function with base a and exponent x**. Its domain is $(-\infty, \infty)$.

In the definition of the exponential function, we rule out the base $a = 1$ because in this case, the function is simply the constant function $f(x) = 1^x$, or $f(x) = 1$. We exclude negative bases so that the domain includes all real numbers. For example, a cannot be -2 because $(-2)^{1/2} = \sqrt{-2}$ is not a real number. Other functions that have variables in the exponent, such as $f(x) = 4 \cdot 3^x$, $g(x) = -5 \cdot 4^{3x-2}$, and $h(x) = c \cdot a^x$ ($a > 0$ and $a \neq 1$), are also called exponential functions.

Evaluate Exponential Functions

We can evaluate exponential functions by using the laws of exponents and/or calculators, as in the next example. (See Section P.2 for the definition of a^x when x is a rational number.)

TECHNOLOGY CONNECTION

You can use either the ∧ or x^y key on your calculator to evaluate exponential functions. For example, to evaluate $4^{3.2}$, press

4 ∧ 3.2 ENTER

or

4 x^y 3.2 =. You will see the following display:

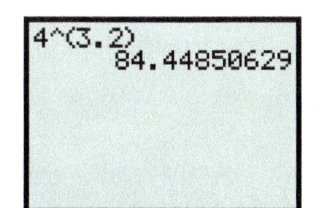

```
4^(3.2)
       84.44850629
```

Here the displayed number is an approximate value, so we should write $4^{3.2} \approx 84.44850629$.

EXAMPLE 1 Evaluating Exponential Functions

a. Let $f(x) = 3^{x-2}$. Find $f(4)$.
b. Let $g(x) = -2 \cdot 10^x$. Find $g(-2)$.
c. Let $h(x) = \left(\dfrac{1}{9}\right)^x$. Find $h\left(-\dfrac{3}{2}\right)$.
d. Let $F(x) = 4^x$. Find $F(3.2)$.

Solution

a. $f(4) = 3^{4-2} = 3^2 = 9$

b. $g(-2) = -2 \cdot 10^{-2} = -2 \cdot \dfrac{1}{10^2} = -2 \cdot \dfrac{1}{100} = -0.02$

c. $h\left(-\dfrac{3}{2}\right) = \left(\dfrac{1}{9}\right)^{-\frac{3}{2}} = (9^{-1})^{-\frac{3}{2}} = 9^{\frac{3}{2}} = (\sqrt{9})^3 = 3^3 = 27$

d. $F(3.2) = 4^{3.2} \approx 84.44850629$ Use a calculator.

Practice Problem 1 Let $f(x) = \left(\dfrac{1}{4}\right)^x$. Find $f(2), f(0), f(-1), f\left(\dfrac{5}{2}\right)$, and $f\left(-\dfrac{3}{2}\right)$.

The domain of an exponential function is $(-\infty, \infty)$. In Example 1, we evaluated exponential functions at some rational numbers. But what is the meaning of a^x when x is an irrational number? For example, what do expressions such as $3^{\sqrt{2}}$ and 2^π mean? It turns out that the definition of a^x (with $a > 0$ and x irrational) requires methods discussed in calculus. However, we can understand the basis for the definition from the following discussion. Suppose we want to define the number 2^π. We use several numbers that approximate π. A calculator shows that $\pi \approx 3.14159265\ldots$. We successively approximate 2^π by using the rational powers shown in Table 3.2.

224 Chapter 3 Exponential and Logarithmic Functions

TABLE 3.2 Values of $f(x) = 2^x$ for Rational Values of x That Approach π

x	3	3.1	3.14	3.141	3.1415
2^x	$2^3 = 8$	$2^{3.1} = 8.5\ldots$	$2^{3.14} = 8.81\ldots$	$2^{3.141} = 8.821\ldots$	$2^{3.1415} = 8.8244\ldots$

It can be shown that the powers $2^3, 2^{3.1}, 2^{3.14}, 2^{3.141}, 2^{3.1415}, \ldots$ approach exactly one number. We define 2^π as that number. Table 3.2 shows that $2^\pi \approx 8.82$ (correct to two decimal places).

For our work with exponential functions, we need the following fundamental facts:

1. Exponential functions $f(x) = a^x$ are defined for all real numbers x.
2. The graph of an exponential function is a continuous (unbroken) curve.
3. The rules of exponents hold for real numbers. To review the rules of exponents, see Section P.1, page 9.

2 Graph exponential functions.

Graphing Exponential Functions

Let's see how to sketch the graph of an exponential function. Although the domain is the set of all real numbers, we usually evaluate the functions only for the integer values of x (for ease of computation). To evaluate a^x for noninteger values of x, use your calculator. Recall that the graph of a function $f(x)$ is the graph of the equation $y = f(x)$. We can use either $f(x) = a^x$ or $y = a^x$ to represent a given function.

EXAMPLE 2 Graphing an Exponential Function with Base $a > 1$

Graph the exponential function $f(x) = 3^x$.

Solution

First, make a table of a few values of x and the corresponding values of y.

x	-3	-2	-1	0	1	2	3
$y = 3^x$	$\dfrac{1}{27}$	$\dfrac{1}{9}$	$\dfrac{1}{3}$	1	3	9	27

Next, plot the points [see Figure 3.1(a)] and draw a smooth curve through them. See Figure 3.1(b).

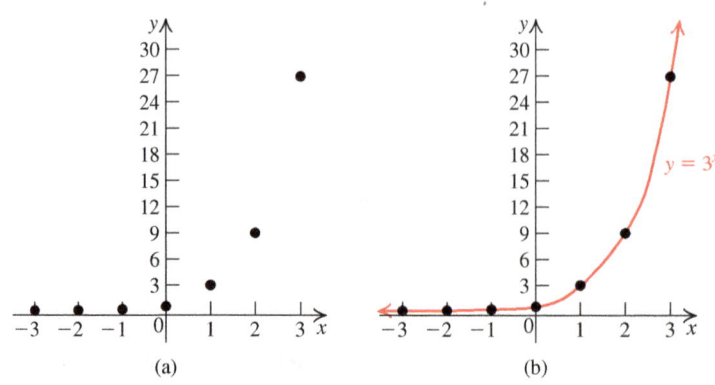

FIGURE 3.1 The graph of $y = a^x$ with $a > 1$

The graph in Figure 3.1(b) is typical of the graphs of exponential functions $f(x) = a^x$ when $a > 1$. Note that the x-axis is the horizontal asymptote of the graph of $y = a^x$.

Practice Problem 2 Sketch the graph of
$$f(x) = 2^x \text{ by making a table of values.}$$

Let's now sketch the graph of an exponential function $f(x) = a^x$ when $0 < a < 1$.

EXAMPLE 3 Graphing an Exponential Function $f(x) = a^x$, with $0 < a < 1$

Sketch the graph of $y = \left(\dfrac{1}{2}\right)^x$.

Solution
Make a table similar to the one in Example 2.

x	-3	-2	-1	0	1	2	3
$y = \left(\dfrac{1}{2}\right)^x$	8	4	2	1	$\dfrac{1}{2}$	$\dfrac{1}{4}$	$\dfrac{1}{8}$

Plotting these points and drawing a smooth curve through them, we get the graph of $y = \left(\dfrac{1}{2}\right)^x$, shown in Figure 3.2. In this graph, as x increases in the positive direction, $y = \left(\dfrac{1}{2}\right)^x$ decreases toward 0. The x-axis is its horizontal asymptote. This is a typical graph of an exponential function $y = a^x$ with $0 < a < 1$.

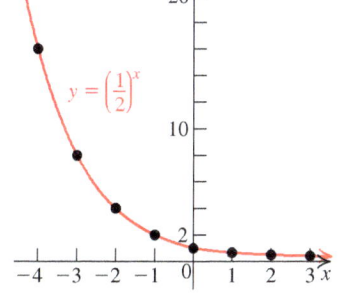

FIGURE 3.2 An exponential function $y = a^x$ with $0 < a < 1$

Practice Problem 3 Sketch the graph of
$$f(x) = \left(\dfrac{1}{3}\right)^x.$$

In general, exponential functions have two basic shapes determined by the base a. If $a > 1$, the graph of $y = a^x$ is rising as in Figure 3.1(b). If $0 < a < 1$, the graph is falling as in Figure 3.2. To get the correct shape for the graph of $f(x) = a^x$, it is helpful to graph the three points:
$$\left(-1, \dfrac{1}{a}\right), (0, 1), \text{ and } (1, a).$$

RECALL
The graph of $y = f(-x)$ is the reflection about the y-axis of the graph of $y = f(x)$.

Because $y = \left(\dfrac{1}{2}\right)^x = (2^{-1})^x = 2^{-x}$, the graph of $y = \left(\dfrac{1}{2}\right)^x$ can also be obtained by reflecting the graph of $y = 2^x$ about the y-axis. The calculator graph in the margin (on page 226) supports this observation.

TECHNOLOGY CONNECTION

Graph $y_1 = 2^x$ and $y_2 = \left(\frac{1}{2}\right)^x$ on the same screen.

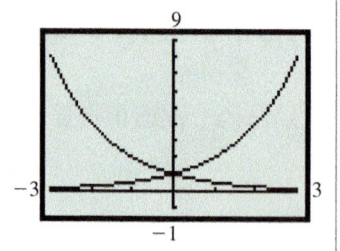

Notice that the graphs of y_1 and y_2 are reflections of each other about the y-axis.

In Figure 3.3, we sketch the graphs of four exponential functions on the same set of axes. These graphs illustrate some general properties of exponential functions.

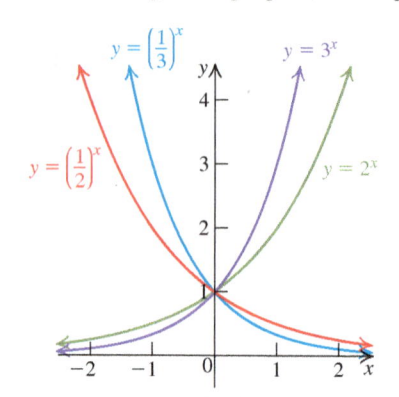

FIGURE 3.3 Graphs of some exponential functions

PROPERTIES OF EXPONENTIAL FUNCTIONS

Let $y = f(x) = a^x, a > 0, a \neq 1$.

1. The domain of f is $(-\infty, \infty)$.
2. The range of f is $(0, \infty)$: the entire graph lies above the x-axis.
3. Because $f(x + 1) = a^{x+1} = a \cdot a^x = af(x)$, the y-values change by a factor of a for each unit increase in x.
4. For $a > 1$, the **growth factor** is a. The y-values *increase* by a factor of a for each unit increase in x,
   (i) f is an increasing function; so the graph rises to the right.
   (ii) as $x \to \infty, y \to \infty$.
   (iii) as $x \to -\infty, y \to 0$.
5. For $0 < a < 1$, the **decay factor** is a. The y-values *decrease* by a factor of a for each unit increase in x,
   (i) f is a decreasing function; so the graph falls to the right.
   (ii) as $x \to -\infty, y \to \infty$.
   (iii) as $x \to \infty, y \to 0$.
6. Each exponential function f is one-to-one. So
   (i) if $a^m = a^n$, then $m = n$.
   (ii) f has an inverse.
7. The graph of f has no x-intercepts, so it never crosses the x-axis. No value of x will cause $f(x) = a^x$ to equal 0.
8. The graph of f is a smooth and continuous curve, and it passes through the points $\left(-1, \frac{1}{a}\right)$, $(0, 1)$, and $(1, a)$.
9. The x-axis is a horizontal asymptote for every exponential function of the form $f(x) = a^x$.
10. The graph of $y = a^{-x}$ is the reflection about the y-axis of the graph of $y = a^x$.

RECALL

A function f is one-to-one if $f(x_1) = f(x_2)$ implies that $x_1 = x_2$.

RECALL

Recall that a line $y = k$ is a horizontal asymptote to the graph of f if $f(x) \to k$ as $x \to \infty$ or as $x \to -\infty$.

EXAMPLE 4 Finding Exponential Functions

Find the exponential function of the form $f(x) = ca^x$ that satisfies the given conditions.

a. $f(0) = 3$ and $f(2) = 75$

b. The graph of f contains the points $(-1, 18)$ and $\left(4, \frac{2}{27}\right)$.

Solution

a. Because $f(0) = 3$, we have $3 = c \cdot a^0 = c \cdot 1 = c$. We now write

$f(x) = 3a^x$ $c = 3$
$75 = 3a^2$ Given $f(2) = 75$
$25 = a^2$ Divide both sides by 3.
$\pm 5 = a$ Solve for a.
$a = 5$ Reject $a = -5$ because the base a is positive.

So $f(x) = 3 \cdot 5^x$.   Replace c with 3 and a with 5.

b. Write the function in the form $y = f(x) = ca^x$.

$18 = ca^{-1}$ $(-1, 18)$ is on the graph, so $f(-1) = 18$.
$18a = c$ Multiply both sides by a; $a^{-1}a = 1$.

Next, $\dfrac{2}{27} = ca^4$ $\left(4, \dfrac{2}{27}\right)$ is on the graph, so $f(4) = \dfrac{2}{27}$.

$\dfrac{2}{27} = 18a \cdot a^4$ Replace c with $18a$.

$\dfrac{2}{27} = 18a^5$ $a \cdot a^4 = a^5$.

$a^5 = \dfrac{1}{243}$ Divide both sides by 18 and simplify.

$a = \sqrt[5]{\dfrac{1}{243}} = \dfrac{1}{3}$ Solve for a.

To find c, we substitute $a = \dfrac{1}{3}$ in the equation $18 = ca^{-1}$. So

$$18 = c\left(\dfrac{1}{3}\right)^{-1} = c(3^{-1})^{-1} = c \cdot 3. \qquad \dfrac{1}{3} = 3^{-1}$$

So $3c = 18$ implies that $c = 6$. Replacing $c = 6$ and $a = \dfrac{1}{3}$ in $f(x) = ca^x$, we have

$$f(x) = 6\left(\dfrac{1}{3}\right)^x.$$

Practice Problem 4 Repeat Example 4(b) with these graph points.

a. $(0, 1)$ and $(2, 49)$ **b.** $(-2, 16)$ and $\left(3, \dfrac{1}{2}\right)$

3 Develop formulas for simple and compound interest.

Simple Interest

Let's first review some terminology concerning simple interest.

Simple Interest

A fee charged for borrowing a lender's money is the **interest**, denoted by I.

The original, or initial, amount of money borrowed is the **principal**, denoted by P.

The period of time during which the borrower pays back the principal plus the interest is the **time**, denoted by t.

The **interest rate** is the percent charged for the use of the principal for the given period. The interest rate, denoted by r, is expressed as a decimal. Unless stated otherwise, the period is assumed to be one year; that is, r is an *annual* rate.

The amount of interest computed only on the principal is called **simple interest**.

RECALL

8% = 8 percent
= 8 per hundred
= $\frac{8}{100}$
= 0.08

When money is deposited with a bank, the bank becomes the borrower. For example, depositing $1000 in an account at 8% interest means that the principal P is $1000 and the interest rate r is 0.08 for the bank as the borrower.

SIMPLE INTEREST FORMULA

The simple interest I on a principal P at a rate r (expressed as a decimal) per year for t years is

$$I = Prt. \quad (1)$$

EXAMPLE 5 **Calculating Simple Interest**

Juanita has deposited $8000 in a bank for five years at a simple interest rate of 6%.

a. How much interest will she receive?
b. How much money will be in her account at the end of five years?

Solution

a. $P = \$8000, r = 0.06,$ and $t = 5$

$I = Prt$ .... Equation (1)
$= \$8000\,(0.06)(5)$ .... Substitute values.
$= \$2400$ .... Simplify.

b. In five years, the amount A she will receive is the original principal plus the interest earned:

$A = P + I$
$= \$8000 + \$2400$
$= \$10,400$

Practice Problem 5 Find the amount that will be in a bank account if $10,000 is deposited at a simple interest rate of 7.5% for two years.

Given P and r, the amount $A(t)$ due in t years and calculated at simple interest is found by using the formula

$$A(t) = P + Prt. \quad (2)$$

Equation (2) is a linear function of t. Simple interest problems are examples of **linear growth**, which take place when the growth of a quantity occurs at a constant rate and so can be modeled by a linear function.

Compound Interest

In the real world, simple interest is rarely used for periods of more than one year. Instead, we use **compound interest**—the interest paid on both the principal and the accrued (previously earned) interest.

To illustrate compound interest, suppose $1000 is deposited in a bank account paying 4% annual interest. At the end of one year, the account will contain the original $1000 plus the 4% interest earned on the $1000:

$$\$1000 + (0.04)(\$1000) = \$1040$$

Similarly, if P represents the initial amount deposited at an interest rate r (expressed as a decimal) per year, then the amount A_1 in the account after one year is

$$A_1 = P + rP \quad \text{Principal } P \text{ plus interest earned}$$
$$= P(1 + r) \quad \text{Factor out } P.$$

During the second year, the account earns interest on the new principal A_1. The amount A_2 in the account after the second year will be equal to A_1 plus the interest on A_1.

$$A_2 = A_1 + rA_1 \quad A_1 \text{ plus interest earned on } A_1$$
$$= A_1(1 + r) \quad \text{Factor out } A_1.$$
$$= P(1 + r)(1 + r) \quad A_1 = P(1 + r)$$
$$= P(1 + r)^2$$

The amount A_3 in the account after the third year is:

$$A_3 = A_2 + rA_2 \quad A_2 \text{ plus interest earned on } A_2$$
$$= A_2(1 + r) \quad \text{Factor out } A_2.$$
$$= P(1 + r)^2(1 + r) \quad A_2 = P(1 + r)^2$$
$$= P(1 + r)^3$$

In general, the amount A in the account after t years is given by

$$A = P(1 + r)^t. \qquad (3)$$

We say that this type of interest is **compounded annually** because it is paid once a year.

EXAMPLE 6 Calculating Compound Interest

Juanita deposits $8000 in a bank at the interest rate of 6% compounded annually for five years.

a. How much money will she have in her account after five years?
b. How much interest will she receive?

Solution
a. Here $P = \$8000$, $r = 0.06$, and $t = 5$; so

$$A = P(1 + r)^t$$
$$= \$8000(1 + 0.06)^5 \quad P = \$8000, r = 0.06, t = 5$$
$$= \$8000(1.06)^5$$
$$= \$10,705.80 \quad \text{Use a calculator.}$$

b. Interest $= A - P = \$10,705.80 - \$8000 = \$2705.80$.

Practice Problem 6 Repeat Example 6 assuming that the bank pays 7.5% interest compounded annually.

Comparing Examples 5 and 6, we see that compounding Juanita's interest made her money grow faster. We expect this because the function $A(t) = P(1 + r)^t$ is an exponential function with base $(1 + r)$.

Banks and other financial institutions usually pay savings account interest more than once a year. They pay a smaller amount of interest more frequently. Suppose the quoted annual interest rate r (also called the **nominal rate**) is compounded n times per year (at equal intervals) instead of annually. Then for each period, the interest rate is $\dfrac{r}{n}$, and there are

$n \cdot t$ periods in t years. Accordingly, we can restate the formula $A(t) = P(1 + r)^t$ as follows:

> **COMPOUND INTEREST FORMULA**
>
> $$A = P\left(1 + \frac{r}{n}\right)^{nt} \quad (4)$$
>
> A = amount after t years
> P = principal
> r = annual interest rate (expressed as a decimal number)
> n = number of times interest is compounded each year
> t = number of years

The total amount accumulated after t years, denoted by A, is also called the **future value** of the investment.

EXAMPLE 7 Using Different Compounding Periods to Compare Future Values

If $100 is deposited in a bank that pays 5% annual interest, find the future value A after one year if the interest is compounded

(i) annually. (ii) semiannually. (iii) quarterly.
(iv) monthly. (v) daily.

DO YOU KNOW?

Compounding that occurs 1, 2, 4, 12, and 365 times a year is known as compounding annually, semiannually, quarterly, monthly, and daily, respectively.

Solution

In the following computations, $P = \$100$, $r = 0.05$, and $t = 1$. Only n, the number of times interest is compounded each year, changes. Because $t = 1$, $nt = n(1) = n$.

(i) Annual Compounding: $\quad A = P\left(1 + \dfrac{r}{n}\right)^{nt} \qquad n = 1; t = 1$

$\qquad A = \$100(1 + 0.05) = \$105.00$

(ii) Semiannual Compounding: $\quad A = P\left(1 + \dfrac{r}{2}\right)^{2} \qquad n = 2; t = 1$

$\qquad A = \$100\left(1 + \dfrac{0.05}{2}\right)^{2}$

$\qquad \approx \$105.06 \qquad$ Use a calculator.

(iii) Quarterly Compounding: $\quad A = P\left(1 + \dfrac{r}{4}\right)^{4} \qquad n = 4; t = 1$

$\qquad A = \$100\left(1 + \dfrac{0.05}{4}\right)^{4}$

$\qquad \approx \$105.09 \qquad$ Use a calculator.

(iv) Monthly Compounding: $\quad A = P\left(1 + \dfrac{r}{12}\right)^{12} \qquad n = 12; t = 1$

$\qquad A = \$100\left(1 + \dfrac{0.05}{12}\right)^{12}$

$\qquad \approx \$105.12 \qquad$ Use a calculator.

SIDE NOTE

You may make errors such as forgetting parentheses when entering complicated expressions in your calculator. Look at your answer to see whether it is reasonable and makes sense.

(v) Daily Compounding: $\quad A = P\left(1 + \dfrac{r}{365}\right)^{365} \qquad n = 365; t = 1$

$\qquad A = \$100\left(1 + \dfrac{0.05}{365}\right)^{365}$

$\qquad \approx \$105.13 \qquad$ Use a calculator.

Practice Problem 7 Repeat Example 7 assuming that $5000 is deposited at a 6.5% annual rate.

The next example shows that we can solve equation (4) for the interest rate r.

EXAMPLE 8 Computing Interest Rate

Carmen has $9000 to invest. She needs $20,000 at the end of 8 years. If the interest is compounded quarterly, find the rate r needed.

Solution

Have $P = 9000, A = 20,000, t = 8$, and $n = 4$. Substituting these values in equation (4), we have

$$20,000 = 9000\left(1 + \frac{r}{4}\right)^{4 \cdot 8}$$

or

$$\left(1 + \frac{r}{4}\right)^{32} = \frac{20,000}{9000} \quad \text{Rewrite, with } 4 \cdot 8 = 32, \text{ and divide by 9000.}$$

$$= \frac{20}{9} \quad \text{Simplify the right side.}$$

Taking the 32nd root or $\frac{1}{32}$ power of both sides, we have

$$1 + \frac{r}{4} = \left(\frac{20}{9}\right)^{1/32}$$

$$\frac{r}{4} = \left(\frac{20}{9}\right)^{1/32} - 1 \quad \text{Subtract 1 from both sides.}$$

$$r = 4\left[\left(\frac{20}{9}\right)^{1/32} - 1\right] \quad \text{Multiply both sides by 4.}$$

$$\approx 0.1010692264 \quad \text{Use a calculator.}$$

$$\approx 10.107\%$$

Carmen needs an interest rate of about 10.107%.

Check: $9000\left(1 + \frac{0.10107}{4}\right)^{32} = 20,000.12073 \approx 20,000.$

Practice Problem 8 Repeat Example 8 assuming that the interest is compounded monthly.

Continuous Compound Interest Formula

Notice in Example 7 that the future value A increases with n, the number of compounding periods. (Of course P, r, and t are fixed.) The question is; If n increases indefinitely (100, 1000, 10,000 times, and so on), does the amount A also increase indefinitely? Let's see why the answer is no. Let $h = \frac{n}{r}$. Then we have

$$A = P\left(1 + \frac{r}{n}\right)^{nt} \quad (4)$$

$$= P\left[\left(1 + \frac{r}{n}\right)^{n/r}\right]^{rt} \quad nt = \frac{n}{r} \cdot rt$$

$$= P\left[\left(1 + \frac{1}{h}\right)^{h}\right]^{rt} \quad (5) \quad h = \frac{n}{r}, \text{ so } \frac{1}{h} = \frac{r}{n}$$

HISTORICAL NOTE

Leonhard Euler (1707–1783)

Leonhard Euler was the son of a Calvinist minister from Switzerland. At 13, Euler entered the University of Basel, pursuing a career in theology. At the university, Euler was tutored by Johann Bernoulli, of the famous family of mathematicians. Euler's interest and skills led him to abandon his theological studies and take up mathematics. In 1741, Euler went to the Berlin Academy, where he stayed until 1766. He then returned to St. Petersburg, where he remained for the rest of his life.

Euler contributed to many areas of mathematics, including number theory, combinatorics, and analysis, as well as applications to areas such as music and naval architecture. He wrote over 1100 books and papers and left so much unpublished work that it took 47 years after he died for all of his work to be published. During Euler's life, his papers accumulated so quickly that he kept a large pile of articles awaiting publication. The Berlin Academy published the papers on top of this pile, so later results often were published before results they depended on or superseded. The project of publishing his collected works, undertaken by the Swiss Society of Natural Science, is still going on and will require more than 75 volumes.

4 Define the number e.

Table 3.3 shows the expression $\left(1 + \dfrac{1}{h}\right)^h$ as h takes on increasingly larger values.

TABLE 3.3

h	$\left(1 + \dfrac{1}{h}\right)^h$
1	2
2	2.25
10	2.59374
100	2.70481
1000	2.71692
10,000	2.71815
100,000	2.71827
1,000,000	2.71828

Table 3.3 suggests that as h gets larger and larger, $\left(1 + \dfrac{1}{h}\right)^h$ gets closer and closer to a fixed number. This observation can be proven, and the fixed number is denoted by e in honor of the famous mathematician Leonhard Euler (pronounced "oiler"). The number e, an irrational number, is sometimes called the **Euler number**.

> The value of e to 15 decimal places is
> $$e \approx 2.718281828459045.$$

TECHNOLOGY CONNECTION

Graphs of $y_1 = e$ and $y_2 = \left(1 + \dfrac{1}{x}\right)^x$ in the window $0 < x < 20, 0 < y < 3$

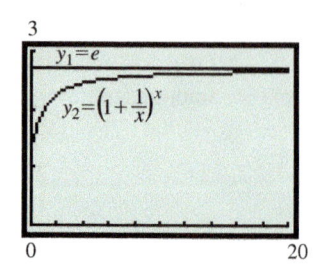

As $x \to \infty$, $\left(1 + \dfrac{1}{x}\right)^x \to e$.

We sometimes write

$$e = \lim_{h \to \infty} \left(1 + \dfrac{1}{h}\right)^h,$$

which means that when h is very large, $\left(1 + \dfrac{1}{h}\right)^h$ has a value very close to e. Note that in equation (5), as n gets very large, the quantity $h = \dfrac{n}{r}$ also gets very large because r is fixed. Therefore, the expression inside the brackets approaches e; so the compounded amount $A = P\left(1 + \dfrac{r}{n}\right)^{nt} = P\left[\left(1 + \dfrac{r}{n}\right)^{n/r}\right]^{rt}$ approaches Pe^{rt}.

Continuous Compounding When interest is compounded continuously, the amount A after t years is given by the following formula:

> **CONTINUOUS COMPOUND INTEREST FORMULA**
> $$A = Pe^{rt} \quad (6)$$
> $A =$ amount after t years
> $P =$ principal
> $r =$ annual interest rate (expressed as a decimal number)
> $t =$ number of years

TECHNOLOGY CONNECTION

Most calculators have the e^x or EXP key for calculating e^x for a given value of x.

On a TI-83 calculator, to evaluate a number such as e^5, you press e^x 5 ENTER and read the display 148.4131591.

EXAMPLE 9 Calculating Continuous Compound Interest

Find the amount when a principal of $8300 is invested at a 7.5% annual rate of interest compounded continuously for eight years and three months.

Solution

We use formula (6), with $P = \$8300$ and $r = 0.075$. We convert eight years and three months to 8.25 years.

$$A = \$8300\, e^{(0.075)(8.25)} \quad \text{Use the formula } A = Pe^{rt}.$$
$$\approx \$15{,}409.83 \quad \text{Use a calculator.}$$

Practice Problem 9 Repeat Example 9 assuming that $9000 is invested at a 6% annual rate.

5 Graph the natural exponential function.

The Natural Exponential Function

The exponential function

$$f(x) = e^x$$

with base e is so prevalent in the sciences that it is often referred to as *the* exponential function or the **natural exponential function**. We use a calculator to find e^x to two decimal places for $x = -2, -1, 0, 1,$ and 2 in Table 3.4.

TABLE 3.4

x	e^x
-2	0.14
-1	0.37
0	1
1	2.72
2	7.39

The graph of $f(x) = e^x$ is sketched in Figure 3.4 by using the ordered pairs in Table 3.4.

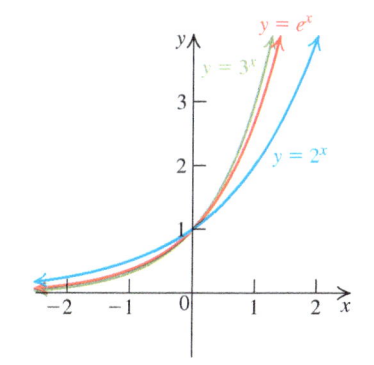

FIGURE 3.4 The graph of $y = e^x$ is between the graphs $y = 2^x$ and $y = 3^x$.

DO YOU KNOW?

Transcendental numbers and transcendental functions.

Numbers that are solutions of polynomial equations with rational coefficients are called **algebraic**: -2 is algebraic because it satisfies the equation $x + 2 = 0$, and $\sqrt{3}$ is algebraic because it satisfies the equation $x^2 - 3 = 0$. Numbers that are not algebraic are called **transcendental**, a term coined by Euler to describe numbers such as e and π that appear to "transcend the power of algebraic methods."

There is a somewhat similar distinction between functions. The exponential and logarithmic functions, which are the main subject of this chapter, are transcendental functions.

Because $2 < e < 3$, the graph of $y = e^x$ lies between the graphs $y = 2^x$ and $y = 3^x$. The function $f(x) = e^x$ has all the properties of exponential functions with base $a > 1$ listed on page 226.

We can apply the transformations from Section 1.5 to the natural exponential function.

EXAMPLE 10 Transformations on $f(x) = e^x$

Use transformations to sketch the graph of

$$g(x) = e^{x-1} + 2.$$

Solution

We start with the natural exponential function $f(x) = e^x$ and shift its graph one unit right to obtain the graph of $y = e^{x-1}$.

234 Chapter 3 Exponential and Logarithmic Functions

We then shift the graph of $y = e^{x-1}$ up two units to obtain the graph of $g(x) = e^{x-1} + 2$. See Figure 3.5.

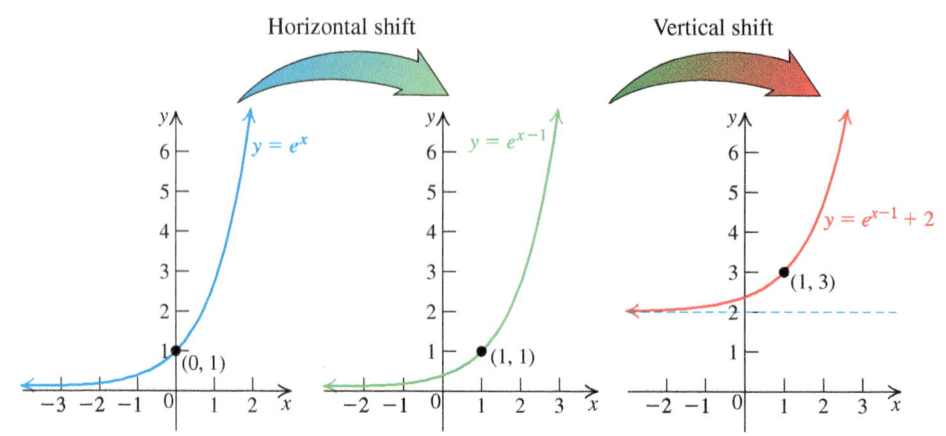

FIGURE 3.5 Graphing $g(x) = e^{x-1} + 2$

Practice Problem 10 Sketch the graph of $g(x) = -e^{x-1} - 2$.

Natural Exponential Growth and Decay

6 Model with exponential functions.

Numerous applications of the exponential function are based on the fact that many different quantities have a growth (or decay) rate proportional to their size. Using calculus, we can show that in such cases, we obtain the following mathematical models:

MODELS FOR NATURAL EXPONENTIAL GROWTH AND DECAY

For $k > 0$,

Exponential growth: $A(t) = A_0 e^{kt}$ and
Exponential decay: $A(t) = A_0 e^{-kt}$, where

$A(t)$ = the amount at time t
$A_0 = A(0)$, the initial amount (the quantity when $t = 0$)
k = relative rate of growth or decay
t = time

EXAMPLE 11 Exponential Growth

In 2000, the human population of the world was approximately 6 billion. Assume the annual rate of growth from 1990 onwards at 2.1%. Using the exponential growth model, estimate the population of the world in the following years.

a. 2030 **b.** 1990

Solution

a. The year 2000 corresponds to $t = 0$. So $A_0 = 6$ billion, $k = 0.021$, and 2030 corresponds to $t = 30$.

$$A(30) = 6e^{(0.021)(30)} \quad A(t) = A_0 e^{kt}$$
$$A(30) \approx 11.265663 \quad \text{Use a calculator.}$$

Thus, the model predicts that if the rate of growth is 2.1% per year, over 11.26 billion people will be in the world in 2030.

b. The year 1990 corresponds to $t = -10$ (because 1990 is ten years prior to 2000). We have

$$A(-10) = 6e^{(0.021)(-10)} \qquad A(t) = A_0 e^{kt}$$
$$= 6e^{(-0.21)} \qquad \text{Simplify.}$$
$$\approx 4.8635055 \qquad \text{Use a calculator.}$$

Thus, the model estimates that the world had over 4.86 billion people in 1990, assuming the growth rate had been 2.1% per year. (The actual population in 1990 was 5.28 billion.)

Practice Problem 11 Repeat Example 11 assuming that the annual rate of growth was 2.3%.

EXAMPLE 12 Exponential Decay

You buy a fishing boat for $22,000. Your boat depreciates (exponentially) at the annual rate of 15% of its value. Find the depreciated value of your boat at the end of 5 years.

Solution
We use the exponential decay model $A(t) = A_0 e^{-kt}$, where $A(t)$ represents the value of the boat after t years.
Here $A_0 = 22{,}000$; $k = 0.15$; and $t = 5$. So

$$A(t) = 22{,}000\, e^{-kt}$$
$$A(5) = 22{,}000\, e^{-(0.15)(5)} \qquad \text{Replace } k \text{ with } 0.15 \text{ and } t \text{ with } 5.$$
$$\approx 10{,}392.06416 \qquad \text{Use a calculator.}$$

So to the nearest cent, the value of your boat at the end of 5 years is $10,392.06.

Practice Problem 12 Repeat Example 12 with $k = 18\%$ and $t = 6$ years.

SECTION 3.1 Exercises

Basic Concepts and Skills

In Exercises 1–6, evaluate each exponential function for the given values. (Use a calculator if necessary.)

1. $f(x) = 5^{x-1}$; $f(3), f(0)$
2. $f(x) = -2^{x+1}$; $f(3), f(-2)$
3. $g(x) = 3^{1-x}$; $g(3.2), g(-1.2)$
4. $g(x) = \left(\dfrac{1}{2}\right)^{x+1}$; $g(2.8), g(-3.5)$
5. $h(x) = \left(\dfrac{2}{3}\right)^{2x-1}$; $h(1.5), h(-2.5)$
6. $f(x) = 3 - 5^x$; $f(2), f(-1)$

In Exercises 7–10, find the function of the form $f(x) = ca^x$ that contains the two given graph points.

7. **a.** $(0, 1)$ and $(2, 16)$ **b.** $(0, 1)$ and $\left(-2, \dfrac{1}{9}\right)$

8. **a.** $(0, 3)$ and $(2, 12)$ **b.** $(0, 5)$ and $(1, 15)$

9. **a.** $(1, 1)$ and $(2, 5)$ **b.** $(1, 1)$ and $\left(2, \dfrac{1}{5}\right)$

10. **a.** $(1, 5)$ and $(2, 125)$ **b.** $(-1, 4)$ and $(1, 16)$

In Exercises 11–18, sketch the graph of the given function by making a table of values. (Use a calculator if necessary.)

11. $f(x) = 4^x$ 12. $g(x) = 10^x$
13. $g(x) = \left(\dfrac{3}{2}\right)^{-x}$ 14. $h(x) = 7^{-x}$
15. $h(x) = \left(\dfrac{1}{4}\right)^x$ 16. $f(x) = \left(\dfrac{1}{10}\right)^x$
17. $f(x) = (1.3)^{-x}$ 18. $g(x) = (0.7)^{-x}$

19. How are the graphs in Exercises 11 and 15 related? Can we obtain the graph of Exercise 15 from that of Exercise 11? If so, how?

20. Repeat Exercise 19 for the graphs in Exercises 12 and 16.

Match each exponential function given in Exercises 21–24 with one of the graphs labeled (a), (b), (c), and (d).

21. $f(x) = 5^x$
22. $f(x) = -5^x$
23. $f(x) = 5^{-x}$
24. $f(x) = 5^{-x} + 1$

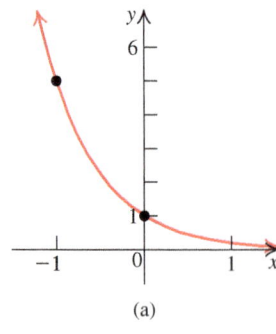

(a)

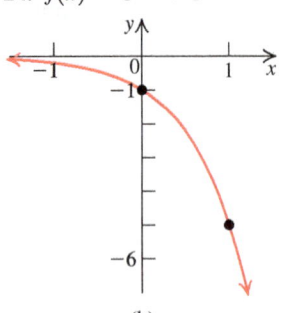

(b)

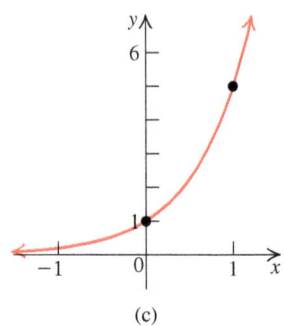

(c)

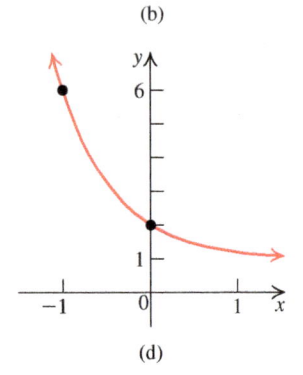

(d)

In Exercises 25–34, start with the graph of the appropriate basic exponential function f and use transformations to sketch the graph of the function g. State the domain and range of g and the horizontal asymptote of its graph.

25. $g(x) = 3^{x-1}$
26. $g(x) = 3^x - 1$
27. $g(x) = 4^{-x}$
28. $g(x) = -4^x$
29. $g(x) = -2.5^{x-1} + 4$
30. $g(x) = \frac{1}{2} \cdot 5^{1-x} - 2$
31. $g(x) = -e^{x-2} + 3$
32. $g(x) = 3 + e^{2-x}$
33. $g(x) = e^{|x|}$
34. $g(x) = -e^{|x|}$

In Exercises 35–38, write an equation of the form $f(x) = a^x + b$ from the given graph. Then compute $f(2)$.

35.

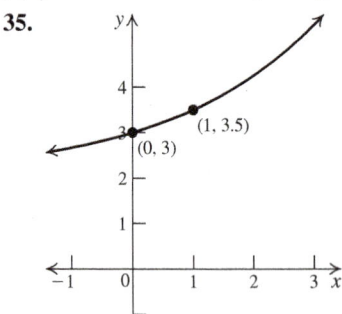

36.

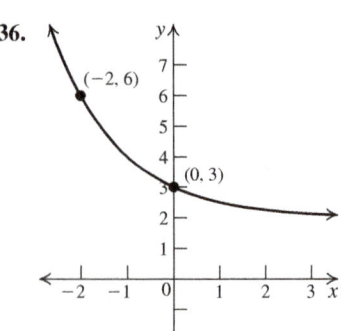

37.

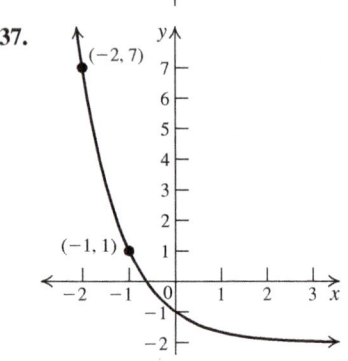

38.

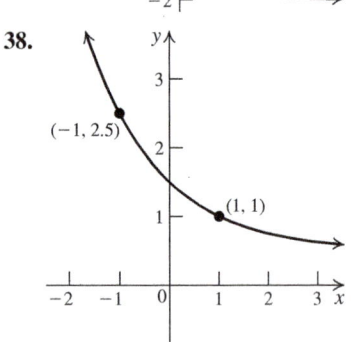

In Exercises 39–42, write an equation of each graph in the final position.

39. The graph of $y = 2^x$ is shifted two units left and then five units up.

40. The graph of $y = 3^x$ is reflected about the y-axis and then shifted three units right.

41. The graph of $y = \left(\frac{1}{2}\right)^x$ is stretched vertically by a factor of 2 and then shifted five units down.

42. The graph of $y = 2^{-x}$ is reflected about the x-axis and then shifted three units up.

In Exercises 43–46, find the simple interest for each value of principal P, rate r per year, and time t.

43. $P = \$5000$, $r = 10\%$, $t = 5$ years
44. $P = \$10{,}000$, $r = 5\%$, $t = 10$ years
45. $P = \$7800$, $r = 6\frac{7}{8}\%$, $t = 10$ years and 9 months
46. $P = \$8670$, $r = 4\frac{1}{8}\%$, $t = 6$ years and 8 months

In Exercises 47–50, find (a) the future value of the given principal P and (b) the interest earned in the given period.

47. $P = \$3500$ at 6.5% compounded annually for 13 years
48. $P = \$6240$ at 7.5% compounded monthly for 12 years
49. $P = \$7500$ at 5% compounded continuously for 10 years
50. $P = \$8000$ at 6.5% compounded daily for 15 years

In Exercises 51–54, find the principal P that will generate the given future value A.

51. $A = \$10,000$ at 8% compounded annually for 10 years
52. $A = \$10,000$ at 8% compounded quarterly for 10 years
53. $A = \$10,000$ at 8% compounded continuously for 10 years
54. $A = \$10,000$ at 8% compounded daily for 10 years

In Exercises 55–62, starting with the graph of $y = e^x$, use transformations to sketch the graph of each function and state its horizontal asymptote.

55. $f(x) = e^{-x}$
56. $f(x) = -e^x$
57. $f(x) = e^{x-2}$
58. $f(x) = e^{2-x}$
59. $f(x) = 1 + e^x$
60. $f(x) = 2 - e^{-x}$
61. $f(x) = -e^{x-2} + 3$
62. $g(x) = 3 + e^{2-x}$

Applying the Concepts

63. **Metal cooling.** Suppose a metal block is cooling so that its temperature T (in °C) is given by $T = 25 + 200 \cdot 4^{-0.1t}$, where t is in hours.
    a. Find the temperature after
       (i) 2 hours.   (ii) 3.5 hours.
    b. How long has the cooling been taking place if the block now has a temperature of 125°C?
    c. Find the eventual temperature ($t \to \infty$).

64. **Ethnic population.** The population (in thousands) of people of East Indian origin in the United States is approximated by the function
$$p(t) = 1600(2)^{0.1047t},$$
where t is the number of years since 2010.
    a. Find the population of this group in
       (i) 2010.   (ii) 2018.
    b. Predict the population in 2025.

65. **Price appreciation.** In 2012, the median price of a house in Miami was $190,000. Assuming a rate of increase of 3% per year, what can we expect the price of such a house to be in 2017?

66. **Investment.** How much should a mother invest at the time her son is born to provide him with $80,000 at age 21? Assume that the interest is 7% compounded quarterly.

67. **Depreciation.** Trans Trucking Co. purchased a truck for $80,000. The company depreciated the truck at the end of each year at the rate of 15% of its current value. What is the value of the truck at the end of the fifth year?

68. **Manhattan Island purchase.** In 1626, Peter Minuit purchased Manhattan Island from the Native Americans for 60 Dutch guilders (about $24). Suppose the $24 was invested in 1626 at a 6% rate. How much money would that investment be worth in 2006 if the interest was
    a. simple interest.
    b. compounded annually.
    c. compounded monthly.
    d. compounded continuously.

69. **Investment.** Ms. Ann Scheiber retired from government service in 1941 with a monthly pension of $83 and $5000 in savings. At the time of her death in January 1995 at the age of 101, Ms. Scheiber had turned the $5000 into $22 million through shrewd investments in the stock market. She bequeathed all of it to Yeshiva University in New York City. What annual rate of return compounded annually would turn $5000 to a whopping $22 million in 54 years?

70. **Population.** The population of Sometown, USA, was 12,000 in 2000 and grew to 15,000 in 2010. Assume that the population will continue to grow exponentially at the same constant rate. What will be the population of Sometown in 2020?
$\left[\text{Hint: Show that } e^k = \left(\dfrac{5}{4}\right)^{1/10}.\right]$

71. **Medicine.** Tests show that a new ointment X helps heal wounds. If A_0 square millimeters is the area of the original wound, then the area A of the wound after n days of application of the ointment X is given by $A = A_0 e^{-0.43n}$. If the area of the original wound was 10 square millimeters, find the area of the wound after ten days of application of the new ointment.

72. **Cooling.** The temperature T (in °C) of coffee at time t minutes after its removal from the microwave is given by the equation
$$T = 25 + 73e^{-0.28t}.$$
Find the temperature of the coffee at each time listed.
    a. $t = 0$ b. $t = 10$
    c. $t = 20$ d. after a long time

73. **Interest rate.** Mary purchases a 12-year bond for $60,000.00. At the end of 12 years, she will redeem the bond for approximately $95,600.00. If the interest is compounded quarterly, what was the interest rate on the bond?

74. **Best deal.** Fidelity Federal offers three types of investments: (a) 9.7% compounded annually, (b) 9.6% compounded monthly, and (c) 9.5% compounded continuously. Which investment is the best deal?

75. **Paper stacking.** Suppose we have a large sheet of paper 0.015 centimeter thick and we tear the paper in half and put the pieces on top of each other. We keep tearing and stacking in this manner, always tearing each piece in half. How high will the resulting pile of paper be if we continue the process of tearing and stacking
    a. 30 times?
    b. 40 times?
    c. 50 times? [*Hint:* Use a calculator.]

76. Doubling. A jar with a volume of 1000 cubic centimeters contains bacteria that doubles in number every minute. If the jar is full in 60 minutes, how long did it take for the container to be half full?

Critical Thinking / Discussion / Writing

77. Discuss why we do not allow a to be 1, 0, or a negative number in the definition of an exponential function of the form $f(x) = a^x$.

78. Consider the function $g(x) = (-3)^x$, which is not an exponential function. What are the possible rational numbers x for which $g(x)$ is defined?

79. Discuss the end-behavior of the graph of $y = \dfrac{b}{1 + ca^x}$, where $a, b, c > 0$ and $a \neq 1$. (Consider the cases $a < 1$ and $a > 1$.)

80. Write a summary of the types of graphs (with sketches) for the exponential functions of the form $y = ca^x$, $c \neq 0$, $a > 0$, and $a \neq 1$.

81. Give a convincing argument to show that the equation $2^x = k$ has exactly one solution for every $k > 0$. Support your argument with graphs.

82. Find all solutions of the equation $2^x = 2x$. How do you know you have found all solutions?

Maintaining Skills

Review the rules of exponents; then simplify the following expressions.

83. 10^0 **84.** 10^{-1}

85. $(-8)^{\frac{1}{3}}$ **86.** $(25)^{\frac{1}{2}}$

87. $\left(\dfrac{1}{7}\right)^{-2}$ **88.** $\left(\dfrac{1}{9}\right)^{-\frac{1}{2}}$

89. $(18)^{\frac{1}{2}}$ **90.** $\left(\dfrac{1}{12}\right)^{-\frac{1}{2}}$

Review the functions and transformations; then find the domain and sketch the graph of each of the following functions:

91. $f(x) = x^2 - 4$ **92.** $s(x) = (x - 1)^3$

93. $h(x) = \sqrt{x} + 3$ **94.** $g(x) = \sqrt{x + 1}$

95. $f(x) = 2\sqrt{1 - x}$ **96.** $s(x) = \sqrt{-x}$

97. $h(x) = \dfrac{1}{x + 1}$ **98.** $g(x) = \dfrac{1}{(x - 1)^2}$

SECTION 3.2 Logarithmic Functions

BEFORE STARTING THIS SECTION, REVIEW

1. Finding inverse of a one-to-one function (Section 1.7, page 142)
2. Polynomial and rational inequalities (Section P.4, page 38)
3. Graphing transformations (Section 1.5, page 110)

OBJECTIVES

1. Define logarithmic functions.
2. Evaluate logarithms.
3. Find the domains of logarithmic functions.
4. Graph logarithmic functions.
5. Use logarithms to solve exponential equations in applications.

◆ The McDonald's Coffee Case

Stella Liebeck of Albuquerque, New Mexico, was in the passenger seat of her grandson's car when she was severely burned by McDonald's coffee in February 1992. Liebeck ordered coffee that was served in a foam cup at the McDonald's drive-through window. While attempting to add cream and sugar to her coffee, Liebeck spilled the entire cup into her lap. A vascular surgeon determined that Liebeck suffered full-thickness burns (third-degree burns) over 6% of her body. She sued McDonald's for damages.

During the preliminary phase of the trial, McDonald's said that on the basis of a consultant's advice, the company brewed its coffee at 195° to 205°F and held it at 180°F to 190°F to maintain optimal taste.

Coffee served at home is generally between 135°F and 140°F. Prior to the Liebeck case, the prestigious Shriners Burn Institute of Cincinnati had published warnings to the franchise food industry that beverages above 130°F were causing serious scald burns. Liebeck's lawsuit was settled out of court, and both parties agreed to keep the award for damages confidential.

Avoiding Further Lawsuits Corporate lawyers informed McDonald's management that further lawsuits from customers who spill their coffee can be avoided if the temperature of the coffee is 125°F or less when it is delivered to the customers. In Example 11, we will learn how long to wait after brewing the coffee before delivering it to customers.

1 Define logarithmic functions.

Logarithmic Functions

Recall from Section 3.1 that every exponential function

$$f(x) = a^x, a > 0, a \neq 1,$$

is a *one-to-one* function and therefore has an inverse function. Let's find the inverse of the function $f(x) = 3^x$. If we think of an exponential function f as "putting an exponent on" a particular base, then the inverse function f^{-1} must "lift the exponent off" to undo the effect of f, as illustrated below.

$$x \xrightarrow{f} 3^x \xrightarrow{f^{-1}} x$$

Let's try to find the inverse of $f(x) = 3^x$ by the procedure outlined in Section 1.7.

Step 1 Replace $f(x)$ in the equation $f(x) = 3^x$ with y: $y = 3^x$.
Step 2 Interchange x and y in the equation in Step 1: $x = 3^y$.
Step 3 Solve the equation in Step 2 for y.

In Step 3, we need to solve the equation $x = 3^y$ for y. Finding y is easy when $3 = 3^y$ or $27 = 3^y$, but what if $5 = 3^y$? There actually is a real number y that solves the equation $5 = 3^y$, but we have not introduced the name for this number yet. (We were in a similar position concerning the solution of $y^3 = 7$ before we introduced the name for the solution, namely, $\sqrt[3]{7}$.) We can say that y is "the exponent on 3 that gives 5." Instead, we use the word *logarithm* (or *log* for short) to describe this exponent. We use the notation $\log_3 5$ (read as "log of 5 with base 3") to express the solution of the equation $5 = 3^y$. This log notation is a new way to write information about exponents: $y = \log_3 x$ means $x = 3^y$. In other words, $\log_3 x$ is the exponent to which 3 must be raised to obtain x. Therefore, the inverse function of the exponential function $f(x) = 3^x$ is written as

$$f^{-1}(x) = \log_3 x \quad \text{or} \quad y = \log_3 x, \text{ where } y = f^{-1}(x).$$

The previous discussion could be repeated for any base a instead of 3 provided that $a > 0$ and $a \neq 1$.

The Logarithmic Function

For $x > 0$, $a > 0$, and $a \neq 1$,

$$y = \log_a x \quad \text{if and only if} \quad x = a^y.$$

The function $f(x) = \log_a x$ is called the **logarithmic function with base a**.

The definition of the logarithmic function says that the two equations

$$y = \log_a x \quad (\textbf{logarithmic form}) \text{ and}$$
$$x = a^y \quad (\textbf{exponential form})$$

are equivalent. For instance, $\log_3 81 = 4$ because the base 3 must be raised to the fourth power to obtain 81.

EXAMPLE 1 Converting from Exponential to Logarithmic Form

Write each exponential equation in logarithmic form.

a. $4^3 = 64$ **b.** $\left(\dfrac{1}{2}\right)^4 = \dfrac{1}{16}$ **c.** $a^{-2} = 7$

Solution

a. $4^3 = 64$ is equivalent to $\log_4 64 = 3$.

b. $\left(\dfrac{1}{2}\right)^4 = \dfrac{1}{16}$ is equivalent to $\log_{1/2} \dfrac{1}{16} = 4$.

c. $a^{-2} = 7$ is equivalent to $\log_a 7 = -2$.

Practice Problem 1 Write each exponential equation in logarithmic form.

a. $2^{10} = 1024$ **b.** $9^{-1/2} = \dfrac{1}{3}$ **c.** $p = a^q$

EXAMPLE 2 Converting from Logarithmic Form to Exponential Form

Write each logarithmic equation in exponential form.

a. $\log_3 243 = 5$ b. $\log_2 5 = x$ c. $\log_a N = x$

Solution

a. $\log_3 243 = 5$ is equivalent to $243 = 3^5$.
b. $\log_2 5 = x$ is equivalent to $5 = 2^x$.
c. $\log_a N = x$ is equivalent to $N = a^x$.

Practice Problem 2 Write each logarithmic equation in exponential form.

a. $\log_2 64 = 6$ b. $\log_v u = w$

Evaluating Logarithms

The technique of converting from logarithmic form to exponential form can be used to evaluate some logarithms by inspection.

EXAMPLE 3 Evaluating Logarithms

Find the value of each of the following logarithms.

a. $\log_5 25$ b. $\log_2 16$ c. $\log_{1/3} 9$

d. $\log_7 7$ e. $\log_6 1$ f. $\log_4 \dfrac{1}{2}$

Solution

Logarithmic Form	Exponential Form	Value
a. $\log_5 25 = y$	$25 = 5^y$ or $5^2 = 5^y$	$y = 2$
b. $\log_2 16 = y$	$16 = 2^y$ or $2^4 = 2^y$	$y = 4$
c. $\log_{1/3} 9 = y$	$9 = \left(\dfrac{1}{3}\right)^y$ or $3^2 = 3^{-y}$	$y = -2$
d. $\log_7 7 = y$	$7 = 7^y$ or $7^1 = 7^y$	$y = 1$
e. $\log_6 1 = y$	$1 = 6^y$ or $6^0 = 6^y$	$y = 0$
f. $\log_4 \dfrac{1}{2} = y$	$\dfrac{1}{2} = 4^y$ or $2^{-1} = 2^{2y}$	$y = -\dfrac{1}{2}$

Practice Problem 3 Evaluate.

a. $\log_3 9$ b. $\log_9 \dfrac{1}{3}$ c. $\log_{1/2} 32$

EXAMPLE 4 Using the Definition of Logarithm

Solve each equation.

a. $\log_5 x = -3$ b. $\log_3 \dfrac{1}{27} = y$

c. $\log_z 1000 = 3$ d. $\log_2 (x^2 - 6x + 10) = 1$

Solution

a. $\log_5 x = -3$

$x = 5^{-3}$ Exponential form

$x = \dfrac{1}{5^3} = \dfrac{1}{125}$ $a^{-n} = \dfrac{1}{a^n}$

b. $\log_3 \dfrac{1}{27} = y$

$\dfrac{1}{27} = 3^y$ Exponential form

$3^{-3} = 3^y$ $\dfrac{1}{27} = \dfrac{1}{3^3} = 3^{-3}$

$-3 = y$ If $a^x = a^y$, then $x = y$.

c. $\log_z 1000 = 3$

$1000 = z^3$ Exponential form

$10^3 = z^3$ $1000 = 10^3$

$10 = z$ Take the cube root of both sides.

d. $\log_2(x^2 - 6x + 10) = 1$

$x^2 - 6x + 10 = 2^1$ Exponential form

$x^2 - 6x + 8 = 0$ $2^1 = 2$; subtract 2 from both sides.

$(x - 2)(x - 4) = 0$ Factor.

$x - 2 = 0$ or $x - 4 = 0$ Zero-product property

$x = 2$ or $x = 4$ Solve for x.

Practice Problem 4 Solve each equation.

a. $\log_2 x = 3$ **b.** $\log_z 125 = 3$ **c.** $\log_3(x^2 - x - 5) = 0$

Domains of Logarithmic Functions

3 Find the domains of logarithmic functions.

Because the exponential function $f(x) = a^x$ has domain $(-\infty, \infty)$ and range $(0, \infty)$, its inverse function $y = \log_a x$ **has domain $(0, \infty)$ and range $(-\infty, \infty)$**. Therefore, the logarithms of 0 and of negative numbers are not defined; so expressions such as $\log_a(-2)$ and $\log_a(0)$ are meaningless.

EXAMPLE 5   Finding the Domain

Find the domain of $f(x) = \log_3(2 - x)$.

Solution

Because the domain of the logarithmic function is $(0, \infty)$, the expression $2 - x$ must be positive. The domain of f is the set of all real numbers x, where

$2 - x > 0$

$2 > x$ Solve for x.

The domain of f is $(-\infty, 2)$.

Practice Problem 5 Find the domain of $f(x) = \log_{10} \sqrt{1 - x}$.

Because $a^1 = a$ and $a^0 = 1$ for any positive base a, expressing these equations in logarithmic notation gives

$\log_a a = 1$ and $\log_a 1 = 0$.

Also, letting $f(x) = a^x$ and $f^{-1}(x) = \log_a x$, then

$$x = f^{-1}(f(x)) = f^{-1}(a^x) = \log_a a^x$$

and

$$x = f(f^{-1}(x)) = f(\log_a x) = a^{\log_a x}.$$

We summarize these relationships as basic properties of logarithms.

RECALL

If f^{-1} is the inverse of f, then

$$f^{-1}(f(x)) = x$$

and

$$f(f^{-1}(x)) = x$$

Basic Properties of Logarithms

For any base $a > 0$, with $a \neq 1$,

1. $\log_a a = 1$.
2. $\log_a 1 = 0$.
3. $\log_a a^x = x$ for any real number x.
4. $a^{\log_a x} = x$ for any $x > 0$.

Graphs of Logarithmic Functions

4 Graph logarithmic functions.

We now look at other properties of logarithmic functions and begin with an example showing how to graph a logarithmic function.

EXAMPLE 6 Sketching a Graph

RECALL

Remember that "$\log_3 x$" means the exponent on 3 that gives x.

Sketch the graph of $y = \log_3 x$.

Plotting points (Method 1) To find selected ordered pairs on the graph of $y = \log_3 x$, we choose the x-values to be powers of 3. We can easily compute the logarithms of these values by using property (3), namely, $\log_3 3^x = x$. We compute the y values in Table 3.5.

TABLE 3.5

x	$y = \log_3 x$	(x, y)
$\dfrac{1}{27} = 3^{-3}$	$y = \log_3 \dfrac{1}{27} = \log_3 3^{-3} = -3$	$\left(\dfrac{1}{27}, -3\right)$
$\dfrac{1}{9} = 3^{-2}$	$y = \log_3 \dfrac{1}{9} = \log_3 3^{-2} = -2$	$\left(\dfrac{1}{9}, -2\right)$
$\dfrac{1}{3} = 3^{-1}$	$y = \log_3 \dfrac{1}{3} = \log_3 3^{-1} = -1$	$\left(\dfrac{1}{3}, -1\right)$
$1 = 3^0$	$y = \log_3 1 = 0$	$(1, 0)$
$3 = 3^1$	$y = \log_3 3 = 1$	$(3, 1)$
$9 = 3^2$	$y = \log_3 9 = \log_3 3^2 = 2$	$(9, 2)$

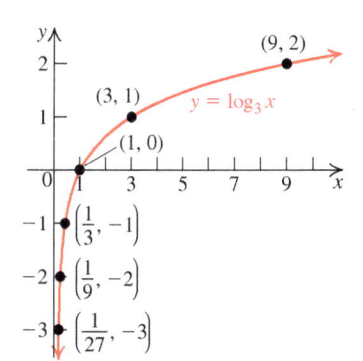

FIGURE 3.6 Graph by plotting points

Plotting these ordered pairs and connecting them with a smooth curve gives us the graph of $y = \log_3 x$, shown in Figure 3.6.

Using the inverse function (Method 2) Because $y = \log_3 x$ is the inverse of the function $y = 3^x$, we first graph the exponential function $y = 3^x$. Then reflect the graph of $y = 3^x$ about the line $y = x$. Both graphs are shown in Figure 3.7. Note that the horizontal asymptote $y = 0$ of the graph of $y = 3^x$ is reflected as the vertical asymptote $x = 0$ for the graph of $y = \log_3 x$.

FIGURE 3.7 Graph $y = \log_3 x$ by using the inverse function $y = 3^x$

Practice Problem 6 Sketch the graph of

$$y = \log_2 x.$$

244 Chapter 3 Exponential and Logarithmic Functions

We can graph $y = \log_{1/3} x$ either by plotting points or by using the graph of its inverse. See Figure 3.8. The graph of $y = \log_3 x$ shown in Figure 3.7 is rising. The graph of $y = \log_{1/3} x$ shown in Figure 3.8, on the other hand, is falling.

In general, logarithmic functions have two basic shapes, determined by the base a. If $a > 1$, the graph of $y = \log_a x$ is rising, as in Figure 3.9(a). If $0 < a < 1$, the graph is falling, as in Figure 3.9(b).

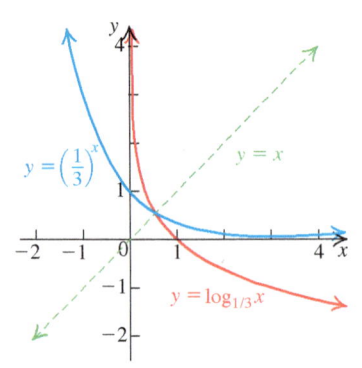

FIGURE 3.8

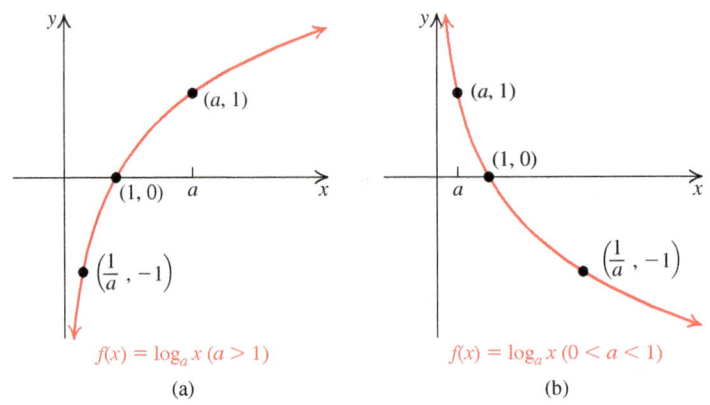

FIGURE 3.9 Basic shapes of $y = \log_a x$

SIDE NOTE

To get the correct shape for a logarithmic function $y = \log_a x$, it is helpful to graph the three points $(a, 1)$, $(1, 0)$, and $\left(\dfrac{1}{a}, -1\right)$.

PROPERTIES OF EXPONENTIAL AND LOGARITHMIC FUNCTIONS

Exponential Function $y = a^x$	Logarithmic Function $y = \log_a x$
1. The domain is $(-\infty, \infty)$, and the range is $(0, \infty)$.	The domain is $(0, \infty)$, and the range is $(-\infty, \infty)$.
2. The y-intercept is 1, and there is no x-intercept.	The x-intercept is 1, and there is no y-intercept.
3. The x-axis $(y = 0)$ is the horizontal asymptote.	The y-axis $(x = 0)$ is the vertical asymptote.
4. The function is one-to-one; that is, $a^u = a^v$ if and only if $u = v$.	The function is one-to-one; that is, $\log_a u = \log_a v$ if and only if $u = v$.
5. The function is increasing if $a > 1$ and decreasing if $0 < a < 1$.	The function is increasing if $a > 1$ and decreasing if $0 < a < 1$.

Once we know the shape of a logarithmic graph, we can shift it vertically or horizontally, stretch it, compress it, check answers with it, and interpret the graph.

EXAMPLE 7   Using Transformations

Start with the graph of $f(x) = \log_3 x$ and use transformations to sketch the graph of each function.

a. $f(x) = \log_3 x + 2$ **b.** $f(x) = \log_3 (x - 1)$
c. $f(x) = -\log_3 x$ **d.** $f(x) = \log_3 (-x)$

State the domain and range and the vertical asymptote for the graph of each function.

Solution

We start with the graph of $f(x) = \log_3 x$ and use the transformations shown in Figure 3.10.

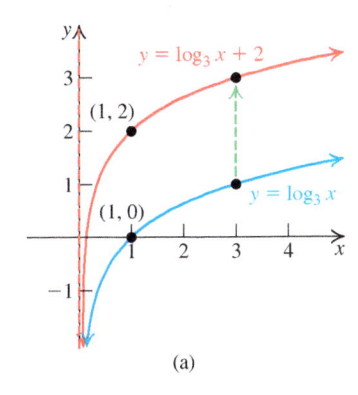

(a)

(b)

a. $f(x) = \log_3 x + 2$

Adding 2 to $\log_3 x$ shifts the graph two units up. Domain: $(0, \infty)$; range: $(-\infty, \infty)$; vertical asymptote: $x = 0$ (y-axis)

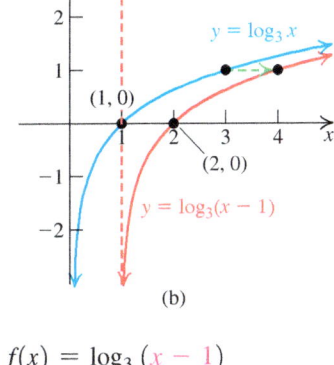

b. $f(x) = \log_3 (x - 1)$

Replacing x with $x - 1$ in $\log_3 x$ shifts the graph one unit right. Domain: $(1, \infty)$; range $(-\infty, \infty)$; vertical asymptote: $x = 1$

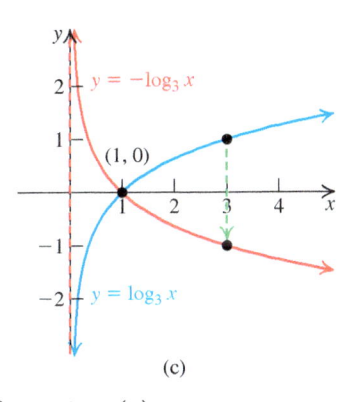

(c)

(d)

c. $f(x) = -\log_3 (x)$

Multiplying $\log_3 x$ by -1 reflects the graph about the x-axis. Domain: $(0, \infty)$; range: $(-\infty, \infty)$; vertical asymptote: $x = 0$

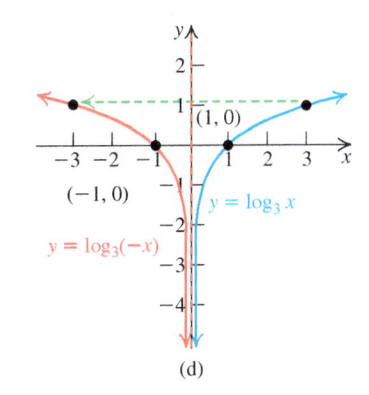

d. $f(x) = \log_3 (-x)$

Replacing x with $-x$ in $\log_3 x$ reflects the graph about the y-axis. Domain: $(-\infty, 0)$; range: $(-\infty, \infty)$; vertical asymptote: $x = 0$

FIGURE 3.10 Transformations on $y = \log_3 x$

Practice Problem 7 Use transformations of the graph of $f(x) = \log_2 x$ to sketch the graph of

$$y = -\log_2 (x - 3).$$

Common Logarithm

The logarithm with base 10 is called the **common logarithm** and is denoted by omitting the base, so

$$\log x = \log_{10} x.$$

COMMON LOGARITHM

$$y = \log x \ (x > 0) \quad \text{if and only if} \quad x = 10^y.$$

TECHNOLOGY CONNECTION

You can find the common logarithms of any positive number with the LOG key on your calculator. The exact key sequence will vary with different calculators. On most calculators, if you want to find log 3.7, press

3.7 LOG ENTER

or

LOG 3.7 ENTER

and see the display 0.5682017. Although the display reads 0.5682017, this number is an approximate value; so you should write $\log_{10} 3.7 \approx 0.5682017$.

BASIC PROPERTIES OF THE COMMON LOGARITHM

Applying the basic properties of logarithms (see page 244) to common logarithms, we have the following:

1. $\log 10 = 1$
2. $\log 1 = 0$
3. $\log 10^x = x$, x any real number
4. $10^{\log x} = x$, $x > 0$

We use property (3) to evaluate the common logarithms of numbers that are powers of 10. For example,

$$\log 1000 = \log 10^3 = 3$$

and $\log 0.01 = \log 10^{-2} = -2.$

We use a calculator to find the common logarithms of numbers that are not powers of 10 by pressing the LOG key.

The graph of $y = \log x$ is similar to the graph in Figure 3.9(a) (on page 245) because the base for log x is understood to be 10.

EXAMPLE 8 Using Transformations to Sketch a Graph

Sketch the graph of

$$y = 2 - \log(x - 2).$$

Solution

We start with the graph of the function $f(x) = \log x$ and use the order of transformations (see page 119) indicated in Figure 3.11 to sketch the graph.

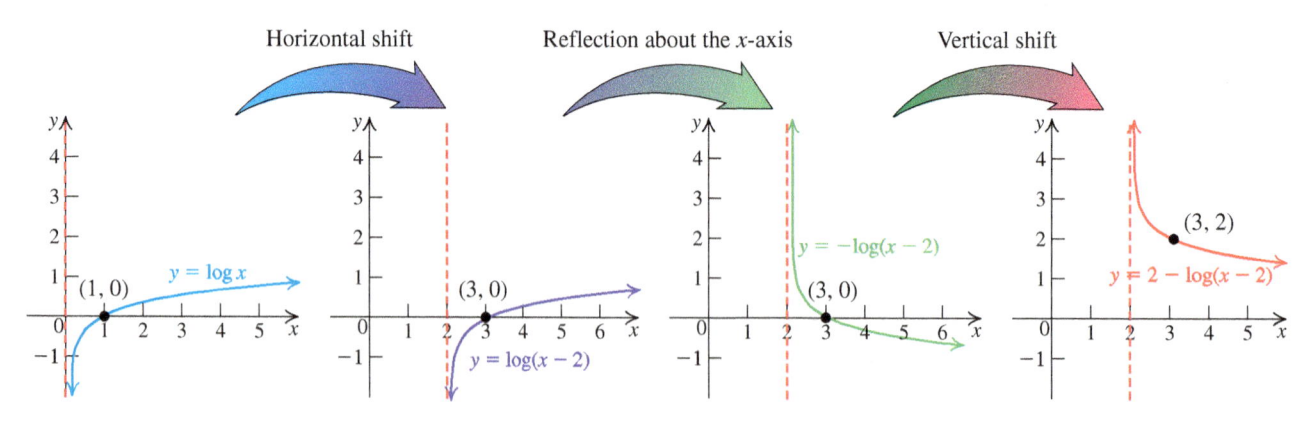

Step 1: Replacing x with $x - 2$ in log x shifts the graph of $y = \log x$ two units right.

Step 2: Multiplying log $(x - 2)$ by -1 reflects the graph of $y = \log(x - 2)$ about the x-axis.

Step 3: Adding 2 to $-\log(x - 2)$ shifts the graph of $y = -\log(x - 2)$ two units up.

FIGURE 3.11 Transformations on $y = \log x$

The domain of $f(x) = 2 - \log(x - 2)$ is $(2, \infty)$, the range is $(-\infty, \infty)$, and the vertical asymptote is the line $x = 2$.

Practice Problem 8 Sketch the graph of

$$y = \log(x - 3) - 2.$$

Natural Logarithm

TECHNOLOGY CONNECTION

To evaluate natural logarithms, we use the LN key on a calculator. To find ln(2.3) on most calculators, press

2.3 LN ENTER

or

LN 2.3 ENTER

and see the display 0.832909123.

In most applications in calculus and the sciences, the convenient base for logarithms is the number e. The logarithm with base e is called the **natural logarithm** and is denoted by ln x (read "ell en x" or "lawn x") so that

$$\ln x = \log_e x.$$

NATURAL LOGARITHM

$y = \ln x \ (x > 0)$ if and only if $x = e^y$.

BASIC PROPERTIES OF THE NATURAL LOGARITHM

Applying the basic properties of logarithms (see page 246) to natural logarithms, we have the following:

1. $\ln e = 1$
2. $\ln 1 = 0$
3. $\ln e^x = x$, x any real number
4. $e^{\ln x} = x$, $x > 0$

We can use property (3) to evaluate the natural logarithms of powers of e.

EXAMPLE 9 Evaluating the Natural Logarithm Function

Evaluate each expression.

a. $\ln e^4$ b. $\ln \dfrac{1}{e^{2.5}}$ c. $\ln 3$

Solution

a. $\ln e^4 = 4$ Property (3)

b. $\ln \dfrac{1}{e^{2.5}} = \ln e^{-2.5} = -2.5$ Property (3)

c. $\ln 3 \approx 1.0986123$ Use a calculator.

Practice Problem 9 Evaluate each expression.

a. $\ln \dfrac{1}{e}$ b. $\ln 2$

Investments

5 Use logarithms to solve exponential equations in applications.

EXAMPLE 10 Doubling Your Money

a. How long will it take to double your money if it earns 6.5% compounded continuously?
b. At what rate of return, compounded continuously, would your money double in 5 years?

Solution

If P dollars is invested and you want to double it, then the final amount $A = 2P$.

a. $A = Pe^{rt}$ Continuous compounding formula
$2P = Pe^{0.065t}$ $A = 2P, r = 0.065$
$2 = e^{0.065t}$ Divide both sides by P.
$\ln 2 = 0.065t$ Logarithmic form of $x = e^y$
$\dfrac{\ln 2}{0.065} = t$ Divide both sides by 0.065.
$t \approx 10.66$ Use a calculator.

It will take approximately 11 years to double your money.

b. $A = Pe^{rt}$ Continuous compounding formula
$2P = Pe^{5r}$ $A = 2P, t = 5$
$2 = e^{5r}$ Divide both sides by P.
$\ln 2 = 5r$ Logarithmic form of $x = e^y$
$\dfrac{\ln 2}{5} = r$ Divide both sides by 5.
$r \approx 0.1386$ Use a calculator.

Your investment will double in 5 years at the approximate rate of 13.86%.

Practice Problem 10 Repeat Example 10 for tripling ($A = 3P$) your money.

Newton's Law of Cooling

When a cool drink is removed from a refrigerator and placed in a room at room temperature, the drink warms to the temperature of the room. When removed from the oven, a pizza baked at a high temperature cools to the temperature of the room.

In situations such as these, the rate at which an object's temperature changes at any given time is proportional to the difference between its temperature and the temperature of the surrounding medium. This observation is called *Newton's Law of Cooling*, although, as in the case of a cool drink, it applies to warming as well.

NEWTON'S LAW OF COOLING

Newton's Law of Cooling states that

$$T(t) = T_S + (T_0 - T_S)e^{-kt},$$

where $T(t)$ is the temperature of the object at time t, T_S is the surrounding temperature, T_0 is the value of $T(t)$ at $t = 0$, and k is a positive constant that depends on the object.

EXAMPLE 11   McDonald's Hot Coffee

The local McDonald's franchise has discovered that when coffee is poured from a coffeemaker whose contents are 180°F into a noninsulated pot, after 1 minute, the coffee cools to 165°F if the room temperature is 72°F. How long should the employees wait before pouring the coffee from this noninsulated pot into cups to deliver it to customers at 125°F?

Solution

We use Newton's Law of Cooling with $T_0 = 180$ and $T_S = 72$.

$T = 72 + (180 - 72)e^{-kt}$
$T = 72 + 108e^{-kt}$ (1)   Simplify.

From the given data, we have $T = 165$ when $t = 1$.

$$165 = 72 + 108e^{-k} \quad \text{Substitute data into equation (1).}$$
$$93 = 108e^{-k} \quad \text{Subtract 72 from both sides.}$$
$$\frac{93}{108} = e^{-k} \quad \text{Solve for } e^{-k}.$$
$$\ln\left(\frac{93}{108}\right) = -k \quad \text{Write in logarithmic form.}$$
$$-0.1495317 = -k \quad \text{Use a calculator.}$$
$$k = 0.1495317$$

With this value of k, equation (1) becomes

$$T = 72 + 108e^{-0.1495317t} \quad (2)$$
$$125 = 72 + 108e^{-0.1495317t} \quad \text{Replace } T \text{ with 125 in equation (2).}$$
$$\frac{125 - 72}{108} = e^{-0.1495317t} \quad \text{Solve for } e^{-0.1495317t}$$
$$\frac{53}{108} = e^{-0.1495317t} \quad \text{Simplify.}$$
$$\ln\left(\frac{53}{108}\right) = -0.1495317t \quad \text{Write in logarithmic form.}$$
$$t = \frac{-1}{0.1495317} \ln\left(\frac{53}{108}\right) \quad \text{Solve for } t.$$
$$t \approx 4.76 \quad \text{Use a calculator.}$$

The employees should wait approximately 4.76 minutes (realistically, about 5 minutes) to deliver the coffee at 125°F to the customers.

Practice Problem 11 Repeat Example 11 assuming that the coffee is to be delivered to the customers at 120°F.

In the next example, we use the mathematical model

$$A(t) = A_0 e^{rt}, \quad (3)$$

where A is the quantity after t units of time, A_0 is the initial (original) quantity when the experiment started $(t = 0)$, and r is the growth or decay rate per period.

EXAMPLE 12 Chemical Toxins in a Lake

In a large lake, one-fifth of the water is replaced by clean water each year. A chemical spill deposits 60,000 cubic meters of soluble toxic waste into the lake.

a. How much of this toxin will be left in the lake after four years?

b. When will the toxic chemical be reduced to 6000 cubic meters?

Solution

Now $A(1) = \frac{4}{5}(60,000) = 48,000$. So, Equation (3) gives

$$48,000 = 60,000e^r \quad A(1) = 48,000;\ A_0 = 60,000;\ t = 1$$
$$\frac{4}{5} = e^r \quad \text{Solve for } e^r$$
$$r = \ln\left(\frac{4}{5}\right) \quad \text{Logarithmic form}$$

So, Equation (3) becomes

$$A(t) = 60,000 e^{\ln\left(\frac{4}{5}\right)t} \quad \text{Equation (4)}$$

250 Chapter 3 Exponential and Logarithmic Functions

a. $A(4) = 60{,}000\, e^{\left(\ln\left(\frac{4}{5}\right)\right)4} = 60{,}000\left(e^{\ln\frac{4}{5}}\right)^4$

$= 60{,}000\left(\frac{4}{5}\right)^4 = 24{,}576\, m^3$ Use a calculator.

After 4 years, $24{,}576\, m^3$ of toxin remain in the lake.

b. $6{,}000 = 60{,}000\, e^{\ln\left(\frac{4}{5}\right)t}$ Replace $A(t)$ with 6,000 in Equation (4)

$0.1 = e^{\ln\left(\frac{4}{5}\right)t}$ Divide both sides by 60,000, simplify.

$\ln(0.1) = t \ln\left(\frac{4}{5}\right)$ Logarithmic form

$\dfrac{\ln(0.1)}{\ln\left(\dfrac{4}{5}\right)} = t$ Solve for t

$t \approx 10.23$ years   Use a calculator

In about 10 years and 3 months the toxin level will be reduced to $6000\, m^3$.

Practice Problem 12 Repeat Example 12 assuming that one-sixth of the water in the lake is replaced by clean water each year.

SECTION 3.2 Exercises

Basic Concepts and Skills

In Exercises 1–12, write each exponential equation in logarithmic form.

1. $5^2 = 25$
2. $81^{1/2} = 9$
3. $(49)^{-\frac{1}{2}} = \dfrac{1}{7}$
4. $\left(\dfrac{1}{16}\right)^{1/2} = \dfrac{1}{4}$
5. $\left(\dfrac{1}{16}\right)^{-\frac{1}{2}} = 4$
6. $(a^2)^2 = a^4$
7. $10^0 = 1$
8. $10^4 = 10{,}000$
9. $(10)^{-1} = 0.1$
10. $3^x = 5$
11. $a^2 = 5$
12. $a^e = \pi$

In Exercises 13–24, write each logarithmic equation in exponential form.

13. $\log_2 32 = 5$
14. $\log_7 49 = 2$
15. $\log_{10} 100 = 2$
16. $\log_{10} 10 = 1$
17. $\log_{10} 1 = 0$
18. $\log_a 1 = 0$
19. $\log_{10} 0.01 = -2$
20. $\log_{1/5} 5 = -1$
21. $\log_8 2 = \dfrac{1}{3}$
22. $\log 1000 = 3$
23. $\ln 2 = x$
24. $\ln \pi = a$

In Exercises 25–34, evaluate each expression without using a calculator.

25. $\log_5 125$
26. $\log_9 81$
27. $\log 10{,}000$
28. $\log_3 \dfrac{1}{3}$
29. $\log_2 \dfrac{1}{8}$
30. $\log_4 \dfrac{1}{64}$
31. $\log_3 \sqrt{27}$
32. $\log_{27} 3$
33. $\log_{16} 2$
34. $\log_5 \sqrt{125}$

In Exercises 35–46, solve each equation.

35. $\log_5 x = 2$
36. $\log_5 x = -2$
37. $\log_5 (x - 2) = 3$
38. $\log_5 (x + 1) = -2$
39. $\log_2\left(\dfrac{1}{16}\right) = x$
40. $\log_x\left(\dfrac{1}{5}\right) = -1$
41. $\log_{16} \sqrt{x - 1} = \dfrac{1}{4}$
42. $\log_{27} \sqrt[3]{1 - x} = \dfrac{1}{3}$
43. $\log_a (x - 1) = 0$
44. $\log_a (x^2 + 5x + 7) = 0$
45. $\log_2 (x^2 - 7x + 14) = 1$
46. $\log (x^2 + 5x + 16) = 1$

In Exercises 47–52, find the domain of each function.

47. $f(x) = \log_2 (x + 1)$
48. $f(x) = \log_2 (x + 3)$
49. $g(x) = \log_3 (x - 5)$
50. $g(x) = \log_3 (x - 8)$
51. $f(x) = \log_3 \sqrt{x - 1}$
52. $g(x) = \log_4 \sqrt{3 - x}$
53. $f(x) = \log (x - 2) + \log (2x - 1)$
54. $g(x) = \ln \sqrt{x + 5} - \ln (x + 1)$

55. $h(x) = \dfrac{\ln(x-1)}{\ln(2-x)}$

56. $f(x) = \ln(x-3) + \ln(2-x)$

57. Match each logarithmic function with one of the graphs labeled (a)–(f).
   a. $f(x) = \log x$
   b. $f(x) = -\log|x|$
   c. $f(x) = -\log(-x)$
   d. $f(x) = \log(x-1)$
   e. $y = (\log x) - 1$
   f. $f(x) = \log(-1-x)$

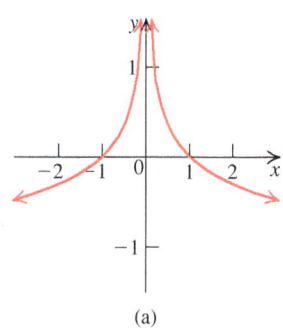

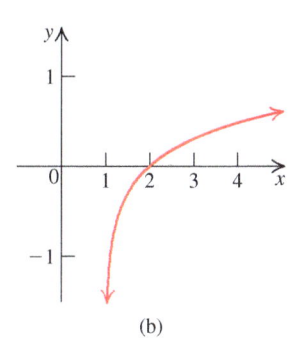

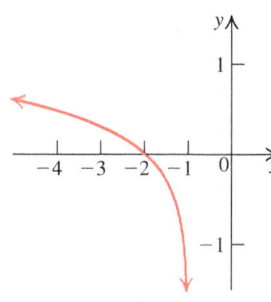

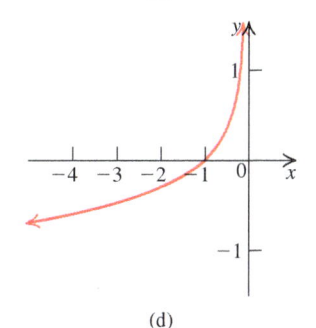

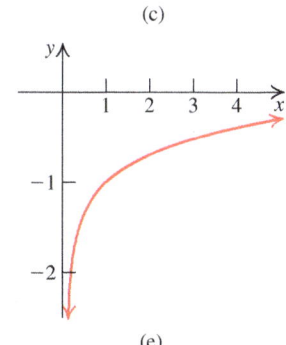

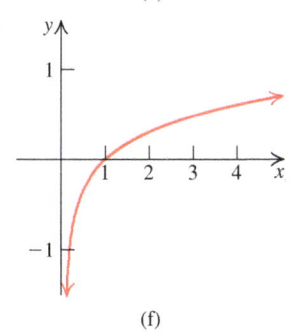

58. For each given graph, find a function of the form $f(x) = \log_a x$ that represents the graph.

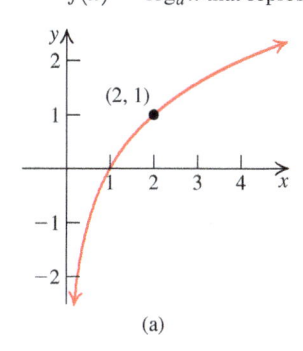

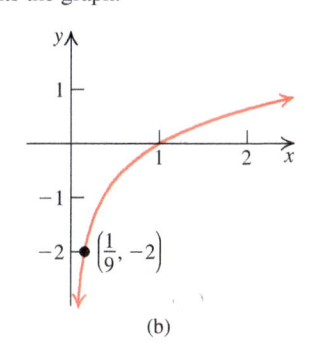

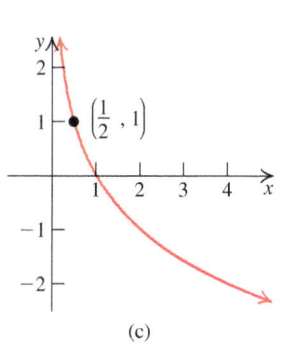

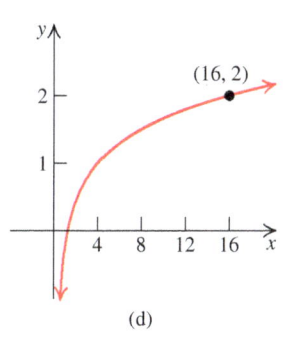

In Exercises 59–66, graph the given function by using transformations on the appropriate basic graph of the form $y = \log_a x$. State the domain and range of the function and the vertical asymptote of the graph.

59. $f(x) = \log_4(x+3)$
60. $g(x) = \log_{1/2}(x-1)$
61. $y = -\log_5 x$
62. $y = 2\log_7 x$
63. $y = \log_{1/5}(-x)$
64. $y = 1 + \log_{1/5}(-x)$
65. $y = |\log_3 x|$
66. $y = \log_3|x|$

In Exercises 67–74, begin with the graph of $f(x) = \log_2 x$ and use transformations to sketch the graph of each function. Find the domain and range of the function and the vertical asymptote of the graph.

67. $y = \log_2(x-1)$
68. $y = \log_2(-x)$
69. $y = \log_2(3-x)$
70. $y = \log_2 x$
71. $y = 2 + \log_2(3-x)$
72. $y = 4 - \log_2(3-x)$
73. $y = \log_2|x|$
74. $y = \log_2 x^2$

In Exercises 75–80, evaluate each expression.

75. $\log_4(\log_3 81)$
76. $\log_4[\log_3(\log_2 8)]$
77. $5^{\log_5 7}$
78. $10^{\log 13}$
79. $e^{\ln 5}$
80. $e^{\ln e^2}$

In Exercises 81–84, solve each equation.

81. $\log x = 2$
82. $\log(x-1) = 1$
83. $\ln x = 1$
84. $\ln x = 0$

In Exercises 85–90, begin with the graph of $y = \ln x$ and use transformations to sketch the graph of each of the given functions.

85. $y = \ln(x+2)$
86. $y = \ln(2-x)$
87. $y = -\ln(2-x)$
88. $y = 2\ln x$
89. $y = 3 - 2\ln x$
90. $y = 1 - \ln(1-x)$

Applying the Concepts

In Exercises 91–94, use the model $A = A_0 e^{kt}$.

91. **Doubling your money.** How long would it take to double your money if you invested P dollars at the rate of 8% compounded continuously?

92. **Rate for doubling your money.** At what annual rate of return, compounded continuously, would your investment double in six years?

93. Population of Canada. The population of Canada in 2010 was 35 million, and it grew 12.33% since 2000.
   a. What was the population of Canada in 2000?
   b. What was the annual rate of growth between 2000 and 2010?
   c. Assuming the same rate of growth, determine the time from 2010 until Canada's population doubles.
   d. Find the time from 2010 until Canada's population reaches 50 million.

94. Toxic chemicals in a lake. In a lake, one-fourth of the water is replaced by clean water every year. Sixteen thousand cubic meters of soluble toxic chemical spill takes place in the lake. Let $T(n)$ represent the amount of toxin left after n years.
   a. Find a formula for $T(n)$.
   b. How much toxin will be left after 12 years?
   c. When will 80% of the toxin be eliminated?

95. The time of murder. A forensic specialist took the temperature of a victim's body lying in a street at 2:10 A.M. and found it to be 85.7°F. At 2:40 A.M., the temperature of the body was 84.8°F. When was the murder committed if the air temperature during the night was 55°F? [Remember, normal body temperature is 98.6°F.]

96. Newton's Law of Cooling. A thermometer is taken from a room at 75°F to the outdoors, where the temperature is 20°F. The reading on the thermometer drops to 50°F after one minute.
   a. Find the reading on the thermometer after
      (i) five minutes.
      (ii) ten minutes.
      (iii) one hour.
   b. How long will it take for the reading to drop to 22°F?

97. Newton's Law of Cooling. The last bit of ice in a picnic cooler has melted ($T_0 = 32°F$). The temperature in the park is 85°F. After 30 minutes, the temperature in the cooler is 40°F. How long will it take the temperature inside the cooler to reach 50°F?

98. Cooking salmon. A salmon filet initially at 50°F is cooked in an oven at a constant temperature of 400°F. After ten minutes, the temperature of the filet rises to 160°F. How long does it take until the salmon is medium rare at 220°F?

Critical Thinking / Discussion / Writing

99. Present value of an investment. Recall that if P dollars is invested in an account at an interest rate r compounded continuously, then the amount A (called the *future value of P*) in the account t years from now will be $A = Pe^{rt}$. Solving the equation for P, we get $P = Ae^{-rt}$. In this formulation, P is called the present value of the investment.
   a. Find the present value of $100,000 at 7% compounded continuously for 20 years.
   b. Find the interest rate r compounded continuously that is needed to have $50,000 be the present value of $75,000 in ten years.

100. Pension. Your uncle is 40 years old, and he wants to have an annual pension of $50,000 each year at age 65. What is the present value of his pension if the money can be invested at all times at a continuously compounded interest rate of
   a. 5%?
   b. 8%?
   c. 10%?

In Exercises 101 and 102, evaluate each expression without using a calculator.

101. $2^{\log_2 3} - 3^{\log_3 2}$

102. $(\log_3 4 + \log_2 9)^2 - (\log_3 4 - \log_2 9)^2$

103. Solve for x: $\log_3[\log_4(\log_2 x)] = 0$

104. Explain why $\log_a x$ is defined only for (a) $a > 0$ and $a \neq 1$ and (b) $x > 0$.

105. Inequalities involving logarithms.
   a. If a is positive, is the statement "$a < b$ if and only if $\log a < \log b$" always true?
   b. What property of logarithmic functions is used whenever the statement in **a** is true?

106. Write a summary of the types of graphs (with sketches) for the logarithmic functions of the form $y = c \log_a x$, $c \neq 0$, $a > 0$, and $a \neq 1$.

Maintaining Skills

In Exercises 107–110, write each expression in the form a^n where n is an integer.

107. $a^2 \cdot a^7$

108. $(a^2)^3$

109. $\sqrt{a^8}$

110. $\sqrt[3]{a^6}$

In Exercises 111–114, solve each equation for x.

111. $2x - (3 - 5x) = -13$

112. $3x(2x - 1) = 9$

113. $Ae^2 = Ae^{x-1}$

114. $A = Ae^{2x}$

115. Find an equation in slope–intercept form for the line through the points $(-1, 4)$ and $(3, 6)$.

SECTION 3.3 Rules of Logarithms

BEFORE STARTING THIS SECTION, REVIEW

1. Definition of logarithm (Section 3.2, page 243)
2. Basic properties of logarithms (Section 3.2, page 246)
3. Rules of exponents (Section P.1, page 9)

OBJECTIVES

1. Learn the rules of logarithms.
2. Estimate a large number.
3. Change the base of a logarithm.
4. Apply logarithms to growth and decay.

◆ The Boy King Tut

Today the most famous pharaoh of Ancient Egypt is King Nebkheperure Tutankhamun, popularly called "King Tut." He was only 9 years old when he became a pharaoh. In 2005, a team of Egyptian scientists headed by Dr. Zahi Hawass determined that the pharaoh was 19 years old when he died (around 1346 B.C.).

Tutankhamun was a short-lived boy king who, unlike the great Egyptian kings Khufu (builder of the Great Pyramid), Amenhotep III (builder of temples throughout Egypt), and Ramesses II (prolific builder and usurper), accomplished nothing significant during his reign. In fact, little was known about him prior to Howard Carter's discovery of his tomb (and the amazing treasures it held) in the Valley of the Kings on November 4, 1922. Carter, with his benefactor Lord Carnarvon at his side, entered the tomb's burial chamber on November 26, 1922. Lord Carnarvon died seven weeks after entering the burial chamber, giving rise to the theory of the "curse" of King Tut.

Work on the tomb continued until 1933. The tomb contained a pristine mummy of an Egyptian king, lying intact in his original burial furniture. He was accompanied by a small slice of the royal world of the pharaohs: golden chariots, statues of gold and ebony, a fleet of miniature ships to accommodate his trip to the hereafter, his throne of gold, bottles of perfume, precious jewelry, and more. The "Treasures of Tutankhamun" exhibition, first shown at London's British Museum in 1972, traveled to many countries, including the United States.

In Example 8, we discuss the age of a work of art found in King Tut's tomb.

King Tut's Golden Mask

1 Learn the rules of logarithms.

Rules of Logarithms

Recall the basic properties of logarithms from Section 3.2 for $a > 0, a \neq 1$:

$$\log_a a = 1 \qquad (1)$$
$$\log_a 1 = 0 \qquad (2)$$
$$\log_a a^x = x, x \text{ real} \qquad (3)$$
$$a^{\log_a x} = x, x > 0 \qquad (4)$$

In this section, we will discuss some important rules of logarithms that are helpful in calculations, simplifications, and applications.

RULES OF LOGARITHMS

Let M, N, and a be positive real numbers with $a \neq 1$ and let r be any real number.

Rule	Description	Examples
1. Product Rule: $\log_a (MN) = \log_a M + \log_a N$	The logarithm of the product of two (or more) numbers is the sum of the logarithms of the numbers.	$\ln (5 \cdot 7) = \ln 5 + \ln 7$ $\log (3x) = \log 3 + \log x$ $\log_2 (5 \cdot 17) = \log_2 5 + \log_2 17$
2. Quotient Rule: $\log_a \left(\dfrac{M}{N}\right) = \log_a M - \log_a N$	The logarithm of the quotient of two numbers is the difference of the logarithms of the numbers.	$\ln \dfrac{5}{7} = \ln 5 - \ln 7$ $\log \dfrac{5}{x} = \log 5 - \log x$
3. Power Rule: $\log_a M^r = r \log_a M$	The logarithm of a number to the power r is r times the logarithm of the number.	$\ln 5^7 = 7 \ln 5$ $\log 5^{3/2} = \dfrac{3}{2} \log 5$ $\log_2 7^{-3} = -3 \log_2 7$

Rules 1, 2, and 3 follow from the corresponding rules of the exponents:

$$a^u \cdot a^v = a^{u+v} \quad \text{Product rule}$$

$$\dfrac{a^u}{a^v} = a^{u-v} \quad \text{Quotient rule}$$

$$(a^u)^r = a^{u\,r} \quad \text{Power rule}$$

For example, to prove the product rule for logarithms, we let

$$\log_a M = u \quad \text{and} \quad \log_a N = v.$$

The corresponding exponential forms of these equations are

$$M = a^u \quad \text{and} \quad N = a^v.$$

Then

$$MN = a^u \cdot a^v$$
$$MN = a^{u+v} \quad \text{Product rule of exponents}$$
$$\log_a MN = u + v \quad \text{Logarithmic form}$$
$$\log_a MN = \log_a M + \log_a N \quad \text{Replace } u \text{ with } \log_a M \text{ and } v \text{ with } \log_a N.$$

This proves the product rule of logarithms. The quotient rule and the power rule can be proved in the same way by using the corresponding rules for exponents.

RECALL

Radicals can also be written as exponents. Recall that

$$\sqrt{x} = x^{1/2}$$
and $\sqrt[3]{x} = x^{1/3}$.

In general,

$$\sqrt[m]{x^n} = x^{n/m}.$$

EXAMPLE 1 Using Rules of Logarithms to Evaluate Expressions

Given that $\log_5 z = 3$ and $\log_5 y = 2$, evaluate each expression.

a. $\log_5 (yz)$ **b.** $\log_5 (125y^7)$ **c.** $\log_5 \sqrt{\dfrac{z}{y}}$ **d.** $\log_5 (z^{1/30} y^5)$

Solution

a. $\log_5 (yz) = \log_5 y + \log_5 z$ Product rule
$\qquad\qquad\quad = 2 + 3 = 5$ Use the given values and simplify.

b. $\log_5 (125y^7) = \log_5 125 + \log_5 y^7$ Product rule
$ = \log_5 5^3 + \log_5 y^7$ $125 = 5^3$
$ = 3 + 7\log_5 y$ Power rule and $\log_a a = 1$
$ = 3 + 7(2) = 17$ Use the given values and simplify.

c. $\log_5 \sqrt{\dfrac{z}{y}} = \log_5 \left(\dfrac{z}{y}\right)^{1/2}$ Rewrite radical as exponent.

$\phantom{\log_5 \sqrt{\dfrac{z}{y}}} = \dfrac{1}{2} \log_5 \dfrac{z}{y}$ Power rule

$\phantom{\log_5 \sqrt{\dfrac{z}{y}}} = \dfrac{1}{2} (\log_5 z - \log_5 y)$ Quotient rule

$\phantom{\log_5 \sqrt{\dfrac{z}{y}}} = \dfrac{1}{2} (3 - 2) = \dfrac{1}{2}$ Use the given values and simplify.

d. $\log_5 (z^{1/30} y^5) = \log_5 z^{1/30} + \log_5 y^5$ Product rule

$\phantom{\log_5 (z^{1/30} y^5)} = \dfrac{1}{30} \log_5 z + 5 \log_5 y$ Power rule

$\phantom{\log_5 (z^{1/30} y^5)} = \dfrac{1}{30}(3) + 5(2)$ Use the given values.

$\phantom{\log_5 (z^{1/30} y^5)} = 0.1 + 10 = 10.1$ Simplify.

Practice Problem 1 Evaluate each expression for the given values in Example 1.
a. $\log_5 (y/z)$ **b.** $\log_5 (y^2 z^3)$

In many applications in more advanced mathematics courses, the rules of logarithms are used in both directions; that is, the rules are read from left to right and from right to left. For example, by the product rule of logarithms, we have

$$\log_2 3x = \log_2 3 + \log_2 x.$$

The expression $\log_2 3 + \log_2 x$ is the *expanded form* of $\log_2 3x$, and $\log_2 3x$ is the *condensed form*, or the *single logarithmic form*, of $\log_2 3 + \log_2 x$.

EXAMPLE 2   Writing Expressions in Expanded Form

Write each expression in expanded form. Assume that all expressions containing variables represent positive numbers.

a. $\log_2 \dfrac{x^2 (x-1)^3}{(2x+1)^4}$ **b.** $\log_c \sqrt{x^3 y^2 z^5}$

Solution

a. $\log_2 \dfrac{x^2 (x-1)^3}{(2x+1)^4} = \log_2 x^2 (x-1)^3 - \log_2 (2x+1)^4$ Quotient rule

$\phantom{\log_2 \dfrac{x^2 (x-1)^3}{(2x+1)^4}} = \log_2 x^2 + \log_2 (x-1)^3 - \log_2 (2x+1)^4$ Product rule

$\phantom{\log_2 \dfrac{x^2 (x-1)^3}{(2x+1)^4}} = 2 \log_2 x + 3 \log_2 (x-1) - 4 \log_2 (2x+1)$ Power rule

b. $\log_c \sqrt{x^3 y^2 z^5} = \log_c (x^3 y^2 z^5)^{1/2}$ $\qquad \sqrt{a} = a^{1/2}$

$= \dfrac{1}{2} \log_c (x^3 y^2 z^5)$ $\qquad$ Power rule

$= \dfrac{1}{2} [\log_c x^3 + \log_c y^2 + \log_c z^5]$ $\qquad$ Product rule

$= \dfrac{1}{2} [3 \log_c x + 2 \log_c y + 5 \log_c z]$ $\qquad$ Power rule

$= \dfrac{3}{2} \log_c x + \log_c y + \dfrac{5}{2} \log_c z$ $\qquad$ Distributive property

Practice Problem 2 Write each expression in expanded form. Assume that all expressions containing variables represent positive numbers.

a. $\ln \dfrac{2x - 1}{x + 4}$ $\qquad$ **b.** $\log \sqrt{\dfrac{4xy}{z}}$

EXAMPLE 3 Writing Expressions in Condensed Form

Write each expression in condensed form.

a. $\log 3x - \log 4y$ $\qquad$ **b.** $2 \ln x + \dfrac{1}{2} \ln (x^2 + 1)$

c. $2 \log_2 5 + \log_2 9 - \log_2 75$ $\qquad$ **d.** $\dfrac{1}{3} [\ln x + \ln (x + 1) - \ln (x^2 + 1)]$

Solution

a. $\log 3x - \log 4y = \log \left(\dfrac{3x}{4y} \right)$ $\qquad$ Quotient rule

b. $2 \ln x + \dfrac{1}{2} \ln (x^2 + 1) = \ln x^2 + \ln (x^2 + 1)^{1/2}$ $\qquad$ Power rule
$\qquad\qquad\qquad\qquad\qquad\quad = \ln (x^2 \sqrt{x^2 + 1})$ $\qquad$ Product rule; $\sqrt{a} = a^{1/2}$

c. $2 \log_2 5 + \log_2 9 - \log_2 75 = \log_2 5^2 + \log_2 9 - \log_2 75$ $\qquad$ Power rule
$\qquad\qquad\qquad\qquad\qquad\qquad\quad = \log_2 (25 \cdot 9) - \log_2 75$ $\qquad$ $5^2 = 25$; Product rule
$\qquad\qquad\qquad\qquad\qquad\qquad\quad = \log_2 \dfrac{25 \cdot 9}{75}$ $\qquad$ Quotient rule
$\qquad\qquad\qquad\qquad\qquad\qquad\quad = \log_2 3$ $\qquad$ $\dfrac{25 \cdot 9}{75} = \dfrac{9}{3} = 3$

d. $\dfrac{1}{3} [\ln x + \ln (x + 1) - \ln (x^2 + 1)] = \dfrac{1}{3} [\ln x(x + 1) - \ln (x^2 + 1)]$ $\qquad$ Product rule

$\qquad\qquad\qquad\qquad\qquad\qquad\qquad\quad = \dfrac{1}{3} \ln \left[\dfrac{x(x + 1)}{x^2 + 1} \right]$ $\qquad$ Quotient rule

$\qquad\qquad\qquad\qquad\qquad\qquad\qquad\quad = \ln \sqrt[3]{\dfrac{x(x + 1)}{x^2 + 1}}$ $\qquad$ Power rule; $a^{1/3} = \sqrt[3]{a}$

Practice Problem 3 Write in condensed form: $\dfrac{1}{2} [\log (x + 1) + \log (x - 1)]$.

Be careful when using the rules of logarithms to simplify expressions. For example, there is no property that allows you to rewrite $\log_a (x + y)$.
In general,

$$\log_a (x + y) \neq \log_a x + \log_a y. \qquad (5)$$

To illustrate this statement, let $x = 100, y = 10$, and $a = 10$. Then the value of the left side of (5) is

$$\log(100 + 10) = \log(110) \approx 2.0414.$$

The value of the right side of (5) is

$$\begin{aligned}\log 100 + \log 10 &= \log 10^2 + \log 10 \\ &= 2 \log 10 + \log 10 \quad \text{Power rule} \\ &= 2(1) + 1 = 3.\end{aligned}$$

Therefore, $\log(100 + 10) \neq \log 100 + \log 10$.

WARNING

To avoid common errors, be aware that in general,

$$\log_a(M + N) \neq \log_a M + \log_a N.$$
$$\log_a(M - N) \neq \log_a M - \log_a N.$$
$$(\log_a M)(\log_a N) \neq \log_a MN.$$
$$\log_a\left(\frac{M}{N}\right) \neq \frac{\log_a M}{\log_a N}.$$
$$\frac{\log_a M}{\log_a N} \neq \log_a M - \log_a N.$$
$$(\log_a M)^r \neq r \log_a M.$$

2 Estimate a large number.

Number of Digits

In estimating the magnitude of a positive number K, we look for the number of digits needed to express K. If K has a fractional part, then "number of digits" means the number of digits to the left of the decimal point. For example, 357.29 and 231.4796 are viewed as 3-digit numbers in this context.

A number K written in the form $K = s \times 10^n$, where $1 \leq s < 10$ and n is an integer, is said to be in scientific notation. For example, $432.1 = 4.321 \times 10^2$ and $0.56 = 5.6 \times 10^{-1}$ are both expressed in scientific notation. The exponent n represents the magnitude of K and is closely related to the common logarithm. In fact, we have

$$\begin{aligned}\log K &= \log(s \times 10^n) &&\text{Take log of both sides of } K = s \times 10^n. \\ &= \log s + \log 10^n &&\text{Product rule} \\ &= \log s + n \log 10 &&\text{Power rule} \\ &= \log s + n &&\log 10 = \log_{10} 10 = 1\end{aligned}$$

Because $1 \leq s < 10$, we have

$$\begin{aligned}\log 1 \leq \log s < \log 10 &\quad \log x \text{ is an increasing function.} \\ 0 \leq \log s < 1 &\quad \log 1 = 0 \text{ and } \log 10 = 1\end{aligned}$$

This says that the exponent n is the largest integer less than or equal to $\log K$.

We note that a number m is a 3-digit number if

$$10^2 = 100 \leq m < 1000 = 10^3$$
$$\text{or} \quad \log 10^2 \leq \log m < \log 10^3$$
$$2 \leq \log m < 3 \qquad \log 10^x = \log_{10} 10^x = x$$

Similarly, a number K has $(n + 1)$ digits if $n \leq \log K < n + 1$.

258 Chapter 3 Exponential and Logarithmic Functions

> **DIGITS AND COMMON LOGARITHMS**
>
> A positive number K has $(n + 1)$ digits if and only if log K is in the interval $[n, n + 1)$.

For example, the number 8765, which has four digits, has its common logarithm between 3 and 4. Conversely, if log N is approximately 57.3, then N is a 58-digit number.

EXAMPLE 4 **Estimating a Large Number**

Write an estimate of the number $K = e^{700}$ in scientific notation.

Solution

$$K = e^{700}$$
$$\log K = \log(e^{700})$$
$$= 700 \log e$$
$$\approx 304.0061373 \quad \text{Use a calculator.}$$

Because log K lies between the integers 304 and 305, the number K requires 305 digits to the left of the decimal point. Also, by definition of the common logarithm, we have

$$K \approx 10^{304.0061373}$$
$$= 10^{0.0061373} \times 10^{304}$$
$$\approx 1.014232 \times 10^{304} \quad \text{Use a calculator.}$$

Practice Problem 4 Repeat Example 4 for $K = (234)^{567}$.

3 Change the base of a logarithm.

Change of Base

Calculators usually come with two types of log keys: a key for natural logarithms (base e, [LN]) and a key for common logarithms (base 10, [LOG]). These two types of logarithms are frequently used in applications. Sometimes, however, we need to evaluate logarithms for bases other than e and 10. The *change-of-base formula* helps us evaluate these logarithms.

Suppose we are given $\log_b x$ and we want to find an equivalent expression in terms of logarithms with base a. We let

$$u = \log_b x. \quad (6)$$

In exponential form, we have

$$x = b^u. \quad (7)$$

Then

$$\log_a x = \log_a b^u \qquad \text{Take } \log_a \text{ of each side of (7).}$$
$$\log_a x = u \log_a b \qquad \text{Power rule}$$
$$u = \frac{\log_a x}{\log_a b} \qquad \text{Solve for } u.$$
$$\log_b x = \frac{\log_a x}{\log_a b} \quad (8) \qquad \text{Substitute for } u \text{ from (6).}$$

Equation (8) is the change-of-base formula. By choosing $a = 10$ and $a = e$, we can state the change-of-base formula as follows:

> **CHANGE-OF-BASE FORMULA**
>
> Let a, b, and x be positive real numbers with $a \neq 1$ and $b \neq 1$. Then $\log_b x$ can be converted to a different base as follows:
>
> $$\log_b x = \underbrace{\frac{\log_a x}{\log_a b}}_{(\text{base } a)} = \underbrace{\frac{\log x}{\log b}}_{(\text{base } 10)} = \underbrace{\frac{\ln x}{\ln b}}_{(\text{base } e)}$$

TECHNOLOGY CONNECTION

To graph a logarithmic function whose base is not e or 10, use the change-of-base formula. For example, graph $y = \log_5 x$ by entering $Y_1 = \text{LN}(X)/\text{LN}(5)$ or $Y_1 = \log(X)/\log(5)$.

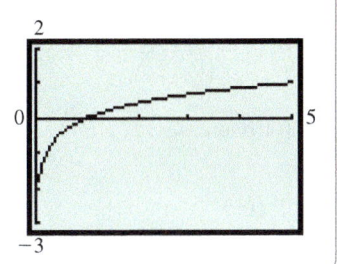

EXAMPLE 5   Using a Change of Base to Compute Logarithms

Compute $\log_5 13$ by changing to

a. common logarithms.   **b.** natural logarithms.

Solution

a. $\log_5 13 = \dfrac{\log 13}{\log 5}$ Change to base 10.   **b.** $\log_5 13 = \dfrac{\ln 13}{\ln 5}$ Change to base e.

$ \approx 1.59369$ Use a calculator. $ \approx 1.59369$ Use a calculator.

Practice Problem 5 Find $\log_3 15$.

EXAMPLE 6   Modeling with Logarithmic Functions

Find

a. an equation of the form $y = c + b \log N$ whose graph contains the points $(2, 3)$ and $(4, -5)$.

b. an equation of the form $y = c + b \log_a x$ whose graph contains the points $(2, 3)$ and $(4, -5)$ where a, b, and c are integers.

Solution

Substitute $(2, 3)$ and $(4, -5)$ in the equation $y = c + b \log x$ to obtain:

$$3 = c + b \log 2 \quad (9)$$
$$-5 = c + b \log 4 \quad (10)$$

a. Find b: Subtract equation (10) from equation (9) to obtain

$$8 = b \log 2 - b \log 4 = b(\log 2 - \log 4) = b\left(\log \frac{2}{4}\right)$$

$$8 = b \log\left(\frac{1}{2}\right) = b \log 2^{-1} = -b \log 2$$

$$b = -\frac{8}{\log 2} \qquad \text{Solve for } b.$$

Find c:

$$c = 3 - b \log 2 \qquad \text{From equation (9)}$$
$$= 3 - \left(-\frac{8}{\log 2}\right) \log 2$$
$$= 3 + 8$$
$$= 11$$

b. Use the change of base formula to find a, b, and c.
Substitute b and c from part **a** in the equation $y = c + b \log x$ to obtain:

$$y = 11 - \frac{8}{\log 2}(\log x) = 11 - 8\left(\frac{\log x}{\log 2}\right)$$

$$y = 11 - 8 \log_2 x \qquad \log_2 x = \frac{\log x}{\log 2}$$

Then $a = 2$, $b = -8$, and $c = 11$.

Practice Problem 6 Repeat Example 6 for the graph that contains the points $(3, 3)$ and $(9, 1)$.

4 Apply logarithms to growth and decay.

Growth and Decay

Exponential growth (or decay) occurs when a quantity grows (or decreases) at a rate proportional to its size. For $k > 0$,

Growth Formula: $A(t) = A_0 e^{kt}$ (11) and

Decay Formula: $A(t) = A_0 e^{-kt}$ (12),

where $A(t) = $ the amount of substance (or population) at time t,

$A_0 = A(0)$ is the initial amount, and

$t = $ time.

Half-Life

The **half-life** of any quantity whose value decreases with time is the time required for the quantity to decay to half its initial value.

Radioactive substances undergo exponential decay. Krypton-85, for example, has a half-life of about ten years; this means that any initial quantity of krypton-85 will dissipate or decay to half that amount in about ten years.

If h is the half-life of a substance undergoing exponential decay at the rate k, then any mass A_0 decays to $\frac{1}{2}A_0$ in time h. See Figure 3.12. This fact leads to a half-life formula.

$$\frac{1}{2}A_0 = A_0 e^{-kh} \qquad A(t) = A_0 e^{kt} \text{ with } t = h \text{ and } A(h) = \frac{1}{2}A_0$$

$$\frac{1}{2} = e^{-kh} \qquad \text{Divide both sides by } A_0.$$

$$\ln\left(\frac{1}{2}\right) = \ln\left(e^{-kh}\right) \qquad \text{Take the natural log of both sides.}$$

$$-\ln 2 = -kh \qquad \ln\left(\frac{1}{2}\right) = \ln(2^{-1}) = -\ln 2; \ln(e^x) = x$$

$$h = \frac{\ln 2}{k} \qquad \text{Solve for } h.$$

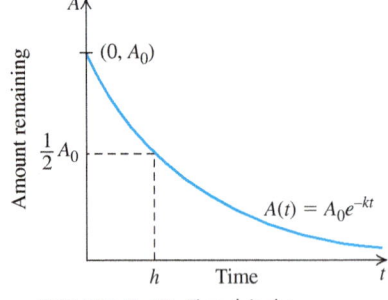

FIGURE 3.12 Time h is the half-life.

HALF-LIFE FORMULA

The half-life h of a substance undergoing exponential decay at a rate k ($k > 0$) is given by the formula

$$h = \frac{\ln 2}{k}.$$

EXAMPLE 7 Finding the Half-Life of a Substance

In an experiment, 18 grams of the radioactive element sodium-24 decayed to 6 grams in 24 hours. Find its half-life to the nearest hour.

Solution

$$A(t) = A_0 e^{-kt} \quad \text{Exponential growth and decay model}$$
$$6 = 18e^{-k(24)} \quad A_0 = 18, A(24) = 6$$
$$\frac{1}{3} = e^{-k(24)} \quad \text{Divide both sides by 18.}$$
$$\ln\left(\frac{1}{3}\right) = \ln\left(e^{-k(24)}\right) \quad \text{Take the natural log of both sides.}$$
$$-\ln(3) = -24k \quad \ln\left(\frac{1}{3}\right) = \ln(3^{-1}) = -\ln 3; \ln(e^x) = x$$
$$k = \frac{\ln 3}{24} \quad \text{Solve for } k.$$

So $A(t) = 18e^{-\frac{\ln 3}{24} t}$.

To find the half-life of sodium-24, we use the formula

$$h = \frac{\ln 2}{k} \quad \text{Half-life formula}$$
$$h = \frac{\ln 2}{\frac{\ln 3}{24}} \quad \text{Replace } k \text{ with } \frac{\ln 3}{24}.$$
$$h = \frac{24 \ln 2}{\ln 3} \approx 15 \text{ hours} \quad \text{Simplify; use a calculator.}$$

Practice Problem 7 In an experiment, 100 grams of the radioactive element strontium-90 decayed to 66 grams in 15 years. Find its half-life to the nearest year.

Radiocarbon Dating

Ordinary carbon, called carbon-12 (^{12}C), is stable and does not decay. However, carbon-14 (^{14}C) is a form of carbon that decays radioactively with a half-life of 5700 years. Sunlight constantly produces carbon-14 in Earth's atmosphere. When a living organism breathes or eats, it absorbs carbon-12 and carbon-14. After the organism dies, no more carbon-14 is absorbed; so the age of its remains can be calculated by determining how much carbon-14 has decayed. The method of radiocarbon dating was developed by the American scientist W. F. Libby.

EXAMPLE 8 King Tut's Treasure

In 1960, a group of specialists from the British Museum in London investigated whether a piece of art containing organic material found in Tutankhamun's tomb had been made during his reign or (as some historians claimed) whether it belonged to an earlier period. We know that King Tut died in 1346 B.C. and ruled Egypt for ten years. What percent of the amount of carbon-14 originally contained in the object should be present in 1960 if the object was made during Tutankhamun's reign?

HISTORICAL NOTE

Willard Frank Libby (1908–1980)

Libby was born in Grand Valley, Colorado. In 1927, he entered the University of California at Berkeley, where he earned his BSc and PhD degrees in 1931 and 1933, respectively. He then became an instructor in the Department of Chemistry at the University of California, Berkeley in 1933. He was awarded a Guggenheim Memorial Foundation Fellowship in 1941, but the fellowship was interrupted when Libby went to Columbia University on the Manhattan District Project upon America's entry into World War II.

Libby was a physical chemist and a specialist in radiochemistry, particularly in hot-atom chemistry, tracer techniques, and isotope tracer work. He became well known at the University of Chicago for his work on natural carbon-14 (a radiocarbon) and its use in dating archaeological artifacts and on natural tritium and its use in hydrology and geophysics. He was awarded the Nobel Prize in Chemistry in 1960.

Solution

Because the half-life of carbon-14 is approximately 5700 years, we can rewrite the equation $A(t) = A_0 e^{-kt}$ as

$$A(5700) = A_0 e^{-5700k}.$$

Now we find the value of k:

$$\frac{1}{2}A_0 = A_0 e^{-5700k} \qquad \text{Replace } A(5700) \text{ with } \frac{1}{2}A_0.$$

$$\frac{1}{2} = e^{-5700k} \qquad \text{Divide both sides by } A_0.$$

$$\ln\left(\frac{1}{2}\right) = -5700k \qquad \text{Logarithmic form}$$

$$k = -\frac{\ln\left(\frac{1}{2}\right)}{5700} \qquad \text{Solve for } k.$$

$$\approx 0.0001216 \qquad \text{Use a calculator.}$$

Substituting this value of k into equation (12), we obtain

$$A(t) = A_0 e^{-0.0001216t} \qquad (13)$$

The time t that elapsed between King Tut's death and 1960 is

$$t = 1960 + 1346 \qquad \text{1346 B.C. to A.D. 1960}$$
$$= 3306$$
$$A(3306) = A_0 e^{-0.0001216(3306)} \qquad \text{Replace } t \text{ with 3306 in (13).}$$
$$\approx 0.66897 A_0$$

Therefore, the percent of the original amount of carbon-14 remaining in the object (after 3306 years) is 66.897%.

Because King Tut began ruling Egypt ten years before he died, the time t_1 that elapsed from the beginning of his reign to 1960 is 3316 years.

$$A(3316) = A_0 e^{-0.0001216(3316)} \qquad \text{Replace } t \text{ with 3316 in (13).}$$
$$\approx 0.66816 A_0$$

Therefore, a piece of art made during King Tut's reign would have between 66.816% and 66.897% of carbon-14 remaining in 1960.

Practice Problem 8 Repeat Example 8 assuming that the art object was made 194 years before King Tut's death.

SECTION 3.3 Exercises

Basic Concepts and Skills

In Exercises 1–12, given that $\log x = 2, \log y = 3, \log 2 \approx 0.3$, and $\log 3 \approx 0.48$, evaluate each expression without using a calculator.

1. $\log 6$
2. $\log 4$
3. $\log 5 \left[\text{Hint: } 5 = \frac{10}{2}\right]$
4. $\log (3x)$
5. $\log\left(\frac{2}{x}\right)$
6. $\log x^2$
7. $\log (2x^2 y)$
8. $\log xy^3$
9. $\log \sqrt[3]{x^2 y^4}$
10. $\log (\log x^2)$
11. $\log \sqrt[3]{48}$
12. $\log_2 3$

In Exercises 13–26, write each expression in expanded form. Assume that all expressions containing variables represent positive numbers.

13. $\ln[x(x-1)]$
14. $\ln\dfrac{x(x+1)}{(x-1)^2}$
15. $\log\dfrac{\sqrt{x^2+1}}{x+3}$
16. $\log_4\left(\dfrac{x^2-9}{x^2-6x+8}\right)^{\frac{2}{3}}$
17. $\log_b x^2 y^3 z$
18. $\log_b \sqrt{xyz}$
19. $\ln\left[\dfrac{x\sqrt{x-1}}{x^2+2}\right]$
20. $\ln\left[\dfrac{\sqrt{x-2}\,\sqrt[3]{x+1}}{x^2+3}\right]$
21. $\ln\left[\dfrac{(x+1)^2}{(x-3)\sqrt{x+4}}\right]$
22. $\ln\left[\dfrac{2x+3}{(x+4)^2(x-3)^4}\right]$
23. $\ln\left[(x+1)\sqrt{\dfrac{x^2+2}{x^2+5}}\right]$
24. $\ln\left[\dfrac{\sqrt[3]{2x+1}\,(x+1)}{(x-1)^2(3x+2)}\right]$
25. $\ln\left[\dfrac{x^3(3x+1)^4}{\sqrt{x^2+1}(x+2)^{-5}(x-3)^2}\right]$
26. $\ln\left[\dfrac{(x+1)^{\frac{1}{2}}(x^2-2)^{\frac{2}{5}}}{(2x-1)^{\frac{3}{2}}(x^2+2)^{-\frac{4}{5}}}\right]$

In Exercises 27–36, write each expression in condensed form.

27. $\log_2 x + \log_2 7$
28. $\log_2 x - \log_2 3$
29. $\dfrac{1}{2}\log x - \log y + \log z$
30. $\dfrac{1}{2}(\log x + \log y)$
31. $\dfrac{1}{5}(\log_2 z + 2\log_2 y)$
32. $\dfrac{1}{3}(\log x - 2\log y + 3\log z)$
33. $\ln x + 2\ln y + 3\ln z$
34. $2\ln x - 3\ln y + 4\ln z$
35. $2\ln x - \dfrac{1}{2}\ln(x^2+1)$
36. $2\ln x + \dfrac{1}{2}\ln(x^2-1) - \dfrac{1}{2}\ln(x^2+1)$

In Exercises 37–42, write an estimate of each number in scientific notation.

37. e^{500}
38. 723^{416}
39. Which is larger, 234^{567} or 567^{234}?
40. Which is larger, 4321^{8765} or 8765^{4321}?
41. Find the number of digits in $17^{200} \cdot 53^{67}$.
42. Find the number of digits in $67^{200} \div 23^{150}$.

In Exercises 43–50, use the change-of-base formula and a calculator to evaluate each logarithm.

43. $\log_2 5$
44. $\log_4 11$
45. $\log_{1/2} 3$
46. $\log_{\sqrt{3}} 12.5$
47. $\log_{\sqrt{5}} \sqrt{17}$
48. $\log_{15} 123$
49. $\log_2 7 + \log_4 3$
50. $\log_2 9 - \log_{\sqrt{2}} 5$

In Exercises 51–58, find the value of each expression without using a calculator.

51. $\log_3 \sqrt{3}$
52. $\log_{1/4} 4$
53. $\log_3(\log_2 8)$
54. $2^{\log_2 2}$
55. $5^{2\log_5 3 + \log_5 2}$
56. $e^{3\ln 2 - 2\ln 3}$
57. $\log 4 + 2\log 5$
58. $\log_2 160 - \log_2 5$

In Exercises 59–66, find an equation of the form $y = c + b\log_a x$ whose graph contains the two given points.

59. $(10, 1)$ and $(1, 2)$
60. $(4, 10)$ and $(2, 12)$
61. $(e, 1)$ and $(1, 2)$
62. $(3, 1)$ and $(9, 2)$
63. $(5, 4)$ and $(25, 7)$
64. $(4, 3)$ and $(8, 5)$
65. $(2, 4)$ and $(3, 9)$
66. $(1, 1)$ and $(5, 7)$

In Exercises 67–70, determine the half-life of each substance that decays from A_0 to A in time t.

67. $A_0 = 50$, $A = 23$, $t = 12$ years
68. $A_0 = 200$, $A = 65$, $t = 10$ years
69. $A_0 = 10.3$, $A = 3.8$, $t = 15$ hours
70. $A_0 = 20.8$, $A = 12.3$, $t = 40$ minutes

Applying the Concepts

In Exercises 71–73, assume the exponential growth model $A(t) = A_0 e^{kt}$ and a world population of 7 billion in 2011.

71. **Rate of population growth.** If the world adds about 90 million people in 2012, what is the rate of growth?
72. **Population.** Use the rate of growth from Exercise 71 to estimate the year in which the world population will be
    a. 12 billion.
    b. 20 billion.
73. **Population growth.** If the population must stay below 20 billion during the next 100 years, what is the maximum acceptable annual rate of growth? Round your answer to six decimal places.
74. **Population decline.** If the world population must decline below 5 billion during the next 25 years, what must be the annual rate of decline? Round your answer to six decimal places.
75. **Continuous compounding.** If $1000 is deposited in a bank at 10% interest compounded continuously, how many years will it take the money to grow to $3500?
76. **Continuous compounding.** At what rate of interest compounded continuously will an investment double in six years?

77. Half-life. Tritium is used in nuclear weapons to increase their power. It decays at the rate of 5.5% per year. Calculate the half-life of tritium.

78. Half-life. Sixty grams of magnesium-28 decayed into 7.5 grams after 63 hours. What is the half-life of magnesium-28?

Critical Thinking / Discussion / Writing

79. What went wrong? Find the error in the following argument:

$$3 < 4$$
$$3 \log \frac{1}{2} < 4 \log \frac{1}{2}$$
$$\log \left(\frac{1}{2}\right)^3 < \log \left(\frac{1}{2}\right)^4$$
$$\log \frac{1}{8} < \log \frac{1}{16}$$
$$\frac{1}{8} < \frac{1}{16}$$
$$2 < 1$$

80. By the power rule for logarithms, $\log x^2 = 2 \log x$. However, the graphs of $f(x) = \log x^2$ and $g(x) = 2 \log x$ are not identical. Explain why.

81. As of 2008, the largest known prime number was $2^{43112609} - 1$. How many digits does this prime number have? [*Hint:* First, argue that if $p = 2^m - 1$ is a prime number, then p and 2^m have the same number of digits.]

Maintaining Skills

In Exercises 82–85, simplify each expression and write your result without exponents.

82. $11 \cdot 3^0$

83. $-4 \cdot 5^x \cdot 5^{-x}$

84. $4^x \cdot 2^{-2x+1}$

85. $(7^x)^2 \cdot (7^2)^{-x}$

In Exercises 86–89, let $t = 5^x$ and write each expression in the form $at^2 + bt + c = 0$, where a, b, and c are real numbers.

86. $5^{2x} - 5^x = -1$

87. $3 \cdot 5^{2x} - 2 \cdot 5^x = 7$

88. $\dfrac{5^x + 3 \cdot 5^{-x}}{5^x} = \dfrac{1}{4}$

89. $\dfrac{5^{-x} + 5^x}{5^{-x} - 5^x} = \dfrac{1}{2}$

In Exercises 90–93, solve for x.

90. $2x - (11 + x) = 8x + (7 + 2x)$

91. $4x - 1 + (6x - 2) = 3 - (5x + 1)$

92. $x^2 + 3x - 1 = 3$

93. $2x^2 - 7x = 3x + 48$

In Exercises 94–97, solve for x.

94. $2x < 7 + x$

95. $5x - 2 \leq 19 - (1 - x)$

96. $12x > 30 - 3x$

97. $4x - 17 \geq 6 - (5 + 2x)$

SECTION 3.4

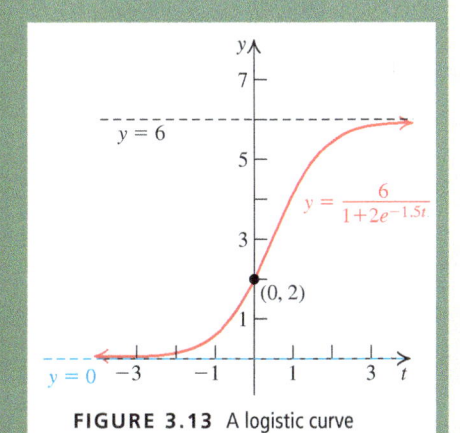

Exponential and Logarithmic Equations and Inequalities

BEFORE STARTING THIS SECTION, REVIEW

1. One-to-one property of exponential functions (Section 3.1, page, 226)
2. One-to-one property of logarithmic functions (Section 3.2, page, 244)
3. Rules of logarithms (Section 3.3, page, 254)

OBJECTIVES

1. Solve exponential equations.
2. Solve applied problems involving exponential equations.
3. Solve logarithmic equations.
4. Use the logistic growth model.
5. Use logarithmic and exponential inequalities.

◆ Logistic Growth Model

In Sections 3.3, we discussed the population growth formula

$$A(t) = A_0 e^{kt}. \quad (1)$$

Equation (1) is a realistic model for a biological population that has plenty of food, enough space to grow, and no threat from predators.

However, most populations are constrained by limitations on resources such as the plants that populations eat. In the 1830s, the Belgian scientist P. F. Verhulst added another constraint to the population growth model: *there is a limited sustainable maximum population M of a species that the habitat can support.* In population biology, M is called the *carrying capacity*.

Verhulst replaced equation (1) with

$$P(t) = \frac{M}{1 + ae^{-kt}}, \quad (2)$$

where k is the rate of growth and a is a positive constant. In 1840, using the data from the first five U.S. censuses, he made a prediction of the U.S. population for the year 1940. As things turned out, it was off by less than 1%.

The mathematical model represented by equation (2) is called the *logistic growth model* or the *Verhulst model*. Verhulst introduced the term *logistique* to refer to the "loglike" qualities of the S-shaped graph of equation (2). See Figure 3.13.

To sketch the graph of equation (2), we notice that the y-intercept is

$$P(0) = \frac{M}{1 + ae^0} = \frac{M}{1 + a}.$$

As $t \to \infty$, the quantity $ae^{-kt} \to 0$ (because k is positive). Consequently, the denominator $1 + ae^{-kt} \to 1$; therefore, $P(t) \to M$ and the line $y = M$ is a horizontal asymptote.

Similarly, as $t \to -\infty$, $e^{-kt} \to \infty$. In this case, the denominator $1 + ae^{-kt} \to \infty$ and $P(t) \to 0$. The x-axis, or $y = 0$, is another horizontal asymptote. See the graph of

$$y = \frac{6}{1 + 2e^{-1.5t}}$$

in Figure 3.13.

In Example 9 we use the logistic function defined by equation (2).

FIGURE 3.13 A logistic curve

Section 3.4 ■ Exponential and Logarithmic Equations and Inequalities **265**

1 Solve exponential equations.

Solving Exponential Equations

An equation containing terms of the form a^x ($a > 0, a \neq 1$) is called an **exponential equation**. Here are some examples of simple exponential equations:

$$2^x = 15$$
$$9^x = 3^{x+1}$$
$$7^x = 13^{2x}$$
$$5 \cdot 2^{x-3} = 17$$

RECALL

The one-to-one property of exponential and logarithmic functions states that
- if $a^u = a^v$, then $u = v$.
- if $\log_a u = \log_a v$, then $u = v$.

If both sides of an equation can be expressed as a power of the same base, then we can use the one-to-one property of exponential functions to solve the equation.

EXAMPLE 1 Solving an Exponential Equation

Solve each equation.

a. $25^x = 125$ **b.** $9^x = 3^{x+1}$

Solution

a.
$(5^2)^x = 5^3$ Rewrite 25 as 5^2 and 125 as 5^3.
$5^{2x} = 5^3$ Power rule for exponents
$2x = 3$ One-to-one property
$x = \dfrac{3}{2}$ Solve for x.

b.
$(3^2)^x = 3^{x+1}$ Rewrite 9 as 3^2.
$3^{2x} = 3^{x+1}$ Power rule of exponents
$2x = x + 1$ One-to-one property
$x = 1$ Solve for x.

Practice Problem 1 Solve each equation.

a. $3^x = 243$ **b.** $8^x = 4$

We can use logarithms to solve those equations for which both sides cannot be readily expressed with the same base. For example, the fact that $\log 2^x = x \log 2$ allows us to solve the equation $2^x = 9$ by first taking the log of both sides.

$2^x = 9$
$\log 2^x = \log 9$ Take log of both sides.
$x \log 2 = \log 9$ Power Rule of logarithms.
$x = \dfrac{\log 9}{\log 2}$ Divide both sides by log 2.

An approximation $x \approx 3.1699$ is found by using a calculator.

The general procedure for solving exponential equations using logarithms is given next.

PROCEDURE IN ACTION

EXAMPLE 2 Solving Exponential Equations Using Logarithms

OBJECTIVE

Solve exponential equations when both sides are not expressed with the same base.

Step 1 Isolate the exponential expression on one side of the equation.

Step 2 Take the common or natural logarithm of both sides.

Step 3 Use the Power Rule, $\log_a M^r = r \log_a M$.

Step 4 Solve for the variable.

EXAMPLE

Solve for x: $5 \cdot 2^{x-3} = 17$.

1. $2^{x-3} = \dfrac{17}{5} = 3.4$ Solve for 2^{x-3}.

2. $\ln 2^{x-3} = \ln(3.4)$ Take ln of both sides.

3. $(x-3)\ln 2 = \ln(3.4)$ Power Rule of logarithms

4. $x - 3 = \dfrac{\ln(3.4)}{\ln 2}$ Divide both sides by ln 2.

$x = \dfrac{\ln(3.4)}{\ln 2} + 3 \approx 4.766$ Solve for x; use a calculator.

Practice Problem 2 Solve for x: $7 \cdot 3^{x+1} = 11$.

EXAMPLE 3 Solving an Exponential Equation with Different Bases

Solve the equation $5^{2x-3} = 3^{x+1}$ and approximate the answer to three decimal places.

Solution

When different bases are involved, we begin with Step 2 of the procedure above.

$\ln 5^{2x-3} = \ln 3^{x+1}$ Take the natural log of both sides.

$(2x - 3)\ln 5 = (x + 1)\ln 3$ Power Rule of logarithms

$2x \ln 5 - 3 \ln 5 = x \ln 3 + \ln 3$ Distributive property

$2x \ln 5 - x \ln 3 = \ln 3 + 3 \ln 5$ Collect terms containing x on one side.

$x(2 \ln 5 - \ln 3) = \ln 3 + 3 \ln 5$ Factor out x on the left side.

$x = \dfrac{\ln 3 + 3 \ln 5}{2 \ln 5 - \ln 3}$ Solve for x to obtain an *exact* solution.

≈ 2.795 Use a calculator.

SIDE NOTE

When using a calculator to evaluate such expressions, make sure you enclose the numerator and denominator in parentheses.

Practice Problem 3 Solve the equation $3^{x+1} = 2^{2x}$.

EXAMPLE 4 An Exponential Equation of Quadratic Form

Solve the equation $3^x - 8 \cdot 3^{-x} = 2$.

Solution

$3^x(3^x - 8 \cdot 3^{-x}) = 2(3^x)$ Multiply both sides by 3^x.

$3^{2x} - 8 \cdot 3^0 = 2 \cdot 3^x$ Distributive property; $3^x \cdot 3^{-x} = 3^0$

$3^{2x} - 8 = 2 \cdot 3^x$ $3^0 = 1$

$3^{2x} - 2 \cdot 3^x - 8 = 0$ Subtract $2 \cdot 3^x$ from both sides.

The last equation is quadratic in form. To see this, we let $y = 3^x$; then $y^2 = (3^x)^2 = 3^{2x}$. So

$$3^{2x} - 2 \cdot 3^x - 8 = 0$$
$$y^2 - 2y - 8 = 0 \quad \text{Substitute: } y = 3^x, y^2 = 3^{2x}.$$
$$(y + 2)(y - 4) = 0 \quad \text{Factor.}$$
$$y + 2 = 0 \quad \text{or} \quad y - 4 = 0 \quad \text{Zero-product property}$$
$$y = -2 \quad \text{or} \quad y = 4 \quad \text{Solve for } y.$$
$$3^x = -2 \quad \text{or} \quad 3^x = 4 \quad \text{Replace } y \text{ with } 3^x.$$

But $3^x = -2$ is not possible because $3^x > 0$ for all real numbers x. Solve $3^x = 4$.

$$\ln 3^x = \ln 4 \quad \text{Take the natural log of both sides.}$$
$$x \ln 3 = \ln 4 \quad \text{Power Rule of logarithms}$$
$$x = \frac{\ln 4}{\ln 3} \quad \text{An exact solution}$$
$$\approx 1.262 \quad \text{Use a calculator.}$$

Practice Problem 4 Solve the equation $e^{2x} - 4e^x - 5 = 0$.

Applications of Exponential Equations

2 Solve applied problems involving exponential equations.

EXAMPLE 5 **Solving a Population Growth Problem**

The following table shows the approximate population and annual growth rate of the United States and Pakistan in 2010.

Country	Population	Annual Population Growth Rate
United States	308 million	0.9%
Pakistan	185 million	2.1%

Source: www.cia.gov

Use the alternate population model $P(t) = P_0(1 + r)^t$, where P_0 is the initial population and t is the time in years since 2010. Assume that the growth rate for each country stays the same.

a. Estimate the population of each country in 2020.
b. In what year will the population of the United States be 350 million?
c. In what year will the population of Pakistan be the same as the population of the United States?

Solution
The given model is

$$P(t) = P_0(1 + r)^t.$$

a. In 2010, the initial year, the U.S. population was $P_0 = 308$. In ten years, 2020, the population of the United States will be

$$308(1 + 0.009)^{10} \approx 336.87 \text{ million} \quad P_0 = 308, r = 0.009, t = 10$$

In the same way, the population of Pakistan in 2020 will be

$$185(1 + 0.021)^{10} \approx 227.73 \text{ million} \quad P_0 = 185, r = 0.021, t = 10$$

b. To find the year when the population of the United States will be 350 million, we solve for t the equation:

$350 = 308(1 + 0.009)^t$ $P(t) = 350, P_0 = 308, r = 0.009$

$\dfrac{350}{308} = (1.009)^t$ Divide both sides by 308.

$\ln\left(\dfrac{350}{308}\right) = \ln(1.009)^t$ Take natural logs of both sides.

$\ln\left(\dfrac{350}{308}\right) = t\ln(1.009)$ Power Rule of logarithms

$t = \dfrac{\ln\left(\dfrac{350}{308}\right)}{\ln(1.009)} \approx 14.27$ Solve for t; use a calculator.

The population of the United States will be 350 million approximately 14.27 years after 2010—that is, sometime in 2025.

c. To find when the population of the two countries will be the same, we solve for t the equation:

$308(1.009)^t = 185(1.021)^t$ Equate populations for both countries.

$\dfrac{308}{185} = \left(\dfrac{1.021}{1.009}\right)^t$ Divide both sides by $185(1.009)^t$ and rewrite.

$\ln\left(\dfrac{308}{185}\right) = \ln\left(\dfrac{1.021}{1.009}\right)^t$ Take natural logs of both sides.

$\ln\left(\dfrac{308}{185}\right) = t\ln\left(\dfrac{1.021}{1.009}\right)$ Power Rule of logarithms

$t = \dfrac{\ln\left(\dfrac{308}{185}\right)}{\ln\left(\dfrac{1.021}{1.009}\right)}$ Solve for t.

≈ 43.12 years   Use a calculator.

Therefore, the two populations will be equal in about 43.12 years, that is, during 2054.

Practice Problem 5 Repeat Example 5 assuming that the rates of growth for the United States and Pakistan are 1.1% and 3.3%, respectively.

Solving Logarithmic Equations

3 Solve logarithmic equations.

Equations that contain terms of the form $\log_a x$ are called **logarithmic equations**. Here are some examples of logarithmic equations.

$$\log_2 x = 4$$
$$\log_3 (2x - 1) = \log_3 (x + 2)$$
$$\log_2 (x - 3) + \log_2 (x - 4) = 1$$

To solve an equation such as $\log_2 x = 4$, we rewrite it in the equivalent exponential form:

$$\log_2 x = 4 \text{ means } x = 2^4 = 16$$

So, $x = 16$ is a solution of the equation $\log_2 x = 4$. Because the domain of logarithmic functions consists of positive numbers, we must check apparent solutions of logarithmic equations in the original equation.

EXAMPLE 6 Solving a Logarithmic Equation

Solve $4 + 3 \log_2 x = 1$.

Solution

$$4 + 3 \log_2 x = 1 \quad \text{Original equation}$$
$$3 \log_2 x = 1 - 4 \quad \text{Subtract 4 from both sides.}$$
$$3 \log_2 x = -3 \quad \text{Simplify.}$$
$$\log_2 x = -1 \quad \text{Divide both sides by 3.}$$
$$x = 2^{-1} \quad \text{Exponential form}$$
$$x = \frac{1}{2} \quad a^{-n} = \frac{1}{a^n}$$

Check: Substitute $x = \frac{1}{2} = 2^{-1}$ into $4 + 3 \log_2 x = 1$.

$$4 + 3 \log_2 2^{-1} \stackrel{?}{=} 1$$
$$4 + 3(-1) \log_2 2 \stackrel{?}{=} 1 \quad \text{Power Rule of logarithms}$$
$$4 - 3 \stackrel{?}{=} 1 \quad \log_2 2 = 1$$
$$1 \stackrel{?}{=} 1 \quad \text{Yes}$$

This check shows that the solution set is $\left\{\frac{1}{2}\right\}$.

Practice Problem 6 Solve $1 + 2 \ln x = 4$.

If each side of an equation can be expressed as a single logarithm with the same base, then we can use the one-to-one property of logarithms to solve the equation.

RECALL

The Product Rule says that $\log_a M + \log_a N = \log_a MN$.

EXAMPLE 7 Using the One-to-One Property of Logarithms

Solve $\log_4 x + \log_4 (x + 1) = \log_4 (x - 1) + \log_4 6$.

Solution

$$\log_4 x + \log_4 (x + 1) = \log_4 (x - 1) + \log_4 6 \quad \text{Original equation}$$
$$\log_4 [x(x + 1)] = \log_4 [6(x - 1)] \quad \text{Product Rule of logarithms}$$
$$x(x + 1) = 6(x - 1) \quad \text{One-to-one property}$$
$$x^2 + x = 6x - 6 \quad \text{Distributive property}$$
$$x^2 - 5x + 6 = 0 \quad \text{Add } -6x + 6 \text{ to both sides.}$$
$$(x - 2)(x - 3) = 0 \quad \text{Factor.}$$
$$x - 2 = 0 \quad \text{or} \quad x - 3 = 0 \quad \text{Zero-product property}$$
$$x = 2 \quad \text{or} \quad x = 3 \quad \text{Solve for } x.$$

Now we check these possible solutions in the original equation.

Check $x = 2$

$$\log_4 2 + \log_4 (2 + 1) \stackrel{?}{=} \log_4 (2 - 1) + \log_4 6$$
$$\log_4 2 + \log_4 3 \stackrel{?}{=} \log_4 1 + \log_4 6$$
$$\log_4 (2 \cdot 3) \stackrel{?}{=} \log_4 (1 \cdot 6) \quad \text{Yes}$$

Check $x = 3$

$$\log_4 3 + \log_4 (3 + 1) \stackrel{?}{=} \log_4 (3 - 1) + \log_4 6$$
$$\log_4 3 + \log_4 4 \stackrel{?}{=} \log_4 2 + \log_4 6$$
$$\log_4 (3 \cdot 4) \stackrel{?}{=} \log_4 (2 \cdot 6) \quad \text{Yes}$$

The solution set is $\{2, 3\}$.

Practice Problem 7 Solve $\ln (x + 5) - \ln (x + 1) = \ln (x - 1)$.

EXAMPLE 8   Using the Product and Quotient Rules

Solve the following equations.

a. $\log_2(x - 3) + \log_2(x - 4) = 1$
b. $\log_2(x + 4) - \log_2(x + 3) = 1$

Solution

a.
$\log_2(x - 3) + \log_2(x - 4) = 1$	Original equation
$\log_2[(x - 3)(x - 4)] = 1$	Product Rule of logarithms
$(x - 3)(x - 4) = 2^1$	Exponential form
$x^2 - 7x + 10 = 0$	Write in standard form.
$(x - 2)(x - 5) = 0$	Factor.
$x - 2 = 0$ or $x - 5 = 0$	Zero-product property
$x = 2$ or $x = 5$	Solve for x.

We check the possible solutions in the original equation.

Check $x = 2$

$\log_2(2 - 3) + \log_2(2 - 4) \stackrel{?}{=} 1$
$\log_2(-1) + \log_2(-2) \stackrel{?}{=} 1$ No

Because logarithms are not defined for negative numbers, $x = 2$ is an extraneous solution of the original equation.

Check $x = 5$

$\log_2(5 - 3) + \log_2(5 - 4) \stackrel{?}{=} 1$
$\log_2 2 + \log_2 1 \stackrel{?}{=} 1$
$1 + 0 \stackrel{?}{=} 1$
$1 \stackrel{?}{=} 1$ Yes

The solution set is $\{5\}$.

> **RECALL**
> Remember that $\log_3 x$ means "the exponent on 3 that gives x."

b.
$\log_2(x + 4) - \log_2(x + 3) = 1$	Original equation
$\log_2 \dfrac{x + 4}{x + 3} = 1$	Quotient rule
$\dfrac{x + 4}{x + 3} = 2^1$	Exponential form
$x + 4 = 2(x + 3)$	Multiply both sides by $x + 3$.
$x + 2 = 2x + 6$	Distributive property
$x = -2$	Solve for x.

We check the possible solution in the original equation.

Check $x = -2$

$\log_2(-2 + 4) - \log_2(-2 + 3) \stackrel{?}{=} 1$
$\log_2 2 - \log_2 1 \stackrel{?}{=} 1$
$1 - 0 \stackrel{?}{=} 1$ Yes

The solution set is $\{-2\}$.

Practice Problem 8 Solve $\log_3(x - 8) + \log_3 x = 2$.

4 Use the logistic growth model.

EXAMPLE 9   Using Logarithms in the Logistic Growth Model

Suppose the carrying capacity M of the human population on Earth is 35 billion. In 1987, the world population was about 5 billion. Use the logistic growth model of P. F. Verhulst to calculate the average rate, k, of growth of the population given that the population was about 6 billion in 2003.

$$P(t) = \frac{M}{1 + ae^{-kt}} \quad (2) \quad \text{From the section introduction}$$

HISTORICAL NOTE

Pierre François Verhulst (1804–1849)

Pierre Verhulst was born and educated in Brussels, Belgium. He received his PhD from the University of Ghent in 1825. Influenced by Lambert Quetelet, he became interested in social statistics. Verhulst's research on the law of population growth is important. Before Quetelet and Verhulst, scientists believed that an increasing population as a function of time was given by

$P(t) = P_0(1 + k)^t$ or $P(t) = P_0 e^{kt}$.

Verhulst showed that forces tend to prevent the population growth according to these laws. He discovered the model given by equation (2).

Solution

If 1987 represents $t = 0$, then $P(0) = 5$ and $M = 35$; so equation (2) becomes

$$5 = \frac{35}{1 + ae^{-k(0)}}$$

$$5 = \frac{35}{1 + a}$$

$5(1 + a) = 35$ Multiply both sides by $1 + a$.

$a = 6$ Solve for a.

Equation (2), with $M = 35$ and $a = 6$, becomes

$$P(t) = \frac{35}{1 + 6e^{-kt}}. \quad (3)$$

We now solve equation (3) for k given that $t = 16$ (for 2003) and $P(16) = 6$.

$$6 = \frac{35}{1 + 6e^{-16k}}$$

$6 + 36e^{-16k} = 35$ Multiply both sides by $1 + 6e^{-16k}$ and distribute.

$e^{-16k} = \dfrac{29}{36}$ Isolate e^{-16k}.

$-16k = \ln\left(\dfrac{29}{36}\right)$ Logarithmic form

$k = -\dfrac{1}{16}\ln\left(\dfrac{29}{36}\right)$ Solve for k.

$k \approx 0.0135 = 1.35\%$ Use a calculator.

The average growth rate of the world population was approximately 1.35%.

Practice Problem 9 Repeat Example 9 assuming that the estimated world population in 2005 was about 6.5 billion.

5 Use logarithmic and exponential inequalities.

Logarithmic and Exponential Inequalities

The properties of logarithms and exponents, together with the same techniques we used to solve polynomials, are used to solve inequalities involving logarithmic and exponential expressions.

The exponential and logarithmic functions with base $a > 1$ are increasing functions. That is, if $x_1 < x_2$, then $a^{x_1} < a^{x_2}$ and if $0 < x_1 < x_2$, then $\log_a x_1 < \log_a x_2$. As their graphs suggest, it is also true for base $a > 1$ that if $a^{x_1} < a^{x_2}$, then $x_1 < x_2$ and if $\log_a x_1 < \log_a x_2$, then $x_1 < x_2$. See Figures 3.14 and 3.15.

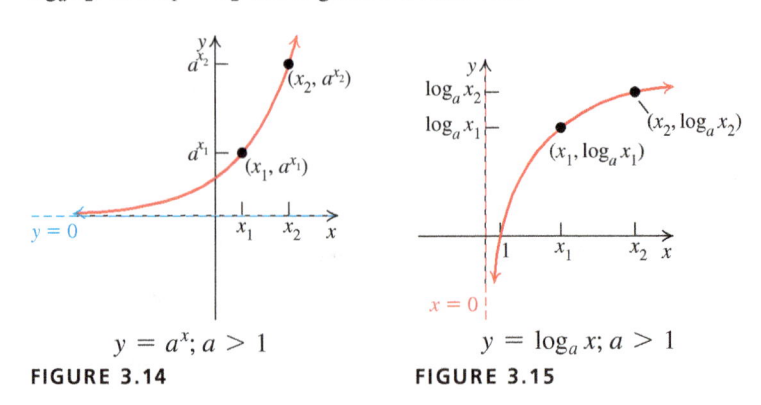

$y = a^x; a > 1$
FIGURE 3.14

$y = \log_a x; a > 1$
FIGURE 3.15

When working with inequalities, it is important to remember that $y = \log_a x, a > 1$, is negative for $0 < x < 1$.

EXAMPLE 10 Solving an Inequality Involving an Exponential Expression

Solve $5(0.7)^x + 3 < 18$.

Solution

$5(0.7)^x + 3 < 18$	Given inequality
$(0.7)^x < 3$	Isolate $(0.7)^x$ on one side.
$\ln (0.7)^x < \ln 3$	Take the natural log of both sides.
$x \ln (0.7) < \ln 3$	Power Rule of logarithms
$x > \dfrac{\ln 3}{\ln (0.7)} \approx -3.080$	Divide both sides by $\ln (0.7)$; reverse the sense of the inequality because $\ln (0.7) < 0$.

RECALL

If $A < B$ and $C < 0$, then $AC > BC$.

Practice Problem 10 Solve $3(0.5)^x + 7 > 19$.

EXAMPLE 11 Solving an Inequality Involving a Logarithmic Expression

Solve $\log (2x - 5) \leq 1$.

Solution

We first identify the domain for $\log (2x - 5)$. Because $2x - 5$ must be positive, we solve $2x - 5 > 0$; so $2x > 5$ or $x > \dfrac{5}{2}$.

Then		
	$\log (2x - 5) \leq 1$	Given inequality
	$10^{\log (2x-5)} \leq 10^1$	If $u < v$, then $10^u < 10^v$.
	$2x - 5 \leq 10$	$a^{\log_a x} = x$
	$x \leq \dfrac{15}{2}$	Solve for x.

For $\log (2x - 5)$ to be a real number, we must have $x > \dfrac{5}{2}$. Combining this with the fact that $x \leq \dfrac{15}{2}$, we find that the solution set for $\log (2x - 5) \leq 1$ is in the interval $\left(\dfrac{5}{2}, \dfrac{15}{2}\right]$.

Practice Problem 11 Solve $\ln (1 - 3x) > 2$.

SECTION 3.4 Exercises

Basic Concepts and Skills

In Exercises 1–14, solve each equation.

1. $2^x = 16$
2. $3^x = 243$
3. $8^x = 32$
4. $5^{x-1} = 1$
5. $4^{|x|} = 128$
6. $9^{|x|} = 243$
7. $5^{-|x|} = 625$
8. $3^{-|x|} = 81$
9. $\ln x = 0$
10. $\ln (x - 1) = 1$
11. $\log_2 x = -1$
12. $\log_2 (x + 1) = 3$
13. $\log_3 |x| = 2$
14. $\log_2 |x + 1| = 3$

In Exercises 15–42, solve each exponential equation and approximate the result correct to three decimal places.

15. $2^x = 3$
16. $3^x = 5$
17. $2^{2x+3} = 15$
18. $3^{2x+5} = 17$

274 Chapter 3 Exponential and Logarithmic Functions

19. $5 \cdot 2^x - 7 = 10$
20. $3 \cdot 5^x + 4 = 11$
21. $3 \cdot 4^{2x-1} + 4 = 14$
22. $2 \cdot 3^{4x-5} - 7 = 10$
23. $2^{1-x} = 3^{4x+6}$
24. $5^{2x+1} = 3^{x-1}$
25. $2 \cdot 3^{x-1} = 5^{x+1}$
26. $5 \cdot 2^{2x+1} = 7 \cdot 3^{x-1}$
27. $(1.065)^t = 2$
28. $(1.0725)^t = 2$
29. $2^{2x} - 4 \cdot 2^x = 21$
30. $4^x - 4^{-x} = 2$
31. $9^x - 6 \cdot 3^x + 8 = 0$
32. $\dfrac{3^x + 5 \cdot 3^{-x}}{3} = 2$
33. $3^{3x} - 4 \cdot 3^{2x} + 2 \cdot 3^x = 8$
34. $2^{3x} + 3 \cdot 2^{2x} - 2^x = 3$
35. $\dfrac{3^x - 3^{-x}}{3^x + 3^{-x}} = \dfrac{1}{4}$
36. $\dfrac{e^x - e^{-x}}{e^x + e^{-x}} = \dfrac{1}{3}$
37. $\dfrac{4}{2 + 3^x} = 1$
38. $\dfrac{7}{2^x - 1} = 3$
39. $\dfrac{17}{5 - 3^x} = 7$
40. $\dfrac{15}{3 + 2 \cdot 5^x} = 4$
41. $\dfrac{5}{2 + 3^x} = 4$
42. $\dfrac{7}{3 + 5 \cdot 2^x} = 4$

In Exercises 43–60, solve each logarithmic equation.

43. $3 + \log(2x + 5) = 2$
44. $1 + \log(3x - 4) = 0$
45. $\log(x^2 - x - 5) = 0$
46. $\log(x^2 - 6x + 9) = 0$
47. $\log_4(x^2 - 7x + 14) = 1$
48. $\log_4(x^2 + 5x + 10) = 1$
49. $\ln(2x - 3) - \ln(x + 5) = 0$
50. $\log(x + 8) + \log(x - 1) = 1$
51. $\log x + \log(x + 9) = 1$
52. $\log_5(3x - 1) - \log_5(2x + 7) = 0$
53. $\log_a(5x - 2) - \log_a(3x + 4) = 0$
54. $\log(x - 1) + \log(x + 2) = 1$
55. $\log_6(x + 2) + \log_6(x - 3) = 1$
56. $\log_2(3x - 2) - \log_2(5x + 1) = 3$
57. $\log_3(2x - 7) - \log_3(4x - 1) = 2$
58. $\log_4 \sqrt{x + 3} - \log_4 \sqrt{2x - 1} = \dfrac{1}{4}$
59. $\log_7 3x + \log_7(2x - 1) = \log_7(16x - 10)$
60. $\log_3(x + 1) + \log_3 2x = \log_3(3x + 1)$

In Exercises 61–68, find a and k and then evaluate the function. Round your answer to three decimal places.

61. Let $f(x) = 20 + a(2^{kx})$ with $f(0) = 50$ and $f(1) = 140$. Find $f(2)$.
62. $f(x) = 40 + a(4^{kx})$ with $f(0) = -216$ and $f(2) = 39$. Find $f(1)$.
63. $f(x) = 16 + a(3^{kx})$ with $f(0) = 21$ and $f(4) = 61$. Find $f(2)$.
64. $f(x) = 50 + a(2^{kx})$ with $f(0) = 34$ and $f(4) = 46$. Find $f(2)$.
65. Let $f(x) = \dfrac{10}{3 + ae^{kx}}$ with $f(0) = 2$ and $f(1) = \dfrac{1}{2}$. Find $f(2)$.
66. Let $f(x) = \dfrac{6}{a + 2e^{kx}}$ with $f(0) = 1$ and $f(1) = 0.8$. Find $f(2)$.
67. Let $f(x) = \dfrac{4}{a + 4e^{kx}}$ with $f(0) = 2$ and $f(1) = 9$. Find $f(2)$.
68. Let $f(x) = \dfrac{a}{1 + 3e^{kx}}$ with $f(0) = -1$ and $f(2) = -2$. Find $f(1)$.

In Exercises 69–78, solve each inequality.

69. $5(0.3)^x + 1 \leq 11$
70. $(0.1)^x - 4 > 15$
71. $-3(1.2)^x + 11 \geq 8$
72. $-7(0.4)^x + 19 < 5$
73. $\log(5x + 15) < 2$
74. $\log(2x + 0.9) > -1$
75. $\ln(x - 5) \geq 1$
76. $\ln(4x + 10) \leq 2$
77. $\log_2(3x - 7) < 3$
78. $\log_2(5x - 4) > 4$

Applying the Concepts

79. **Investment.** Find the time required for an investment of $10,000 to grow to $18,000 at an annual interest rate of 6% if the interest is compounded
    a. yearly.
    b. quarterly.
    c. monthly.
    d. daily.
    e. continuously.

80. **Investment.** How long will it take for an investment of $100 to double in value if the annual rate of interest is 7.2% compounded
    a. yearly.
    b. quarterly.
    c. monthly.
    d. continuously.

81. Doubling time. An amount P dollars is deposited in a regular bank account. How long does it take to double the initial investment (called the **doubling time**) if
 a. the interest rate is r compounded annually.
 b. the interest rate is r (per year) compounded continuously.

82. Epidemic outbreak. The number of people in a community who became infected in an epidemic t weeks after its outbreak is given by the function

$$f(t) = \frac{20{,}000}{1 + ae^{-kt}},$$

where 20,000 people of the community are susceptible to the disease. Assume that 1000 people were infected initially and 8999 had been infected by the end of the fourth week.
 a. Find the number of people infected after 8 weeks.
 b. After how many weeks will 12,400 people be infected?

83. Biological growth. In his laboratory experiment in 1934, G. F. Gause placed paramecia (unicellular microorganisms) in 5 cubic centimeters of a saline (salt) solution with a constant amount of food and measured their growth on a daily basis. He found that the population $P(t)$ after t days was approximated by

$$P(t) = \frac{4490}{1 + e^{5.4094 - 1.0255t}}, \quad t \geq 0.$$

 a. What was the initial population of the paramecia?
 b. What was the carrying capacity of the medium?

84. Facebook users. In 2004, there were 1 million Facebook users. In 2012, there were 845 million users. Use the model $f(t) = \dfrac{20{,}000}{20 + ae^{-kt}}$, where t is the number of years since 2004 and $f(t)$ is the number of users in millions, to estimate
 a. the numbers a and k.
 b. the number of users in 2015.
 c. the year in which the number of users will be 2 billion.

Richter scale. In Exercises 85–88, use the following information. The energy E (measured in joules) released by an earthquake of magnitude M on the Richter scale is given by the equation

$$\log E = 4.4 + 1.5 M.$$

85. The Great China Earthquake of 1920 registered 8.6 on the Richter scale. Let $I_0 = 1$.
 a. What was the intensity of this earthquake?
 b. How many joules of energy were released?

86. Suppose one earthquake registers one more point on the Richter scale than another.
 a. How are their corresponding intensities related?
 b. How are their released energies related?

87. If one earthquake is 150 times as intense as another, what is the difference in the Richter scale readings of the two earthquakes?

88. If the energy released by one earthquake is 150 times that of another, what is the difference in the Richter scale readings of the two earthquakes?

Sound. In Exercises 89–92, use the following information. The *loudness* (level of sound) of a sound is measured in *decibels* (dB). It is one-tenth of a *bel* (named after Alexander Graham Bell). The loudness L of a sound is related to its intensity I by the equation

$$L = 10 \log\left(\frac{I}{I_0}\right),$$

where $I_0 = 10^{-12}$ watts per square meter (W/m^2) is the intensity of the faintest sound the human ear can detect.

89. Find the dB level L of a TV at average volume from 12 feet away if it has an intensity of 250×10^{-7} (W/m^2).

90. Compare the intensity of a 65 dB sound to that of a 42 dB sound.

91. Find the intensity of a 73 dB sound.

92. The intensity of one sound is 5000 times that of another sound. Find the difference in the dB levels of the two sounds.

Critical Thinking / Discussion / Writing

In Exercises 93–97, solve each equation for x.

93. $(\log_2 x)(\log_2 8x) = 10$

94. $(\log_3 x)(\log_3 3x) = 2$

95. $(\log x)^2 = \log x$

96. $\log_2 x + \log_4 x = 6$

97. $\log_4 x^2 (x - 1)^2 - \log_2 (x - 1) = 1$

98. Logistic function. For $f(t) = \dfrac{P}{1 + ae^{-kt}}$, it can be shown that the maximum rate of growth occurs when $f(t) = \dfrac{P}{2}$. Find the time when the rate of growth is maximum.

99. Solve each of the following equations for x.
 a. $\log_4 (x - 1)^2 = 3$
 b. $2 \log_4 (x - 1) = 3$
 c. $2 \log_4 |x - 1| = 3$
 Explain why these three equations do not have identical solutions.

Maintaining Skills

In Exercises 100–104, find the center and radius of each circle.

100. $x^2 + y^2 = 1$
101. $(x + 2)^2 + (y - 3)^2 = 16$
102. $x^2 + y^2 - 2x - 3 = 0$
103. $x^2 + y^2 + 4y = 0$
104. $x^2 + y^2 - 2x + 4y - 4 = 0$

SUMMARY Definitions, Concepts, and Formulas

3.1 Exponential Functions

i. A function $f(x) = a^x$, with $a > 0$ and $a \neq 1$, is called an exponential function with base a.

Rules of exponents: $a^x a^y = a^{x+y}, \dfrac{a^x}{a^y} = a^{x-y}, (a^x)^y = a^{xy}$,

$$a^0 = 1, a^{-x} = \dfrac{1}{a^x}$$

ii. Exponential functions are one-to-one: If $a^x = a^y$, then $x = y$.

iii. If $a > 1$, then $f(x) = a^x$ is an increasing function; $f(x) \to \infty$ as $x \to \infty$ and $f(x) \to 0$ as $x \to -\infty$.

iv. If $0 < a < 1$, then $f(x) = a^x$ is a decreasing function; $f(x) \to 0$ as $x \to \infty$ and $f(x) \to \infty$ as $x \to -\infty$.

v. The graph of $f(x) = a^x$ has y-intercept 1, and the x-axis is a horizontal asymptote.

vi. **Simple interest formula.** If P dollars is invested at an interest rate r per year for t years, then the simple interest is given by the formula $I = Prt$. The future value $A(t) = P + Prt$.

vii. **Compound interest.** P dollars invested at annual rate r compounded n times per year for t years amounts to

$$A_n(t) = P\left(1 + \dfrac{r}{n}\right)^{nt}.$$

viii. The Euler constant $e = \lim\limits_{h \to \infty}\left(1 + \dfrac{1}{h}\right)^h \approx 2.718$.

ix. **Continuous compounding.** P dollars invested at an annual rate r compounded continuously for t years amounts to $A = Pe^{rt}$.

x. The function $f(x) = e^x$ is the natural exponential function.

3.2 Logarithmic Functions

i. For $a > 0$ and $a \neq 1$, $y = \log_a x$ if and only if $x = a^y$.

ii. Basic properties: $\log_a a = 1, \log_a 1 = 0$,

$$\log_a a^x = x, a^{\log_a x} = x \quad \text{Inverse properties}$$

iii. The domain of $\log_a x$ is $(0, \infty)$, the range is $(-\infty, \infty)$, and the y-axis is a vertical asymptote. The x-intercept is 1.

iv. Logarithmic functions are one-to-one: if $\log_a x = \log_a y$, then $x = y$.

v. If $a > 1$, then $f(x) = \log_a x$ is an increasing function; $f(x) \to \infty$ as $x \to \infty$ and $f(x) \to -\infty$ as $x \to 0^+$.

vi. If $0 < a < 1$, then $f(x) = \log_a x$ is a decreasing function; $f(x) \to -\infty$ as $x \to \infty$ and $f(x) \to \infty$ as $x \to 0^+$.

vii. The common logarithmic function is $y = \log x$ (base 10); the natural logarithmic function is $y = \ln x$ (base e).

3.3 Rules of Logarithms

i. Rules of logarithms:

$\log_a MN = \log_a M + \log_a N$ — Product rule

$\log_a \dfrac{M}{N} = \log_a M - \log_a N$ — Quotient rule

$\log_a M^r = r \log_a M$ — Power rule

ii. **Change-of-base formula:**

$$\log_b x = \dfrac{\log_a x}{\log_a b} = \dfrac{\log x}{\log b} = \dfrac{\ln x}{\ln b}$$

$\quad\quad$ (base a) $\quad$ (base 10) $\quad$ (base e)

3.4 Exponential and Logarithmic Equations and Inequalities

An *exponential equation* is an equation in which a variable occurs in one or more exponents.

A *logarithmic equation* is an equation that involves the logarithm of a function of the variable.

Exponential and logarithmic equations are solved by using some or all of the following techniques:

i. Using the one-to-one property of exponential and logarithmic functions

ii. Converting from exponential to logarithmic form or vice versa

iii. Using the Product, Quotient, and Power rules for exponents and logarithms

iv. Using similar techniques to solve logarithmic and exponential inequalities

REVIEW EXERCISES

Basic Concepts and Skills

In Exercises 1–10, state whether the given statement is true or false.

1. The function $f(x) = a^x$ is exponential if $a > 0$.
2. The graph of $f(x) = 4^x$ approaches the x-axis as $x \to -\infty$.
3. The domain of $f(x) = \log(2 - x)$ is $(-\infty, 2]$.
4. The equation $u = 10^v$ means that $\log u = v$.
5. The inverse of $f(x) = \ln x$ is $g(x) = e^x$.
6. The graph of $y = a^x$ ($a > 0, a \neq 1$) always contains the points $(0, 1)$ and $(1, a)$.
7. $\ln(M + N) = \ln M + \ln N$
8. $\log \sqrt{300} = 1 + \dfrac{1}{2} \log 3$
9. The functions $f(x) = 2^{-x}$ and $g(x) = \left(\dfrac{1}{2}\right)^x$ have the same graph.
10. $\ln u = \dfrac{\log u}{\log e}$

In Exercises 11–18, match the function with its graph in (a)–(h).

11. $f_1(x) = \log_2 x$
12. $f_2(x) = 2^x$
13. $f_3(x) = \log_2(3 - x)$
14. $f_4(x) = \log_{1/2} x$
15. $f_5(x) = -\log_2 x$
16. $f_6(x) = \left(\dfrac{1}{2}\right)^x$
17. $f_7(x) = 3 - 2^{-x}$
18. $f_8(x) = \dfrac{6}{1 + 2e^{-x}}$

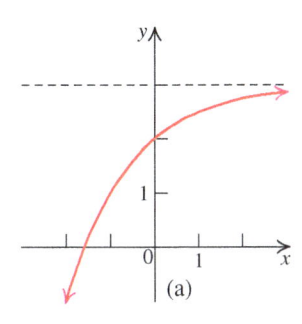

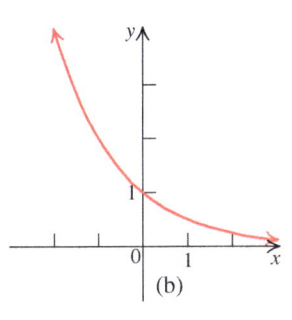

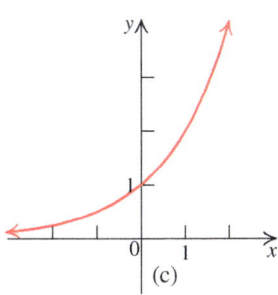

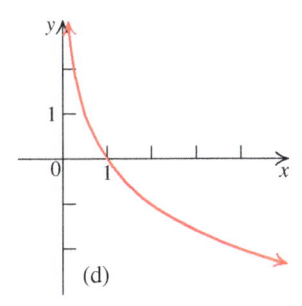

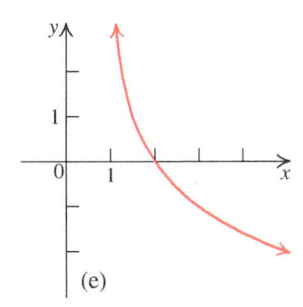

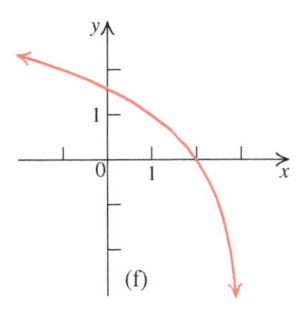

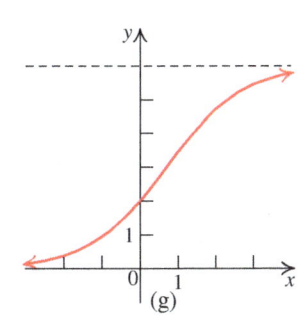

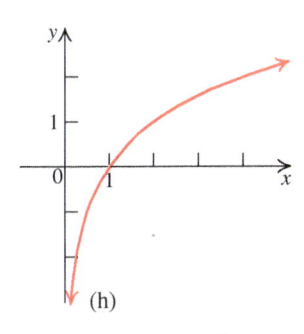

(a) (b) (c) (d) (e) (f) (g) (h)

In Exercises 19–30, graph each function using transformations on an appropriate graph. Determine the domain, range, and asymptotes (if any).

19. $f(x) = 2^{-x}$
20. $g(x) = 2^{-0.5x}$
21. $h(x) = 3 + 2^{-x}$
22. $f(x) = e^{|x|}$
23. $g(x) = e^{-|x|}$
24. $h(x) = e^{-x+1}$
25. $f(x) = \ln(-x)$
26. $g(x) = 2\ln|x|$
27. $h(x) = 2\ln(x - 1)$
28. $f(x) = 2 - \left(\dfrac{1}{2}\right)^x$
29. $g(x) = 2 - \ln(-x)$
30. $h(x) = 3 + 2\ln(5 + x)$

In Exercises 31–34, sketch the graph of the given function using these two steps:
a. Find the intercepts.
b. Find the end behavior of f.

31. $f(x) = 3 - 2e^{-x}$
32. $f(x) = \dfrac{5}{2 + 3e^{-x}}$
33. $f(x) = e^{-x^2}$
34. $f(x) = 3 - \dfrac{6}{1 + 2e^{-x}}$

In Exercises 35–38, find a and k and then evaluate the function.

35. Let $f(x) = a(2^{kx})$ with $f(0) = 10$ and $f(3) = 640$. Find $f(2)$.
36. Let $f(x) = 50 - a(5^{kx})$ with $f(0) = 10$ and $f(2) = 0$. Find $f(1)$.
37. Let $f(x) = \dfrac{4}{1 + ae^{-kx}}$ with $f(0) = 1$ and $f(3) = \dfrac{1}{2}$. Find $f(4)$.
38. Let $f(x) = 6 - \dfrac{3}{1 + ae^{-kx}}$ with $f(0) = 5$ and $f(4) = 4$. Find $f(10)$.

In Exercises 39 and 40, find an exponential function of the form $f(x) = ca^x$ with the given graph.

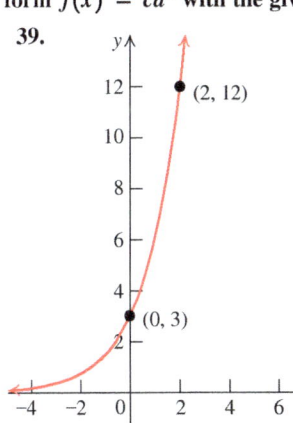

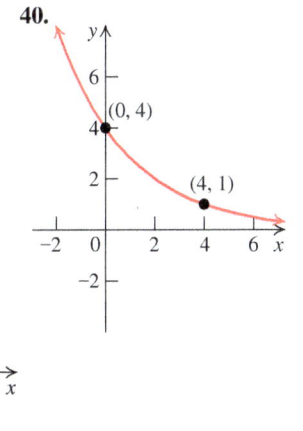

39. (0, 3), (2, 12)
40. (0, 4), (4, 1)

In Exercises 41 and 42, find a logarithmic function of the form $y = \log_a(x - c)$ with the given graph.

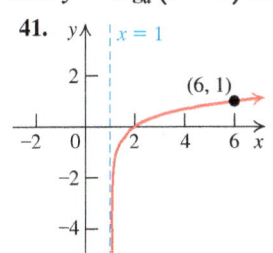

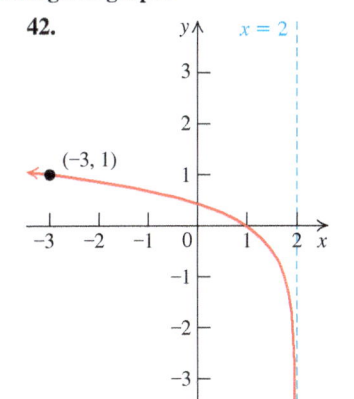

41. $x = 1$; (6, 1)
42. $x = 2$; (-3, 1)

43. Start with the graph of $y = 2^x$. Find an equation of the graph that results when you
    a. shift the graph right one unit, shift up three units, and reflect about the x-axis.
    b. shift the graph right one unit, reflect about the x-axis, and shift up three units.

44. Start with the graph of $y = \ln x$. Find an equation of the graph that results when you
    a. shift the graph left one unit, stretch horizontally by a factor of 2, and compress vertically by a factor of 3.
    b. compress horizontally to $\frac{1}{3}$, shift left by one unit, and reflect about the y-axis.

In Exercises 45–48, write each logarithm in expanded form.

45. $\ln (xy^2z^3)$
46. $\log (x^3\sqrt{y-1})$
47. $\ln \left[\dfrac{x\sqrt{x^2+1}}{(x^2+3)^2}\right]$
48. $\ln \sqrt{\dfrac{x^3+5}{x^3-7}}$

In Exercises 49–54, write y as a function of x.

49. $\ln y = \ln x + \ln 3$
50. $\ln y = \ln (C) + kx$; C and k are constants.
51. $\ln y = \ln (x-3) - \ln (y+2)$
52. $\ln y = \ln x - \ln (x^2y) - 2\ln y$
53. $\ln y = \dfrac{1}{2}\ln (x-1) + \dfrac{1}{2}\ln (x+1) - \ln (x^2+1)$
54. $\ln (y-1) = \dfrac{1}{x} + \ln (y)$

In Exercises 55–78, solve each equation.

55. $3^x = 81$
56. $5^{x-1} = 625$
57. $2^{x^2+2x} = 16$
58. $3^{x^2-6x+8} = 1$
59. $3^x = 23$
60. $2^{x-1} = 5.2$
61. $273^x = 19$
62. $27 = 9^x \cdot 3^{x^2}$
63. $3^{2x} = 7^x$
64. $2^{x+4} = 3^{x+1}$
65. $(1.7)^{3x} = 3^{2x-1}$
66. $3(2^{x+5}) = 5(7^{2x-3})$
67. $\log_3 (x+2) - \log_3 (x-1) = 1$
68. $\log_3 (x+12) - \log_3 (x+4) = 2$
69. $\log_2 (x+2) + \log_2 (x+4) = 3$
70. $\log_5 (3x+7) + \log_5 (x-5) = 2$
71. $\log_5 (x^2-5x+6) - \log_5 (x-2) = 1$
72. $\log_3 (x^2-x-6) - \log_3 (x-3) = 1$
73. $\log_6 (x-2) + \log_6 (x+1) = \log_6 (x+4) + \log_6 (x-3)$
74. $\ln (x-2) - \ln (x+2) = \ln (x-1) - \ln (2x+1)$
75. $2\ln 3x = 3\ln x$
76. $2\log x = \ln e$
77. $2^x - 8 \cdot 2^{-x} - 7 = 0$
78. $3^x - 24 \cdot 3^{-x} = 10$

In Exercises 79–82, solve each inequality.

79. $3(0.2)^x + 5 \leq 20$
80. $-4(0.2)^x + 15 < 6$
81. $\log (2x+7) < 2$
82. $\ln (3x+5) \leq 1$

Applying the Concepts

83. **Doubling your money.** How much time is required for a $1000 investment to double in value if interest is earned at the rate of 6.25% compounded annually?

84. **Tripling your money.** How much time (to the nearest month) is required to triple your money if the interest rate is 100% compounded continuously?

85. **Half-life.** The half-life of a certain radioactive substance is 20 hours. How long does it take for this substance to fall to 25% of its original value?

86. **Plutonium-210.** Find the half-life of plutonium-210 assuming that its decay equation is $Q = Q_0 e^{-5 \cdot 10^{-3}t}$, where t is in days.

87. **Comparing rates.** You have $7000 to invest for seven years. Which investment will provide the greater return, 5% compounded yearly or 4.75% compounded monthly?

88. **Investment growth.** How long will it take $8000 to grow to $20,000 if the rate of interest is 7% compounded continuously?

89. **Population.** The formula $P(t) = 33e^{0.003t}$ models the population of Canada, in millions, t years after 2007.
    a. Estimate the population of Canada in 2017.
    b. According to this model, when will the population of Canada be 60 million?

90. **House appreciation.** The formula $C(t) = 100 + 25e^{0.03t}$ models the average cost of a house in Sometown, USA, t years after 2000. The cost is expressed in thousands of dollars.
    a. Sketch the graph of $y = C(t)$.
    b. Estimate the average cost in 2010.
    c. According to this model, when will the average cost of a house in Sometown be $250,000?

91. **Drug concentration.** An experimental drug was injected into the bloodstream of a rat. The concentration $C(t)$ of the drug (in micrograms per milliliter of blood) after t hours was modeled by the function

    $$C(t) = 0.3e^{-0.47t}, \ 0.5 \leq t \leq 10.$$

    a. Graph the function $y = C(t)$ for $0.5 \leq t \leq 10$.
    b. When will the concentration of the drug be 0.029 micrograms per milliliter? (1 microgram = 10^{-6} gram)

92. **X-ray intensity.** X-ray technicians are shielded by a wall insulated with lead. The equation $x = \dfrac{1}{152}\log\left(\dfrac{I_0}{I}\right)$ measures the thickness x (in centimeters) of the lead insulation required to reduce the initial intensity I_0 of X-rays to the desired intensity I.
    a. What thickness of lead is required to reduce the intensity of X-rays to one-tenth their initial intensity?
    b. What thickness of lead is required to reduce the intensity of X-rays to $\dfrac{1}{40}$ their initial intensity?
    c. How much is the intensity reduced if the lead insulation is 10 centimeters thick?

93. **Cooling tea.** Chai (tea) is made by adding boiling water (212°F) to the chai mix. Suppose you make chai in a room with the air temperature at 75°F. According to Newton's Law of Cooling, the temperature of the chai t minutes after it is boiled is given by a function of the form $f(t) = 75 + ae^{-kt}$. After one minute, the temperature of the chai falls to 200°F. How long will it take for the chai to be drinkable at 150°F?

94. **Spread of influenza.** Approximately $P(t) = \dfrac{6}{1 + 5e^{-0.7t}}$ thousand people caught a new form of influenza within t weeks of its outbreak.
   a. Sketch the graph of $y = P(t)$.
   b. How many people had the disease initially?
   c. How many people contracted the disease within four weeks?
   d. If the trend continues, how many people in all will contract the disease?

95. **Population density.** The population density x miles from the center of a town called Greenville is approximated by the equation $D(x) = 5e^{0.08x}$, in thousands of people per square mile.
   a. What is the population density at the center of Greenville?
   b. What is the population density 5 miles from the center of Greenville?
   c. Approximately how far from the center of Greenville would the density be 15,000 people per square mile?

96. **Population.** In 2000, the population of the United States was 280 million and the number of vehicles was 200 million. If the population of the United States is growing at the rate of 1% per year while the number of vehicles is growing at the rate of 3%, in what year will there be an average of one vehicle per person?

97. **Light intensity.** The *Bouguer–Lambert Law* states that the intensity I of sunlight filtering down through water at a depth x (in meters) decreases according to the exponential decay function $I = I_0 e^{-kx}$, where I_0 is the intensity at the surface and $k > 0$ is an absorption constant that depends on the murkiness of the water. Suppose the absorption constant of Carrollwood Lake was experimentally determined to be $k = 0.73$. How much light has been absorbed by the water at a depth of 2 meters?

98. **Signal strength.** The strength of a TV signal usually fades due to a damping effect of cable lines. If I_0 is the initial strength of the signal, then its strength I at a distance x miles is measured by the formula $I = I_0 e^{-kx}$, where $k > 0$ is a damping constant that depends on the type of wire used. Suppose the damping constant has been measured experimentally to be $k = 0.003$. What percent of the signal is lost at a distance of 10 miles? 20 miles?

99. **Walking speed in a city.** In 1976, Marc and Helen Bernstein discovered that in a city with population p, the average speed s (in feet per second) that a person walks on main streets can be approximated by the formula

$$s(p) = 0.04 + 0.86 \ln p.$$

   a. What is the average walking speed of pedestrians in Tampa (population 336,000)?
   b. What is the average walking speed of pedestrians in Bowman, Georgia (population 962)?
   c. What is the estimated population of a town in which the estimated average walking speed is 4.6 feet per second?

100. **Drinking and driving.** Just after Eric had his last drink, the alcohol level in his bloodstream was 0.26 (milligram of alcohol per milliliter of blood). After one-half hour, his alcohol level was 0.18. The alcohol level $A(t)$ in a person follows the exponential decay law

$$A(t) = A_0 e^{-kt}, \ (t \text{ in hours})$$

where $k > 0$ depends on the individual.
   a. What is the value of A_0 for Eric?
   b. What is the value of k for Eric?
   c. If the legal driving limit for alcohol level is 0.08, how long should Eric wait (after his last drink) before he can drive legally?

101. **Bacteria culture.** The mass $m(t)$, in grams, of a bacteria culture grows according to the logistic growth model

$$m(t) = \dfrac{6}{1 + 5e^{-0.7t}},$$

where t is time measured in days.
   a. What is the initial mass of the culture?
   b. What happens to the mass in the long run?
   c. When will the mass reach 5 grams?

102. **A learning model.** The number of units $n(t)$ produced per day after t days of training is given by

$$n(t) = 60(1 - e^{-kt}),$$

where $k > 0$ is a constant that depends on the individual.
   a. Estimate k for Rita, who produced 20 units after one day of training.
   b. How many units will Rita produce per day after ten days of training?
   c. How many days should Rita be trained if she is expected to produce 40 units per day?

EXERCISES FOR CALCULUS

1. If $a > 0$ and $a \neq 1$, find a number k such that $a^x = e^{kx}$ for all real numbers, x.

2. a. Use a calculator to fill in the following table. Round your answers to five decimal places in columns 2 and 4.

	e^x	$x + 0.001$	$e^{x+0.001}$
-1.5			
-1			
-0.5			
0			
0.5			
1			
1.5			

   b. Comparing the values for e^x and $e^{x+0.001}$, what can you say about the slope of the secant line connecting the graph points (x, e^x) and $(x + 0.001, e^{x+0.001})$ on the graph of $y = e^x$?

3. The population of bacteria in a culture at a particular time t, was 1000, and two hours later the population was 8000. Assuming exponential growth, express the population after t hours as a function of t, in hours.

4. The equation $e^x = 1 + 2x$ has zero as one root; graph $y = e^x$ and $y = 1 + 2x$ for visual evidence that the equation has a positive real root as well. Then replace e^x in this equation with the approximation $e^x \approx 1 + x + \dfrac{x^2}{2} + \dfrac{x^3}{6}$ to approximate this real root.

5. Show that if $x > \ln(10)$, then $e^{-x} < \dfrac{1}{10}$ and that if $x > \ln(100)$, then $e^{-x} < \dfrac{1}{100}$. Find a real number, a, such that if $x > \ln(a)$, then $e^{-x} < \dfrac{1}{10,000}$.

6. Expand $\log_2(x^2(1-x))$

7. Expand $\log(x^4(2x-3)^3(x+7)^5)$

8. Expand $\ln\left(\dfrac{x^2(x-1)^5}{\sqrt{x^2+1}}\right)$

9. A goat is growing exponentially. If it weighs 3 pounds at birth 5 pounds one month later, how old will it be when it weighs 90 pounds?

10. Explain why the number e may be fairly described as the amount that accumulates in one year if \$1 is invested at an interest rate of 100% compounded continuously.

PRACTICE TEST A

1. Solve the equation $5^{-x} = 125$.
2. Solve the equation $\log_2 x = 5$.
3. State the range of $y = -e^x + 1$ and find the asymptote of its graph.
4. Evaluate $\log_2 \dfrac{1}{8}$.
5. Solve the exponential equation $\left(\dfrac{1}{4}\right)^{2-x} = 4$.
6. Evaluate $\log 0.001$ without using a calculator.
7. Rewrite the expression $\ln 3 + 5 \ln x$ in condensed form.
8. Solve the equation $2^{x+1} = 5$.
9. Solve the equation $e^{2x} + e^x - 6 = 0$.
10. Rewrite the expression $\ln \dfrac{2x^3}{(x+1)^5}$ in expanded logarithmic form.
11. Evaluate $\ln e^{-5}$.
12. Give the equation for the graph obtained by shifting the graph of $y = \ln x$ three units up and one unit right.
13. Sketch the graph of $y = 3^{x-1} + 2$.
14. State the domain of the function $f(x) = \ln(-x) + 4$.
15. Write $3 \ln x + \ln(x^3 + 2) - \dfrac{1}{2} \ln(3x^2 + 2)$ in condensed form.
16. Solve the equation $\log x = \log 6 - \log(x - 1)$.
17. Find x if $\log_x 9 = 2$.
18. Rewrite the expression $\ln \sqrt{2x^3 y^2}$ in expanded logarithmic form.
19. Suppose \$15,000 is invested in a savings account paying 7% interest per year. Write the formula for the amount in the account after t years if the interest is compounded quarterly.
20. Suppose the number of Hispanic people living in the United States is approximated by $H = 15{,}000 e^{0.2t}$, where $t = 0$ represents 1960. According to this model, in which year did the Hispanic population in the United States reach 1.5 million?

PRACTICE TEST B

1. Solve the equation $3^{-x} = 9$.
   a. $\{2\}$ b. $\{-2\}$ c. $\left\{\dfrac{1}{2}\right\}$ d. $\left\{-\dfrac{1}{2}\right\}$ e. $\{\ln 2\}$

2. Solve the equation $\log_5 x = 2$.
   a. $\{10\}$ b. $\{25\}$ c. $\left\{\dfrac{5}{2}\right\}$ d. $\left\{\dfrac{2}{5}\right\}$ e. $\{2\}$

3. State the range and asymptote of $y = e^{-x} - 1$.
   a. $(0, \infty); y = -1$
   b. $(-1, \infty); y = -1$
   c. $(-\infty, \infty); y = -1$
   d. $(-1, \infty); y = 1$
   e. $(\infty, 1); y = 1$

4. Evaluate $\log_4 64$.
   a. 16 b. 8 c. 2 d. 3 e. 4

5. Find the solution of the exponential equation $\left(\dfrac{1}{3}\right)^{1-x} = 3$.
   a. $\left\{-\dfrac{1}{3}\right\}$ b. $\left\{\dfrac{1}{3}\right\}$ c. $\{1\}$ d. $\{2\}$

6. Evaluate $\log 0.01$.
   a. -1.99999999
   b. -2
   c. 2
   d. 100

7. Which of the following expressions is equivalent to $\ln 7 + 2 \ln x$?
   a. $\ln(7 + 2x)$
   b. $\ln(7x^2)$
   c. $\ln(9x)$
   d. $\ln(14x)$

8. Solve the equation $2^{x^2} = 3^x$.
   a. $\{0\}$
   b. $\left\{\dfrac{\ln 3}{\ln 2}\right\}$
   c. $\left\{0, \dfrac{\ln 3}{\ln 2}\right\}$
   d. $\{0, \ln 3 - \ln 2\}$

9. Solve the equation $e^{2x} - e^x - 6 = 0$.
   a. $\{-\ln 2\}$
   b. $\{-\ln 3\}$
   c. $\{\ln 6\}$
   d. $\{\ln 3\}$

10. Rewrite the expression $\ln \dfrac{3x^2}{(x+1)^{10}}$ in expanded logarithmic form.
    a. $\ln 6x - 10 \ln x + 1$
    b. $2 \ln 3x - 10 \ln(x + 1)$
    c. $2 \ln 3 + 2 \ln x - 10 \ln(x + 1)$
    d. $\ln 3 + 2 \ln x - 10 \ln(x + 1)$

11. Find $\ln e^{3x}$.
    a. 3 b. $3x$ c. $3 + x$ d. x

12. The equation for the graph obtained by shifting the graph of $y = \log_3 x$ two units up and three units left is
    a. $y = \log_3(x - 3) + 2$
    b. $y = \log_3(x + 3) - 2$
    c. $y = \log_3(x - 3) - 2$
    d. $y = \log_3(x + 3) + 2$

13. Which of the following graphs is the graph of $y = 5^{x+1} - 4$?

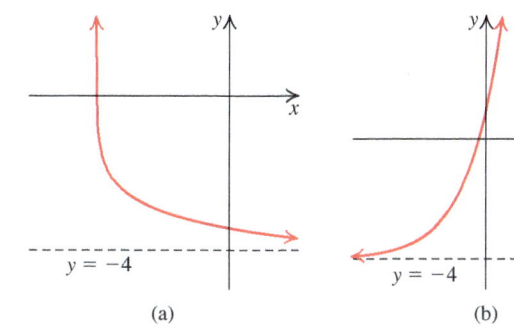

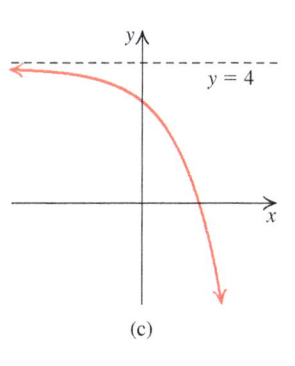

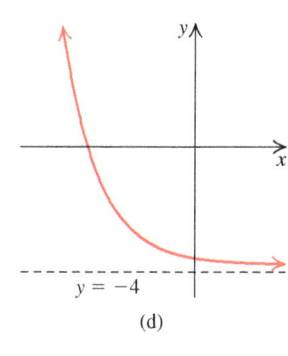

    (a) (b) (c) (d)

14. Find the domain of the function $f(x) = \ln(1 - x) + 3$.
    a. $(-\infty, 1)$ b. $(1, \infty)$ c. $(-\infty, -1)$
    d. $(-\infty, 3)$ e. $(3, \infty)$

15. Write $\ln x - 2 \ln(x^2 + 1) + \dfrac{1}{2} \ln(x^4 + 1)$ in condensed form.
    a. $\ln \dfrac{x\sqrt{x^4 + 1}}{(x^2 + 1)^2}$
    b. $\ln \dfrac{x}{x^2 + 1}$
    c. $\ln(x - (x^2 + 1)^2 + (x^4 + 1)^{1/2})$
    d. $2 \ln \dfrac{x(1 + x^4)}{x^2 + 1}$

16. Solve the equation $\log x = \log 12 - \log(x + 1)$.
    a. $\left\{\dfrac{11}{2}\right\}$ b. $\left\{\dfrac{13}{2}\right\}$ c. $\{3, -4\}$
    d. $\{3\}$ e. $\left\{\dfrac{2}{12}\right\}$

17. Find x if $\log_x 16 = 4$.
    a. 4 b. 2 c. 64 d. $\dfrac{1}{4}$ e. 16

18. Rewrite the expression $\ln \sqrt[3]{5x^2y^3}$ in expanded logarithmic form.
    a. $\dfrac{1}{3}(\ln 5x^2 + \ln y^3)$
    b. $\dfrac{1}{3}(2 \ln 5x + 3 \ln y)$
    c. $\dfrac{1}{3}(\ln 5 + 2 \ln x + 3 \ln y)$
    d. $\dfrac{1}{3} \ln 5 + 2 \ln x + 3 \ln y$
    e. $\dfrac{1}{3} \ln 5 + 2 + \ln x + 3 + \ln y$

19. Suppose $12,000 is invested in a savings account paying 10.5% interest per year. Write the formula for the amount in the account after t years assuming that the interest is compounded monthly.
    a. $A = 12{,}000(1.105)^t$
    b. $A = 12{,}000(1.525)^{2t}$
    c. $A = 12{,}000(1.2625)^{4t}$
    d. $A = 12{,}000(1.00875)^{12t}$

20. The population of a certain city is growing according to the model $P = 10{,}000 \log_5(t + 5)$, where t is time in years. If $t = 0$ corresponds to the year 2000, what will the population of the city be in 2020?
    a. 30,000
    b. 20,000
    c. 50,000
    d. 10,000

CHAPTER 4

Trigonometric Functions

TOPICS

4.1 Angles and Their Measure

4.2 The Unit Circle; Trigonometric Functions

4.3 Graphs of the Sine and Cosine Functions

4.4 Graphs of the Other Trigonometric Functions

4.5 Inverse Trigonometric Functions

4.6 Right-Triangle Trigonometry

4.7 Trigonometric Identities

4.8 Sum and Difference Formulas

Angles are vital for activities ranging from taking photographs to designing layouts for city streets. Scientists use angles and basic trigonometry to measure remote distances such as the distance between mountain peaks and the distances between planets. Angles are universally used to specify locations on Earth via longitude and latitude. In this chapter, we begin the study of trigonometry and its many uses.

SECTION 4.1

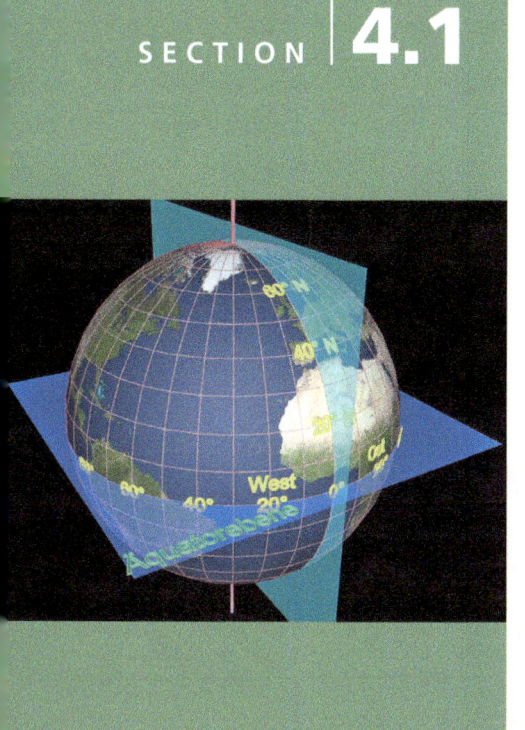

Angles and Their Measure

BEFORE STARTING THIS SECTION, REVIEW

1. Circumference and area of a circle

OBJECTIVES

1. Learn the vocabulary associated with angles.
2. Use degree and radian measure.
3. Convert between degree and radian measure.
4. Find complements and supplements.
5. Find the length of an arc of a circle.

◆ Latitude and Longitude

Any location on Earth can be described with two numbers, *latitude* and *longitude*. To understand how these two location numbers are assigned, we think of Earth as a perfect sphere.

Lines of latitude are parallel circles of different size around the sphere representing Earth. The longest circle is the equator, with latitude 0, and at the poles, the circles shrink to a point. Lines of longitude (also called *meridians*) are circles of identical size that pass through the North Pole and the South Pole as they circle the globe. Each of these circles crosses the equator. The equator is divided into 360 degrees, and the longitude of a location is the number of degrees where the meridian through that location meets the equator. For historical reasons, the meridian near the old Royal Astronomical Observatory in Greenwich, England, is chosen as 0 longitude.

Today this prime meridian is marked with a band of brass that stretches across the yard of the Observatory. Longitude is measured from the prime meridian, with positive values going east (0 to 180) and negative values going west (0 to −180). Both ±180-degree longitudes share the same line, in the middle of the Pacific Ocean. At the equator, the distance on Earth's surface for each one degree of latitude or longitude is just over 69 miles.

Every location on Earth is then identified by the meridian and the latitude lines that pass through it. In Example 6, we explain how latitude values are assigned and how they can be used to compute distances between cities having the same longitude.

1 Learn the vocabulary associated with angles.

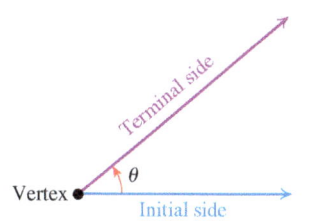

FIGURE 4.1 An angle

Angles

A **ray** is a part of a line made up of a point, called the **endpoint**, and all of the points on one side of the endpoint. An **angle** is formed by rotating a ray about its endpoint. The angle's **initial side** is the ray's original position, while the angle's **terminal side** is the ray's position after the rotation. The endpoint is called the **vertex** of the angle. A curved arrow drawn near the angle's vertex indicates both the direction and amount of rotation from the initial side to the terminal side. See Figure 4.1.

The Greek letters α (alpha), β (beta), γ (gamma), and θ (theta) are often used to name angles. If the rotation is counterclockwise, the result is a **positive angle**; if the rotation is

284 Chapter 4 Trigonometric Functions

clockwise, the result is a **negative angle**. See Figure 4.2. Rotation in both directions is unrestricted. Angles that have the same initial and terminal sides are called **coterminal angles**. In Figure 4.2(c), α and β are different angles but are coterminal.

Angle θ is positive.   Angle θ is negative.   Angles α and β are coterminal.
(a)   (b)   (c)

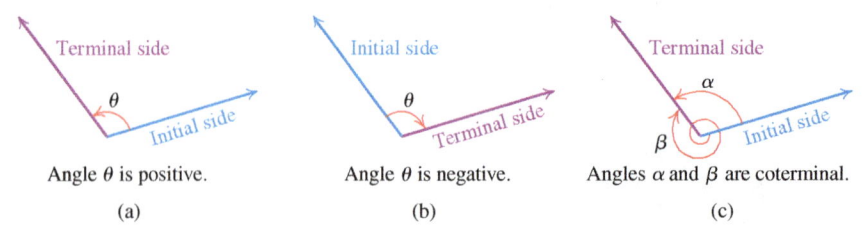

FIGURE 4.2 Positive, negative, and coterminal angles

An angle in a rectangular coordinate system is in **standard position** if its vertex is at the origin and its initial side lies on the positive x-axis. All of the angles in Figure 4.3 are in standard position. An angle in standard position is **quadrantal** if its terminal side lies on a coordinate axis; it is said to **lie in a quadrant** if its terminal side lies in that quadrant. See Figure 4.3.

We measure angles by determining the amount of rotation from the initial side to the terminal side, indicated by a curved arrow that also shows direction. Two units of measurement for angles are *degrees* and *radians*.

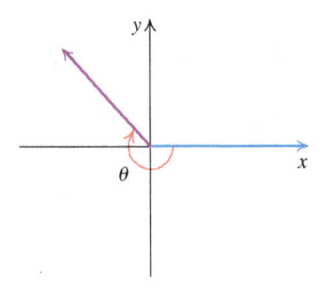

Angle θ is negative and lies in quadrant II.
(a)

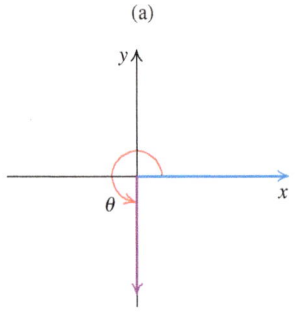

Angle θ is positive and is a quadrantal angle.
(b)

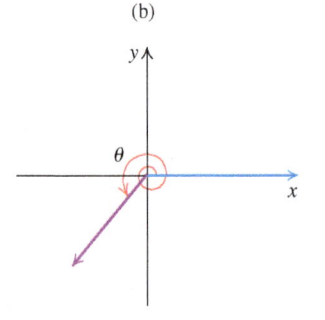

Angle θ is positive and lies in quadrant III.
(c)

FIGURE 4.3 Angles in standard position

Measuring Angles Using Degrees

A measure of **one degree** (denoted by 1°) is assigned to an angle resulting from a rotation $\frac{1}{360}$ of a complete revolution counterclockwise about the vertex. An angle formed by rotating the initial side counterclockwise one full rotation so that the terminal and initial sides coincide has measure 360 degrees, written as 360°.

Angles are classified according to their measures. As illustrated in Figure 4.4, an **acute angle** has measure between 0° and 90°; a **right angle** has measure 90°, or one-fourth of a revolution; an **obtuse angle** has measure between 90° and 180°; and a **straight angle** has measure 180°, or half of a revolution. Figure 4.4 also shows several angles measured in degrees.

2 Use degree and radian measure.

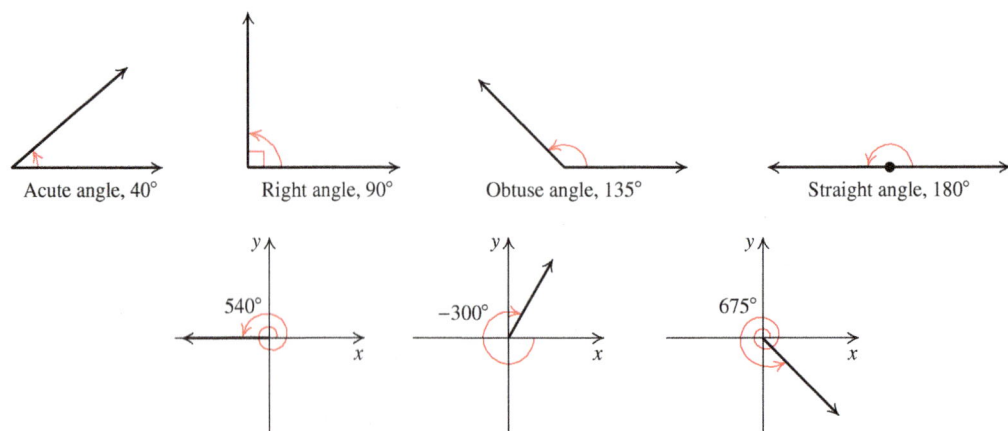

Acute angle, 40°   Right angle, 90°   Obtuse angle, 135°   Straight angle, 180°

540°   −300°   675°

FIGURE 4.4 Degree measure of various angles

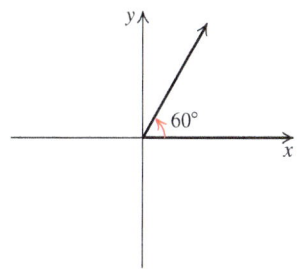

Section 4.1 ■ Angles and Their Measure   285

The measure of an angle has no numerical limit because the terminal side can be rotated without limitation.

EXAMPLE 1   Drawing an Angle in Standard Position

Draw each angle in standard position.

a. 60°   **b.** 135°   **c.** −240°   **d.** 405°

Solution

a. Because $60 = \frac{2}{3}(90)$, a 60° angle is $\frac{2}{3}$ of a 90° angle. See Figure 4.5.

b. Because $135 = 90 + 45$, a 135° angle is a counterclockwise rotation of 90°, followed by half of a 90° counterclockwise rotation. See Figure 4.6.

c. Because $-240 = -180 - 60$, a −240° angle is a clockwise rotation of 180°, followed by a clockwise rotation of 60°. See Figure 4.7.

d. Because $405 = 360 + 45$, a 405° angle is one complete counterclockwise rotation, followed by half of a 90° counterclockwise rotation. See Figure 4.8.

Practice Problem 1   Draw a 225° angle in standard position.

Today it is common to divide degrees into fractional parts using *decimal degree notation* such as 30.5°. Traditionally, however, degrees were expressed in terms of *minutes* and *seconds*. One **minute**, denoted 1′, is defined as $\frac{1}{60}(1°)$, and one **second**, denoted 1″, is defined as $\frac{1}{60}(1')$. An angle measuring 27 degrees, 14 minutes, and 39 seconds is written as 27°14′39″. Because $\frac{1}{60} \cdot \frac{1}{60} = \frac{1}{3600}$, we have

$$1'' = \frac{1}{60}(1') = \frac{1}{60}\left[\frac{1}{60}(1°)\right] = \frac{1}{3600}(1°).$$

FIGURE 4.5

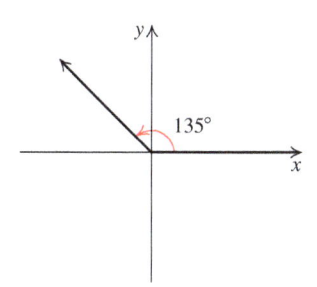

FIGURE 4.6

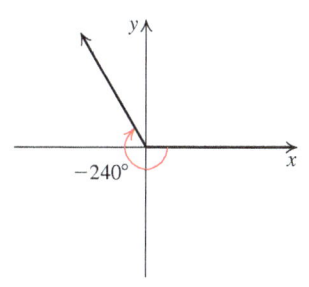

FIGURE 4.7

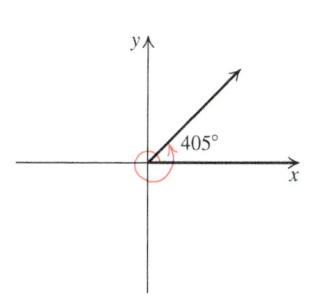

FIGURE 4.8

RELATIONSHIP AMONG DEGREES, MINUTES, AND SECONDS

$$1' = \frac{1}{60}(1°) = \left(\frac{1}{60}\right)° \qquad 1° = 60' \qquad 1° = (3600)''$$

$$1'' = \frac{1}{3600}(1°) = \left(\frac{1}{3600}\right)° \qquad 1'' = \frac{1}{60}(1') \qquad 1' = 60''$$

TECHNOLOGY CONNECTION

Many calculators can perform conversions between angle measurements in decimal degree form and in degrees, minutes, and seconds (DMS) form. From the ANGLE menu, you get the following:

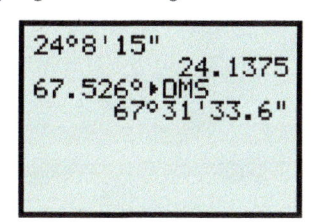

```
24°8'15"
          24.1375
67.526°▶DMS
          67°31'33.6"
```

Let's use these relationships to convert between degree, minute, second (DMS) notation and decimal degree notation. For example,

$$30' = 30 \cdot 1' = 30 \cdot \frac{1}{60}(1°) = 0.5(1°) = 0.5°.$$

$$45'' = 45 \cdot 1'' = 45 \cdot \frac{1}{3600}(1°) = 0.0125°$$

So $18°30'45'' = 18° + 30' + 45'' = 18° + 0.5° + 0.0125° = 18.5125°.$

Because $0.2° = \frac{2}{10}(1°) = \frac{2}{10}(60') = 12'$

and $0.02° = \frac{2}{100}(1°) = \frac{2}{100}(3600'') = 72'' = 1' + 12''$

then $39.22° = 39° + 0.2° + 0.02° = 39° + 12' + (1' + 12'') = 39°13'12''.$

286 Chapter 4 Trigonometric Functions

Radian Measure

An angle whose vertex is at the center of a circle is called a **central angle**. A central angle intercepts the arc of the circle from the initial side to the terminal side. The smallest positive central angle that intercepts an arc of the circle of length equal to the radius of the circle is said to have measure **1 radian**. See Figure 4.9.

Because the circumference of a circle of radius r is $2\pi r \approx 6.28r$, six arcs of length r can be consecutively marked off on a circle of radius r, leaving only a small portion of the circumference uncovered. See Figure 4.10.

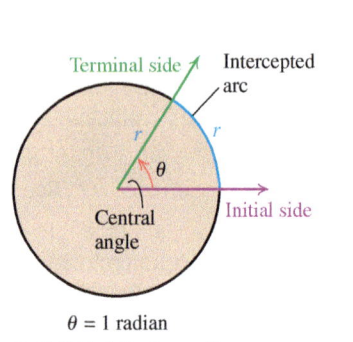

$\theta = 1$ radian
FIGURE 4.9 One radian

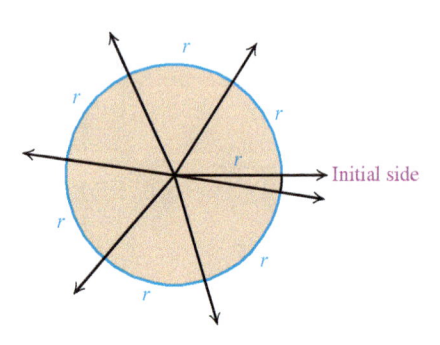

Circumference $= 2\pi r \approx 6.28r$
FIGURE 4.10

Each of the positive central angles in Figure 4.10 that intercepts an arc of length r has measure 1 radian; so the measure of one complete revolution is a little more than 6 radians. In this book, we assume (unless otherwise specified) that all central angles are positive. From geometry, we know that the measure of a central angle, θ ($0 \leq \theta \leq 2\pi$), is proportional to the length, s, of the arc it intercepts. This fact leads to the following relation.

SIDE NOTE

Since both s and r are in units of length, the units "cancel" in

$$\theta = \frac{s \text{ units}}{r \text{ units}} = \frac{s}{r}$$

making radians a unitless measure and consequently a real number.

RADIAN MEASURE OF A CENTRAL ANGLE

The radian measure θ of a central angle that intercepts an arc of length s on a circle of radius r is given by

$$\theta = \frac{s}{r} \text{ radians.}$$

Notice that when $s = r$, we have $\theta = \frac{r}{r} = 1$ radian.

3 Convert between degree and radian measure.

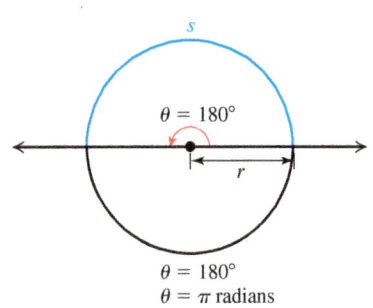

Relationship Between Degrees and Radians

To see the relationship between degrees and radians, consider the central angle in Figure 4.11 with degree measure $\theta = 180°$ in a circle of radius r. The length of the arc, s, that is intercepted by this angle is half the circumference of the circle. So

$$s = \frac{1}{2}(\text{circumference})$$

$$s = \frac{1}{2}(2\pi r) = \pi r \qquad \text{Circumference} = 2\pi r$$

$\theta = 180°$
$\theta = \pi$ radians
FIGURE 4.11

We can now find the radian measure for $\theta = 180°$.

$$180° = \theta = \frac{s}{r} \text{ radians} \quad \text{Radian measure of a central angle}$$

$$180° = \frac{\pi r}{r} \text{ radians} \quad \text{Replace } s \text{ with } \pi r.$$

$$180° = \pi \text{ radians} \quad \text{Simplify.}$$

CONVERTING BETWEEN DEGREES AND RADIANS

The relationship
$$180° = \pi \text{ radians}$$
allows us to convert between the degree and the radian measure of an angle.

Degrees to Radians:

$$180° = \pi \text{ radians} \quad \text{Note that } 180° = (180) \cdot 1°.$$

$$1° = \frac{\pi}{180} \text{ radian} \quad \text{Divide both sides by 180.}$$

$$\theta° = \frac{\theta \pi}{180} \text{ radians} \quad \text{Multiply both sides by } \theta.$$

Radians to Degrees:

$$\pi \text{ radians} = 180°$$

$$1 \text{ radian} = \left(\frac{180}{\pi}\right)° \quad \text{Divide both sides by } \pi.$$

$$\theta \text{ radians} = \left(\frac{180\,\theta}{\pi}\right)° \quad \text{Multiply both sides by } \theta.$$

TECHNOLOGY CONNECTION

A graphing calculator can convert degrees to radians from the ANGLE menu. The calculator should be set in *radian* mode.

```
180°
         3.141592654
30°
         .5235987756
55°
         .9599310886
```

EXAMPLE 2 **Converting Degrees to Radians**

Convert each angle from degrees to radians.

a. 30° b. 90°
c. −225° d. 55°

Solution

To convert degrees to radians, multiply degrees by $\frac{\pi}{180°}$.

a. $30° = 30° \cdot \frac{\pi}{180°} = \frac{30\pi}{180} = \frac{\pi}{6}$ radians

b. $90° = 90° \cdot \frac{\pi}{180°} = \frac{90\pi}{180} = \frac{\pi}{2}$ radians

c. $-225° = -225° \cdot \frac{\pi}{180°} = \frac{-225\pi}{180} = -\frac{5\pi}{4}$ radians

d. $55° = 55° \cdot \frac{\pi}{180°} = \frac{55\pi}{180} = \frac{11\pi}{36}$ radians

Practice Problem 2 Convert −45° to radians.

288 Chapter 4 Trigonometric Functions

If an angle is a fraction of a complete revolution, we usually write it in radian measure as a fractional multiple of π, rather than as a decimal. For example, we write 30° as $\dfrac{\pi}{6}$ radian, which is the exact value, instead of writing the approximation 30° ≈ 0.52 radian.

EXAMPLE 3 Converting Radians to Degrees

Convert each angle in radians to degrees.

a. $\dfrac{\pi}{3}$ radians   **b.** $-\dfrac{3\pi}{4}$ radians   **c.** 1 radian

Solution

To convert radians to degrees, multiply radians by $\dfrac{180°}{\pi}$.

a. $\dfrac{\pi}{3}$ radians $= \dfrac{\pi}{3} \cdot \dfrac{180°}{\pi} = \left(\dfrac{180}{3}\right)° = 60°$

b. $-\dfrac{3\pi}{4}$ radians $= -\dfrac{3\pi}{4} \cdot \dfrac{180°}{\pi} = \left(-\dfrac{3}{4}\right)180° = -135°$

c. 1 radian $= 1 \cdot \dfrac{180°}{\pi} \approx 57.3°$

Practice Problem 3 Convert $\dfrac{3\pi}{2}$ radians to degrees.

TECHNOLOGY CONNECTION

A graphing calculator can convert radians to degrees from the ANGLE menu. The calculator should be set in *degree* mode.

```
(π/3)ʳ
               60
(-3π/4)ʳ
             -135
1ʳ
       57.29577951
■
```

Figure 4.12 includes the degree and radian measures of some commonly used angles. With practice, you should be able to make these conversions without using the figure.

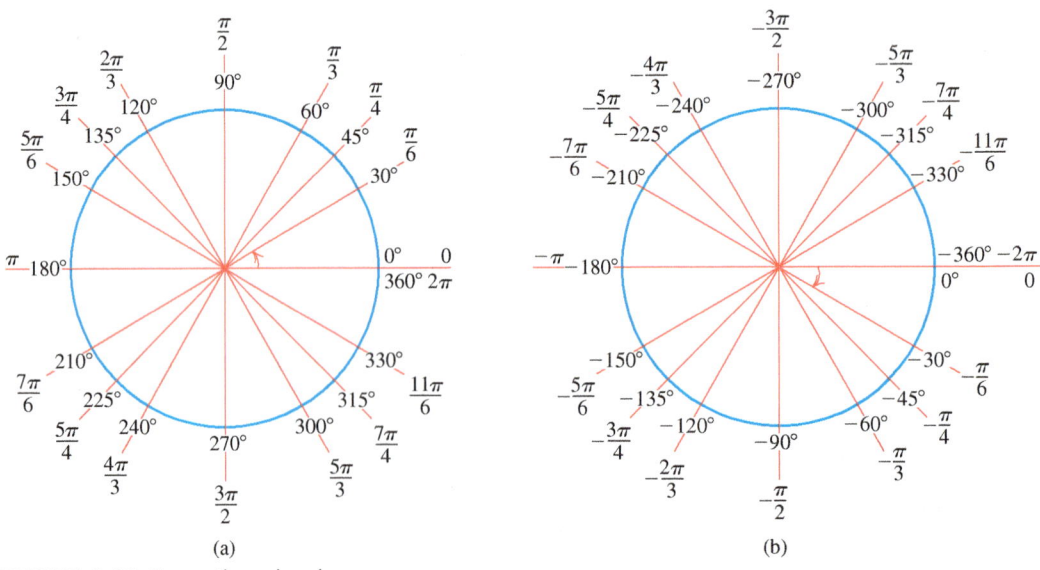

FIGURE 4.12 Frequently used angles

Complements and Supplements

4 Find complements and supplements.

Two positive angles are **complements** (or **complementary angles**) if the sum of their measures is 90° $\left(\dfrac{\pi}{2} \text{ radians}\right)$. So two angles with measures 50° and 40° are complements because 50° + 40° = 90°. Each angle is the complement of the other. Two positive angles are **supplements** (or **supplementary angles**) if the sum of their measures is 180°. So two angles with measures 120° and 60° are supplements because 120° + 60° = 180° (π radians). Each angle is the supplement of the other.

EXAMPLE 4   Finding Complements and Supplements

Find the complement and the supplement of the given angle or explain why the angle has no complement or supplement.

a. 73°   **b.** 110°

Solution

a. If θ represents the complement of 73°, then $\theta + 73° = 90°$; so $\theta = 90° - 73° = 17°$. The complement of 73° is 17°.
If α represents the supplement for 73°, then $\alpha + 73° = 180°$; so $\alpha = 180° - 73° = 107°$. The supplement of 73° is 107°.

b. There is no complement of a 110° angle because if θ represents the complement of 110°, then $\theta = 90° - 110° = -20°$, which is not a positive angle.
If β represents the supplement of 110°, then $\beta = 180° - 110° = 70°$. The supplement of 110° is 70°.

Practice Problem 4 Find the complement and the supplement for 67° or explain why 67° does not have a complement or a supplement.

> **SIDE NOTE**
> A positive angle with measure ≥90° does not have a complement. A positive angle with measure ≥180° does not have a supplement.

Length of an Arc of a Circle

5 Find the length of an arc of a circle.

Recall that the formula for the radian measure θ of a central angle that intercepts an arc of length s is $\theta = \dfrac{s}{r}$. See Figure 4.13. Multiplying both sides of this equation by r yields $s = r\theta$.

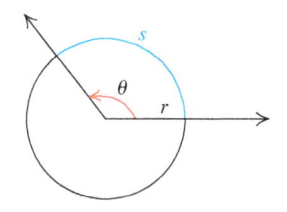

FIGURE 4.13

> **ARC LENGTH FORMULA**
>
> The length s of the arc intercepted by a central angle with radian measure θ in a circle of radius r is given by
>
> $$s = r\theta.$$

EXAMPLE 5   Finding Arc Length of a Circle

A circle has a radius of 18 inches. Find the length of the arc intercepted by a central angle with measure 210°.

Solution

We must first convert the central angle measure from degrees to radians.

$s = r\theta$ Arc length formula

$s = 18\left(\dfrac{7\pi}{6}\right)$ $\theta = 210° = 210°\left(\dfrac{\pi}{180°}\right) = \dfrac{7\pi}{6}$ radians

$= 21\pi \approx 65.97$ inches   Simplify and use a calculator.

Practice Problem 5 A circle has a radius of 2 meters. Find the length of the arc intercepted by a central angle with measure 225°.

EXAMPLE 6 Finding the Distance Between Cities

We discussed longitude and latitude in the introduction to this section. We determine the latitude of a location L by first finding the point of intersection, P, between the meridian through L and the equator. The latitude of L is the angle formed by rays drawn from the center of Earth to points L and P, with the ray through P being the initial ray. See Figure 4.14(a).

Billings, Montana, is due north of Grand Junction, Colorado. Find the distance between Billings (latitude 45°48′ N) and Grand Junction (latitude 39° 7′ N). Use 3960 miles as the radius of Earth. The N in 45°48′ N means that the location is north of the equator.

Solution

Because Billings is due north of Grand Junction, the same meridian passes through both cities. The distance between the cities is the length of the arc, s, on this meridian intercepted by the central angle, θ, that is the difference in their latitudes. See Figure 4.14(b).

The measure of angle θ is $45°48′ - 39°7′ = 6°41′$. To use the arc length formula, we must convert this angle to radians.

$$\theta = 6°41′ \approx 6.6833° = 6.6833\left(\frac{\pi}{180}\right) \text{ radian} \approx 0.117 \text{ radian}$$

We use this value of θ and $r = 3960$ miles in the arc length formula:

$$s \approx (3960)(0.117) \text{ miles} \approx 463 \text{ miles} \qquad s = r\theta$$

The distance between Billings and Grand Junction is about 463 miles.

Practice Problem 6 Chicago, Illinois, is due north of Pensacola, Florida. Find the distance between Chicago (latitude 41°51′ N) and Pensacola (latitude 30°25′ N). Use $r = 3960$ miles.

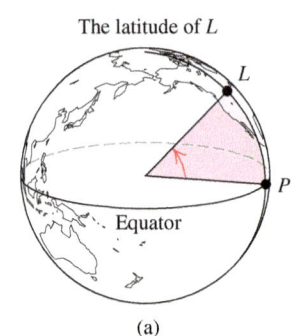

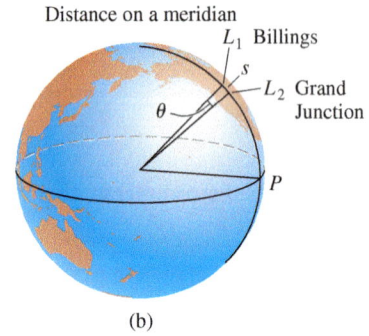

FIGURE 4.14 Distance between cities

SECTION 4.1 Exercises

Basic Concepts and Skills

In Exercises 1–8, draw each angle in standard position.

1. 30°
2. 150°
3. −120°
4. −330°
5. $\frac{5\pi}{3}$
6. $\frac{11\pi}{6}$
7. $-\frac{4\pi}{3}$
8. $-\frac{13\pi}{4}$

In Exercises 9–18, convert each angle from degrees to radians. Express each answer as a multiple of π.

9. 20°
10. 40°
11. −180°
12. −210°
13. 315°
14. 330°
15. 480°
16. 450°
17. −510°
18. −420°

In Exercises 19–28, convert each angle from radians to degrees.

19. $\frac{\pi}{12}$
20. $\frac{3\pi}{8}$
21. $-\frac{5\pi}{9}$
22. $-\frac{3\pi}{10}$
23. $\frac{5\pi}{3}$
24. $\frac{11\pi}{6}$
25. $\frac{5\pi}{2}$
26. $\frac{17\pi}{6}$
27. $-\frac{11\pi}{4}$
28. $-\frac{7\pi}{3}$

In Exercises 29–32, convert each angle from degrees to radians. Round your answers to two decimal places.

29. 12°
30. 127°
31. −84°
32. −175°

In Exercises 33–36, convert each angle from radians to degrees. Round your answers to two decimal places.

33. 0.94
34. 5
35. −8.21
36. −6.28

In Exercises 37–42, find the complement and the supplement of the given angle or explain why the angle has no complement or supplement.

37. 47°
38. 75°
39. 120°
40. 160°
41. 210°
42. −50°

In Exercises 43–50, use the following notations: θ = central angle of a circle, r = radius of a circle, and s = length of the intercepted arc. In each case, find the missing quantity. Round your answers to three decimal places.

43. $r = 25$ inches, $s = 7$ inches, $\theta = ?$
44. $r = 5$ feet, $s = 6$ feet, $\theta = ?$
45. $r = 10.5$ centimeters, $s = 22$ centimeters, $\theta = ?$
46. $r = 60$ meters, $s = 120$ meters, $\theta = ?$
47. $r = 3$ m, $\theta = 25°$, $s = ?$
48. $r = 0.7$ m, $\theta = 357°$, $s = ?$
49. $r = 6.5$ m, $\theta = 12$ radians, $s = ?$
50. $r = 6$ m, $\theta = \dfrac{\pi}{6}$ radians, $s = ?$

Applying the Concepts

51. **Wheel ratios.** An automobile is pushed so that its wheels turn three-quarters of a revolution. If the tires have a radius of 15 inches, how many inches does the car move?

52. **Nautical miles.** A nautical mile is the length of an arc of the equator intercepted by a central angle of 1′. Using 3960 miles for the value of the radius of Earth, express 1 nautical mile in terms of land (**statute**) miles.

53. **Angles on a clock.** What is the radian measure of the smaller central angle made by the hands of a clock at 4:00? Express your answer as a rational multiple of π.

54. **Diameter of a pizza.** You are told that a slice of pizza whose edges form a 25° angle with an outer crust edge 4 inches long was found in a gym locker. What was the diameter of the original pizza? Round your answer to the nearest inch.

55. **Arc of a pendulum.** How far does the tip of a 5-foot pendulum travel as it swings through an angle of 30°? Round your answer to two decimal places.

56. **Security cameras.** A security camera rotates through an angle of 120°. To the nearest foot, what is the arc width of the field of view 40 feet from the camera?

In Exercises 57–61, the latitude of any location on Earth is the angle formed by the two rays drawn from the center of Earth to the location and to the point of intersection of the meridian for the location with the equator. The ray through the location is the initial ray. Use 3960 miles as the radius of the Earth.

57. **Distance between cities.** Indianapolis, Indiana, is due north of Montgomery, Alabama. Find the distance between Indianapolis (latitude 39°44′ N) and Montgomery (latitude 32°23′ N).

58. **Distance between cities.** Pittsburgh, Pennsylvania, is due north of Charleston, South Carolina. Find the distance between Pittsburgh (latitude 40°30′ N) and Charleston (latitude 32°54′ N).

59. **Distance between cities.** Amsterdam, Netherlands, is due north of Lyon, France. Find the distance between Amsterdam (latitude 52°23′ N) and Lyon (latitude 45°42′ N).

60. **Distance between cities.** Adana, Turkey (north latitude 36°59′ N), is due north of Jerusalem, Israel (latitude 31°47′ N). Find the distance between Adana and Jerusalem.

61. **Difference in latitudes.** Miles City, Montana, is 440 miles due north of Boulder, Colorado. Find the difference in the latitudes of these two cities.

Linear and angular speed. Use the following definitions in Exercises 62–66. If an object travels along the circumference of a circle of radius r through a central angle θ radians, arc length s, in time t, then $v = \dfrac{s}{t}$ is called the *linear speed* of the object and $\omega = \dfrac{\theta}{t}$ is called the *angular speed* of the object.

62. Show that $v = r\omega$.

63. **Linear speed at the equator.** Earth rotates on an axis that goes through both the North and South Poles. It makes one complete revolution in 24 hours. Find the linear speed (in miles per hour) of a location on the equator.

64. **Ferris wheel speed.** A Ferris wheel in Vienna, Austria, has a diameter of approximately 61 meters. Assume that it takes 90 seconds for the Ferris wheel to make one complete revolution.
    a. Find the angular speed of the Ferris wheel. Round your answer to two decimal places.
    b. Find the linear speed of the Ferris wheel in meters per second. Round your answer to two decimal places.

65. **Bicycle speed.** A bicycle's wheels are 24 inches in diameter. If the bike is traveling at a rate of 25 miles per hour, find the angular speed of the wheels.

66. **Bicycle speed.** A bicycle's wheels are 30 inches in diameter. If the angular speed of the wheels is 11 radians per second, find the linear speed of the bicycle in inches per second.

Sector of a circle. In Exercises 67–70, use the following definition: a *sector* of a circle is a region bounded by the two sides of a central angle and the intercepted arc.

67. Use the proportionality property from geometry

$$\dfrac{\text{Area of a sector}}{\text{Total area of the circle}} = \dfrac{\text{Length of the intercepted arc}}{\text{Circumference of the circle}}$$

to show that the area A of a sector of radius r formed by a central angle of θ radians is $A = \dfrac{1}{2}r^2\theta$.

68. How many square inches of pizza have you eaten if you eat a slice of an 18-inch diameter pizza whose edges form a 30° angle? Round your answer to the nearest square inch.

69. If the area A of a sector of a circle of radius 10 inches is 60 square inches, find the central angle θ to the nearest degree.

70. A slice of pizza with an angle of 45° has an area of 19 square inches. Find the diameter of this pizza. Round your answer to the nearest inch.

Critical Thinking / Discussion / Writing

71. Use the fact that $\theta + \pi$ and $\theta - \pi$ are coterminal angles to prove that $\frac{5\pi}{3}$ and $-\frac{\pi}{3}$ are coterminal.

72. Find the smallest positive integer n so that θ and $\theta - \frac{n\pi}{2}$ are coterminal.

73. Which line of latitude has a smaller radius, the line of latitude through a city of latitude N 30°40′13″ or the line of latitude through a city of latitude N 40°30′13″?

74. **Earth circumference.** In about 250 B.C., the philosopher Eratosthenes estimated the radius of Earth based on what might appear to be scant information. He knew that the city of Syene (modern Aswan) is directly south of the city of Alexandria and that Syene and Alexandria are (in modern units) about 500 miles apart. He also knew that in Syene at noon on a midsummer day, an upright rod casts no shadow but makes a 7°12′ angle with a vertical rod at Alexandria and that then the sun's rays at Syene are effectively parallel to the sun's rays at Alexandria. By measuring the length of the shadow in Alexandria at noon on the summer solstice when there was no shadow in Syene, explain how Eratosthenes could measure the circumference of Earth.

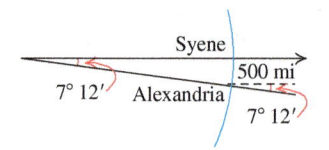

Maintaining Skills

In Exercises 75–80, determine whether the given point (x, y) satisfies the equation of the unit circle $x^2 + y^2 = 1$.

75. $\left(\dfrac{3}{5}, \dfrac{4}{5}\right)$

76. $\left(-\dfrac{12}{13}, \dfrac{5}{13}\right)$

77. $\left(\dfrac{3}{4}, -\dfrac{\sqrt{7}}{4}\right)$

78. $\left(-\dfrac{\sqrt{13}}{7}, -\dfrac{6}{7}\right)$

79. $\left(\dfrac{1}{2}, \dfrac{1}{2}\right)$

80. $\left(\dfrac{5}{4}, -\dfrac{1}{4}\right)$

SECTION 4.2

The Unit Circle; Trigonometric Functions

BEFORE STARTING THIS SECTION, REVIEW

1. Equation of a circle (Section 1.1, page 60)
2. Arc length formula (Section 4.1, page 289)
3. Symmetry (Section 1.1, page 57)
4. Equation of a line (Section 1.2, page 71)

OBJECTIVES

1. Review properties of the unit circle.
2. Define trigonometric functions of real numbers.
3. Find exact trigonometric function values using a terminal point on the unit circle.
4. Define trigonometric function of angles.
5. Find trigonometric function values of an angle in standard position.
6. Approximate trigonometric function values using a calculator.
7. Determine the signs of the trigonometric functions in each quadrant.
8. Find a reference angle.

◆ Blood Pressure

In 1628, William Harvey wrote in his book *De Motu Cordis (On the Motion of the Heart)*, "Just as the king is the first and highest authority in the state, so the heart governs the whole body!" He thereby announced to the world that the heart pumped blood through the entire body. There was a time, however, when even the fact that blood circulated at all was not accepted: the arteries were thought to carry air, not blood.

The first measurements of blood pressure were made in the early eighteenth century by Stephen Hales, an English botanist, physiologist, and clergyman. He found that pressure from a horse's heart filled a vertical glass tube with the horse's blood to a height of 8 feet 3 inches. He accomplished this by inserting a narrow brass pipe directly into an artery and fitting a 9-foot-long vertical glass to the pipe.

Fortunately, today a much safer, more convenient, and simplified method for measuring blood pressure is available everywhere. In Exercise 73, we use trigonometric or circular functions to find blood pressure.

1 Review properties of the unit circle.

The Unit Circle

Recall the definition of the unit circle (see Section 1.1).

THE UNIT CIRCLE

The **unit circle** is the circle of radius 1 with its center at the origin. In the xy-plane, it is the set of points (x, y) that satisfy the equation

$$x^2 + y^2 = 1.$$

294 Chapter 4 Trigonometric Functions

We list some properties of the unit circle that are useful in the development of the trigonometric functions.

1. **Circumference.** The circumference of the unit circle is 2π. This means that the arc length for one revolution around the unit circle is 2π units.
2. **Symmetry.** If $P = (x, y)$ is a point on the unit circle, then the following *symmetric* images of P are also on the unit circle:

   (i) $Q = (x, -y)$ about the x-axis   (ii) $R = (-x, y)$ about the y-axis
   (iii) $S = (-x, -y)$ about the origin   (iv) $T = (y, x)$ about the line $y = x$

See Figure 4.15.

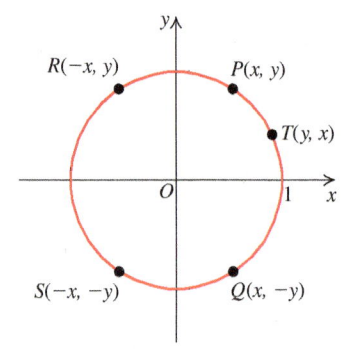

FIGURE 4.15

EXAMPLE 1 Finding Points on the Unit Circle

Find all points on the unit circle whose y-coordinate is $\dfrac{3}{5}$.

Solution

We need to find the x-coordinates of all points (if any) of the form $\left(x, \dfrac{3}{5}\right)$ that satisfy the equation $x^2 + y^2 = 1$.

We let $y = \dfrac{3}{5}$ in the equation $x^2 + y^2 = 1$ and solve for x.

$$x^2 + y^2 = 1 \qquad \text{The unit circle equation}$$

$$x^2 + \left(\frac{3}{5}\right)^2 = 1 \qquad \text{Replace } y \text{ with } \frac{3}{5}.$$

$$x^2 + \frac{9}{25} = 1$$

$$x^2 = 1 - \frac{9}{25} = \frac{16}{25}$$

$$x = \pm\sqrt{\frac{16}{25}} = \pm\frac{4}{5}$$

The points on the unit circle whose y-coordinate is $\dfrac{3}{5}$ are $\left(\dfrac{4}{5}, \dfrac{3}{5}\right)$ and $\left(-\dfrac{4}{5}, \dfrac{3}{5}\right)$. These are the points of intersection of the unit circle and the line $y = \dfrac{3}{5}$. See Figure 4.16.

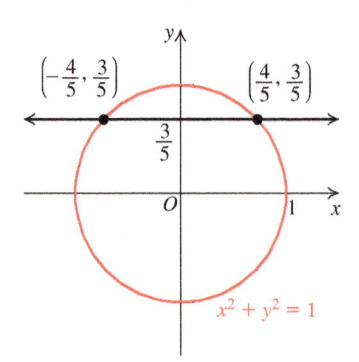

FIGURE 4.16

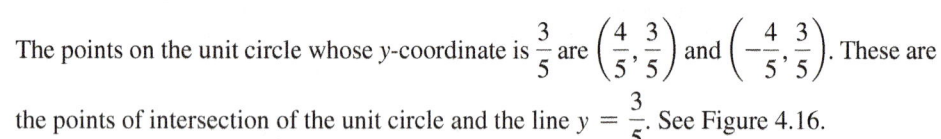

Practice Problem 1 Find all points on the unit circle whose x-coordinate is $-\dfrac{2}{3}$.

2 Define trigonometric functions of real numbers.

Trigonometric Functions of Real Numbers

In Figure 4.17(a), the number line is drawn tangent to the unit circle at the point $(1, 0)$. The unit length on the number line is the same as on the coordinate axes. For a real number t on the tangent number line, we "wrap" the segment from 0 to t around the unit circle in a counterclockwise direction if t is positive or in a clockwise direction if t is negative. See Figures 4.17(b) and (c).

In this way, each real number t corresponds to a unique point $P(t) = (x, y)$ on the unit circle and is often called the **terminal point** associated with the real number t. Because the circumference of the unit circle is 2π, if $t > 2\pi$, you must go around the unit circle more than once to arrive at the point $P(t)$. Because the arc length between t and $t \pm 2\pi$ is 2π, the points corresponding to $P(t)$ and $P(t \pm 2\pi)$ coincide on the unit circle. In fact, for any integer n and for any real number t, the points corresponding to $P(t)$ and $P(t + 2n\pi)$ coincide on the unit circle.

RECALL

A line is tangent to a circle if it intersects the circle at exactly one point.

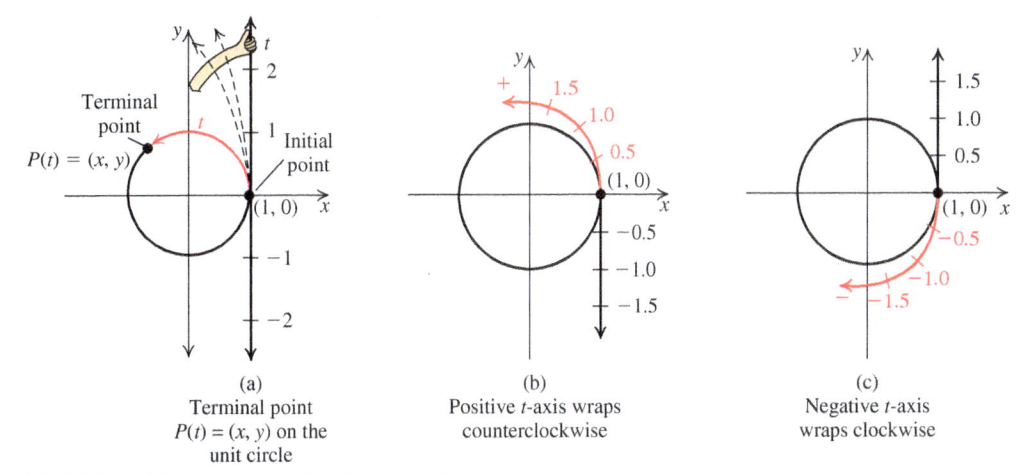

FIGURE 4.17 Wrapping *t*-axis on the unit circle

(a) Terminal point $P(t) = (x, y)$ on the unit circle
(b) Positive *t*-axis wraps counterclockwise
(c) Negative *t*-axis wraps clockwise

The correspondence between real numbers and terminal points of arcs on the unit circle is used to define six important functions, called the **trigonometric functions.** The full names of these functions are **sine, cosine, tangent, cosecant, secant,** and **cotangent.** The customary abbreviations for the values of these functions at a real number *t* are **sin *t*, cos *t*, tan *t*, csc *t*, sec *t*,** and **cot *t*,** respectively. Because the definitions of trigonometric functions are based on the unit circle, trigonometric functions are often called **circular functions.**

Unit Circle Definitions of the Trigonometric Functions of Real Numbers

Let *t* be any real number and let $P(t) = (x, y)$ be the terminal point on the unit circle associated with *t*. Then

$$\sin t = y \qquad \csc t = \frac{1}{y} \quad (y \neq 0)$$

$$\cos t = x \qquad \sec t = \frac{1}{x} \quad (x \neq 0)$$

$$\tan t = \frac{y}{x} \quad (x \neq 0) \qquad \cot t = \frac{x}{y} \quad (y \neq 0)$$

Note that each function in the second column is the *reciprocal* of the corresponding function in the first column. Here are the "**reciprocal identities**":

$$\csc t = \frac{1}{y} = \frac{1}{\sin t} \qquad \sec t = \frac{1}{x} = \frac{1}{\cos t} \qquad \cot t = \frac{x}{y} = \frac{1}{y/x} = \frac{1}{\tan t}.$$

We also note the following "**quotient identities**":

$$\tan t = \frac{y}{x} = \frac{\sin t}{\cos t} \quad \text{and} \quad \cot t = \frac{x}{y} = \frac{\cos t}{\sin t}$$

From the definitions of the tangent and secant functions, we see that these functions are not defined when $x = 0$. Similarly, the cotangent and cosecant functions are not defined when $y = 0$.

EXAMPLE 2 Using the Definition to Write Trigonometric Function Values

For the point $P(t) = (x, y)$ on the unit circle, write the values of all trigonometric functions at t.

a. $P(t) = \left(\dfrac{5}{13}, -\dfrac{12}{13}\right)$ **b.** $P(t) = \left(-\dfrac{3}{7}, -\dfrac{2\sqrt{10}}{7}\right)$

Solution

a. For $P(t) = (x, y) = \left(\dfrac{5}{13}, -\dfrac{12}{13}\right)$,

$\sin t = y = -\dfrac{12}{13},$ $\cos t = x = \dfrac{5}{13},$ $\tan t = \dfrac{y}{x} = \dfrac{-12/13}{5/13} = -\dfrac{12}{5},$

$\csc t = -\dfrac{13}{12},$ $\sec t = \dfrac{13}{5},$ $\cot t = -\dfrac{5}{12}.$

b. For $P(t) = \left(-\dfrac{3}{7}, -\dfrac{2\sqrt{10}}{7}\right)$

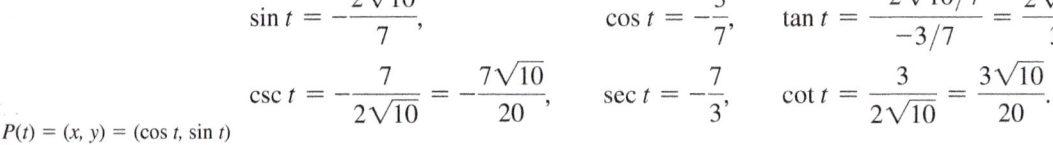

$\sin t = -\dfrac{2\sqrt{10}}{7},$ $\cos t = -\dfrac{3}{7},$ $\tan t = \dfrac{-2\sqrt{10}/7}{-3/7} = \dfrac{2\sqrt{10}}{3},$

$\csc t = -\dfrac{7}{2\sqrt{10}} = -\dfrac{7\sqrt{10}}{20},$ $\sec t = -\dfrac{7}{3},$ $\cot t = \dfrac{3}{2\sqrt{10}} = \dfrac{3\sqrt{10}}{20}.$

Practice Problem 2 Repeat Example 2 for

a. $P(t) = \left(-\dfrac{4}{5}, \dfrac{3}{5}\right).$ **b.** $P(t) = \left(\dfrac{1}{3}, \dfrac{2\sqrt{2}}{3}\right).$

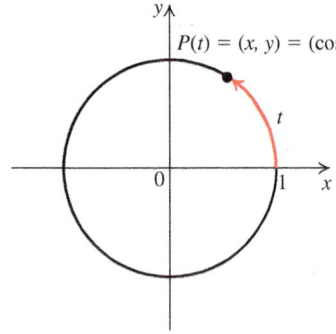

FIGURE 4.18 Unit circle definitions of $\cos t$ and $\sin t$

> **COSINE AND SINE**
>
> The terminal point $P(t) = (x, y)$ on the unit circle associated with a real number t has coordinates $(\cos t, \sin t)$ because $x = \cos t$ and $y = \sin t$. See Figure 4.18.

3 Find exact trigonometric function values using a terminal point on the unit circle.

Finding Exact Trigonometric Function Values

EXAMPLE 3 Evaluating Trigonometric Functions

Find the values (if any) of the six trigonometric functions at each value of t.

a. $t = 0$ **b.** $t = \dfrac{3\pi}{2}$ **c.** $t = -3\pi$

Solution

For each value of t, find the corresponding point $P(t) = (x, y)$ on the unit circle. See Figure 4.19. Then use the definitions of the trigonometric functions. Note that the length of the unit circle (one wrap) is 2π.

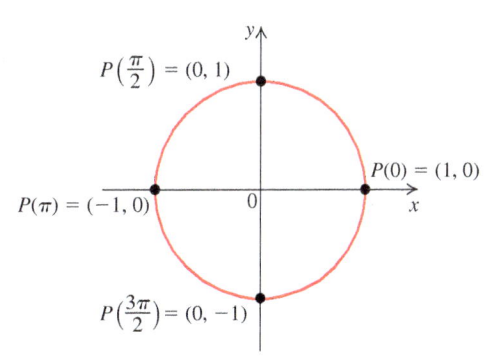

FIGURE 4.19 Points corresponding to various values of t

a. $t = 0$ corresponds to the point $(x, y) = (1, 0)$. So

$$\sin 0 = y = 0 \qquad \csc 0 = \frac{1}{y} = \frac{1}{0} \text{ is undefined}$$

$$\cos 0 = x = 1 \qquad \sec 0 = \frac{1}{x} = \frac{1}{1} = 1$$

$$\tan 0 = \frac{y}{x} = \frac{0}{1} = 0 \qquad \cot 0 = \frac{x}{y} = \frac{1}{0} \text{ is undefined}$$

b. Length of $\frac{3}{4}$ (one wrap) $= \frac{3}{4}(2\pi) = \frac{3\pi}{2}$, so $t = \frac{3\pi}{2}$ corresponds to the terminal point $(x, y) = (0, -1)$.

$$\sin \frac{3\pi}{2} = y = -1 \qquad \csc \frac{3\pi}{2} = \frac{1}{y} = \frac{1}{-1} = -1$$

$$\cos \frac{3\pi}{2} = x = 0 \qquad \sec \frac{3\pi}{2} = \frac{1}{x} = \frac{1}{0} \text{ is undefined}$$

$$\tan \frac{3\pi}{2} = \frac{y}{x} = \frac{-1}{0} \text{ is undefined} \qquad \cot \frac{3\pi}{2} = \frac{x}{y} = \frac{0}{-1} = 0$$

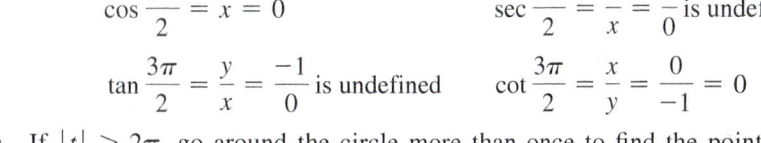

c. If $|t| > 2\pi$, go around the circle more than once to find the point corresponding to t. From Figure 4.20, we see that $t = -3\pi$ corresponds to exactly the same point $(-1, 0)$ as does $t = \pi$. So

$$\sin(-3\pi) = \sin \pi = 0 \qquad \csc(-3\pi) = \csc \pi \text{ is undefined}$$

$$\cos(-3\pi) = \cos \pi = -1 \qquad \sec(-3\pi) = \sec \pi = -1$$

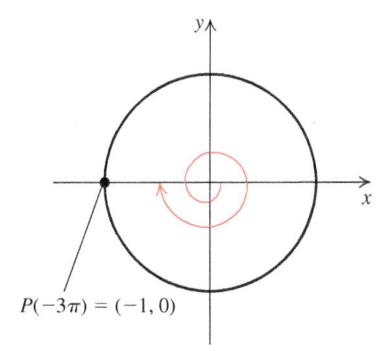

$$\tan(-3\pi) = \tan \pi = 0 \qquad \cot(-3\pi) = \cot \pi \text{ is undefined}$$

FIGURE 4.20

Practice Problem 3 Repeat Example 3 for $t = \frac{5\pi}{2}$.

Trigonometric Function Values for $t = \frac{\pi}{4}$. Let $P(t) = (x, y)$ be the terminal point on the unit circle associated with a real number t. For $t = \frac{\pi}{4}$, the point $P\left(\frac{\pi}{4}\right) = (x, y)$ bisects the arc from $(1, 0)$ to $(0, 1)$ on the unit circle. See Figure 4.21. Because the unit circle is symmetric about the line $y = x$, we conclude that the point $P\left(\frac{\pi}{4}\right) = (x, y)$ lies on the line $y = x$.

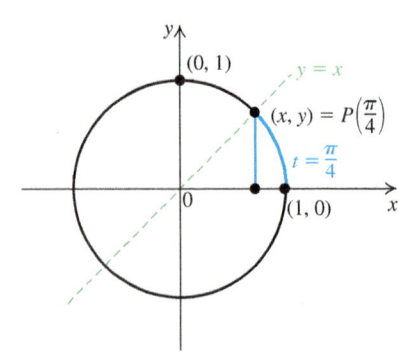

FIGURE 4.21 Terminal point
$P\left(\dfrac{\pi}{4}\right) = \left(\dfrac{\sqrt{2}}{2}, \dfrac{\sqrt{2}}{2}\right)$

That is for $t = \dfrac{\pi}{4}$, $y = x$. We then have:

$$x^2 + y^2 = 1 \qquad \text{Equation for the unit circle}$$
$$x^2 + x^2 = 1 \qquad y = x \text{ when } t = \dfrac{\pi}{4}$$
$$2x^2 = 1 \qquad \text{Simplify.}$$
$$x^2 = \dfrac{1}{2} \qquad \text{Divide both sides by 2.}$$

$$x = \sqrt{\dfrac{1}{2}} = \dfrac{1}{\sqrt{2}} \cdot \dfrac{\sqrt{2}}{\sqrt{2}} = \dfrac{\sqrt{2}}{2} \qquad x \text{ is positive because } (x, y) \text{ is in quadrant I.}$$

So $y = x = \dfrac{\sqrt{2}}{2}$ when $t = \dfrac{\pi}{4}$.

Therefore, $(x, y) = \left(\dfrac{\sqrt{2}}{2}, \dfrac{\sqrt{2}}{2}\right)$ is the point on the unit circle associated with $t = \dfrac{\pi}{4}$.

We have:

$$\sin\dfrac{\pi}{4} = y = \dfrac{\sqrt{2}}{2} \qquad \cos\dfrac{\pi}{4} = x = \dfrac{\sqrt{2}}{2} \qquad \tan\dfrac{\pi}{4} = \dfrac{y}{x} = 1$$

$$\csc\dfrac{\pi}{4} = \dfrac{1}{y} = \sqrt{2} \qquad \sec\dfrac{\pi}{4} = \dfrac{1}{x} = \sqrt{2} \qquad \cot\dfrac{\pi}{4} = \dfrac{x}{y} = 1$$

Trigonometric Function Values for $t = \dfrac{\pi}{6}$ and $t = \dfrac{\pi}{3}$. In Exercises 71 and 72, we ask you to show that the terminal points $P(t) = (x, y)$ associated with arcs $t = \dfrac{\pi}{6}$ and $t = \dfrac{\pi}{3}$ are $P\left(\dfrac{\pi}{6}\right) = \left(\dfrac{\sqrt{3}}{2}, \dfrac{1}{2}\right)$ and $P\left(\dfrac{\pi}{3}\right) = \left(\dfrac{1}{2}, \dfrac{\sqrt{3}}{2}\right)$. We then have:

$$\sin\dfrac{\pi}{6} = y = \dfrac{1}{2} \qquad \cos\dfrac{\pi}{6} = x = \dfrac{\sqrt{3}}{2} \qquad \tan\dfrac{\pi}{6} = \dfrac{y}{x} = \dfrac{1}{\sqrt{3}} = \dfrac{\sqrt{3}}{3}$$

$$\csc\dfrac{\pi}{6} = \dfrac{1}{y} = 2 \qquad \sec\dfrac{\pi}{6} = \dfrac{1}{x} = \dfrac{2}{\sqrt{3}} = \dfrac{2\sqrt{3}}{3} \qquad \cot\dfrac{\pi}{6} = \dfrac{x}{y} = \sqrt{3}$$

$$\sin\dfrac{\pi}{3} = y = \dfrac{\sqrt{3}}{2} \qquad \cos\dfrac{\pi}{3} = x = \dfrac{1}{2} \qquad \tan\dfrac{\pi}{3} = \dfrac{y}{x} = \dfrac{\frac{\sqrt{3}}{2}}{\frac{1}{2}} = \sqrt{3}$$

$$\csc\dfrac{\pi}{3} = \dfrac{1}{y} = \dfrac{2}{\sqrt{3}} = \dfrac{2\sqrt{3}}{3} \qquad \sec\dfrac{\pi}{3} = \dfrac{1}{x} = 2 \qquad \cot\dfrac{\pi}{3} = \dfrac{x}{y} = \dfrac{1}{\sqrt{3}} = \dfrac{\sqrt{3}}{3}$$

Figure 4.22 and Table 4.1 identify the terminal points $P(t) = (x, y)$ on the unit circle for some special values of t in the interval $[0, 2\pi]$. In Table 4.1, we also list the corresponding values of $\sin t$, $\cos t$, and $\tan t$.

Section 4.2 ■ The Unit Circle; Trigonometric Functions 299

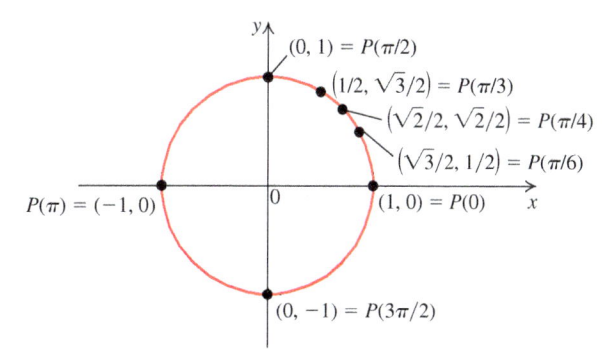

FIGURE 4.22 Terminal points

TABLE 4.1

Real Number t	Terminal Point $P(t) = (x, y)$	$\sin t$	$\cos t$	$\tan t$
0	(1, 0)	0	1	0
$\dfrac{\pi}{6}$	$\left(\dfrac{\sqrt{3}}{2}, \dfrac{1}{2}\right)$	$\dfrac{1}{2}$	$\dfrac{\sqrt{3}}{2}$	$\dfrac{1}{\sqrt{3}} = \dfrac{\sqrt{3}}{3}$
$\dfrac{\pi}{4}$	$\left(\dfrac{\sqrt{2}}{2}, \dfrac{\sqrt{2}}{2}\right)$	$\dfrac{\sqrt{2}}{2}$	$\dfrac{\sqrt{2}}{2}$	1
$\dfrac{\pi}{3}$	$\left(\dfrac{1}{2}, \dfrac{\sqrt{3}}{2}\right)$	$\dfrac{\sqrt{3}}{2}$	$\dfrac{1}{2}$	$\sqrt{3}$
$\dfrac{\pi}{2}$	(0, 1)	1	0	undefined
π	(−1, 0)	0	−1	0
$\dfrac{3\pi}{2}$	(0, −1)	−1	0	undefined
2π	(1, 0)	0	1	0

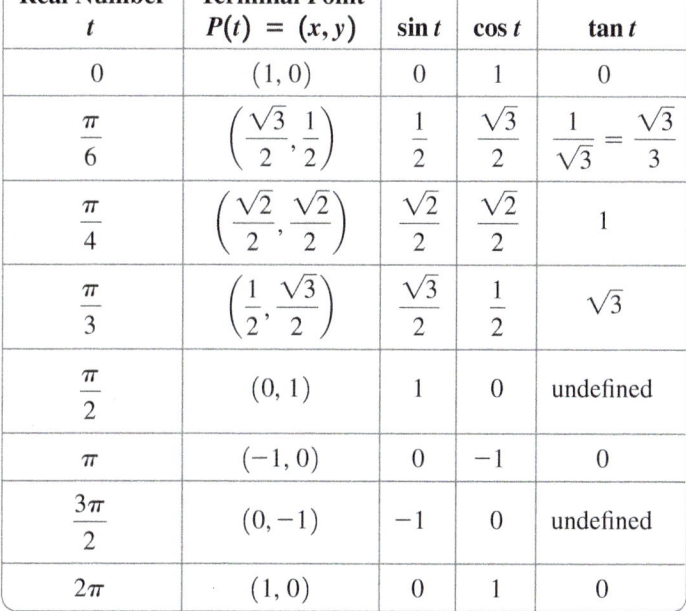

We can use Figure 4.22 and the symmetry of the unit circle to find the terminal points for values of t that are integer multiples of $\dfrac{\pi}{6}$ and $\dfrac{\pi}{4}$.

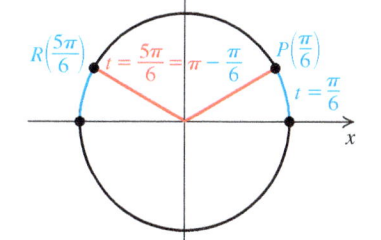

EXAMPLE 4 Finding Terminal Points by Symmetry

Use Table 4.1 and symmetry to find $\sin t$, $\cos t$, and $\tan t$ for each value of t.

a. $t = -\dfrac{\pi}{3}$ **b.** $t = \dfrac{5\pi}{6}$ **c.** $t = \dfrac{5\pi}{4}$

Solution

a. The terminal point $Q\left(-\dfrac{\pi}{3}\right)$ is the symmetric image about the x-axis of the point $P\left(\dfrac{\pi}{3}\right) = \left(\dfrac{1}{2}, \dfrac{\sqrt{3}}{2}\right)$. See Figure 4.23(a). So $Q\left(-\dfrac{\pi}{3}\right) = \left(\dfrac{1}{2}, -\dfrac{\sqrt{3}}{2}\right)$. Then $\sin\left(-\dfrac{\pi}{3}\right) = -\dfrac{\sqrt{3}}{2}$, $\cos\left(-\dfrac{\pi}{3}\right) = \dfrac{1}{2}$, and $\tan\left(-\dfrac{\pi}{3}\right) = \dfrac{-\sqrt{3}/2}{1/2} = -\sqrt{3}$.

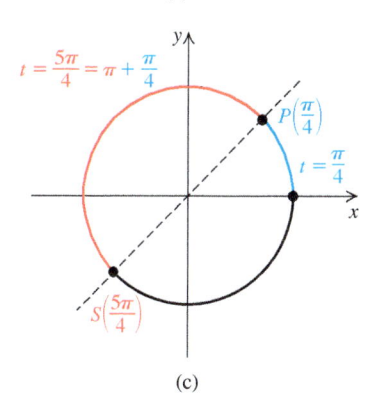

FIGURE 4.23

b. The terminal point $R\left(\dfrac{5\pi}{6}\right)$ is the symmetric image about the y-axis of the point

$P\left(\dfrac{\pi}{6}\right) = \left(\dfrac{\sqrt{3}}{2}, \dfrac{1}{2}\right)$. See Figure 4.23(b). So $R\left(\dfrac{5\pi}{6}\right) = \left(-\dfrac{\sqrt{3}}{2}, \dfrac{1}{2}\right)$. Then

$\sin\left(\dfrac{5\pi}{6}\right) = \dfrac{1}{2}, \cos\left(\dfrac{5\pi}{6}\right) = -\dfrac{\sqrt{3}}{2}$, and $\tan\left(\dfrac{5\pi}{6}\right) = \dfrac{1/2}{-\sqrt{3}/2} = -\dfrac{1}{\sqrt{3}} = -\dfrac{\sqrt{3}}{3}$.

c. The terminal point $S\left(\dfrac{5\pi}{4}\right)$ is symmetric about the origin to the point

$P\left(\dfrac{\pi}{4}\right) = \left(\dfrac{\sqrt{2}}{2}, \dfrac{\sqrt{2}}{2}\right)$. See Figure 4.23(c). So $S\left(\dfrac{5\pi}{4}\right) = \left(-\dfrac{\sqrt{2}}{2}, -\dfrac{\sqrt{2}}{2}\right)$.

Then $\sin\left(\dfrac{5\pi}{4}\right) = -\dfrac{\sqrt{2}}{2}, \cos\left(\dfrac{5\pi}{4}\right) = -\dfrac{\sqrt{2}}{2}$, and $\tan\left(\dfrac{5\pi}{4}\right) = \dfrac{-\sqrt{2}/2}{-\sqrt{2}/2} = 1$.

Practice Problem 4 Find terminal points (x, y) associated with each value of t.

a. $t = -\dfrac{5\pi}{6}$. **b.** $t = \dfrac{2\pi}{3}$. **c.** $t = \dfrac{7\pi}{6}$.

EXAMPLE 5   Finding Terminal Points for $|t| > 2\pi$

Find $\sin t$, $\cos t$, and $\tan t$ for each value of t.

a. $t = \dfrac{33\pi}{4}$ **b.** $t = \dfrac{17\pi}{3}$ **c.** $t = -\dfrac{19\pi}{6}$

Solution
Recall that for any integer n, $P(t) = P(t + 2n\pi)$. See page 298.

a. $t = \dfrac{33\pi}{4} = \dfrac{\pi}{4} + \dfrac{32\pi}{4} = \dfrac{\pi}{4} + 4(2\pi)$. So the terminal point associated with

$t = \dfrac{33\pi}{4}$ coincides with the terminal point associated with $t = \dfrac{\pi}{4}$, $P\left(\dfrac{\pi}{4}\right)$. That

is, $P\left(\dfrac{33\pi}{4}\right) = P\left(\dfrac{\pi}{4}\right) = \left(\dfrac{\sqrt{2}}{2}, \dfrac{\sqrt{2}}{2}\right)$. See Table 4.1. Then

$\sin\left(\dfrac{33\pi}{4}\right) = \dfrac{\sqrt{2}}{2}, \cos\left(\dfrac{33\pi}{4}\right) = \dfrac{\sqrt{2}}{2}$, and $\tan\left(\dfrac{33\pi}{4}\right) = 1$.

b. $t = \dfrac{17\pi}{3} = 6\pi - \dfrac{\pi}{3} = -\dfrac{\pi}{3} + 3(2\pi)$. So the terminal point associated with

$t = \dfrac{17\pi}{3}$ coincides with the terminal point associated with $t = -\dfrac{\pi}{3}$. From

Example 4a, $P = \left(\dfrac{17\pi}{3}\right) = Q\left(-\dfrac{\pi}{3}\right) = \left(\dfrac{1}{2}, -\dfrac{\sqrt{3}}{2}\right)$. Then

$\sin\left(\dfrac{17\pi}{3}\right) = -\dfrac{\sqrt{3}}{2}, \cos\left(\dfrac{17\pi}{3}\right) = \dfrac{1}{2}$, and $\tan\left(\dfrac{17\pi}{3}\right) = \dfrac{-\sqrt{3}/2}{1/2} = -\sqrt{3}$.

c. Similarly, $t = -\dfrac{19\pi}{6} = -4\pi + \dfrac{5\pi}{6}$. So from Example 4b,

$P\left(-\dfrac{19\pi}{6}\right) = P\left(\dfrac{5\pi}{6}\right) = \left(-\dfrac{\sqrt{3}}{2}, \dfrac{1}{2}\right)$. Then

$\sin\left(-\dfrac{19\pi}{6}\right) = \dfrac{1}{2}, \cos\left(-\dfrac{19\pi}{6}\right) = -\dfrac{\sqrt{3}}{2}$, and

$\tan\left(-\dfrac{19\pi}{6}\right) = -\dfrac{1}{\sqrt{3}} = -\dfrac{\sqrt{3}}{3}$.

Practice Problem 5 Find terminal points on the unit circle associated with each value of t.

a. $t = \dfrac{31\pi}{6}$ **b.** $t = -\dfrac{28\pi}{3}$

Trigonometric Functions of an Angle

4 Define trigonometric functions of angles.

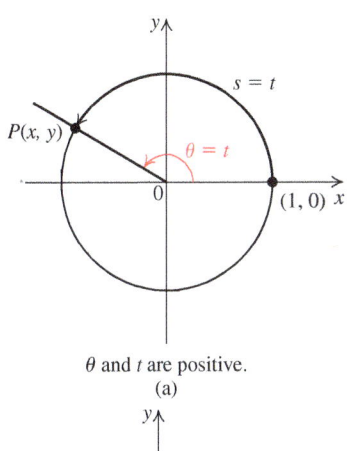

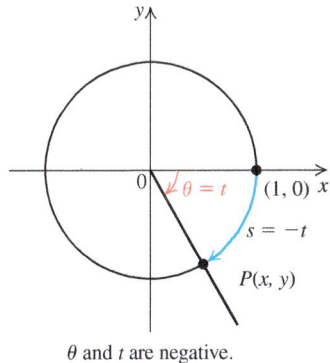

So far, we have defined the trigonometric functions of real numbers. That is, the domains of these functions consist of real numbers. We now define new functions whose domains are angles. These definitions are based on the following relationship between the real numbers and angles measured in radians.

Given an angle θ in standard position, let $P(x, y)$ be the point where the terminal side of θ intersects the unit circle. Because $r = 1$ for the unit circle, the arc length formula $s = r|\theta|$ gives us $s = 1 \cdot |\theta| = |\theta|$, where θ is measured in radians. Because θ is in standard position, we have $s = |t|$. If the measure of the central angle θ indicates direction of rotation, then the arc from $(1, 0)$ to $P(x, y)$ has length and direction; so $\theta = t$. See Figure 4.24.

For example, an angle $\theta = \dfrac{\pi}{3}$ radians in standard position intercepts an arc on the unit circle from $(1, 0)$ to the point $P(t) = P\left(\dfrac{\pi}{3}\right) = \left(\dfrac{1}{2}, \dfrac{\sqrt{3}}{2}\right)$. See page 299. By definition, we know that

$$\cos t = \cos \frac{\pi}{3} = \frac{1}{2} \quad \text{and} \quad \sin t = \sin \frac{\pi}{3} = \frac{\sqrt{3}}{2}.$$

real number

It seems reasonable to define

$$\cos \theta = \cos \frac{\pi}{3} = \frac{1}{2} \quad \text{and} \quad \sin \theta = \sin \frac{\pi}{3} = \frac{\sqrt{3}}{2}.$$

angle in radians

This leads to the following definition of trigonometric functions of an angle.

FIGURE 4.24
(a) θ and t are positive.
(b) θ and t are negative.

Trigonometric Function of an Angle

If θ is an angle with radian measure t, then

$\sin \theta = \sin t \qquad \cos \theta = \cos t \qquad \tan \theta = \tan t$
$\csc \theta = \csc t \qquad \sec \theta = \sec t \qquad \cot \theta = \cot t$

If θ is given in degrees, convert θ to radians before using these equations.

EXAMPLE 6   **Finding the Trigonometric Function Values of a Quadrantal Angle**

Find the trigonometric function values of $90°$.

Solution

Because $90° = \dfrac{\pi}{2}$ radians, we use Figure 4.22 and the definition on page 301 for $t = \dfrac{\pi}{2}$ and $\theta = 90°$.

$$\sin 90° = \sin\left(\dfrac{\pi}{2} \text{ radians}\right) = \sin\left(\dfrac{\pi}{2}\right) = 1$$

$$\cos 90° = \cos\left(\dfrac{\pi}{2} \text{ radians}\right) = \cos\left(\dfrac{\pi}{2}\right) = 0$$

$$\tan 90° = \tan\left(\dfrac{\pi}{2} \text{ radians}\right) = \tan\left(\dfrac{\pi}{2}\right) \text{ is undefined}$$

$$\csc 90° = \csc\left(\dfrac{\pi}{2} \text{ radians}\right) = \csc\left(\dfrac{\pi}{2}\right) = 1$$

$$\sec 90° = \sec\left(\dfrac{\pi}{2} \text{ radians}\right) = \sec\left(\dfrac{\pi}{2}\right) \text{ is undefined}$$

$$\cot 90° = \cot\left(\dfrac{\pi}{2} \text{ radians}\right) = \cot\left(\dfrac{\pi}{2}\right) = 0$$

Practice Problem 6 Find the trigonometric function values of $-270°$.

If an angle θ is measured in degrees, we use the degree symbol to indicate a trigonometric function value, such as $\cos 60°$ and $\tan 28°$. If θ is measured in radians, we do not use a symbol (for example, $\sin \dfrac{\pi}{4}$, $\sec \pi$, and $\cos 16$). Notice that the 0 coordinate of a point on the terminal side of a quadrantal angle cannot appear in a denominator. See Figure 4.25(a), where the terminal side is on the x-axis. The y-coordinate is 0; so the cotangent $\left(\cot \theta = \dfrac{x}{y}\right)$ and the cosecant $\left(\csc \theta = \dfrac{1}{y}\right)$ functions are undefined. If the terminal side is on the y-axis, as in Figure 4.25(b), then the x-coordinate is 0 and the tangent $\left(\tan \theta = \dfrac{y}{x}\right)$ and the secant $\left(\sec \theta = \dfrac{1}{x}\right)$ functions are undefined.

The definitions of trigonometric functions together with Table 4.1 (page 299) lead to the following useful table.

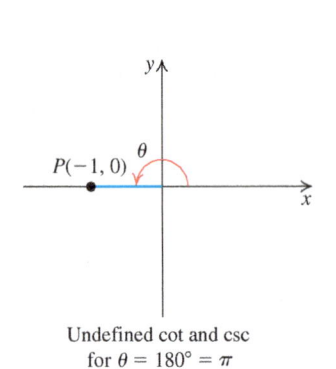

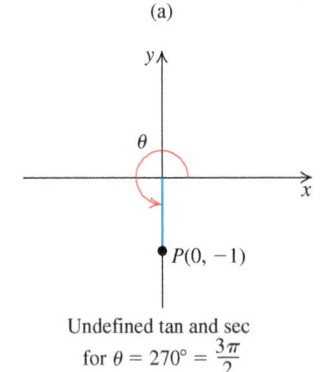

FIGURE 4.25 Undefined trigonometric function values

(a) Undefined cot and csc for $\theta = 180° = \pi$, with $P(-1, 0)$

(b) Undefined tan and sec for $\theta = 270° = \dfrac{3\pi}{2}$, with $P(0, -1)$

TABLE 4.2 Trigonometric Function Values of Common Angles

Real Number t	Angle $t = \theta$ Radians	Angle θ in Degrees	$\sin \theta$	$\cos \theta$	$\tan \theta$	$\cot \theta$	$\sec \theta$	$\csc \theta$
0	0	0°	0	1	0	undefined	1	undefined
$\dfrac{\pi}{6}$	$\dfrac{\pi}{6}$	30°	$\dfrac{1}{2}$	$\dfrac{\sqrt{3}}{2}$	$\dfrac{\sqrt{3}}{3}$	$\sqrt{3}$	$\dfrac{2\sqrt{3}}{3}$	2
$\dfrac{\pi}{4}$	$\dfrac{\pi}{4}$	45°	$\dfrac{\sqrt{2}}{2}$	$\dfrac{\sqrt{2}}{2}$	1	1	$\sqrt{2}$	$\sqrt{2}$
$\dfrac{\pi}{3}$	$\dfrac{\pi}{3}$	60°	$\dfrac{\sqrt{3}}{2}$	$\dfrac{1}{2}$	$\sqrt{3}$	$\dfrac{\sqrt{3}}{3}$	2	$\dfrac{2\sqrt{3}}{3}$
$\dfrac{\pi}{2}$	$\dfrac{\pi}{2}$	90°	1	0	undefined	0	undefined	1
π	π	180°	0	-1	0	undefined	-1	undefined
$\dfrac{3\pi}{2}$	$\dfrac{3\pi}{2}$	270°	-1	0	undefined	0	undefined	-1
2π	2π	360°	0	1	0	undefined	1	undefined

Trigonometric Function Values of an Angle θ

5 Find trigonometric function values of an angle in standard position.

Suppose the terminal ray of an angle θ in standard position intersects the unit circle at the point $P_1(x_1, y_1)$, as illustrated for an acute angle θ in Figure 4.26(a). Let $P(x, y)$ be any point on the terminal side of θ and $r = \sqrt{x^2 + y^2}$, as shown in Figure 4.26(b). The triangles POQ and P_1OQ_1 are similar because they have equal angles; so the ratios of their corresponding sides are equal. We have $\sin \theta = y_1 = \dfrac{y_1}{1} = \dfrac{y}{r}$, $\cos \theta = x_1 = \dfrac{x_1}{1} = \dfrac{x}{r}$, $\tan \theta = \dfrac{y_1}{x_1} = \dfrac{y}{x}$, and so on. We can now evaluate the trigonometric functions for any angle θ using any point on its terminal side.

(a) $x_1 = \cos \theta,\; y_1 = \sin \theta$

(b) $\dfrac{x_1}{1} = \dfrac{x}{r},\; \dfrac{y_1}{1} = \dfrac{y}{r}$

FIGURE 4.26

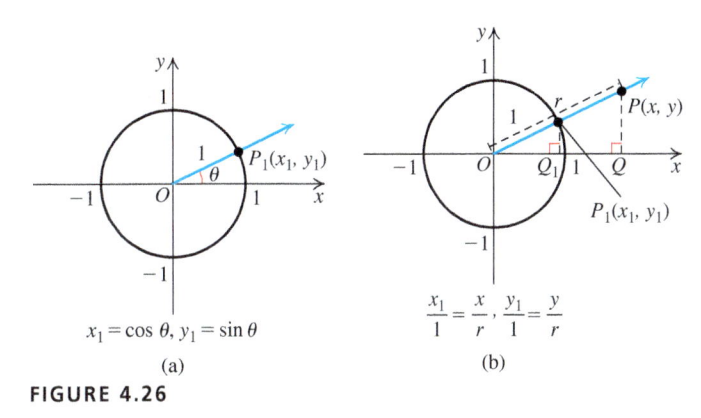

Figure 4.27 shows the connection between the degree measure and radian measure of common angles and the coordinates of the terminal points of arcs on the unit circle.

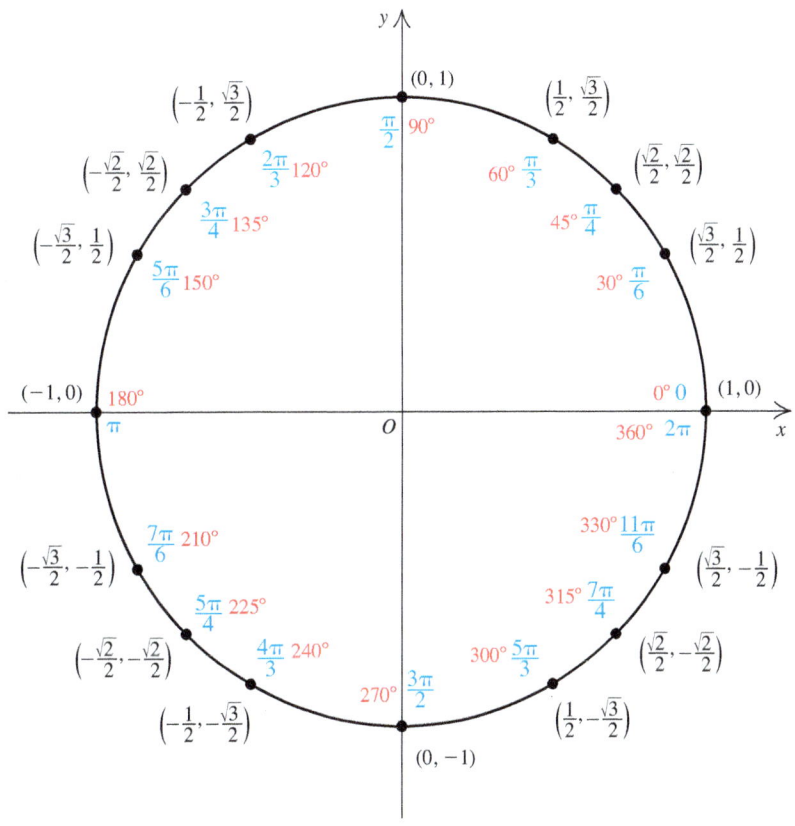

FIGURE 4.27 Degrees, radians, ordered pairs

> **VALUES OF THE TRIGONOMETRIC FUNCTIONS OF AN ANGLE θ**
>
> Let $P(x, y)$ be any point on the terminal ray of an angle θ in standard position (other than the origin) and let $r = \sqrt{x^2 + y^2}$. Then $r > 0$ and
>
> $$\sin \theta = \frac{y}{r} \qquad \csc \theta = \frac{r}{y} \, (y \neq 0)$$
>
> $$\cos \theta = \frac{x}{r} \qquad \sec \theta = \frac{r}{x} \, (x \neq 0)$$
>
> $$\tan \theta = \frac{y}{x} \, (x \neq 0) \qquad \cot \theta = \frac{x}{y} \, (y \neq 0)$$

EXAMPLE 7 Finding Trigonometric Function Values

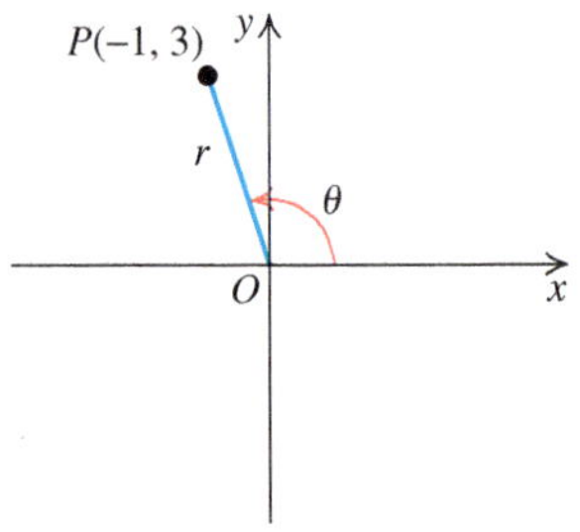

Suppose θ is an angle whose terminal side contains the point $P(-1, 3)$. Find the exact values of the six trigonometric functions of θ.

Solution

Because $x = -1$ and $y = 3$ (see Figure 4.28), we have

$$r = \sqrt{x^2 + y^2} \qquad \text{Definition of } r$$
$$= \sqrt{(-1)^2 + 3^2} = \sqrt{10} \qquad \text{Substitute and simplify.}$$

Replacing x with -1, y with 3, and r with $\sqrt{10}$ in the definition of the trigonometric functions, we have

$$\sin \theta = \frac{y}{r} = \frac{3}{\sqrt{10}} = \frac{3\sqrt{10}}{10} \qquad \csc \theta = \frac{r}{y} = \frac{\sqrt{10}}{3}$$

$$\cos \theta = \frac{x}{r} = \frac{-1}{\sqrt{10}} = -\frac{\sqrt{10}}{10} \qquad \sec \theta = \frac{r}{x} = \frac{\sqrt{10}}{-1} = -\sqrt{10}$$

$$\tan \theta = \frac{y}{x} = \frac{3}{-1} = -3 \qquad \cot \theta = \frac{x}{y} = \frac{-1}{3} = -\frac{1}{3}$$

Practice Problem 7 Suppose θ is an angle whose terminal side contains the point $P(2, -5)$. Find the exact values of the six trigonometric functions of θ.

FIGURE 4.28

6 Approximate trigonometric function values using a calculator.

Evaluating Trigonometric Functions Using a Calculator

Suppose we want to find the trigonometric function values of $20° = \frac{\pi}{9}$ radians. Unfortunately, the coordinates $(x, y) = \left(\cos \frac{\pi}{9}, \sin \frac{\pi}{9} \right)$ of the terminal point $P\left(\frac{\pi}{9} \right)$ on the unit circle do not have a nice form. Neither coordinate is a rational number or a radical of a rational number. In such cases, we use a calculator to find their approximate values.

When you are using a calculator to find the values of the trigonometric functions, the first step is to set the *mode* of measurement. If you are working with an angle given in degrees, set the mode to *degrees*; otherwise, set the mode to *radians*.

Your calculator has keys labeled SIN, COS, and TAN but no keys for directly evaluating the cosecant, secant, or cotangent functions. However, these functions are the

TECHNOLOGY CONNECTION

The first screen shows how sin 71° is displayed on a graphing calculator in *degree* mode.

```
sin(71)
        .9455185756
```

This screen shows how $\tan\frac{5\pi}{7}$ and sec 1.3 are displayed on a graphing calculator in *radian* mode.

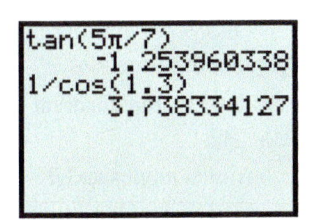

```
tan(5π/7)
       -1.253960338
1/cos(1.3)
        3.738334127
```

reciprocal functions of sine, cosine, and tangent functions, respectively. Therefore, to evaluate these functions, use the $\boxed{x^{-1}}$ key with the appropriate function.

EXAMPLE 8 Approximating Trigonometric Function Values Using a Calculator

Use a calculator to find the approximate value of each expression. Round your answers to two decimal places.

a. sin 71°

b. $\tan\frac{5\pi}{7}$

c. sec 1.3

Solution

a. Set the MODE to degrees. $\sin 71° \approx 0.9455185756 \approx 0.95$

b. Set the MODE to radians. $\tan\frac{5\pi}{7} \approx -1.253960338 \approx -1.25$

c. Set the MODE to radians. $\sec 1.3 = \frac{1}{\cos 1.3} \approx 3.738334127 \approx 3.74$

Practice Problem 8 Repeat Example 8 for each expression.

a. cos 114°   b. cot 3.6

⚠ **WARNING** The calculator keys $\boxed{\text{SIN}^{-1}}$, $\boxed{\text{COS}^{-1}}$, and $\boxed{\text{TAN}^{-1}}$ do not represent the reciprocal functions for the sine, cosine, and tangent functions, respectively.

Signs of the Trigonometric Functions

7 Determine the signs of the trigonometric functions in each quadrant.

II	I
$(-, +)$	$(+, +)$
$\sin\theta > 0$,	All positive
$\csc\theta > 0$	
Others < 0	
$(-, -)$	$(+, -)$
$\tan\theta > 0$,	$\cos\theta > 0$,
$\cot\theta > 0$	$\sec\theta > 0$
Others < 0	Others < 0
III	IV

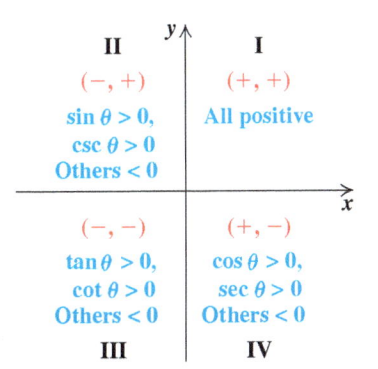

FIGURE 4.29 Signs of the trigonometric functions

Suppose an angle θ is not quadrantal and its terminal side contains the point (x, y). (Notice that $r = \sqrt{x^2 + y^2}$ is always *positive*.) The signs of x and y determine the signs of the trigonometric functions.

If θ lies in quadrant I, then both x and y are positive; so all six trigonometric function values are positive. However, if θ lies in quadrant II, then x is negative and y is positive; so only $\sin\theta = \frac{y}{r}$ and $\csc\theta = \frac{r}{y}$ are positive. If θ lies in quadrant III, then x and y are both negative; so only $\tan\theta = \frac{y}{x}$ and $\cot\theta = \frac{x}{y}$ are positive. If θ lies in quadrant IV, then x is positive and y is negative; so only $\cos\theta = \frac{x}{r}$ and $\sec\theta = \frac{r}{x}$ are positive. Figure 4.29 summarizes the signs of the trigonometric functions.

EXAMPLE 9 Evaluating Trigonometric Functions

Given that $\tan\theta = \frac{3}{2}$ and $\cos\theta < 0$, find the exact values of $\sin\theta$ and $\sec\theta$.

Solution

Because $\tan\theta > 0$ and $\cos\theta < 0$, θ lies in quadrant III. We want to identify a point (x, y) in quadrant III that is on the terminal side of θ. We have

$$\tan\theta = \frac{y}{x} = \frac{3}{2},$$

and because the point (x, y) is in quadrant III, both x and y must be negative. (See Figure 4.30.) If we choose $x = -2$ and $y = -3$, then

$$\tan \theta = \frac{y}{x} = \frac{-3}{-2} = \frac{3}{2}$$

and $r = \sqrt{x^2 + y^2} = \sqrt{(-2)^2 + (-3)^2} = \sqrt{4 + 9} = \sqrt{13}$.

Using $x = -2$, $y = -3$, and $r = \sqrt{13}$, we can find $\sin \theta$ and $\sec \theta$.

$$\sin \theta = \frac{y}{r} = \frac{-3}{\sqrt{13}} = -\frac{3\sqrt{13}}{13} \qquad \sec \theta = \frac{r}{x} = \frac{\sqrt{13}}{-2} = -\frac{\sqrt{13}}{2}$$

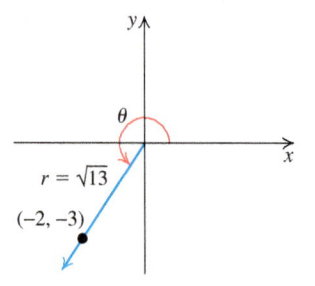

FIGURE 4.30

SIDE NOTE

The phrase *All Students Take Calculus* can be useful in remembering which trigonometric functions are positive in quadrants I–IV. Use the first letter of each word in the phrase

A ll functions (Q I)
S ine (Q II)
T angent (Q III)
C osine (Q IV)

and recall that a and $\frac{1}{a}$ have the same sign.

Practice Problem 9 Given that $\tan \theta = -\frac{4}{5}$ and $\cos \theta > 0$, find the exact values of $\sin \theta$ and $\sec \theta$.

Because the value of each trigonometric function of an angle in standard position is completely determined by the position of the terminal side, the following statements are true.

1. Coterminal angles are assigned identical values by the six trigonometric functions.
2. The signs of the values of the trigonometric functions are determined by the quadrant containing the terminal side.
3. For any integer n, θ and $\theta + n360°$ (in degree measure) are coterminal angles and θ and $\theta + 2\pi n$ (in radian measure) are coterminal angles.

For example, if α is any angle (in degrees) with $\alpha \geq 360°$ or $\alpha < 0°$, we can obtain its coterminal angle θ with $0° \leq \theta < 360°$ as follows:

(i) If $\alpha \geq 360°$, *subtract* appropriate multiples of 360° to obtain θ.

(ii) If $\alpha < 0°$, *add* appropriate multiples of 360° to obtain θ.

For α in radians, replace 360° with 2π.

Some frequently used consequences of these observations are given next.

TRIGONOMETRIC FUNCTION VALUES OF COTERMINAL ANGLES

θ in degrees

$\sin \theta = \sin (\theta + n360°)$

$\cos \theta = \cos (\theta + n360°)$

θ in radians

$\sin \theta = \sin (\theta + 2\pi n)$

$\cos \theta = \cos (\theta + 2\pi n)$

These equations hold for any integer n.

---

8 Find a reference angle.

Reference Angle

For any angle θ, there is a corresponding acute angle called its *reference angle*, whose trigonometric function values are either equal to or opposite in sign to that of θ.

Reference Angle

Let θ be an angle in standard position that is not a quadrantal angle. The **reference angle** for θ is the acute angle θ' ("theta prime") formed by the terminal side of θ and the x-axis.

If θ is measured in radians and $0 < \theta < 2\pi$, we have the following values of θ' that correspond to θ in each quadrant.

Angle θ	Reference angle θ'	Example
θ in Q I	$\theta' = \theta$	If $\theta = \dfrac{\pi}{3}$, then $\theta' = \dfrac{\pi}{3}$.
θ in Q II	$\theta' = \pi - \theta$	If $\theta = \dfrac{3\pi}{5}$, then $\theta' = \pi - \dfrac{3\pi}{5} = \dfrac{2\pi}{5}$
θ in Q III	$\theta' = \theta - \pi$	If $\theta = \dfrac{5\pi}{4}$, then $\theta' = \dfrac{5\pi}{4} - \pi = \dfrac{\pi}{4}$
θ in Q IV	$\theta' = 2\pi - \theta$	If $\theta = 5.75$, then $\theta' = 2\pi - 5.75 \approx 6.28 - 5.75 = 0.53$

If θ is measured in degrees and $0° < \theta < 360°$, replace π with $180°$ in these calculations.

We next give a three-step procedure for using reference angles to find trigonometric function values.

PROCEDURE IN ACTION

EXAMPLE 10 Finding Trigonometric Function Values Using the Reference Angles

OBJECTIVE
Find the value of any trigonometric function of an angle θ.

EXAMPLE
Find $\sin 1320°$.

Step 1 If $\theta \geq 360°$ or $\theta < 0°$, find a coterminal angle between $0°$ and $360°$. Otherwise go to Step 2.

1. Because $1320° - 3(360°) = 240°$, $240°$ is coterminal with $1320°$.

Step 2 Find the reference angle θ' for the angle resulting from Step 1. Write the trigonometric function of θ'.

2. Because $240°$ is in quadrant III, its reference angle θ' is
$$\theta' = 240° - 180° = 60°$$
and
$$\sin \theta' = \sin 60° = \dfrac{\sqrt{3}}{2} \quad \text{See page 305.}$$

Step 3 Choose the correct sign for θ based on the quadrant in which it lies.

3. Angle θ and its coterminal angle, $240°$, lie in quadrant III, where the sine is *negative*. So,
$$\sin 1320° = \sin 240° = -\sin 60° = -\dfrac{\sqrt{3}}{2}$$

Practice Problem 10 Find the exact value of $\cos 1035°$.

SECTION 4.2 Exercises

Basic Concepts and Skills

In Exercises 1–6, find all numbers u (if any) so that the given point (u, y) or (x, u) is on the unit circle.

1. $\left(u, \dfrac{1}{2}\right)$
2. $\left(u, -\dfrac{1}{3}\right)$
3. $\left(-\dfrac{3}{4}, u\right)$
4. $\left(-\dfrac{2}{5}, u\right)$
5. $\left(u, \dfrac{3}{2}\right)$
6. $\left(-\dfrac{4}{3}, u\right)$

In Exercises 7–12, $P(t) = (x, y)$ is the terminal point on the unit circle that corresponds to the real number t. Find the values of the six trigonometric functions of t.

7. $\left(\dfrac{2\sqrt{2}}{3}, \dfrac{1}{3}\right)$
8. $\left(\dfrac{1}{2}, \dfrac{\sqrt{3}}{2}\right)$
9. $\left(-\dfrac{1}{3}, \dfrac{2\sqrt{2}}{3}\right)$
10. $\left(-\dfrac{\sqrt{3}}{2}, \dfrac{1}{2}\right)$
11. $\left(\dfrac{1}{2}, -\dfrac{\sqrt{3}}{2}\right)$
12. $\left(-\dfrac{2\sqrt{2}}{3}, -\dfrac{1}{3}\right)$

In Exercises 13–18, find each trigonometric function value.

13. $\tan(4\pi)$
14. $\sec(7\pi)$
15. $\cos(-5\pi)$
16. $\sin(-2\pi)$
17. $\cos\left(-\dfrac{3\pi}{2}\right)$
18. $\sin\left(-\dfrac{\pi}{2}\right)$

In Exercises 19–26, use Table 4.1 and symmetry to find $\sin t$, $\cos t$, and $\tan t$ for the given value of t.

19. $t = -\dfrac{\pi}{6}$
20. $t = -\dfrac{\pi}{4}$
21. $t = \dfrac{4\pi}{3}$
22. $t = \dfrac{7\pi}{6}$
23. $t = \dfrac{11\pi}{6}$
24. $t = \dfrac{5\pi}{3}$
25. $t = -\dfrac{7\pi}{6}$
26. $t = -\dfrac{7\pi}{4}$

In Exercises 27–32, find $\sin t$, $\cos t$, and $\tan t$ for the given value of t.

27. $t = \dfrac{13\pi}{6}$
28. $t = \dfrac{9\pi}{4}$
29. $t = -\dfrac{23\pi}{6}$
30. $t = -\dfrac{15\pi}{4}$
31. $t = \dfrac{29\pi}{3}$
32. $t = \dfrac{31\pi}{3}$

In Exercises 33–38, find the exact value of each expression. Do not use a calculator.

33. $\sin 180° - \cos 90°$
34. $\cos 180° - \sin 90°$
35. $\sin\dfrac{\pi}{4} - \cos\pi$
36. $\sec\pi + \sin\dfrac{\pi}{6}$
37. $\sin\dfrac{3\pi}{2}\tan\dfrac{\pi}{4}$
38. $\cos\dfrac{\pi}{2}\sec\pi$

In Exercises 39–46, a point on the terminal side of an angle θ is given. Find the exact values of the six trigonometric functions of θ.

39. $(-4, 3)$
40. $(-3, 5)$
41. $(-\sqrt{3}, -1)$
42. $(-1, -2)$
43. $(3, 3)$
44. $(-2, -2)$
45. $(12, -5)$
46. $(7, -2)$

In Exercises 47–56, use a calculator to find the approximate value of each expression. Round your answers to two decimal places.

47. $\cos 14°$
48. $\sin 20°$
49. $\sec 34°$
50. $\csc 72°$
51. $\tan\dfrac{2\pi}{9}$
52. $\cot\dfrac{3\pi}{7}$
53. $\sin(-41°)$
54. $\cos(-54°)$
55. $\sin(-17.3)$
56. $\cos(-38.6)$

In Exercises 57–62, find the exact values of the remaining trigonometric functions of θ from the given information.

57. $\cos\theta = -\dfrac{5}{13}$, θ in quadrant III
58. $\tan\theta = -\dfrac{3}{4}$, θ in quadrant IV
59. $\cot\theta = -\dfrac{3}{4}$, θ in quadrant II
60. $\sec\theta = \dfrac{4}{\sqrt{7}}$, θ in quadrant IV
61. $\sin\theta = \dfrac{3}{5}$, $\tan\theta < 0$
62. $\sec\theta = 3$, $\sin\theta < 0$

In Exercises 63–70, find the reference angle for each angle.

63. $120°$
64. $275°$
65. $500°$
66. $420°$
67. $\dfrac{19\pi}{4}$
68. $\dfrac{28\pi}{6}$
69. $\dfrac{31\pi}{6}$
70. $\dfrac{5\pi}{6}$

71. Find the terminal point $P(t)$ associated with $t = \dfrac{\pi}{6}$.

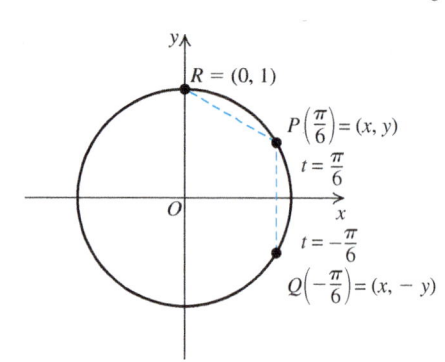

[Hint: Use the figure and the fact from geometry that equal arcs of a circle subtend equal chords. Use the distance formula for $d(P, Q) = d(P, R)$ to show that $4y^2 = x^2 + y^2 - 2y + 1$. Use this equation along with $x^2 + y^2 = 1$ to find x and y.]

72. Find the terminal point $P(t)$ associated with $t = \dfrac{\pi}{3}$.

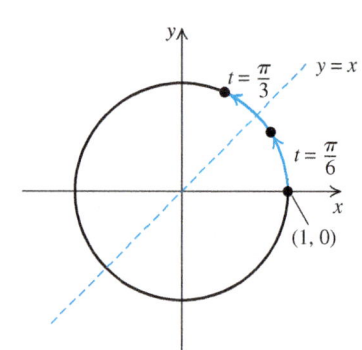

[Hint: Use $P\left(\dfrac{\pi}{6}\right)$ from Exercise 71 and symmetry about the line $y = x$ to find x and y. See the figure.]

Applying the Concepts

73. Blood pressure. Maurice's blood pressure (in millimeters of mercury) while resting is given by the function $R(t) = 25 \sin(2\pi t) + 100$, where t is time in seconds. Find his blood pressure after
  a. 1 second.
  b. 1.75 seconds.

74. Deer population. The number of deer in a region is modeled by the equation

$$N(t) = 450 \sin\left(\frac{\pi}{6}t\right) + 1550,$$

where t is measured in years and $t = 0$ represents 2010. Find the deer population in
  a. 2012
  b. 2018
  c. 2021

75. Tide pattern. The depth of water, d feet, in a channel t hours after midnight is given by

$$d = 3 \cos\left(\frac{\pi}{6}t\right) + 10.$$

Find the channel depth at
  a. noon.
  b. 6 P.M.

76. Pollution levels. On a typical day, a particular city's air pollution level (suspended particle levels in mg/m³) is modeled by the equation

$$P(t) = 0.5 + 15t - 0.4t^2 + 10 \sin\left(\frac{\pi}{12}t\right),$$

where t is measured in hours and $t = 0$ represents 6 A.M. Find the pollution level at
  a. noon.
  b. 6 P.M.
  c. midnight.

Critical Thinking / Discussion / Writing

77. In the adjoining figure, by definition, $P = (\cos \theta, \sin \theta)$ and
$$Q = \left(\cos\left(\theta + \frac{\pi}{2}\right), \sin\left(\theta + \frac{\pi}{2}\right)\right).$$

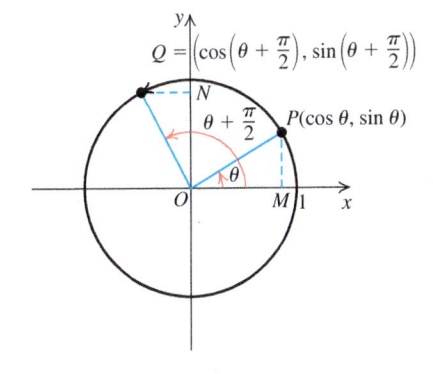

  a. Show that triangles OMP and ONQ are congruent.
  b. Use the result from part (a) to show that $Q = (-\sin \theta, \cos \theta)$.
  c. Use initial definitions and the results from part (b) to show that $\cos\left(\theta + \frac{\pi}{2}\right) = -\sin \theta$ and $\sin\left(\theta + \frac{\pi}{2}\right) = \cos \theta$.
  d. Use quotient and reciprocal identities to find
  $\tan\left(\theta + \frac{\pi}{2}\right)$, $\cot\left(\theta + \frac{\pi}{2}\right)$, $\sec\left(\theta + \frac{\pi}{2}\right)$, and $\csc\left(\theta + \frac{\pi}{2}\right)$ in terms of the trigonometric functions of θ.

78. Use the adjoining figure to show that $\cos(\pi - \theta) = -\cos \theta$, $\sin(\pi - \theta) = \sin \theta$, and $\tan(\pi - \theta) = -\tan \theta$.

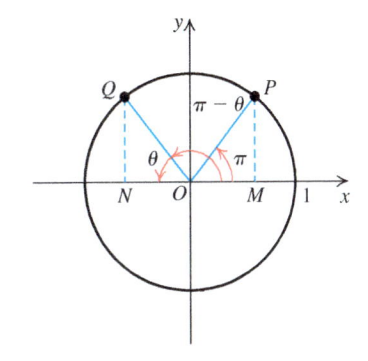

79. Make a figure for angles θ and $\pi + \theta$. Then show that $\cos(\pi + \theta) = -\cos \theta$, $\sin(\pi + \theta) = -\sin \theta$, and $\tan(\pi + \theta) = \tan \theta$.

80. Make a figure for angles θ and $-\theta$. Then show that $\cos(-\theta) = \cos \theta$, $\sin(-\theta) = -\sin \theta$, and $\tan(-\theta) = -\tan \theta$.

81. Find a formula for the reference angle θ', of a negative angle θ, with $-180° < \theta < -90°$.

82. Find a positive number x for which
  a. $\cos(\ln x) = 0$.
  b. $\sin(e^x) = 0$.

Maintaining Skills

In Exercises 83–88, determine whether each function is even, odd, or neither.

83. $f(x) = 2x^3 - 4x$
84. $g(x) = \sqrt{1 - x^2}$
85. $h(x) = x^3 - x^2$
86. $p(x) = \sqrt[3]{x^5}$
87. $g(x) = 3x^4 + 2x^2 - 17$
88. $f(x) = 5x + 3$

SECTION 4.3

Graphs of the Sine and Cosine Functions

BEFORE STARTING THIS SECTION, REVIEW

1. Equation of a circle (Section 1.1, page 60)
2. Transformations of functions (Section 1.5, page 109)
3. Unit circle definitions of the sine and cosine (Section 4.2, page 295)

OBJECTIVES

1. Discuss properties of the sine and cosine functions.
2. Graph the sine and cosine functions.
3. Find the amplitude and period of sinusoidal curves.
4. Find the phase shifts and graph sinusoidal functions of the forms $y = a \sin[b(x - c)]$ and $y = a \cos[b(x - c)]$.

◆ Length of Days

During winter, because the Northern Hemisphere tilts away from the sun, it receives less direct solar radiation than does the Southern Hemisphere, which tilts toward the sun. In the Northern Hemisphere, the summer solstice (the longest day of the year) occurs on or around June 21 and the winter solstice (the shortest day of the year) occurs on or about December 21. In the Southern Hemisphere, the seasons are reversed. The number of hours of daylight varies at different latitudes. The length of a day is one of the many phenomena that can be described by the trigonometric functions we study in this section. In Example 9, we see how to model the number of daylight hours that the city of Paris enjoys throughout the year.

1 Discuss properties of the sine and cosine functions.

Properties of Sine and Cosine

We first discuss some properties of the sine and cosine functions that will be useful in sketching the graphs of these functions.

Domain and Range of Sine and Cosine

Each real number t determines a terminal point $P(t) = (x, y)$ on the unit circle. Because $x = \cos t$ and $y = \sin t$, both functions are defined for each real number t. So the domain of both the cosine and sine functions is the set of real numbers: $(-\infty, \infty)$ in interval notation.

For each real number t, the point $P(t) = (\cos t, \sin t)$ lies on the unit circle $x^2 + y^2 = 1$. Indeed substituting $x = \cos t$ and $y = \sin t$ in the equation, we have

$$(\cos t)^2 + (\sin t)^2 = x^2 + y^2 = 1 \text{ for every real number } t.$$

Because $(\cos t)^2$ and $(\sin t)^2$ are both nonnegative, we have

$$(\cos t)^2 \leq 1 \text{ and } (\sin t)^2 \leq 1.$$

310 Chapter 4 Trigonometric Functions

These equations state that the values of both cos t and sin t lie between -1 and 1. That is,

$$|\cos t| \leq 1 \text{ and } |\sin t| \leq 1, \text{ or}$$
$$-1 \leq \cos t \leq 1 \text{ and } -1 \leq \sin t \leq 1 \text{ for every number } t.$$

Furthermore, let c be a real number in the interval $[-1, 1]$. Then c is the x-coordinate of some terminal point $P(t) = (\cos t, \sin t)$ on the unit circle. See Figure 4.31. So $c = \cos t$. This shows that the interval $[-1, 1]$ is the range of the cosine function. A similar argument shows that the interval $[-1, 1]$ is also the range of the sine function.

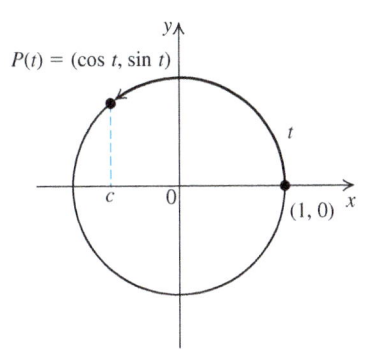

FIGURE 4.31

> **DOMAIN AND RANGE OF SINE AND COSINE**
>
> The *domain* of both the sine and cosine functions is $(-\infty, \infty)$.
> The *range* of both the sine and cosine functions is $[-1, 1]$.

Zeros of Sine and Cosine

We know that $\cos t = 0$ precisely when the terminal point $P(t) = (\cos t, \sin t)$ lies on the y-axis. This happens when $t = \dfrac{\pi}{2}, -\dfrac{\pi}{2}, \dfrac{3\pi}{2}, -\dfrac{3\pi}{2}, \ldots$, in other words, when t is an odd integer multiple of $\dfrac{\pi}{2}$. That is, the zeros of $\cos t$ are given by

$$t = (2n + 1)\frac{\pi}{2} = \frac{\pi}{2} + n\pi \text{ for any integer } n.$$

Similarly, $\sin t = 0$ only if the terminal point $P(t) = (\cos t, \sin t)$ lies on the x-axis. This occurs when $t = 0, \pi, -\pi, 2\pi, -2\pi, \ldots$, that is, when t is an integer multiple of π. So the zeros of $\sin t$ are given by

$$t = n\pi \text{ for any integer } n.$$

Even–Odd Properties

Let $P(x, y)$ be the terminal point on the unit circle associated with a real number t. Then from Figure 4.32, we see that the point $Q(x, -y)$ is the terminal point associated with the number $-t$. Then

$$\cos(-t) = x = \cos t \quad \text{and}$$
$$\sin(-t) = -y = -\sin t.$$

Because $\sin(-t) = -\sin t$, the sine function is an *odd* function; so its graph is symmetric about the origin. Similarly, because $\cos(-t) = \cos t$, the cosine function is an *even* function; so its graph is symmetric about the y-axis. See Exercise 80 in Section 4.2.

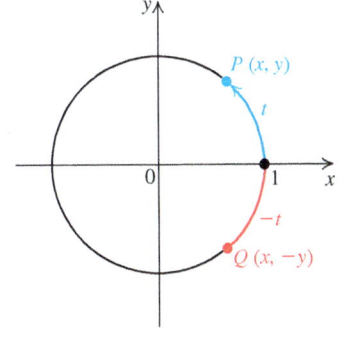

FIGURE 4.32

Periodic Functions

Periodic Function

> A function f is said to be **periodic** if there is a positive number p such that
>
> $$f(x + p) = f(x)$$
>
> for every x in the domain of f.
> The smallest value of p (if there is one) for which $f(x + p) = f(x)$ is called the **period** of f. The graph of f over any interval of length p is called one **cycle** of the graph.

We note that the values of a periodic function f of period p repeat themselves every p units. This means that we may draw only one cycle of the graph of f because the rest of the graph is generated by repeated copies of it. Figure 4.33 shows examples of periodic functions.

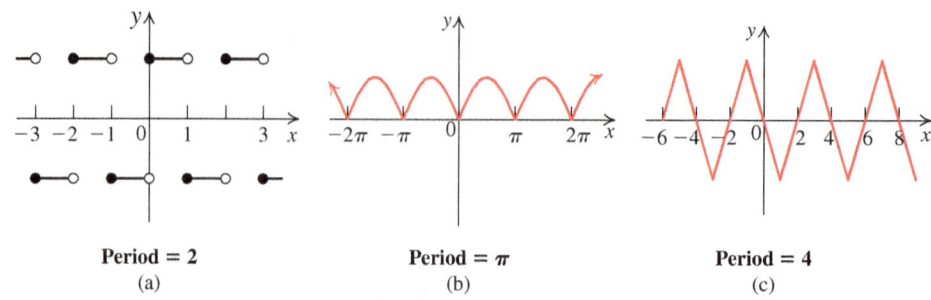

Period = 2
(a)

Period = π
(b)

Period = 4
(c)

FIGURE 4.33 Periodic functions

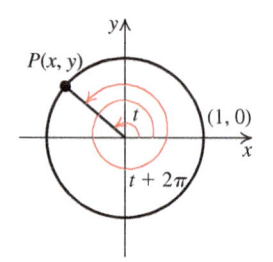

FIGURE 4.34

Let $P(x, y)$ be a point on the unit circle that corresponds to angle t, measured in radians. If we add 2π to t, we obtain the same point P on the unit circle. See Figure 4.34. Then $\sin(t + 2\pi) = \sin t = y$ and $\cos(t + 2\pi) = \cos t = x$. Therefore, $\sin t$ and $\cos t$ are periodic functions. The period for each of the sine and cosine functions is 2π. See Exercise 55.

PERIOD OF THE SINE AND COSINE FUNCTIONS

For every real number t,

$$\sin(t + 2\pi) = \sin t \quad \text{and} \quad \cos(t + 2\pi) = \cos t.$$

The sine and cosine functions are periodic with **period 2π**.

We summarize the properties of the sine and cosine functions.

PROPERTIES OF SINE AND COSINE

$y = \sin t$

1. Domain: $(-\infty, \infty)$
2. Range: $[-1, 1]$
3. Zeros at $t = n\pi$, n any integer
4. Odd: $\sin(-t) = -\sin t$
5. Period: 2π

$y = \cos t$

1. Domain: $(-\infty, \infty)$
2. Range: $[-1, 1]$
3. Zeros at $t = \dfrac{\pi}{2} + n\pi$, n any integer
4. Even: $\cos(-t) = \cos t$
5. Period: 2π

---

2 Graph the sine and cosine functions.

Graphs of Sine and Cosine Functions

To graph the sine and cosine functions on an xy-coordinate plane, we use x as the independent variable and y as the dependent variable. So we write $y = \sin x$ and $y = \cos x$ rather than $y = \sin t$ and $y = \cos t$, respectively.

In calculus, it can be shown that the trigonometric functions are continuous on their domains. That is, they are continuous wherever they are defined.

Graph of the Sine Function

To sketch the graph of one cycle of $y = \sin x$ on the interval $[0, 2\pi]$, we use Table 4.3. This table consists of values of $y = \sin x$ for common values of x. In other words, x is a real number that is the radian measure of a common angle. Because the common angles x (in radians) are expressed in multiples of π, we mark off the x-axis in multiples of π. However, the range of y is between -1 (minimum value) and 1 (maximum value). We mark off the y-axis in integers.

TABLE 4.3

x	0	$\dfrac{\pi}{6}$	$\dfrac{\pi}{4}$	$\dfrac{\pi}{3}$	$\dfrac{\pi}{2}$	$\dfrac{2\pi}{3}$	$\dfrac{3\pi}{4}$	$\dfrac{5\pi}{6}$	π	$\dfrac{7\pi}{6}$	$\dfrac{5\pi}{4}$	$\dfrac{4\pi}{3}$	$\dfrac{3\pi}{2}$	$\dfrac{5\pi}{3}$	$\dfrac{7\pi}{4}$	$\dfrac{11\pi}{6}$	2π
$\sin x$	0	$\dfrac{1}{2}$	$\dfrac{1}{\sqrt{2}}$	$\dfrac{\sqrt{3}}{2}$	1	$\dfrac{\sqrt{3}}{2}$	$\dfrac{1}{\sqrt{2}}$	$\dfrac{1}{2}$	0	$-\dfrac{1}{2}$	$-\dfrac{1}{\sqrt{2}}$	$-\dfrac{\sqrt{3}}{2}$	-1	$-\dfrac{\sqrt{3}}{2}$	$-\dfrac{1}{\sqrt{2}}$	$-\dfrac{1}{2}$	0

Connecting the points $(x, \sin x)$ with a smooth curve gives us the graph in Figure 4.35. We have shown the points $(x, \sin x)$, where x is a multiple of $\dfrac{\pi}{4}$ or $\dfrac{\pi}{6}$.

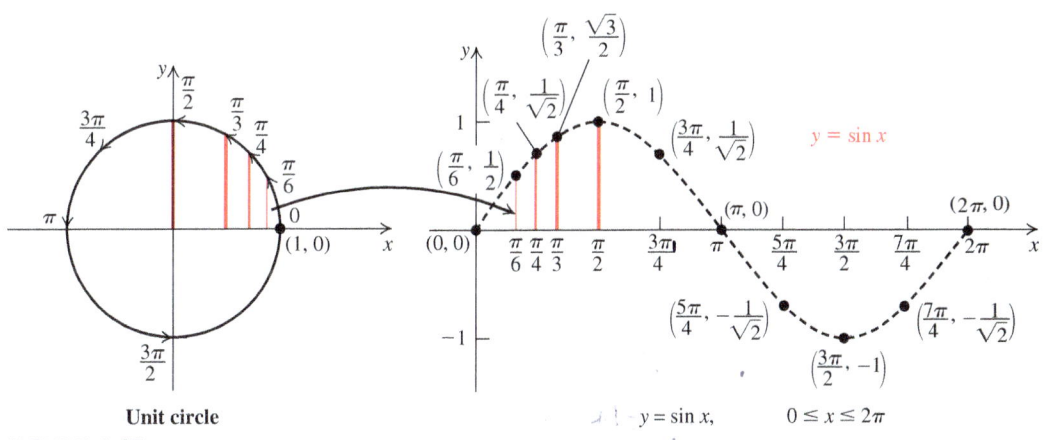

FIGURE 4.35

Because the sine function is periodic with period 2π, we can sketch the complete graph of $y = \sin x$ by extending the graph in Figure 4.35 indefinitely to the left and to the right in every successive interval of length 2π. See Figure 4.36.

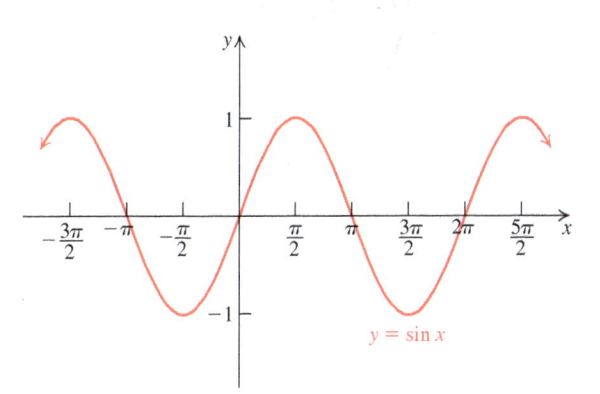

FIGURE 4.36 The sine curve

Graph of the Cosine Function

We can use the same method for graphing $y = \cos x$ that we used for $y = \sin x$.

Table 4.4 gives the values for $y = \cos x$, where x has the same values as in Table 4.3.

TABLE 4.4

x	0	$\dfrac{\pi}{6}$	$\dfrac{\pi}{4}$	$\dfrac{\pi}{3}$	$\dfrac{\pi}{2}$	$\dfrac{2\pi}{3}$	$\dfrac{3\pi}{4}$	$\dfrac{5\pi}{6}$	π	$\dfrac{7\pi}{6}$	$\dfrac{5\pi}{4}$	$\dfrac{4\pi}{3}$	$\dfrac{3\pi}{2}$	$\dfrac{5\pi}{3}$	$\dfrac{7\pi}{4}$	$\dfrac{11\pi}{6}$	2π
$\cos x$	1	$\dfrac{\sqrt{3}}{2}$	$\dfrac{1}{\sqrt{2}}$	$\dfrac{1}{2}$	0	$-\dfrac{1}{2}$	$-\dfrac{1}{\sqrt{2}}$	$-\dfrac{\sqrt{3}}{2}$	-1	$-\dfrac{\sqrt{3}}{2}$	$-\dfrac{1}{\sqrt{2}}$	$-\dfrac{1}{2}$	0	$\dfrac{1}{2}$	$\dfrac{1}{\sqrt{2}}$	$\dfrac{\sqrt{3}}{2}$	1

Connecting the points $(x, \cos x)$ with a smooth curve gives the graph of $y = \cos x$, $0 \le x \le 2\pi$, shown in Figure 4.37.

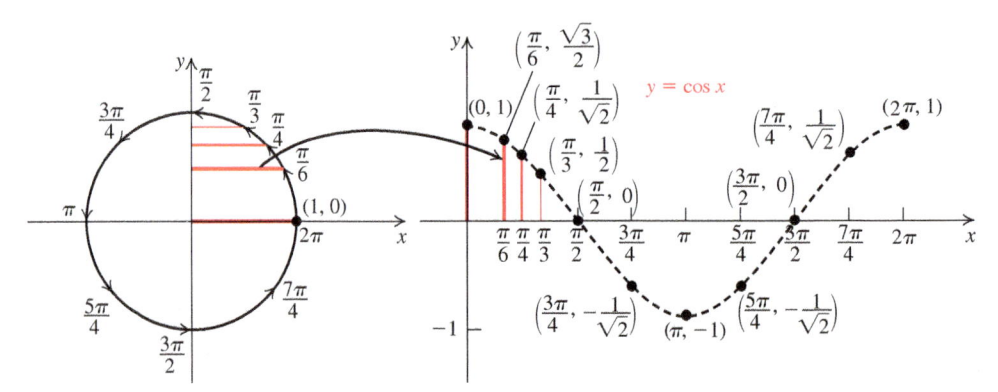

FIGURE 4.37 $y = \cos x, 0 \le x \le 2\pi$

As with the sine function, the values of x in $y = \cos x$ are not restricted, because the domain of the cosine function is $(-\infty, \infty)$. We can draw the complete graph of $y = \cos x$ by extending the graph in Figure 4.37 indefinitely to the left and right in every successive interval of length 2π. See Figure 4.38.

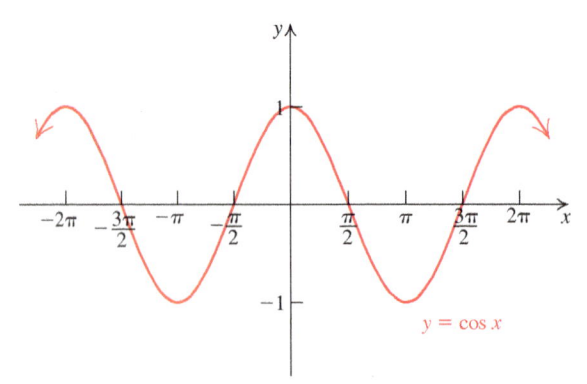

FIGURE 4.38 The cosine curve

Five Key Points

There are five *key points* in every cycle of the graphs of the sine and cosine functions. The key points consist of the x-intercepts, high points, and low points. See Figure 4.39. The key points are very helpful for graphing the sine and cosine functions and their transformations. Note that the x-coordinates of the five key points divide the period into four congruent intervals.

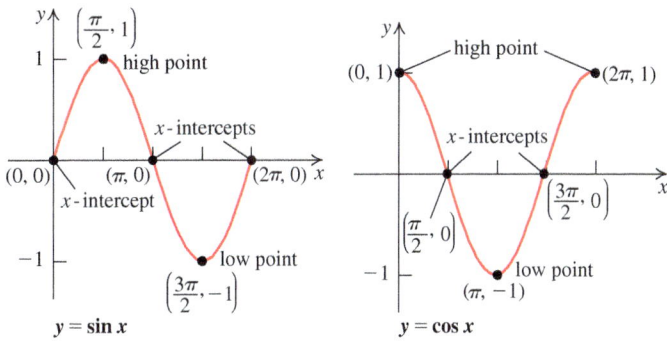

FIGURE 4.39 Key points of the sine and cosine graphs

3 Find the amplitude and period of sinusoidal curves.

Amplitude and Period

The graphs of the sine and cosine functions and their transformations are called **sinusoidal graphs** or **sinusoidal curves**. In this section, we study sinusoidal curves of the form $y = af[b(x - c)] + d$, where f is a sine or cosine function. We first consider transformations of the form $y = af(bx)$.

We start with the effect of the multiplier a in $y = af(bx)$. If the maximum (or minimum) value of $f(x)$ is M and $a > 0$, then the maximum (or minimum) value of $af(x)$ is aM. For example, the maximum value of $y = 3 \sin x$ is $3(1) = 3$ and its minimum value is $3(-1) = -3$. If the graph of $y = f(x)$ is "centered" about the x-axis; then the maximum value of $f(x)$ is the *amplitude* of $f(x)$. In general, we have the following definition of the amplitude of a periodic function.

Amplitude

Let f be a periodic function. Suppose M is the maximum value of f and m is the minimum value of f. The amplitude of f is defined by

$$\text{Amplitude} = \frac{1}{2}(M - m).$$

SIDE NOTE

If a function f does not have both a maximum and a minimum value, we say that f has no amplitude or that its amplitude is undefined.

Because the maximum value of both functions $y = a \sin x$ and $y = a \cos x$ is $|a|$ and their minimum value is $-|a|$, the amplitude of both functions is $\frac{1}{2}[|a| - (-|a|)] = |a|$.

AMPLITUDES OF SINE AND COSINE

The functions $y = a \sin x$ and $y = a \cos x$ have **amplitude** $= |a|$ and range $= [-|a|, |a|]$.

EXAMPLE 1 Graphing $y = a \sin x$

Graph $y = 3 \sin x$, $y = \frac{1}{3} \sin x$, and $y = \sin x$ on the same coordinate system over the interval $[-2\pi, 2\pi]$.

Solution

We begin with the graph of $y = \sin x$ and multiply the y-coordinate of each point (including the key points) on this graph by 3 to get the graph of $y = 3 \sin x$. This results in stretching the graph of $y = \sin x$ vertically by a factor of 3. Because $3(0) = 0$, the x-intercepts are unchanged. However, the maximum y value is $3(1) = 3$ and the minimum y value is $3(-1) = -3$. See Figure 4.40. Similarly, we multiply the y-coordinate of the graph by $\frac{1}{3}$ to get the graph of $y = \frac{1}{3} \sin x$. This results in compressing the graph of $y = \sin x$ vertically by a factor of $\frac{1}{3}$. As before, the x-intercepts are unchanged; but here the maximum y value is $\frac{1}{3}$ and the minimum y value is $-\frac{1}{3}$. See Figure 4.40.

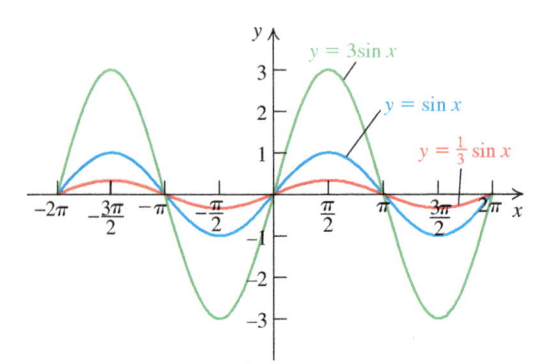

FIGURE 4.40 Varying a in $y = a \sin x$

Practice Problem 1 Graph $y = 5 \sin x$, $y = \frac{1}{5} \sin x$, and $y = \sin x$ on the same coordinate system over the interval $[-2\pi, 2\pi]$.

EXAMPLE 2   Graphing $y = a \cos x$

Graph $y = -2 \cos x$ over the interval $[-2\pi, 2\pi]$. Find the amplitude and range of the function.

Solution

Begin with the graph of $y = \cos x$. Multiply the y-coordinate of each point of that graph by 2 to stretch the graph vertically by a factor of 2 to get the graph of $y = 2 \cos x$. Pay particular attention to the key points. Notice that the x-intercepts are unchanged. Next, reflect the graph of $y = 2 \cos x$ about the x-axis to produce the graph of $y = -2 \cos x$. See Figure 4.41. The amplitude is $|-2| = 2$, the largest value attained by $y = -2 \cos x$; the smallest function value attained is -2. Therefore, the range of $y = -2 \cos x$ is $[-2, 2]$.

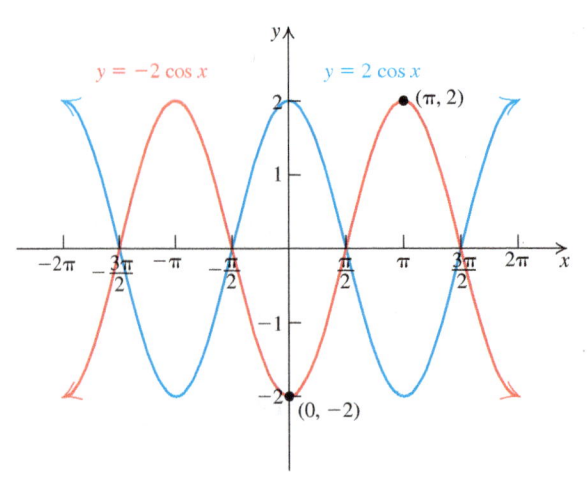

FIGURE 4.41 Graph of $y = -2 \cos x$

Practice Problem 2 Graph $y = 5 \cos x$ over the interval $[-2\pi, 2\pi]$. Find the amplitude and range of the function.

Both the sine and cosine functions have period 2π. For $b > 0$, $0 \leq bx \leq 2\pi$ is equivalent to $0 \leq x \leq \frac{2\pi}{b}$. Therefore, as x takes on the values from 0 to $\frac{2\pi}{b}$, the expression bx takes on the values from 0 to 2π.

The graphs of $y = \sin bx$ and $y = \cos bx$ have similar shapes to those of $y = \sin x$ and $y = \cos x$, respectively, but they are compressed or expanded horizontally so that they complete one cycle over any interval of length $\frac{2\pi}{b}$. Consequently, the functions $y = \sin bx$ and $y = \cos bx$ both have period $\frac{2\pi}{b}$. When $b < 0$, we can use the facts that $\sin(-x) = -\sin x$ and $\cos(-x) = \cos x$ to rewrite the given function with $b > 0$. For example, $y = \sin(-3x) = -\sin 3x$.

EXAMPLE 3 Graphing $y = \sin bx$

Sketch one cycle of the graphs of $y = \sin 3x$, $y = \sin \frac{1}{3}x$, and $y = \sin x$ on the same coordinate system.

Solution

We begin with the graph of $y = \sin x$ and adjust its period, 2π, by dividing 2π by 3 to find the period $\frac{2\pi}{3}$ of $y = \sin 3x$. We then compress the graph of $y = \sin x$ horizontally so that one cycle of $y = \sin 3x$ is completed on the interval $\left[0, \frac{2\pi}{3}\right]$. Divide the interval $0 \leq x \leq \frac{2\pi}{3}$ into four equal parts to find the x-coordinates for the key points: $0, \frac{\pi}{6}, \frac{\pi}{3}, \frac{\pi}{2}$, and $\frac{2\pi}{3}$. Because the amplitude is 1, the y values of the key points are unchanged. See Figure 4.42.

Similarly, to find the period of $y = \sin \frac{1}{3}x$, we divide 2π by $\frac{1}{3}$, resulting in $\frac{2\pi}{\frac{1}{3}} = 6\pi$.

Therefore, one cycle for $y = \sin \frac{1}{3}x$ is completed on the interval $[0, 6\pi]$. Divide the interval $0 \leq x \leq 6\pi$ into four equal parts to get the x-coordinates for the key points: $0, \frac{3\pi}{2}, 3\pi, \frac{9\pi}{2}$, and 6π. Again, the y values of the key points are unchanged. See Figure 4.42.

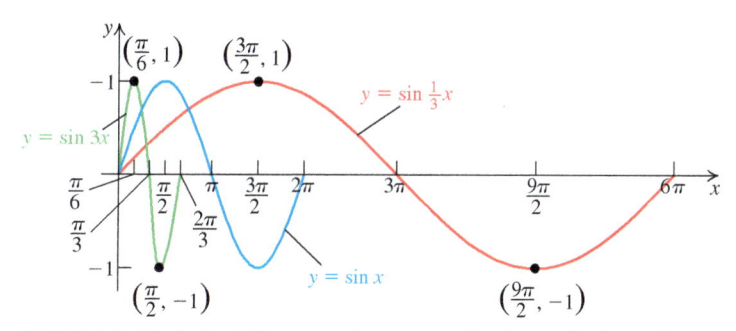

FIGURE 4.42 Horizontally compressing or expanding the graph of $y = \sin x$

Practice Problem 3 Sketch one cycle of the graphs of $y = \sin \frac{1}{2}x$ and $y = \sin x$ on the same coordinate system.

The next box summarizes information used to graph $y = a \sin bx$ and $y = a \cos bx$.

CHANGING THE AMPLITUDE AND PERIOD OF THE SINE AND COSINE FUNCTIONS

The functions $y = a \sin bx$ and $y = a \cos bx$ $(b > 0)$ have amplitude $|a|$ and period $\frac{2\pi}{b}$. If $a > 0$, the graphs of $y = a \sin bx$ and $y = a \cos bx$ are similar to the graphs of $y = \sin x$ and $y = \cos x$, respectively, with three changes.

1. The range is $[-a, a]$.
2. One cycle is completed over the interval $\left[0, \frac{2\pi}{b}\right]$.
3. The x-coordinates for the key points are $0, \frac{\pi}{2b}, \frac{\pi}{b}, \frac{3\pi}{2b},$ and $\frac{2\pi}{b}$. The y-values at each of these numbers is one of $-a$, 0, or a.

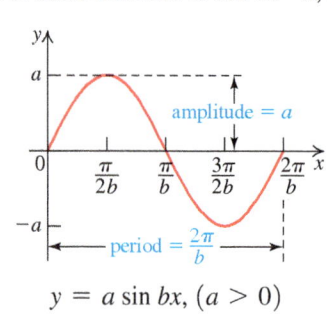

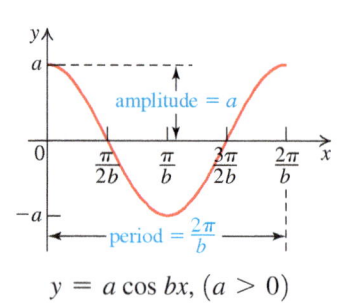

$y = a \sin bx,\ (a > 0)$ $y = a \cos bx,\ (a > 0)$

If $a < 0$, the graphs are the reflections of the graph of $y = |a| \sin bx$ and $y = |a| \cos bx$, respectively, about the x-axis.

If $b < 0$, the graphs of $y = a \cos bx$ and $y = a \cos(|b|x)$ are identical, but the graph of $y = a \sin bx$ is the reflection of the graph of $y = a \sin(|b|x)$ about the x-axis.

TECHNOLOGY CONNECTION

To graph $y = 3 \cos \frac{1}{2}x$ with a graphing calculator, choose Radian mode and these WINDOW settings:

X min = 0
X max = 4π
X scl = $\pi/2$
Y min = -4
Y max = 4
Y scl = 1

EXAMPLE 4   Graphing $y = a \cos bx$

Graph $y = 3 \cos \frac{1}{2}x$ over a one-period interval.

Solution

We begin with the graph of $y = \cos x$ and find the period of $y = \cos \frac{1}{2}x$ by dividing 2π by $\frac{1}{2}$ to get

$$\frac{2\pi}{\frac{1}{2}} = 4\pi \qquad \text{Replace } b \text{ with } \frac{1}{2} \text{ in } \frac{2\pi}{b}.$$

The period of $y = \cos \frac{1}{2}x$ is 4π. So the graph of $y = \cos \frac{1}{2}x$ is obtained by horizontally stretching the graph of $y = \cos x$ by a factor of 2.

Next, we multiply the y-coordinate of each point on the graph of $y = \cos \frac{1}{2}x$ by 3 to stretch this graph vertically by a factor of 3. For help in sketching the graphs, divide the period,

4π, into four equal parts, starting at 0, to find the x-coordinates for the key points. These x-coordinates $(0, \pi, 2\pi, 3\pi,$ and $4\pi)$ correspond to the highest points, the lowest point, and the x-intercepts of the graph. The key points on the graph are $(0, 3)$, $(\pi, 0)$, $(2\pi, -3)$, $(3\pi, 0)$, and $(4\pi, 3)$. See Figure 4.43. The graph can then be extended indefinitely to the left and right. The amplitude is 3; so the largest value of y is 3. The range is $[-3, 3]$.

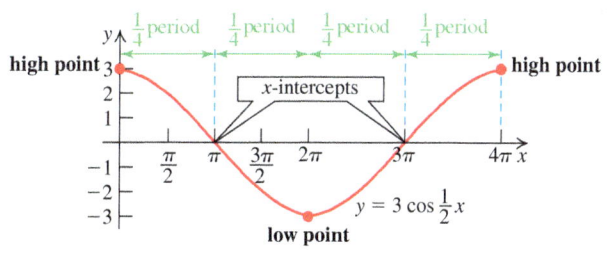

FIGURE 4.43 One cycle of $y = 3\cos\dfrac{1}{2}x$

Practice Problem 4 Graph $y = 2\cos 2x$ over a one-period interval.

Phase Shift

4 Find the phase shifts and graph sinusoidal functions of the forms $y = a\sin[b(x - c)]$ and $y = a\cos[b(x - c)]$.

We now consider sinusoidal graphs with horizontal shifts.

EXAMPLE 5 Graphing $y = a \sin(x - c)$

Graph $y = \sin\left(x - \dfrac{\pi}{2}\right)$ over a one-period interval.

Solution

Recall that for any function f, the graph of $y = f(x - c)$ is the graph of $y = f(x)$ shifted right c units if $c > 0$ and left $|c|$ units if $c < 0$. Comparing the equation $y = \sin\left(x - \dfrac{\pi}{2}\right)$ with $y = \sin x$, we see that $y = \sin\left(x - \dfrac{\pi}{2}\right)$ has the form $y = \sin(x - c)$, where $c = \dfrac{\pi}{2}$. Therefore, the graph of $y = \sin\left(x - \dfrac{\pi}{2}\right)$ is the graph of $y = \sin x$ shifted right $\dfrac{\pi}{2}$ units. See Figure 4.44.

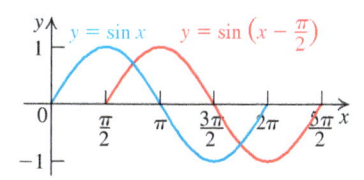

FIGURE 4.44 Graph of $y = \sin\left(x - \dfrac{\pi}{2}\right)$

Practice Problem 5 Graph $y = \cos\left(x - \dfrac{\pi}{4}\right)$ over a one-period interval.

Before describing the general procedure for graphing the sinusoidal functions $y = a\sin[b(x - c)]$ and $y = a\cos[b(x - c)]$, with $b > 0$, we find the value of x for which $b(x - c) = 0$. This number is called the **phase shift**. Let

$$b(x - c) = 0$$
$$x - c = 0 \quad \text{Divide both sides by } b.$$
$$x = c \quad \text{Add } c \text{ to both sides.}$$

The phase shift is the amount by which the graphs of $y = a\sin bx$ or $y = a\cos bx$ are shifted horizontally. These graphs are shifted right if $c > 0$ and left if $c < 0$. To graph a cycle of either function, we start the cycle at $x = c$ and then divide the period into four equal parts. We describe this procedure next.

PROCEDURE IN ACTION

EXAMPLE 6 Graphing $y = a \sin [b(x - c)]$ and $y = a \cos [b(x - c)]$ with $b > 0$

OBJECTIVE
Graph a function of the form $y = a \sin [b(x - c)]$ or $y = a \cos [b(x - c)]$, with $b > 0$, by finding the amplitude, period, and phase shift.

EXAMPLE
Find the amplitude, period, and phase shift and sketch the graph of $y = 3 \sin \left(2x - \dfrac{\pi}{2} \right)$ over a one-period interval.

Step 1 If necessary, rewrite the equation in the form $y = a \sin [b(x - c)]$ or $y = a \cos [b(x - c)]$. Find the amplitude, period, and phase shift.

amplitude $= |a|$

period $= \dfrac{2\pi}{b}$

phase shift $= c$

If $c > 0$, shift to the right.
If $c < 0$, shift to the left.

1. $y = 3 \sin \left[2 \left(x - \dfrac{\pi}{4} \right) \right]$ Rewrite

$a = 3, b = 2, c = \dfrac{\pi}{4}$

amplitude $= |3| = 3$

period $= \dfrac{2\pi}{2} = \pi$

phase shift $= \dfrac{\pi}{4}$ Shift right because $\dfrac{\pi}{4} > 0$.

Step 2 The cycle begins at $x = c$. One complete cycle occurs over the interval $\left[c, c + \dfrac{2\pi}{b} \right]$.

2. Begin the cycle at $x = \dfrac{\pi}{4}$ and graph one cycle over $\left[\dfrac{\pi}{4}, \dfrac{\pi}{4} + \pi \right] = \left[\dfrac{\pi}{4}, \dfrac{5\pi}{4} \right]$.

Step 3 Divide the interval $\left[c, c + \dfrac{2\pi}{b} \right]$ into four congruent parts, each of length $\dfrac{1}{4}(\text{period}) = \dfrac{1}{4}\left(\dfrac{2\pi}{b} \right)$. This gives the x-coordinates for the five key points:

$c, c + \dfrac{1}{4}\left(\dfrac{2\pi}{b} \right), c + \dfrac{1}{2}\left(\dfrac{2\pi}{b} \right),$

$c + \dfrac{3}{4}\left(\dfrac{2\pi}{b} \right)$, and $c + \dfrac{2\pi}{b}$.

3. Divide the interval $\left[\dfrac{\pi}{4}, \dfrac{5\pi}{4} \right]$ into four equal parts, each having length $\dfrac{1}{4}(\text{period}) = \dfrac{1}{4}(\pi) = \dfrac{\pi}{4}$.

The x-coordinates of the five key points are

$\dfrac{\pi}{4}, \dfrac{\pi}{4} + \dfrac{\pi}{4} = \dfrac{\pi}{2}, \dfrac{\pi}{4} + \dfrac{\pi}{2} = \dfrac{3\pi}{4},$

$\dfrac{\pi}{4} + \dfrac{3\pi}{4} = \pi$, and $\dfrac{\pi}{4} + \pi = \dfrac{5\pi}{4}$.

Step 4 If $a > 0$, for $y = a \sin [b(x - c)]$, sketch one cycle of the sine curve through the key points $(c, 0)$, $\left(c + \dfrac{\pi}{2b}, a \right)$, $\left(c + \dfrac{\pi}{b}, 0 \right)$, $\left(c + \dfrac{3\pi}{2b}, -a \right)$, and $\left(c + \dfrac{2\pi}{b}, 0 \right)$.

For $y = a \cos [b(x - c)]$, sketch one cycle of the cosine curve through the key points (c, a), $\left(c + \dfrac{\pi}{2b}, 0 \right)$, $\left(c + \dfrac{\pi}{b}, -a \right)$, $\left(c + \dfrac{3\pi}{2b}, 0 \right)$, and $\left(c + \dfrac{2\pi}{b}, a \right)$.

If $a < 0$, reflect the graph of $y = |a| \sin [b(x - c)]$ or $y = |a| \cos [b(x - c)]$ about the x-axis.

4. Sketch one cycle of the sine curve through the five key points: $\left(\dfrac{\pi}{4}, 0 \right)$, $\left(\dfrac{\pi}{2}, 3 \right)$, $\left(\dfrac{3\pi}{4}, 0 \right)$, $(\pi, -3)$, and $\left(\dfrac{5\pi}{4}, 0 \right)$.

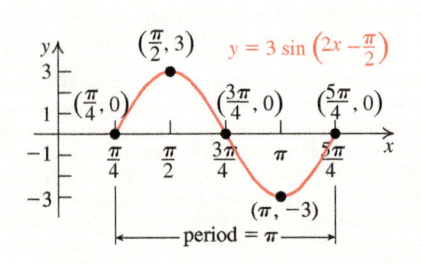

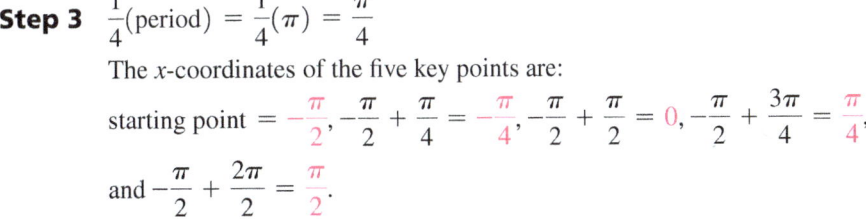

Practice Problem 6 Find the amplitude, period, and phase shift and sketch the graph of the function

$$y = \frac{1}{3} \cos \left[\frac{2}{5} \left(x + \frac{\pi}{6} \right) \right].$$

Graphing When b < 0. The equations $\sin(-x) = -\sin x$ and $\cos(-x) = \cos x$ allow us to graph $y = a \sin[b(x - c)]$ and $y = a \cos[b(x - c)]$ when $b < 0$. For example, to graph $y = 3 \sin \left[(-2) \left(x - \frac{\pi}{2} \right) \right]$, we rewrite and graph the equation as $y = -3 \sin \left[2 \left(x - \frac{\pi}{2} \right) \right]$.

EXAMPLE 7 Graphing $y = a \cos[b(x - c)]$

Graph $y = -\cos \left[2 \left(x + \frac{\pi}{2} \right) \right]$ over a one-period interval.

Solution
Use the steps in Example 6 for the procedure for graphing $y = a \cos[b(x - c)]$.

Step 1 Here $y = -\cos \left[2 \left(x + \frac{\pi}{2} \right) \right] = (-1) \cos \left[2 \left(x - \left(-\frac{\pi}{2} \right) \right) \right]$

$a = -1 \quad b = 2 \quad c = -\frac{\pi}{2}$

amplitude $= |a| = |-1| = 1 \quad$ period $= \frac{2\pi}{b} = \frac{2\pi}{2} = \pi$

phase shift $= c = -\frac{\pi}{2} \quad$ Shift left because $c < 0$.

Step 2 The cycle begins at $x = -\frac{\pi}{2}$.

One cycle is graphed over the interval $\left[-\frac{\pi}{2}, -\frac{\pi}{2} + \pi \right] = \left[-\frac{\pi}{2}, \frac{\pi}{2} \right]$.

Step 3 $\frac{1}{4}$(period) $= \frac{1}{4}(\pi) = \frac{\pi}{4}$

The x-coordinates of the five key points are:

starting point $= -\frac{\pi}{2}, -\frac{\pi}{2} + \frac{\pi}{4} = -\frac{\pi}{4}, -\frac{\pi}{4} + \frac{\pi}{4} = 0, -\frac{\pi}{2} + \frac{3\pi}{4} = \frac{\pi}{4},$

and $-\frac{\pi}{2} + \frac{2\pi}{2} = \frac{\pi}{2}$.

Step 4 Because $a = -1 < 0$, we first graph $y = |-1| \cos \left[2 \left(x + \frac{\pi}{2} \right) \right]$ by sketching one cycle, beginning at $\left(-\frac{\pi}{2}, 1 \right)$ and going through the points $\left(-\frac{\pi}{4}, 0 \right), (0, -1), \left(\frac{\pi}{4}, 0 \right),$ and $\left(\frac{\pi}{2}, 1 \right)$. We then reflect this graph about the x-axis to obtain the graph of $y = -\cos \left[2 \left(x + \frac{\pi}{2} \right) \right]$. See Figure 4.45.

Practice Problem 7 Graph $y = -2 \cos[3(x + \pi)]$ over a one-period interval.

TECHNOLOGY CONNECTION

To graph $y = -\cos 2 \left(x + \frac{\pi}{2} \right)$ with a graphing calculator, choose **radian** mode and enter these WINDOW settings:

X min $= -\pi/2$
X max $= \pi/2$
X scl $= \pi/4$
Y min $= -2$
Y max $= 2$
Y scl $= 1$

FIGURE 4.45

322 Chapter 4 Trigonometric Functions

Vertical Shifts

Recall that the graph of $y = f(x) + d$ results from shifting the graph of $y = f(x)$ vertically d units up if $d > 0$ and $|d|$ units down if $d < 0$. The next example shows this effect on a sinusoidal graph.

EXAMPLE 8 Graphing $y = a\cos[b(x - c)] + d$

Graph $y = \cos\left[2\left(x + \dfrac{\pi}{2}\right)\right] + 3$ over a one-period interval.

Solution

In Example 7, we graphed the function $y = \cos\left[2\left(x + \dfrac{\pi}{2}\right)\right]$ in order to graph $y = -\cos\left[2\left(x + \dfrac{\pi}{2}\right)\right]$. To graph $y = \cos\left[2\left(x + \dfrac{\pi}{2}\right)\right] + 3$, shift the graph of $y = \cos\left[2\left(x + \dfrac{\pi}{2}\right)\right]$ up three units. See Figure 4.46.

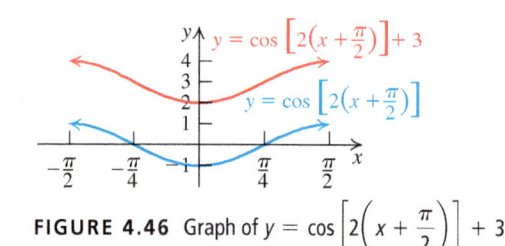

FIGURE 4.46 Graph of $y = \cos\left[2\left(x + \dfrac{\pi}{2}\right)\right] + 3$

Practice Problem 8 Graph $y = 4\sin\left(3x - \dfrac{\pi}{3}\right) - 2$ over a one-period interval.

EXAMPLE 9 Modeling the Number of Daylight Hours in Paris

Table 4.5 gives the average number of daylight hours in Paris each month. Let y represent the number of daylight hours in Paris in month x to find a function of the form $y = a\sin[b(x - c)] + d$ that models the hours of daylight throughout the year.

Solution

Plot the values given in Table 4.5, representing the months by the integers 1 through 12 (Jan. = 1, ..., Dec. = 12). Then sketch a function of the form $y = a\sin[b(x - c)] + d$ that models the points just graphed. See Figure 4.47.

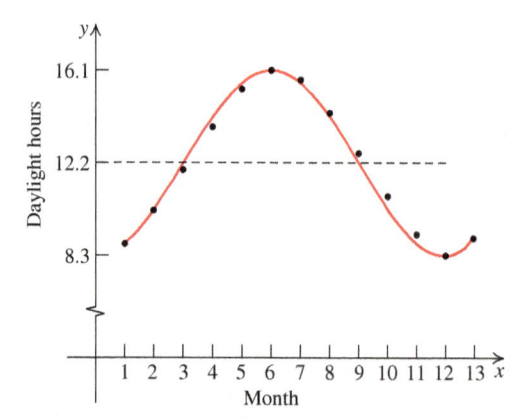

TABLE 4.5

Daylight Hours in Paris	
January	8.8
February	10.2
March	11.9
April	13.7
May	15.3
June	16.1
July	15.7
August	14.3
September	12.6
October	10.8
November	9.2
December	8.3

FIGURE 4.47 Hours of daylight in Paris each month

We want to find the values for the constants a, b, c, and d that will produce the graph shown in Figure 4.48. From Table 4.5, the range of the function is $[8.3, 16.1]$; so

$$\text{Amplitude} = a = \frac{\text{highest value} - \text{lowest value}}{2} = \frac{(16.1 - 8.3)}{2} = 3.9.$$

The sine graph has been shifted vertically by $\frac{1}{2}$ (highest value + lowest value), the average of the highest and lowest values. Thus, this average value gives

$$d = \frac{1}{2}(16.1 + 8.3) = 12.2.$$

The weather repeats every 12 months; so the period is 12. Because the period $= \frac{2\pi}{b}$, we can write

$$12 = \frac{2\pi}{b} \qquad \text{Replace period with 12.}$$

$$b = \frac{2\pi}{12} = \frac{\pi}{6} \qquad \text{Solve for } b.$$

So far, we can write $y = 3.9 \sin\left[\left(\frac{\pi}{6}\right)(x - c)\right] + 12.2$. The horizontally "unshifted" graph of $y = 3.9 \sin\left(\frac{\pi}{6}x\right) + 12.2$ is superimposed on Figure 4.48. This graph must be shifted to the right to fit the plotted data. To determine the phase shift, find the value of c when y has the greatest value. This value is $y = 16.1$ when $x = 6$ (in June); so we now have

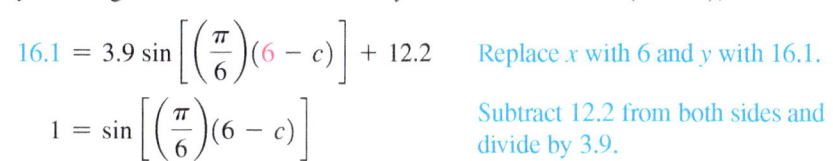

$$16.1 = 3.9 \sin\left[\left(\frac{\pi}{6}\right)(6 - c)\right] + 12.2 \qquad \text{Replace } x \text{ with 6 and } y \text{ with 16.1.}$$

$$1 = \sin\left[\left(\frac{\pi}{6}\right)(6 - c)\right] \qquad \begin{array}{l}\text{Subtract 12.2 from both sides and}\\ \text{divide by 3.9.}\end{array}$$

Now $\sin\left[\left(\frac{\pi}{6}\right)(6 - c)\right]$ will first equal 1 when the argument of sine is $\frac{\pi}{2}$. So

$$\frac{\pi}{6}(6 - c) = \frac{\pi}{2} \qquad \sin t = 1 \text{ when } t = \frac{\pi}{2}.$$

$$c = 3 \qquad \text{Solve for } c.$$

Note that this shifts the graph of $y = 3.9 \sin\left(\frac{\pi}{6}x\right) + 12.2$ three units right. The equation describing the hours of daylight in Paris *in month x* is

$$y = 3.9 \sin\left[\frac{\pi}{6}(x - 3)\right] + 12.2.$$

Practice Problem 9 Rework Example 9 for the city of Fargo, North Dakota, using the values given in Table 4.6.

Simple Harmonic Motion

Trigonometric functions can often describe or model motion caused by vibration, rotation, or oscillation. The motions associated with sound waves, radio waves, alternating electric current, a vibrating guitar string, or the swing of a pendulum are examples of such motion. Another example is the movement of a ball suspended by a spring. If the ball is pulled

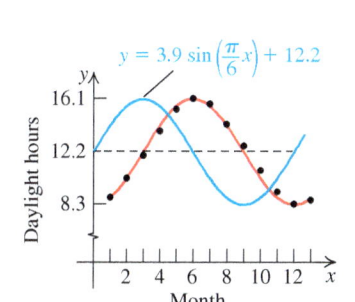

FIGURE 4.48

TABLE 4.6

Daylight Hours in Fargo	
January	9
February	10.3
March	11.9
April	13.6
May	15.1
June	15.8
July	15.4
August	14.2
September	12.5
October	10.9
November	9.4
December	8.6

downward and then released (ignoring the effects of friction and air resistance), the ball will repeatedly move up and down past its rest, or equilibrium, position. See Figure 4.49.

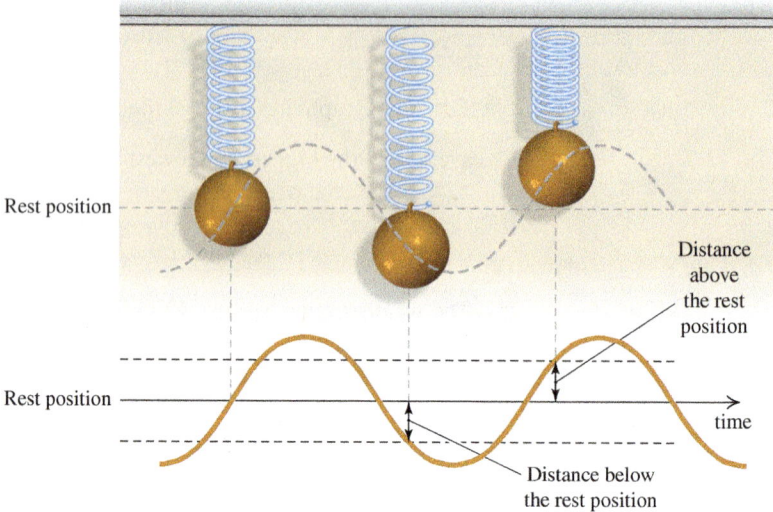

FIGURE 4.49 Simple harmonic motion

Simple Harmonic Motion

An object whose position relative to an equilibrium position at time t can be described by either

$$y = a \sin \omega t \quad \text{or} \quad y = a \cos \omega t \quad (\omega > 0)$$

is in **simple harmonic motion.**

The *amplitude*, $|a|$, is the maximum distance the object reaches from its equilibrium position. The *period* of the motion, $\frac{2\pi}{\omega}$, is the time it takes the object to complete one full cycle. The **frequency** of the motion is $\frac{\omega}{2\pi}$ and gives the number of cycles completed per unit of time.

If the distance above and below the rest position of the ball is graphed as a function of time, a sine or a cosine curve results. The amplitude of the curve is the maximum distance above the rest position the ball travels, and one cycle is completed as the ball travels from its highest position to its lowest position and back to its highest position.

EXAMPLE 10 Simple Harmonic Motion of a Ball Attached to a Spring

Suppose a ball attached to a spring is pulled down 6 inches and released and the resulting simple harmonic motion has a period of eight seconds. Write an equation for the ball's simple harmonic motion.

Solution

We must first choose between an equation of the form $y = a \sin \omega t$ or $y = a \cos \omega t$. We start tracking the motion of the ball when $t = 0$. Note that for $t = 0$,

$$y = a \sin(\omega \cdot 0) = a \cdot \sin 0 = 0 \quad \text{and} \quad y = a \cos(\omega \cdot 0) = a \cdot \cos 0 = a \cdot 1 = a.$$

If we choose to start tracking the ball's motion when we release it after pulling it down 6 inches, we should choose $a = -6$ and $y = -6 \cos \omega t$. Because we pulled the ball *down* in order to start, a is negative. We now have the form of the equation of motion.

$$y = -6 \cos \omega t \quad \text{When } t = 0, y = -6.$$

$$\text{period} = \frac{2\pi}{\omega} = 8 \quad \text{The period is given as eight seconds.}$$

$$\omega = \frac{2\pi}{8} = \frac{\pi}{4} \quad \text{Solve for } \omega.$$

Replacing ω with $\frac{\pi}{4}$ in $y = -6 \cos \omega t$ yields the equation of the ball's simple harmonic motion:

$$y = -6 \cos \left(\frac{\pi}{4} t\right)$$

Practice Problem 10 Rework Example 10 with the motion period of three seconds and the ball pulled down 4 inches and released.

SECTION 4.3 Exercises

Basic Concepts and Skills

In Exercises 1–18, sketch the graph of each given equation over the interval $[-2\pi, 2\pi]$.

1. $y = 2 \sin x$
2. $y = 4 \cos x$
3. $y = -\frac{1}{2} \sin x$
4. $y = -2 \sin x$
5. $y = \cos 2x$
6. $y = \sin 4x$
7. $y = \cos \frac{2}{3} x$
8. $y = \sin \frac{4}{3} x$
9. $y = \cos \left(x + \frac{\pi}{2}\right)$
10. $y = \sin \left(x + \frac{\pi}{4}\right)$
11. $y = \cos \left(x - \frac{\pi}{3}\right)$
12. $y = \sin (x - \pi)$
13. $y = 2 \cos \left(x - \frac{\pi}{2}\right)$
14. $y = 2 \sin \left(x + \frac{\pi}{3}\right)$
15. $y = (\sin x) + 1$
16. $y = (\cos x) - 2$
17. $y = (-\cos x) + 1$
18. $y = (\sin x) - 3$

In Exercises 19–26, find the amplitude, period, and phase shift of each given function.

19. $y = 5 \cos (x - \pi)$
20. $y = 3 \sin \left(x - \frac{\pi}{8}\right)$
21. $y = 7 \cos \left[9\left(x + \frac{\pi}{6}\right)\right]$
22. $y = 11 \sin \left[8\left(x + \frac{\pi}{3}\right)\right]$
23. $y = -6 \cos \left[\frac{1}{2}(x + 2)\right]$
24. $y = -8 \sin \left[\frac{1}{5}(x + 9)\right]$
25. $y = 0.9 \sin \left[0.25\left(x - \frac{\pi}{4}\right)\right]$
26. $y = \sqrt{5} \cos [\pi(x + 1)]$

In Exercises 27–34, graph each function over a one-period interval.

27. $y = -4 \cos \left(x + \frac{\pi}{6}\right)$
28. $y = -3 \sin \left(x - \frac{\pi}{6}\right)$
29. $y = \frac{5}{2} \sin \left[2\left(x - \frac{\pi}{4}\right)\right]$
30. $y = \frac{3}{2} \cos \left[2\left(x + \frac{\pi}{3}\right)\right]$
31. $y = -5 \cos \left[4\left(x - \frac{\pi}{6}\right)\right]$
32. $y = -3 \sin \left[4\left(x + \frac{\pi}{6}\right)\right]$
33. $y = \frac{1}{2} \sin \left[4\left(x + \frac{\pi}{4}\right)\right] + 2$
34. $y = -\frac{1}{2} \cos \left[2\left(x - \frac{\pi}{2}\right)\right] - 3$

In Exercises 35–44, write each function in the form $y = a \sin [b(x - c)]$ or $y = a \cos [b(x - c)]$. Find each period and phase shift.

35. $y = 4 \cos \left(2x + \frac{\pi}{3}\right)$
36. $y = 5 \sin \left(3x + \frac{\pi}{2}\right)$
37. $y = -\frac{3}{2} \sin (2x - \pi)$
38. $y = -2 \cos \left(5x - \frac{\pi}{4}\right)$
39. $y = 3 \cos \pi x$
40. $y = \sin \frac{\pi x}{3}$
41. $y = \frac{1}{2} \cos \left(\frac{\pi x}{4} + \frac{\pi}{4}\right)$
42. $y = -\sin \left(\frac{\pi x}{6} + \frac{\pi}{6}\right)$
43. $y = 2 \sin (\pi x + 3)$
44. $y = -\cos \left(\pi x - \frac{1}{4}\right)$

Applying the Concepts

45. Pulse and blood pressure. Blood pressure is given by two numbers written as a fraction: $\dfrac{\text{systolic}}{\text{diastolic}}$. The *systolic* reading is the maximum pressure in an artery, and the *diastolic* reading is the minimum pressure in an artery. As the heart beats, the systolic measurement is taken; when the heart rests, the diastolic measurement is taken. A reading of $\dfrac{120}{80}$ is considered normal. Your pulse is the number of heartbeats per minute. Suppose Desmond's blood pressure after t minutes is given by

$$p(t) = 20 \sin(140\pi t) + 122,$$

where $p(t)$ is the pressure in millimeters of mercury.
  a. Find the period and explain how it relates to pulse.
  b. Graph the function p over one period.
  c. What is Desmond's blood pressure?

46. Oscillating ball. Suppose a ball attached to a spring is pulled down 5 inches and released and the resulting simple harmonic motion has a period of ten seconds. Write an equation of the ball's simple harmonic motion.

47. Kangaroo population. The kangaroo population in a certain region is given by the function

$$N(t) = 650 + 150 \sin 2t,$$

where the time t is measured in years.
  a. What is the largest number of kangaroos present in the region at any time?
  b. What is the smallest number of kangaroos present in the region at any time?
  c. How much time elapses between occurrences of the largest and the smallest kangaroo population?

48. Daylight hours in London. The table gives the average number of daylight hours in London each month, where $x = 1$ represents January. Let y represent the number of daylight hours in London in month x. Find a function of the form $y = a \sin [b(x - c)] + d$ that models the hours of daylight throughout the year.

Jan	Feb	Mar	Apr	May	June
8.4	10	11.9	13.9	15.6	16.6

July	Aug	Sept	Oct	Nov	Dec
16.1	14.6	12.6	10.7	8.9	7.9

Critical Thinking / Discussion / Writing

49. Suppose a function of the form $y = a \sin [b(x - c)] + d$, with $a > 0$, has period $= 20$.
  a. In a one-period interval, how many units are between the value of x at which the maximum value is attained and the value of x at which the minimum value is attained?
  b. In a one-period interval, how many units are between the value of x at which y has value d and the value of x at which the maximum value is attained?

50. Suppose the minimum value of a function of the form $y = a \cos [b(x - c)] + d$, with $a > 0$, occurs at a value of x that is five units from the value of x at which the function has the maximum value. What is the period of the function?

In Exercises 51–54, write an equation for each sinusoidal graph.

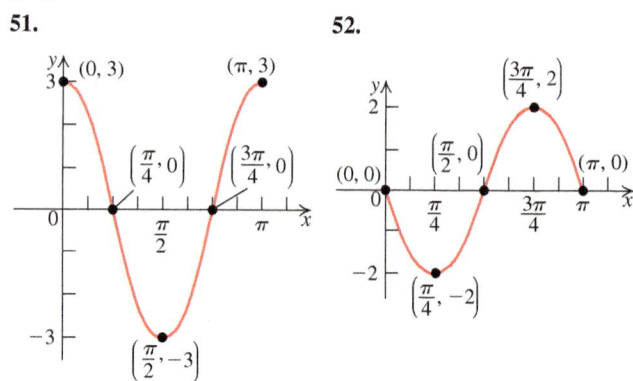

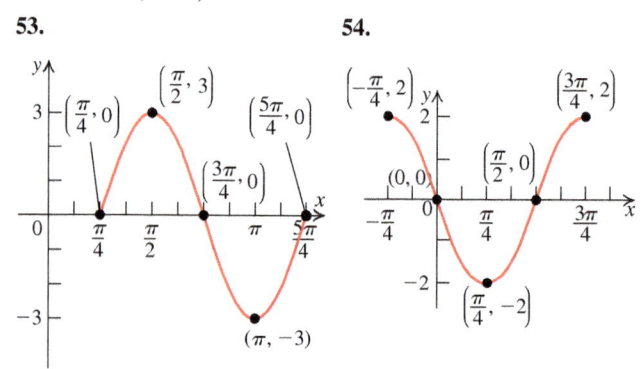

51.

52.

53.

54.

55. We know that $\sin(x + 2\pi) = \sin x$ for all real numbers x. Consequently, if 2π is not the period for the sine function, then $\sin(x + p) = \sin x$ for some real number p, with $0 < p < 2\pi$, and all real numbers x.
  a. By evaluating $\sin(x + p) = \sin x$ for $x = 0$, show that if the period is not 2π, then $p = \pi$.
  b. Find a value of x for which $\sin(x + \pi) \neq \sin x$ and conclude that the period of the sine function is 2π.

Maintaining Skills

56. In slope–intercept form, find the equation of the line that passes through the point $(2, -3)$ and has slope $m = \dfrac{1}{2}$.

57. Find the domain and the range of $f(x) = |x - 1|$.

58. Find the zeros and horizontal and vertical asymptotes of the graph of $f(x) = \dfrac{x^2 + 5x - 14}{2x^2 + 5x - 3}$.

59. Find the period of $f(x) = 2 \sin(3x)$.

SECTION 4.4

Graphs of the Other Trigonometric Functions

BEFORE STARTING THIS SECTION, REVIEW

1. Vertical asymptotes (Section 2.5, page 200)
2. Transformations of functions (Section 1.5, page 109)

OBJECTIVES

1. Discuss properties of the tangent function.
2. Graph the tangent function.
3. Graph the cosecant, secant, and cotangent functions.

◆ The U.S.S. Enterprise and Mach Numbers

Star Trek fans probably know quite a bit about the warp speeds attainable by the Starfleet's U.S.S. Enterprise. Near Earth, Mach numbers provide a more relevant measure of speed.

The Mach number (named after the Austrian physicist Ernst Mach) is the ratio of the speed of the aircraft to the speed of sound. For example, an aircraft traveling at Mach 2 is traveling at twice the speed of sound. An airplane flying at less than the speed of sound is traveling at *subsonic* speed; at the speed of sound (Mach 1), the plane is traveling at *transonic* speed; at speeds greater than the speed of sound, it is traveling at *supersonic* speeds; and at speeds greater than 5 times the speed of sound, it is traveling at *hypersonic* speeds. In Example 5, we see how at supersonic and hypersonic speeds, Mach numbers are given as values of the cosecant function and we graph a range of Mach numbers that are attainable by various military aircraft.

1 Discuss properties of the tangent function.

Tangent Function

Recall that if $P(x, y)$ is a point on the terminal side of an angle θ in standard position, then (see Figure 4.50)

$$\tan \theta = \frac{y}{x}, x \neq 0$$

The initial point of the line OP is $(0, 0)$, so the slope of the line OP is $\frac{y - 0}{x - 0} = \frac{y}{x} = \tan \theta$. Because parallel lines have equal slopes (see page 72), we have the following geometric interpretation of the tangent of an angle.

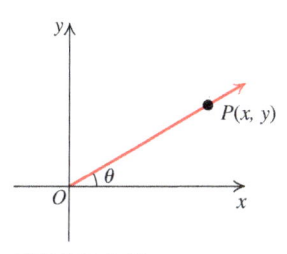

FIGURE 4.50

TANGENT AS SLOPE

If a line makes an angle $\theta \left(0 \leq \theta \leq \pi, \theta \neq \dfrac{\pi}{2}\right)$ with the positive x-axis, then the slope m of the line is given by $m = \tan \theta$.

EXAMPLE 1 Equation of a Line

A line ℓ makes an angle $\theta = 60°$ with positive x-axis and passes through the point $P = (7, 5)$. Find the equation of ℓ in slope–intercept form.

Solution
We have $m = \tan 60° = \sqrt{3}$ (page 302). So the slope of the line ℓ is $\sqrt{3}$. Then

$$y - 5 = \sqrt{3}(x - 7) \qquad \text{Point–slope form: } y - y_1 = m(x - x_1)$$
$$y - 5 = \sqrt{3}x - 7\sqrt{3} \qquad \text{Distributive property}$$
$$y = \sqrt{3}x + (5 - 7\sqrt{3}) \qquad \text{Slope–intercept form: } y = mx + b$$

Practice Problem 1 Repeat Example 1 with $\theta = 30°$ and $P = (2, 3)$.

Domain of tan x From the quotient identity (page 295), we have

$$\tan x = \frac{\sin x}{\cos x}.$$

The function $y = \tan x$ is defined for all real numbers x except where $\cos x = 0$. Because $\cos x = 0$ for $x = \frac{\pi}{2} + n\pi$ for any integer n, we conclude that the domain of $\tan x$ is $(-\infty, \infty)$ with $x \neq \frac{\pi}{2} + n\pi$. In other words, $\tan x$ is undefined at $x = \pm\frac{\pi}{2}, \pm\frac{3\pi}{2}, \pm\frac{5\pi}{2}, \ldots$.

Range of tan x We note that the tangent of an angle θ in the standard position is the slope of the corresponding terminal side. Because every real number is the slope of some line through the origin, every number is the tangent of some angle in standard position. This means that the range of the tangent function is $(-\infty, \infty)$.

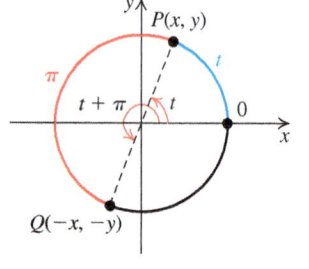

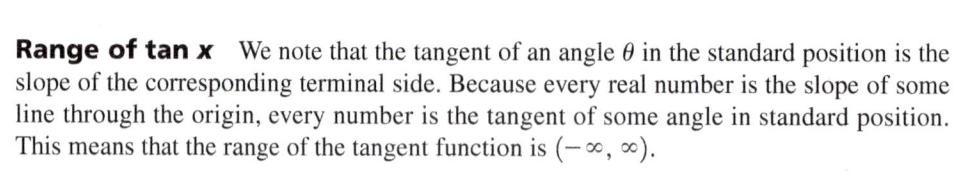

Zeros of tan x From $\tan x = \frac{\sin x}{\cos x}$, we have $\tan x = 0$ whenever $\sin x = 0$. Because $\sin x = 0$ for $x = n\pi$ for any integer n, we conclude that the zeros of $y = \tan x$ are $x = n\pi$ (Note that $\cos x \neq 0$ if $\sin x = 0$ because $\cos^2 x + \sin^2 x = 1$.)

Period of tan x The tangent function has period π. Figure 4.51 illustrates that $\tan(t + \pi) = \tan t$. Therefore, the values of the tangent function repeat every π units.

$\tan(t + \pi) = \frac{-y}{-x} = \frac{y}{x} = \tan t$

FIGURE 4.51 $\tan(t + \pi) = \tan t$

Even–Odd Property

$$\text{We have } \tan(-x) = \frac{\sin(-x)}{\cos(-x)} \qquad \text{Quotient identity}$$
$$= \frac{-\sin x}{\cos x} \qquad \text{Sine is odd, and cosine is even.}$$
$$= -\tan x \qquad \text{Quotient identity}$$

So $\tan(-x) = -\tan x$. Therefore, $y = \tan x$ is an odd function: its graph is symmetric about the origin.

Section 4.4 ■ Graphs of the Other Trigonometric Functions   329

2 Graph the tangent function.

Graph of $y = \tan x$

We first summarize the main facts about $y = \tan x$.

MAIN FACTS ABOUT TANGENT FUNCTION

1. Domain: All real numbers except odd multiples of $\dfrac{\pi}{2}$
2. Range: $(-\infty, \infty)$
3. Zeros or x-intercepts: All integer multiples of $\pi = n\pi$
4. Period: π
5. Vertical asymptotes: At $x =$ odd integer multiples of $\dfrac{\pi}{2}$, $x = (2n + 1)\dfrac{\pi}{2}$ for any integer
6. Symmetry: Odd function, so the graph is symmetric about the origin

TECHNOLOGY CONNECTION

To graph the tangent, cotangent, secant, or cosecant functions on a graphing calculator, use *dot* mode and appropriate window settings.

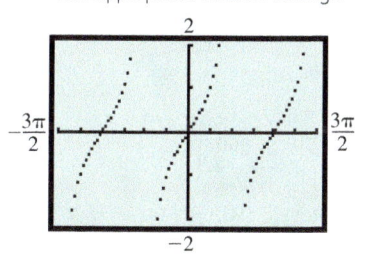

$Y_1 = \tan x$, dot mode

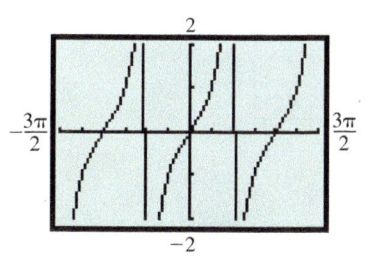

$Y_1 = \tan x$, Connected mode

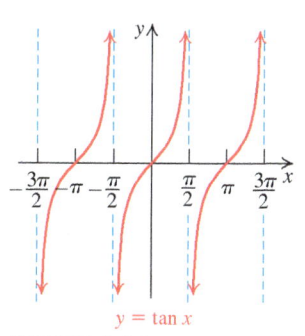

FIGURE 4.53

To graph $y = \tan x$, we use the approximate values of $\tan x$ for x in the interval $\left[0, \dfrac{\pi}{2}\right)$. See Table 4.7.

TABLE 4.7

x	0	$\dfrac{\pi}{6}$	$\dfrac{\pi}{4}$	$\dfrac{\pi}{3}$	$\dfrac{7\pi}{18}$	$\dfrac{4\pi}{9}$	$\dfrac{17\pi}{36}$
$y = \tan x$	0	0.6	1	1.7	2.7	5.7	11.4

As x approaches $\dfrac{\pi}{2}$, the values of $\tan x$ continue to increase; in fact, they increase indefinitely. The decimal value of $\dfrac{\pi}{2}$ is approximately 1.5707963. When $x = \dfrac{89\pi}{180} \approx 1.55$, $\tan x > 57$; when $x = 1.57$, $\tan x > 1255$; and when $x = 1.5707$, $\tan x > 10,000$. The graph of $y = \tan x$ has a vertical asymptote at $x = \dfrac{\pi}{2}$. See Figure 4.52(a).

(a)

(b)

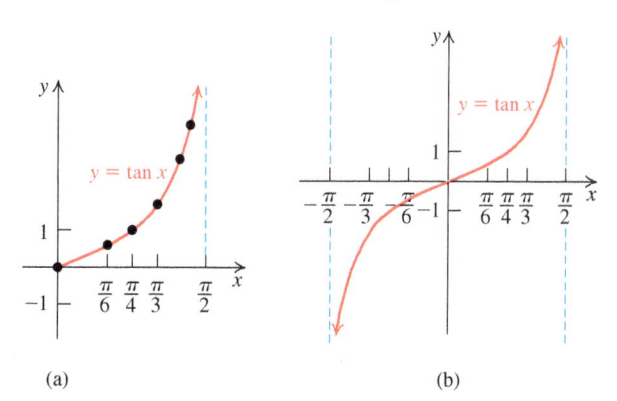

FIGURE 4.52 Graphing $y = \tan x$

We can extend the graph of $y = \tan x$ to the interval $\left(-\dfrac{\pi}{2}, \dfrac{\pi}{2}\right)$ by using the fact that the graph is symmetric with respect to the origin. See Figure 4.52(b).

Because the period of the tangent function is π and we have graphed $y = \tan x$ over an interval of length π, we can draw the complete graph of $y = \tan x$ by repeating the graph in Figure 4.52(b) indefinitely to the left and right over intervals of length π. Figure 4.53 shows three cycles of the graph.

Graphing $y = a \tan [b(x - c)]$ The procedure for graphing $y = a \tan [b(x - c)]$ is based on the essential features of the graph of $y = \tan x$.

PROCEDURE IN ACTION

EXAMPLE 2 Graphing $y = a \tan(bx - k)$

OBJECTIVE

Graph a function of the form $y = a \tan(bx - k)$, where $b > 0$, by finding the period and phase shift.

Step 1 If necessary, rewrite the equation in the form $y = a \tan[b(x - c)], c = \dfrac{k}{b}$.

Find the following:

vertical stretch factor $= |a|$

period $= \dfrac{\pi}{b}$

phase shift $= c$

Step 2 Locate two adjacent vertical asymptotes by solving the following equations for x:

$$b(x - c) = -\dfrac{\pi}{2} \quad \text{and} \quad b(x - c) = \dfrac{\pi}{2}$$

Step 3 Divide the interval on the x-axis between the two vertical asymptotes into four equal parts, each of length $\dfrac{1}{4}\left(\dfrac{\pi}{b}\right)$.

Step 4 Evaluate the function at the three x values found in Step 3 that are the division points of the interval. These values (from left to right) are: $-a$, 0, and a.

Step 5 Sketch the vertical asymptotes using the values found in Step 2. Connect the points in Step 4 with a smooth curve in the standard shape of a cycle for the tangent function. Repeat the graph to the left and right over intervals of length $\dfrac{\pi}{b}$.

EXAMPLE

Graph $y = 3 \tan\left(2x - \dfrac{\pi}{2}\right)$.

1. $y = 3 \tan\left[2\left(x - \dfrac{\pi}{4}\right)\right]$ Rewrite.

$a = 3 \quad b = 2 \quad c = \dfrac{\pi}{4}$

vertical stretch factor $= |a| = |3| = 3$

period $= \dfrac{\pi}{b} = \dfrac{\pi}{2}$ phase shift $= c = \dfrac{\pi}{4}$

2. $2\left(x - \dfrac{\pi}{4}\right) = -\dfrac{\pi}{2} \qquad\qquad 2\left(x - \dfrac{\pi}{4}\right) = \dfrac{\pi}{2}$

$x - \dfrac{\pi}{4} = -\dfrac{\pi}{4} \qquad\qquad x - \dfrac{\pi}{4} = \dfrac{\pi}{4}$

$x = -\dfrac{\pi}{4} + \dfrac{\pi}{4} \qquad\qquad x = \dfrac{\pi}{4} + \dfrac{\pi}{4}$

$x = 0 \qquad\qquad\qquad\quad x = \dfrac{\pi}{2}$

3. The interval $\left(0, \dfrac{\pi}{2}\right)$ has length $\dfrac{\pi}{2}$, and $\dfrac{1}{4}\left(\dfrac{\pi}{2}\right) = \dfrac{\pi}{8}$.

The division points of the interval $\left(0, \dfrac{\pi}{2}\right)$ are

$0 + \dfrac{\pi}{8} = \dfrac{\pi}{8}, \; 0 + 2\left(\dfrac{\pi}{8}\right) = \dfrac{\pi}{4}, \text{ and } 0 + 3\left(\dfrac{\pi}{8}\right) = \dfrac{3\pi}{8}.$

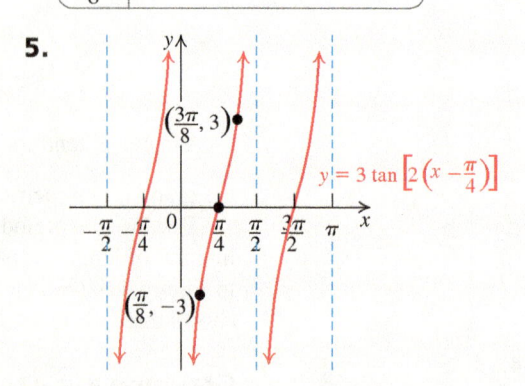

4.

x	$y = 3 \tan\left[2\left(x - \dfrac{\pi}{4}\right)\right]$
$\dfrac{\pi}{8}$	-3
$\dfrac{\pi}{4}$	0
$\dfrac{3\pi}{8}$	3

5.

Practice Problem 2 Graph $y = -3 \tan \left(x + \dfrac{\pi}{4} \right)$.

The fact that the tangent function is an odd function allows us to graph $y = a \tan [b(x - c)]$ when $b < 0$. For example, to graph $y = 3 \tan \left[(-2)\left(x - \dfrac{\pi}{2} \right) \right]$, we rewrite the equation as $y = -3 \tan \left[2\left(x - \dfrac{\pi}{2} \right) \right]$. Graph $y = 3 \tan \left[2\left(x - \dfrac{\pi}{2} \right) \right]$ and reflect it about the x-axis.

3 Graph the cosecant, secant, and cotangent functions.

Graphs of the Reciprocal Functions

The cosecant, secant, and cotangent functions are reciprocals of the sine, cosine, and tangent functions, respectively. We first consider some relationships between the properties of any pair of these reciprocal functions. We use these relationships to sketch the graph of the reciprocal function of a trigonometric function $g(x)$.

> **THE GRAPH OF THE RECIPROCAL OF A TRIGONOMETRIC FUNCTION**
>
> Let $f(x)$ be the reciprocal of $g(x)$: $f(x) = \dfrac{1}{g(x)}$, where g is any trigonometric function.
>
> - **Periodicity** If $g(x)$ has period p, then $f(x)$ also has period p.
> - **Zeros** If $g(c) = 0$, then $f(c)$ is undefined. So if c is an x-intercept of the graph of g, then the line $x = c$ is a vertical asymptote of the graph of f. Conversely, if $g(d)$ is undefined, then $f(d) = 0$.
> - **Even–Odd**
>   a. If $g(x)$ is odd, then $f(x)$ is odd.
>   b. If $g(x)$ is even, then $f(x)$ is even.
> - **Special Values**
>   a. If $g(x_1) = 1$, then $f(x_1) = 1$. Both graphs pass through the point $(x_1, 1)$.
>   b. If $g(x_2) = -1$, then $f(x_2) = -1$. Both graphs pass through the point $(x_2, -1)$.
> - **Sign**
>   a. If $g(x) > 0$ on an interval (a, b), then $f(x) > 0$ on the interval (a, b). Both graphs are above the x-axis on the interval (a, b).
>   b. If $g(x) < 0$ on an interval (c, d), then $f(x) < 0$ on the interval (c, d). Both graphs are below the x-axis on the interval (c, d).
> - **Increasing-Decreasing**
>   a. If $g(x)$ is increasing on an interval (a, b), then $f(x)$ is decreasing on the interval (a, b).
>   b. If $g(x)$ is decreasing on an interval (c, d), then $f(x)$ is increasing on the interval (c, d).
> - **Magnitude**
>   a. If $|g(x)|$ is small, then $|f(x)|$ is large.
>   b. If $|g(x)|$ is large, then $|f(x)|$ is small. If c is in the domain of f and $|g(x)|$ approaches ∞ as x approaches c, then $f(c) = 0$.

We use the properties of the reciprocal functions to sketch the graphs of the cosecant, secant, and cotangent functions from the graphs of the sine, cosine, and tangent functions. Two or more cycles of these graphs along with their reciprocals are shown in the following box.

PROPERTIES OF THE COSECANT, SECANT, AND COTANGENT FUNCTIONS

	$y = \csc x = \dfrac{1}{\sin x}$	$y = \sec x = \dfrac{1}{\cos x}$	$y = \cot x = \dfrac{1}{\tan x}$
For any integer n:			
Domain	All real numbers $x \neq n\pi$	All real numbers $x \neq \dfrac{\pi}{2} + n\pi$	All real numbers $x \neq n\pi$
Range	$(-\infty, -1] \cup [1, \infty)$	$(-\infty, -1] \cup [1, \infty)$	$(-\infty, \infty)$
Period	2π	2π	π
x-intercepts	No x-intercepts	No x-intercepts	$x = \dfrac{\pi}{2} + n\pi$
Even–Odd	Odd	Even	Odd
Vertical Asymptotes	$x = n\pi$	$x = \dfrac{\pi}{2} + n\pi$	$x = n\pi$
Graph	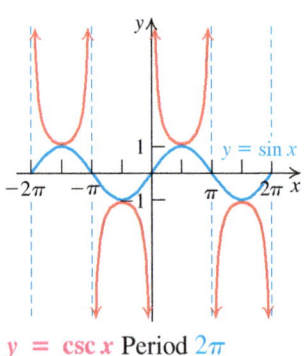 $y = \csc x$ Period 2π	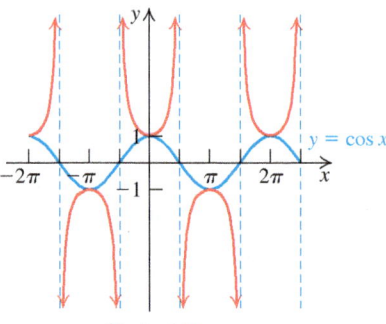 $y = \sec x$ Period 2π	$y = \cot x$ Period π

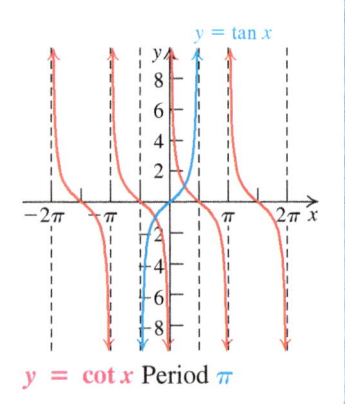

To sketch the graph of $y = a \cot(bx - k)$, we use the steps that are similar to those used in Example 2.

EXAMPLE 3   Graphing $y = a\cot(bx - k)$

Graph $y = -4\cot\left(x - \dfrac{\pi}{2}\right)$ over the interval $[-\pi, 2\pi]$.

Solution

Step 1   For $y = -4\cot\left(x - \dfrac{\pi}{2}\right)$, we have $a = -4$, $b = 1$, and $c = \dfrac{k}{b} = \dfrac{\pi}{2}$.

Therefore, the vertical stretch factor $= |-4| = 4$, the period $= \dfrac{\pi}{1} = \pi$,

and the phase shift $= \dfrac{\pi}{2}$.

Step 2   Locate two adjacent asymptotes for a cotangent function. Solve the equations:

$$x - \dfrac{\pi}{2} = 0 \quad \text{and} \quad x - \dfrac{\pi}{2} = \pi$$

$$x = \dfrac{\pi}{2} \qquad\qquad x = \dfrac{3\pi}{2}$$

Section 4.4 ■ Graphs of the Other Trigonometric Functions   333

Step 3 The interval $\left(\dfrac{\pi}{2}, \dfrac{3\pi}{2}\right)$ has length π, the period of the cotangent function.

The division points of $\left(\dfrac{\pi}{2}, \dfrac{3\pi}{2}\right)$ are $\dfrac{\pi}{2} + \dfrac{1}{4}(\pi) = \dfrac{3\pi}{4}$,

$\dfrac{\pi}{2} + \dfrac{2}{4}(\pi) = \pi$, and $\dfrac{\pi}{2} + \dfrac{3}{4}(\pi) = \dfrac{5\pi}{4}$.

Step 4 Evaluate the function at $\dfrac{3\pi}{4}$, π, and $\dfrac{5\pi}{4}$.

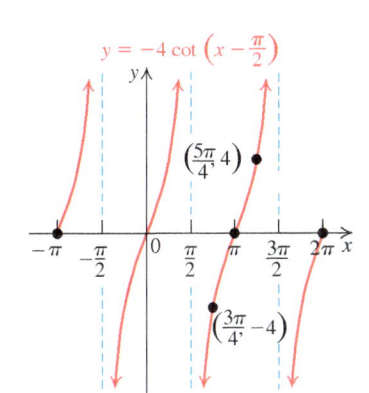

x	$y = -4\cot\left(x - \dfrac{\pi}{2}\right)$
$\dfrac{3\pi}{4}$	-4
π	0
$\dfrac{5\pi}{4}$	4

FIGURE 4.54 Graph of $y = -4\cot\left(x - \dfrac{\pi}{2}\right)$

Step 5 Sketch the vertical asymptotes at $x = \dfrac{\pi}{2}$ and $x = \dfrac{3\pi}{2}$. Draw the cycle for the cotangent function through the three graph points found in Step 4. Repeat the graph to the left and right over intervals of length π. See Figure 4.54.

Practice Problem 3 Graph $y = -3\cot\left(x + \dfrac{\pi}{4}\right)$ over the interval $\left(-\dfrac{3\pi}{4}, \dfrac{5\pi}{4}\right)$.

Graphing $y = a\csc(bx - k) + d$ and $y = a\sec(bx - k) + d$

The procedures for graphing the functions

$$y = a\csc(bx - k) + d \quad \text{and} \quad y = a\sec(bx - k) + d$$

rely on graphing the corresponding functions

$$y = a\sin(bx - k) + d \text{ and } y = a\cos(bx - k) + d,$$

respectively.

EXAMPLE 4 Graphing $y = a\csc(bx - k) + d$

Graph one cycle of $y = 2\csc(x - \pi) + 1$.

Solution

Step 1 Graph one cycle of the corresponding sine function on the interval $[\pi, 3\pi]$.

$$y = 2\sin(x - \pi) + 1 \quad \text{Replace csc with sin.}$$

See the blue graph in Figure 4.55.

Step 2 Recall that $y = \csc x$ has vertical asymptotes at places where $\sin x = 0$. For a function of the form $y = a\csc(bx - k) + d$, the vertical asymptotes will occur where $\sin(bx - k) = 0$. For the graph of $y = 2\csc(x - \pi) + 1$, on the interval $[\pi, 3\pi]$, the vertical asymptotes will occur at $x = \pi$, $x = 2\pi$, and $x = 3\pi$. See Figure 4.55.

334 Chapter 4 Trigonometric Functions

FIGURE 4.55

Step 3 To obtain the graph of $y = 2 \csc(x - \pi) + 1$

a. take the portion of the graph of $y = 2\sin(x - \pi) + 1$ that lies above the line $y = 1$, reflect it about the horizontal line $y = 3$, and stretch it vertically up. See the red graph on the interval $[\pi, 2\pi]$.

b. take the portion of the graph of $y = 2\sin(x - \pi) + 1$ that lies below the line $y = 1$, reflect it about the horizontal line $y = -1$, and stretch it vertically down. See the red graph on the interval $[2\pi, 3\pi]$.

Practice Problem 4 Graph $y = 2\sec 3x$ over a two-period interval.

EXAMPLE 5   Graphing a Range of Mach Numbers

When a plane travels at supersonic and hypersonic speeds, small disturbances in the atmosphere are transmitted downstream within a cone. The cone intersects the ground. Figure 4.56 shows the edge of the cone's intersection with the ground. The sound waves strike the edge of the cone at a right angle. The speed of the sound wave is represented by leg s of the right triangle in Figure 4.56. The plane is moving at speed v, which is represented by the hypotenuse of the right triangle in this figure.

The Mach number, M, is given by

$$M = M(x) = \frac{\text{speed of the aircraft}}{\text{speed of sound}} = \frac{v}{s} = \csc\left(\frac{x}{2}\right),$$

where x is the angle at the vertex of the cone. Graph the Mach number function, $M(x)$, as the angle at the vertex of the cone varies. What is the range of Mach numbers associated with the interval $\left[\dfrac{\pi}{4}, \pi\right)$?

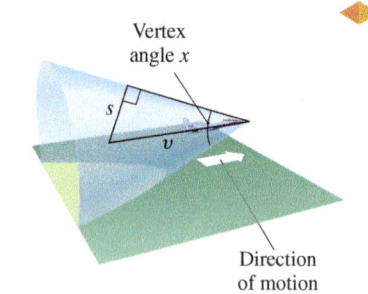

FIGURE 4.56 Sonic cone

Solution

Because $\csc\dfrac{x}{2} = \dfrac{1}{\sin\dfrac{x}{2}}$, first graph $y = \sin\dfrac{x}{2}$. For convenience, we have sketched the graph of $y = \sin\dfrac{x}{2}$ over the interval $[0, 2\pi]$ in Figure 4.57. The graph of $y = \csc\dfrac{x}{2}$ is sketched over the interval $(0, 2\pi)$ using the reciprocal relationship between the sine graph and the cosecant graph.

For $x = \dfrac{\pi}{4}$, $y = \csc\dfrac{x}{2} = \csc\dfrac{\pi}{8} \approx 2.6$.

For $x = \pi$, $y = \csc\dfrac{x}{2} = \csc\dfrac{\pi}{2} = 1$.

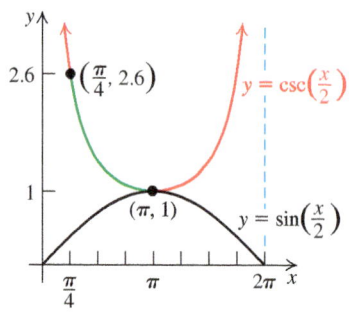

FIGURE 4.57 Mach numbers

The range of Mach numbers associated with the interval $\left[\dfrac{\pi}{4}, \pi\right)$ is $(1, 2.6]$.

Practice Problem 5 In Example 5, what is the range of Mach numbers associated with the interval $\left[\dfrac{\pi}{8}, \dfrac{\pi}{4}\right]$?

SECTION 4.4 Exercises

Basic Concepts and Skills

In Exercises 1–4, find the slope-intercept form of the equation of each line that passes through the point P and makes angle θ with the positive x-axis.

1. $P = (-2, 3), \theta = 45°$
2. $P = (3, -1), \theta = 60°$
3. $P = (-3, -2), \theta = 120°$
4. $P = (2, 5), \theta = 135°$

In Exercises 5–18, graph each function over a one-period interval.

5. $y = \tan\left(x - \dfrac{\pi}{4}\right)$
6. $y = \cot\left(x + \dfrac{\pi}{4}\right)$
7. $y = \tan 2x$
8. $y = \tan \dfrac{x}{2}$
9. $y = \cot \dfrac{x}{2}$
10. $y = \cot 2x$
11. $y = 3 \tan x$
12. $y = 3 \cot x$
13. $y = \sec 2x$
14. $y = \sec \dfrac{x}{2}$
15. $y = \csc 3x$
16. $y = \csc \dfrac{x}{3}$
17. $y = \sec(x - \pi)$
18. $y = \csc(x - \pi)$

In Exercises 19–36, graph each function over a two-period interval.

19. $y = \tan\left[2\left(x + \dfrac{\pi}{2}\right)\right]$
20. $y = \tan\left[2\left(x - \dfrac{\pi}{2}\right)\right]$
21. $y = \cot\left[2\left(x - \dfrac{\pi}{2}\right)\right]$
21. $y = \cot\left[2\left(x + \dfrac{\pi}{2}\right)\right]$
23. $y = \tan\left[\dfrac{1}{2}(x + 2\pi)\right]$
24. $y = \tan\left[\dfrac{1}{2}(x - 2\pi)\right]$
25. $y = \cot\left[\dfrac{1}{2}(x - 2\pi)\right]$
26. $y = \cot\left[\dfrac{1}{2}(x + 2\pi)\right]$
27. $y = \sec 4\left(x - \dfrac{\pi}{4}\right)$
28. $y = \sec \dfrac{1}{2}\left(x - \dfrac{\pi}{2}\right)$
29. $y = 3 \csc\left(x + \dfrac{\pi}{2}\right)$
30. $y = 3 \sec\left[2\left(x - \dfrac{\pi}{6}\right)\right]$
31. $y = \tan\left[\dfrac{2}{3}\left(x - \dfrac{\pi}{2}\right)\right]$
32. $y = 2 \cot\left[2\left(x - \dfrac{\pi}{6}\right)\right]$
33. $y = -5 \tan\left[2\left(x + \dfrac{\pi}{3}\right)\right]$
34. $y = -3 \cot\left[\dfrac{1}{2}\left(x - \dfrac{\pi}{3}\right)\right]$
35. $y = \dfrac{1}{3} \cot[2(x - \pi)]$
36. $y = \dfrac{1}{2} \tan\left[4\left(x - \dfrac{\pi}{6}\right)\right]$

Applying the Concepts

37. **Prison searchlight.** A dual-beam rotating light on a movie set is positioned as a spotlight shining on a prison wall. The light is 20 feet from the wall and rotates clockwise. The light shines on point P on the wall when first turned on ($t = 0$). After t seconds, the distance (in feet) from the beam on the wall to the point P is given by the function

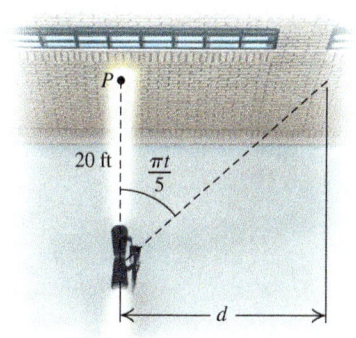

$$d(t) = 20 \tan \dfrac{\pi t}{5}.$$

When the light beam is to the right of P, the value of d is positive, and when the beam is to the left of P, the value of d is negative.
a. Graph d over the interval $0 \le t \le 5$.
b. Because $d(t)$ is undefined for $t = 2.5$, where is the rotating light pointing when $t = 2.5$?

38. **Prison searchlight.** In Exercise 37, find the value for b assuming that the light beam is to sweep the entire wall in 10 seconds and $d(t) = 20 \tan bt$.

Critical Thinking / Discussion / Writing

39. For what number k, with $-2\pi < k < 0$, is $x = k$ a vertical asymptote for $y = \cot\left[\dfrac{1}{2}\left(x - \dfrac{\pi}{4}\right)\right]$?

40. For what number b does $y = \sec bx$ have period $\dfrac{\pi}{3}$?

41. Write an equation in the form $y = a \csc[b(x - c)]$ that has the same graph as $y = -2 \sec x$.

In Exercises 42–45, write an equation for each graph.

42.

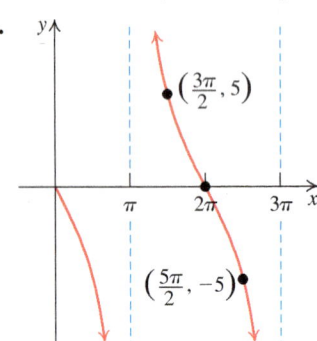

43.

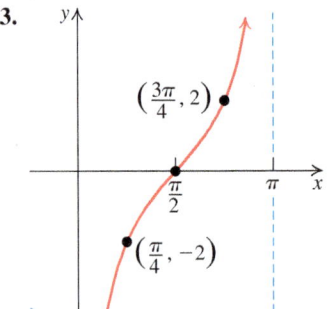

44.

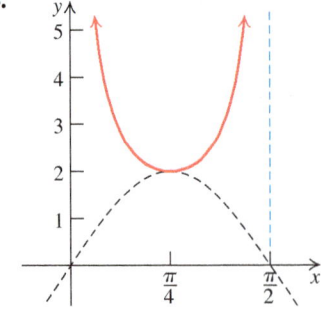

45.

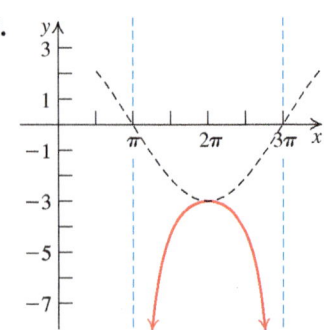

Modeling with tangent function. In Exercises 46 and 47, model the given data by using the function $y = a \tan[b(x - c)]$. Use your equation to find y when $x = -1.5$ and $x = 1.5$. Round your answers to two decimal places.

46.

x	y
-5	undef. $(-\infty)$
-3	-7
-1	0
1	7
3	undef. $(+\infty)$

47.

x	y
-3	undef. $(+\infty)$
-0.5	6
2	0
4.5	-6
7	undef. $(-\infty)$

Maintaining Skills

In Exercises 48–50, state whether the given function is one-to-one. For each one-to-one function, find its inverse.

48. $f(x) = 2x - 3$
49. $g(x) = x^2 + 1$
50. $f(x) = 2x^2, x \geq 0$
51. **True or False.** If f is one-to-one, then it has an inverse function.
52. **True or False.** For a one-to-one function f, the graph of f^{-1} is the reflection of the graph of f about the line $y = x$.

SECTION 4.5

Inverse Trigonometric Functions

BEFORE STARTING THIS SECTION, REVIEW

1. Inverse functions (Section 1.7, page 136)
2. Composition of functions (Section 1.6, page 126)
3. Exact values of the trigonometric functions (Section 4.2, page 302)

OBJECTIVES

1. Graph and apply the inverse sine function.
2. Graph and apply the inverse cosine function.
3. Graph and apply the inverse tangent function.
4. Evaluate inverse trigonometric functions.
5. Find exact values of composite functions involving the inverse trigonometric functions.

◆ Retail Theft

In 2005, security cameras at Filene's Basement store in Boston filmed a theft coordinated by a thief and an accomplice. The accomplice distracted the salesperson so that the thief could steal a $16,000 necklace. Retail theft is a major concern for all retail outlets.

The National Retail Federation, the industry's largest trade group, and the Retail Industry Leaders Association have instituted password-protected national crime databases online. These databases allow retailers to share information about thefts and determine whether they have been a target of individual shoplifters who steal for themselves or a target of organized crime. In addition to participating in the shared databases, many large retailers have their own organized anti-crime squads. One estimate puts loss to organized theft at over $30 billion annually. In Example 12, we see how methods in this section can be used in an attempt to prevent loss from theft.

The Inverse Sine Function

1 Graph and apply the inverse sine function.

Recall that a function f has an inverse function if no horizontal line intersects the graph of f in more than one point. Because every horizontal line $y = b$, where $-1 \leq b \leq 1$, intersects the graph of $y = \sin x$ at more than one point, the sine function fails the horizontal line test; so it is not one-to-one and consequently has no inverse.

The solid portion of the sine graph shown in Figure 4.58 is the graph of $y = \sin x$ for $-\frac{\pi}{2} \leq x \leq \frac{\pi}{2}$. If we restrict the domain of $y = \sin x$ to the interval $\left[-\frac{\pi}{2}, \frac{\pi}{2}\right]$, the resulting function

$$y = \sin x, -\frac{\pi}{2} \leq x \leq \frac{\pi}{2}$$

is one-to-one (it passes the horizontal line test); so it has an inverse. Notice that the restricted function takes on all values in the range of $y = \sin x$, which is $[-1, 1]$, and that each of

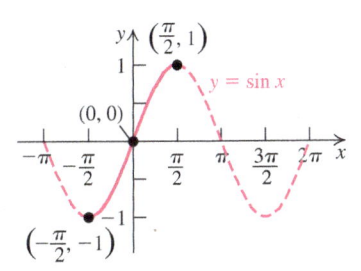

FIGURE 4.58 $y = \sin x$, $-\frac{\pi}{2} \leq x \leq \frac{\pi}{2}$

these y-values corresponds to exactly one x-value in the restricted domain $\left[-\frac{\pi}{2}, \frac{\pi}{2}\right]$. The inverse function for $y = \sin x$, $-\frac{\pi}{2} \leq x \leq \frac{\pi}{2}$, is called the **inverse sine**, or **arcsine** function and is denoted by $\sin^{-1} x$ or $\arcsin x$.

Inverse Sine Function

The equation $y = \sin^{-1} x$ means $\sin y = x$, where $-1 \leq x \leq 1$ and $-\frac{\pi}{2} \leq y \leq \frac{\pi}{2}$.

Read $y = \sin^{-1} x$ as "y equals inverse sine at x."

The range of $y = \sin x$ is $[-1, 1]$; so the domain of $y = \sin^{-1} x$ is $[-1, 1]$. The domain of the restricted sine function is $\left[-\frac{\pi}{2}, \frac{\pi}{2}\right]$; so the range of $y = \sin^{-1} x$ is $\left[-\frac{\pi}{2}, \frac{\pi}{2}\right]$. We can graph $y = \sin^{-1} x$ by reflecting the graph of $y = \sin x$, for $-\frac{\pi}{2} \leq x \leq \frac{\pi}{2}$, about the line $y = x$. See Figure 4.59.

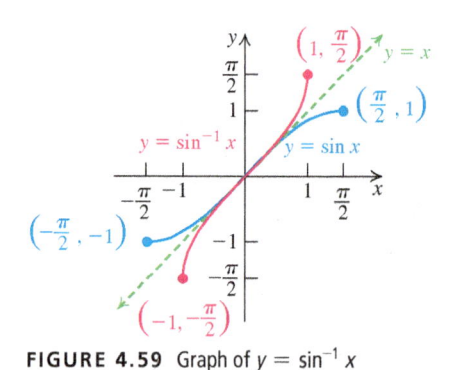

FIGURE 4.59 Graph of $y = \sin^{-1} x$

RECALL

If two functions are inverses, their graphs are symmetric with respect to the line $y = x$.

EXAMPLE 1 Finding the Exact Values for $y = \sin^{-1} x$

Find the exact values of y.

a. $y = \sin^{-1} \frac{\sqrt{3}}{2}$ **b.** $y = \sin^{-1}\left(-\frac{1}{2}\right)$ **c.** $y = \sin^{-1} 3$

Solution

a. The equation $y = \sin^{-1} \frac{\sqrt{3}}{2}$ means $\sin y = \frac{\sqrt{3}}{2}$ and $-\frac{\pi}{2} \leq y \leq \frac{\pi}{2}$.

Because $\sin \frac{\pi}{3} = \frac{\sqrt{3}}{2}$ and $-\frac{\pi}{2} \leq \frac{\pi}{3} \leq \frac{\pi}{2}$, we have $y = \sin^{-1} \frac{\sqrt{3}}{2} = \frac{\pi}{3}$.

b. The equation $y = \sin^{-1}\left(-\frac{1}{2}\right)$ means $\sin y = -\frac{1}{2}$ and $-\frac{\pi}{2} \leq y \leq \frac{\pi}{2}$.

Because $\sin\left(-\frac{\pi}{6}\right) = -\frac{1}{2}$ and $-\frac{\pi}{2} \leq -\frac{\pi}{6} \leq \frac{\pi}{2}$, we have $y = -\frac{\pi}{6}$.

c. Since 3 is not in the domain of the inverse sine function $[-1, 1]$, $\sin^{-1} 3$ does not exist.

SIDE NOTE

You can also read $y = \sin^{-1} x$ as "y is the number in the interval $\left[-\frac{\pi}{2}, \frac{\pi}{2}\right]$ whose sine is x."

Practice Problem 1 Find the exact values of y.

a. $y = \sin^{-1}\left(-\frac{\sqrt{3}}{2}\right)$ **b.** $y = \sin^{-1}(-1)$

The Inverse Cosine Function

2 Graph and apply the inverse cosine function.

When we restrict the domain of $y = \cos x$ to the interval $[0, \pi]$, the resulting function, $y = \cos x$ (with $0 \leq x \leq \pi$), is one-to-one. No horizontal line intersects the graph of $y = \cos x$, with $0 \leq x \leq \pi$, in more than one point. See Figure 4.60. Consequently, the restricted cosine function has an inverse function.

The inverse function for $y = \cos x$, $0 \leq x \leq \pi$, is called the **inverse cosine**, or **arccosine** function and is denoted by $\cos^{-1} x$, or $\arccos x$.

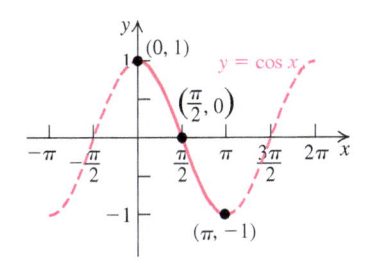

FIGURE 4.60 $y = \cos x, 0 \leq x \leq \pi$

Inverse Cosine Function

The equation $y = \cos^{-1} x$ means $\cos y = x$ where $-1 \leq x \leq 1$ and $0 \leq y \leq \pi$. Read $y = \cos^{-1} x$ as "y equals inverse cosine at x."

SIDE NOTE

You can also read $y = \cos^{-1} x$ as "y is the number in the interval $[0, \pi]$ whose cosine is x."

Reflecting the graph of $y = \cos x$, for $0 \leq x \leq \pi$, about the line, $y = x$ produces the graph of $y = \cos^{-1} x$, shown in Figure 4.61.

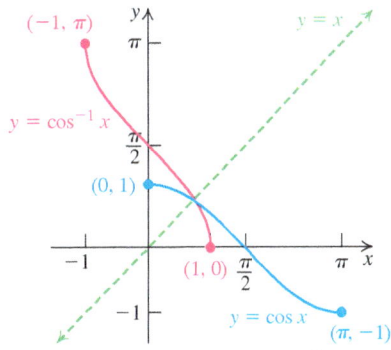

FIGURE 4.61 Graph of $y = \cos^{-1} x$

EXAMPLE 2 Finding an Exact Value for $\cos^{-1} x$

Find the exact value of y.

a. $y = \cos^{-1} \dfrac{\sqrt{2}}{2}$ **b.** $y = \cos^{-1}\left(-\dfrac{1}{2}\right)$

Solution

a. The equation $y = \cos^{-1} \dfrac{\sqrt{2}}{2}$ means $\cos y = \dfrac{\sqrt{2}}{2}$ and $0 \leq y \leq \pi$.

Because $\cos \dfrac{\pi}{4} = \dfrac{\sqrt{2}}{2}$ and $0 \leq \dfrac{\pi}{4} \leq \pi$, we have $y = \dfrac{\pi}{4}$.

b. The equation $y = \cos^{-1}\left(-\dfrac{1}{2}\right)$ means $\cos y = -\dfrac{1}{2}$ and $0 \leq y \leq \pi$.

Because $\cos \dfrac{2\pi}{3} = -\dfrac{1}{2}$ and $0 \leq \dfrac{2\pi}{3} \leq \pi$, we have $y = \dfrac{2\pi}{3}$.

Practice Problem 2 Find the exact value of y.

a. $y = \cos^{-1}\left(-\dfrac{\sqrt{2}}{2}\right)$ **b.** $y = \cos^{-1} \dfrac{1}{2}$

3 Graph and apply the inverse tangent function.

The Inverse Tangent Function

The *inverse tangent function* results from restricting the domain of the tangent function to the interval $\left(-\dfrac{\pi}{2}, \dfrac{\pi}{2}\right)$ to obtain a one-to-one function. The inverse of this restricted tangent function is the **inverse tangent**, or **arctangent** function and is denoted by $\tan^{-1} x$, or $\arctan x$.

> **SIDE NOTE**
>
> You can also read $y = \tan^{-1} x$ as "y is the number in the interval $\left(-\dfrac{\pi}{2}, \dfrac{\pi}{2}\right)$ whose tangent is x."

Inverse Tangent Function

The equation $y = \tan^{-1} x$ means $\tan y = x$, where $-\infty < x < \infty$ and $-\dfrac{\pi}{2} < y < \dfrac{\pi}{2}$.

Read $y = \tan^{-1} x$ as "y equals the inverse tangent at x."

The graph of $y = \tan^{-1} x$ is obtained by reflecting the graph of $y = \tan x$, with $-\dfrac{\pi}{2} < x < \dfrac{\pi}{2}$, about the line $y = x$. Figure 4.62 shows the graph of the restricted tangent function. Figure 4.63 shows the graph of $y = \tan^{-1} x$.

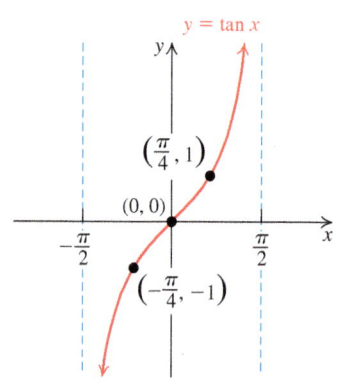

FIGURE 4.62 $y = \tan x, -\dfrac{\pi}{2} < x < \dfrac{\pi}{2}$

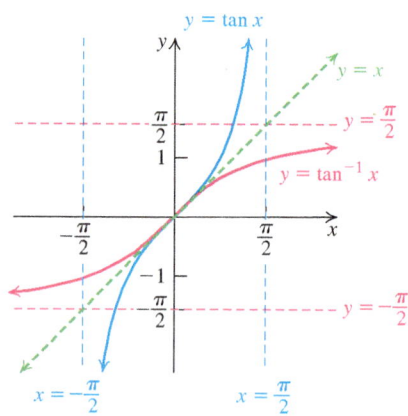

FIGURE 4.63 Graph of $y = \tan^{-1} x$

EXAMPLE 3 Finding the Exact Value for $\tan^{-1} x$

Find the exact value of y.

a. $y = \tan^{-1} 0$ **b.** $y = \tan^{-1} (-\sqrt{3})$

Solution

a. Because $\tan 0 = 0$ and $-\dfrac{\pi}{2} < 0 < \dfrac{\pi}{2}$, we have $y = 0$.

b. Because $\tan\left(-\dfrac{\pi}{3}\right) = -\sqrt{3}$ and $-\dfrac{\pi}{2} < -\dfrac{\pi}{3} < \dfrac{\pi}{2}$, we have $y = -\dfrac{\pi}{3}$.

Practice Problem 3 Find the exact value of $y = \tan^{-1} \dfrac{\sqrt{3}}{3}$.

Other Inverse Trigonometric Functions

Sometimes the ranges of the *inverse secant* and *inverse cosecant* functions differ from those used in this text. Always check the definitions of the domains of these two functions when they are used outside this course.

Inverse Cotangent, Cosecant, Secant Functions

Inverse cotangent $y = \cot^{-1} x$ means $\cot y = x$,
where $-\infty < x < \infty$ and $0 < y < \pi$.

Inverse cosecant $y = \csc^{-1} x$ means $\csc y = x$,
where $|x| \geq 1$ and $-\dfrac{\pi}{2} \leq y \leq \dfrac{\pi}{2}, y \neq 0$.

Inverse secant $y = \sec^{-1} x$ means $\sec y = x$,
where $|x| \geq 1$ and $0 \leq y \leq \pi, y \neq \dfrac{\pi}{2}$.

EXAMPLE 4 Finding the Exact Value for $\csc^{-1} x$

Find the exact value for $y = \csc^{-1} 2$.

Solution

Because $\csc \dfrac{\pi}{6} = 2$ and $-\dfrac{\pi}{2} \leq \dfrac{\pi}{6} \leq \dfrac{\pi}{2}$, we have $y = \csc^{-1} 2 = \dfrac{\pi}{6}$.

Practice Problem 4 Find the exact value of $y = \sec^{-1} 2$.

SUMMARY OF MAIN FACTS

Inverse Trigonometric Functions

Inverse Function	Equivalent to	Domain	Range
$y = \sin^{-1} x$	$\sin y = x$	$[-1, 1]$	$\left[-\dfrac{\pi}{2}, \dfrac{\pi}{2}\right]$
$y = \cos^{-1} x$	$\cos y = x$	$[-1, 1]$	$[0, \pi]$
$y = \tan^{-1} x$	$\tan y = x$	$(-\infty, \infty)$	$\left(-\dfrac{\pi}{2}, \dfrac{\pi}{2}\right)$
$y = \cot^{-1} x$	$\cot y = x$	$(-\infty, \infty)$	$(0, \pi)$
$y = \csc^{-1} x$	$\csc y = x$	$(-\infty, -1] \cup [1, \infty)$	$\left[-\dfrac{\pi}{2}, 0\right) \cup \left(0, \dfrac{\pi}{2}\right]$
$y = \sec^{-1} x$	$\sec y = x$	$(-\infty, -1] \cup [1, \infty)$	$\left[0, \dfrac{\pi}{2}\right) \cup \left(\dfrac{\pi}{2}, \pi\right]$

4 Evaluate inverse trigonometric functions.

Evaluating Inverse Trigonometric Functions

In Section 4.2, we defined the six trigonometric functions of *real numbers*, and in this section, we have defined the corresponding six inverse trigonometric functions of real numbers. For example,

$$\sin \dfrac{\pi}{4} = \dfrac{\sqrt{2}}{2}$$

and

$$\sin^{-1} \dfrac{\sqrt{2}}{2} = \dfrac{\pi}{4}.$$

TECHNOLOGY CONNECTION

The secondary functions on your calculator, labeled SIN⁻¹, COS⁻¹, and TAN⁻¹, are associated with the keys labeled [SIN], [COS], and [TAN], respectively. Consult your manual to learn how to access these secondary functions. The screen shows several values for the trigonometric inverse functions on a calculator set to *radian* mode.

```
cos⁻¹(3/4)
           .723
sin⁻¹(-0.86)
         -1.035
tan⁻¹(-6.25)
         -1.412
```

However, because we also defined the trigonometric functions of *angles* in Section 4.2, it is meaningful when working with angles in degree measure to write a statement such as

$$\sin^{-1}\frac{\sqrt{2}}{2} = 45°.$$

We state some useful identities involving inverse trigonometric functions of $\frac{1}{x}$ and $-x$ in terms of the inverse trigonometric functions of x.

INVERSE TRIGONOMETRIC IDENTITIES

For $\frac{1}{x}$	For $-x$
$\sin^{-1}\left(\frac{1}{x}\right) = \csc^{-1} x, \ |x| \geq 1$	$\sin^{-1}(-x) = -\sin^{-1} x, \ |x| \leq 1$
$\csc^{-1}\left(\frac{1}{x}\right) = \sin^{-1} x, \ |x| \leq 1$	$\csc^{-1}(-x) = -\csc^{-1} x, \ |x| \geq 1$
$\cos^{-1}\left(\frac{1}{x}\right) = \sec^{-1} x, \ |x| \geq 1$	$\cos^{-1}(-x) = \pi - \cos^{-1} x, \ |x| \leq 1$
$\sec^{-1}\left(\frac{1}{x}\right) = \cos^{-1} x, \ |x| \leq 1$	$\sec^{-1}(-x) = \pi - \sec^{-1} x, \ |x| \geq 1$
$\tan^{-1}\left(\frac{1}{x}\right) = \begin{cases} \cot^{-1} x, & x > 0 \\ -\pi + \cot^{-1} x, & x < 0 \end{cases}$	$\tan^{-1}(-x) = -\tan^{-1} x, \ -\infty < x < \infty$
$\cot^{-1}\left(\frac{1}{x}\right) = \begin{cases} \tan^{-1} x, & x > 0 \\ \pi + \tan^{-1} x, & x < 0 \end{cases}$	$\cot^{-1}(-x) = \pi - \cot^{-1} x, \ -\infty < x < \infty$

Note that the equation $\tan^{-1}\left(\frac{1}{x}\right) = \cot^{-1} x$ is true for $x > 0$, because the values of both sides are in the interval $\left(0, \frac{\pi}{2}\right)$ and $\tan\left[\tan^{-1}\left(\frac{1}{x}\right)\right] = \frac{1}{x}$, $\tan(\cot^{-1} x) = \frac{1}{\cot(\cot^{-1} x)} = \frac{1}{x}$. But when $x < 0$, $\tan^{-1}\left(\frac{1}{x}\right)$ is negative and lies in the interval $\left(-\frac{\pi}{2}, 0\right)$ and $\cot^{-1} x$ is positive and lies in the interval $\left(\frac{\pi}{2}, \pi\right)$. So for $x < 0$, the correct equation is $\tan^{-1}\left(\frac{1}{x}\right) = -\pi + \cot^{-1} x$. In other words, for $x < 0$, we have $\cot^{-1} x = \pi + \tan^{-1}\left(\frac{1}{x}\right)$.

EXAMPLE 5 Verifying an Inverse Trigonometric Identity

a. Show that $\cos^{-1}(-x) = \pi - \cos^{-1} x$ for $|x| \leq 1$.

b. Use part **a** to evaluate $\sin\left[\cos^{-1}\left(-\frac{1}{2}\right)\right]$.

Solution

a. For $|x| \leq 1$, let $\theta = \cos^{-1} x$ with θ in the interval $[0, \pi]$. Then

$$\cos \theta = x \qquad \text{Definition of } \cos^{-1} x$$

For θ in $[0, \pi]$, $(\pi - \theta)$ is also in $[0, \pi]$ and

$$\cos(\pi - \theta) = -\cos\theta \qquad \text{See page 309 Exercise 78.}$$
$$\cos(\pi - \theta) = -x \qquad \text{Replace } \cos\theta \text{ with } x.$$
$$\pi - \theta = \cos^{-1}(-x) \qquad \text{Definition of } \cos^{-1} x$$
$$\pi - \cos^{-1} x = \cos^{-1}(-x) \qquad \text{Replace } \theta \text{ with } \cos^{-1} x.$$

b.
$$\sin\left[\cos^{-1}\left(-\frac{1}{2}\right)\right] = \sin\left[\pi - \cos^{-1}\left(\frac{1}{2}\right)\right] \qquad \text{By part a}$$
$$= \sin\left[\pi - \frac{\pi}{3}\right] \qquad \cos^{-1}\left(\frac{1}{2}\right) = \frac{\pi}{3}$$
$$= \sin\left(\frac{\pi}{3}\right) \qquad \sin(\pi - \theta) = \sin\theta. \text{ See page 309.}$$
$$= \frac{\sqrt{3}}{2} \qquad \text{See page 302.}$$

Practice Problem 5 **a.** Show that $\sin^{-1}(-x) = -\sin^{-1} x$ for $|x| \leq 1$.

b. Use part **a** to evaluate $\sin\left[\frac{\pi}{3} - \sin^{-1}\left(-\frac{1}{2}\right)\right]$.

Using a Calculator with Inverse Functions When using the inverse trigonometric functions on a calculator to find a real number (or equivalently, an angle measured in radians), make sure you set your calculator to Radian mode.

When using a calculator to find $\csc^{-1} x$ or $\sec^{-1} x$, find $\sin^{-1}\frac{1}{x}$ and $\cos^{-1}\frac{1}{x}$, respectively. For example, if $\csc^{-1} 5 = \theta$, then $\csc\theta = 5$, or $\frac{1}{\sin\theta} = 5$. So $\sin\theta = \frac{1}{5}$ and $\theta = \sin^{-1}\left(\frac{1}{5}\right)$. However, to find $\cot^{-1} x$, begin by finding $\tan^{-1}\frac{1}{x}$; this gives you a value in the interval $\left(-\frac{\pi}{2}, \frac{\pi}{2}\right)$. If $x > 0$, this is the correct value, but for $x < 0$, $\cot^{-1}(x) = \pi + \tan^{-1}\frac{1}{x}$; so that $\cot^{-1}(x)$ is in the interval $\left(\frac{\pi}{2}, \pi\right)$.

When using a calculator to find an unknown angle measure in degrees, make sure you set your calculator to degree measure.

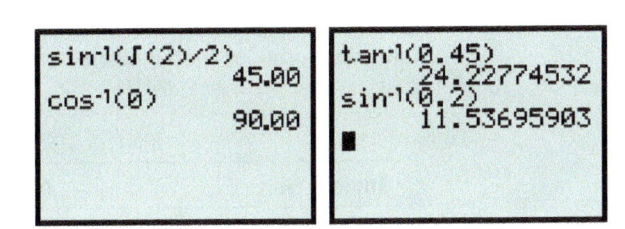

```
sin-1(√(2)/2)
           45.00
cos-1(0)
           90.00
```

```
tan-1(0.45)
      24.22774532
sin-1(0.2)
      11.53695903
```

EXAMPLE 6 Using a Calculator to Find the Values of Inverse Functions

Use a calculator to find the value of y in radians rounded to four decimal places.

a. $y = \sin^{-1} 0.75$ **b.** $y = \cot^{-1} 2.8$ **c.** $y = \cot^{-1}(-2.3)$

Solution

Set your calculator to Radian mode.

a. $y = \sin^{-1} 0.75 \approx 0.8481$

b. $y = \cot^{-1} 2.8 = \tan^{-1}\left(\dfrac{1}{2.8}\right) \approx 0.3430$

c. $y = \cot^{-1}(-2.3) = \pi + \tan^{-1}\left(-\dfrac{1}{2.3}\right) \approx 2.7315$

Practice Problem 6 Use a calculator to find the value of y in radians rounded to four decimal places.

a. $y = \cos^{-1} 0.22$ **b.** $y = \csc^{-1} 3.5$ **c.** $y = \cot^{-1}(-4.7)$

EXAMPLE 7 Using a Calculator to Find the Values of Inverse Functions

Use a calculator to find the value of y in degrees rounded to four decimal places.

a. $y = \tan^{-1} 0.99$ **b.** $y = \sec^{-1} 25$ **c.** $y = \cot^{-1}(-1.3)$

Solution

Set your calculator to Degree mode.

a. $y = \tan^{-1} 0.99 \approx 44.7121°$

b. $y = \sec^{-1} 25 = \cos^{-1}\dfrac{1}{25} \approx 87.7076°$

c. $y = \cot^{-1}(-1.3) = 180° + \tan^{-1}\left(-\dfrac{1}{1.3}\right) \approx 142.4314°$

Practice Problem 7 Repeat Example 7 for each expression.

a. $y = \cot^{-1} 0.75$ **b.** $y = \csc^{-1} 13$ **c.** $y = \tan^{-1}(-12)$

5 Find exact values of composite functions involving the inverse trigonometric functions.

Composition of Trigonometric and Inverse Trigonometric Functions

Recall that if f is a one-to-one function with inverse f^{-1}, then $f^{-1}[f(x)] = x$ for every x in the domain of f and $f[f^{-1}(x)] = x$ for every x in the domain of f^{-1}. This leads to the following formulas for the inverse sine, cosine, and tangent functions.

INVERSE FUNCTION PROPERTIES

Inverse Sine	Inverse Cosine	Inverse Tangent
$\sin^{-1}(\sin x) = x$, $-\dfrac{\pi}{2} \le x \le \dfrac{\pi}{2}$	$\cos^{-1}(\cos x) = x$, $0 \le x \le \pi$	$\tan^{-1}(\tan x) = x$, $-\dfrac{\pi}{2} < x < \dfrac{\pi}{2}$
$\sin(\sin^{-1} x) = x$, $-1 \le x \le 1$	$\cos(\cos^{-1} x) = x$, $-1 \le x \le 1$	$\tan(\tan^{-1} x) = x$, $-\infty < x < \infty$

EXAMPLE 8 Finding the Exact Value of $\sin^{-1}(\sin x)$ and $\cos^{-1}(\cos x)$

Find the exact value of

a. $\sin^{-1}\left[\sin\left(-\dfrac{\pi}{8}\right)\right].$ **b.** $\cos^{-1}\left(\cos\dfrac{5\pi}{4}\right).$

Solution

a. Because $-\dfrac{\pi}{2} \leq -\dfrac{\pi}{8} \leq \dfrac{\pi}{2}$, we have $\sin^{-1}\left[\sin\left(-\dfrac{\pi}{8}\right)\right] = -\dfrac{\pi}{8}.$

b. We cannot use the formula $\cos^{-1}(\cos x) = x$ for $x = \dfrac{5\pi}{4}$ because $\dfrac{5\pi}{4}$ is not in the interval $[0, \pi]$. However, $\cos\dfrac{5\pi}{4} = \cos\left(2\pi - \dfrac{5\pi}{4}\right) = \cos\dfrac{3\pi}{4}$ and $\dfrac{3\pi}{4}$ is in the interval $[0, \pi]$. Therefore,

$$\cos^{-1}\left(\cos\dfrac{5\pi}{4}\right) = \cos^{-1}\left(\cos\dfrac{3\pi}{4}\right) = \dfrac{3\pi}{4}.$$

Practice Problem 8 Find the exact value of $\sin^{-1}\left(\sin\dfrac{3\pi}{2}\right).$

To find the exact values of expressions involving the composition of a trigonometric function and the inverse of a *different* trigonometric function, we use points on the terminal side of an angle in standard position.

EXAMPLE 9 Finding the Exact Value of a Composite Trigonometric Expression

Find the exact value of

a. $\cos\left(\tan^{-1}\dfrac{2}{3}\right).$ **b.** $\sin\left[\cos^{-1}\left(-\dfrac{1}{4}\right)\right].$

Solution

a. Let θ represent the radian measure of the angle in the interval $\left(-\dfrac{\pi}{2}, \dfrac{\pi}{2}\right)$, with $\tan\theta = \dfrac{2}{3}$. Then because $\tan\theta$ is positive, θ must be positive. We have

$$\theta = \tan^{-1}\dfrac{2}{3} \quad \text{and} \quad 0 < \theta < \dfrac{\pi}{2}.$$

Figure 4.64 shows θ in standard position. If (x, y) is a point on the terminal side of θ, then $\tan\theta = \dfrac{y}{x}$.

Consequently, we can choose the point with coordinates $(3, 2)$ to determine the terminal side of θ. Then $x = 3$, $y = 2$ and we have

$$\tan\theta = \dfrac{2}{3} \quad \text{and} \quad \cos\theta = \dfrac{x}{r} = \dfrac{3}{r}, \text{ where}$$

$$r = \sqrt{x^2 + y^2} = \sqrt{3^2 + 2^2} = \sqrt{9 + 4} = \sqrt{13}. \text{ So}$$

$$\cos\left(\tan^{-1}\dfrac{2}{3}\right) = \cos\theta = \dfrac{3}{r} = \dfrac{3}{\sqrt{13}} = \dfrac{3\sqrt{13}}{13}.$$

$\theta = \tan^{-1}\dfrac{2}{3}$

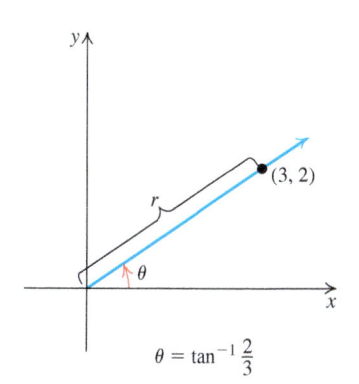

FIGURE 4.64

b. Let θ represent the radian measure of the angle in $[0, \pi]$, with $\cos \theta = -\dfrac{1}{4}$.

Then because $\cos \theta$ is negative, θ is in quadrant II; so

$$\theta = \cos^{-1}\left(-\dfrac{1}{4}\right) \quad \text{and} \quad \dfrac{\pi}{2} < \theta < \pi.$$

Figure 4.65 shows θ in standard position. If (x, y) is a point on the terminal side of θ and r is the distance between (x, y) and the origin, then $\sin \theta = \dfrac{y}{r}$. We choose the point with coordinates $(-1, y)$, a distance of $r = 4$ units from the origin, on the terminal side of θ. Then

$$\cos \theta = -\dfrac{1}{4} \quad \text{and} \quad \sin \theta = \dfrac{y}{4}, \text{ where}$$

$$\begin{aligned} r = \sqrt{x^2 + y^2} = \sqrt{(-1)^2 + y^2} \text{ or } \quad r^2 &= 1 + y^2 \\ 4^2 &= 1 + y^2 &&\text{Replace } r \text{ with } 4. \\ 15 &= y^2 &&\text{Simplify.} \\ \sqrt{15} &= y &&y \text{ is positive.} \end{aligned}$$

Thus,

$$\sin\left[\cos^{-1}\left(-\dfrac{1}{4}\right)\right] = \sin \theta = \dfrac{y}{r} = \dfrac{\sqrt{15}}{4}.$$

Practice Problem 9 Find the exact value of $\cos\left[\sin^{-1}\left(-\dfrac{1}{3}\right)\right]$.

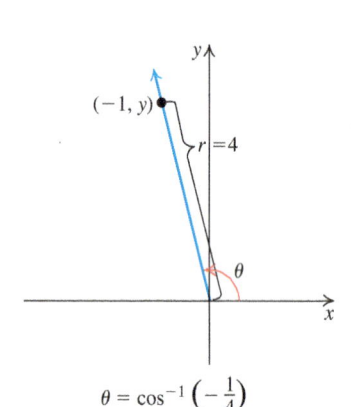

FIGURE 4.65 $\theta = \cos^{-1}\left(-\dfrac{1}{4}\right)$

EXAMPLE 10 Converting Composite Trigonometric Expressions to Algebraic Expressions

Write algebraic expression for $y = \tan\left(\sin^{-1}\dfrac{t}{4}\right)$ for $|t| < 4$.

Solution

Let $\theta = \sin^{-1}\dfrac{t}{4}$. Then $\sin \theta = \dfrac{t}{4}$ and θ is in the interval $\left(-\dfrac{\pi}{2}, \dfrac{\pi}{2}\right)$.

Figure 4.66 shows angle θ in standard position with a point $P(x, t)$ on the terminal side of θ. Then by the Pythagorean Theorem, we have

$$\begin{aligned} x^2 + t^2 &= 4^2 \\ x^2 &= 16 - t^2 \\ x &= \sqrt{16 - t^2} &&\text{For } \theta \text{ in } \left(-\dfrac{\pi}{2}, \dfrac{\pi}{2}\right), x > 0. \end{aligned}$$

The expression $y = \tan\left(\sin^{-1}\dfrac{t}{4}\right)$ becomes

$$\begin{aligned} y &= \tan \theta &&\theta = \sin^{-1}\dfrac{t}{4} \\ y &= \dfrac{t}{x} &&\text{Definition of } \tan \theta \\ y &= \dfrac{t}{\sqrt{16 - t^2}} &&\text{Replace } x \text{ with } \sqrt{16 - t^2}. \end{aligned}$$

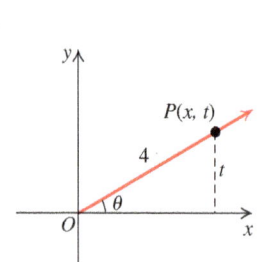

FIGURE 4.66

Practice Problem 10 Write an algebraic expression for

$$y = \tan\left(\cos^{-1}\dfrac{x}{3}\right) \text{ for } 0 < x < 3.$$

In many applications of calculus, we use trigonometric substitutions to simplify algebraic expressions.

EXAMPLE 11 Using Trigonometry to Simplify Algebraic Expressions

Simplify $y = \dfrac{\sqrt{5 - x^2}}{x}$ using the substitution $x = \sqrt{5} \sin \theta$ for $0 < \theta < \dfrac{\pi}{2}$.

Solution

$$y = \frac{\sqrt{5 - (\sqrt{5} \sin \theta)^2}}{\sqrt{5} \sin \theta} \qquad \text{Substitute for } x.$$

$$= \frac{\sqrt{5 - 5 \sin^2 \theta}}{\sqrt{5} \sin \theta} \qquad (\sqrt{5} \sin \theta)^2 = 5 \sin^2 \theta$$

$$= \frac{\sqrt{5(1 - \sin^2 \theta)}}{\sqrt{5} \sin \theta} \qquad \text{Factor.}$$

$$= \frac{\sqrt{5 \cos^2 \theta}}{\sqrt{5} \sin \theta} \qquad 1 - \sin^2 \theta = \cos^2 \theta$$

$$= \frac{\sqrt{5} \cos \theta}{\sqrt{5} \sin \theta} \qquad \sqrt{5 \cos^2 \theta} = \sqrt{5}\sqrt{\cos^2 \theta} = \sqrt{5} \cos \theta \text{ for } 0 < \theta < \frac{\pi}{2}$$

$$= \cot \theta \qquad \text{Quotient identity}$$

Practice Problem 11 Simplify $y = \dfrac{x}{\sqrt{7 - x^2}}$ using the substitution $x = \sqrt{7} \cos \theta$ for $0 < \theta < \dfrac{\pi}{2}$.

EXAMPLE 12 Finding the Rotation Angle for a Security Camera

A security camera is to be installed 20 feet away from the center of a jewelry counter. The counter is 30 feet long. What angle, to the nearest degree, should the camera rotate through so that it scans the entire counter? See Figure 4.67.

Solution

The counter center C, the camera A, and a counter end B form a right triangle. The angle at vertex A is $\dfrac{\theta}{2}$, where θ is the angle through which the camera rotates. Note that

$$\tan \frac{\theta}{2} = \frac{15}{20} = \frac{3}{4}$$

$$\frac{\theta}{2} = \tan^{-1} \frac{3}{4} \approx 36.87° \qquad \text{Use a calculator in Degree mode.}$$

$$\theta \approx 73.74° \qquad \text{Multiply both sides by 2.}$$

Set the camera to rotate through 74° to scan the entire counter.

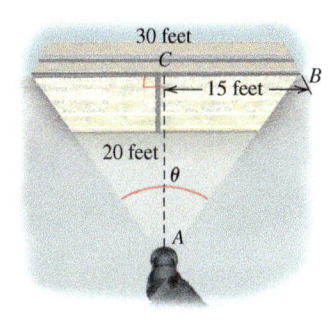

FIGURE 4.67

Practice Problem 12 Rework Example 12 for a counter that is 20 feet long and a camera set 12 feet from the center of the counter.

SECTION 4.5 Exercises

Basic Concepts and Skills

In Exercises 1–18, find each exact value of y or state that y is undefined.

1. $y = \sin^{-1} 0$
2. $y = \cos^{-1} 0$
3. $y = \sin^{-1}\left(-\dfrac{1}{2}\right)$
4. $y = \cos^{-1}\left(-\dfrac{\sqrt{3}}{2}\right)$
5. $y = \cos^{-1}(-1)$
6. $y = \sin^{-1} \dfrac{1}{2}$
7. $y = \cos^{-1} \dfrac{\pi}{2}$
8. $y = \sin^{-1} \pi$
9. $y = \tan^{-1} \sqrt{3}$
10. $y = \tan^{-1} 1$
11. $y = \tan^{-1}(-1)$
12. $y = \tan^{-1}\left(-\dfrac{\sqrt{3}}{3}\right)$
13. $y = \cot^{-1}(-1)$
14. $y = \sin^{-1}\left(-\dfrac{\sqrt{2}}{2}\right)$
15. $y = \cos^{-1}(-2)$
16. $y = \sin^{-1} \sqrt{3}$
17. $y = \sec^{-1}(-2)$
18. $y = \csc^{-1}(-2)$

In Exercises 19–28, find each exact value of y or state that y is undefined.

19. $y = \sin\left(\sin^{-1} \dfrac{1}{8}\right)$
20. $y = \cos\left(\cos^{-1} \dfrac{1}{5}\right)$
21. $y = \tan^{-1}\left(\tan \dfrac{\pi}{7}\right)$
22. $y = \tan^{-1}\left(\tan \dfrac{\pi}{4}\right)$
23. $y = \tan(\tan^{-1} 247)$
24. $y = \tan(\tan^{-1} 7)$
25. $y = \sin^{-1}\left(\sin \dfrac{4\pi}{3}\right)$
26. $y = \cos^{-1}\left(\cos \dfrac{5\pi}{3}\right)$
27. $y = \tan^{-1}\left(\tan \dfrac{2\pi}{3}\right)$
28. $y = \tan\left(\tan^{-1} \dfrac{2\pi}{3}\right)$

In Exercise 29–44, use the identities on page 344 to find the exact value of each expression.

29. $\cot^{-1}\left(\dfrac{1}{\sqrt{3}}\right)$
30. $\sec^{-1}(\sqrt{2})$
31. $\csc^{-1}(2)$
32. $\csc^{-1}\left(\dfrac{2\sqrt{3}}{3}\right)$
33. $\cot^{-1}\left(-\dfrac{1}{\sqrt{3}}\right)$
34. $\cot^{-1}(-\sqrt{3})$
35. $\sin^{-1}\left(-\dfrac{\sqrt{3}}{2}\right)$
36. $\csc^{-1}\left(-\dfrac{2}{\sqrt{3}}\right)$
37. $\cos^{-1}\left(-\dfrac{1}{2}\right)$
38. $\sec^{-1}\left(-\dfrac{2}{\sqrt{3}}\right)$
39. $\sin\left[\dfrac{\pi}{3} - \sin^{-1}\left(-\dfrac{1}{2}\right)\right]$
40. $\cos\left[\dfrac{\pi}{6} + \cos^{-1}\left(-\dfrac{\sqrt{3}}{2}\right)\right]$
41. $\sin\left[\dfrac{\pi}{2} - \cos^{-1}(-1)\right]$
42. $\tan\left[\dfrac{\pi}{6} + \cot^{-1}\left(-\dfrac{1}{\sqrt{3}}\right)\right]$
43. $\sin\left[\tan^{-1}(-\sqrt{3}) + \cos^{-1}\left(-\dfrac{\sqrt{3}}{2}\right)\right]$
44. $\cos\left[\cot^{-1}(-\sqrt{3}) + \sin^{-1}\left(-\dfrac{1}{2}\right)\right]$

In Exercises 45–54, use a calculator to find each value of y in degrees rounded to two decimal places.

45. $y = \cos^{-1} 0.6$
46. $y = \sin^{-1} 0.23$
47. $y = \sin^{-1}(-0.69)$
48. $y = \cos^{-1}(-0.57)$
49. $y = \sec^{-1}(3.5)$
50. $y = \csc^{-1}(6.8)$
51. $y = \tan^{-1} 14$
52. $y = \tan^{-1} 50$
53. $y = \tan^{-1}(-42.147)$
54. $y = \tan^{-1}(-0.3863)$

In Exercises 55–66, use a sketch to find each exact value of y.

55. $y = \cos\left(\sin^{-1} \dfrac{2}{3}\right)$
56. $y = \sin\left(\cos^{-1} \dfrac{3}{4}\right)$
57. $y = \sin\left[\cos^{-1}\left(-\dfrac{4}{5}\right)\right]$
58. $y = \cos\left(\sin^{-1} \dfrac{3}{5}\right)$
59. $y = \cos\left(\tan^{-1} \dfrac{5}{2}\right)$
60. $y = \sin\left(\tan^{-1} \dfrac{13}{5}\right)$
61. $y = \tan\left(\cos^{-1} \dfrac{4}{5}\right)$
62. $y = \tan\left[\sin^{-1}\left(-\dfrac{3}{4}\right)\right]$
63. $y = \sin(\tan^{-1} 4)$
64. $y = \cos(\tan^{-1} 3)$
65. $y = \tan(\sec^{-1} 2)$
66. $y = \tan\left[\csc^{-1}(-2)\right]$

In Exercises 67–72, write an algebraic expression for each composite trigonometric expression.

67. $y = \sin\left(\tan^{-1} \dfrac{x}{2}\right); x > 0$
68. $y = \sin\left(\cot^{-1} \dfrac{2x}{3}\right); x > 0$
69. $y = \cot\left(\sin^{-1} \dfrac{3x}{5}\right); 0 < x < \dfrac{5}{3}$
70. $y = \sin\left(\sec^{-1} \dfrac{x}{3}\right); x > 3$
71. $y = \cot\left(\cos^{-1} \dfrac{x}{2}\right); -2 < x < 0$
72. $y = \sin\left(\cot^{-1} \dfrac{x}{5}\right); x < 0$

In Exercises 73–78, simplify the algebraic expression by using the given trigonometric substitution. Assume that $0 < \theta < \dfrac{\pi}{2}$.

73. $y = \dfrac{x}{\sqrt{9 - x^2}}$; $x = 3 \cos \theta$

74. $y = \dfrac{\sqrt{16 - x^2}}{x}$; $x = 4 \sin \theta$

75. $y = \dfrac{x}{\sqrt{4 + x^2}}$; $x = 2 \tan \theta$

76. $y = \dfrac{\sqrt{9 + x^2}}{x}$; $x = 3 \cot \theta$

77. $y = \dfrac{x}{\sqrt{x^2 - 25}}$; $x = 5 \sec \theta$

78. $y = \dfrac{\sqrt{x^2 - 36}}{x}$; $x = 6 \csc \theta$

Applying the Concepts

79. Sprinkler rotation. A sprinkler rotates back and forth through an angle θ, as shown in the figure. At a distance of 5 feet from the sprinkler, the rays that form the sides of angle θ are 6 feet apart. Find θ.

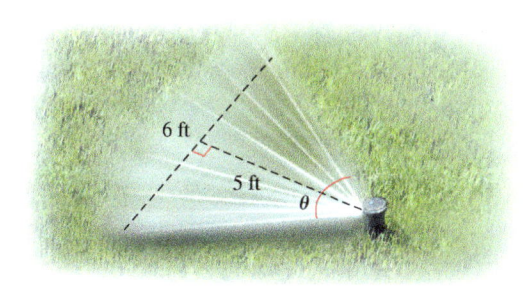

80. Irradiating flowers. A tray of flowers is being irradiated by a beam from a rotating lamp, as shown in the figure. If the tray is 8 feet long and the lamp is 2 feet from the center of the tray, through what angle should the lamp rotate to irradiate the full length of the tray?

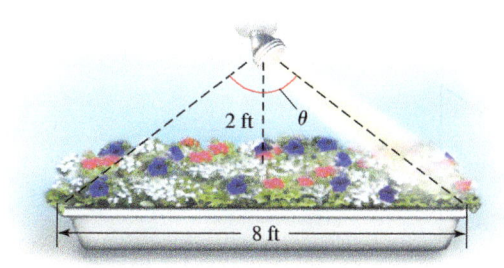

81. Motorcycle racing. A video camera is set up 110 feet away from and at a right angle to a straight quarter-mile racetrack, as shown in the figure. The starting line is to the left, and the finish line is to the right. Through what angle must the camera rotate to film the entire race?

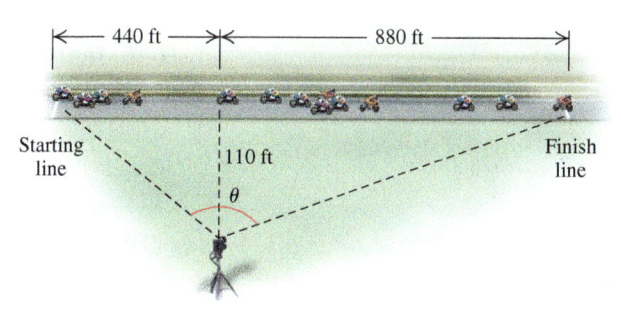

82. Camera's viewing angle. The viewing angle for the 35-millimeter camera is given (in degrees) by
$$\theta = 2 \tan^{-1} \dfrac{18}{x},$$
where x is the focal length of the lens. The focal length on most adjustable cameras is marked in millimeters on the lens mount.
  a. Find the viewing angle, in degrees, if the focal length is 50 millimeters.
  b. Find the viewing angle, in degrees, if the focal length is 200 millimeters.

Critical Thinking / Discussion / Writing

83. Show that $\sin^{-1} x = \begin{cases} \cos^{-1}(\sqrt{1 - x^2}); & 0 \leq x \leq 1 \\ -\cos^{-1}(\sqrt{1 - x^2}); & -1 \leq x < 0 \end{cases}$.

84. Show that $\sin^{-1} x + \cos^{-1} x = \dfrac{\pi}{2}$; $|x| \leq 1$.

Maintaining Skills

In Exercises 85–86, perform the indicated operations and simplify the result.

85. $\dfrac{1}{x - 1} - \dfrac{1}{x + 1}$

86. $\dfrac{1}{x - 1} - \dfrac{1}{x^2 - 1}$

87. Let $f(x) = x^2 + 3$. Find and simplify $\dfrac{f(x) - f(a)}{x - a}, x \neq a$.

88. If $x^2 + y^2 = 1$, simplify $\dfrac{x}{1 + y} - \dfrac{1 - y}{x}$.

SECTION 4.6

Right-Triangle Trigonometry

BEFORE STARTING THIS SECTION, REVIEW

1. Rationalizing the denominator (Section P.2, page 15)
2. Acute angle definition (Section 4.1, page 284)
3. Pythagorean Theorem (Section 1.1, page 53)
4. Functions (Section 1.3, page 77)

OBJECTIVES

1. Express the trigonometric functions using a right triangle.
2. Evaluate trigonometric functions of angles in a right triangle.
3. Solve right triangles.
4. Use right-triangle trigonometry in applications.

◆ Measuring Mount Kilimanjaro

Trigonometry developed as a result of attempts to solve practical problems in astronomy, navigation, and land measurement. The word *trigonometry*, coined by Pitiscus in 1594, is derived from two Greek words, *trigonon* (triangle) and *metron* (measure), and means "triangle measurement."

Measuring objects that are generally inaccessible is one of the many successes of trigonometry. In Example 7, we use trigonometry to approximate the height of Mount Kilimanjaro, Tanzania. Mount Kilimanjaro is the highest mountain in Africa.

Mount Kilimanjaro

1 Express the trigonometric functions using a right triangle.

RECALL

In a right triangle, the side opposite the right angle is called the hypotenuse and the other two sides are called the legs.

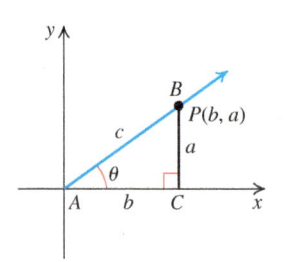

FIGURE 4.68

Trigonometric Ratios and Functions

Consider a right triangle with one of its acute angles labeled θ. In Figure 4.67, the capital letters A, B, and C designate the vertices of the triangle and the lowercase letters a, b, and c, respectively, represent the lengths of the sides opposite these vertices.

a = length of the side opposite A
b = length of the side adjacent to A
c = length of the hypotenuse

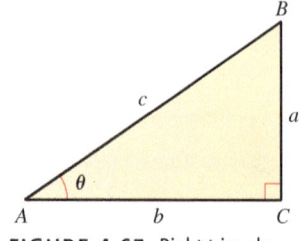

FIGURE 4.67 Right triangle

Now place the angle θ in standard position, as shown in Figure 4.68. The point corresponding to the vertex B is $P(b, a)$. Because P is on the terminal side of θ, we can use it to express the trigonometric functions of θ as ratios of the sides of a right triangle. We use the words *opposite* for the length of the leg opposite θ, *adjacent* for the length of the leg adjacent to θ, and *hypotenuse* for the length of the hypotenuse.

350 Chapter 4 Trigonometric Functions

SIDE NOTE

The mnemonic SOHCAHTOA is helpful in remembering that **S**ine is **O**pposite over **H**ypotenuse (SOH); **C**osine is **A**djacent over **H**ypotenuse (CAH); and **T**angent is **O**pposite over **A**djacent (TOA).

TRIGONOMETRIC FUNCTIONS OF AN ANGLE θ IN A RIGHT TRIANGLE

$$\sin \theta = \frac{\text{opposite}}{\text{hypotenuse}} = \frac{a}{c} \qquad \csc \theta = \frac{\text{hypotenuse}}{\text{opposite}} = \frac{c}{a}$$

$$\cos \theta = \frac{\text{adjacent}}{\text{hypotenuse}} = \frac{b}{c} \qquad \sec \theta = \frac{\text{hypotenuse}}{\text{adjacent}} = \frac{c}{b}$$

$$\tan \theta = \frac{\text{opposite}}{\text{adjacent}} = \frac{a}{b} \qquad \cot \theta = \frac{\text{adjacent}}{\text{opposite}} = \frac{b}{a}$$

Because any two right triangles having acute angle θ are similar, the ratio of any two sides in one of them is the same as the ratio of the corresponding sides in the other. Thus, the ratios of corresponding sides depend only on the angle θ, not on the size of the triangle. Three similar right triangles with acute angle θ are shown in Figure 4.69.

RECALL

Two triangles are *similar* if their corresponding angles are congruent and their corresponding sides are proportional. Two similar triangles have the same shape but not necessarily the same size.

$\sin \theta = \frac{3}{5}$ $\sin \theta = \frac{9}{15} = \frac{3}{5}$ $\sin \theta = \frac{30}{50} = \frac{3}{5}$

$\cos \theta = \frac{4}{5}$ $\cos \theta = \frac{12}{15} = \frac{4}{5}$ $\cos \theta = \frac{40}{50} = \frac{4}{5}$

$\tan \theta = \frac{3}{4}$ $\tan \theta = \frac{9}{12} = \frac{3}{4}$ $\tan \theta = \frac{30}{40} = \frac{3}{4}$

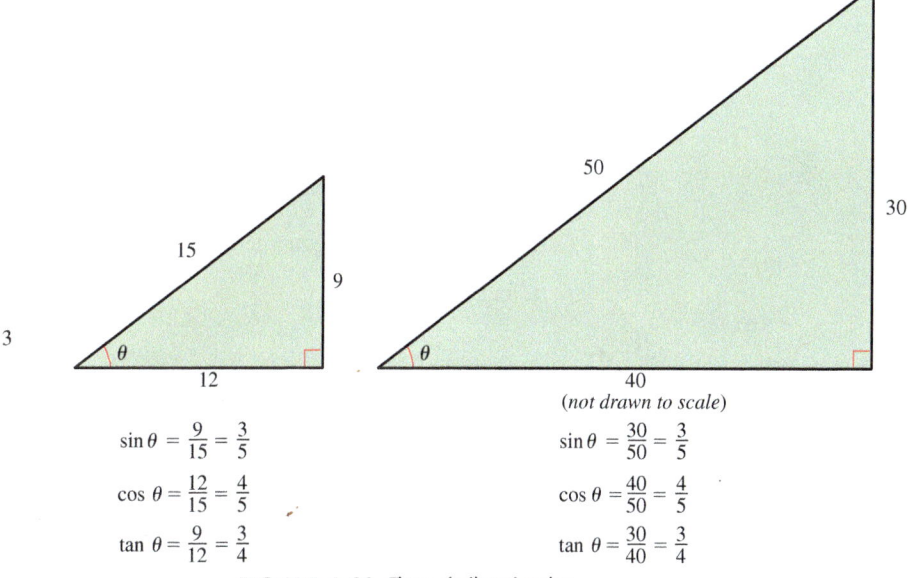

FIGURE 4.69 Three similar triangles

Notice that the value of each of the six trigonometric functions is independent of the triangle used to find it. Recall that $\csc \theta$, $\sec \theta$, and $\cot \theta$ are the reciprocals of $\sin \theta$, $\cos \theta$, and $\tan \theta$, respectively. Consequently, for any of the triangles in Figure 4.69, we have

$$\csc \theta = \frac{5}{3}, \; \sec \theta = \frac{5}{4}, \; \text{and} \; \cot \theta = \frac{4}{3}.$$

2 Evaluate trigonometric functions of angles in a right triangle.

Evaluating Trigonometric Functions

EXAMPLE 1   Finding the Values of Trigonometric Functions

Find the exact values for the six trigonometric functions of the angle θ in Figure 4.70.

Solution

To find the values for the six trigonometric functions of θ, we must first find the value of c, the length of the hypotenuse.

FIGURE 4.70

352 Chapter 4 Trigonometric Functions

$$a^2 + b^2 = c^2 \quad \text{Pythagorean Theorem}$$
$$(3)^2 + (\sqrt{7})^2 = c^2 \quad \text{Replace } a \text{ with 3 and } b \text{ with } \sqrt{7}.$$
$$9 + 7 = c^2$$
$$16 = c^2$$
$$4 = c \quad c \text{ is a positive number.}$$

Now with $c = 4$, $a = 3$, and $b = \sqrt{7}$, we have

$$\sin \theta = \frac{\text{opposite}}{\text{hypotenuse}} = \frac{3}{4} \qquad \csc \theta = \frac{\text{hypotenuse}}{\text{opposite}} = \frac{4}{3}$$

$$\cos \theta = \frac{\text{adjacent}}{\text{hypotenuse}} = \frac{\sqrt{7}}{4} \qquad \sec \theta = \frac{\text{hypotenuse}}{\text{adjacent}} = \frac{4}{\sqrt{7}} = \frac{4\sqrt{7}}{7}$$

$$\tan \theta = \frac{\text{opposite}}{\text{adjacent}} = \frac{3}{\sqrt{7}} = \frac{3\sqrt{7}}{7} \qquad \cot \theta = \frac{\text{adjacent}}{\text{opposite}} = \frac{\sqrt{7}}{3}$$

These are exact values; using a calculator would yield approximate values.

Practice Problem 1 Find the exact values for the six trigonometric functions of the acute angle θ in a right triangle if the length of the leg opposite θ is 4 and the length of the hypotenuse is 5.

EXAMPLE 2 Finding the Remaining Trigonometric Function Values from a Given Value

Find the other five trigonometric function values of θ, given that θ is an acute angle of a right triangle with $\sin \theta = \frac{2}{5}$.

Solution
Because

$$\sin \theta = \frac{2}{5} = \frac{\text{opposite}}{\text{hypotenuse}}$$

we draw a right triangle with hypotenuse of length 5 and the side opposite θ of length 2. See Figure 4.71. Then

$$5^2 = 2^2 + b^2 \quad \text{Pythagorean Theorem}$$
$$b^2 = 5^2 - 2^2 \quad \text{Subtract } 2^2 \text{ from both sides; switch sides.}$$
$$b^2 = 21 \quad \text{Simplify.}$$
$$b = \sqrt{21} \quad b \text{ is a positive number.}$$

For this triangle we have:

$$c = \text{length of the hypotenuse} = 5$$
$$a = \text{length of the opposite side} = 2$$
$$b = \text{length of the adjacent side} = \sqrt{21}$$

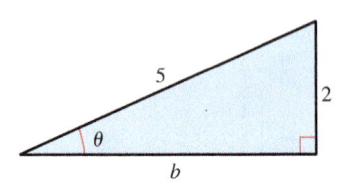

FIGURE 4.71

So

$$\sin \theta = \frac{\text{opposite}}{\text{hypotenuse}} = \frac{2}{5} \qquad \csc \theta = \frac{\text{hypotenuse}}{\text{opposite}} = \frac{5}{2}$$

$$\cos \theta = \frac{\text{adjacent}}{\text{hypotenuse}} = \frac{\sqrt{21}}{5} \qquad \sec \theta = \frac{\text{hypotenuse}}{\text{adjacent}} = \frac{5}{\sqrt{21}} = \frac{5\sqrt{21}}{21}$$

$$\tan \theta = \frac{\text{opposite}}{\text{adjacent}} = \frac{2}{\sqrt{21}} = \frac{2\sqrt{21}}{21} \qquad \cot \theta = \frac{\text{adjacent}}{\text{opposite}} = \frac{\sqrt{21}}{2}$$

Practice Problem 2 Find the other five trigonometric function values of θ, given that θ is an acute angle of a right triangle with $\cos \theta = \dfrac{1}{3}$.

Complements

Because θ and $90° - \theta$ are acute angles in the same right triangle, the same six possible ratios of the lengths of the sides are used to compute the values of the trigonometric functions for both angles. Notice that the leg opposite θ is the leg adjacent to $90° - \theta$; so

$$\sin \theta = \frac{\text{opposite to } \theta}{\text{hypotenuse}} = \frac{\text{adjacent to } (90° - \theta)}{\text{hypotenuse}} = \cos(90° - \theta).$$

Similarly, $\cos \theta = \sin(90° - \theta)$ and $\tan \theta = \cot(90° - \theta)$.

Taking reciprocals of these values extends this relationship to the remaining three trigonometric functions.

The prefix *co* (in *cosine*, *cotangent*, and *cosecant*) stands for *complement*. In general, for any acute angle θ in a right triangle, the other acute angle is its complement, $90° - \theta$. See Figure 4.72. The pairs of functions—sine and cosine, tangent and cotangent, secant and cosecant—are cofunctions of each other.

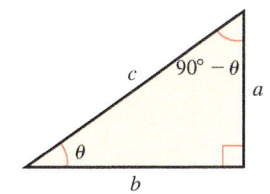

FIGURE 4.72 *a* is opposite θ and adjacent to $90° - \theta$.
b is adjacent to θ and opposite $90° - \theta$.

SIDE NOTE

Here is a scheme for remembering the cofunction identities for acute angles A and B with $A + B = 90°$, or $A + B = \dfrac{\pi}{2}$:

complementary angles
$\sin A = \cos B$
cofunctions

The sine and cosine may be replaced with any other cofunction pairs.

COMPLEMENTARY RELATIONSHIPS

The value of any trigonometric function of an acute angle θ is equal to the cofunction value of the complement of θ. This is true whether θ is measured in degrees or in radians.

θ in degrees

$\sin \theta = \cos(90° - \theta)$ $\cos \theta = \sin(90° - \theta)$
$\tan \theta = \cot(90° - \theta)$ $\cot \theta = \tan(90° - \theta)$
$\sec \theta = \csc(90° - \theta)$ $\csc \theta = \sec(90° - \theta)$

If θ is measured in radians, replace $90°$ with $\dfrac{\pi}{2}$.

EXAMPLE 3 Finding Trigonometric Function Values of a Complementary Angle

a. Given that $\cot 68° \approx 0.4040$, find $\tan 22°$.
b. Given that $\cos 72° \approx 0.3090$, find $\sin 18°$.

Solution

a. Note that $68° = 90° - 22°$. So
$$\tan 22° = \cot(90° - 22°) = \cot 68° \approx 0.4040.$$

b. Here $72° = 90° - 18°$. So
$$\sin 18° = \cos(90° - 18°) = \cos 72° \approx 0.3090.$$

Practice Problem 3

a. Given that $\csc 21° \approx 2.7904$, find $\sec 69°$.
b. Given that $\tan 75° \approx 3.7321$, find $\cot 15°$.

3 Solve right triangles.

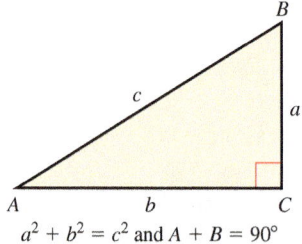

Solving Right Triangles

We use right-triangle trigonometry to find unknown lengths in triangles and unknown distances in applications. Using inverse trigonometric functions, we also find unknown angle measures in right triangles as well as unknown angle measures in applied problems. To **solve a triangle** means to find all unknown side lengths and angle measures. In this section, we discuss how to solve right triangles. When solving right triangles, we solve each triangle ABC, where $\angle C$ is always the right angle and $\angle A$ and $\angle B$ are the acute angles. (We frequently use A as a shorthand way to write measure of $\angle A$.) The side lengths are a, b, and c, where a is the length of the side opposite $\angle A$, and so on. Because $\angle C$ is always the right angle, the length of the hypotenuse is always c and the lengths of the legs are a and b. See Figure 4.73. With this notation, $a^2 + b^2 = c^2$ and $A + B = 90°$.

$a^2 + b^2 = c^2$ and $A + B = 90°$

FIGURE 4.73 Labeling a right triangle

In applications, degree measure is traditionally used for the angles. Here are some useful facts about solving right triangles:

1. The measure of one angle (the right angle) is automatically given: $C = 90°$.
2. Using the Pythagorean Theorem, the length of the third side of any right triangle can be found if two of the side lengths are known.
3. Knowing the measure of one acute angle allows us to find the other because the two acute angles of any right triangle are complementary.
4. We can solve a right triangle if we are given *either* of the following combinations of measurements:
   a. The lengths of any two sides
   b. The length of any one side and the measure of either of the acute angles

EXAMPLE 4 Solving a Right Triangle, Given One Acute Angle and One Side

Solve right triangle ABC if $A = 23°$ and $c = 5.8$.

Solution

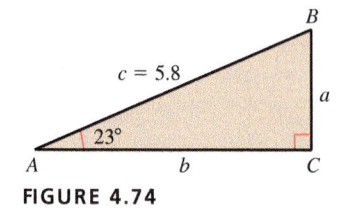

Sketch the triangle. See Figure 4.74.

To find a, find a trigonometric function that involves a and the given parts A and c.

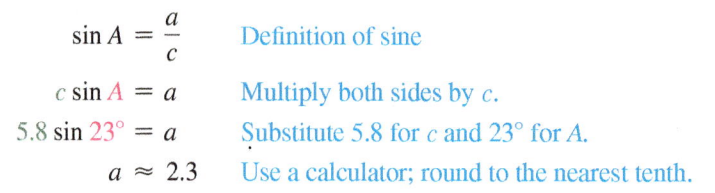

$\sin A = \dfrac{a}{c}$ Definition of sine

$c \sin A = a$ Multiply both sides by c.

$5.8 \sin 23° = a$ Substitute 5.8 for c and 23° for A.

$a \approx 2.3$ Use a calculator; round to the nearest tenth.

FIGURE 4.74

To find b, find a trigonometric function that involves b and the given parts A and c.

$\cos A = \dfrac{b}{c}$ Definition of cosine

$c \cos A = b$ Multiply both sides by c.

$5.8 \cos 23° = b$ Substitute 5.8 for c and 23° for A.

$b \approx 5.3$ Use a calculator; round to the nearest tenth.

Once a and b have been found, the Pythagorean Theorem can be used as a check.
Finally, because $\angle A$ and $\angle B$ are complementary, $B = 90° - 23° = 67°$.

Practice Problem 4 Solve right triangle ABC if $c = 25.8$ and $A = 56°$.

EXAMPLE 5 Solving a Right Triangle, Given Two Sides

Solve right triangle ABC if $a = 9.5$ and $b = 3.4$.

Solution

Sketch the triangle. See Figure 4.75.
To find A, use a trigonometric function that involves A and the given parts a and b.

$$\tan A = \frac{a}{b} \qquad \text{Definition of tangent}$$

$$\tan A = \frac{9.5}{3.4} \qquad \text{Substitute 9.5 for } a \text{ and 3.4 for } b.$$

$$A = \tan^{-1}\left(\frac{9.5}{3.4}\right) \approx 70.3° \qquad \text{Use a calculator; round to one decimal place.}$$

To find c, using the Pythagorean Theorem, we have:

$$c = \sqrt{a^2 + b^2}$$
$$c = \sqrt{(9.5)^2 + (3.4)^2} \qquad \text{Substitute 9.5 for } a \text{ and 3.4 for } b.$$
$$c \approx 10.1 \qquad \text{Use a calculator; round to the nearest tenth.}$$

All three sides have been found.
Finally, $B = 90° - A$; so $B \approx 90° - 70.3°$, or $B \approx 19.7°$.

Practice Problem 5 Solve right triangle ABC if $b = 24.5$ and $c = 36.9$.

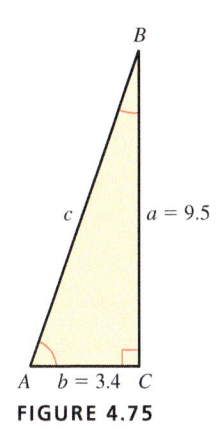

FIGURE 4.75

Applications

4 Use right-triangle trigonometry in applications.

Angles that are measured between a line of sight and a horizontal line occur in many applications. If the line of sight is *above* the horizontal line, the angle between these two lines is called the **angle of elevation**. If the line of sight is *below* the horizontal line, the angle between the two lines is called the **angle of depression**.

For example, suppose you are taking a picture of a friend who has climbed to the top of the Arc de Triomphe in Paris. See Figure 4.76. The angle your eyes rotate through as your gaze changes from looking straight ahead to looking at your friend is the angle of elevation.

The angle your friend's eyes rotate through as she changes from looking straight ahead to looking down at you (no pun intended) is the angle of depression.

SIDE NOTE

Remember that an angle of elevation or an angle of depression is the angle between the line of sight and a *horizontal* line.

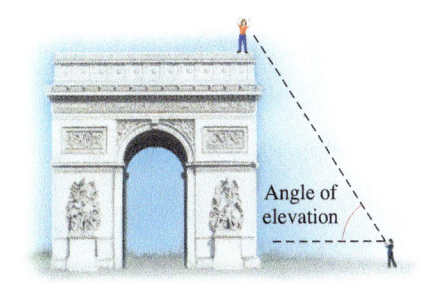

FIGURE 4.76 Angles of elevation and depression

356 Chapter 4 Trigonometric Functions

EXAMPLE 6 Finding the Height of a Cloud

To determine the height of a cloud at night, a farmer shines a spotlight straight up to a spot on the cloud. The angle of elevation to this same spot on the cloud from a point located 150 feet horizontally from the spotlight is 68.2°. Find the height of the cloud.

Solution

In Figure 4.77, h represents the height of the cloud, in feet. We have

$\tan 68.2° = \dfrac{h}{150}$ Definition of tangent

$h = 150 \tan 68.2°$ Multiply both sides by 150.

$h \approx 375$ Use a calculator.

The height of the cloud is approximately 375 feet.

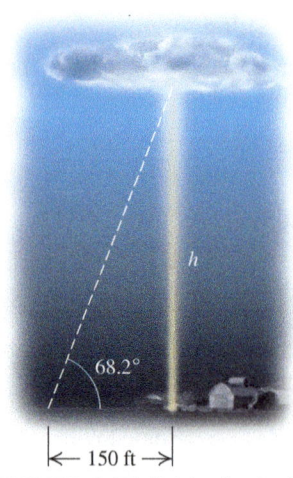

FIGURE 4.77 Height of a cloud

Practice Problem 6 From an observation deck 425 feet high on the top of a lighthouse, the angle of depression to a ship at sea is 4.2°. How many miles is the ship from a point at sea level directly below the observation deck?

TECHNOLOGY CONNECTION

When solving a triangle, to find **Length**, use the SIN, COS, and TAN keys; **Angles**, use the SIN⁻¹, COS⁻¹, and TAN⁻¹ keys.

EXAMPLE 7 Measuring the Height of Mount Kilimanjaro

A surveyor wants to measure the height of Mount Kilimanjaro by using the known height of a nearby mountain. The nearby location is at an altitude of 8720 feet, the distance between that location and Mount Kilimanjaro's peak is 4.9941 miles, and the angle of elevation from the lower location is 23.75°. See Figure 4.78. Use this information to find the approximate height of Mount Kilimanjaro.

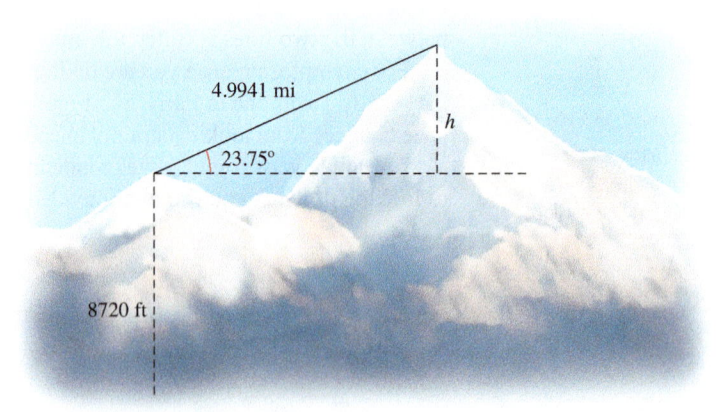

FIGURE 4.78 Height of Mount Kilimanjaro

Solution

The sum of the side length h and the location height of 8720 feet gives the approximate height of Mount Kilimanjaro. Let h be measured in miles. Use the definition of $\sin \theta$ for $\theta = 23.75°$.

$\sin \theta = \dfrac{\text{opposite}}{\text{hypotenuse}} = \dfrac{h}{4.9941}$

$h = (4.9941) \sin \theta$ Multiply both sides by 4.9941.

$= (4.9941) \sin 23.75°$ Replace θ with 23.75°.

$h \approx 2.0114$ Use a calculator.

Because 1 mile = 5280 feet,

$$2.0114 \text{ miles} = (2.0114)(5280)$$
$$\approx 10{,}620 \text{ feet.}$$

Thus, the height of Mount Kilimanjaro

$$\approx 10{,}620 + 8720 = 19{,}340 \text{ feet.}$$

Practice Problem 7 The height of the nearby location to Mount McKinley, Alaska, is 12,870 feet, its distance to Mount McKinley's peak is 3.3387 miles, and the angle of elevation θ is 25°. What is the approximate height of Mount McKinley?

EXAMPLE 8   Finding the Width of a River

To find the width of a river, a surveyor sights straight across the river from a point A on her side to a point B on the opposite side. See Figure 4.79. She then walks 200 feet upstream to a point C. The angle θ that the line of sight from point C to point B makes with the riverbank is 58°. How wide is the river to the nearest foot?

Solution

The points A, B, and C are the vertices of a right triangle with acute angle $\theta = 58°$.
Let w represent the width of the river. From Figure 4.79, we have

$$\frac{w}{200} = \tan 58° \qquad \tan \theta = \frac{\text{opposite}}{\text{adjacent}}$$

$$w = 200 \tan 58° \qquad \text{Multiply both sides by 200.}$$

$$w \approx 320.07 \text{ feet} \qquad \text{Use a calculator.}$$

The river is about 320 feet wide at the point A.

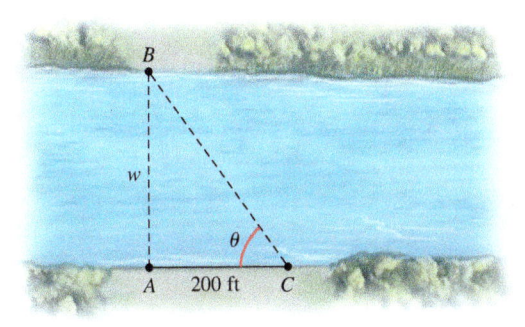

FIGURE 4.79 Width of a river

Practice Problem 8 From a point on the edge of one of two parallel rooftop sides, a point directly opposite is identified on the other rooftop. From a second point 25 feet from the first point along the rooftop edge, a second sighting to the point on the opposite rooftop is made, and the angle θ that this line of sight makes with the rooftop is 60°. How far apart are the two buildings? Round your answer to the nearest foot.

EXAMPLE 9   Finding the Rotation Angle for a Security Camera

A security camera is to be installed 40 feet from the center of a jewelry counter. The counter is 30 feet long. Through what angle, to the nearest degree, should the camera rotate so that it scans the entire counter? See Figure 4.80.

FIGURE 4.80 Security camera

Solution

The counter center B, the camera A, and one end of the counter C form a right triangle. The angle at vertex A is $\frac{\theta}{2}$, where θ is the angle through which the camera rotates. See Figure 4.80. We have

$\tan \frac{\theta}{2} = \frac{15}{40} = \frac{3}{8}$ Definition of tangent

$\frac{\theta}{2} = \tan^{-1}\left(\frac{3}{8}\right) \approx 20.556°$ Use a calculator in Degree mode.

$\theta \approx 41.11°$ Multiply both sides by 2.

Set the camera to rotate through 42° to scan the entire counter.

Practice Problem 9 Rework Example 9 for a counter that is 25 feet long and a camera set 15 feet from the center of the counter.

SECTION 4.6 Exercises

Basic Concepts and Skills

In Exercises 1–6, find the exact values for the six trigonometric functions of the angle θ in each figure. Rationalize the denominator where necessary.

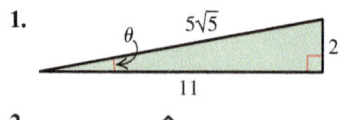

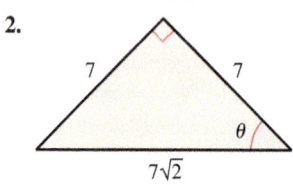

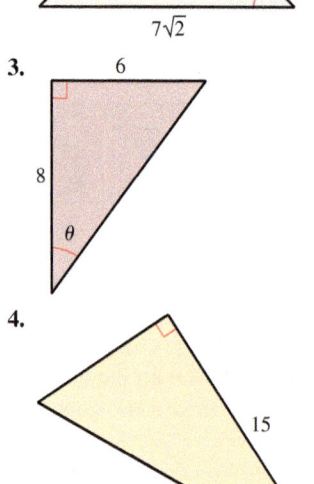

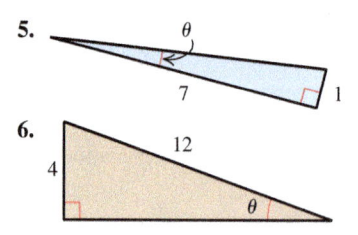

1.
2.
3.
4.
5.
6.

In Exercises 7–12, use each given trigonometric function value of θ to find the five other trigonometric function values of the acute angle θ. Rationalize the denominators where necessary.

7. $\cos \theta = \frac{2}{3}$
8. $\sin \theta = \frac{3}{4}$
9. $\tan \theta = \frac{5}{3}$
10. $\cot \theta = \frac{6}{11}$
11. $\sec \theta = \frac{13}{12}$
12. $\csc \theta = 4$

In Exercises 13–18, find the trigonometric function values of each corresponding complementary angle.

13. Given that $\sin 58° \approx 0.8480$, find $\cos 32°$.
14. Given that $\cos 37° \approx 0.7986$, find $\sin 53°$.
15. Given that $\tan 27° \approx 0.5095$, find $\cot 63°$.
16. Given that $\cot 49° \approx 0.8693$, find $\tan 41°$.
17. Given that $\sec 65° \approx 2.3662$, find $\csc 25°$.
18. Given that $\csc 78° \approx 1.0223$, find $\sec 12°$.

In Exercises 19–26, solve each right triangle; angle C is the right angle. Round to the nearest tenth where necessary.

19. $A = 50°, c = 9.2$
20. $B = 28°, c = 10.5$
21. $A = 36°, a = 12.0$
22. $B = 74°, b = 6.3$
23. $a = 12.5, b = 6.2$
24. $a = 4.3, b = 8.1$
25. $b = 9.4, c = 14.5$
26. $a = 6.2, c = 18.7$

In Exercises 27–34, use the figure and the given values to find each specified side length, angle, and trigonometric function value. Round your answer to three decimal places.

27. $a = 8, b = 10$ Find c, $\sin \theta$, and $\tan \theta$.
28. $a = 18, b = 3$ Find c, $\sin \theta$, and $\cos \theta$.
29. $a = 23, b = 7$ Find $\cos \theta$ and $\tan \theta$.
30. $a = 19, b = 27$ Find c, $\cos \theta$, and $\tan \theta$.
31. $\theta = 30°, a = 9$ Find $\sin \theta$, b, and c.
32. $\theta = 30°, a = 5$ Find b and c.
33. $\theta = 60°, b = 7$ Find α, a, and c.
34. $\theta = 60°, b = 10$ Find α, a, and c.

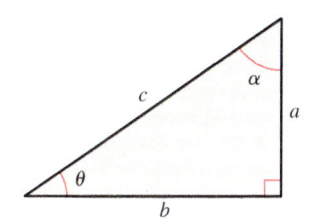

Applying the Concepts

35. **Positioning a ladder.** An 18-foot ladder is resting against a wall of a building in such a way that the top of the ladder is 14 feet above the ground. How far is the foot of the ladder from the base of the building?

36. **Positioning a ladder.** The ladder from Exercise 35 is placed against a wall so that the foot of the ladder is 7 feet from the base of the wall. How high up the wall will the ladder reach?

In Exercises 37 and 38, use the following definitions: The *vertical rise* of a ski lift is the vertical distance traveled when a lift car goes from the bottom terminal to the top terminal. The *slope length* is the linear distance a lift car travels as it moves from the bottom terminal to the top terminal. We treat the slope length as the hypotenuse and the vertical rise as the side opposite the angle formed by the horizontal line at the top terminal and the lift cables.

37. **Ski lift dimensions.** Find the horizontal distance a lift car passes over if the vertical rise is 295 meters and the slope length is 1070 meters.

38. **Ski lift dimensions.** Find the vertical rise (to the nearest foot) for a ski lift with slope length 2050 feet if the angle formed by a horizontal line at the top terminal and the lift cable is 16°.

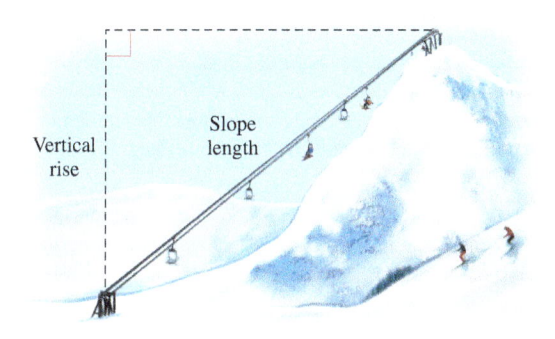

39. **Ski lift height.** If a skier travels 5000 feet up a ski lift whose angle of inclination is 30°, how high above his starting level is he?

40. **Shadow length.** At a time when the sun's rays make a 50° angle with the horizontal, how long is the shadow cast by a 6-foot man standing on flat ground?

41. **Height of a tree.** The angle relative to the horizontal from the top of a tree to a point 110 feet from its base (on flat ground) is 15°. Find the height of the tree to the nearest foot.

42. **Height of a cliff.** The angle of elevation from a rowboat moored 75 feet from a cliff is 73.8°. Find the height of the cliff to the nearest foot.

43. **Height of a flagpole.** A flagpole is supported by a 30-foot tension wire attached to the flagpole and to a stake in the ground. If the ground is flat and the angle the wire makes with the ground is 40°, how high up the flagpole is the wire attached? Round your answer to the nearest foot.

44. **Flying a kite.** A girl 5 feet tall is flying a kite. How long must the string be in order for her to raise the kite 200 feet above the ground if the string makes a 28°4′ angle with the ground?

45. **Width of a river.** Sal and his friend are on opposite sides of a river. To find the width of the river, each of them hammered a stake on his bank of the river in such a way that the distance between the two stakes approximates the width of the river. Sal walks 10 feet away from his stake along the river and finds that the line of sight from his new

position to his friend's stake across the river and the line of sight to his own stake form a 73°34′ angle. How wide is the river?

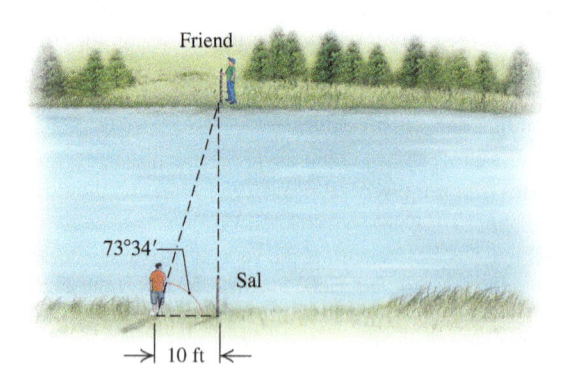

46. Measuring the Empire State Building. From a neighboring building's thirteenth-floor window that is 128 meters above the ground, the angle of elevation of the top of the Empire State Building is 59°20′, while the angle of depression of the base is 40°29′. How far away from the neighboring building and how tall is the Empire State Building?

47. Sighting a mortar station. The angle of depression from a helicopter 3000 feet directly above an allied mortar station to an enemy supply depot is 13°. How far is the depot from the mortar station?

48. Distance between Earth and the moon. If the radius of Earth is 3960 miles and angle A (the latitude of C) is 89°3′, find the distance between the center of Earth and the center of the moon.

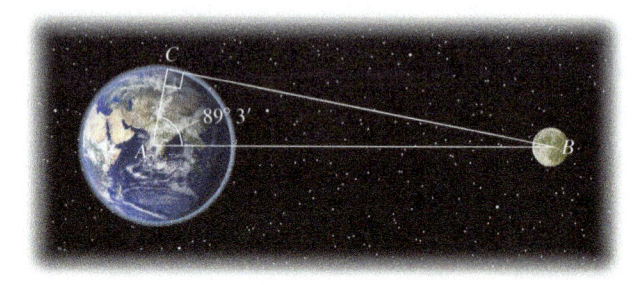

49. Sprinkler rotation. A sprinkler rotates back and forth through an angle θ, as shown in the figure. At a distance of 5 feet from the sprinkler, the rays that form the sides of angle θ are 6 feet apart. Find θ.

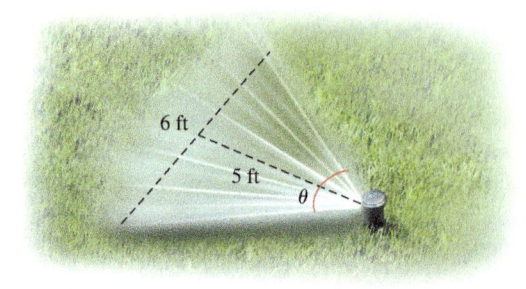

50. Irradiating flowers. A tray of flowers is being irradiated by a beam from a rotating lamp, as shown in the figure. If the tray is 8 feet long and the lamp is 2 feet from the center of the tray, through what angle should the lamp rotate to irradiate the full length of the tray?

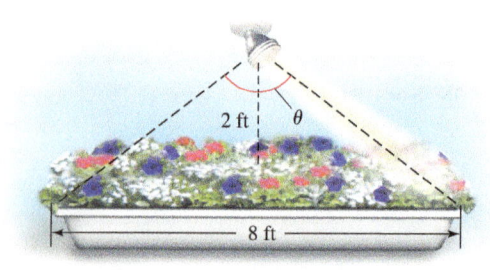

In Exercises 51 and 52, use the figure to find the radius r for the given latitudes (values of θ). Use 3960 miles for the radius of Earth.

51. Latitude. Find the radius of the 50th parallel (of latitude).

52. Latitude. Find the radius of the 53rd parallel (of latitude).

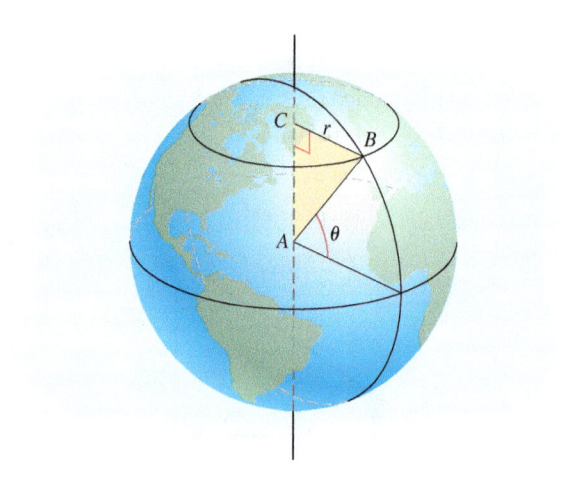

53. Motorcycle racing. A video camera is set up 110 feet at a right angle to a straight quarter-mile racetrack, as shown in the figure. The starting line is to the left, and the finish line is to the right. Through what angle must the camera rotate to film the entire race?

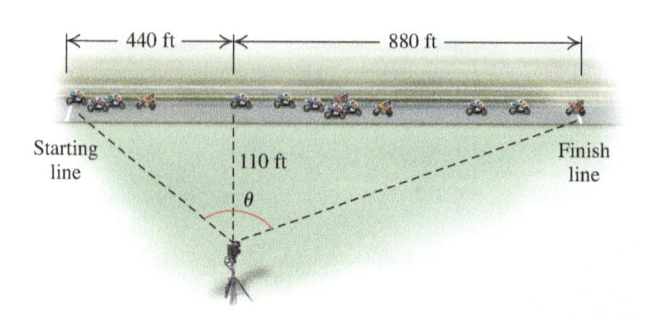

54. Camera's viewing angle. The viewing angle for the 35-millimeter camera is given (in degrees) by $\theta = 2\tan^{-1}\dfrac{18}{x}$, where x is the focal length of the lens. The focal length on most adjustable cameras is marked in millimeters on the lens mount.
  a. Find the viewing angle, in degrees, if the focal length is 50 millimeters.
  b. Find the viewing angle, in degrees, if the focal length is 200 millimeters.

Critical Thinking / Discussion / Writing

55. Draw a right triangle with acute angle θ. Then use the definitions of $\sin \theta$ and $\cos \theta$ together with the Pythagorean Theorem to show that $\sin^2 \theta + \cos^2 \theta = 1$.

56. In a right triangle with acute angle θ, what value must the adjacent side have for the length of the opposite side to represent $\tan \theta$?

57. Use the result of Exercise 56 to compare $\tan \alpha$ and $\tan \beta$ if α and β are acute angles and the measure of α is less than the measure of β.

58. In the figure, show that $h = \dfrac{d}{\cot \alpha - \cot \beta}$.

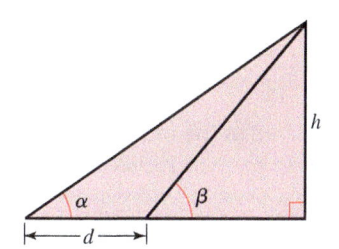

Maintaining Skills

In Exercises 59–64, solve each inequality. Write the answer in interval notation.

59. $|x - 5| < 3$
60. $|2x + 3| < 7$
61. $|x - 1| \geq 3$
62. $|3x + 4| \geq 2$
63. $|x + 2| < |x - 4|$
64. $|2x - 3| < |x + 1|$

SECTION 4.7

Trigonometric Identities

BEFORE STARTING THIS SECTION, REVIEW

1. Signs of the trigonometric functions (Section 4.2, page 305)

OBJECTIVES

1. Use fundamental trigonometric identities.
2. Prove that a given equation is not an identity.
3. Verify a trigonometric identity.

Hipparchus of Rhodes (190–120 B.C.) Hipparchus was born in Nicaea (now called Iznik) in Bithynia (northwest Turkey) but spent much of his life in Rhodes (Greece). Many historians consider him the founder of trigonometry. He introduced trigonometric functions in the form of a chord table—the ancestor of the sine table found in ancient Indian astronomical works. With his chord table, Hipparchus could solve the height–distance problems of plane trigonometry.

◆ Visualizing Trigonometric Functions

Trigonometric functions are commonly defined as the ratios of two sides corresponding to an angle in a right triangle. The Greek mathematician Hipparchus was the first person to relate trigonometric functions to a circle. His work allows us to visualize the trigonometric functions of an acute angle θ as the length of various segments associated with a unit circle. See Figure 4.81, showing several similar triangles. For example, triangles PCT and OCP are similar, so

$$\frac{CT}{PC} = \frac{PC}{OC} \quad \text{Sides are proportional.}$$

From Figure 4.81, we use the definitions of trigonometric functions to get

$$PC = \sin \theta, \quad OC = \cos \theta, \quad \text{and} \quad CT = OT - OC = \sec \theta - \cos \theta.$$

Substituting these values into the equation $\dfrac{CT}{PC} = \dfrac{PC}{OC}$, we obtain

$$\frac{\sec \theta - \cos \theta}{\sin \theta} = \frac{\sin \theta}{\cos \theta}.$$

In Example 4, we verify that this equation is an identity. Recall that an *identity* is an equation that is true for all numbers for which both sides are defined.

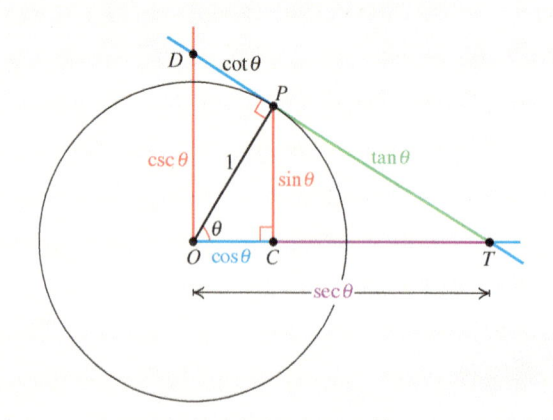

FIGURE 4.81 The unit circle and trigonometric functions

---

1 Use fundamental trigonometric identities.

Fundamental Trigonometric Identities

Because the equation for the unit circle is

$$x^2 + y^2 = 1, \quad \text{The center is } (0, 0); \text{ the radius is 1.}$$

we have the following identity:

$$(\cos t)^2 + (\sin t)^2 = 1 \quad \text{Replace } x \text{ with } \cos t \text{ and } y \text{ with } \sin t. \text{ See page 310.}$$

362 Chapter 4 Trigonometric Functions

By agreement, we write $\cos^2 t$ instead of $(\cos t)^2$ and $\sin^2 t$ instead of $(\sin t)^2$. A similar agreement holds for powers of all of the trigonometric functions.

The equation $x^2 + y^2 = 1$ yields two other useful identities.

$$\frac{x^2}{x^2} + \frac{y^2}{x^2} = \frac{1}{x^2}, x \neq 0 \quad \text{Divide both sides by } x^2.$$

$$1 + \left(\frac{y}{x}\right)^2 = \left(\frac{1}{x}\right)^2 \quad \text{Simplify.}$$

$$1 + \left(\frac{\sin t}{\cos t}\right)^2 = \left(\frac{1}{\cos t}\right)^2 \quad \text{Replace } x \text{ with } \cos t \text{ and } y \text{ with } \sin t.$$

$$1 + \tan^2 t = \sec^2 t \quad \frac{\sin t}{\cos t} = \tan t, \frac{1}{\cos t} = \sec t$$

Dividing both sides of the equation $x^2 + y^2 = 1$ by y^2, and then replacing x with $\cos t$ and y with $\sin t$, leads to the identity

$$1 + \cot^2 t = \csc^2 t.$$

The three identities resulting from the equation $x^2 + y^2 = 1$ are called the **Pythagorean identities** because the Pythagorean Theorem is the basis for this equation.

We use the definitions of the trigonometric functions and properties of the unit circle to establish reciprocal, quotient, Pythagorean, and even–odd identities. These are called **fundamental trigonometric identities**.

FUNDAMENTAL TRIGONOMETRIC IDENTITIES

1. **Reciprocal Identities**

$$\csc x = \frac{1}{\sin x} \qquad \sec x = \frac{1}{\cos x} \qquad \cot x = \frac{1}{\tan x}$$

$$\sin x = \frac{1}{\csc x} \qquad \cos x = \frac{1}{\sec x} \qquad \tan x = \frac{1}{\cot x}$$

2. **Quotient Identities**

$$\tan x = \frac{\sin x}{\cos x} \qquad \cot x = \frac{\cos x}{\sin x}$$

3. **Pythagorean Identities**

$$\sin^2 x + \cos^2 x = 1 \qquad 1 + \tan^2 x = \sec^2 x \qquad 1 + \cot^2 x = \csc^2 x$$

4. **Even–Odd Identities**

$$\sin(-x) = -\sin x \qquad \cos(-x) = \cos x \qquad \tan(-x) = -\tan x$$

$$\csc(-x) = -\csc x \qquad \sec(-x) = \sec x \qquad \cot(-x) = -\cot x$$

Evaluating Trigonometric Functions

The next example shows how fundamental trigonometric identities are used to evaluate the remaining trigonometric functions from one specified trigonometric function value and the quadrant in which the angle lies.

EXAMPLE 1 Using the Fundamental Trigonometric Identities

If $\cot x = \frac{3}{4}$ and $\pi < x < \frac{3\pi}{2}$, find the values of the remaining trigonometric functions.

Solution

When given cot x, we can find csc x by using the identity:

$$\csc^2 x = 1 + \cot^2 x \qquad \text{Pythagorean identity}$$

$$= 1 + \left(\frac{3}{4}\right)^2 \qquad \text{Replace cot } x \text{ with } \frac{3}{4}.$$

$$\csc^2 x = 1 + \frac{9}{16} = \frac{25}{16} \qquad \text{Simplify.}$$

$$\csc x = -\frac{5}{4} \qquad \text{Square root property: because } x \text{ is in quadrant III, csc } x \text{ is negative.}$$

$$\sin x = -\frac{4}{5} \qquad \sin x = \frac{1}{\csc x}$$

Multiplying both sides of the quotient identity $\dfrac{\cos x}{\sin x} = \cot x$ by sin x, we have

$$\cos x = \cot x \sin x$$

$$= \left(\frac{3}{4}\right)\left(-\frac{4}{5}\right) \qquad \text{Substitute values of cot } x \text{ and sin } x.$$

$$= -\frac{3}{5} \qquad \text{Simplify.}$$

We have the following values of the trigonometric functions:

$$\sin x = -\frac{4}{5} \qquad \cos x = -\frac{3}{5} \qquad \tan x = \frac{4}{3}$$

$$\csc x = -\frac{5}{4} \qquad \sec x = -\frac{5}{3} \qquad \cot x = \frac{3}{4} \qquad \text{Use reciprocal identities.}$$

Practice Problem 1 If $\tan x = -\dfrac{1}{2}$ and $\dfrac{\pi}{2} < x < \pi$, find the values of the remaining trigonometric functions.

Trigonometric Equations and Identities

2 Prove that a given equation is not an identity.

To verify that an equation is a trigonometric identity, you must *prove* (*verify*) that both sides of the equation are equal for all values of the variable for which both sides are defined.

Consider the following equation:

$$\cos x = \sqrt{1 - \sin^2 x} \qquad (1)$$

Equation (1) is true for all values of x in the interval $\left[-\dfrac{\pi}{2}, \dfrac{\pi}{2}\right]$. See Figure 4.82(a).

x in $\left[-\dfrac{\pi}{2}, \dfrac{\pi}{2}\right]$ \qquad x in $[-\pi, \pi]$

$Y_1 = \cos x$
$Y_2 = \sqrt{1 - \sin^2 x}$

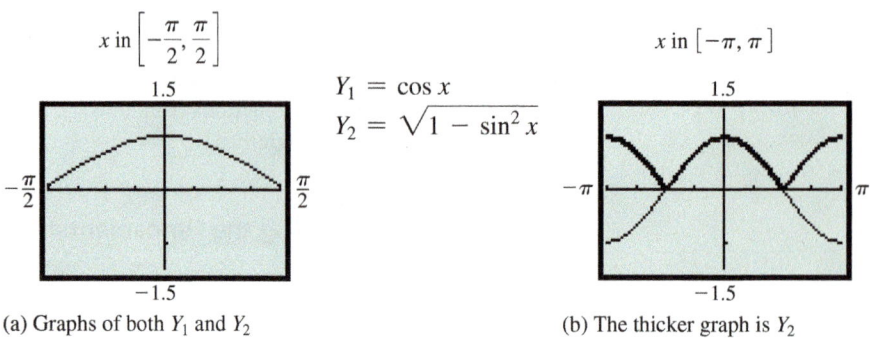

(a) Graphs of both Y_1 and Y_2 \qquad (b) The thicker graph is Y_2

FIGURE 4.82

Because the domain of the sine and cosine functions is the interval $(-\infty, \infty)$, both sides of equation (1) are defined for all real numbers. But for any value of x in the interval $\left(\dfrac{\pi}{2}, \pi\right]$, the left side of equation (1) has a negative value (because $\cos x$ is negative in quadrant II), and the right side of equation (1) has a positive value. Therefore, equation (1) is *not* an identity. Figure 4.82(b) illustrates that a graphing calculator can help verify that a given equation is *not* an identity. Figure 4.82(a), however, shows that a graphing calculator cannot prove that a given equation *is* an identity because you might happen to use a viewing window where the graphs coincide.

EXAMPLE 2 Proving That an Equation Is Not an Identity

Prove that the following equation is not an identity.
$$(\sin x - \cos x)^2 = \sin^2 x - \cos^2 x$$

Solution

To prove that the equation is not an identity we look for a **counterexample**. That is, we find at least one value of x that results in both sides being defined but not equal.

Let $x = 0$. We know that $\sin 0 = 0$ and $\cos 0 = 1$.

The left side $= (\sin x - \cos x)^2 = (0 - 1)^2 = 1$ Replace x with 0.
The right side $= \sin^2 x - \cos^2 x = (0)^2 - (1)^2 = -1$ Replace x with 0.

For $x = 0$, the equation's two sides are not equal; so it is not an identity.

Practice Problem 2 Prove that the equation $\cos x = 1 - \sin x$ is not an identity.

> **SIDE NOTE**
>
> A **counterexample** is an example that disproves a statement.

3 Verify a trigonometric identity.

Verifying Trigonometric Identities

Verifying a trigonometric identity differs from solving an equation. In verifying an identity, we are given an equation and want to show that it is true for *all* values of the variable for which both sides of the equation are defined. We will use the following method.

> **VERIFYING TRIGONOMETRIC IDENTITIES**
>
> To verify that an equation is an identity, transform one side of the equation into the other side by a sequence of steps, each of which produces an identity. The steps involved can be algebraic manipulations or can use known identities. Note that in verifying an identity, we *do not* just perform the same operation on both sides of the equation.

Methods of Verifying Trigonometric Identities

Verifying trigonometric identities requires practice and experience; there is no set procedure. In the following examples, we suggest five guidelines for verifying trigonometric identities, with the first two being more widely used.

1. **Start with the more complicated side and transform it to the simpler side.**

EXAMPLE 3 Verifying an Identity

Verify the identity:
$$\frac{\csc^2 x - 1}{\cot x} = \cot x$$

TECHNOLOGY CONNECTION

In Example 3, we use a graphing calculator to graph
$Y_1 = \dfrac{\csc^2 x - 1}{\cot x}$ and $Y_2 = \cot x$.
Although not a definitive method of proof, the graphs appear identical; so the equation in Example 3 *appears* to be an identity.

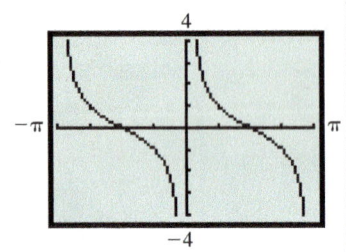

Graph of Y_1

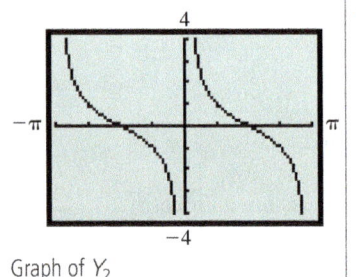

Graph of Y_2

Solution

We start with the left side of the equation because it appears more complicated than the right side.

$$\dfrac{\csc^2 x - 1}{\cot x} = \dfrac{\cot^2 x}{\cot x} \qquad \csc^2 x = 1 + \cot^2 x, \text{ so } \csc^2 x - 1 = \cot^2 x.$$

$$= \cot x \qquad \text{Remove the common factor, } \cot x.$$

We have shown that the left side of the equation is equal to the right side, which verifies the identity.

Practice Problem 3 Verify the identity:

$$\dfrac{1 - \sin^2 x}{\cos x} = \cos x$$

2. Stay focused on the final expression.

While working on one side of the equation, stay focused on your goal of converting it to the form on the other side. This often helps in deciding what your next step should be.

EXAMPLE 4 Verifying an Identity

Verify the following identity proposed in the introduction to this section:

$$\dfrac{\sec \theta - \cos \theta}{\sin \theta} = \dfrac{\sin \theta}{\cos \theta}$$

Solution

We start with the more complicated left side.

$$\dfrac{\sec \theta - \cos \theta}{\sin \theta} = \dfrac{\dfrac{1}{\cos \theta} - \cos \theta}{\sin \theta} \qquad \sec \theta = \dfrac{1}{\cos \theta}$$

$$= \dfrac{\left(\dfrac{1}{\cos \theta} - \cos \theta\right) \cos \theta}{\sin \theta \cos \theta} \qquad \text{Multiply numerator and denominator by } \cos \theta.$$

$$= \dfrac{\dfrac{1}{\cos \theta} \cdot \cos \theta - \cos^2 \theta}{\sin \theta \cos \theta} \qquad \text{Distributive property}$$

$$= \dfrac{1 - \cos^2 \theta}{\sin \theta \cos \theta} \qquad \text{Simplify.}$$

$$= \dfrac{\sin^2 \theta}{\sin \theta \cos \theta} \qquad \sin^2 \theta + \cos^2 \theta = 1$$

$$= \dfrac{\sin \theta}{\cos \theta} \qquad \text{Remove the common factor } \sin \theta.$$

The left side is identical to the right side; the given equation is an identity.

Practice Problem 4 In Figure 4.81 on page 362, triangles *OPD* and *TCP* are similar. So, $\dfrac{OD}{OP} = \dfrac{PT}{CT}$ leads to

$$\dfrac{\csc \theta}{1} = \dfrac{\tan \theta}{\sec \theta - \cos \theta}.$$

Verify that this equation is an identity.

3. Convert to sines and cosines.

It may be helpful to rewrite all trigonometric functions in the equation in terms of sines and cosines and then simplify. This approach is recommended if an identity involves three or more functions.

EXAMPLE 5 Verifying by Rewriting with Sines and Cosines

Verify the identity $\cot^4 x + \cot^2 x = \cot^2 x \csc^2 x$.

Solution

We start with the more complicated left side.

$$\cot^4 x + \cot^2 x = \frac{\cos^4 x}{\sin^4 x} + \frac{\cos^2 x}{\sin^2 x} \qquad \cot x = \frac{\cos x}{\sin x}$$

$$= \frac{\cos^4 x}{\sin^4 x} + \frac{\cos^2 x}{\sin^2 x} \cdot \frac{\sin^2 x}{\sin^2 x} \qquad \text{The LCD is } \sin^4 x.$$

$$= \frac{\cos^4 x + \cos^2 x \sin^2 x}{\sin^4 x} \qquad \text{Add rational expressions.}$$

$$= \frac{\cos^2 x (\cos^2 x + \sin^2 x)}{\sin^4 x} \qquad \text{Factor out } \cos^2 x.$$

$$= \frac{\cos^2 x (1)}{\sin^4 x} \qquad \cos^2 x + \sin^2 x = 1$$

$$= \frac{\cos^2 x}{\sin^2 x} \cdot \frac{1}{\sin^2 x} \qquad \text{Factor to obtain the form on the right side.}$$

$$= \cot^2 x \csc^2 x \qquad \cot x = \frac{\cos x}{\sin x}, \csc x = \frac{1}{\sin x}$$

Because the left side is identical to the right side, the given equation is an identity.

Practice Problem 5 Verify the identity $\tan^4 x + \tan^2 x = \tan^2 x \sec^2 x$.

You could also verify the identity in Example 5 as follows:

$$\cot^4 x + \cot^2 x = \cot^2 x (\cot^2 x + 1) \qquad \text{Distributive property}$$
$$= \cot^2 x \csc^2 x \qquad \text{Pythagorean identity}$$

This approach shows that rewriting the expression using only sines and cosines is not always the quickest way to verify an identity. It is a useful approach when you are stuck, however.

4. Work on both sides.

Sometimes it is helpful to work separately on both sides of the equation. To verify the identity $P(x) = Q(x)$, for example, we transform the left side $P(x)$ into $R(x)$ by using algebraic manipulations and known identities. Then $P(x) = R(x)$ is an identity. Next, we transform the right side, $Q(x)$, into $R(x)$ so that the equation $Q(x) = R(x)$ is an identity. It then follows that $P(x) = Q(x)$ is an identity. See the next example.

EXAMPLE 6 Verifying an Identity by Transforming Both Sides Separately

Verify the identity:

$$\frac{1}{1 - \sin x} - \frac{1}{1 + \sin x} = \frac{\tan^2 x + \sec^2 x + 1}{\csc x}$$

Solution

We start with the left side of the equation.

$$\frac{1}{1-\sin x} - \frac{1}{1+\sin x}$$ 　　Begin with the left side.

$$= \frac{1}{1-\sin x} \cdot \frac{1+\sin x}{1+\sin x} - \frac{1}{1+\sin x} \cdot \frac{1-\sin x}{1-\sin x}$$ 　　Rewrite using the LCD, $(1-\sin x)(1+\sin x)$.

$$= \frac{(1+\sin x) - (1-\sin x)}{(1-\sin x)(1+\sin x)}$$ 　　Subtract fractions.

$$= \frac{1+\sin x - 1 + \sin x}{1-\sin^2 x}$$ 　　Rewrite the numerator; multiply in the denominator.

$$= \frac{2\sin x}{\cos^2 x}$$ 　　Simplify; $1 - \sin^2 x = \cos^2 x$

We now try to convert the right side to the form $\dfrac{2\sin x}{\cos^2 x}$.

$$\frac{\tan^2 x + \sec^2 x + 1}{\csc x}$$

$$= \frac{(\tan^2 x + 1) + \sec^2 x}{\csc x}$$ 　　Regroup terms.

$$= \frac{\sec^2 x + \sec^2 x}{\csc x}$$ 　　$\tan^2 x + 1 = \sec^2 x$

$$= \frac{2\sec^2 x}{\csc x}$$ 　　Simplify.

$$= 2\left(\frac{1}{\csc x}\right)\sec^2 x$$ 　　Rewrite.

$$= 2\sin x\left(\frac{1}{\cos^2 x}\right)$$ 　　Reciprocal identities

$$= \frac{2\sin x}{\cos^2 x}$$ 　　Multiply.

From the original equation, we see that

$$\underbrace{\frac{1}{1-\sin x} - \frac{1}{1+\sin x}}_{\text{Left-Hand Side}} = \underbrace{\frac{2\sin x}{\cos^2 x}}_{\text{Third Expression}} = \underbrace{\frac{\tan^2 x + \sec^2 x + 1}{\csc x}}_{\text{Right-Hand Side}}$$

Because both sides of the original equation are equal to $\dfrac{2\sin x}{\cos^2 x}$, the identity is verified.

Practice Problem 6 Verify the identity:

$$\tan\theta + \sec\theta = \frac{\csc\theta + 1}{\cot\theta}$$

5. Use conjugates.

It may be helpful to multiply both the numerator and denominator of a fraction by the same factor. Recall that the sum $a + b$ and the difference $a - b$ are the *conjugates* of each other and that $(a+b)(a-b) = a^2 - b^2$.

SIDE NOTE

You cannot verify a trigonometric identity that involves rational expressions by cross multiplication, because you *do not know* that the expressions are equal. In fact, equality is what you are trying to verify.

EXAMPLE 7   **Verifying an Identity by Using a Conjugate**

Verify the identity:

$$\frac{\cos x}{1 + \sin x} = \frac{1 - \sin x}{\cos x}$$

Solution

We start with the left side of the equation.

$$\frac{\cos x}{1 + \sin x} = \frac{\cos x (1 - \sin x)}{(1 + \sin x)(1 - \sin x)}$$ Multiply the numerator and the denominator by $1 - \sin x$, the conjugate of $1 + \sin x$.

$$= \frac{\cos x (1 - \sin x)}{1 - \sin^2 x}$$ $(a + b)(a - b) = a^2 - b^2$

$$= \frac{\cos x (1 - \sin x)}{\cos^2 x}$$ $1 - \sin^2 x = \cos^2 x$

$$= \frac{1 - \sin x}{\cos x}$$ Remove the common factor $\cos x$.

Because the left side is identical to the right side, the given equation is an identity.

Practice Problem 7   Verify the identity $\dfrac{\tan x}{\sec x + 1} = \dfrac{\sec x - 1}{\tan x}$.

SUMMARY OF MAIN FACTS

Guidelines for Verifying Trigonometric Identities

Algebra Operations

Review the procedure for combining fractions by finding the least common denominator.

Fundamental Trigonometric Identities

Review the fundamental trigonometric identities summarized on page 363. Look for an opportunity to apply the fundamental trigonometric identities when working on either side of the identity to be verified. Become thoroughly familiar with alternative forms of fundamental identities. For example, $\sin^2 x = 1 - \cos^2 x$ and $\sec^2 x - \tan^2 x = 1$ are alternative forms of the fundamental identities $\sin^2 x + \cos^2 x = 1$ and $\sec^2 x = 1 + \tan^2 x$, respectively.

1. **Start with the more complicated side.**
   It is generally helpful to start with the more complicated side of an identity and simplify it until it becomes identical to the other side. See Example 3.

2. **Stay focused on the final expression.**
   While working on one side of the identity, stay focused on your goal of converting it to the form on the other side. See Example 4.

   The following three techniques are *sometimes* helpful.

3. **Option: Convert to sines and cosines.**
   Rewrite one side of the identity in terms of sines and cosines. See Example 5.

4. **Option: Work on both sides.**
   Transform each side separately to the same equivalent expression. See Example 6.

5. **Option: Use conjugates.**
   In expressions containing $1 + \sin x$, $1 - \sin x$, $1 + \cos x$, $1 - \cos x$, $\sec x + \tan x$, and so on, multiply the numerator and the denominator by its conjugate and then use the appropriate Pythagorean identity. See Example 7.

In calculus, it is sometimes helpful to convert algebraic expressions to trigonometric ones. This technique is called *trigonometric substitution* and is illustrated in the next example.

EXAMPLE 8 Trigonometric Substitution

Replace x with $\cos \theta$ in the expression $\sqrt{1 - x^2}$ and simplify. Assume that $0 \leq \theta \leq \pi/2$.

Solution

$$\sqrt{1 - x^2} = \sqrt{1 - \cos^2 \theta} \quad \text{Replace } x \text{ with } \cos \theta.$$
$$= \sqrt{\sin^2 \theta} \quad \text{Pythagorean identity}$$
$$= \sin \theta \quad \sin \theta > 0 \text{ because } 0 < \theta < \frac{\pi}{2}.$$

Practice Problem 8 Replace x with $3 \sin \theta$ in the expression $\sqrt{9 - x^2}$ and simplify. Assume that $0 \leq \theta \leq \frac{\pi}{2}$.

SECTION 4.7 Exercises

Basic Concepts and Skills

In Exercises 1–12, use the fundamental identities and the given information to find the exact values of the remaining trigonometric functions of x.

1. $\cos x = -\dfrac{3}{5}$ and $\sin x = \dfrac{4}{5}$
2. $\cos x = \dfrac{3}{5}$ and $\sin x = -\dfrac{4}{5}$
3. $\sin x = \dfrac{1}{\sqrt{3}}$ and $\cos x = \sqrt{\dfrac{2}{3}}$
4. $\sin x = \dfrac{1}{\sqrt{3}}$ and $\cos x = -\sqrt{\dfrac{2}{3}}$
5. $\tan x = \dfrac{1}{2}$ and $\sec x = -\dfrac{\sqrt{5}}{2}$
6. $\tan x = \dfrac{1}{2}$ and $\sec x = \dfrac{\sqrt{5}}{2}$
7. $\csc x = 3$ and $\cot x = 2\sqrt{2}$
8. $\csc x = 3$ and $\cot x = -2\sqrt{2}$
9. $\cot x = \dfrac{12}{5}$ and $\sin x = -\dfrac{5}{13}$
10. $\cot x = -\dfrac{12}{5}$ and $\sin x = -\dfrac{5}{13}$
11. $\sec x = 3$ and $\dfrac{3\pi}{2} < x < 2\pi$
12. $\tan x = -2$ and $\dfrac{\pi}{2} < x < \pi$

In Exercises 13–22, use the fundamental identities and appropriate algebraic operations to simplify each expression.

13. $(1 + \tan x)(1 - \tan x) + \sec^2 x$
14. $(\sec x - 1)(\sec x + 1) - \tan^2 x$
15. $(\sec x + \tan x)(\sec x - \tan x)$
16. $\dfrac{\sec^2 x - 4}{\sec x - 2}$
17. $\csc^4 x - \cot^4 x$
18. $\sin x \cos x (\tan x + \cot x)$
19. $\dfrac{\sec x \csc x (\sin x + \cos x)}{\sec x + \csc x}$
20. $\dfrac{1}{\csc x + 1} - \dfrac{1}{\csc x - 1}$
21. $\dfrac{\tan^2 x - 2 \tan x - 3}{\tan x + 1}$
22. $\dfrac{\tan^2 x + \sec x - 1}{\sec x - 1}$

In Exercises 23–28, prove that the given equation is not an identity by finding a value of x for which the two sides have different values.

23. $\sin x = 1 - \cos x$
24. $\tan x = \sec x - 1$
25. $\cos x = \sqrt{1 - \sin^2 x}$
26. $\sec x = \sqrt{1 + \tan^2 x}$
27. $\sin^2 x = (1 - \cos x)^2$
28. $\cot^2 x = (\csc x + 1)^2$

In Exercises 29–64, verify each identity.

29. $\sin x \tan x + \cos x = \sec x$
30. $\cos x \cot x + \sin x = \csc x$
31. $\dfrac{1 - 4\cos^2 x}{1 - 2\cos x} = 1 + 2\cos x$
32. $\dfrac{9 - 16\sin^2 x}{3 + 4\sin x} = 3 - 4\sin x$
33. $(\cos x - \sin x)(\cos x + \sin x) = 1 - 2\sin^2 x$
34. $(\sin x - \cos x)(\sin x + \cos x) = 1 - 2\cos^2 x$
35. $\sin^2 x \cot^2 x + \sin^2 x = 1$
36. $\tan^2 x - \sin^2 x = \sin^4 x \sec^2 x$
37. $\sin^3 x - \cos^3 x = (\sin x - \cos x)(1 + \sin x \cos x)$
38. $\sin^3 x + \cos^3 x = (\sin x + \cos x)(1 - \sin x \cos x)$
39. $\cos^4 x - \sin^4 x = 1 - 2\sin^2 x$
40. $\cos^4 x - \sin^4 x = 2\cos^2 x - 1$
41. $\dfrac{1}{1 - \sin x} + \dfrac{1}{1 + \sin x} = 2\sec^2 x$
42. $\dfrac{1}{1 - \cos x} + \dfrac{1}{1 + \cos x} = 2\csc^2 x$
43. $\dfrac{1}{\csc x - 1} - \dfrac{1}{\csc x + 1} = 2\tan^2 x$
44. $\dfrac{1}{\sec x - 1} + \dfrac{1}{\sec x + 1} = 2\sec x \cot^2 x$
45. $\sec^2 x + \csc^2 x = \sec^2 x \csc^2 x$
46. $\cot^2 x + \tan^2 x = \sec^2 x \csc^2 x - 2$
47. $\dfrac{1}{\sec x - \tan x} + \dfrac{1}{\sec x + \tan x} = \dfrac{2}{\cos x}$
48. $\dfrac{1}{\csc x + \cot x} + \dfrac{1}{\csc x - \cot x} = \dfrac{2}{\sin x}$
49. $\dfrac{\sin x}{1 + \cos x} = \csc x - \cot x$
50. $\dfrac{\sin x}{1 - \cos x} = \csc x + \cot x$
51. $(\sin x + \cos x)^2 = 1 + 2\sin x \cos x$
52. $(\sin x - \cos x)^2 = 1 - 2\sin x \cos x$
53. $(1 + \tan x)^2 = \sec^2 x + 2\tan x$
54. $(1 - \cot x)^2 = \csc^2 x - 2\cot x$
55. $\dfrac{\tan x \sin x}{\tan x + \sin x} = \dfrac{\tan x - \sin x}{\tan x \sin x}$
56. $\dfrac{\cot x \cos x}{\cot x + \cos x} = \dfrac{\cot x - \cos x}{\cot x \cos x}$
57. $(\tan x + \cot x)^2 = \sec^2 x + \csc^2 x$
58. $(1 + \cot^2 x)(1 + \tan^2 x) = \dfrac{1}{\sin^2 x \cos^2 x}$
59. $\dfrac{\sin^2 x - \cos^2 x}{\sec^2 x - \csc^2 x} = \sin^2 x \cos^2 x$
60. $\left(\tan x + \dfrac{1}{\cot x}\right)\left(\cot x + \dfrac{1}{\tan x}\right) = 4$
61. $\dfrac{\tan x}{1 + \sec x} + \dfrac{1 + \sec x}{\tan x} = 2\csc x$
62. $\dfrac{\cot x}{1 + \csc x} + \dfrac{1 + \csc x}{\cot x} = 2\sec x$
63. $\dfrac{\sin x + \tan x}{\cos x + 1} = \tan x$
64. $\dfrac{\sin x}{1 + \tan x} = \dfrac{\cos x}{1 + \cot x}$

In Exercises 65–70, make the indicated trigonometric substitution in the given algebraic expression and simplify. Assume that $0 < \theta < \dfrac{\pi}{2}$.

65. $\sqrt{1 + x^2},\ x = \tan\theta$
66. $\sqrt{4 - x^2},\ x = 2\cos\theta$
67. $\sqrt{x^2 - 1},\ x = \sec\theta$
68. $\sqrt{x^2 - 4},\ x = 2\sec\theta$
69. $\dfrac{x}{\sqrt{1 - x^2}},\ x = \cos\theta$
70. $\dfrac{x}{\sqrt{x^2 - 1}},\ x = \sec\theta$

Applying the Concepts

71. **Length of a ladder.** A ladder x feet long makes an angle θ with the horizontal and reaches a height of 20 feet. Then $x = \dfrac{20}{\sin\theta}$. Use a reciprocal identity to rewrite this formula.

72. **Distance from a building.** From a distance of x feet to a 60-foot-high building, the angle of elevation is θ degrees. Then $x = \dfrac{60}{\tan\theta}$. Use a reciprocal identity to rewrite this formula.

73. **Intersecting lines.** If two nonvertical lines with slopes m_1 and $m_2\,(m_1 > m_2)$ intersect, then the acute angle θ between the lines satisfies the equation $m_1\cos\theta - \sin\theta = m_2\cos\theta + m_1 m_2 \sin\theta$. Rewrite this equation in terms of a single trigonometric function.

74. **Tower height.** The angles to the top of a tower measured from two locations d units apart are α and β as shown in the figure. The height h of the tower satisfies the equation
$$h = \dfrac{d\sin\alpha\sin\beta}{\cos\alpha\sin\beta - \sin\alpha\cos\beta}.$$
Rewrite this equation in terms of cotangents.

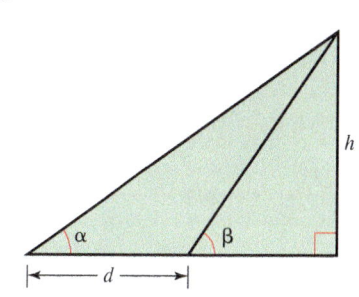

Critical Thinking / Discussion / Writing

75. If x and y are real numbers, explain why $\sec\theta$ cannot be equal to $\dfrac{xy}{x^2+y^2}$.

76. Find values of t for which $\sin\theta = \dfrac{1+t^2}{1-t^2}$ is possible.

77. Find values of a and b for which $\cos^2\theta = \dfrac{a^2+b^2}{2ab}$ is possible.

78. Find values of a and b for which $\csc^2\theta = \dfrac{2ab}{a^2+b^2}$ is possible.

79. Prove that the following inequalities are true for $0 < \theta < \dfrac{\pi}{2}$.
  a. $\sin^2\theta + \csc^2\theta \geq 2$
  b. $\cos^2\theta + \sec^2\theta \geq 2$
  c. $\sec^2\theta + \csc^2\theta \geq 4$

Maintaining Skills

80. Find the distance between the points $(1, 0)$ and $(3, -4)$.

81. If $x^2 + y^2 = 1$, simplify $(x-y)^2 + (x+y)^2$.

82. Simplify: $\dfrac{\dfrac{1}{x-1} + \dfrac{1}{x+1}}{\dfrac{1}{x-1} - \dfrac{1}{x+1}}$

SECTION 4.8 Sum and Difference Formulas

BEFORE STARTING THIS SECTION, REVIEW

1. Distance formula (Section 1.1, page 54)
2. Fundamental trigonometric identities (Section 4.7, page 363)
3. Trigonometric functions of common angles (Section 4.2, page 302)

OBJECTIVES

1. Use the sum and difference formulas for cosine.
2. Use cofunction identities.
3. Use the sum and difference formulas for sine.
4. Use the sum and difference formulas for tangent.
5. Use double-angle formulas.
6. Use power-reducing formulas.
7. Use half-angle formulas.

◆ Cost of Using an Electric Blanket

Three basic properties of electricity—*voltage, current*, and *power*—together with the rate charged by your electric company, are used to find the cost of using an electric device.

Voltage (*V*) is measured in *volts*. It is the force that pushes electricity through a wire. *Current* (*I*), measured in *amperes* (amps), measures how much electricity is moving through the device per second. *Power* (*P*), measured in *watts*, gives the energy consumed per second by an electric device and is defined by the equation

$$P = VI \quad \text{Power} = (\text{Voltage})(\text{Current})$$

Your electric company bills you by the kilowatt-hour. When you turn on an electric device that consumes 1000 watts for one hour, it consumes 1 kilowatt-hour. Suppose your electric company charges 8¢ per kilowatt-hour. An electric blanket might use 250 watts (depending on the setting). If you turn it on for ten hours, it will consume $250 \times 10 = 2500 = 2.5$ kilowatt-hours. This will cost you $(2.5)(8) = 20$¢.

In Exercises 57–60, we use trigonometric identities to compute the *wattage rating* of some electric appliances.

1 Use the sum and difference formulas for cosine.

Sum and Difference Formulas for Cosine

In this section, we develop identities involving the sum and difference of two variables, u and v. The variables represent any two real numbers, or angles, in radian or degree measure.

> **SUM AND DIFFERENCE FORMULAS FOR COSINE**
> $$\cos(u + v) = \cos u \cos v - \sin u \sin v$$
> $$\cos(u - v) = \cos u \cos v + \sin u \sin v$$

Section 4.8 ■ Sum and Difference Formulas 373

To prove the second formula, we assume that $0 < v < u < 2\pi$, although this identity is true for all real numbers u and v. Figure 4.83(a) shows points P and Q on the unit circle on the terminal sides of angles u and v. The definition of the circular functions tells us that $P = (\cos u, \sin u)$ and $Q = (\cos v, \sin v)$. In Figure 4.83(b), we rotated angle $(u - v)$ to standard position and labeled point B on the unit circle and the terminal side of the angle $(u - v)$.

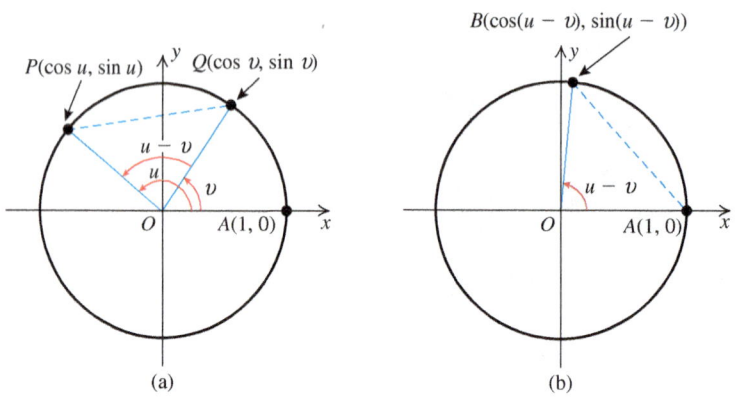

FIGURE 4.83

RECALL

The distance $d(P, Q)$ between two points $P(x_1, y_1)$ and $Q(x_2, y_2)$ is given by
$$d(P, Q) = \sqrt{(x_2 - x_1)^2 + (y_2 - y_1)^2}.$$

We know that triangle POQ in Figure 4.83(a) is congruent to triangle BOA in Figure 4.83(b) by SAS (side-angle-side) congruence; so

$$d(P, Q) = d(A, B)$$
$$[d(P, Q)]^2 = [d(A, B)]^2 \quad (1) \qquad \text{Square both sides.}$$

Now use the distance formula to simplify the left side of equation (1).

$$[d(P, Q)]^2 = (\cos u - \cos v)^2 + (\sin u - \sin v)^2 \qquad \text{Distance formula}$$
$$= \cos^2 u - 2 \cos u \cos v + \cos^2 v \qquad (a - b)^2 = a^2 - 2ab + b^2$$
$$\quad + \sin^2 u - 2 \sin u \sin v + \sin^2 v$$
$$= (\cos^2 u + \sin^2 u) + (\cos^2 v + \sin^2 v) \qquad \text{Combine terms.}$$
$$\quad - 2(\cos u \cos v + \sin u \sin v)$$
$$= 1 + 1 - 2(\cos u \cos v + \sin u \sin v) \qquad \text{Pythagorean identity}$$
$$[d(P, Q)]^2 = 2 - 2(\cos u \cos v + \sin u \sin v) \qquad \text{Simplify.}$$

Now simplify the right side of equation (1).

$$[d(A, B)]^2 = [\cos(u - v) - 1]^2 + [\sin(u - v) - 0]^2 \qquad \text{Distance formula}$$
$$= \cos^2(u - v) - 2\cos(u - v) + 1 + \sin^2(u - v) \qquad \text{Expand binomial.}$$
$$= [\cos^2(u - v) + \sin^2(u - v)] + 1 - 2\cos(u - v) \qquad \text{Combine terms.}$$
$$= 1 + 1 - 2\cos(u - v) \qquad \text{Pythagorean identity}$$
$$[d(A, B)]^2 = 2 - 2\cos(u - v) \qquad \text{Simplify.}$$

Because $[d(A, B)]^2 = [d(P, Q)]^2$,

$$2 - 2\cos(u - v) = 2 - 2(\cos u \cos v + \sin u \sin v) \qquad \text{Substitute in each side.}$$
$$\mathbf{\cos(u - v) = \cos u \cos v + \sin u \sin v} \qquad \text{Simplify.}$$

We have proved the formula for the cosine of the difference of two angles or two real numbers.

Section 4.8 ■ Sum and Difference Formulas   375

EXAMPLE 1 Using the Difference Formula for Cosine

Find the exact value of $\cos \dfrac{\pi}{12}$ by using $\dfrac{\pi}{12} = \dfrac{\pi}{3} - \dfrac{\pi}{4}$.

Solution

We use the exact values of the functions sine and cosine at $\dfrac{\pi}{3}$ and $\dfrac{\pi}{4}$ as well as the difference formula for cosine.

$$\cos \dfrac{\pi}{12} = \cos\left(\dfrac{\pi}{3} - \dfrac{\pi}{4}\right) \qquad\qquad \dfrac{\pi}{3} - \dfrac{\pi}{4} = \dfrac{4\pi}{12} - \dfrac{3\pi}{12} = \dfrac{\pi}{12}$$

$$= \cos \dfrac{\pi}{3} \cos \dfrac{\pi}{4} + \sin \dfrac{\pi}{3} \sin \dfrac{\pi}{4} \qquad \text{Formula for } \cos(u - v)$$

$$= \dfrac{1}{2} \cdot \dfrac{\sqrt{2}}{2} + \dfrac{\sqrt{3}}{2} \cdot \dfrac{\sqrt{2}}{2} \qquad \text{Use exact values. See page 302.}$$

$$= \dfrac{\sqrt{2} + \sqrt{6}}{4} \qquad \text{Multiply and add.}$$

Practice Problem 1 Find the exact value of $\cos 15°$ by using $15° = 45° - 30°$.

RECALL

The cosine function is even: $\cos(-x) = \cos(x)$.
The sine function is odd: $\sin(-x) = -\sin x$.

The difference formula for $\cos(u - v)$ is true for all real numbers and angles u and v. We can use this formula and the odd–even identities to prove the formula for $\cos(u + v)$.

$$\cos(u + v) = \cos[u - (-v)] \qquad u + v = u - (-v)$$
$$= \cos u \cos(-v) + \sin u \sin(-v) \qquad \text{Difference formula for cosine}$$
$$= \cos u \cos v + \sin u (-\sin v) \qquad \cos(-v) = \cos v;\ \sin(-v) = -\sin v$$
$$\mathbf{\cos(u + v) = \cos u \cos v - \sin u \sin v} \qquad \text{Simplify.}$$

We have proved the formula for the cosine of the sum of two angles or two real numbers.

EXAMPLE 2 Using the Sum Formula for Cosine

Find the exact value of $\cos 75°$ by using $75° = 45° + 30°$.

Solution

$$\cos 75° = \cos(45° + 30°) \qquad 75° = 45° + 30°$$
$$= \cos 45° \cos 30° - \sin 45° \sin 30° \qquad \text{Sum formula for cosine}$$
$$= \dfrac{\sqrt{2}}{2} \cdot \dfrac{\sqrt{3}}{2} - \dfrac{\sqrt{2}}{2} \cdot \dfrac{1}{2} \qquad \text{Use exact values.}$$
$$= \dfrac{\sqrt{6}}{4} - \dfrac{\sqrt{2}}{4} \qquad \text{Multiply.}$$
$$= \dfrac{\sqrt{6} - \sqrt{2}}{4} \qquad \text{Simplify.}$$

Practice Problem 2 Find the exact value of $\cos \dfrac{7\pi}{12}$ by using $\dfrac{7\pi}{12} = \dfrac{\pi}{3} + \dfrac{\pi}{4}$.

2 Use cofunction identities.

Cofunction Identities

Two trigonometric functions f and g are called **cofunctions** if

$$f\left(\dfrac{\pi}{2} - x\right) = g(x) \quad \text{and} \quad g\left(\dfrac{\pi}{2} - x\right) = f(x).$$

We derive two *cofunction identities*.

$$\cos(u - v) = \cos u \cos v + \sin u \sin v \quad \text{Difference formula for cosine}$$

$$\cos\left(\frac{\pi}{2} - v\right) = \cos\frac{\pi}{2}\cos v + \sin\frac{\pi}{2}\sin v \quad \text{Replace } u \text{ with } \frac{\pi}{2}.$$

$$= 0 \cdot \cos v + 1 \cdot \sin v = \sin v \quad \cos\frac{\pi}{2} = 0; \sin\frac{\pi}{2} = 1$$

The result is a cofunction identity that holds for any real number v or angle v in radian measure:

$$\cos\left(\frac{\pi}{2} - v\right) = \sin v$$

If we replace v with $\frac{\pi}{2} - v$ in the identity $\cos\left(\frac{\pi}{2} - v\right) = \sin v$, we have

$$\cos\left[\frac{\pi}{2} - \left(\frac{\pi}{2} - v\right)\right] = \sin\left(\frac{\pi}{2} - v\right) \quad \text{Replace } v \text{ with } \frac{\pi}{2} - v.$$

$$\cos v = \sin\left(\frac{\pi}{2} - v\right) \quad \text{Simplify.}$$

So, $$\sin\left(\frac{\pi}{2} - v\right) = \cos v.$$

BASIC COFUNCTION IDENTITIES

If v is any real number or angle measured in radians, then

$$\cos\left(\frac{\pi}{2} - v\right) = \sin v$$

$$\sin\left(\frac{\pi}{2} - v\right) = \cos v$$

If angle v is measured in degrees, then replace $\frac{\pi}{2}$ with $90°$ in these identities.

EXAMPLE 3 Using Cofunction Identities

Prove that for any real number x, $\tan\left(\frac{\pi}{2} - x\right) = \cot x$.

Solution

$$\tan\left(\frac{\pi}{2} - x\right) = \frac{\sin\left(\frac{\pi}{2} - x\right)}{\cos\left(\frac{\pi}{2} - x\right)} \quad \text{Quotient identity}$$

$$= \frac{\cos x}{\sin x} \quad \text{Use cofunction identities.}$$

$$= \cot x \quad \text{Quotient identity}$$

Practice Problem 3 Prove that for any real number x, $\sec\left(\frac{\pi}{2} - x\right) = \csc x$.

3 Use the sum and difference formulas for sine.

Sum and Difference Formulas for Sine

To prove the difference formula for the sine function, we start with a cofunction identity.

$$\sin(u - v) = \cos\left[\frac{\pi}{2} - (u - v)\right] \qquad \text{Cofunction identity}$$

$$= \cos\left[\left(\frac{\pi}{2} - u\right) + v\right] \qquad \frac{\pi}{2} - (u - v) = \left(\frac{\pi}{2} - u\right) + v$$

$$= \cos\left(\frac{\pi}{2} - u\right)\cos v - \sin\left(\frac{\pi}{2} - u\right)\sin v \qquad \text{Sum formula for cosine}$$

$$= \sin u \cos v - \cos u \sin v \qquad \text{Cofunction identities}$$

We have the difference formula for sine:

$$\sin(u - v) = \sin u \cos v - \cos u \sin v,$$

which holds for all real numbers u and v. If we replace v with $-v$, we derive the sum formula for sine.

$$\sin(u + v) = \sin[u - (-v)] \qquad u + v = u - (-v)$$

$$= \sin u \cos(-v) - \cos u \sin(-v) \qquad \text{Difference formula for sine}$$

$$= \sin u \cos v - \cos u (-\sin v) \qquad \cos(-v) = \cos v; \sin(-v) = -\sin v$$

$$= \sin u \cos v + \cos u \sin v \qquad \text{Simplify.}$$

This proves the sum formula for sine:

$$\sin(u + v) = \sin u \cos v + \cos u \sin v$$

EXAMPLE 4 Using the Sum and Difference Formulas for Sine

Prove the identity $\sin(\pi - x) = \sin x$.

Solution

$$\sin(\pi - x) = \sin \pi \cos x - \cos \pi \sin x \qquad \text{Difference formula for sine}$$

$$= (0) \cos x - (-1) \sin x \qquad \sin \pi = 0; \cos \pi = -1$$

$$= \sin x \qquad \text{Simplify.}$$

Practice Problem 4 Prove the identity $\sin(\pi + x) = -\sin x$.

EXAMPLE 5 Using the Sum and Difference Formulas for Sine

Find the exact value of $\sin 63° \cos 27° + \cos 63° \sin 27°$ without using a calculator.

Solution

The given expression is the right side of the sum formula for sine:

$$\sin(u + v) = \sin u \cos v + \cos u \sin v$$

$$\sin 63° \cos 27° + \cos 63° \sin 27° = \sin(63° + 27°) \qquad u = 63°, v = 27°$$

$$= \sin 90° = 1$$

Practice Problem 5 Find the exact value of $\sin 43° \cos 13° - \cos 43° \sin 13°$ without using a calculator.

4 Use the sum and difference formulas for tangent.

Sum and Difference Formulas for Tangent

We use the quotient identity $\tan x = \dfrac{\sin x}{\cos x}$ and the formulas for $\sin(u - v)$ and $\cos(u - v)$ to derive a difference formula for tangent.

$$\tan(u - v) = \dfrac{\sin(u - v)}{\cos(u - v)} \qquad \text{Quotient identity}$$

$$= \dfrac{\sin u \cos v - \cos u \sin v}{\cos u \cos v + \sin u \sin v} \qquad \text{Formulas for } \sin(u - v) \text{ and } \cos(u - v)$$

$$= \dfrac{\dfrac{\sin u \cos v}{\cos u \cos v} - \dfrac{\cos u \sin v}{\cos u \cos v}}{\dfrac{\cos u \cos v}{\cos u \cos v} + \dfrac{\sin u \sin v}{\cos u \cos v}} \qquad \text{Divide numerator and denominator by } \cos u \cos v.$$

$$= \dfrac{\dfrac{\sin u}{\cos u} - \dfrac{\sin v}{\cos v}}{1 + \dfrac{\sin u \sin v}{\cos u \cos v}} \qquad \text{Simplify.}$$

$$= \dfrac{\tan u - \tan v}{1 + \tan u \tan v} \qquad \text{Quotient identity}$$

We have derived the difference formula for the tangent:

$$\tan(u - v) = \dfrac{\tan u - \tan v}{1 + \tan u \tan v}$$

Replacing v with $-v$ in the difference formula, we have

$$\tan(u + v) = \tan[u - (-v)] \qquad u + v = u - (-v)$$

$$= \dfrac{\tan u - \tan(-v)}{1 + \tan u \tan(-v)} \qquad \text{Formula for } \tan(u - v)$$

$$= \dfrac{\tan u - (-\tan v)}{1 + \tan u (-\tan v)} \qquad \tan(-v) = -\tan v$$

$$= \dfrac{\tan u + \tan v}{1 - \tan u \tan v} \qquad \text{Simplify.}$$

We now have the sum formula for tangent:

$$\tan(u + v) = \dfrac{\tan u + \tan v}{1 - \tan u \tan v}$$

EXAMPLE 6 Verifying an Identity

Verify the identity $\tan(\pi - x) = -\tan x$.

Solution

Apply the difference formula for the tangent to $\tan(\pi - x)$.

$$\tan(\pi - x) = \dfrac{\tan \pi - \tan x}{1 + \tan \pi \tan x} \qquad \begin{array}{l}\text{Replace } u \text{ with } \pi \text{ and } v \text{ with } x \text{ in} \\ \text{the formula for } \tan(u - v).\end{array}$$

$$= \dfrac{0 - \tan x}{1 + 0 \cdot \tan x} \qquad \tan \pi = 0$$

$$= -\tan x \qquad \text{Simplify.}$$

Therefore, the given equation is an identity.

Practice Problem 6 Verify the identity $\tan(\pi + x) = \tan x$.

Double-Angle Formulas

5 Use double-angle formulas.

Suppose an angle measures x (radians or degrees); then $2x$ is double the measure of x. *Double-angle formulas* express trigonometric functions of $2x$ in terms of functions of x.

To find the double-angle formula for the sine, replace u and v with x in the sum formula for $\sin(u + v)$.

$\sin(u + v) = \sin u \cos v + \cos u \sin v$ Sum formula for sine
$\sin(x + x) = \sin x \cos x + \cos x \sin x$ Replace both u and v with x.
$\sin 2x = 2 \sin x \cos x$ Simplify.

The identity $\sin 2x = 2 \sin x \cos x$

is the double-angle formula for the sine function. We derive double-angle formulas for the cosine and tangent functions by replacing both u and v with x in the sum formulas $\cos(u + v)$ and $\tan(u + v)$, respectively. We get

$$\cos 2x = \cos^2 x - \sin^2 x \qquad \tan 2x = \frac{2 \tan x}{1 - \tan^2 x}$$

To derive two other useful forms for $\cos 2x$, first replace $\cos^2 x$ with $1 - \sin^2 x$ and then replace $\sin^2 x$ with $1 - \cos^2 x$ in the formula for $\cos 2x$.

$\cos 2x = \cos^2 x - \sin^2 x$ $\cos 2x = \cos^2 x - \sin^2 x$
$\quad\quad = (1 - \sin^2 x) - \sin^2 x$ $\quad\quad = \cos^2 x - (1 - \cos^2 x)$
$\cos 2x = 1 - 2 \sin^2 x$ $\cos 2x = 2 \cos^2 x - 1$

DOUBLE-ANGLE FORMULAS

$\sin 2x = 2 \sin x \cos x \qquad \cos 2x = \cos^2 x - \sin^2 x$

$\tan 2x = \dfrac{2 \tan x}{1 - \tan^2 x} \qquad \cos 2x = 1 - 2 \sin^2 x$

$\cos 2x = 2 \cos^2 x - 1$

EXAMPLE 7   Using Double-Angle Formulas

If $\cos \theta = -\dfrac{3}{5}$ and θ is in quadrant II, find the exact value of each expression.

a. $\sin 2\theta$ **b.** $\cos 2\theta$ **c.** $\tan 2\theta$

Solution

We use identities to find $\sin \theta$ and $\tan \theta$.

$\sin \theta = \sqrt{1 - \cos^2 \theta}$ θ is in quadrant II.

$\quad\quad = \sqrt{1 - \left(-\dfrac{3}{5}\right)^2}$ $\cos \theta = -\dfrac{3}{5}$

$\quad\quad = \dfrac{4}{5}$ Simplify.

$\tan \theta = \dfrac{\sin \theta}{\cos \theta} = \dfrac{4/5}{-3/5} = -\dfrac{4}{3}$ $\cos \theta = -\dfrac{3}{5}$ is given.

a. $\sin 2\theta = 2 \sin \theta \cos \theta$ Double-angle formula for sine

$\quad\quad\quad = 2\left(\dfrac{4}{5}\right)\left(-\dfrac{3}{5}\right)$ Replace $\sin \theta$ with $\dfrac{4}{5}$ and $\cos \theta$ with $-\dfrac{3}{5}$.

$\quad\quad\quad = -\dfrac{24}{25}$ Simplify.

b. $\cos 2\theta = \cos^2 \theta - \sin^2 \theta$ Double-angle formula for cosines

$= \left(-\dfrac{3}{5}\right)^2 - \left(\dfrac{4}{5}\right)^2$ Replace $\cos \theta$ with $-\dfrac{3}{5}$ and $\sin \theta$ with $\dfrac{4}{5}$.

$= \dfrac{9}{25} - \dfrac{16}{25} = -\dfrac{7}{25}$ Simplify.

c. $\tan 2\theta = \dfrac{2 \tan \theta}{1 - \tan^2 \theta}$ Double-angle formula for tangent

$= \dfrac{2\left(-\dfrac{4}{3}\right)}{1 - \left(-\dfrac{4}{3}\right)^2}$ Replace $\tan \theta$ with $-\dfrac{4}{3}$.

$= \dfrac{-\dfrac{8}{3}}{1 - \dfrac{16}{9}} = \dfrac{24}{7}$ Simplify.

You can also find $\tan 2\theta$ by using parts **a** and **b**:

$$\tan 2\theta = \dfrac{\sin 2\theta}{\cos 2\theta} = \dfrac{-\dfrac{24}{25}}{-\dfrac{7}{25}} = \dfrac{24}{7}$$

Practice Problem 7 If $\sin x = \dfrac{12}{13}$ and $\dfrac{\pi}{2} < x < \pi$, find the exact value of each expression.

a. $\sin 2x$ **b.** $\cos 2x$ **c.** $\tan 2x$

EXAMPLE 8 **Using Double-Angle Formulas**

Find the exact value of each expression.

a. $1 - 2\sin^2\left(\dfrac{\pi}{12}\right)$ **b.** $\dfrac{2 \tan 22.5°}{1 - \tan^2 22.5°}$

Solution

a. The given expression is the right side of the following formula, where $\theta = \dfrac{\pi}{12}$.

$\cos 2\theta = 1 - 2\sin^2 \theta$ A double-angle formula for cosine

$1 - 2\sin^2\left(\dfrac{\pi}{12}\right) = \cos\left[2\left(\dfrac{\pi}{12}\right)\right]$ Replace θ with $\dfrac{\pi}{12}$; interchange sides.

$= \cos \dfrac{\pi}{6} = \dfrac{\sqrt{3}}{2}$

b. The given expression is the right side of the following formula, where $\theta = 22.5°$.

$\tan 2\theta = \dfrac{2 \tan \theta}{1 - \tan^2 \theta}$ Double-angle formula for tangent

$\dfrac{2 \tan(22.5°)}{1 - \tan^2(22.5°)} = \tan[2(22.5°)]$ Replace θ with $22.5°$; interchange sides.

$= \tan 45° = 1$

Practice Problem 8 Find the exact value of each expression.

a. $2\cos^2\left(\dfrac{\pi}{12}\right) - 1$ **b.** $\cos^2 22.5° - \sin^2 22.5°$

We can use the double-angle formulas to express trigonometric functions of $4\theta, 6\theta$, and 8θ in terms of $2\theta, 3\theta$, and 4θ, respectively. For example, the identities

$$\cos 4\theta = 2\cos^2 2\theta - 1, \quad \sin 6\theta = 2\sin 3\theta \cos 3\theta, \quad \text{and} \quad \tan 8\theta = \dfrac{2\tan 4\theta}{1 - \tan^2 4\theta}$$

can be directly derived from the appropriate double-angle formulas.

Power-Reducing Formulas

6 Use power-reducing formulas.

The purpose of *power-reducing formulas* is to express $\sin^2 x, \cos^2 x$, and $\tan^2 x$ in terms of trigonometric functions with powers not greater than 1. These formulas are useful in calculus.

POWER-REDUCING FORMULAS

$$\sin^2 x = \dfrac{1 - \cos 2x}{2} \qquad \cos^2 x = \dfrac{1 + \cos 2x}{2} \qquad \tan^2 x = \dfrac{1 - \cos 2x}{1 + \cos 2x}$$

We can derive the first power-reducing formula by using the appropriate formula for $\cos 2x$.

$\cos 2x = 1 - 2\sin^2 x$ Double-angle formula for $\cos 2x$ in terms of sine
$2\sin^2 x = 1 - \cos 2x$ Add $2\sin^2 x - \cos 2x$ to both sides and simplify.
$\sin^2 x = \dfrac{1 - \cos 2x}{2}$ Divide both sides by 2.

Similarly, we can derive the second formula.

$\cos 2x = 2\cos^2 x - 1$ Double-angle formula for $\cos 2x$ in terms of cosine
$1 + \cos 2x = 2\cos^2 x$ Add 1 to both sides.
$\dfrac{1 + \cos 2x}{2} = \cos^2 x$ Divide both sides by 2.

We begin with the quotient identity to prove the third formula.

$\tan^2 x = \dfrac{\sin^2 x}{\cos^2 x}$ Quotient identity

$= \dfrac{\dfrac{1 - \cos 2x}{2}}{\dfrac{1 + \cos 2x}{2}}$ Power-reducing formulas for $\sin^2 x$ and $\cos^2 x$

$= \dfrac{1 - \cos 2x}{1 + \cos 2x}$ Multiply numerator and denominator by 2. Simplify.

EXAMPLE 9 Using Power-Reducing Formulas

Write an equivalent expression for $\cos^4 x$ that contains only first powers of cosines of multiple angles.

Solution

We use the power-reducing formulas repeatedly.

$$\cos^4 x = (\cos^2 x)^2 \qquad\qquad a^4 = (a^2)^2$$

$$= \left(\frac{1 + \cos 2x}{2}\right)^2 \qquad\qquad \text{Power-reducing formula}$$

$$= \frac{1}{4}(1 + 2\cos 2x + \cos^2 2x) \qquad\qquad \text{Expand the binomial.}$$

$$= \frac{1}{4}\left(1 + 2\cos 2x + \frac{1 + \cos 4x}{2}\right) \qquad\qquad \text{Power-reducing formula for } \cos^2 x; \text{ replace } x \text{ with } 2x.$$

$$= \frac{1}{4}\left(1 + 2\cos 2x + \frac{1}{2} + \frac{1}{2}\cos 4x\right) \qquad\qquad \frac{a+b}{c} = \frac{a}{c} + \frac{b}{c}$$

$$= \frac{1}{4} + \frac{2}{4}\cos 2x + \frac{1}{8} + \frac{1}{8}\cos 4x \qquad\qquad \text{Distributive property}$$

$$= \frac{3}{8} + \frac{1}{2}\cos 2x + \frac{1}{8}\cos 4x \qquad\qquad \text{Simplify.}$$

Practice Problem 9 Write an equivalent expression for $\sin^4 x$ that contains only first powers of cosines of multiple angles.

7 Use half-angle formulas.

Half-Angle Formulas

If an angle measures θ, then $\frac{\theta}{2}$ is half the measure of θ. Half-angle formulas express trigonometric functions of $\frac{\theta}{2}$ in terms of functions of θ. To derive the half-angle formulas, we replace x with $\frac{\theta}{2}$ in the power-reducing formulas and then take the square root of both sides. For example,

$$\cos^2 x = \frac{1 + \cos 2x}{2} \qquad\qquad \text{Power-reducing formula for cosines}$$

$$\cos^2 \frac{\theta}{2} = \frac{1 + \cos\left[2\left(\frac{\theta}{2}\right)\right]}{2} \qquad\qquad \text{Replace } x \text{ with } \frac{\theta}{2}.$$

$$\cos^2 \frac{\theta}{2} = \frac{1 + \cos\theta}{2} \qquad\qquad \text{Simplify.}$$

$$\cos \frac{\theta}{2} = \pm\sqrt{\frac{1 + \cos\theta}{2}} \qquad\qquad \text{Square root property}$$

We call the last equation a *half-angle formula* for cosine. The sign + or − depends on the quadrant in which $\frac{\theta}{2}$ lies. We can derive half-angle formulas for sine and tangent in a similar manner.

HALF-ANGLE FORMULAS

$$\sin\frac{\theta}{2} = \pm\sqrt{\frac{1 - \cos\theta}{2}} \qquad \cos\frac{\theta}{2} = \pm\sqrt{\frac{1 + \cos\theta}{2}} \qquad \tan\frac{\theta}{2} = \pm\sqrt{\frac{1 - \cos\theta}{1 + \cos\theta}}$$

The sign + or − depends on the quadrant in which $\frac{\theta}{2}$ lies.

Section 4.8 ■ Sum and Difference Formulas 383

EXAMPLE 10 Finding the Exact Value

Given that $\sin \theta = -\dfrac{5}{13}$, $\pi < \theta < \dfrac{3\pi}{2}$, find the exact value of each expression.

a. $\sin \dfrac{\theta}{2}$ **b.** $\tan \dfrac{\theta}{2}$

Solution

Because θ lies in quadrant III, $\cos \theta$ is negative; so

$\cos \theta = -\sqrt{1 - \sin^2 \theta}$ Pythagorean identity

$\cos \theta = -\sqrt{1 - \left(-\dfrac{5}{13}\right)^2}$ Replace $\sin \theta$ with $-\dfrac{5}{13}$.

$= -\sqrt{1 - \dfrac{25}{169}} = -\sqrt{\dfrac{169 - 25}{169}} = -\dfrac{12}{13}$ Simplify.

a. Because $\pi < \theta < \dfrac{3\pi}{2}$, $\dfrac{\theta}{2}$ lies in quadrant II $\left(\text{if } \pi < \theta < \dfrac{3\pi}{2}, \text{ then } \dfrac{\pi}{2} < \dfrac{\theta}{2} < \dfrac{3\pi}{4}\right)$.

The half-angle formula gives

$\sin \dfrac{\theta}{2} = \sqrt{\dfrac{1 - \cos \theta}{2}}$ In quadrant II, $\sin \dfrac{\theta}{2}$ is positive.

$= \sqrt{\dfrac{1 - \left(-\dfrac{12}{13}\right)}{2}}$ Replace $\cos \theta$ with $-\dfrac{12}{13}$.

$= \sqrt{\dfrac{\dfrac{25}{13}}{2}} = \dfrac{5}{\sqrt{26}} = \dfrac{5\sqrt{26}}{26}$ Simplify.

b. From the half-angle formula, we have

$\tan \dfrac{\theta}{2} = -\sqrt{\dfrac{1 - \cos \theta}{1 + \cos \theta}}$ In quadrant II, $\tan \dfrac{\theta}{2}$ is negative.

$= -\sqrt{\dfrac{1 - \left(-\dfrac{12}{13}\right)}{1 + \left(-\dfrac{12}{13}\right)}}$ $\cos \theta = -\dfrac{12}{13}$

$= -\sqrt{\dfrac{25/13}{1/13}} = -5$ Simplify.

Practice Problem 10 For θ in Example 10, find $\cos \dfrac{\theta}{2}$.

EXAMPLE 11 Verifying an Identity Containing Half-Angles

Verify the identity $\sin x \cos \dfrac{x}{2} = \sin \dfrac{x}{2}(1 + \cos x)$.

Solution

The left side contains $\sin x$, and the right side contains $\sin \frac{x}{2}$. We replace x with $\frac{x}{2}$ in the double-angle identity for sine.

$$\sin 2x = 2 \sin x \cos x \quad \text{Double-angle identity for sine}$$

$$\sin x = 2 \sin \frac{x}{2} \cos \frac{x}{2} \quad \text{Replace } x \text{ with } \frac{x}{2}; \sin 2\left(\frac{x}{2}\right) = \sin x.$$

Begin with the left side of the original equation and use this expression for $\sin x$.

$$\sin x \cos \frac{x}{2} = 2 \sin \frac{x}{2} \cos \frac{x}{2} \cos \frac{x}{2} \quad \text{Replace } \sin x \text{ with } 2 \sin \frac{x}{2} \cos \frac{x}{2}.$$

$$= 2 \sin \frac{x}{2} \cos^2 \frac{x}{2}$$

$$= 2 \left(\sin \frac{x}{2} \right) \left(\frac{1 + \cos \left[2 \left(\frac{x}{2} \right) \right]}{2} \right) \quad \text{Power-reducing formula for cosine}$$

$$= \sin \frac{x}{2} (1 + \cos x) \quad \cos \left[2\left(\frac{x}{2}\right) \right] = \cos x; \text{ simplify.}$$

Because the left side is identical to the right side of the equation, the identity is verified.

Practice Problem 11 Verify the identity $\sin \frac{x}{2} \sin x = \cos \frac{x}{2} (1 - \cos x)$.

SUMMARY OF MAIN FACTS

Difference and Sum Formulas

$$\cos(u - v) = \cos u \cos v + \sin u \sin v \qquad \cos(u + v) = \cos u \cos v - \sin u \sin v$$

$$\sin(u - v) = \sin u \cos v - \cos u \sin v \qquad \sin(u + v) = \sin u \cos v + \cos u \sin v$$

$$\tan(u - v) = \frac{\tan u - \tan v}{1 + \tan u \tan v} \qquad \tan(u + v) = \frac{\tan u + \tan v}{1 - \tan u \tan v}$$

Cofunction Identities

$$\sin\left(\frac{\pi}{2} - x\right) = \cos x \qquad \cos\left(\frac{\pi}{2} - x\right) = \sin x \qquad \tan\left(\frac{\pi}{2} - x\right) = \cot x$$

$$\csc\left(\frac{\pi}{2} - x\right) = \sec x \qquad \sec\left(\frac{\pi}{2} - x\right) = \csc x \qquad \cot\left(\frac{\pi}{2} - x\right) = \tan x$$

Double-Angle Formulas

$$\sin 2x = 2 \sin x \cos x \qquad \cos 2x = \cos^2 x - \sin^2 x$$

$$\tan 2x = \frac{2 \tan x}{1 - \tan^2 x} \qquad \cos 2x = 2 \cos^2 x - 1$$

$$\cos 2x = 1 - 2 \sin^2 x$$

Power-Reducing Formulas

$$\sin^2 x = \frac{1 - \cos 2x}{2} \qquad \cos^2 x = \frac{1 + \cos 2x}{2} \qquad \tan^2 x = \frac{1 - \cos 2x}{1 + \cos 2x}$$

Half-Angle Formulas

$$\sin \frac{\theta}{2} = \pm \sqrt{\frac{1 - \cos \theta}{2}} \qquad \cos \frac{\theta}{2} = \pm \sqrt{\frac{1 + \cos \theta}{2}} \qquad \tan \frac{\theta}{2} = \pm \sqrt{\frac{1 - \cos \theta}{1 + \cos \theta}}$$

The sign + or − depends on the quadrant in which $\frac{\theta}{2}$ lies.

SECTION 4.8 Exercises

Basic Concepts and Skills

In Exercises 1–12, find the exact value of each expression.

1. $\sin\left(\dfrac{\pi}{6} + \dfrac{\pi}{4}\right)$
2. $\cos\left(\dfrac{\pi}{3} - \dfrac{\pi}{4}\right)$
3. $\tan\left(\dfrac{\pi}{4} - \dfrac{\pi}{6}\right)$
4. $\cot\left(\dfrac{\pi}{3} - \dfrac{\pi}{4}\right)$
5. $\sec\left(\dfrac{\pi}{3} + \dfrac{\pi}{4}\right)$
6. $\csc\left(\dfrac{\pi}{4} - \dfrac{\pi}{3}\right)$
7. $\cos\dfrac{-5\pi}{12}$
8. $\sin\dfrac{7\pi}{12}$
9. $\tan\dfrac{19\pi}{12}$
10. $\sec\dfrac{\pi}{12}$
11. $\tan\dfrac{17\pi}{12}$
12. $\csc\dfrac{11\pi}{12}$

In Exercises 13–20, verify each identity.

13. $\sin\left(x + \dfrac{\pi}{2}\right) = \cos x$
14. $\cos\left(x + \dfrac{\pi}{2}\right) = -\sin x$
15. $\sin\left(x - \dfrac{\pi}{2}\right) = -\cos x$
16. $\cos\left(x - \dfrac{\pi}{2}\right) = \sin x$
17. $\tan\left(x + \dfrac{\pi}{2}\right) = -\cot x$
18. $\tan\left(x - \dfrac{\pi}{2}\right) = -\cot x$
19. $\csc(x + \pi) = -\csc x$
20. $\sec(x + \pi) = -\sec x$

In Exercises 21–24, find the exact value of each expression without using a calculator.

21. $\sin\dfrac{7\pi}{12}\cos\dfrac{3\pi}{12} - \cos\dfrac{7\pi}{12}\sin\dfrac{3\pi}{12}$
22. $\cos\dfrac{5\pi}{12}\cos\dfrac{\pi}{12} - \sin\dfrac{5\pi}{12}\sin\dfrac{\pi}{12}$
23. $\dfrac{\tan\dfrac{5\pi}{12} - \tan\dfrac{2\pi}{12}}{1 + \tan\dfrac{5\pi}{12}\tan\dfrac{2\pi}{12}}$
24. $\dfrac{\tan\dfrac{5\pi}{12} + \tan\dfrac{7\pi}{12}}{1 - \tan\dfrac{5\pi}{12}\tan\dfrac{7\pi}{12}}$

In Exercises 25–28, find the exact value of each expression, given that $\tan u = \dfrac{3}{4}$, with u in quadrant III, and $\sin v = \dfrac{5}{13}$, with v in quadrant II.

25. $\sin(u - v)$
26. $\cos(u + v)$
27. $\tan(u + v)$
28. $\tan(u - v)$

In Exercises 29–34, use the given information about the angle θ to find the exact value of
a. $\sin 2\theta$.   b. $\cos 2\theta$.   c. $\tan 2\theta$.

29. $\sin\theta = \dfrac{3}{5}$, θ in quadrant II
30. $\cos\theta = -\dfrac{5}{13}$, θ in quadrant III
31. $\tan\theta = 4$, $\sin\theta < 0$
32. $\sec\theta = -\sqrt{3}$, $\sin\theta > 0$
33. $\tan\theta = -2$, $\dfrac{\pi}{2} < \theta < \pi$
34. $\cot\theta = -7$, $\dfrac{3\pi}{2} < \theta < 2\pi$

In Exercises 35–40, use a double-angle formula to find the exact value of each expression.

35. $1 - 2\sin^2 75°$
36. $2\cos^2 105° - 1$
37. $\dfrac{2\tan 165°}{1 - \tan^2 165°}$
38. $1 - 2\sin^2\dfrac{\pi}{8}$
39. $2\cos^2\left(-\dfrac{\pi}{8}\right) - 1$
40. $1 - 2\sin^2\left(-\dfrac{7\pi}{12}\right)$

In Exercises 41–44, verify each identity.

41. $\cos^4 x - \sin^4 x = \cos 2x$
42. $1 + \cos 2x + 2\sin^2 x = 2$
43. $\dfrac{\cos 2x}{\sin 2x} + \dfrac{\sin x}{\cos x} = \csc 2x$
44. $\dfrac{\sin 2x}{\sin x} - \dfrac{\cos 2x}{\cos x} = \sec x$

In Exercises 45–50, use half-angle formulas to find the exact value of each expression.

45. $\sin\dfrac{\pi}{12}$
46. $\sin\dfrac{\pi}{8}$
47. $\cos\dfrac{\pi}{8}$
48. $\cos\left(-\dfrac{3\pi}{8}\right)$
49. $\tan 112.5°$
50. $\sin(-75°)$

In Exercises 51–56, use the information about the angle θ $0 < \theta < 2\pi$ to find the exact value of
a. $\sin\dfrac{\theta}{2}$.   b. $\cos\dfrac{\theta}{2}$.   c. $\tan\dfrac{\theta}{2}$.

51. $\sin\theta = \dfrac{4}{5}$, $\dfrac{\pi}{2} < \theta < \pi$
52. $\cos\theta = -\dfrac{12}{13}$, $\pi < \theta < \dfrac{3\pi}{2}$
53. $\tan\theta = -\dfrac{2}{3}$, $\dfrac{\pi}{2} < \theta < \pi$
54. $\sin\theta = \dfrac{1}{5}$, $\cos\theta < 0$
55. $\sec\theta = \sqrt{5}$, $\sin\theta > 0$
56. $\csc\theta = \sqrt{7}$, $\tan\theta < 0$

Applying the Concepts

In Exercises 57–60, assume that the voltage of the current is given by $V = 170 \sin(120\pi t)$, where t is in seconds. Find the wattage rating $\left(\dfrac{\text{maximum wattage}}{\sqrt{2}}\right)$ of each electric device if the current flowing through the device is I amps.

57. **Lightbulb.** $I = 0.832 \sin(120\pi t)$.
58. **Microwave oven.** $I = 7.487 \sin(120\pi t)$.
59. **Toaster.** $I = 9.983 \sin(120\pi t)$.
60. **Refrigerator.** $I = 6.655 \sin(120\pi t)$.
61. **Analytic geometry.** Suppose two nonvertical lines l_1 and l_2 lie in a plane. The slope of l_1 is m_1, and the slope of l_2 is m_2. Note that $m_1 = \tan \theta_1$ and $m_2 = \tan \theta_2$. Show that the tangent of the smallest nonnegative angle α between the lines l_1 and l_2 is given by $\tan \alpha = \left|\dfrac{m_2 - m_1}{1 + m_1 m_2}\right|$.

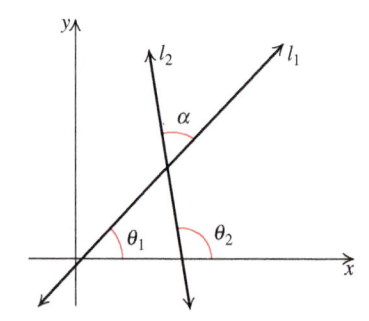

In Exercises 62–64, use Exercise 61 to find the smallest nonnegative angle between the lines l_1 and l_2.

62. **Angle between lines.** $l_1: y = x + 5$, $l_2: y = 4x + 2$.
63. **Angle between lines.** $l_1: y = 2x + 5$, $l_2: 6x - 3y = 21$.
64. **Angle between lines.** $l_1: y = 2x - 4$, $l_2: x + 2y = 7$.
65. **Throwing a football.** A quarterback throws a ball with an initial velocity of v_0 feet per second at an angle θ with the horizontal. The horizontal distance x in feet that the ball is thrown is modeled by the equation $x = \dfrac{v_0^2}{16} \sin \theta \cos \theta$. For a fixed v_0, find the angle that produces the maximum distance x.

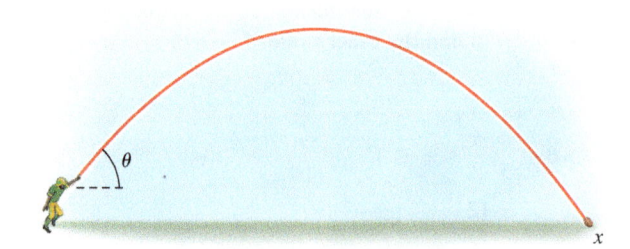

66. **Angle of elevation.** A, B, and C are three points on the same horizontal line. The segment CT represents a vertical tower. The angle of elevation of T from the point B is twice that from the point A. The lengths of the segments AB and BC are 80 meters and 50 meters, respectively. Show that the angle of elevation θ from A in the figure is given by $\sin^{-1}\left(\dfrac{\sqrt{3}}{4}\right)$.

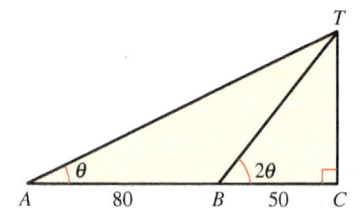

Critical Thinking / Discussion / Writing

67. Use the figure to find the exact value of each expression.
    a. $\sin \theta$ b. $\cos \theta$ c. $\sin 2\theta$
    d. $\cos 2\theta$ e. $\tan 2\theta$ f. $\sin \dfrac{\theta}{2}$
    g. $\cos \dfrac{\theta}{2}$ h. $\tan \dfrac{\theta}{2}$

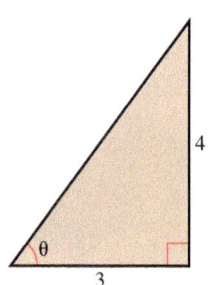

68. Explain why we cannot use the formula for $\tan(u - v)$ to verify that $\tan\left(\dfrac{\pi}{2} - x\right) = \cot x$.

69. Find a formula for $\cot(x - y)$ in terms of $\cot x$ and $\cot y$.

70. Verify the following **product-to-sum** identities:
    (i) $\cos x \cos y = \dfrac{1}{2}[\cos(x - y) + \cos(x + y)]$
    (ii) $\sin x \sin y = \dfrac{1}{2}[\cos(x - y) - \cos(x + y)]$
    (iii) $\sin x \cos y = \dfrac{1}{2}[\sin(x + y) + \sin(x - y)]$
    (iv) $\cos x \sin y = \dfrac{1}{2}[\sin(x + y) - \sin(x - y)]$

    (*Hint:* Start with the right side of each identity and then use the appropriate sum or difference identity.)

71. Verify the following **sum-to-product** identities:

    (i) $\cos x + \cos y = 2 \cos\left(\dfrac{x+y}{2}\right) \cos\left(\dfrac{x-y}{2}\right)$

    (ii) $\cos x - \cos y = -2 \sin\left(\dfrac{x+y}{2}\right) \sin\left(\dfrac{x-y}{2}\right)$

    (iii) $\sin x + \sin y = 2 \sin\left(\dfrac{x+y}{2}\right) \cos\left(\dfrac{x-y}{2}\right)$

    (iv) $\sin x - \sin y = 2 \sin\left(\dfrac{x-y}{2}\right) \cos\left(\dfrac{x+y}{2}\right)$

    (*Hint:* Start with the right side of each identity and then use the appropriate product-to-sum identity of Exercise 70.)

Maintaining Skills

In Exercises 72–76, state whether the given function is even, odd, or neither.

72. $f(x) = 2 \sin x + \tan x - 3x$

73. $g(x) = 3 \cos x - 7 \sin x + x^2$

74. $h(x) = 5x^2 - 4 \cos x + \sec x$

75. $f(x) = \dfrac{\sin x}{x}$

76. $f(x) = \dfrac{1 - \cos x}{x^2}$

SUMMARY Definitions, Concepts, and Formulas

4.1 Angles and Their Measure

i. An **angle** is formed by rotating a ray around its endpoint.

ii. An angle in a rectangular coordinate system is in **standard position** if its vertex is at the origin and its initial side coincides with the positive *x*-axis. (See pages 283–284 for positive and negative angles.)

iii. If the terminal side of an angle in standard position lies on the *x*-axis or the *y*-axis, the angle is a **quadrantal angle**.

iv. Angles can be measured in **degrees**, where $1° = \dfrac{1}{360}$ of a complete revolution.

v. An **acute angle** is an angle with measure between 0° and 90°, a **right angle** is an angle with measure 90°, an **obtuse angle** is an angle with measure between 90° and 180°, and a **straight angle** is an angle with measure 180°.

vi. Angles that have the same initial and terminal sides are **coterminal** angles.

vii. Angles can also be measured in **radians**. The radian measure of a central angle θ that intercepts an arc of length *s* on a circle of radius *r* is $\theta = \dfrac{s}{r}$ radians.

viii. To convert from degrees to radians, multiply degrees by $\dfrac{\pi}{180°}$. To convert from radians to degrees, multiply radians by $\dfrac{180°}{\pi}$.

ix. Two positive angles are **complements** if their sum is 90°. Two positive angles are **supplements** if their sum is 180°.

x. The formula for the length *s* of the arc intercepted by a central angle θ in a circle of radius *r* is $s = r\theta$.

4.2 The Unit Circle; Trigonometric Functions of an Angle

i. A **number line** is wrapped around the unit circle ($x^2 + y^2 = 1$). See Figure 4.17. For any real number *t*, let $P(t) = (x, y)$ be the **terminal point** on the unit circle associated with *t*. We define

$$\sin t = y, \quad \cos t = x, \quad \tan t = \dfrac{y}{x},$$

$$\csc t = \dfrac{1}{y}, \quad \sec t = \dfrac{1}{x}, \quad \text{and} \quad \cot t = \dfrac{x}{y}.$$

ii. If the terminal side of an angle θ (in standard position) intersects the unit circle at the point (x, y), then $\sin \theta = y$,

$$\cos \theta = x, \tan \theta = \dfrac{y}{x}, \cot \theta = \dfrac{x}{y}, \sec \theta = \dfrac{1}{x}, \text{ and } \csc \theta = \dfrac{1}{y}.$$

iii. Trigonometric function values of coterminal angles

θ in degrees	θ in radians
$\sin \theta = \sin(\theta + n360°)$	$\sin \theta = \sin(\theta + 2\pi n)$
$\cos \theta = \cos(\theta + n360°)$	$\cos \theta = \cos(\theta + 2\pi n)$

These equations hold for any integer *n*.

TABLE 4.1

$\theta°$	θ radians	$\sin \theta$	$\cos \theta$	$\tan \theta$	$\csc \theta$	$\sec \theta$	$\cot \theta$
30°	$\dfrac{\pi}{6}$	$\dfrac{1}{2}$	$\dfrac{\sqrt{3}}{2}$	$\dfrac{1}{\sqrt{3}}$	2	$\dfrac{2}{\sqrt{3}}$	$\sqrt{3}$
45°	$\dfrac{\pi}{4}$	$\dfrac{1}{\sqrt{2}}$	$\dfrac{1}{\sqrt{2}}$	1	$\sqrt{2}$	$\sqrt{2}$	1
60°	$\dfrac{\pi}{3}$	$\dfrac{\sqrt{3}}{2}$	$\dfrac{1}{2}$	$\sqrt{3}$	$\dfrac{2}{\sqrt{3}}$	2	$\dfrac{1}{\sqrt{3}}$

iv. *Signs of the trigonometric functions*

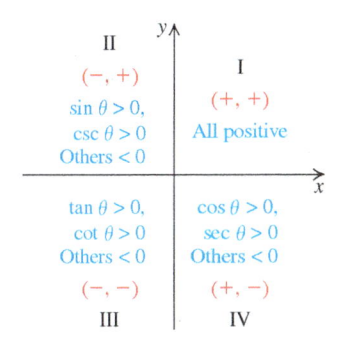

II   (−, +)
$\sin \theta > 0$,
$\csc \theta > 0$
Others < 0

I   (+, +)
All positive

III   (−, −)
$\tan \theta > 0$,
$\cot \theta > 0$
Others < 0

IV   (+, −)
$\cos \theta > 0$,
$\sec \theta > 0$
Others < 0

4.3 Graphs of the Sine and Cosine Functions

i. Sine and cosine function graphs and properties

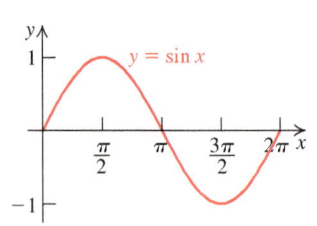

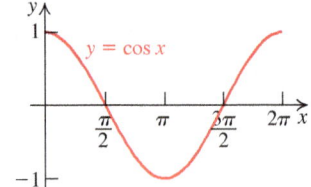

Sine Function

1. Period: 2π
2. Domain: $(-\infty, \infty)$
3. Range: $[-1, 1]$
4. Odd: $\sin(-t) = -\sin t$
5. Zeros at integer multiples of π

Cosine Function

1. Period: 2π
2. Domain: $(-\infty, \infty)$
3. Range: $[-1, 1]$
4. Even: $\cos(-t) = \cos t$
5. Zeros at odd integer multiples of $\frac{\pi}{2}$.

ii. The functions $y = a \sin [b(x - c)] + d$ and $y = a \cos [b(x - c)] + d$ ($b > 0$) have amplitude $|a|$, period $\frac{2\pi}{b}$, phase shift c, and vertical shift d.

4.4 Graphs of the Other Trigonometric Functions

i. If a line makes an angle θ with the positive x-axis, its **slope** is $m = \tan \theta$.

ii. Tangent and cotangent function graphs and properties

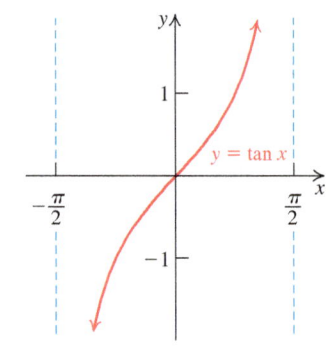

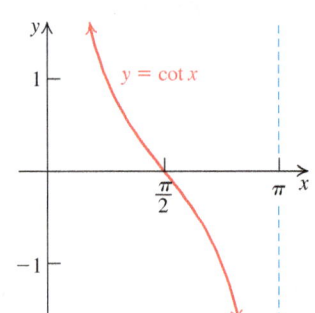

Tangent Function

1. Period: π
2. Domain: All real numbers except odd multiples of $\frac{\pi}{2}$
3. Range: $(-\infty, \infty)$
4. Odd: $\tan(-x) = -\tan x$
5. Zeros at integer multiples of π
6. Vertical asymptotes at odd integer multiples of $\frac{\pi}{2}$

Cotangent Function

1. Period: π
2. Domain: All real numbers except integer multiples of π
3. Range: $(-\infty, \infty)$
4. Odd: $\cot(-x) = -\cot x$
5. Zeros at odd integer multiples of $\frac{\pi}{2}$
6. Vertical asymptotes at integer multiples of π

iii. Secant and cosecant function graphs and properties

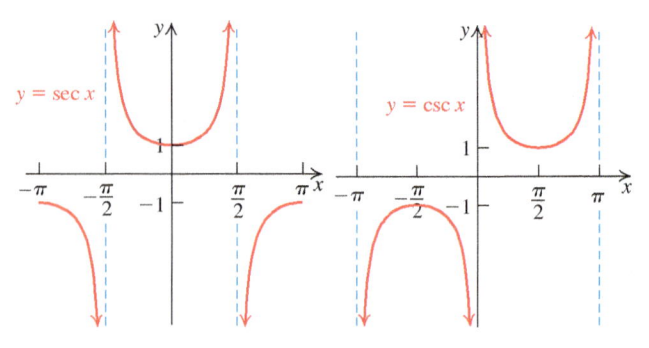

Secant Function

1. Period: 2π
2. Domain: All real numbers except for odd multiples of $\frac{\pi}{2}$
3. Range: $(-\infty, -1] \cup [1, \infty)$
4. Even: $\sec(-x) = \sec x$
5. Vertical asymptotes at odd integer multiples of $\frac{\pi}{2}$

Cosecant Function

1. Period: 2π
2. Domain: All real numbers except for integer multiples of π
3. Range: $(-\infty, -1] \cup [1, \infty)$
4. Odd: $\csc(-x) = -\csc x$
5. Vertical asymptotes at integer multiples of π

4.5 Inverse Trigonometric Functions

i.

Inverse Trigonometric Functions

Inverse Function	Equivalent to	Domain	Range
$y = \sin^{-1} x$	$\sin y = x$	$[-1, 1]$	$\left[-\frac{\pi}{2}, \frac{\pi}{2}\right]$
$y = \cos^{-1} x$	$\cos y = x$	$[-1, 1]$	$[0, \pi]$
$y = \tan^{-1} x$	$\tan y = x$	$(-\infty, \infty)$	$\left(-\frac{\pi}{2}, \frac{\pi}{2}\right)$
$y = \cot^{-1} x$	$\cot y = x$	$(-\infty, \infty)$	$(0, \pi)$
$y = \csc^{-1} x$	$\csc y = x$	$(-\infty, -1] \cup [1, \infty)$	$\left[-\frac{\pi}{2}, 0\right) \cup \left(0, \frac{\pi}{2}\right]$
$y = \sec^{-1} x$	$\sec y = x$	$(-\infty, -1] \cup [1, \infty)$	$\left[0, \frac{\pi}{2}\right) \cup \left(\frac{\pi}{2}, \pi\right]$

ii. Points on the terminal sides of angles in the standard position are used to find exact values of the composition of a trigonometric function and a different inverse trigonometric function.

iii. (a) $\sin^{-1}(\sin x) = x$ for $-\frac{\pi}{2} \leq x \leq \frac{\pi}{2}$ and
$\sin(\sin^{-1} x) = x$ for $-1 \leq x \leq 1$
(b) $\cos^{-1}(\cos x) = x$ for $0 \leq x \leq \pi$ and
$\cos(\cos^{-1} x) = x$ for $-1 \leq x \leq 1$
(c) $\tan^{-1}(\tan x) = x$ for $-\frac{\pi}{2} < x < \frac{\pi}{2}$ and
$\tan(\tan^{-1} x) = x$ for $-\infty < x < \infty$

4.6 Right-Triangle Trigonometry

i. The right-triangle definitions for trigonometric functions of an acute angle θ are given by

$$\sin \theta = \frac{\text{opposite side}}{\text{hypotenuse}} = \frac{a}{c} \qquad \csc \theta = \frac{\text{hypotenuse}}{\text{opposite side}} = \frac{c}{a}$$

$$\cos \theta = \frac{\text{adjacent side}}{\text{hypotenuse}} = \frac{b}{c} \qquad \sec \theta = \frac{\text{hypotenuse}}{\text{adjacent side}} = \frac{c}{b}$$

$$\tan \theta = \frac{\text{opposite side}}{\text{adjacent side}} = \frac{a}{b} \qquad \cot \theta = \frac{\text{adjacent side}}{\text{opposite side}} = \frac{b}{a}$$

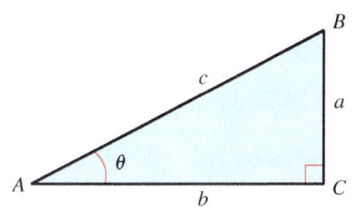

ii. Two acute angles are complements if their sum is 90° or $\frac{\pi}{2}$.

The value of any trigonometric function of an acute angle θ is equal to the cofunction value of the complement of θ.

iii. To solve a right triangle means to find the measurements of missing angles and the sides.

4.7 Trigonometric Identities

Reciprocal Identities

$$\csc x = \frac{1}{\sin x} \qquad \sec x = \frac{1}{\cos x} \qquad \cot x = \frac{1}{\tan x}$$

$$\sin x = \frac{1}{\csc x} \qquad \cos x = \frac{1}{\sec x} \qquad \tan x = \frac{1}{\cot x}$$

Quotient Identities

$$\tan x = \frac{\sin x}{\cos x} \qquad \cot x = \frac{\cos x}{\sin x}$$

Pythagorean Identities

$$\sin^2 x + \cos^2 x = 1$$
$$1 + \tan^2 x = \sec^2 x$$
$$1 + \cot^2 x = \csc^2 x$$

Even–Odd Identities

$\sin(-x) = -\sin x \quad \cos(-x) = \cos x \quad \tan(-x) = -\tan x$
$\csc(-x) = -\csc x \quad \sec(-x) = \sec x \quad \cot(-x) = -\cot x$

See page 367 for *guidelines for verifying trigonometric identities*.

4.8 Sum and Difference Formulas

Sum and Difference Formulas

$$\cos(u - v) = \cos u \cos v + \sin u \sin v$$
$$\cos(u + v) = \cos u \cos v - \sin u \sin v$$
$$\sin(u - v) = \sin u \cos v - \cos u \sin v$$
$$\sin(u + v) = \sin u \cos v + \cos u \sin v$$
$$\tan(u - v) = \frac{\tan u - \tan v}{1 + \tan u \tan v}$$
$$\tan(u + v) = \frac{\tan u + \tan v}{1 - \tan u \tan v}$$

Cofunction Identities

$$\sin\left(\frac{\pi}{2} - x\right) = \cos x \qquad \cos\left(\frac{\pi}{2} - x\right) = \sin x$$

$$\tan\left(\frac{\pi}{2} - x\right) = \cot x \qquad \csc\left(\frac{\pi}{2} - x\right) = \sec x$$

$$\sec\left(\frac{\pi}{2} - x\right) = \csc x \qquad \cot\left(\frac{\pi}{2} - x\right) = \tan x$$

Double-Angle Formulas

$$\sin 2x = 2 \sin x \cos x$$
$$\cos 2x = \cos^2 x - \sin^2 x$$
$$\cos 2x = 1 - 2 \sin^2 x$$
$$\cos 2x = 2 \cos^2 x - 1$$
$$\tan 2x = \frac{2 \tan x}{1 - \tan^2 x}$$

Power-Reducing Formulas

$$\sin^2 x = \frac{1 - \cos 2x}{2}$$
$$\cos^2 x = \frac{1 + \cos 2x}{2}$$
$$\tan^2 x = \frac{1 - \cos 2x}{1 + \cos 2x}$$

Half-Angle Formulas

$$\sin \frac{\theta}{2} = \pm\sqrt{\frac{1 - \cos \theta}{2}}$$
$$\cos \frac{\theta}{2} = \pm\sqrt{\frac{1 + \cos \theta}{2}}$$
$$\tan \frac{\theta}{2} = \pm\sqrt{\frac{1 - \cos \theta}{1 + \cos \theta}},$$

where the sign + or − depends on the quadrant in which $\frac{\theta}{2}$ lies.

REVIEW EXERCISES

Basic Concepts and Skills

In Exercises 1–3, convert each angle from degrees to radians. Express each answer as a multiple of π.

1. $20°$
2. $36°$
3. $-60°$

In Exercises 4–6, convert each angle from radians to degrees.

4. $\dfrac{7\pi}{10}$
5. $\dfrac{5\pi}{18}$
6. $-\dfrac{4\pi}{9}$

In Exercises 7 and 8, find the length of the arc on each circle of radius r intercepted by a central angle θ. Round your answers to three decimal places.

7. $r = 2$ meters, $\theta = 36°$
8. $r = 0.9$ meter, $\theta = 12°$

In Exercises 9–12, a point on the terminal side of an angle θ is given. For each angle, find the exact values of the six trigonometric functions. Rationalize the denominators where necessary.

9. $(2, 8)$
10. $(-3, 7)$
11. $(-\sqrt{5}, 2)$
12. $(-\sqrt{3}, -\sqrt{6})$

In Exercises 13–16, find the exact values of the remaining trigonometric functions of θ from the given information.

13. $\cos\theta = -\dfrac{4}{5}$, θ in quadrant III
14. $\tan\theta = -\dfrac{5}{12}$, θ in quadrant IV
15. $\sin\theta = \dfrac{3}{5}$, θ in quadrant II
16. $\csc\theta = -\dfrac{5}{4}$, θ in quadrant III

In Exercises 17–20, find the amplitude or vertical stretch factor, period, and phase shift of each given function.

17. $y = 14\sin\left(2x + \dfrac{\pi}{7}\right)$
18. $y = 21\cos\left(8x + \dfrac{\pi}{9}\right)$
19. $y = 6\tan\left(2x + \dfrac{\pi}{5}\right)$
20. $y = -11\cot\left(6x + \dfrac{\pi}{12}\right)$

In Exercises 21–26, graph each function over a two-period interval.

21. $y = -2\cos\left(x + \dfrac{2\pi}{3}\right)$
22. $y = 3\cos(x - \pi)$
23. $y = 5\tan\left[2\left(x - \dfrac{\pi}{4}\right)\right]$
24. $y = -\cot\left[2\left(x + \dfrac{\pi}{4}\right)\right]$
25. $y = \dfrac{1}{2}\sec\dfrac{x}{2}$
26. $y = -3\csc\dfrac{x}{2}$

In Exercises 27–34, find the exact value of y or state that y is undefined.

27. $y = \cos^{-1}\left(\cos\dfrac{5\pi}{8}\right)$
28. $y = \sin^{-1}\left(\sin\dfrac{7\pi}{6}\right)$
29. $y = \tan^{-1}\left[\tan\left(-\dfrac{2\pi}{3}\right)\right]$
30. $y = \sin\left(\cos^{-1}\dfrac{1}{2}\right)$
31. $y = \cos\left(\sin^{-1}\dfrac{\sqrt{2}}{2}\right)$
32. $y = \cos\left(\tan^{-1}\dfrac{3}{4}\right)$
33. $y = \tan\left[\cos^{-1}\left(-\dfrac{1}{2}\right)\right]$
34. $y = \tan\left[\sin^{-1}\left(-\dfrac{\sqrt{3}}{2}\right)\right]$

In Exercises 35–38, solve right triangle ABC with right angle at C. Round each answer to the nearest tenth.

35. $A = 30°$, $a = 6$
36. $B = 35°$, $b = 5$
37. $a = 3$, $b = 5$
38. $a = 4$, $c = 6$

In Exercises 39–50, verify each identity.

39. $(\sin x + \cos x)^2 + (\sin x - \cos x)^2 = 2$
40. $(1 - \tan x)^2 + (1 + \tan x)^2 = 2\sec^2 x$
41. $\dfrac{1 - \tan^2\theta}{1 + \tan^2\theta} = \cos^2\theta - \sin^2\theta$
42. $\dfrac{\sin\theta}{1 + \cos\theta} + \dfrac{\sin\theta}{1 - \cos\theta} = 2\csc\theta$
43. $\tan^2 x \sin^2 x = \tan^2 x - \sin^2 x$
44. $\dfrac{\sin\theta}{1 - \cot\theta} + \dfrac{\cos\theta}{1 - \tan\theta} = \cos\theta + \sin\theta$
45. $\dfrac{\tan\theta - \sin\theta}{\sin^3\theta} = \dfrac{\sec\theta}{1 + \cos\theta}$
46. $\dfrac{\tan x}{\sec x - 1} + \dfrac{\tan x}{\sec x + 1} = 2\csc x$
47. $\dfrac{1}{\csc x - \cot x} - \dfrac{1}{\cot x + \csc x} = 2\cot x$
48. $\dfrac{1 + \sin\theta}{1 - \sin\theta} = (\sec\theta + \tan\theta)^2$
49. $\dfrac{\sec x - \tan x}{\sec x + \tan x} = (\sec x - \tan x)^2$
50. $\dfrac{\csc x + \cot x}{\csc x - \cot x} = (\csc x + \cot x)^2$

In Exercises 51–58, find the exact value of each expression.

51. $\cos 15°$
52. $\sin 105°$
53. $\csc 75°$
54. $\tan 75°$
55. $\sin 41° \cos 49° + \cos 41° \sin 49°$
56. $\cos 50° \cos 10° - \sin 50° \sin 10°$
57. $\dfrac{\tan 69° + \tan 66°}{1 - \tan 69° \tan 66°}$
58. $2\cos 75° \cos 15°$

In Exercises 59–62, let $\sin u = \dfrac{4}{5}$, $\cos v = \dfrac{5}{13}$, for $0 < u \le \dfrac{\pi}{2}$, $0 < v \le \dfrac{\pi}{2}$. Find the value of each expression.

59. $\sin(u - v)$ **60.** $\cos(u + v)$
61. $\cos(u - v)$ **62.** $\tan(u - v)$

In Exercises 63–70, verify each identity.

63. $\sin(x - y)\cos y + \cos(x - y)\sin y = \sin x$

64. $\cos(x - y)\cos y - \sin(x - y)\sin y = \cos x$

65. $\dfrac{\sin(u + v)}{\sin(u - v)} = \dfrac{\tan u + \tan v}{\tan u - \tan v}$

66. $\dfrac{\tan(x + y)}{\cot(x - y)} = \dfrac{\tan^2 x - \tan^2 y}{1 - \tan^2 x \tan^2 y}$

67. $\dfrac{\sin 4x}{\sin 2x} - \dfrac{\cos 4x}{\cos 2x} = \sec 2x$

68. $\dfrac{\sin 3x}{\sin x} - \dfrac{\cos 3x}{\cos x} = 2$

69. $\sin\left(x - \dfrac{\pi}{6}\right) + \cos\left(x + \dfrac{\pi}{3}\right) = 0$

70. $\cos 2x = \cos^4 x - \sin^4 x$

71. If $x = r\cos\theta$ and $y = r\sin\theta$, show that $x^2 + y^2 = r^2$.

72. If $x = a\cos\theta$ and $y = b\sin\theta$, show that $\dfrac{x^2}{a^2} + \dfrac{y^2}{b^2} = 1$.

73. If $a = x\sin\theta$ and $b = y\tan\theta$, show that $\dfrac{x^2}{a^2} - \dfrac{y^2}{b^2} = 1$.

74. If $x = a\cos^3\theta$ and $y = b\sin^3\theta$, show that $\left(\dfrac{x}{a}\right)^{2/3} + \left(\dfrac{y}{b}\right)^{2/3} = 1$.

75. If $x = a\sin A\cos B$, $y = r\sin A\sin B$, and $z = r\cos A$, show that $x^2 + y^2 + z^2 = r^2$.

76. If $x = r\cos\alpha\sin\beta$, $y = r\cos\alpha\cos\beta$, and $z = r\sin\alpha$, show that $x^2 + y^2 + z^2 = r^2$.

Applying the Concepts

77. Distance. A tractor rolls backward so that its wheels turn a quarter of a revolution. If the tires have a radius of 28 inches, how many inches has the tractor moved?

78. Angle measure. What is the radian measure of the smaller central angle made by the hands of a clock at 5:00? Express your answer as a rational multiple of π.

79. Distance. New Orleans, Louisiana, is due south of Dubuque, Iowa. Find the distance between New Orleans (north latitude 29°59′ N) and Dubuque (north latitude 42°31′ N). Use 3960 miles as the value of the radius of Earth.

80. Daylight hours. The table gives the average number of daylight hours in Santa Fe, New Mexico, each month. Let y represent the number of daylight hours in Santa Fe in month x and find a function of the form $y = a\sin[b(x - c)] + d$ that models the hours of daylight throughout the year, where $x = 1$ represents January.

Jan	Feb	Mar	Apr	May	June
10.1	9.9	12.0	12.7	14.1	14.1

July	Aug	Sept	Oct	Nov	Dec
14.3	13.5	12.0	11.3	10.0	9.8

81. Rabbit population. The rabbit population of an area is described by $N(t) = 2400 + 500\sin\pi t$, where $N(t)$ gives the number of rabbits in the area after t years. When will the rabbit population first reach 1900?

82. Voltage. The voltage from an alternating current generator is given by $V(t) = 160\cos 120\pi t$, where t is measured in seconds. Find the time t when the voltage first equals 80.

EXERCISES FOR CALCULUS

1. Difference quotient. Let $f(x) = \sin x$. Show that
$$\dfrac{f(x + h) - f(x)}{h} = \sin x\left(\dfrac{\cos h - 1}{h}\right) + \cos x\left(\dfrac{\sin h}{h}\right)$$

2. Difference quotient. Let $f(x) = \cos x$. Show that
$$\dfrac{f(x + h) - f(x)}{h} = \cos x\left(\dfrac{\cos h - 1}{h}\right) - \sin x\left(\dfrac{\sin h}{h}\right)$$

Circumscribed polygon. A regular circumscribed n-gon about a circle of radius r has n vertices and its sides are tangent to the circumference of the circle. The adjoining figure is a circumscribed hexagon ($n = 6$) about a circle of radius r. Use this figure for Exercises 3–6.

3. a. In the figure, show that $B = \left(r, r\tan\dfrac{\pi}{6}\right)$ and conclude that the length of each side of the hexagon is $2r\tan\dfrac{\pi}{6}$.

b. Use part (a) to find the perimeter of the hexagon.

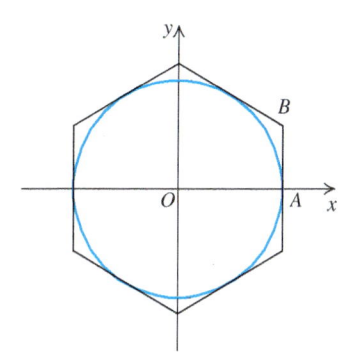

4. a. Use the figure to show that the area of the triangle AOB is $\dfrac{1}{2}r^2\tan\dfrac{\pi}{6}$.

b. Use part (a) to find the area of the circumscribed hexagon.

5. a. Show that the perimeter P of a regular circumscribed n-gon about a circle of radius r is

$$P = 2nr \tan \frac{\pi}{n}.$$

b. What is the circumference of a disc of radius $r = 10$?
c. Calculate P in part (a) with $r = 10$ and $n = 4, 10, 50$, and 100. Round your answer to four decimal places. What do you observe?

6. a. Show that the area A of a regular n-gon circumscribed about a circle of radius r is

$$A = nr^2 \tan \frac{\pi}{n}.$$

b. What is the area of a circle of radius $r = 10$?
c. Calculate A by using the formula in part (a) with $r = 10$ and $n = 4, 10, 50$, and 100. What do you observe?

Exercises 7–10 deal with a regular polygon. For $n \geq 3$, a regular n-gon is an n-sided polygon, all of whose sides are equal in length and all of whose angles are equal in measure.

Inscribed polygon. A regular inscribed n-gon has n vertices that are equally spaced points on a circle of radius r. The adjoining figure is an inscribed octagon ($n = 8$) on a circle of radius r.

Use this figure for Exercises 7–10.

7. a. Write the coordinates of the vertices of the inscribed octagon with one of the vertices at $A_1 = (r, 0)$.
b. Find the length of the side $A_1 A_2$.
c. Use part (b) to find the perimeter of the octagon.

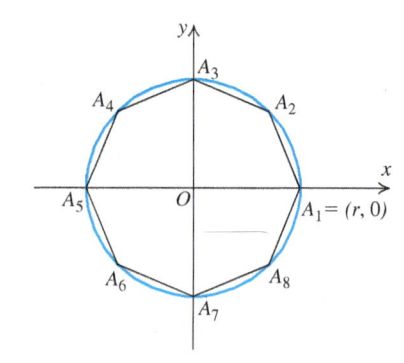

8. a. In the figure, find the area of triangle $A_1 O A_2$.
b. Use part (a) to find the area of the inscribed octagon.

9. a. Show that, the perimeter P of a regular n-gon inscribed in a circle of radius r is

$$P = nr\sqrt{2 - 2\cos\frac{2\pi}{n}}.$$

b. What is the circumference (perimeter) of a circle of radius $r = 5$?
c. Calculate P in part (a) with $r = 5$ and $n = 4, 10, 50$, and 100. Round your answer to four decimal places. What do you observe?

10. a. Show that the area A of a regular n-gon inscribed in a circle of radius r is

$$A = \frac{nr^2}{2} \sin \frac{2\pi}{n}.$$

b. What is the area of a disc of radius $r = 5$?
c. Calculate A from part (a) with $r = 5$ and $n = 4, 10, 50$, and 100. What do you observe?

PRACTICE TEST A

1. Convert $140°$ to radians.

2. Convert $\dfrac{7\pi}{5}$ to degrees.

3. If $(2, -5)$ is a point on the terminal side of an angle θ, find $\cos \theta$.

4. If $\tan \theta < 0$ and $\csc \theta > 0$, find the quadrant in which θ lies.

5. If $\cot \theta = -\dfrac{5}{12}$ and θ is in quadrant IV, find $\sec \theta$.

6. What is the reference angle for $217°$?

7. If $\cos \dfrac{3\pi}{7} = 0.223$, find $\sin \dfrac{\pi}{14}$.

8. Find the amplitude, period, and phase shift for $y = 19 \sin 12(x + 3\pi)$.

9. Find the exact value of $y = \sin^{-1}\left(-\dfrac{\sqrt{3}}{2}\right)$.

10. Find the exact value of $y = \cos^{-1}\left[\cos \dfrac{4\pi}{3}\right]$.

11. Find the exact value of $y = \tan\left[\sin^{-1}\left(-\dfrac{1}{3}\right)\right]$.

In Problems 12–14, verify each identity.

12. $\dfrac{1 - \sin^2 x}{\sin^2 x} = \cot^2 x$

13. $\sin x \sin\left(\dfrac{\pi}{2} - x\right) = \dfrac{\sin 2x}{2}$

14. $\dfrac{\sin 2x}{2(\cos x + \sin x)} = \dfrac{\sin x}{1 + \tan x}$

15. Use the information in the figure to solve for x.

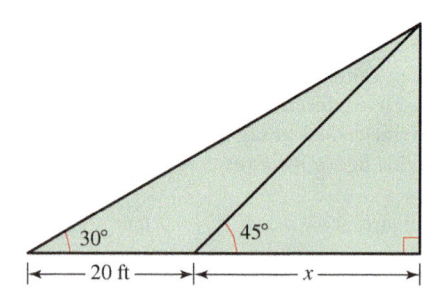

16. The angle of depression from the top of a tree to the point 85 feet from its base (on the ground) is $12°$. Find the height of the tree to the nearest foot.

17. If $\tan \theta = \dfrac{3}{4}$, find the exact value of $\sin 2\theta$.

18. Determine whether the function $y = \cos\left(\dfrac{\pi}{2} - x\right) + \tan x$ is odd, even, or neither.

19. Determine whether $\sin x + \sin y = \sin(x + y)$ is an identity.

20. A golf ball is hit on a level fairway with an initial velocity of $v_0 = 128$ ft/sec on an initial angle of flight θ (in degrees). The range (in feet) of the ball is given by the expression $\dfrac{v_0^2 \sin 2\theta}{32}$. Find the value(s) of θ if the range is 512 feet.

PRACTICE TEST B

Multiple choice—select the correct answer.

1. Convert $160°$ to radians.
   a. $\dfrac{7\pi}{9}$ b. $\dfrac{8\pi}{9}$ c. $\dfrac{4\pi}{3}$ d. $\dfrac{7\pi}{3}$

2. Convert $\dfrac{13\pi}{5}$ to degrees.
   a. $36°$ b. $72°$ c. $108°$ d. $468°$

3. If $(-3, 1)$ is a point on the terminal side of an angle θ, find $\cos \theta$.
   a. $-\dfrac{3\sqrt{10}}{10}$ b. $-\dfrac{\sqrt{10}}{10}$ c. $\dfrac{\sqrt{10}}{10}$ d. $\dfrac{3\sqrt{10}}{10}$

4. If $\sin \theta < 0$ and $\sec \theta > 0$, find the quadrant in which θ lies.
   a. I   b. II   c. III   d. IV

5. If $\tan \theta = \dfrac{12}{5}$ and θ is in quadrant III, find $\csc \theta$.
   a. $-\dfrac{5}{12}$ b. $-\dfrac{13}{12}$ c. $\dfrac{5}{13}$ d. $\dfrac{5}{12}$

6. What is the reference angle for $640°$?
   a. $10°$ b. $60°$ c. $80°$ d. $280°$

7. In which quadrant is it true that both $\cos \theta > 0$ and $\csc \theta > 0$?
   a. I   b. II   c. III   d. IV

8. Which of the following statements is *false*?
   a. $\sin(-x) = -\sin x$ b. $\cos(-x) = -\sin x$
   c. $\tan(-x) = -\tan x$ d. $\cot(-x) = -\cot x$

9. This is the graph of which function?

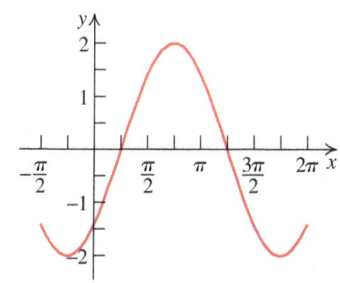

   a. $y = -2 \sin\left(x + \dfrac{\pi}{2}\right)$ b. $y = 2 \sin\left(x + \dfrac{\pi}{2}\right)$
   c. $y = -2 \cos\left(x + \dfrac{\pi}{4}\right)$ d. $y = 2 \cos\left(x + \dfrac{\pi}{4}\right)$

10. Find the exact value of $y = \cos^{-1}\left(\cos \dfrac{7\pi}{4}\right)$.
    a. $\dfrac{2\pi}{3}$ b. $\dfrac{-2\pi}{3}$ c. $\dfrac{7\pi}{4}$ d. $\dfrac{\pi}{4}$

11. Find the exact value of $y = \cos\left[\tan^{-1}\left(\dfrac{5}{3}\right)\right]$.
    a. $\dfrac{3}{5}$ b. $\dfrac{5}{34}$ c. $\dfrac{5\sqrt{34}}{34}$ d. $\dfrac{3\sqrt{34}}{34}$

12. Which expression is equal to $\dfrac{\sin x}{1 + \cos x} + \dfrac{\sin x}{1 - \cos x}$?
    a. $2 \sin x$ b. $2 \cos x$ c. $2 \sec x$ d. $2 \csc x$

13. Which expression is equal to $\dfrac{\sin 2x}{1 + \cos 2x}$?
    a. $\cot x$ b. $\tan x$ c. $\tan 2x$ d. $\cot 2x$

14. Which expression is equal to $\dfrac{1 - \cos 2x}{1 + \cos 2x}$?
    a. $\sin^2 x$ b. $\cos^2 x$ c. $\tan^2 x$ d. $\cot^2 x$

15. A 3-foot shrub is 4.5 feet away from a 12-foot streetlight. What length shadow is cast by the shrub?
    a. 1.5 ft   b. 2 ft   c. 2.67 ft   d. 4 ft

16. In a right triangle ABC, with $a = 12$ and $c = 13$, which is closest to the correct value of B?
    a. $67.38°$ b. $42.71°$ c. $22.62°$ d. $47.29°$

17. If $\sin \theta = \dfrac{2}{3}$, find the exact value of $\cos 2\theta$.
    a. $\dfrac{1}{3}$ b. $\dfrac{1}{9}$ c. $\dfrac{14}{9}$ d. $\dfrac{\sqrt{5}}{3}$

18. If $\tan \theta = \dfrac{4}{3}$, find the exact value of $\sin 2\theta$.
    a. $\dfrac{24}{25}$ b. $\dfrac{4}{5}$ c. $\dfrac{3}{5}$ d. $\dfrac{8}{5}$

19. Which pair of values results in a false statement for $\tan(x + y) = \tan x + \tan y$?
    a. $x = 0, y = \pi$ b. $x = \dfrac{\pi}{4}, y = -\dfrac{\pi}{4}$
    c. $x = \dfrac{\pi}{4}, y = \dfrac{5\pi}{4}$ d. $x = \dfrac{\pi}{4}, y = \dfrac{3\pi}{4}$

20. A golf ball is hit on a level fairway with an initial velocity of $v_0 = 128$ ft/sec and an initial angle of flight θ (in degrees). The range (in feet) of the ball is given by the expression $\dfrac{v_0^2 \sin 2\theta}{32}$. Find the value(s) of θ if the range is $256\sqrt{3}$ feet.
    a. $45°$ b. $30°, 60°$
    c. $45°, 135°$ d. $150°, 175°$

CHAPTER 5

Applications of Trigonometric Functions

TOPICS

5.1 The Law of Sines and the Law of Cosines

5.2 Areas of Polygons Using Trigonometry

5.3 Polar Coordinates

5.4 Parametric Equations

Applications of trigonometry permeate virtually every field, from architecture and bionics to medicine and physics. In this chapter, we introduce polar coordinates and investigate the many applications of trigonometry to real-world problems.

SECTION 5.1

The Law of Sines and the Law of Cosines

BEFORE STARTING THIS SECTION, REVIEW

1. Trigonometric functions (Section 4.2, page 301)
2. Right-triangle trigonometry (Section 4.6, page 350)
3. Inverse trigonometric functions (Section 4.5, page 337)

OBJECTIVES

1. Learn vocabulary and conventions for solving triangles.
2. Use the Law of Sines.
3. Use the Law of Cosines.

Mount Everest
In 1852, an unsung hero Radhanath Sickdhar managed to *calculate* the height of Peak XV, an icy peak in the Himalayas. Peak XV, the highest mountain in the world, was later named Mount Everest. It stands at 29,002 feet. Sickdhar's calculations used triangulations as well as the phenomenon called refraction—the bending of light rays by the density of Earth's atmosphere.

◆ The Great Trigonometric Survey of India

Triangulation is the process of finding the coordinates and the distance to a point by using the Law of Sines. Triangulation was used in surveying the Indian subcontinent. The survey, which was called "one of the most stupendous works in the whole history of science," was begun by William Lambton, a British army officer, in 1802. It started in the south of India and extended north to Nepal, a distance of approximately 1600 miles. Lasting several decades and employing thousands of workers, the survey was named the Great Trigonometric Survey (GTS). In 1818, George Everest (1790–1866), a Welsh surveyor and geographer, was appointed assistant to Lambton. Everest, who succeeded Lambton in 1823 and was later promoted as surveyor general of India, completed the GTS. Mount Everest was surveyed and named after Everest by his successor Andrew Waugh. In Example 2, we use the Law of Sines to estimate the height of a mountain.

1 Learn vocabulary and conventions for solving triangles.

RECALL

Two triangles are *similar* if they have equal angles and their corresponding sides are proportional.

Two triangles are *congruent* if corresponding sides are congruent (equal in length) and corresponding angles are congruent (equal in measure). An *included side* is the side between two angles, and an *included angle* is the angle between two sides.

Solving Oblique Triangles

The process of finding the unknown side lengths and angle measures in a triangle is called *solving a triangle*. In Section 4.6, you learned techniques of solving right triangles. We now discuss triangles that are not necessarily right triangles. A triangle with no right angle is called an **oblique triangle**. From geometry, we know that two triangles with any of the following sets of equal parts will always be **congruent**. In other words, the following sets of given parts determine a unique triangle:

ASA: Two angles and the included side
AAS: Two angles and a nonincluded side
SAS: Two sides and the included angle
SSS: All three sides

Although **SSA** (two sides and a nonincluded angle) does not guarantee a unique triangle, there can be, at most, two triangles with the given parts. On the other hand,

396 Chapter 5 Applications of Trigonometric Functions

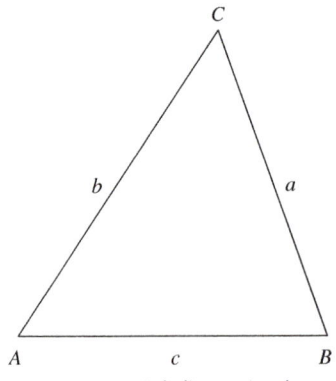

FIGURE 5.1 Labeling a triangle

AAA (or AA) guarantees only *similar* triangles, not congruent ones; so there are infinitely many triangles with the same three angle measures.

For any triangle *ABC*, we let each letter, such as *A*, stand for both the vertex *A* and the measure of the angle at *A*. As usual, the angles of a triangle *ABC* are labeled *A*, *B*, and *C* and the lengths of their opposite sides are labeled, *a*, *b*, and *c*, respectively. See Figure 5.1.

To solve an oblique triangle, given at least one side and two other measures, we consider four possible cases:

Case 1. Two angles and a side are known (**AAS** and **ASA** triangles).
Case 2. Two sides and an angle opposite one of the given sides are known (**SSA** triangles).
Case 3. Two sides and their included angle are known (**SAS** triangles).
Case 4. All three sides are known (**SSS** triangles).

We use the *Law of Sines* to analyze triangles in Cases 1 and 2, and we use the *Law of Cosines* to solve triangles in Cases 3 and 4.

2 Use the Law of Sines.

The Law of Sines

The Law of Sines states that in a triangle *ABC* with sides of length *a*, *b*, and *c*,

$$\frac{\sin A}{a} = \frac{\sin B}{b} = \frac{\sin C}{c}.$$

We can derive the Law of Sines by using the right triangle definition of the sine of an acute angle θ:

$$\sin \theta = \frac{\text{opposite}}{\text{hypotenuse}}$$

RECALL

An *altitude* of a triangle is a segment drawn from any vertex of the triangle perpendicular to the opposite side or to an extension of the opposite side.

We begin with an oblique triangle *ABC*. See Figures 5.2 and 5.3. Draw altitude *CD* from the vertex *C* to the side *AB* (or its extension, as in Figure 5.3). Let *h* be the length of segment *CD*.

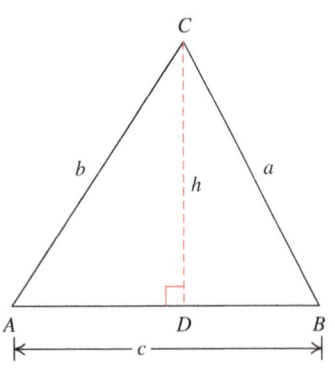

FIGURE 5.2 Altitude *CD* with acute $\angle B$

In right triangle *CDA*,

$$\frac{h}{b} = \sin A \quad \text{or}$$

$$h = b \sin A.$$

In right triangle *CDB*,

$$\frac{h}{a} = \sin B \quad \text{or}$$

$$h = a \sin B.$$

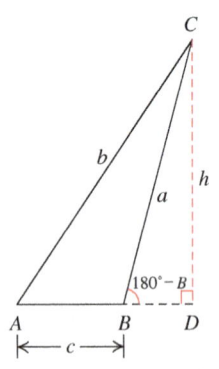

FIGURE 5.3 Altitude *CD* with obtuse $\angle B$

In right triangle *CDA*,

$$\frac{h}{b} = \sin A \quad \text{or}$$

$$h = b \sin A.$$

In right triangle *CDB*,

$$\frac{h}{a} = \sin(180° - B) = \sin B \quad \text{or}$$

$$h = a \sin B.$$

Because in each figure $h = b \sin A$ and $h = a \sin B$, we have

$$b \sin A = a \sin B$$

$$\frac{b \sin A}{ab} = \frac{a \sin B}{ab} \quad \text{Divide both sides by } ab.$$

$$\frac{\sin A}{a} = \frac{\sin B}{b} \quad \text{Simplify.}$$

Similarly, by drawing altitudes from the other two vertices, we can show that

$$\frac{\sin B}{b} = \frac{\sin C}{c} \quad \text{and} \quad \frac{\sin C}{c} = \frac{\sin A}{a}.$$

We have proved the Law of Sines.

THE LAW OF SINES

In any triangle ABC, with sides of length a, b, and c,

$$\frac{\sin A}{a} = \frac{\sin B}{b}, \quad \frac{\sin B}{b} = \frac{\sin C}{c}, \quad \text{and} \quad \frac{\sin C}{c} = \frac{\sin A}{a}.$$

We can rewrite these relations in compact notation:

$$\frac{\sin A}{a} = \frac{\sin B}{b} = \frac{\sin C}{c} \quad \text{or equivalently,} \quad \frac{a}{\sin A} = \frac{b}{\sin B} = \frac{c}{\sin C}$$

Solving AAS and ASA Triangles

We next use the Law of Sines to solve AAS and ASA triangles of Case 1, where two angles and a side are given. Note that if two angles of a triangle are given, then you also know the third angle (because **the sum of the angles of a triangle is 180°**). So in Case 1, all three angles and one side of a triangle are known.

PROCEDURE
IN ACTION

EXAMPLE 1 Solving the AAS and ASA Triangles

OBJECTIVE
Solve a triangle given two angles and a side.

EXAMPLE
Solve triangle ABC with $A = 62°$, $c = 14$ feet, $B = 74°$. Round side lengths to the nearest tenth.

Step 1 Find the third angle. Find the measure of the third angle by subtracting the measures of the known angles from 180°.

1. $C = 180° - A - B$
   $= 180° - 62° - 74°$
   $= 44°$

Step 2 Make a chart. Make a chart of the six parts of the triangle, including the known and the unknown parts. Sketch the triangle.

2.
$A = 62°$	$a = ?$
$B = 74°$	$b = ?$
$C = 44°$	$c = 14$

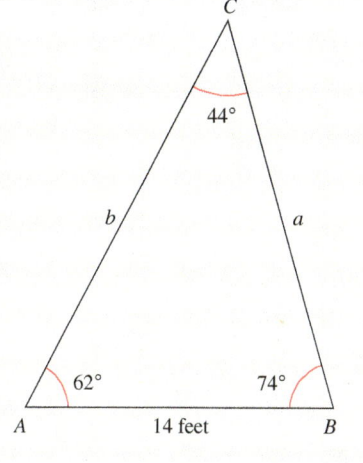

(continued)

Step 3 Apply the Law of Sines. Select two ratios from the Law of Sines in which three of the four quantities are known. Solve for the fourth quantity. Use the form of the Law of Sines in which the unknown quantity is in the numerator.

3. $\dfrac{b}{\sin B} = \dfrac{c}{\sin C}$ $\quad\quad$ $\dfrac{a}{\sin A} = \dfrac{c}{\sin C}$

$\dfrac{b}{\sin 74°} = \dfrac{14}{\sin 44°}$ $\quad\quad$ $\dfrac{a}{\sin 62°} = \dfrac{14}{\sin 44°}$

$b = \dfrac{14 \sin 74°}{\sin 44°}$ $\quad\quad$ $a = \dfrac{14 \sin 62°}{\sin 44°}$

$b \approx 19.4$ feet $\quad\quad$ $a \approx 17.8$ feet

Step 4 Show the solution. Show the solution by completing the chart.

4.
$A = 62°$	$a \approx 17.8$ feet
$B = 74°$	$b \approx 19.4$ feet
$C = 44°$	$c = 14$

Practice Problem 1 Solve triangle ABC with $A = 70°$, $B = 65°$, and $a = 16$ inches. Round side lengths to the nearest tenth.

EXAMPLE 2 Height of a Mountain

From a point on a level plain at the foot of a mountain, a surveyor finds the angle of elevation of the peak of the mountain to be 20°. She walks 3465 meters closer (on a direct line between the first point and the base of the mountain) and finds the angle of elevation to be 23°. Estimate the height of the mountain to the nearest meter. See Figure 5.4.

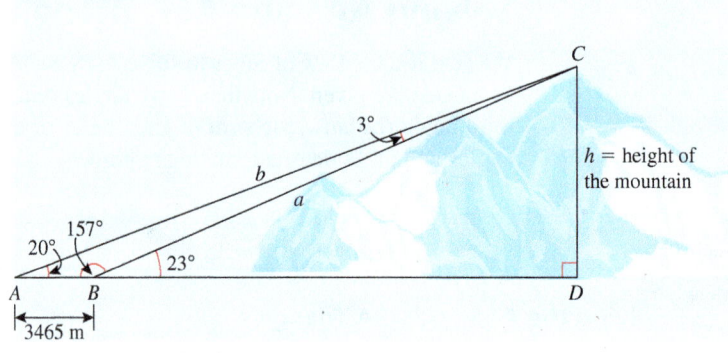

FIGURE 5.4 Height of a mountain

SIDE NOTE
When using your calculator to solve triangles, make sure it is in *degree* mode.

Solution

In Figure 5.4, consider triangle ABC.

$\angle ABC = 180° - 23° = 157°$ $\quad$ $\angle ABC + \angle CBD = 180°$

Find the third angle in triangle ABC.

$C = 180° - 20° - 157° = 3°$

Apply the Law of Sines.

$$\dfrac{a}{\sin 20°} = \dfrac{3465}{\sin 3°}$$

$$a = \dfrac{3465 \sin 20°}{\sin 3°} \approx 22{,}644 \text{ meters}$$

Now in triangle BCD, $\sin 23° = \dfrac{h}{a}$. So $h = a \sin 23° \approx 22{,}644 \sin 23° \approx 8848$ meters.

The mountain is approximately 8848 meters high.

Practice Problem 2 From a point on a level plain at the foot of a mountain, the angle of elevation of the peak is 40°. If you move 2500 feet farther (on a direct line extending from the first point and the base of the mountain), the angle of elevation of the peak is 35°. How high above the plain is the peak? Round your answer to the nearest foot.

Solving SSA Triangles—the Ambiguous Case

In Case 1, where we know two angles and a side, we obtain a unique triangle. However, if we know the lengths of two sides and the measure of the angle opposite one of these sides, then we could have a result that is (1) not a triangle, (2) exactly one triangle, or (3) two different triangles. For this reason, Case 2 is called the **ambiguous case**.

Suppose we want to draw triangle *ABC* with given measures of *A*, *a*, and *b*. We draw angle *A* with a horizontal initial side, and we place the point *C* on the terminal side of *A* so that the length of segment *AC* is *b*. We need to locate point *B* on the initial side of *A* so that the measure of the segment *CB* is *a*. See Figure 5.5.

The number of possible triangles depends upon the length *h* of the altitude from *C* to the initial side of angle *A*. Because $\sin A = \dfrac{h}{b}$, we have $h = b \sin A$. Various possibilities for forming triangle *ACB* are illustrated in Figure 5.6 (where *A* is an acute angle) and Figure 5.7 (where *A* is an obtuse angle).

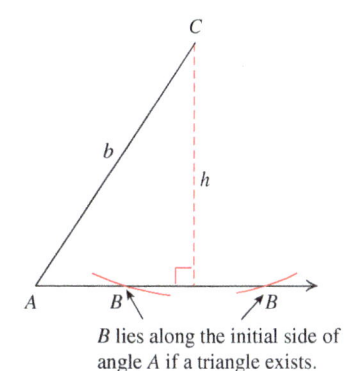

B lies along the initial side of angle *A* if a triangle exists.
FIGURE 5.5 Need to locate *B*

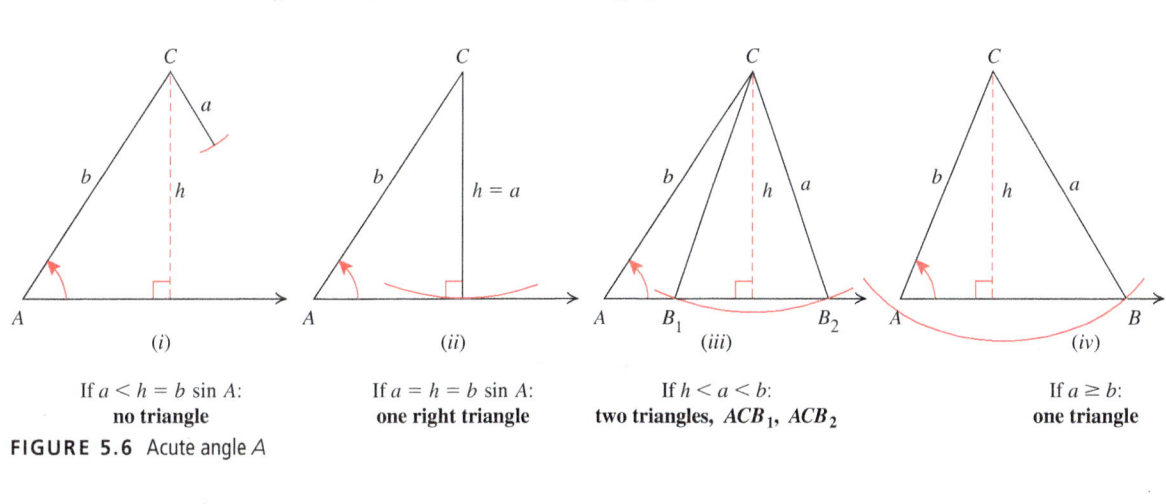

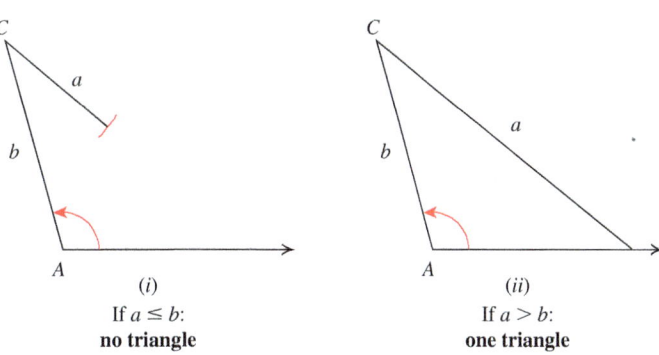

(i) If $a < h = b \sin A$: no triangle
(ii) If $a = h = b \sin A$: one right triangle
(iii) If $h < a < b$: two triangles, ACB_1, ACB_2
(iv) If $a \geq b$: one triangle

FIGURE 5.6 Acute angle *A*

(i) If $a \leq b$: no triangle
(ii) If $a > b$: one triangle

FIGURE 5.7 Obtuse angle *A*

In general in an SSA triangle, we are given:

(i) the known angle θ,
(ii) opposite side (side opposite to angle θ), and
(iii) adjacent side (side adjacent to angle θ).

Then **altitude = (adjacent side) sin θ**.

We summarize the situations from Figures 5.6 and 5.7 in Tables 5.1 and 5.2, respectively

TABLE 5.1 If θ is an acute angle

Case	Condition on Opposite Side	Number of Triangles
1	opposite side < altitude	None
2	opposite side = altitude	One right triangle
3	altitude < opposite side < adjacent side	Two triangles
4	opposite side ≥ adjacent side	One triangle

TABLE 5.2 If θ is an obtuse angle

Case	Condition on Opposite Side	Number of Triangles
1	opposite side ≤ adjacent side	None
2	opposite side > adjacent side	One triangle

EXAMPLE 3 Finding the Number of Triangles

How many triangles can be drawn with $A = 42°$, $b = 2.4$, and $a = 2.1$?

Solution

$A = 42°$ is an acute angle, and opposite side $= a = 2.1 < 2.4 = b =$ adjacent side. We calculate

$$\text{altitude} = b \sin A = 2.4 \sin (42°) \quad \text{Replace } b \text{ with 2.4 and } A \text{ with } 42°.$$
$$\approx 1.6 \quad \text{Use a calculator.}$$

We have altitude < opposite side < adjacent side. So by Case 3 of Table 5.1 two triangles can be drawn with the given measurements.

Practice Problem 3 How many triangles can be drawn with $A = 35°$, $b = 6.5$, and $a = 3.5$?

EXAMPLE 4 Finding the Number of Triangles

How many triangles can be drawn with $A = 98°$, $b = 4.8$, and $a = 5.1$?

Solution

$A = 98°$ is an obtuse angle, and opposite side $= a = 5.1 > 4.8 = b =$ adjacent side. So by Case 2 of Table 5.2 exactly one triangle can be drawn with the given measurements.

Practice Problem 4 How many triangles can be drawn with $A = 105°$, $b = 12.5$ and $a = 11.2$?

We strongly suggest that you begin with a tentative sketch of the triangle ABC.

PROCEDURE IN ACTION

EXAMPLE 5 Solving the SSA Triangles

OBJECTIVE
Solve a triangle if two sides and an angle opposite one of them is given.

EXAMPLE
Solve triangle ABC with $B = 32°$, $b = 100$ feet, and $c = 150$ feet. Round each answer to the nearest tenth.

Step 1 Make a chart of the six parts. Identify θ, opposite and adjacent sides.

1. $\theta = B$, opposite side $= b$, adjacent side $= c$

$A = ?$	$a = ?$
$B = 32°$	$b = 100$
$C = ?$	$c = 150$

Three quantities are known.

Step 2 Count the solutions.
There may be none, one, or two solutions.
If the known angle
(i) θ is acute—use Table 5.1
(ii) θ is obtuse—use Table 5.2
(iii) $\theta = 90°$—use procedures in Section 4.6.
If there are no solutions, you are done.
Otherwise continue with Step 3.

Step 3 Apply the Law of Sines to the two ratios in which two sides and one angle are known (see the chart in Step 1). Solve for the sine of the unknown angle. Use the form of the Law of Sines in which the unknown angle is in the numerator.

Step 4 Find the second angle(s)
If you determined in step 2 that:
(i) there are two solutions, find two angles; one acute and one obtuse.
(ii) there is one solution, find one acute angle.

Step 5 Find the third angle of the triangle(s).

Step 6 Use the Law of Sines to find the remaining side(s).

Step 7 Show the solution(s).

2. Here $\theta = B = 32°$ is an acute angle, so we use Table 6.1.

$$\text{Altitude} = (\text{adjacent side}) \sin B$$
$$= 150 \sin 32° \quad \text{Substitute values}$$
$$\approx 79.488 \quad \text{Use a calculator}$$

We have:
altitude < opposite side < adjacent side, so there will be two solutions.

3. $\dfrac{\sin C}{c} = \dfrac{\sin B}{b}$ The Law of Sines

$\sin C = \dfrac{c \sin B}{b}$ Multiply both sides by c.

$= \dfrac{(150) \sin 32°}{100}$ Substitute values.

$\sin C \approx 0.7949$ Use a calculator.

4. Since there are two solutions, we find two angles:

$$C_1 \approx \sin^{-1}(0.7949) \approx 52.6°$$
$$C_2 \approx 180° - 52.6° = 127.4°$$

5. $\angle BAC_1 \approx 180° - 32° - 52.6° = 95.4°$
$\angle BAC_2 \approx 180° - 32° - 127.4° = 20.6°$

6. $\dfrac{a_1}{\sin \angle BAC_1} = \dfrac{b}{\sin B}$ $\dfrac{a_2}{\sin \angle BAC_2} = \dfrac{b}{\sin B}$

$a_1 = \dfrac{b \sin \angle BAC_1}{\sin B}$ $a_2 = \dfrac{b \sin \angle BAC_2}{\sin B}$

$a_1 = \dfrac{100 \sin 95.4°}{\sin 32°}$ $a_2 = \dfrac{(100) \sin 20.6°}{\sin 32°}$

$a_1 \approx 187.9$ feet $a_2 \approx 66.4$ feet

7.

$\triangle BAC_1$

$\angle BAC_1 \approx 95.4°$	$a_1 \approx 187.9$
$B = 32°$	$b = 100$
$C_1 \approx 52.6°$	$c = 150$

$\triangle BAC_2$

$\angle BAC_2 \approx 20.6°$	$a_2 \approx 66.4$
$B = 32°$	$b = 100$
$C_2 \approx 127.4°$	$c = 150$

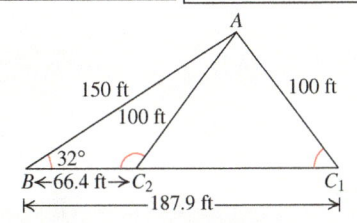

Practice Problem 5 Solve triangle ABC with $C = 35°$, $b = 15$ feet, and $c = 12$ feet. Round each answer to the nearest tenth.

EXAMPLE 6   Solving an SSA Triangle (No Solution)

Solve triangle ABC with $A = 50°$, $a = 8$ inches, and $b = 15$ inches.

Solution
Step 1 Make a chart.

402 Chapter 5 Applications of Trigonometric Functions

$A = 50°$	$a = 8$
$B = ?$	$b = 15$
$C = ?$	$c = ?$

Three quantities are known.

Step 2 **Count the solutions.** Here $\theta = A = 50°$ is acute, we use Table 5.1. The opposite side $a = 8$ is not greater than the adjacent side $b = 15$, so we need to compute the altitude.

$$\text{altitude} = (\text{adjacent side}) \sin A$$
$$= 15 \sin (50°) \quad \text{Replace: adjacent side with 15, } A = 50°$$
$$\approx 11.491. \quad \text{Use a calculator}$$

We have: opposite side $a = 8 < 11.491 = $ altitude, so there is no triangle.

Practice Problem 6 Solve triangle ABC with $A = 65°$, $a = 16$ meters, and $b = 30$ meters.

EXAMPLE 7 Solving an SSA Triangle (One Solution)

Solve triangle ABC with $C = 40°$, $c = 20$ meters, and $a = 15$ meters. Round each answer to the nearest tenth.

Solution

Step 1 Make a chart.

$A = ?$	$a = 15$
$B = ?$	$b = ?$
$C = 40°$	$c = 20$

Three quantities are known.

Step 2 **Find the number of triangles.** Here $\theta = C = 40°$ is acute, we use Table 5.1. The opposite side $c = 20$ is greater than the adjacent side $a = 15$, so there is one triangle.

Step 3 Apply the Law of Sines.

$$\frac{\sin A}{a} = \frac{\sin C}{c} \quad \text{The Law of Sines}$$

$$\sin A = \frac{a \sin C}{c} \quad \text{Multiply both sides by } a.$$

$$= \frac{(15) \sin 40°}{20} \quad \text{Substitute values.}$$

$$\sin A \approx 0.4821 \quad \text{Use a calculator.}$$

Step 4 **Find the second angle(s).** Since there is one solution, we find one angle.

$$A = \sin^{-1}(0.4821) \approx 28.8°.$$

Step 5 The third angle at B has measure $\approx 180° - 40° - 28.8° = 111.2°$.

Step 6 **Find the remaining side length.**

$$\frac{b}{\sin B} = \frac{c}{\sin C} \quad \text{The Law of Sines}$$

$$b = \frac{c \sin B}{\sin C} \quad \text{Multiply both sides by } \sin B.$$

$$= \frac{(20) \sin (111.2°)}{\sin 40°} \quad \text{Substitute values.}$$

$$b \approx 29.0 \text{ m} \quad \text{Use a calculator.}$$

Section 5.1 ■ The Law of Sines and the Law of Cosines   403

Step 7 Show the solution. See Figure 5.8.

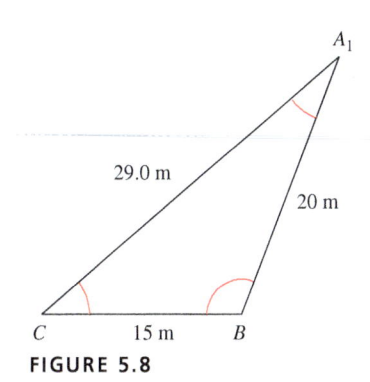

$A_1 \approx 28.8°$	$a = 15$ meters
$B \approx 111.2°$	$b \approx 29.0$ meters
$C = 40°$	$c = 20$ meters

FIGURE 5.8

Practice Problem 7 Solve triangle ABC with $C = 60°$, $c = 50$ feet, and $a = 30$ feet.

In some special fields, such as navigation and surveying, it is traditional to measure directions directly from a meridian by using *bearings*.

Bearings

A **bearing** is the measure of an acute angle from due north or due south. The bearing N 40° E means 40° to the east of due north, whereas S 30° W means 30° to the west of due south. See Figure 5.9.

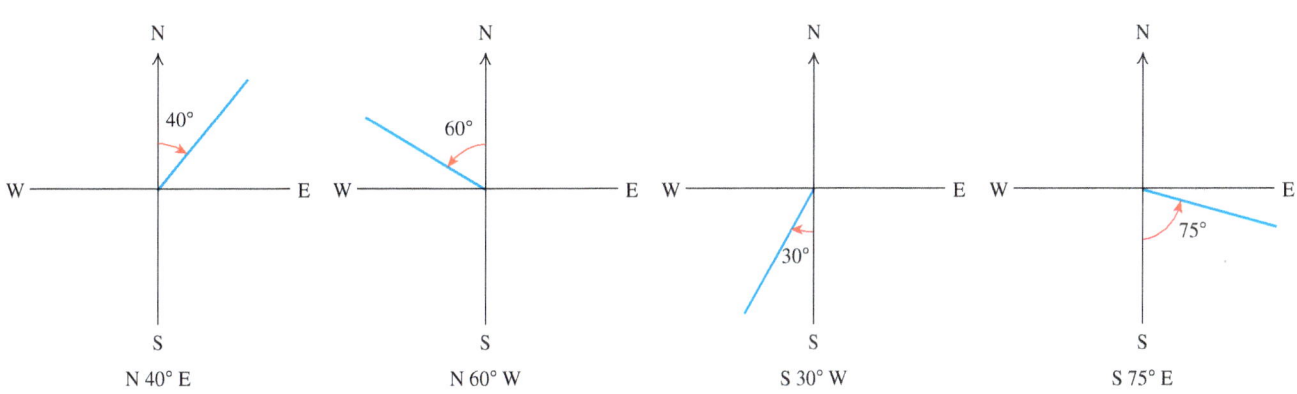

FIGURE 5.9 Bearings

EXAMPLE 8   Navigation Using Bearings

A ship sailing due west at 20 miles per hour records the bearing of an oil rig at N 55.4° W. An hour and a half later the bearing of the same rig is N 66.8° E.

a. How far is the ship from the oil rig the second time?

b. How close did the ship pass to the oil rig?

Solution

a. In an hour and a half, the ship travels $(1.5)(20) = 30$ miles due west. The oil rig (C), the starting point for the ship (A), and the position of the ship after an hour and a half (B) form the vertices of the triangle ABC shown in Figure 5.10.

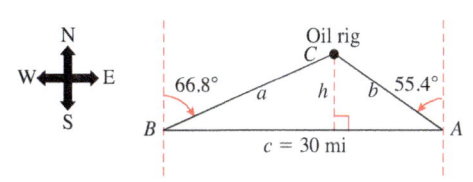

FIGURE 5.10 Navigation

Then $A = 90° - 55.4° = 34.6°$ and $B = 90° - 66.8° = 23.2°$. Therefore, $C = 180° - 34.6° - 23.2° = 122.2°$.

404 Chapter 5 Applications of Trigonometric Functions

$$\frac{a}{\sin A} = \frac{c}{\sin C} \qquad \text{The Law of Sines}$$

$$a = \frac{c \sin A}{\sin C} \qquad \text{Multiply both sides by } \sin A.$$

$$= \frac{30 \sin 34.6°}{\sin 122.2°} \approx 20 \text{ miles} \qquad \text{Substitute values and use a calculator.}$$

The ship is approximately 20 miles from the oil rig when the second bearing is taken.

b. The shortest distance between the ship and the oil rig is the length of the segment h in triangle ABC.

$$\sin B = \frac{h}{a} \qquad \text{Right-triangle definition of the sine}$$

$$h = a \sin B \qquad \text{Multiply both sides by } a.$$

$$h \approx 20 \sin 23.2° \qquad \text{Substitute values.}$$

$$\approx 7.9 \qquad \text{Use a calculator.}$$

The ship passes within 7.9 miles of the oil rig.

Practice Problem 8 Rework Example 8 assuming that the second bearing is taken after two hours and is N 60.4° E and the ship is traveling due west at 25 miles per hour.

3 Use the Law of Cosines.

The Law of Cosines

The Pythagorean Theorem states that the relationship $c^2 = a^2 + b^2$ holds in a right triangle ABC, where c represents the length of the hypotenuse. This relationship is not true if the triangle is not a right triangle. The **Law of Cosines** is a generalization of the Pythagorean Theorem that is true in any triangle. This law will be used to solve triangles in which two sides and the included angle are known (SAS triangles), as well as those triangles in which all three sides are known (SSS triangles).

FIGURE 5.11

> ### THE LAW OF COSINES
>
> In a triangle ABC with sides of lengths a, b, and c (as in Figure 5.11),
>
> $$a^2 = b^2 + c^2 - 2bc \cos A,$$
> $$b^2 = c^2 + a^2 - 2ca \cos B,$$
> $$c^2 = a^2 + b^2 - 2ab \cos C.$$
>
> In words, the square of any side of a triangle is equal to the sum of the squares of the length of the other two sides, less twice the product of the lengths of the other sides and the cosine of their included angle.

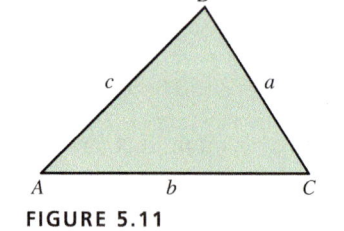

Derivation of the Law of Cosines

To derive the Law of Cosines, place triangle ABC in a rectangular coordinate system with the vertex A at the origin and the side c along the positive x-axis. See Figure 5.12.

Angle A is an acute angle.
(a)

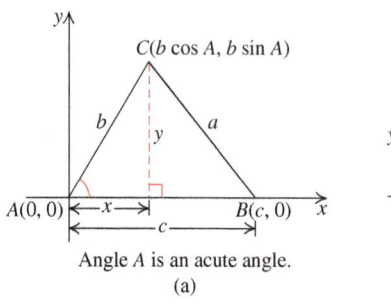

Angle A is an obtuse angle.
(b)

FIGURE 5.12 Two cases for an angle A

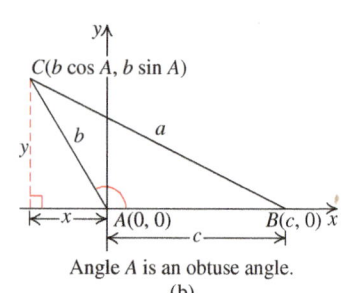

Label the coordinates of the vertices as shown in Figure 5.12. In Figures 5.12(a) and (b), the point $C(x, y)$ on the terminal side of angle A has coordinates $(b \cos A, b \sin A)$, which are found using the cosine and sine definitions.

$$\cos A = \frac{x}{b} \qquad \sin A = \frac{y}{b}$$
$$x = b \cos A \qquad y = b \sin A$$

The point B has coordinates $(c, 0)$. Applying the distance formula to the line segment joining $B(c, 0)$ and $C(b \cos A, b \sin A)$, we have

$a = d(C, B)$	
$a^2 = [d(C, B)]^2$	Square both sides.
$a^2 = (b \cos A - c)^2 + (b \sin A - 0)^2$	Distance formula
$a^2 = b^2 \cos^2 A - 2bc \cos A + c^2 + b^2 \sin^2 A$	Expand binomial.
$a^2 = b^2(\sin^2 A + \cos^2 A) + c^2 - 2bc \cos A$	Regroup terms.
$a^2 = b^2 + c^2 - 2bc \cos A$	$\sin^2 A + \cos^2 A = 1$

The last equation is one of the forms of the Law of Cosines. Similarly, by placing the vertex B and then the vertex C at the origin, we obtain the other two forms. Notice that the Law of Cosines becomes the Pythagorean Theorem if the included angle is $90°$ because $\cos 90° = 0$.

Solving SAS Triangles

Let's solve SAS triangles (Case 3 in Section 3.1).

PROCEDURE
IN ACTION

EXAMPLE 9 Solving the SAS Triangles

OBJECTIVE

Solve triangles in which the measures of two sides and the included angle are known.

Step 1 Use the appropriate form of the Law of Cosines to find the side opposite the given angle.

EXAMPLE

Solve triangle ABC with $a = 15$ inches, $b = 10$ inches, and $C = 60°$. Round each answer to the nearest tenth.

1. Find side c opposite angle C.

$c^2 = a^2 + b^2 - 2ab \cos C$	The Law of Cosines
$c^2 = (15)^2 + (10)^2 - 2(15)(10) \cos 60°$	Substitute values.
$c^2 = 225 + 100 - 2(15)(10)\left(\frac{1}{2}\right)$	$\cos 60° = \frac{1}{2}$
$c^2 = 175$	Simplify.
$c = \sqrt{175} \approx 13.2$	Use a calculator.

Step 2 Use the Law of Sines to find the angle opposite the shorter of the two given sides. Note that this angle is always an acute angle.

2. Find angle B.

$\dfrac{\sin B}{b} = \dfrac{\sin C}{c}$	The Law of Sines
$\sin B = \dfrac{b \sin C}{c}$	Multiply both sides by b.
$\sin B = \dfrac{10 \sin 60°}{\sqrt{175}}$	Substitute values.
$B = \sin^{-1}\left(\dfrac{10 \sin 60°}{\sqrt{175}}\right) \approx 40.9°$	$0° < B < 90°$

(continued)

Step 3 Use the angle sum formula to find the third angle.

3. $A \approx 180° - 60° - 40.9° \approx 79.1°$ $A + B + C = 180°$

Step 4 Write the solution.

4. | $A \approx 79.1°$ | $a = 15$ inches |
|---|---|
| $B \approx 40.9°$ | $b = 10$ inches |
| $C = 60°$ | $c \approx 13.2$ inches |

Practice Problem 9 Solve triangle ABC with $c = 25$, $a = 15$, and $B = 60°$. Round each answer to the nearest tenth.

Solving SSS Triangles

Let's solve SSS triangles (Case 4 in Section 3.1).

PROCEDURE
IN ACTION

EXAMPLE 10 Solving SSS Triangles

OBJECTIVE
Solve triangles in which the measures of the three sides are known.

EXAMPLE
Solve triangle ABC with $a = 3.1$ feet, $b = 5.4$ feet, and $c = 7.2$ feet. Round each answer to the nearest tenth.

Step 1 Use the Law of Cosines to find the angle opposite the longest side.

1. Because c is the longest side, we first find angle C.

$c^2 = a^2 + b^2 - 2ab \cos C$ The Law of Cosines
$2ab \cos C = a^2 + b^2 - c^2$ Add $2ab \cos C - c^2$ to both sides.
$\cos C = \dfrac{a^2 + b^2 - c^2}{2ab}$ Solve for cos C.
$\cos C = \dfrac{(3.1)^2 + (5.4)^2 - (7.2)^2}{2(3.1)(5.4)}$ Substitute values.
$\cos C \approx -0.39$ Use a calculator.
$C \approx \cos^{-1}(-0.39) \approx 113.0°$ $0° < C < 180°$

Step 2 Use the Law of Sines to find either of the two remaining acute angles.

2. Find angle B.

$\dfrac{\sin B}{b} = \dfrac{\sin C}{c}$ The Law of Sines
$\sin B = \dfrac{b \sin C}{c}$ Multiply both sides by b.
$B = \sin^{-1}\left(\dfrac{b \sin C}{c}\right)$ $0° < B < 90°$
$B = \sin^{-1}\left(\dfrac{5.4 \sin 113°}{7.2}\right)$ Substitute values.
$B \approx 43.7°$ Use a calculator.

Step 3 Use the angle sum formula to find the third angle.

3. $A \approx 180° - 43.7° - 113°$ $A + B + C = 180°$
$A \approx 23.3°$ Simplify.

Step 4 Write the solution.

4. | $A \approx 23.3°$ | $a = 3.1$ feet |
|---|---|
| $B = 43.7°$ | $b = 5.4$ feet |
| $C \approx 113.0°$ | $c = 7.2$ feet |

Practice Problem 10 Solve triangle ABC with $a = 4.5$, $b = 6.7$, and $c = 5.3$. Round each answer to the nearest tenth.

EXAMPLE 11 Solving an SSS Triangle (No Solution)

Solve triangle ABC with $a = 2$ meters, $b = 9$ meters, and $c = 5$ meters. Round each answer to the nearest tenth.

Solution

Step 1 We first find B, the angle opposite the longest side.

$$b^2 = c^2 + a^2 - 2ca \cos B \quad \text{The Law of Cosines}$$

$$\cos B = \frac{c^2 + a^2 - b^2}{2ca} \quad \text{Solve for } \cos B.$$

$$\cos B = \frac{5^2 + 2^2 - 9^2}{2(5)(2)} \quad \text{Substitute values.}$$

$$\cos B = -2.6 \quad \text{Simplify.}$$

RECALL

The Triangle Inequality states that in any triangle, the sum of the lengths of any two sides of a triangle is greater than the length of the third side.

Because the range of the cosine function is $[-1, 1]$, there is no angle B for which $\cos B = -2.6$. This means that a triangle with the given information cannot exist. Because $2 + 5 < 9$, you can also use the Triangle Inequality from geometry to see that there is no such triangle.

Practice Problem 11 Solve triangle ABC with $a = 2$ inches, $b = 3$ inches, and $c = 6$ inches.

SECTION 5.1 Exercises

Basic Concepts and Skills

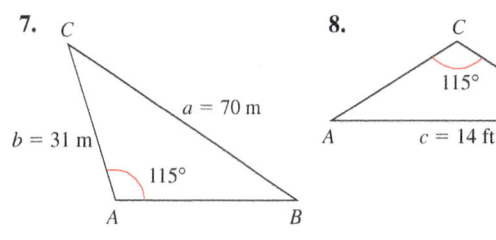

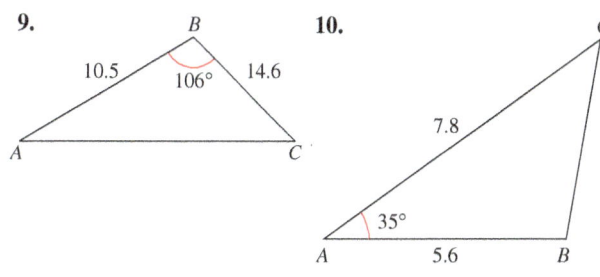

In Exercises 1–12, solve each triangle. Round each answer to the nearest tenth.

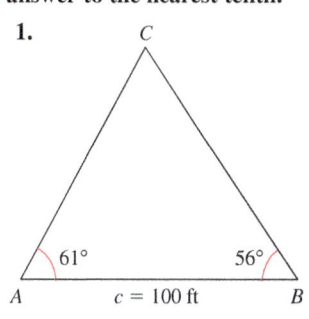

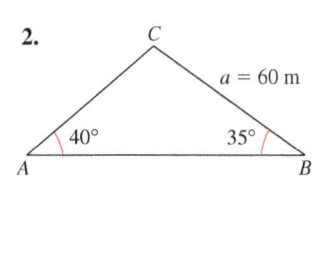

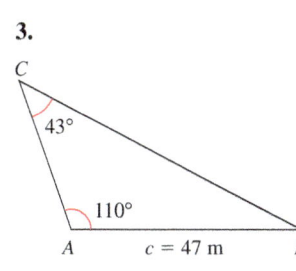

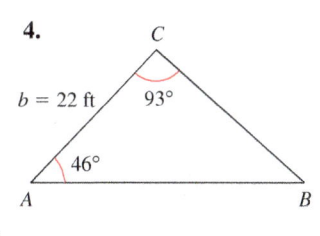

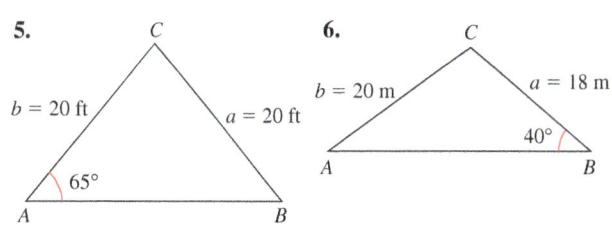

1. Triangle with $A = 61°$, $B = 56°$, $c = 100$ ft
2. Triangle with $A = 40°$, $B = 35°$, $a = 60$ m
3. Triangle with $A = 110°$, $C = 43°$, $c = 47$ m
4. Triangle with $A = 46°$, $C = 93°$, $b = 22$ ft
5. Triangle with $A = 65°$, $a = 20$ ft, $b = 20$ ft
6. Triangle with $B = 40°$, $a = 18$ m, $b = 20$ m
7. Triangle with $C = 115°$, $a = 70$ m, $b = 31$ m
8. Triangle with $C = 115°$, $a = 12$ ft, $c = 14$ ft
9. Triangle with $B = 106°$, sides 10.5 and 14.6
10. Triangle with $A = 35°$, side 5.6, side 7.8

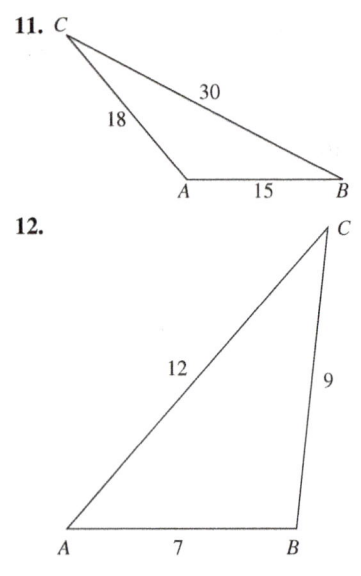

408 Chapter 5 Applications of Trigonometric Functions

11. [Triangle with vertices A, B, C; side from C to A = 18, side from C to B = 30, side AB = 15]

12. [Triangle with vertices A, B, C; side from A to C = 12, side from C to B = 9, side AB = 7]

In Exercises 13–24, solve triangle ABC. Round each answer to the nearest tenth.

13. $A = 40°, B = 35°, a = 100$ meters
14. $A = 80°, B = 20°, a = 100$ meters
15. $A = 46°, C = 55°, a = 75$ centimeters
16. $A = 35°, C = 98°, a = 75$ centimeters
17. $A = 35°, C = 47°, c = 60$ feet
18. $A = 44°, C = 76°, c = 40$ feet
19. $B = 43°, C = 67°, b = 40$ inches
20. $B = 95°, C = 35°, b = 100$ inches
21. $B = 110°, C = 46°, c = 23.5$ feet
22. $B = 67°, C = 63°, c = 16.8$ feet
23. $A = 35.7°, B = 45.8°, c = 30$ meters
24. $A = 64.5°, B = 54.3°, c = 40$ meters

In Exercises 25–44, solve each SSA triangle. Indicate whether the given measurements result in no triangle, one triangle, or two triangles. Solve each resulting triangle. Round each answer to the nearest tenth.

25. $A = 40°, a = 23, b = 20$
26. $A = 36°, a = 30, b = 24$
27. $A = 30°, a = 25, b = 50$
28. $A = 60°, a = 20\sqrt{3}, b = 40$
29. $A = 40°, a = 10, b = 20$
30. $A = 62°, a = 30, b = 40$
31. $A = 95°, a = 18, b = 21$
32. $A = 110°, a = 37, b = 41$
33. $A = 100°, a = 40, b = 34$
34. $A = 105°, a = 70, b = 30$
35. $B = 50°, b = 22, c = 40$
36. $B = 64°, b = 45, c = 60$
37. $B = 46°, b = 35, c = 40$
38. $B = 32°, b = 50, c = 60$
39. $B = 97°, b = 27, c = 30$
40. $B = 110°, b = 19, c = 21$
41. $A = 42°, a = 55, c = 62$
42. $A = 34°, a = 6, c = 8$
43. $C = 40°, a = 3.3, c = 2.1$
44. $C = 62°, a = 50, c = 100$

In Exercises 45–60, solve each triangle ABC. Round each answer to the nearest tenth. All sides are measured in feet.

45. $a = 15, b = 9, C = 120°$
46. $a = 14, b = 10, C = 75°$
47. $b = 10, c = 12, A = 62°$
48. $b = 11, c = 16, A = 110°$
49. $c = 12, a = 15, b = 11$
50. $c = 16, a = 11, b = 13$
51. $a = 9, b = 13, c = 18$
52. $a = 14, b = 6, c = 10$
53. $a = 2.5, b = 3.7, c = 5.4$
54. $a = 4.2, b = 2.9, c = 3.6$
55. $b = 3.2, c = 4.3, A = 97.7°$
56. $b = 5.4, c = 3.6, A = 79.2°$
57. $c = 4.9, a = 3.9, B = 68.3°$
58. $c = 7.8, a = 9.8, B = 95.6°$
59. $a = 2.3, b = 2.8, c = 3.7$
60. $a = 5.3, b = 2.9, c = 4.6$

Applying the Concepts

61. Finding distance. Angela wants to find the distance from point A to her friend Carmen's house at point C on the other side of the river. She knows the distance from A to Betty's house at B is 540 feet. See the figure. The measurement of angles A and B are 57° and 46°, respectively. Calculate the distance from A to C. Round to the nearest foot.

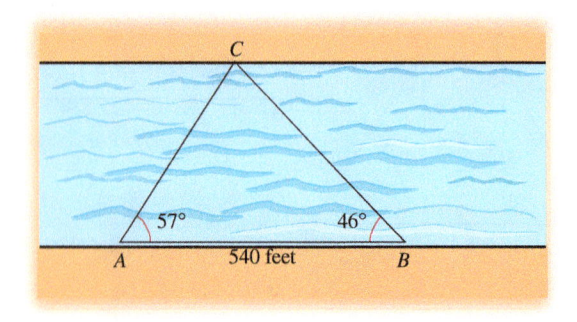

[Figure: Triangle ABC across a river with angle at A = 57°, angle at B = 46°, AB = 540 feet]

62. Height of a flagpole. Two surveyors stand 200 feet apart with a flagpole between them. Suppose the "transit" at each location is 5 feet high. The transit measures the angles of elevation of the top of the flagpole at the two locations to be 30° and 25°. Find the height of the flagpole. Round to the nearest foot.

63. Radio beacon. A ship sailing due east at the rate of 16 miles per hour records the bearing of a radio beacon at N 36.5° E. Two hours later the bearing of the same beacon is N 55.7° W. Round each answer to the nearest tenth of a mile.
  a. How far is the ship from the beacon the second time?
  b. How close to the beacon did the ship pass?

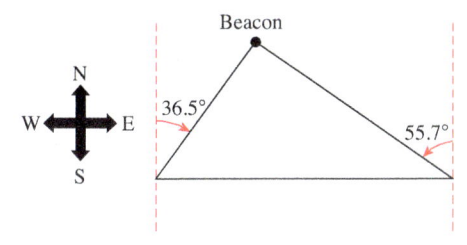

64. Lighthouse. A boat sailing due north at the rate of 14 miles per hour records the bearing of a lighthouse as N 8.4° E. Two hours later the bearing of the same lighthouse is N 30.9° E. Round each answer to the nearest tenth of a mile.
  a. How far is the ship from the lighthouse the second time?
  b. If the boat follows the same course and speed, how close to the lighthouse will it approach?

65. Distance between planets. The angle subtended at Earth by the lines joining Earth to Venus and the sun is 31°. If Venus is 67 million miles from the sun and Earth is 93 million miles from the sun, what is the distance (to the nearest million) between Earth and Venus? (*Hint:* There are two possible answers.)

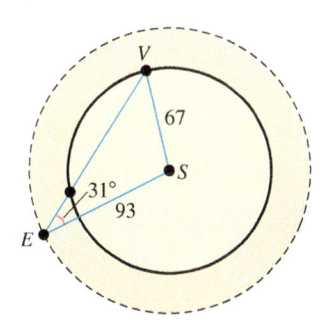

66. Navigation. A point on an island is located 24 miles southwest of a dock. A ship leaves the dock at 1 P.M. traveling west at 12 mph. At what time(s) to the nearest minute is the ship 20 miles from the point?

67. Hikers. Two hikers, Sonia and Tony, leave the same point at the same time. Sonia walks due east at the rate of 3 miles per hour, and Tony walks 45° northeast at the rate of 4.3 miles per hour. How far apart are the hikers after three hours?

68. Distance between ships. Two ships leave the same port—Ship A at 1.00 P.M. and ship B at 3:30 P.M. Ship A sails on a bearing of S 37° E at 18 miles per hour, and B sails on a bearing of N 28° E at 20 miles per hour. How far apart are the ships at 8.00 P.M.?

69. Navigation. A ship is traveling due north. At two different points A and B, the navigator of the ship sites a lighthouse at the point C, as shown in the accompanying figure.
  a. Determine the distance from B to C.
  b. How much farther due north must the ship travel to reach the point closest to the lighthouse?

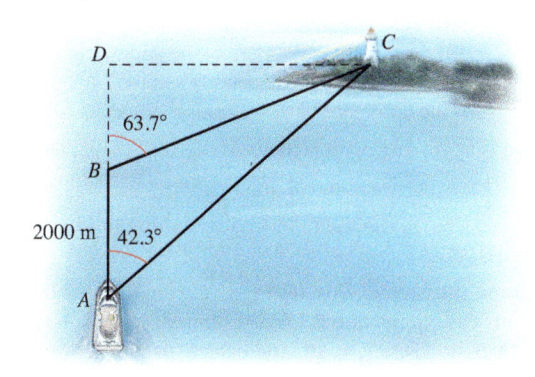

70. Tangential circles. Three circles of radii 1.2 inches, 2.2 inches, and 3.1 inches are tangent to each other externally. Find the angles of the triangle formed by joining the centers of the circles.

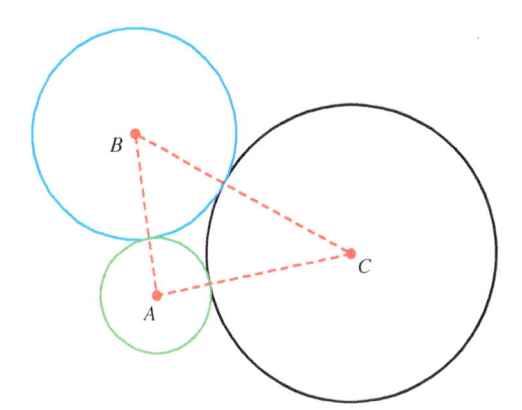

Critical Thinking / Discussion / Writing

71. Mollweide's formula. In a triangle ABC, prove that

$$\frac{b-c}{a} = \frac{\sin\frac{B-C}{2}}{\cos\frac{A}{2}}.$$

(*Hint:* Let $\frac{a}{\sin A} = \frac{b}{\sin B} = \frac{c}{\sin C} = k$. Write an expression for $\frac{b-c}{a}$ in terms of sines and then use Half-Angle and Sum-to-Product formulas (see Exercise 71, page 387. Note that because the formula contains all six parts of a triangle, it can be used as a check for the solutions of a triangle.)

72. Mollweide's formula. In a triangle ABC, prove that

$$\frac{b+c}{a} = \frac{\cos\frac{B-C}{2}}{\sin\frac{A}{2}}.$$

73. Law of Tangents. In a triangle ABC, prove that

$$\frac{b-c}{b+c} = \frac{\tan\left(\frac{B-C}{2}\right)}{\tan\left(\frac{B+C}{2}\right)}.$$

74. Angle measure. Two sides of a triangle are $\sqrt{3}+1$ feet and $\sqrt{3}-1$ feet, and the measure of the angle between these sides is $60°$. Use the Law of Tangents (Exercise 73) to find the measure of the difference of the remaining angles.

75. Solve triangle ABC with vertices $A(-2,1)$, $B(5,3)$, and $C(3,6)$.

In Exercises 76–78, use the following technique to solve each triangle ABC:

You can use the Law of Cosines to solve triangle ABC in the ambiguous case of Section 5.1. For example, to solve triangle ABC with $B = 150°$, $b = 10$, and $c = 6$, use the Law of Cosines to write $b^2 = a^2 + c^2 - 2ac \cos B$.

Substitute values of b, c, and B in this equation to obtain a quadratic equation in a. Find the roots of this equation and interpret your results. Again use the Law of Cosines to find A. Then find $C = 180° - A - B$.

76. $B = 150°$, $b = 10$, and $c = 6$

77. $A = 30°$, $a = 6$, and $b = 10$

78. $A = 60°$, $a = 12$, and $c = 15$

Maintaining Skills

79. The length of a rectangle is three times its width. If the perimeter of the rectangle is 120 feet, find the dimensions of the rectangle.

80. The length of a rectangle is 48 inches, and its diagonal is 50 inches. Find the width of the rectangle.

81. Convert $162°$ to radians.

82. Find the area of the sector subtended by a central angle of $70°$ in a circle of radius 3 feet.

SECTION 5.2

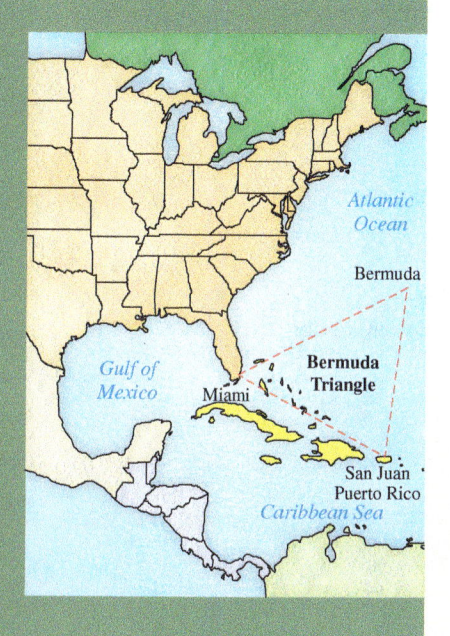

Areas of Polygons Using Trigonometry

BEFORE STARTING THIS SECTION, REVIEW

1. Congruent triangles (Section 5.1, page 395)
2. Solving SSA triangles (Section 5.1, page 399)

OBJECTIVES

1. Use geometry formulas.
2. Find the area of SAS triangles.
3. Find the area of AAS and ASA triangles.
4. Find the area of SSS triangles.

◆ The Bermuda Triangle

The Bermuda Triangle, or Devil's Triangle, is the expanse of the Atlantic Ocean between Florida, Bermuda, and Puerto Rico. This mysterious stretch of sea has an unusually high occurrence of disappearing ships and planes.

For example, on Halloween in 1991, pilot John Verdi and his copilot were flying a Grumman Cougar jet over the triangle. They radioed the nearest tower to get permission to increase their altitude. The tower agreed and watched as the jet began the ascent and then disappeared off the radar. The jet didn't fly out of range of the radar, didn't descend, and didn't radio a mayday (distress call). It just vanished, and the plane and crew were never recovered. Popular culture has attributed such disappearances to extraterrestrial forces. However, documented evidence indicates that the nature of disappearances in the region is similar to that in any region of the ocean. In Exercise 33, you are asked to find the area of the Bermuda Triangle.

1 Use geometry formulas.

Geometry Formulas

Although the area of a rectangle is just the product of its length and width, the general concept of area (and the related concepts of perimeter and volume) is developed with the use of calculus. Fortunately, the formulas for calculating these quantities require only algebra. Figure 5.13 lists some useful formulas from geometry.

Formulas for area (K), circumference (C), perimeter (P), volume (V), and surface area (S)

Plane Figures

Square
$K = s^2$
$P = 4s$

Rectangle
$K = lw$
$P = 2l + 2w$

Triangle
$K = \frac{1}{2}bh$
$P = a + b + c$

Trapezoid
$K = \frac{1}{2}h(a + b)$
$P = a + b + c + d$

Circle
$K = \pi r^2$
$C = 2\pi r$

Solids

Cube
$V = s^3$
$S = 6s^2$

Rectangular Solid
$V = lwh$
$S = 2(lw + wh + hl)$

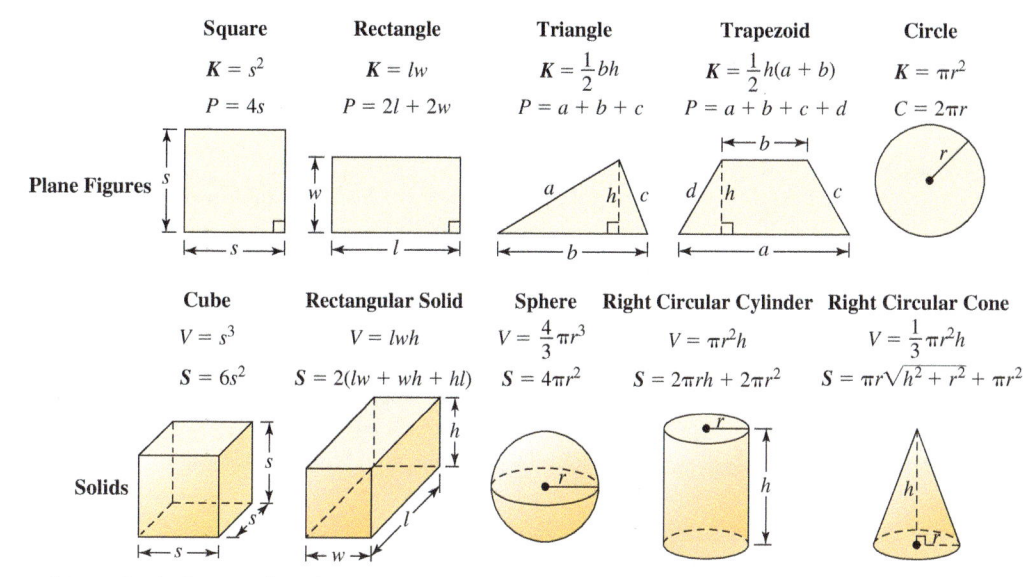

Sphere
$V = \frac{4}{3}\pi r^3$
$S = 4\pi r^2$

Right Circular Cylinder
$V = \pi r^2 h$
$S = 2\pi rh + 2\pi r^2$

Right Circular Cone
$V = \frac{1}{3}\pi r^2 h$
$S = \pi r \sqrt{h^2 + r^2} + \pi r^2$

FIGURE 5.13 Geometry formulas

EXAMPLE 1 Finding Perimeter and Area

Figure 5.14 shows a rectangle with a semicircle on the top. Find

a. its perimeter.   **b.** the shaded area enclosed by the figure.

Solution

a. The figure consists of three sides of a rectangle and a semicircle. Because the radius shown in the figure is 2 ft, its diameter is $2(2) = 4$ ft. So the sum of the lengths of the three sides of the rectangle is $6 + 4 + 6 = 16$ ft. The circumference (perimeter) of the semicircle is $\frac{1}{2}(\text{circumference}) = \frac{1}{2}(2\pi r) = \frac{1}{2}(2\pi \cdot 2) = 2\pi$ ft.

The perimeter of the figure $= 16 \text{ ft} + 2\pi \text{ ft}$

$\qquad\qquad = (16 + 2\pi) \text{ ft}$ Exact answer

$\qquad\qquad \approx 22.2832 \text{ ft}$ Use a calculator.

b. Area enclosed by the figure

$\qquad = $ Area of the rectangle $+$ area of the semicircle

$\qquad = l \cdot w + \frac{1}{2}(\pi r^2)$

$\qquad = (4)(6) + \frac{1}{2}[\pi(2)^2]$ $l = 4, w = 6, r = 2$

$\qquad = (24 + 2\pi) \text{ ft}^2$ Simplify. Exact answer

$\qquad \approx 30.2832 \text{ ft}^2$ Use a calculator.

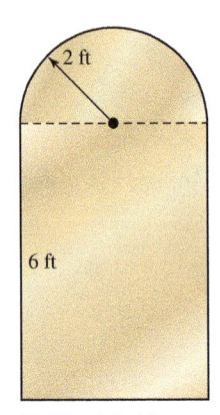

FIGURE 5.14

Practice Problem 1 One side of a rectangular plot is 48 feet, and its diagonal is 50 feet. Find

a. its perimeter.   **b.** its area.

We recall from Figure 5.13 the following formula.

Area of a Triangle

The area K of a triangle is

$$K = \frac{1}{2}(\text{base})(\text{height})$$

or

$$K = \frac{1}{2}bh,$$

where b is the base and h is the height (the length of the altitude to the base).

If a base (b) and the height (h) of a triangle are given, then we use this formula to calculate the area of the triangle. In this section, we discover alternative formulas when both b and h are not known.

2 Find the area of SAS triangles.

Area of SAS Triangles

We first derive a formula for the area of a triangle in which two sides and the included angle are known.

Let ABC be a triangle with known sides b and c and included angle $A = \theta$. Let h be the altitude from B to the side b.

FIGURE 5.15 Acute angle A

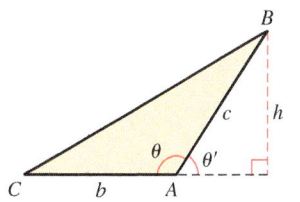

FIGURE 5.16 Obtuse angle A

If $A = \theta$ is an acute angle in a triangle (see Figure 5.15), then $\sin A = \dfrac{h}{c}$; so $h = c \sin A$. The area of this triangle is given by

$$K = \frac{1}{2}bh = \frac{1}{2}bc \sin A \qquad \text{Replace } h \text{ with } c \sin A.$$

The angle $A = \theta$ shown in Figure 5.16 is an obtuse angle. Here $\theta' = 180° - \theta$ and $\sin \theta = \sin \theta' = \dfrac{h}{c}$.

So

$$h = c \sin \theta' = c \sin \theta = c \sin A$$

$$K = \frac{1}{2}bh = \frac{1}{2}bc \sin A \qquad \text{Replace } h \text{ with } c \sin A.$$

In both cases (A is acute or obtuse), the area K of the triangle is $\dfrac{1}{2}bc \sin A$. By dropping altitudes from A and C, we have the following result.

Area of an SAS Triangle

The area K of a triangle ABC with sides a, b, and c is

$$K = \frac{1}{2}bc \sin A \qquad K = \frac{1}{2}ca \sin B \qquad K = \frac{1}{2}ab \sin C.$$

In words: The area K of a triangle is one-half the product of two of its sides and the sine of the included angle.

EXAMPLE 2 Finding the Area of a Triangle

Find the area of the triangle ABC in Figure 5.17.

Solution

We are given the angle of measure $62°$ between the sides of lengths 36 feet and 29 feet. The area K of the triangle ABC is given by

$$K = \frac{1}{2}bc \sin \theta \qquad \text{Area formula}$$

$$= \frac{1}{2}(36)(29) \sin 62° \qquad b = 36, c = 29, \theta = 62°$$

$$\approx 460.9 \text{ square feet} \qquad \text{Use a calculator.}$$

FIGURE 5.17

Practice Problem 2 Find the area of triangle ABC assuming that the lengths of the sides AC and BC are 27 feet and 38 feet, respectively, and the measure of the angle between these sides is $47°$.

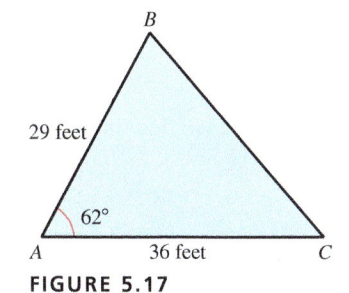

EXAMPLE 3 The Ambiguous Case

A triangle ABC has area $K = 20 \text{ ft}^2$. The lengths of two of its sides are $a = 10$ ft and $b = 8$ ft. Find the measure of the angle between these sides.

Solution

Let θ be the angle between the sides of lengths a and b. Then

$$\frac{1}{2}ab \sin \theta = K \quad \text{Area formula}$$

$$\frac{1}{2}(10)(8) \sin \theta = 20 \quad a = 10, b = 8, K = 20$$

$$\sin \theta = \frac{1}{2} \quad \text{Solve for } \sin \theta \text{ and simplify.}$$

Then $\theta = \sin^{-1}\left(\frac{1}{2}\right) = 30°$ is an angle between the sides of lengths a and b. However, because

$$\sin(180° - \theta) = \sin \theta, \quad \text{See page 309, Exercise 78.}$$

we have

$$\sin 150° = \sin(180° - 30°) = \sin 30° = \frac{1}{2}.$$

So the angle between the sides of lengths a and b is also $150°$. Figure 5.18 shows two possible triangles with the given information.

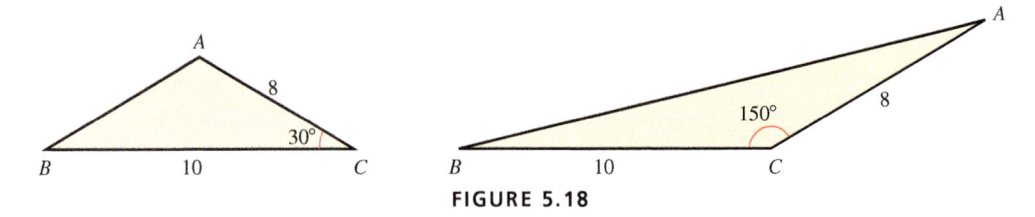

FIGURE 5.18

Practice Problem 3 Repeat Example 3 with $K = 6, a = 4,$ and $b = 3$.

3 Find the area of AAS and ASA triangles.

Area of AAS and ASA Triangles

Suppose we are given two angles A and C and the side a of triangle ABC. We find the third angle B by the angle sum formula. So

$$B = 180° - A - C.$$

We find b by the Law of Sines.

$$\frac{b}{\sin B} = \frac{a}{\sin A}$$

$$b = \frac{a \sin B}{\sin A} \quad \text{Multiply both sides by } \sin B.$$

The area, K, of triangle ABC is

$$K = \frac{1}{2}ab \sin C \quad \text{SAS formula}$$

$$= \frac{1}{2}a\left(\frac{a \sin B}{\sin A}\right) \sin C \quad \text{Replace } b \text{ with } \frac{a \sin B}{\sin A}.$$

$$= \frac{a^2 \sin B \sin C}{2 \sin A}.$$

We obtain similar results if any two angles and the side b or the side c is given.

Area of AAS and ASA Triangles

The area K of an AAS triangle ABC with side a, b, or c is:

$$K = \frac{a^2 \sin B \sin C}{2 \sin A}, \quad K = \frac{b^2 \sin C \sin A}{2 \sin B}, \quad \text{or} \quad K = \frac{c^2 \sin A \sin B}{2 \sin C}.$$

EXAMPLE 4   **Area of an AAS Triangle**

Find the area of triangle ABC with

$$B = 36°, C = 69°, \text{ and } b = 15 \text{ ft}.$$

Solution

First, we find the third angle of the triangle. We have

$$\begin{aligned} A &= 180° - B - C &&\text{Angle sum formula} \\ &= 180° - 36° - 69° &&\text{Substitute for } B \text{ and } C. \\ &= 75°. \end{aligned}$$

Because we are given side b, we use the area formula for an AAS triangle that contains side b. Then

$$\begin{aligned} K &= \frac{b^2 \sin C \sin A}{2 \sin B} &&\text{Area of an AAS triangle} \\ &= \frac{(15)^2 \sin 69° \sin 75°}{2 \sin 36°} &&\text{Substitute values for } b, C, A, \text{ and } B. \\ &\approx 172.6 \text{ ft}^2 &&\text{Use a calculator.} \end{aligned}$$

Practice Problem 4 Find the area of triangle ABC with $A = 63°$, $B = 74°$, and $c = 18$ in.

4 Find the area of SSS triangles.

Area of SSS Triangles

The Law of Cosines can be used to derive a formula for the area of a triangle if the lengths of the three sides (SSS triangles) are known. The formula is called **Heron's formula**.

HERON'S FORMULA FOR SSS TRIANGLES

The area K of an SSS triangle with sides of lengths a, b, and c is given by:

$$K = \sqrt{s(s-a)(s-b)(s-c)},$$

where $s = \frac{1}{2}(a + b + c)$ is the **semiperimeter**.

The derivation of Heron's formula is given in Exercise 37.

EXAMPLE 5   **Using Heron's Formula**

Find the area of triangle ABC with $a = 29$ inches, $b = 25$ inches, and $c = 40$ inches. Round your answer to the nearest tenth.

Solution

We first find s:

$$s = \frac{a + b + c}{2}$$

$$= \frac{29 + 25 + 40}{2} = 47$$

Area of the triangle $= \sqrt{s(s-a)(s-b)(s-c)}$ Heron's formula
$= \sqrt{47(47-29)(47-25)(47-40)}$ Substitute values.
≈ 360.9 square inches   Use a calculator.

Practice Problem 5 Find the area of triangle ABC with $a = 11$ meters, $b = 17$ meters, and $c = 20$ meters. Round your answer to the nearest tenth.

EXAMPLE 6   Using Heron's Formula

A triangular swimming pool has side lengths 23 feet, 17 feet, and 26 feet. How many gallons of water will fill the pool to a depth of 5 feet? Round your answer to the nearest whole number.

Solution

To calculate the volume of water, we first calculate the area of the triangular surface. We have $a = 23$, $b = 17$, and $c = 26$. So $s = \frac{1}{2}(a + b + c) = 33$.

By Heron's formula, the area K of the triangular surface is

$K = \sqrt{s(s-a)(s-b)(s-c)}$
$= \sqrt{33(33-23)(33-17)(33-26)}$ Substitute values for a, b, c, and s.
≈ 192.2498 square feet.

The volume of water = surface area × depth
$\approx 192.2498 \times 5$
≈ 961.25 cubic feet

One cubic foot contains approximately 7.5 gallons of water. So, $961.25 \times 7.5 \approx 7209$ gallons of water will fill the pool.

Practice Problem 6 Repeat Example 6, assuming that the swimming pool has side lengths of 25 feet, 30 feet, and 33 feet and the depth of the pool is 5.5 feet.

SECTION 5.2   Exercises

Basic Concepts and Skills

In Exercises 1–8, find the area of each SAS triangle ABC. Round your answers to the nearest tenth.

1. $A = 57°$, $b = 30$ in., $c = 52$ in.
2. $B = 110°$, $a = 20$ cm, $c = 27$ cm
3. $C = 46°$, $a = 15$ km, $b = 22$ km
4. $A = 146.7°$, $b = 16.7$ ft, $c = 18$ ft
5. $B = 38.6°$, $a = 12$ mm, $c = 16.7$ mm
6. $B = 112.5°$, $a = 151.6$ ft, $c = 221.8$ ft
7. $C = 107.3°$, $a = 271$ ft, $b = 194.3$ ft
8. $A = 131.8°$, $b = 15.7$ mm, $c = 18.2$ mm

In Exercises 9–16, find the area of each AAS or ASA triangle ABC. Round your answers to the nearest tenth.

9. $a = 16$ ft, $B = 57°$, $C = 49°$
10. $a = 12$ ft, $A = 73°$, $C = 64°$
11. $b = 15.3$ yd, $A = 64°$, $B = 38°$
12. $b = 10$ m, $B = 53.4°$, $C = 65.6°$

13. $c = 16.3$ cm, $B = 55°$, $C = 37.5°$
14. $c = 20.5$ ft, $A = 64°$, $B = 84.2°$
15. $a = 65.4$ ft, $A = 62°15'$, $B = 44°30'$
16. $b = 24.3$ m, $B = 56°18'$, $C = 37°36'$

In Exercises 17–24, find the area of each SSS triangle ABC. Round your answers to the nearest tenth.

17. $a = 2, b = 3, c = 4$
18. $a = 50, b = 100, c = 130$
19. $a = 50, b = 50, c = 75$
20. $a = 100, b = 100, c = 125$
21. $a = 7.5, b = 4.5, c = 6.0$
22. $a = 8.5, b = 5.1, c = 4.5$
23. $a = 3.7, b = 5.1, c = 4.2$
24. $a = 9.8, b = 5.7, c = 6.5$

In Exercises 25–28, find an angle θ between the sides a and b of a triangle ABC with given area K.

25. $K = 12, a = 6, b = 8$
26. $K = 6, a = 4, b = 2\sqrt{3}$
27. $K = 30, a = 12, b = 5$
28. $K = 15, a = 5\sqrt{2}, b = 6$

In Exercises 29–32, use the following figure of a parallelogram (not drawn to scale).

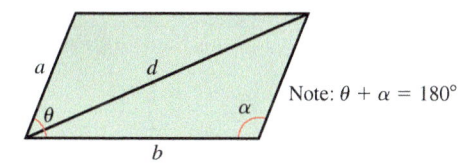

Note: $\theta + \alpha = 180°$

29. Show that the area of the parallelogram is $ab \sin \theta = ab \sin \alpha$.
30. Find the area of the parallelogram if $a = 8$ cm, $b = 13$ cm, and $\theta = 60°$.
31. Find the area of the parallelogram if $a = 12$ cm, $b = 16$ cm, and $d = 22$ cm.
32. If the area of parallelogram is 10 cm² with $b = 4$ cm and $\alpha = \dfrac{5\pi}{6}$, find a.

Applying the Concepts

33. **Area of the Bermuda Triangle.** The angle formed by the lines of sight from Fort Lauderdale, Florida, to Bermuda and to San Juan, Puerto Rico, is approximately 67°. The distance from Fort Lauderdale to Bermuda is 1026 miles, and the distance from Fort Lauderdale to San Juan is 1046 miles. Find the area of the Bermuda Triangle, which is formed with those cities as vertices.

34. **Landscaping.** A triangular region between the three streets shown in the figure will be landscaped for $30 a square foot. How much will the landscaping cost?

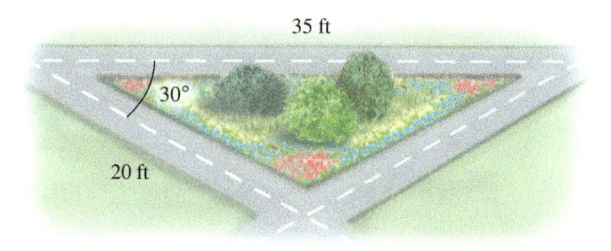

35. **Real estate.** You have inherited a commercial triangular-shaped lot of side lengths 400 feet, 250 feet, and 274 feet. The neighboring property is selling for $1 million dollars per acre. How much is your lot worth? (*Hint:* 1 acre = 43,560 square feet)

36. **Swimming pool.** Find the number of gallons of water in a triangular swimming pool with sides 11 feet, 16 feet, and 19 feet and a depth of 5 feet. Round your answer to the nearest whole number. (Recall: Approximately 7.5 gallons of water are in 1 cubic foot.)

Critical Thinking / Discussion / Writing

37. Derive Heron's formula by justifying each step.

In a triangle ABC,

$$K = \frac{1}{2}ab \sin C$$
$$4K^2 = a^2b^2(1 - \cos^2 C)$$
$$4K^2 = a^2b^2\left[1 - \left(\frac{a^2 + b^2 - c^2}{2ab}\right)^2\right]$$
$$16K^2 = 4a^2b^2 - (a^2 + b^2 - c^2)^2$$
$$16K^2 = [2ab + (a^2 + b^2 - c^2)][2ab - (a^2 + b^2 - c^2)]$$
$$16K^2 = [(a + b)^2 - c^2][c^2 - (a - b)^2]$$
$$K^2 = \left(\frac{a + b + c}{2}\right)\left(\frac{a + b - c}{2}\right)\left(\frac{c + a - b}{2}\right)\left(\frac{c + b - a}{2}\right)$$
$$K = \sqrt{s(s - a)(s - b)(s - c)}, \quad s = \frac{1}{2}(a + b + c)$$

38. In a triangle ABC, show that

$$\sin A = \frac{2}{bc}\sqrt{s(s - a)(s - b)(s - c)}.$$

39. Use Heron's formula to show that the area of an equilateral triangle whose sides have length a is $K = \dfrac{\sqrt{3}}{4}a^2$.

40. A **segment** of a circle is a part of a circle bounded by a chord and the arc cut by the chord. See the shaded region in the figure.

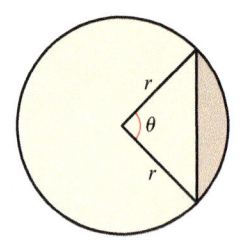

   a. Show that the area of the segment of a circle of radius r subtended (formed) by a central angle of θ radians is $\frac{1}{2}r^2(\theta - \sin\theta)$.
   b. Find the area of the segment subtended by a central angle of $60°$ in a circle of radius 6 inches.

41. **Circumscribed circle.** Perpendicular bisectors of the sides of a triangle meet at a point O, and the distance from O to each of the vertices is the same number (say, R). The circle with center at O and radius R is called the **circumscribed circle** of the triangle. See the figure.

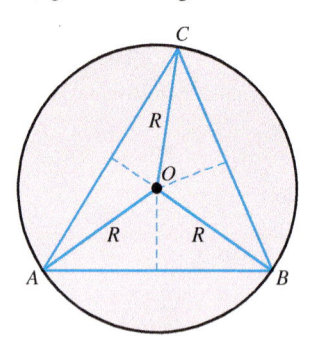

   Show that the radius R of the circumscribed circle of triangle ABC is given by
   $$R = \frac{a}{2\sin A} = \frac{b}{2\sin B} = \frac{c}{2\sin C}.$$
   (*Hint:* Use a theorem of geometry: $\angle BOC = 2\angle A$.)

42. Use Exercises 38 and 41 to show that
   $$R = \frac{abc}{4K},$$
   where K is the area of the triangle ABC.

43. **Inscribed circle.** The bisectors of the angles of a triangle meet at a point I, and the perpendicular distance from I to each of the sides is the same number (say, r). The circle with center I and radius r is called the **inscribed circle** of the triangle. See the figure.

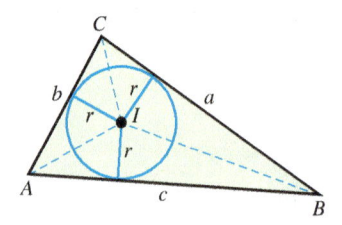

   Show that the radius r of the inscribed circle of triangle ABC is given by
   $$r = \frac{\sqrt{s(s-a)(s-b)(s-c)}}{s}.$$
   (*Hint:* Show that $rs = K$.)

44. From Exercises 42 and 43, conclude that
   $$rR = \frac{abc}{2(a+b+c)}.$$

45. In triangle ABC, let p_1, p_2, and p_3 be the lengths of the perpendiculars from the vertices A, B, and C to the sides a, b, and c, respectively. Show that
   $$\frac{1}{p_1} + \frac{1}{p_2} + \frac{1}{p_3} = \sqrt{\frac{s}{(s-a)(s-b)(s-c)}}.$$
   $$\left(\text{Hint: } \frac{1}{2}ap_1 = \frac{1}{2}bp_2 = \frac{1}{2}cp_3 = K\right)$$

46. From Exercises 44 and 45, conclude that
   $$\frac{1}{p_1} + \frac{1}{p_2} + \frac{1}{p_3} = \frac{1}{r}.$$

Maintaining Skills

In Exercises 47–49, test each equation for symmetry with respect to the x-axis, the y-axis, and the origin.

47. $x^3 + y^2 = 5$
48. $x^3 + y^3 - 3xy^2 = 0$
49. $2x^2 + 3y^2 = 6$
50. Find the x- and y-intercepts of the graph of $x^2 + y^2 + 6x - 4y - 12 = 0$.

SECTION 5.3 Polar Coordinates

BEFORE STARTING THIS SECTION, REVIEW

1. Coordinate plane (Section 1.1, page 52)
2. Degree and radian measure of angles (Section 4.1, pages 284–287)
3. Trigonometric functions (Section 4.2, page 301)

OBJECTIVES

1. Plot points using polar coordinates.
2. Convert points between polar and rectangular forms.
3. Convert equations between rectangular and polar forms.
4. Graph polar equations.

◆ Bionics

On large construction sites and maintenance projects, it is common to see mechanical systems that lift, dig, and scoop just like biological systems having arms, legs, and hands. The technique of solving mechanical problems with our knowledge of biological systems is called **bionics**. Bionics is used to make robotic devices for the remote-control handling of radioactive materials and the machines that are sent to the moon or to Mars to scoop soil samples, bring them into the space vehicle, perform analysis on them, and radio the results back to Earth. In Example 4, we use polar coordinates to find the reach of a robotic arm.

Up to now, we have used a rectangular coordinate system to identify points in the plane. In this section, we introduce another coordinate system, called a **polar coordinate system**, that allows us to describe many curves using rather simple equations. The polar coordinate system is believed to have been introduced by Jakob Bernoulli (1654–1705).

1 Plot points using polar coordinates.

Polar Coordinates

In a rectangular coordinate system, a point in the plane is described in terms of the horizontal and vertical directed distances from the origin. In a polar coordinate system, we draw a horizontal ray, called the **polar axis**, in the plane; its endpoint is called the **pole**. A point P in the plane is described by an ordered pair of numbers (r, θ), the **polar coordinates** of P described below.

Figure 5.19 shows the point $P(r, \theta)$ in the polar coordinate system, where r is the "directed distance" (see Figures 5.19 and 5.20) from the pole O to the point P and θ is a directed angle from the polar axis to the line segment OP. The angle θ may be measured in degrees or radians. As usual, positive angles are measured counterclockwise from the polar axis and negative angles are measured clockwise from the polar axis.

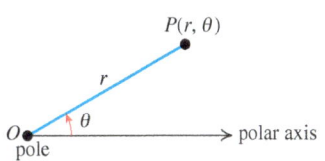

FIGURE 5.19 Polar coordinates (r, θ) of a point

Section 5.3 ■ Polar Coordinates **419**

420 Chapter 5 Applications of Trigonometric Functions

To locate a point with polar coordinates (5, 40°), we draw an angle $\theta = 40°$ counterclockwise from the polar axis. We then measure five units along the terminal side of θ. In Figure 5.20, we have plotted the points with polar coordinates (5, 40°), (3, 220°), (4, −30°), and (2, 110°).

> **SIDE NOTE**
>
> When plotting a point in polar coordinates, measure the angle θ first. Then measure a distance of r units from the pole along the terminal ray of θ. Notice that in working with polar coordinates, you use the second coordinate first.

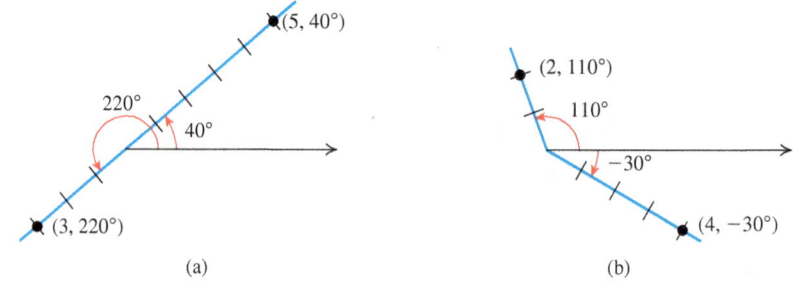

FIGURE 5.20 Plotting points using polar coordinates

Unlike rectangular coordinates, the polar coordinates of a point are not unique. For example, the polar coordinates (3, 60°), (3, 420°), and (3, −300°) represent the same point. See Figure 5.21. In general, if a point P has polar coordinates (r, θ), then for any integer n,

$$(r, \underbrace{\theta + n \cdot 360°}_{\theta \text{ in degrees}}) \quad \text{or} \quad (r, \underbrace{\theta + 2n\pi}_{\theta \text{ in radians}})$$

are also polar coordinates of P.

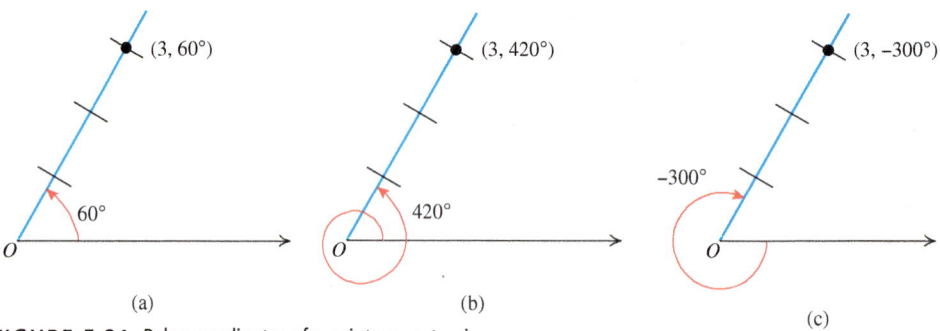

FIGURE 5.21 Polar coordinates of a point are not unique.

When equations are graphed in polar coordinates, it is convenient to allow r to be negative. That's why we defined r as *a directed distance*. If r is negative, then instead of the point being on the terminal side of the angle, it is $|r|$ units on the extension of the terminal side, which is the ray from the pole that extends in the direction opposite the terminal side of θ. For example, the point $P(-4, -30°)$ shown in Figure 5.22 is the same point as (4, 150°). Some other polar coordinates for the same point P are (−4, 330°), (4, −210°), and (4, 510°).

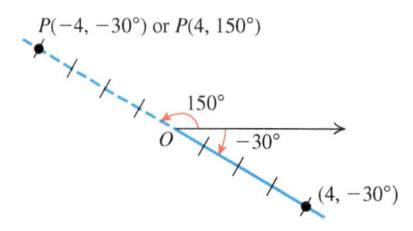

FIGURE 5.22 Negative r

Section 5.3 ■ Polar Coordinates 421

EXAMPLE 1 Finding Different Polar Coordinates

a. Plot the point P with polar coordinates $(3, 225°)$.

Find another pair of polar coordinates of P with the following properties.

b. $r < 0$ and $0° < \theta < 360°$

c. $r < 0$ and $-360° < \theta < 0°$

d. $r > 0$ and $-360° < \theta < 0°$

Solution

The point $P(3, 225°)$ is plotted by first drawing $\theta = 225°$ counterclockwise from the polar axis. Because $r = 3 > 0$, the point P is on the terminal side of the angle, three units from the pole. See Figure 5.23(a).

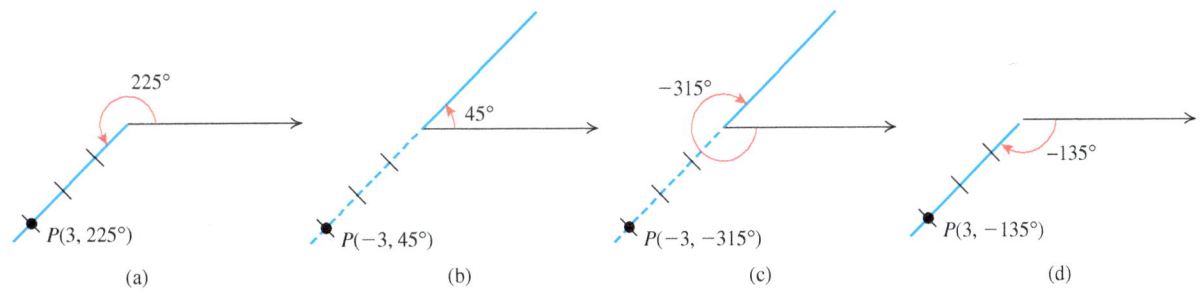

FIGURE 5.23 Plotting points

The answer to parts, **b**, **c**, and **d** are, respectively, $(-3, 45°)$, $(-3, -315°)$, and $(3, -135°)$. See Figures 5.23(b)–(d).

Practice Problem 1

a. Plot the point P with polar coordinates $(2, -150°)$.

Find another pair of polar coordinates of P with the following properties.

b. $r < 0$ and $0° < \theta < 360°$

c. $r > 0$ and $0° < \theta < 360°$

d. $r < 0$ and $-360° < \theta < 0°$

2 Convert points between polar and rectangular forms.

Converting Between Polar and Rectangular Forms

Frequently, we need to use polar and rectangular coordinates in the same problem. To do this, we let the positive x-axis of the rectangular coordinate system serve as the polar axis and the origin serve as the pole. Each point P then has polar coordinates (r, θ) and rectangular coordinates (x, y), as shown in Figure 5.24.

From Figure 5.24, we see the following relationships: $x^2 + y^2 = r^2$, $\sin \theta = \dfrac{y}{r}$, $\cos \theta = \dfrac{x}{r}$, and $\tan \theta = \dfrac{y}{x}$. These relationships hold whether $r > 0$ or $r < 0$, and they are used to convert a point's coordinates between rectangular and polar forms.

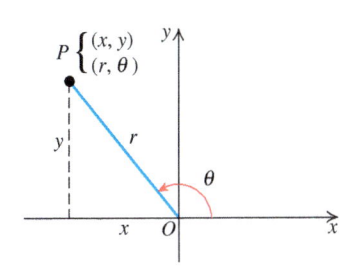

FIGURE 5.24 $P = (x, y)$ or $P = (r, \theta)$

CONVERTING FROM POLAR TO RECTANGULAR COORDINATES

To convert the polar coordinates (r, θ) of a point to rectangular coordinates, use the equations

$$x = r \cos \theta \quad \text{and} \quad y = r \sin \theta.$$

TECHNOLOGY CONNECTION

A graphing calculator can be used to check conversions from polar to rectangular coordinates. Choose the appropriate options from the ANGLE menu. The x- and y-coordinates must be calculated separately. The screen shown here is from a calculator set in *degree* mode.

```
P▶Rx(2,-30)
           1.732050808
√(3)
           1.732050808
P▶Ry(2,-30)
                    -1
■
```

EXAMPLE 2   Converting from Polar to Rectangular Coordinates

Convert the polar coordinates of each point to rectangular coordinates.

a. $(2, -30°)$ **b.** $\left(-4, \dfrac{\pi}{3}\right)$

Solution

a. $x = r \cos \theta$ $y = r \sin \theta$ Conversion equations

$x = 2 \cos(-30°)$ $y = 2 \sin(-30°)$ Replace r with 2 and θ with $-30°$.

$x = 2 \cos 30°$ $y = -2 \sin 30°$ $\cos(-\theta) = \cos \theta$,
$\sin(-\theta) = -\sin \theta$

$x = 2\left(\dfrac{\sqrt{3}}{2}\right) = \sqrt{3}$ $y = -2\left(\dfrac{1}{2}\right) = -1$ $\cos 30° = \dfrac{\sqrt{3}}{2}, \sin 30° = \dfrac{1}{2}$

The rectangular coordinates of $(2, -30°)$ are $(\sqrt{3}, -1)$.

b. $x = r \cos \theta$ $y = r \sin \theta$ Conversion equations

$x = -4 \cos \dfrac{\pi}{3}$ $y = -4 \sin \dfrac{\pi}{3}$ Replace r with -4 and θ with $\dfrac{\pi}{3}$.

$x = -4\left(\dfrac{1}{2}\right) = -2$ $y = -4\left(\dfrac{\sqrt{3}}{2}\right) = -2\sqrt{3}$ $\cos \dfrac{\pi}{3} = \dfrac{1}{2}, \sin \dfrac{\pi}{3} = \dfrac{\sqrt{3}}{2}$

The rectangular coordinates of $\left(-4, \dfrac{\pi}{3}\right)$ are $(-2, -2\sqrt{3})$.

Practice Problem 2 Convert the polar coordinates of each point to rectangular coordinates.

a. $(-3, 60°)$ **b.** $\left(2, -\dfrac{\pi}{4}\right)$

When converting from rectangular coordinates (x, y) of a point to polar coordinates (r, θ), remember that infinitely many representations are possible. We will find the polar coordinates (r, θ) of the point (x, y) that have $r > 0$ and $0° \leq \theta < 360°$ (or $0 \leq \theta < 2\pi$).

CONVERTING FROM RECTANGULAR TO POLAR COORDINATES

To convert the rectangular coordinates (x, y) of a point to polar coordinates:

1. Find the quadrant in which the given point (x, y) lies.
2. Use $r = \sqrt{x^2 + y^2}$ to find r.
3. Find θ by using $\tan \theta = \dfrac{y}{x}$ and choose θ so that it lies in the same quadrant as the point (x, y).

EXAMPLE 3   Converting from Rectangular to Polar Coordinates

Find polar coordinates (r, θ) of the point P with $r > 0$ and $0 \leq \theta < 2\pi$ whose rectangular coordinates are $(x, y) = (-2, 2\sqrt{3})$.

TECHNOLOGY CONNECTION

A graphing calculator can also be used to check conversions from rectangular to polar coordinates. Choose the appropriate options on the ANGLE menu. The r and θ coordinates must be calculated separately. The screen shown here is from a calculator set in *radian* mode

```
R▶Pr(-2,2√(3))
                    4
R▶Pθ(-2,2√(3))
         2.094395102
2π/3
         2.094395102
```

Solution

1. The point $P(-2, 2\sqrt{3})$ lies in quadrant II with $x = -2$ and $y = 2\sqrt{3}$.
2. $r = \sqrt{x^2 + y^2}$ Formula for computing r
   $= \sqrt{(-2)^2 + (2\sqrt{3})^2}$ Substitute values of x and y.
   $= \sqrt{4 + 12} = \sqrt{16} = 4$ Simplify.
3. $\tan \theta = \dfrac{y}{x}$ Definition of $\tan \theta$

   $\tan \theta = \dfrac{2\sqrt{3}}{-2} = -\sqrt{3}$ Substitute values and simplify.

Because the tangent is negative in quadrants II and IV, the equation $\tan \theta = -\sqrt{3}$ gives $\theta = \pi - \dfrac{\pi}{3} = \dfrac{2\pi}{3}$ or $\theta = 2\pi - \dfrac{\pi}{3} = \dfrac{5\pi}{3}$. Because P lies in quadrant II, we choose $\theta = \dfrac{2\pi}{3}$. So the required polar coordinates are $\left(4, \dfrac{2\pi}{3}\right)$.

Practice Problem 3 Find the polar coordinates of the point P with $r > 0$ and $0 \leq \theta < 2\pi$ whose rectangular coordinates are $(x, y) = (-1, -1)$.

EXAMPLE 4   Positioning a Robotic Hand

For the robotic arm of Figure 5.25, find the polar coordinates of the hand relative to the shoulder. Round all answers to the nearest tenth.

Solution

FIGURE 5.25 Robotic arm

FIGURE 5.26

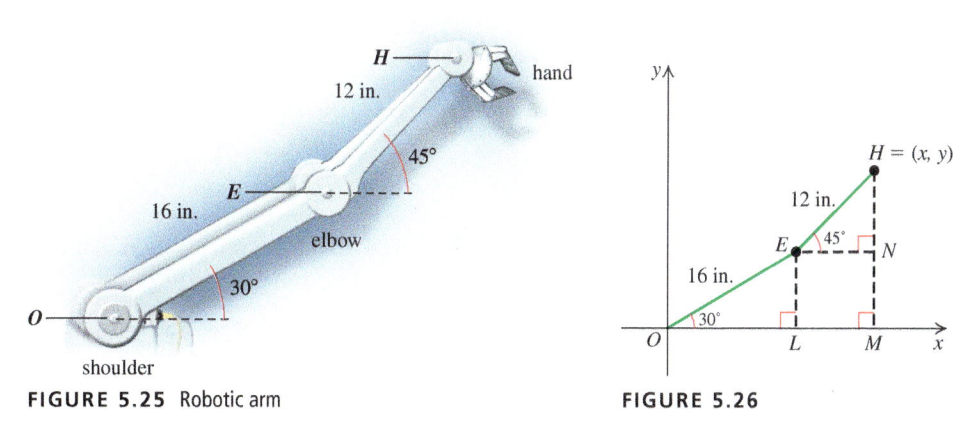

We need to find the polar coordinates (r, θ) of the hand H, with the shoulder O as the pole. We construct a geometric figure (see Figure 5.26) in the Cartesian coordinate system associated with Figure 5.25. From the right triangles OEL and EHN in Figure 5.26, we have

$$OL = 16 \cos 30°, LE = 16 \sin 30°, EN = 12 \cos 45°, \text{ and } NH = 12 \sin 45°.$$

Let (x, y) be the rectangular coordinates of H. Then

$$x = OM = OL + LM = OL + EN = 16 \cos 30° + 12 \cos 45° \text{ and}$$
$$y = MH = MN + NH = LE + NH = 16 \sin 30° + 12 \sin 45°.$$

We now convert the coordinates (x, y) of H to polar coordinates (r, θ).

$$\begin{aligned} r &= \sqrt{x^2 + y^2} & \text{Formula for } r \\ &= \sqrt{(16 \cos 30° + 12 \cos 45°)^2 + (16 \sin 30° + 12 \sin 45°)^2} & \text{Substitute values for } x \text{ and } y. \\ &\approx 27.8 \text{ inches} & \text{Use a calculator.} \\ \theta &= \tan^{-1}\left(\frac{y}{x}\right) & \text{Definition of inverse tangent} \\ &= \tan^{-1}\left(\frac{16 \sin 30° + 12 \sin 45°}{16 \cos 30° + 12 \cos 45°}\right) \approx 36.4° & \text{Substitute values and use a calculator.} \end{aligned}$$

The polar coordinates of H relative to the pole O are $(r, \theta) = (27.8, 36.4°)$.

Practice Problem 4 Repeat Example 4 with $OE = 15$ inches and $EH = 10$ inches.

3 Convert equations between rectangular and polar forms.

Converting Equations Between Rectangular and Polar Forms

An equation that has the rectangular coordinates x and y as variables is called a **rectangular** (or **Cartesian**) **equation**. Similarly, an equation where the polar coordinates r and θ are the variables is called a **polar equation**. Some examples of polar equations are

$$r = \sin \theta, \quad r = 1 + \cos \theta, \quad \text{and} \quad r = \theta.$$

To convert a rectangular equation to a polar equation, we simply replace x with $r \cos \theta$ and y with $r \sin \theta$ and then simplify where possible.

EXAMPLE 5 Converting an Equation from Rectangular to Polar Form

Convert the equation $x^2 + y^2 - 3x + 4 = 0$ to polar form.

Solution

$$\begin{aligned} x^2 + y^2 - 3x + 4 &= 0 & \text{Given equation} \\ (r \cos \theta)^2 + (r \sin \theta)^2 - 3(r \cos \theta) + 4 &= 0 & x = r \cos \theta, y = r \sin \theta \\ r^2 \cos^2 \theta + r^2 \sin^2 \theta - 3r \cos \theta + 4 &= 0 & (ab)^2 = a^2 b^2 \\ r^2(\cos^2 \theta + \sin^2 \theta) - 3r \cos \theta + 4 &= 0 & \text{Factor out } r^2 \text{ from the first two terms} \\ r^2 - 3r \cos \theta + 4 &= 0 & \cos^2 \theta + \sin^2 \theta = 1 \end{aligned}$$

The equation $r^2 - 3r \cos \theta + 4 = 0$ is a polar form of the given rectangular equation.

Practice Problem 5 Convert the equation $x^2 + y^2 - 2y + 3 = 0$ to polar form.

Converting an equation from polar to rectangular form will frequently require some ingenuity in order to use the substitutions:

$$r^2 = x^2 + y^2, \quad r \cos \theta = x, \quad r \sin \theta = y, \quad \text{and} \quad \tan \theta = \frac{y}{x}$$

Section 5.3 ■ Polar Coordinates   425

FIGURE 5.27 $x^2 + y^2 = 3^2$ or $r = 3$

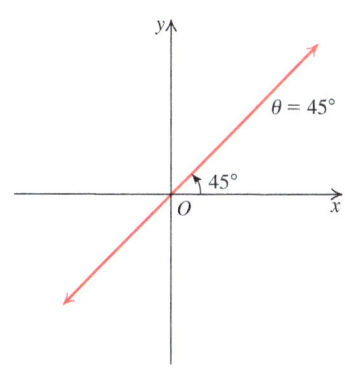

FIGURE 5.28 $y = x$ or $\theta = 45°$

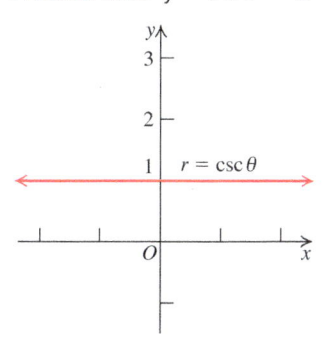

FIGURE 5.29 $y = 1$ or $r \sin \theta = 1$

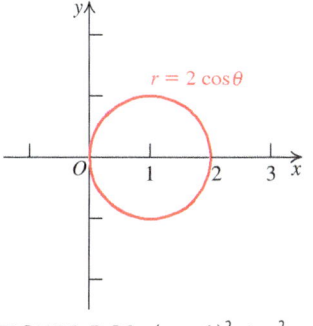

FIGURE 5.30 $(x - 1)^2 + y^2 = 1$ or $r = 2 \cos \theta$

EXAMPLE 6 Converting an Equation from Polar to Rectangular Form

Convert each polar equation to a rectangular equation and identify its graph.

a. $r = 3$
b. $\theta = 45°$
c. $r = \csc \theta$
d. $r = 2 \cos \theta$

Solution

a. $r = 3$ Given polar equation
    $r^2 = 9$ Square both sides.
    $x^2 + y^2 = 9$ Replace r^2 with $x^2 + y^2$.

The graph of $r = 3$ (or $x^2 + y^2 = 3^2$) is a circle with center at the pole and radius = 3 units. See Figure 5.27.

b. $\theta = 45°$ Given polar equation
    $\tan \theta = \tan 45°$ If $x = y$, then $\tan x = \tan y$.
    $\dfrac{y}{x} = 1$ $\tan \theta = \dfrac{y}{x}, x \neq 0$, and $\tan 45° = 1$
    $y = x$ Multiply both sides by x.

The graph of $\theta = 45°$ (or $y = x$) is a line through the origin, making an angle of 45° with the positive x-axis. See Figure 5.28.

c. $r = \csc \theta$ Given polar equation
    $r = \dfrac{1}{\sin \theta}$ Reciprocal identity
    $r \sin \theta = 1$ Multiply both sides by $\sin \theta$.
    $y = 1$ Replace $r \sin \theta$ with y.

The graph of $r = \csc \theta$ (or $y = 1$) is a horizontal line. See Figure 5.29.

d. $r = 2 \cos \theta$ Given polar equation
    $r^2 = 2r \cos \theta$ Multiply both sides by r.
    $x^2 + y^2 = 2x$ Replace r^2 with $x^2 + y^2$ and $r \cos \theta$ with x.
    $x^2 - 2x + y^2 = 0$ Subtract $2x$ from both sides and simplify.
    $(x^2 - 2x + 1) + y^2 = 1$ Add 1 to both sides to complete the square.
    $(x - 1)^2 + y^2 = 1$ Factor perfect square trinomial.

The graph of $r = 2 \cos \theta$ [or $(x - 1)^2 + y^2 = 1$] is a circle with center at $(1, 0)$ and radius 1 unit. See Figure 5.30.

Practice Problem 6 Convert each polar equation to a rectangular equation and identify its graph.

a. $r = 5$ b. $\theta = -\dfrac{\pi}{3}$
c. $r = \sec \theta$ d. $r = 2 \sin \theta$

4 Graph polar equations.

The Graph of a Polar Equation

Many important polar equations do not convert "nicely" to rectangular equations. To graph these equations, we plot points in polar coordinates. The graph of a polar equation

$$r = f(\theta)$$

is the set of all points $P(r, \theta)$ that have at least one polar coordinate representation that satisfies the equation. We make a table of several ordered pair solutions (r, θ) of the equation, plot these points, and join them with a smooth curve.

426 Chapter 5 Applications of Trigonometric Functions

EXAMPLE 7 Sketching a Graph by Plotting Points

Sketch the graph of the polar equation $r = \theta$ ($\theta \geq 0$) by plotting points.

Solution

We choose values of θ that are integer multiples of $\dfrac{\pi}{2}$. Because $r = \theta$, the polar coordinates of points on the curve are as follows:

$$(0, 0), \left(\frac{\pi}{2}, \frac{\pi}{2}\right), (\pi, \pi), \left(\frac{3\pi}{2}, \frac{3\pi}{2}\right), (2\pi, 2\pi), \left(\frac{5\pi}{2}, \frac{5\pi}{2}\right), \ldots$$

Plot these points on a polar grid and join them in order of increasing θ to obtain the graph of $r = \theta$ shown in Figure 5.31. This curve is called the **spiral of Archimedes**.

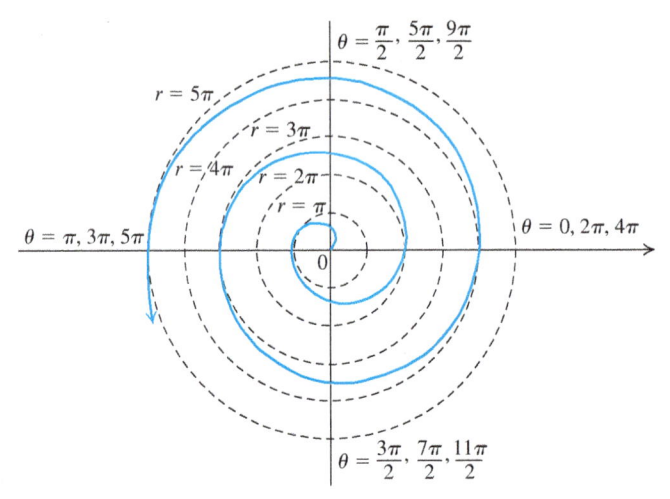

FIGURE 5.31 Spiral of Archimedes $r = \theta$

Practice Problem 7 Sketch the graph of the polar equation $r = 2\theta$ ($\theta \geq 0$) by plotting points.

As with Cartesian equations, it is helpful to determine symmetry in the graph of a polar equation. If the graph of a polar equation has symmetry, then we can sketch its graph by plotting fewer points.

TESTS FOR SYMMETRY IN POLAR COORDINATES

The graph of a polar equation has the symmetry described below if an equivalent equation results when making the replacements shown.

Symmetry with respect to the polar axis

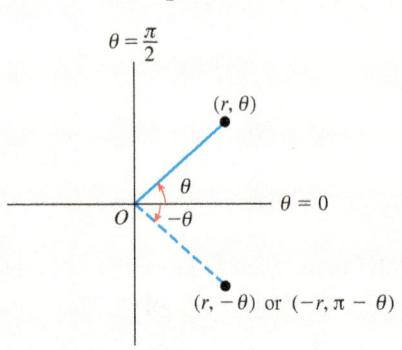

Replace (r, θ) with $(r, -\theta)$ or $(-r, \pi - \theta)$.

Symmetry with respect to the line $\theta = \dfrac{\pi}{2}$

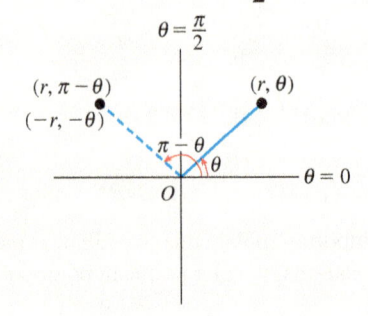

Replace (r, θ) with $(r, \pi - \theta)$ or $(-r, -\theta)$.

Symmetry with respect to the pole

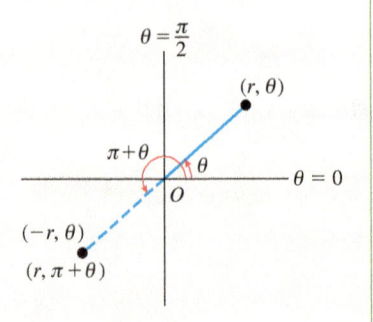

Replace (r, θ) with $(r, \pi + \theta)$ or $(-r, \theta)$.

Note that if a polar equation has two of these symmetries, then it must have the third symmetry.

TECHNOLOGY CONNECTION

A graphing calculator can be used to graph polar equations. The calculator must be set in *polar* (Pol) mode. Either *radian* or *degree* mode may be used, but make sure your WINDOW settings are appropriate for your choice. See your owner's manual for details on polar graphing. The screen shows the graph of $r = 2(1 + \cos \theta)$.

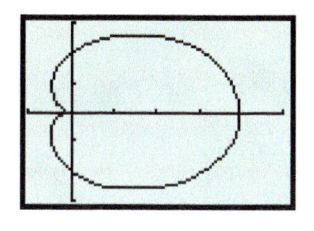

EXAMPLE 8 Sketching the Graph of a Polar Equation

Sketch the graph of the polar equation $r = 2(1 + \cos \theta)$.

Solution

Because $\cos \theta$ is an even function (that is, $\cos(-\theta) = \cos \theta$), the graph of $r = 2(1 + \cos \theta)$ is symmetric about the polar axis. Thus, to graph this equation, it is sufficient to compute values for $0 \leq \theta \leq \pi$. We construct Table 5.1 by substituting values for θ at increments of $\dfrac{\pi}{6}$, calculating the corresponding values of r.

TABLE 5.1

θ	0	$\dfrac{\pi}{6}$	$\dfrac{\pi}{3}$	$\dfrac{\pi}{2}$	$\dfrac{2\pi}{3}$	$\dfrac{5\pi}{6}$	π
$r = 2(1 + \cos \theta)$	4	$2 + \sqrt{3} \approx 3.73$	3	2	1	$2 - \sqrt{3} \approx 0.27$	0

We plot the points (r, θ), draw a smooth curve through them. Because the graph is symmetric about the polar axis, we reflect this curve about the polar axis.

The graph of the equation $r = 2(1 + \cos \theta)$ is shown in Figure 5.32. This type of curve is called a *cardioid* because it resembles a heart.

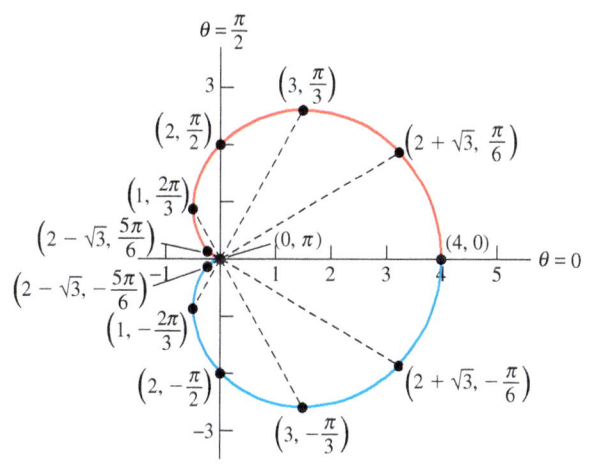

FIGURE 5.32 $r = 2(1 + \cos \theta)$

Practice Problem 8 Sketch the graph of the polar equation $r = 2(1 + \sin \theta)$.

PROCEDURE IN ACTION

EXAMPLE 9 Graphing a Polar Equation

OBJECTIVE

Sketch the graph of a polar equation $r = f(\theta)$, where f is a periodic function.

Step 1 Test for symmetry.

EXAMPLE

Sketch the graph of $r = \cos 2\theta$.

1. Replace θ with $-\theta$,

$$\cos(-2\theta) = \cos 2\theta \quad (1)$$

Replace θ with $\pi - \theta$,

$$\cos[2(\pi - \theta)] = \cos(2\pi - 2\theta) = \cos 2\theta \quad (2)$$

From (1) and (2), we conclude that the graph is symmetric about the polar axis and about the line $\theta = \dfrac{\pi}{2}$. It is also symmetric about the pole.

(continued)

428 Chapter 5 Applications of Trigonometric Functions

Step 2 Analyze the behavior of r, symmetries from Step 1, and then find selected points.

2. To get a sense of how r behaves for different values of θ, we make the following table:

As θ goes from:	r varies from:
0 to $\dfrac{\pi}{4}$	1 to 0
$\dfrac{\pi}{4}$ to $\dfrac{\pi}{2}$	0 to -1
$\dfrac{\pi}{2}$ to $\dfrac{3\pi}{4}$	-1 to 0
$\dfrac{3\pi}{4}$ to π	0 to 1

Because the curve is symmetric about $\theta = 0$ and $\theta = \dfrac{\pi}{2}$, we find selected points for $0 \leq \theta \leq \dfrac{\pi}{2}$:

θ	0	$\dfrac{\pi}{6}$	$\dfrac{\pi}{4}$	$\dfrac{\pi}{3}$	$\dfrac{\pi}{2}$
$r = \cos 2\theta$	1	$\dfrac{1}{2}$	0	$-\dfrac{1}{2}$	-1

Step 3 Sketch the graph of $r = f(\theta)$ using the points (r, θ) found in Step 2.

3.

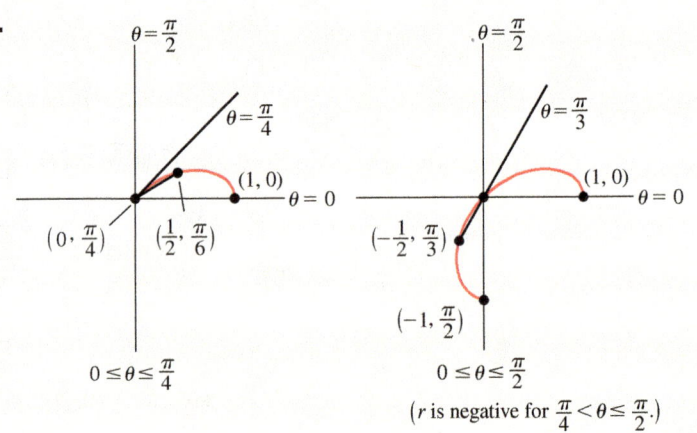

(r is negative for $\dfrac{\pi}{4} < \theta \leq \dfrac{\pi}{2}$.)

Step 4 Use symmetries to complete the graph.

4.

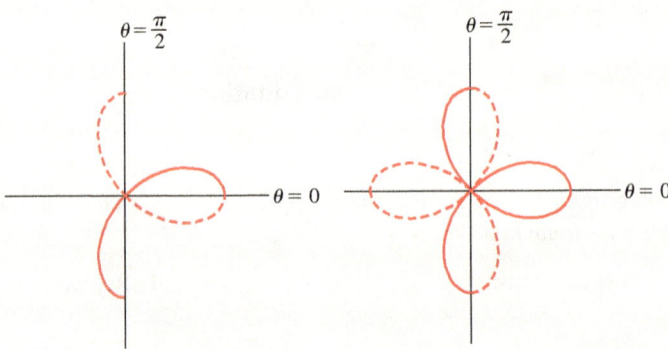

Using symmetry about the polar axis   Using symmetry about the line $\theta = \dfrac{\pi}{2}$

The curve of $r = \cos 2\theta$ is called a *four-leafed rose*.

Practice Problem 9 Sketch the graph of $r = \sin 2\theta$.

Section 5.3 ■ Polar Coordinates 429

SIDE NOTE

If a curve $r = f(\theta)$ is symmetric about the line $\theta = \frac{\pi}{2}$ (y-axis), you can first sketch the portion of the graph for $-\frac{\pi}{2} \le \theta \le \frac{\pi}{2}$ or for $0 \le \theta \le \frac{\pi}{2}$ and $\pi \le \theta \le \frac{3\pi}{2}$, then complete the graph by symmetry.

EXAMPLE 10 Graphing a Polar Equation

Sketch a graph of $r = 1 + 2 \sin \theta$.

Solution

Step 1 **Symmetry.** Replacing (r, θ) with $(r, \pi - \theta)$, we obtain an equivalent equation. The graph is symmetric about the line $\theta = \frac{\pi}{2}$.

Step 2 **Analyze and make a table of values.** We first analyze the behavior of r:

As θ goes from $-\frac{\pi}{2}$ to $-\frac{\pi}{6}$, r varies from -1 to 0.

As θ goes from $-\frac{\pi}{6}$ to 0, r varies from 0 to 1.

As θ goes from 0 to $\frac{\pi}{2}$, r varies from 1 to 3.

As θ goes from $\frac{\pi}{2}$ to π, r varies from 3 to 1.

From Step 1 it is sufficient to first sketch the graph for $-\frac{\pi}{2} \le \theta \le \frac{\pi}{2}$. Table 5.2 gives the coordinates of some points on the graph.

TABLE 5.2

θ	$-\frac{\pi}{2}$	$-\frac{\pi}{3}$	$-\frac{\pi}{4}$	$-\frac{\pi}{6}$	0	$\frac{\pi}{6}$	$\frac{\pi}{4}$	$\frac{\pi}{3}$	$\frac{\pi}{2}$
r	-1	$1 - \sqrt{3}$	$1 - \sqrt{2}$	0	1	2	$1 + \sqrt{2}$	$1 + \sqrt{3}$	3

Step 3 Sketch the graph for $-\frac{\pi}{2} \le \theta \le \frac{\pi}{2}$ using the points (r, θ) found in Step 2. See Figure 5.33.

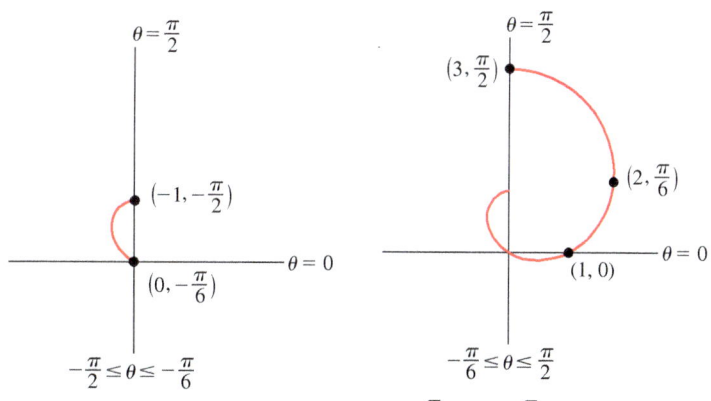

FIGURE 5.33 Graph of $r = 1 + 2 \sin \theta$ for $-\frac{\pi}{2} \le \theta \le \frac{\pi}{2}$

FIGURE 5.34 Graph of $r = 1 + 2 \sin \theta$

Step 4 Complete the graph by symmetry about the line $\theta = \frac{\pi}{2}$. The sketch is shown in Figure 5.34. This curve is called a *limaçon*.

Practice Problem 10 Sketch a graph of $r = 1 + 2 \cos \theta$.

The graph of an equation of the form

$$r = a \pm b \cos \theta \quad \text{or} \quad r = a \pm b \sin \theta$$

is called a **limaçon of Pascal**. If $b > a$, as in Example 10, the limaçon has a loop. If $b = a$, the limaçon is a **cardioid**, a heart-shaped curve. The curve in Example 8 is a cardioid. If $b < a$, the limaçon has a shape similar to the one shown in Example 11.

430 Chapter 5 Applications of Trigonometric Functions

> **SIDE NOTE**
> The symmetry for the curve $r = 3 + 2\sin\theta$ about $\theta = \dfrac{\pi}{2}$ cannot be decided from the failure of the test: replace (r, θ) with $(-r, -\theta)$. But the other option of replacing (r, θ) with $(r, \pi - \theta)$ tells you that the curve is symmetric about $\theta = \pi/2$.

EXAMPLE 11   Graphing a Limaçon

Sketch a graph of $r = 3 + 2 \sin\theta$.

Solution

Step 1   The graph is symmetric about the line $\theta = \dfrac{\pi}{2}$ because if (r, θ) is replaced with $(r, \pi - \theta)$, an equivalent equation is obtained.

Step 2   Analyze the behavior of $r = 3 + 2\sin\theta$. As θ goes from 0 to $\dfrac{\pi}{2}$, r increases from 3 to 5; as θ goes from $\dfrac{\pi}{2}$ to π, r decreases from 5 to 3; as θ goes from π to $\dfrac{3\pi}{2}$, r decreases from 3 to 1; and as θ goes from $\dfrac{3\pi}{2}$ to 2π, r increases from 1 to 3. Because of symmetry about the line $\theta = \dfrac{\pi}{2}$, we find selected points for $0 \leq \theta \leq \dfrac{\pi}{2}$ and $\pi \leq \theta \leq \dfrac{3\pi}{2}$ (see Side Note on page 429). Table 5.3 gives polar coordinates of some points in these intervals on the graph.

TABLE 5.3

θ	0	$\dfrac{\pi}{6}$	$\dfrac{\pi}{3}$	$\dfrac{\pi}{2}$	π	$\dfrac{7\pi}{6}$	$\dfrac{4\pi}{3}$	$\dfrac{3\pi}{2}$
r	3	4	$3 + \sqrt{3}$	5	3	2	$3 - \sqrt{3}$	1

Step 3   Sketch the portion of the graph for $0 \leq \theta \leq \dfrac{\pi}{2}$ and $\pi \leq \theta \leq \dfrac{3\pi}{2}$. See Figure 5.35.

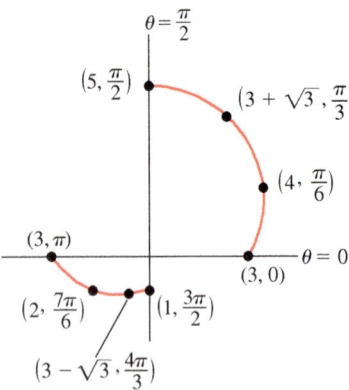

FIGURE 5.35 $r = 3 + 2\sin\theta$; $0 \leq \theta \leq \dfrac{\pi}{2}$ and $\pi \leq \theta \leq \dfrac{3\pi}{2}$

Step 4   Complete the graph by symmetry about the line $\theta = \dfrac{\pi}{2}$. See Figure 5.36.

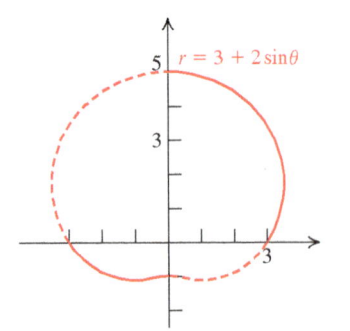

FIGURE 5.36   Graph of $r = a + b\cos\theta$ with $a > b$

Practice Problem 11   Sketch a graph of $r = 2 - \cos\theta$.

Note that even if a polar equation fails a test for a given symmetry, its graph nevertheless may exhibit that symmetry. See the margin Side Note. In other words, the tests for polar coordinate symmetry are *sufficient* but not *necessary* for showing symmetry.

The following graphs have simple polar equations.

LIMAÇONS

The graphs of
$r = a \pm b \cos \theta$,
$r = a \pm b \sin \theta$
$(a > 0, b > 0)$
are called **limaçons**
(French for "snail").

| Inner loop if $\dfrac{a}{b} < 1$ | Cardioid if $\dfrac{a}{b} = 1$ | Dimpled if $1 < \dfrac{a}{b} < 2$ | Not Dimpled if $\dfrac{a}{b} \geq 2$ |

ROSE CURVES

The graphs of
$r = a \cos n\theta$,
$r = a \sin n\theta$
$(a > 0)$ are called
rose curves.
If n is odd, the rose has n petals. If n is even, the rose has $2n$ petals.

| $r = a \cos 3\theta$ | $r = a \sin 5\theta$ | $r = a \sin 2\theta$ | $r = a \cos 4\theta$ |

CIRCLES AND LEMNISCATES

$(a > 0)$

| $r = a \cos \theta$ Circle | $r = a \sin \theta$ Circle | $r^2 = a^2 \sin 2\theta$ Lemniscate | $r^2 = a^2 \cos 2\theta$ Lemniscate |

SPIRALS

$(a > 0)$

$r = a\theta,\ \theta \geq 0$

$r = a^{k\theta},\ \theta \geq 0,\ a \neq 1, k > 0$

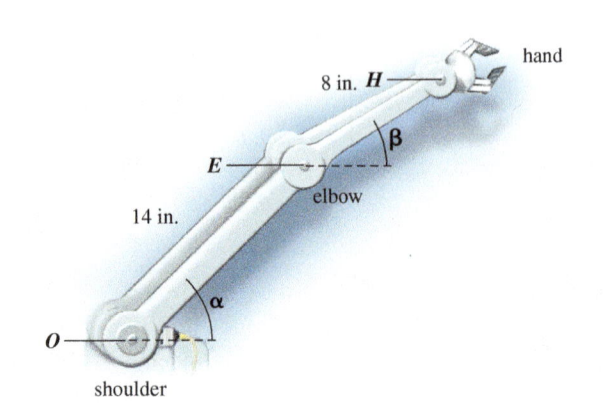

SECTION 5.3 Exercises

Basic Concepts and Skills

In Exercises 1–8, plot the point having the given polar coordinates. Then find different polar coordinates (r, θ) for the same point for which (a) $r < 0$ and $0° \leq \theta < 360°$; (b) $r < 0$ and $-360° < \theta < 0°$; (c) $r > 0$ and $-360° < \theta < 0°$.

1. $(4, 45°)$
2. $(5, 150°)$
3. $(3, 90°)$
4. $(2, 240°)$
5. $(3, 60°)$
6. $(4, 210°)$
7. $(6, 300°)$
8. $(2, 270°)$

In Exercises 9–16, plot the point having the given polar coordinates. Then give two different pairs of polar coordinates of the same point, (a) one with the given value of r and (b) one with r having the opposite sign of the given value of r.

9. $\left(2, -\dfrac{\pi}{3}\right)$
10. $\left(\sqrt{2}, -\dfrac{\pi}{4}\right)$
11. $\left(-3, \dfrac{\pi}{6}\right)$
12. $\left(-2, \dfrac{3\pi}{4}\right)$
13. $\left(-2, -\dfrac{\pi}{6}\right)$
14. $(-2, -\pi)$
15. $\left(4, \dfrac{7\pi}{6}\right)$
16. $(2, 3)$

In Exercises 17–24, convert the given polar coordinates of each point to rectangular coordinates.

17. $(3, 60°)$
18. $(-2, -30°)$
19. $(5, -60°)$
20. $(-3, 90°)$
21. $(3, \pi)$
22. $\left(\sqrt{2}, -\dfrac{\pi}{4}\right)$
23. $\left(-2, -\dfrac{5\pi}{6}\right)$
24. $\left(-1, \dfrac{7\pi}{6}\right)$

In Exercises 25–32, convert the rectangular coordinates of each point to polar coordinates (r, θ) with $r > 0$ and $0 \leq \theta < 2\pi$.

25. $(1, -1)$
26. $(-\sqrt{3}, 1)$
27. $(3, 3)$
28. $(-4, 0)$
29. $(3, -3)$
30. $(2\sqrt{3}, 2)$
31. $(-1, \sqrt{3})$
32. $(-2, -2\sqrt{3})$

In Exercises 33–40, convert each rectangular equation to polar form.

33. $x^2 + y^2 = 16$
34. $x + y = 1$
35. $y^2 = 4x$
36. $x^3 = 3y^2$
37. $y^2 = 6y - x^2$
38. $x^2 - y^2 = 1$
39. $xy = 1$
40. $x^2 + y^2 - 4x + 6y = 12$

In Exercises 41–48, convert each polar equation to rectangular form. Identify each curve.

41. $r = 2$
42. $r = -3$
43. $\theta = \dfrac{3\pi}{4}$
44. $\theta = -\dfrac{\pi}{6}$
45. $r = 4 \cos \theta$
46. $r = 4 \sin \theta$
47. $r = -2 \sin \theta$
48. $r = -3 \cos \theta$

In Exercises 49–56, sketch the graph of each polar equation by transforming it to rectangular coordinates.

49. $r = -2$
50. $r \cos \theta = -1$
51. $r \sin \theta = 2$
52. $r = \sin \theta$
53. $r = 2 \sec \theta$
54. $r + 3 \cos \theta = 0$
55. $r = \dfrac{1}{\cos \theta + \sin \theta}$
56. $r = \dfrac{6}{2 \cos \theta + 3 \sin \theta}$

In Exercises 57–72, sketch the graph of each polar equation. Identify the curve.

57. $r = \sin \theta$
58. $r = \cos \theta$
59. $r = 1 - \cos \theta$
60. $r = 1 + \sin \theta$
61. $r = \cos 3\theta$
62. $r = \sin 3\theta$
63. $r = \sin 4\theta$
64. $r = \cos 4\theta$
65. $r = 1 - 2 \sin \theta$
66. $r = 2 - 4 \sin \theta$
67. $r = 2 + 4 \cos \theta$
68. $r = 2 - 4 \cos \theta$
69. $r = 3 - 2 \sin \theta$
70. $r = 5 + 3 \sin \theta$
71. $r = 4 + 3 \cos \theta$
72. $r = 5 + 2 \cos \theta$

Applying the Concepts

In Exercises 73–76, determine the position of the hand relative to the shoulder by considering the robotic arm illustrated in the figure. Round all answers to the nearest tenth.

73. $\alpha = 45°, \beta = 30°$
74. $\alpha = -30°, \beta = 60°$
75. $\alpha = -70°, \beta = 0°$
76. $\alpha = 47°, \beta = 17°$

Critical Thinking / Discussion / Writing

Determine whether each statement is true or false. Explain your reasoning.

77. The points (r, θ) and $(-r, \theta)$ are symmetric with respect to the y-axis.

78. The points (r, θ) and $(-r, -\theta)$ are symmetric with respect to the y-axis.

79. The points (r, θ) and $(r, -\theta)$ are symmetric with respect to the x-axis.

80. The polar coordinates $(r, -\theta)$ and $(-r, \pi - \theta)$ represent the same point in the plane.

81. Prove that the distance between the points $P(r_1, \theta_1)$ and $Q(r_2, \theta_2)$ is given by
$$d(P, Q) = \sqrt{r_1^2 + r_2^2 - 2r_1 r_2 \cos(\theta_1 - \theta_2)}.$$

82. Prove that the equation $r = a \sin \theta + b \cos \theta$ represents a circle. Find its center and radius.

83. Prove that the area K of the triangle whose polar coordinates are $(0, 0)$, (r_1, θ_1), and (r_2, θ_2) is given by
$$K = \frac{1}{2} r_1 r_2 \sin(\theta_2 - \theta_1).$$
(Assume that $0 \leq \theta_1 < \theta_2 \leq \pi$ and $r_1 > 0, r_2 > 0$.)

84. Show that the polar equation of a circle with center (a, θ_0) and radius a is given by $r = 2a \cos(\theta - \theta_0)$. Discuss the cases $\theta_0 = 0$ and $\theta_0 = \frac{\pi}{2}$.

Maintaining Skills

85. Find the slope–intercept form the equation of the line that passes through the points $(2, 5)$ and $(-3, 7)$.

86. Find in slope–intercept form the equation of the line that passes through the point $(-3, 2)$ and makes an angle of $120°$ with the positive x-axis.

87. Find the acute angle θ between the lines
$$y = 3x - 10 \text{ and } y = -2x + 5.$$
$\left(\text{Hint: Recall that } \tan \theta = \left|\dfrac{m_1 - m_2}{1 + m_1 m_2}\right|.\right)$

88. Find the general form of the equation of a circle with center $(-5, 2)$ and radius 3.

89. Find the center and the radius of the circle with equation
$$x^2 + y^2 - 4x + 2y = 0.$$

SECTION 5.4 Parametric Equations

BEFORE STARTING THIS SECTION, REVIEW

1. Fundamental trigonometric identities (Section 4.7, page 363)
2. Radian measure and arc length (Section 4.1, page 289)
3. Conversion of polar equations to Cartesian equations (Section 5.3, page 427)

OBJECTIVES

1. Graph plane curves described by parametric equations.
2. Eliminate the parameter.
3. Find parametric equations of a curve.

◆ The Brachistochrone Problem

Brachistochrone is a combination of two Greek words, *brachistos* meaning "shortest" and *chronos* meaning "time." In 1692, Johann Bernoulli posed the *brachistochrone problem* as a challenge to other mathematicians: Find the shape of a wire (or slide) along which a bead might slide in the least amount of time if it starts from a point A with zero speed and ends at a point B (not directly below A) under the influence of gravity and ignoring friction. The problem was solved by Isaac Newton, Jakob Bernoulli (Johann's brother), Gottfried Leibniz, Guillaume de l' Hôpital, and Johann Bernoulli.

You might think that the wire would form a straight line because a straight line is the shortest distance between the two points A and B. However, the correct shape is half of one arch of an inverted *cycloid*, shown in Figure 5.37. The brachistochrone curve (or the curve of fastest descent) is the cycloid, which is investigated in Example 6.

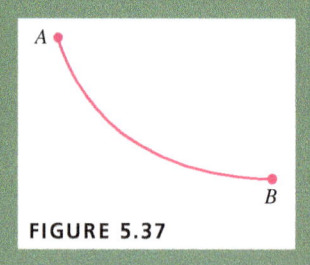

FIGURE 5.37

1 Graph plane curves described by parametric equations.

Parametric Equations

So far, we have used **Cartesian** or **rectangular equations** such as $x^2 + y^2 = 16$ and $y^2 = 8x$ to represent curves in the plane. We have also represented curves using **polar equations** such as $r = 2 \sin \theta$ and $r = 1 - \cos \theta$. When describing the path of a particle moving in the plane, it is sometimes more helpful and informative to express each of the x- and y-coordinates of the path as a function of a third variable, called a **parameter**.

Parametric Equations of a Plane Curve

Suppose $f(t)$ and $g(t)$ are functions defined on an interval I. A **plane curve** C is the set of points (x, y) where

$$x = f(t) \text{ and } y = g(t).$$

The equations $x = f(t)$ and $y = g(t)$ are called the **parametric equations** for the curve C, and t is called the **parameter** for C.

If I is the closed interval $[a, b]$, then $a \le t \le b$. The point $(f(a), g(a))$ is the **initial point** of the curve C, and the point $(f(b), g(b))$ is the **terminal point** of C. In applications involving the motion of a particle, the parameter t often represents time, but we also use other variables, such as θ (for an angle) and s (for distance), as parameters.

Graphing a Plane Curve

To sketch the graph of a plane curve C represented by the parametric equations $x = f(t)$ and $y = g(t)$, you plot points in the xy-plane corresponding to successive values of t. You then connect these points with a smooth curve. The path of a particle moving along the curve C with increasing values of t describes the **direction of increasing parameter** or the **orientation** of the curve C.

PROCEDURE IN ACTION

EXAMPLE 1 Graphing a Plane Curve

OBJECTIVE

Sketch the graph of a plane curve represented by parametric equations.

EXAMPLE

Sketch the graph represented by the parametric equations

$$x = t^2 - 3 \quad y = \frac{1}{2}t, \text{ where } -2 \leq t \leq 3.$$

Step 1 Select values of t. Select some values of t (in increasing order) in the given interval.

1. Select integer values $t = -2, -1, 0, 1, 2,$ and 3 for t in the interval $[-2, 3]$.

Step 2 Make a table. For each value of t in Step 1, use the given parametric equations to compute the x and y coordinates of the points (x, y) in the plane.

2.

t	$x = t^2 - 3$	$y = \frac{1}{2}t$	(x, y)
-2	$(-2)^2 - 3 = 1$	$\frac{1}{2}(-2) = -1$	$(1, -1)$
-1	$(-1)^2 - 3 = -2$	$\frac{1}{2}(-1) = -\frac{1}{2}$	$\left(-2, -\frac{1}{2}\right)$
0	$0^2 - 3 = -3$	$\frac{1}{2}(0) = 0$	$(-3, 0)$
1	$1^2 - 3 = -2$	$\frac{1}{2}(1) = \frac{1}{2}$	$\left(-2, \frac{1}{2}\right)$
2	$2^2 - 3 = 1$	$\frac{1}{2}(2) = 1$	$(1, 1)$
3	$3^2 - 3 = 6$	$\frac{1}{2}(3) = \frac{3}{2}$	$\left(6, \frac{3}{2}\right)$

Step 3 Sketch the curve. Plot the points (x, y) from Step 2 in an xy-plane. Connect the points by the order of increasing values of t with a smooth curve. Show the orientation by placing arrowheads on the curve.

3.

Practice Problem 1 Sketch the graph of $x = 2t, y = t^2 - 3, -2 \leq t \leq 2$.

436 Chapter 5 Applications of Trigonometric Functions

2 Eliminate the parameter.

Eliminating the Parameter

In Example 1, we sketched the graph of a curve represented by parametric equations by plotting points. Sometimes we can identify or sketch the curve by first finding its Cartesian representation. We do this by eliminating the parameter in the given parametric equations. If the parametric equations involve trigonometric functions, we often can use trigonometric identities (see Example 4) to eliminate the parameter.

PROCEDURE IN ACTION

EXAMPLE 2 Eliminating the Parameter

OBJECTIVE
Convert parametric equations with parameter t to a Cartesian equation by eliminating the parameter. Sketch the graph and indicate its orientation.

EXAMPLE
Identify the parametric curve
$$x = \sqrt{t+1} \quad y = \sqrt{t}, \text{ where } t \geq 0,$$
by converting it to a Cartesian equation. Sketch its graph and show the orientation.

Step 1 Solve for t. Solve one of the parametric equations for the parameter.

1. $y = \sqrt{t}$ A parametric equation
   $y^2 = t$ Square both sides.

Step 2 Substitute. Substitute the expression for the parameter obtained in Step 1 into the other parametric equation.

2. $x = \sqrt{t+1}$ The other parametric equation
   $x = \sqrt{y^2 + 1}$ Substitute y^2 for t from Step 1.

Step 3 Simplify. Simplify the resulting Cartesian equation.

3. $x^2 = y^2 + 1$ Square both sides.
   $x^2 - y^2 = 1$ Subtract y^2 from both sides.

Step 4 Graph. Sketch the graph of the Cartesian equation in Step 3. Sketch the portion of the graph for the given values of the parameter and indicate its orientation.

4. The graph of $x^2 - y^2 = 1$ is a "**hyperbola**" shown in the figure. Because $t \geq 0$, we have $x = \sqrt{t+1} \geq 1$, $y = \sqrt{t} \geq 0$. The curve starts at the initial point $(1, 0)$ when $t = 0$ and rises along the hyperbola into the first quadrant as t increases. The parameter interval is $[0, \infty)$, so there is no terminal point. The graph, along with its orientation, is the part of the hyperbola shown by a solid curve in the figure.

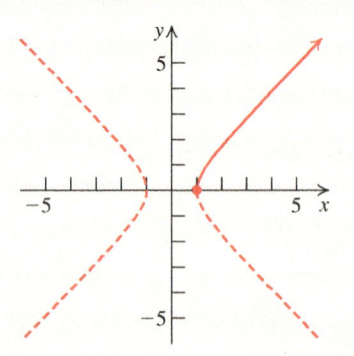

Practice Problem 2 Identify the parametric curve $x = t + 1, y = 3 + t^2, t \leq 0$ by eliminating the parameter and indicate the orientation of the graph.

EXAMPLE 3 Graphing a Plane Curve

Sketch a graph of the curve represented by the parametric equations

$$x = 4\cos\theta \quad y = 3\sin\theta, \text{ where } 0 \le \theta \le 2\pi.$$

Eliminate the parameter θ and write the Cartesian equation of the curve.

Solution

Step 1 We select common values of θ that are integer multiples of $\dfrac{\pi}{6}$ and $\dfrac{\pi}{4}$.

Step 2 We make a table of approximate values

θ	$x = 4\cos\theta$	$y = 3\sin\theta$	(x, y)	θ	$x = 4\cos\theta$	$y = 3\sin\theta$	(x, y)
0	4	0	(4, 0)				
$\dfrac{\pi}{6}$	3.5	1.5	(3.5, 1.5)	$\dfrac{7\pi}{6}$	−3.5	−1.5	(−3.5, −1.5)
$\dfrac{\pi}{4}$	2.8	2.1	(2.8, 2.1)	$\dfrac{5\pi}{4}$	−2.8	−2.1	(−2.8, −2.1)
$\dfrac{\pi}{3}$	2.0	2.6	(2, 2.6)	$\dfrac{4\pi}{3}$	−2.0	−2.6	(−2, −2.6)
$\dfrac{\pi}{2}$	0	3.0	(0, 3)	$\dfrac{3\pi}{2}$	0	−3.0	(0, −3)
$\dfrac{2\pi}{3}$	−2.0	2.6	(−2, 2.6)	$\dfrac{5\pi}{3}$	2.0	−2.6	(2, −2.6)
$\dfrac{3\pi}{4}$	−2.8	2.1	(−2.8, 2.1)	$\dfrac{7\pi}{4}$	2.8	−2.1	(2.8, −2.1)
$\dfrac{5\pi}{6}$	−3.5	1.5	(−3.5, 1.5)	$\dfrac{13\pi}{6}$	3.5	−1.5	(3.5, −1.5)
π	−4.0	0	(−4, 0)	2π	4.0	0	(4, 0)

Step 3 Plot the points (x, y) from Step 2 and connect them by the order of increasing values of θ with a smooth curve.

The oval shaped curve in Figure 5.38 is called an **ellipse**.

Step 4 We eliminate the parameter θ by using the identity $\cos^2\theta + \sin^2\theta = 1$. From the parametric equations we have

$$\frac{x}{4} = \cos\theta \quad \frac{y}{3} = \sin\theta$$

$$\frac{x^2}{16} + \frac{y^2}{9} = \cos^2\theta + \sin^2\theta \quad \text{Square both sides and add.}$$

or $\quad \dfrac{x^2}{16} + \dfrac{y^2}{9} = 1 \quad\quad \cos^2\theta + \sin^2\theta = 1$

The last equation is the Cartesian equation of the ellipse shown in Figure 5.38.

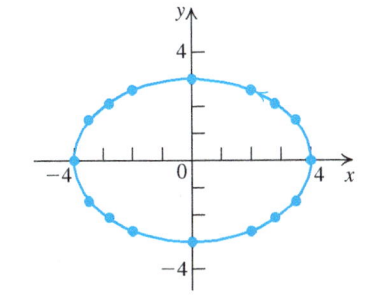

FIGURE 5.38

Practice Problem 3 Repeat Example 3 for the curve represented by the parametric equation

$$x = 4\sec\theta \quad y = 3\tan\theta, \text{ where } 0 \le \theta \le 2\pi.$$

(*Hint:* $\sec^2\theta - \tan^2\theta = 1$)

438 Chapter 5 Applications of Trigonometric Functions

EXAMPLE 4 Eliminating the Parameter

Identify the parametric curve and indicate its orientation.

a. $x = \sin t, y = \cos^2 t, 0 \leq t \leq \dfrac{\pi}{2}$ **b.** $x = \sin t, y = \cos^2 t, t \geq 0$

Solution

$$\begin{aligned} y &= \cos^2 t \\ &= 1 - \sin^2 t \quad \text{Pythagorean identity} \\ &= 1 - x^2 \quad \sin t = x \end{aligned}$$

The graph of the Cartesian equation $y = 1 - x^2$ is shown in Figure 5.39(a).

a. When $t = 0$, $x = \sin 0 = 0$ and $y = \cos^2 0 = 1$. When $t = \dfrac{\pi}{2}$, $x = \sin \dfrac{\pi}{2} = 1$ and $y = \cos^2 \dfrac{\pi}{2} = 0$. As t ranges from 0 to $\dfrac{\pi}{2}$, x values increase from 0 to 1, while y values decrease from 1 to 0. Thus, the parametric equations describe the parabolic arc $y = 1 - x^2$ with initial point $(0, 1)$ and terminal point $(1, 0)$. See Figure 5.39(b) for the oriented graph.

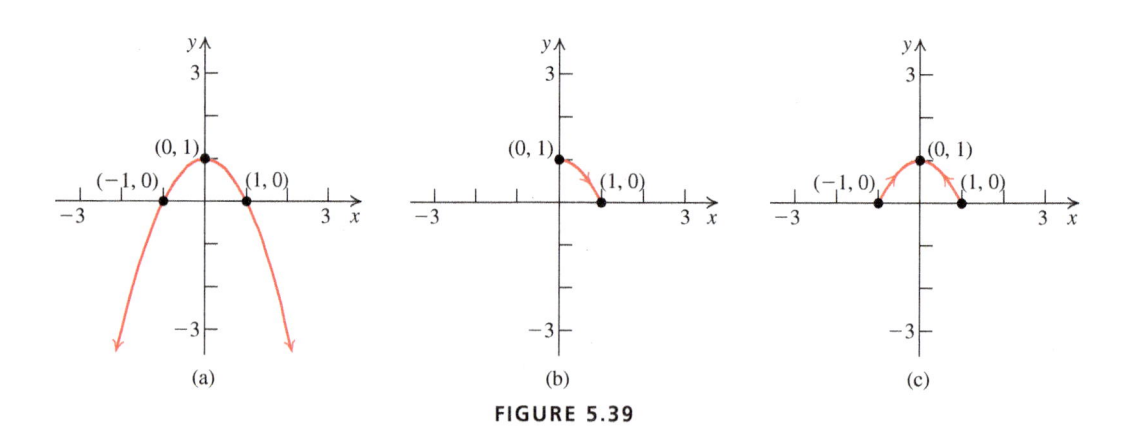

> **SIDE NOTE**
> Be careful that the process of elimination does not include more (or fewer) points than those given by the parametric equations.

FIGURE 5.39

b. Note that for any value of t, we have $-1 \leq x \leq 1$ and $0 \leq y \leq 1$. Thus, as t ranges over the interval $[0, \infty)$, the parabolic arc $y = 1 - x^2$ shown in Figure 5.39(c) is traced back and forth an infinite number of times. The initial point is $(0, 1)$, and there is no terminal point.

Practice Problem 4 Identify the parametric curve and indicate its orientation.

a. $x = \cos t, y = \sin t, 0 \leq t \leq \pi$ **b.** $x = \cos t, y = \sin t, t \geq 0$

3 Find parametric equations of a curve.

Finding Parametric Equations

We now consider the reverse problem of finding parametric equations for a given curve. We call the parametric equations together with a parameter interval I a **parameterization** of the curve. A given curve can be parameterized in infinitely many ways.

The easiest way to write the graph of any Cartesian equation $y = f(x)$ in parametric form is $x = t, y = f(t)$, with t in the domain of f. For example, a parameterization of the parabola $y = x^2 + 1$ is

$$x = t, \quad y = t^2 + 1$$

with $I = (-\infty, \infty)$.

You can convert any polar equation $r = g(\theta)$ to a parametric form as follows: Convert the polar coordinates to rectangular coordinates (see page 421):

$$x = r\cos\theta, \qquad y = r\sin\theta$$
$$x = g(\theta)\cos\theta \qquad y = g(\theta)\sin\theta \qquad \text{Replace } r \text{ with } g(\theta).$$

The equations $x = g(\theta)\cos\theta$, $y = g(\theta)\sin\theta$ give the parametric equations of $r = g(\theta)$ with parameter θ. For example, a parameterization of the polar equation $r = 1 + \cos\theta$ of cardioid is

$$x = (1 + \cos\theta)\cos\theta$$
$$y = (1 + \cos\theta)\sin\theta$$

with $I = [0, 2\pi]$.

The next example illustrates that the parametric equations of a curve are *not unique*.

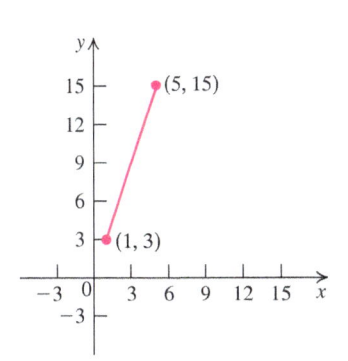

EXAMPLE 5 Parameterizing a Line Segment

The equation $y = 3x$, $1 \le x \le 5$ represents the line segment joining the points $(1, 3)$ and $(5, 15)$. See Figure 5.40. Find a set of parametric equations and interpret the motion of a particle moving along this segment by using the following parameters.

a. $x = t$ **b.** $x = 1 + 4t$ **c.** $x = 5 - t$

Solution

a. With $x = t$, we obtain the following parameterization of the line segment $y = 3x$, $1 \le x \le 5$:

$$x = t, \quad y = 3t, \quad 1 \le t \le 5$$

At time $t = 1$, the particle is at $(1, 3)$. It travels up the line segment and reaches the point $(5, 15)$ in $t = 5$ time units.

b. If $x = 1 + 4t$, then $y = 3x$, $1 \le x \le 5$ becomes

$$y = 3(1 + 4t), \quad 1 \le 1 + 4t \le 5 \qquad \text{Replace } x \text{ with } 1 + 4t.$$
$$y = 3 + 12t, \qquad 0 \le t \le 1 \qquad \text{Solve } 1 \le 1 + 4t \le 5 \text{ for } t.$$

The parametric equations are $x = 1 + 4t$, $y = 3 + 12t$; $0 \le t \le 1$. At time $t = 0$, the particle is at $(1, 3)$. It travels the same line segment upward and reaches the point $(5, 15)$ in $t = 1$ time unit.

c. If $x = 5 - t$, then $y = 3x$, $1 \le x \le 5$ becomes

$$y = 3(5 - t), 1 \le 5 - t \le 5 \qquad \text{Replace } x \text{ with } 5 - t.$$
$$= 15 - 3t, -4 \le -t \le 0 \qquad \text{Subtract 5 from all parts of } 1 \le 5 - t \le 5.$$
$$= 15 - 3t, 4 \ge t \ge 0 \qquad \text{Multiply the inequality } -4 \le -t \le 0 \text{ by } -1$$
$$\qquad\qquad\qquad\qquad\qquad \text{and reverse the inequality signs.}$$
$$y = 15 - 3t, 0 \le t \le 4 \qquad \text{Rewrite.}$$

The parametric equations are $x = 5 - t$, $y = 15 - 3t$; $0 \le t \le 4$. At time $t = 0$, the particle is at $(5, 15)$. It travels the same line segment (in the opposite direction) and reaches the point $(1, 3)$ in $t = 4$ time units.

Practice Problem 5 The graph of the equation $y = \sqrt{1 - x^2}$, $-1 \le x \le 1$ is the upper semicircle of radius 1 shown in Figure 5.41. Find a set of parametric equations and interpret the motion of a particle moving along this arc using the following parameters.

a. $x = \cos t$ **b.** $x = \cos 2t$ **c.** $x = -\cos t$

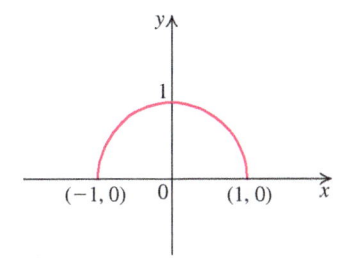

Remark. Example 5 shows the advantages of parametric equations over Cartesian equations in describing the motion of a particle. A Cartesian equation only indicates the path on which a particle is traveling. However, parametric equations of a curve give not

440 Chapter 5 Applications of Trigonometric Functions

only the path, but also information about the location of the particle at any specific time as well as the direction in which the particle is traveling.

EXAMPLE 6 Parametric Equations of a Cycloid

If a wheel (circle) of radius a rolls along a straight line, then a point P on the rim of the wheel traces a curve called a **cycloid**. Find parametric equations for a cycloid.

Solution

Suppose the wheel rolls on the x-axis and the point P touches the x-axis at the origin. We let $t = 0$ when P is at the origin. Let $P(x, y)$ denote the location of the point P after the wheel rolls through an angle of t radians. In Figure 5.42, the measure of $\angle PCT$ is t radians.

When the wheel in Figure 5.42 rolls the distance OT, the radial line from the center C to the point P rotates through angle t. We treat t as a positive angle. So

$OT = \widehat{PT}$ The distance OT equals the arc length from P to T.
$ = at$ arc length = (radius)(angle measured in radians)

We now compute the coordinates of the point $P(x, y)$ in terms of the parameter t.

$y = PM$
$ = QT$ Opposite sides of a rectangle
$ = CT - CQ$ Figure 5.42
$ = a - a\cos t$ In triangle PCQ, $\cos t = \dfrac{CQ}{PC} = \dfrac{CQ}{a}$.
$ = a(1 - \cos t)$ Factor out a.

$x = OM$
$ = OT - MT$ Figure 5.42
$ = at - PQ$ $OT = at, MT = PQ$
$ = at - a\sin t$ In triangle PCQ, $\sin t = \dfrac{PQ}{PC} = \dfrac{PQ}{a}$.
$ = a(t - \sin t)$

Thus, the parametric equations of a cycloid generated by a circle of radius a are given by

$$x = a(t - \sin t), \; y = a(1 - \cos t), \; -\infty < t < \infty.$$

Practice Problem 6 A wheel of radius a rolls along a horizontal straight line; then a point P that is b units from the center with $b < a$ traces a curve called a **curtate cycloid**. See Figure 5.43. Find parametric equations for a curtate cycloid.

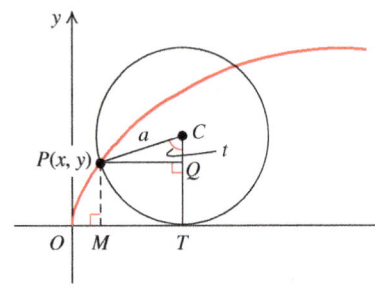

FIGURE 5.42 Graph of a cycloid

TECHNOLOGY CONNECTION

Graphing calculators must be set in *radian* and *parametric* modes; the viewing window requires minimum and maximum values of x, y, and t. The graph of

$x = t - \sin(t)$
$y = 1 - \cos(t)$

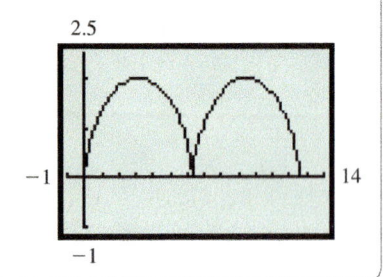

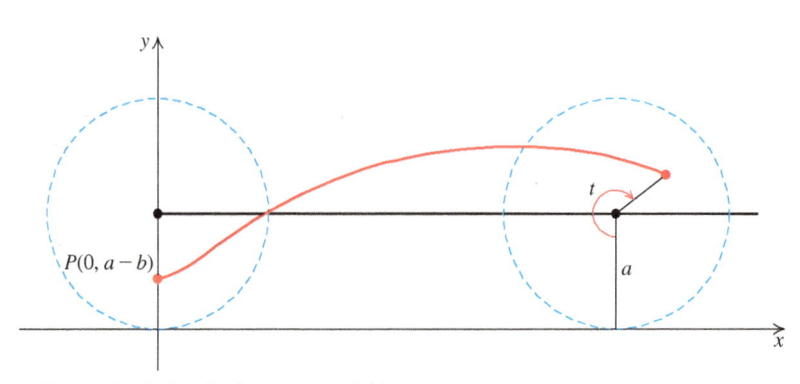

with $T_{min} = 0$, $T_{max} = 4\pi$ is shown.

FIGURE 5.43 Graph of a curtate cycloid

SECTION 5.4 Exercises

Basic Skills and Concepts

In Exercises 1–4, sketch the graph of each curve by plotting points. Indicate the points on the curve that correspond to the integer values of t in the given interval.

1. $x = t - 1, y = 2t - 1, -2 \leq t \leq 2$
2. $x = t + 1, y = 3t + 2, -2 \leq t \leq 3$
3. $x = \frac{1}{2}t, y = t^2 + 1, -2 \leq t \leq 3$
4. $x = 2t^2 - 1, y = 2t, -2 \leq t \leq 2$

In Exercises 5–8, sketch the graph of each curve by plotting points. Indicate the points on the curve that correspond to $\theta = 0, \frac{\pi}{4}, \frac{\pi}{2}, \frac{3\pi}{4}, \pi$.

5. $x = \cos\theta, y = \sin\theta, 0 \leq \theta \leq \pi$
6. $x = \sin\theta, y = \cos\theta, 0 \leq \theta \leq \pi$
7. $x = 2\cos\theta, y = \sin\theta, 0 \leq \theta \leq \pi$
8. $x = 2\sin\theta, y = 3\cos\theta, 0 \leq \theta \leq \pi$

In Exercises 9–20, eliminate the parameter to obtain a Cartesian equation of the curve. Graph the curve and indicate its orientation.

9. $x = t, y = 2t, -\infty < t < \infty$
10. $x = 3t, y = t, -\infty < t < \infty$
11. $x = 2t, y = 6t, 0 \leq t \leq 4$
12. $x = 2t + 1, y = -3t + \frac{3}{2}, -1 \leq t \leq \frac{1}{2}$
13. $x = \sqrt{t}, y = 3 - t, 0 \leq t \leq 4$
14. $x = t - 2, y = \sqrt{t}, 0 \leq t \leq 6$
15. $x = \sqrt{t}, y = \sqrt{4 - t}, 0 \leq t \leq 4$
16. $x = \sqrt{9 - t}, y = -\sqrt{t}, 0 \leq t \leq 9$
17. $x = 2\sqrt{2 - t}, y = 3\sqrt{t - 1}, 1 \leq t \leq 2$
18. $x = 3\sqrt{t - 2}, y = 4\sqrt{3 - t}, 2 \leq t \leq 3$
19. $x = t + 2, y = \sqrt{t^2 + 4t}, t \geq 0$
20. $x = 3\sqrt{t - 1}, y = 2\sqrt{t}, t \geq 1$

In Exercises 21–36, eliminate the parameter to obtain a Cartesian equation of the curve. Graph the curve and indicate its orientation.

21. $x = \sin t, y = \cos t, 0 \leq t \leq 2\pi$
22. $x = \cos t, y = \sin t, 0 \leq t \leq 2\pi$
23. $x = \cos^2 t, y = \sin^2 t, 0 \leq t \leq \frac{\pi}{2}$
24. $x = \sin^2 t, y = \cos^2 t, 0 \leq t \leq \pi$
25. $x = 3\cos t, y = 2\sin t, 0 \leq t \leq 2\pi$
26. $x = 3\sin t, y = 2\cos t, 0 \leq t \leq 4\pi$
27. $x = \cos\theta, y = \cos^2 2\theta, -\infty < \theta < \infty$
28. $x = \sin\theta, y = \cos^2 2\theta, -\infty < \theta < \infty$
29. $x = 1 - \cos t, y = 2 - \sin t, 0 \leq t \leq 2\pi$
30. $x = 3 - 2\cos t, y = -1 + 5\sin t, 0 \leq t \leq 2\pi$
31. $x = 2\tan t - 1, y = 3\sec t + 2, 0 \leq t < \frac{\pi}{2}$
32. $x = 3\csc t + 2, y = 5\cot t - 1, 0 < t < \pi$
33. $x = e^t, y = e^{2t}, t \geq 0$
34. $x = 4e^{2t}, y = -e^t, t \geq 0$
35. $x = t^2, y = 2\ln t, t \geq 1$
36. $x = 5\ln t, y = t^5, t \geq 1$

In Exercises 37–40, consider the Cartesian equation $y = 3x + 2$ of a line. Find a set of parametric equations and interpret the motion of a particle moving along this line by using the parameter t with $-\infty < t < \infty$.

37. $x = t$
38. $x = \sin t$
39. $x = e^t$
40. $x = e^{-t}$

In Exercises 41–44, consider the Cartesian equation $y = x^2 - 4$ of a parabola. Find a set of parametric equations and interpret the motion of a particle moving along this parabola by using the parameter t with $-\infty < t < \infty$.

41. $x = -t$
42. $x = t^2$
43. $x = \cos t$
44. $x = e^t$

In Exercises 45–48, consider the Cartesian equation $x = \sqrt{1 - y^2}, -1 \leq y \leq 1$ of a right semicircle of radius 1. Find a set of parametric equations and interpret the motion of a particle moving along this semicircle by using the following parameters.

45. $x = \sin t$
46. $x = \sin 2t$
47. $x = -\sin t$
48. $x = -\sin 4t$

Applying the Concepts

Projectile motion. Suppose an object is launched upward with an initial speed v_0 at an angle of θ with the horizontal from a height h above the ground. The resulting motion is called *projectile motion*. It can be shown that the parametric equations of the path of a projectile are

$$x = (v_0 \cos\theta)t, \quad y = (v_0 \sin\theta)t - \frac{1}{2}gt^2 + h,$$

where t is the time and g is the constant acceleration due to gravity ($g \approx 32$ ft/sec^2 or 9.8 m/sec^2).

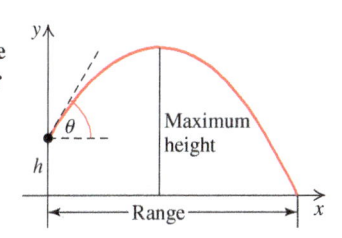

In Exercises 49–52, a golf ball is hit at an angle of 30° with an initial speed of 64 ft/sec. (Here $h = 0$.)

49. Write parametric equations that model the path of the golf ball.
50. What is the location of the ball 2 seconds later?
51. When does the ball land on the ground?
52. What is the range of the ball?

In Exercises 53–56, a gun is fired from a tower 96 feet above the ground. The angle of elevation of the gun is 60°, and its muzzle speed is 1600 ft/sec.

53. Write parametric equations that model the path of the bullet.
54. When does the bullet hit the ground?
55. What is the range for the gun?
56. Find a Cartesian equation for the path of the bullet. Identify the curve.

Critical Thinking / Discussion / Writing

57. **a.** Verify that the parametric equations
$$x = x_1 + (x_2 - x_1)t, \quad y = y_1 + (y_2 - y_1)t, \quad -\infty < t < \infty$$
represent the line passing through the points (x_1, y_1) and (x_2, y_2).

    b. Find parametric equations of the line segment from $(2, -3)$ to $(-1, 4)$.

58. **a.** Verify that the parametric equations
$$x = h + a\cos t, \quad y = k + a\sin t, \quad 0 \le t \le 2\pi$$
represent the curve that starts at the point $(h + a, k)$ and goes counterclockwise once around the circle with center (h, k) and radius a.

    b. Find a parametric equation of the curve that starts at the point $(3, 0)$ and goes counterclockwise once around the circle with center at $(1, 0)$ and radius 2.

59. Which of the following parametric equations represent the line $y = x + 1$? Explain.
    a. $x = 2t - 1, y = 2t$
    b. $x = t^2 - 1, y = t^2$
    c. $x = t^3 + 1, y = t^3 + 3$
    d. $x = e^t - 1, y = e^t$

60. **Projectile motion.** Eliminate the parameter t from the parametric equations $x = (v_0 \cos\theta)t$, $y = (v_0 \sin\theta)t - \frac{1}{2}gt^2$ to obtain a Cartesian equation.
    a. Show that the trajectory of a projectile is a parabola.
    b. Find the vertex of the parabola.
    c. Find the time of flight of the projectile.
    d. Find the range.
    e. Show that for a fixed initial speed v_0, the maximum range is obtained when $\theta = 45°$. Find a formula for this maximum range.
    f. Show that doubling the initial speed of a projectile has the effect of multiplying both the range and maximum height by a factor of four.

Maintaining Skills

61. Let $f(x) = 3x - 1$. Find $f(0), f(1), f(2),$ and $f(7)$.

62. Let $g(x) = \dfrac{x-3}{x+1}$. Find $g(0), g(1), g(2),$ and $g(7)$.

63. Let $f(x) = \dfrac{8(1-x^4)}{1-x}$. Find $f\left(\dfrac{1}{2}\right)$.

64. The ratio of female and male employees in a factory is 5:2. If there are 115 female employees, find the number of male employees.

SUMMARY Definitions, Concepts, and Formulas

5.1 The Law of Sines and the Law of Cosines

i. **The Law of Sines:** In a triangle ABC with sides of length $a, b,$ and c, $\dfrac{\sin A}{a} = \dfrac{\sin B}{b} = \dfrac{\sin C}{c}$, or equivalently, $\dfrac{a}{\sin A} = \dfrac{b}{\sin B} = \dfrac{c}{\sin C}$.

The Law of Sines is used to solve AAS, ASA, and SSA triangles. The SSA case is called the *ambiguous case* because the information provided may lead to two triangles, one triangle, or no triangle.

ii. **The Law of Cosines:** In a triangle ABC with sides of lengths $a, b,$ and c,
$$a^2 = b^2 + c^2 - 2bc\cos A,$$
$$b^2 = c^2 + a^2 - 2ca\cos B,$$
$$c^2 = a^2 + b^2 - 2ab\cos C.$$

The Law of Cosines is used to solve SAS and SSS triangles.

5.2 Areas of Polygons Using Trigonometry

i. **Area of an SAS triangle**
The area K of a triangle ABC with sides $a, b,$ and c is
$$K = \frac{1}{2}bc\sin A = \frac{1}{2}ca\sin B = \frac{1}{2}ab\sin C.$$
Area K is one-half the product of two of its sides and the sine of the included angle.

ii. **Area of AAS and ASA triangles**
The area K of a triangle ABC with sides $a, b,$ and c is
$$K = \frac{a^2 \sin B \sin C}{2 \sin A} = \frac{b^2 \sin C \sin A}{2 \sin B} = \frac{c^2 \sin A \sin B}{2 \sin C}.$$

iii. **Area of SSS triangles**
Heron's formula The area K of a triangle with sides of lengths $a, b,$ and c is given by
$$K = \sqrt{s(s-a)(s-b)(s-c)},$$
where $s = \dfrac{1}{2}(a + b + c)$ is the *semiperimeter*.

5.3 Polar Coordinates

i. The ordered pair (r, θ) represents a point P in the plane that is a "directed distance" of r units from the origin (pole) O, where θ is the directed angle between the polar axis (positive x-axis) and the line segment OP. The ordered pair (r, θ) gives the polar coordinates of P.

ii. The polar coordinates of a point $P(r, \theta)$ are not unique. In fact, if n is any integer, then

$$(r, \theta), (r, \theta + 2n\pi), \text{ and } (-r, \theta + \pi + 2n\pi)$$

are all polar coordinates of the same point.

iii. To convert a point from polar coordinates (r, θ) to rectangular coordinates, use the equations $x = r \cos \theta$ and $y = r \sin \theta$.

iv. To convert a point from rectangular coordinates (x, y) to polar coordinates (r, θ), use the equations on page 423.

v. A polar equation is an equation whose variables are r and θ.

5.4 Parametric Equations

i. Suppose $f(t)$ and $g(t)$ are functions defined on interval I. A **plane curve** C is a set of points (x, y), where $x = f(t)$ and $y = g(t)$. The equations $x = f(t)$ and $y = g(t)$ are the **parametric equations** for the curve C, and t is the **parameter**. The path of a particle moving along the curve C with increasing values of t describes the **orientation** (direction) of C.

ii. We can graph a plane curve represented by parametric equations by plotting points. (A procedure is described on page 435.)

iii. We can eliminate the parameter t in the parametric equations to convert to a Cartesian equation of the curve. It may be necessary to change the domain of the Cartesian equation to be consistent with the domain of the parameter t. (See page 438.)

iv. The easiest way to convert any Cartesian equation $y = f(x)$ in parametric form is $x = t$ and $y = f(t)$, with t in the domain of f. A polar equation $r = f(\theta)$ in parametric form is $x = f(\theta) \cos \theta$ and $y = f(\theta) \sin \theta$ with parameter θ.

REVIEW EXERCISES

Basic Concepts and Skills

In Exercises 1–8, use the Law of Sines to solve each triangle ABC. Round each answer to the nearest tenth.

1. $A = 40°, B = 35°, c = 100$
2. $B = 30°, C = 80°, a = 100$
3. $A = 45°, a = 25, b = 75$
4. $B = 36°, a = 12.5, b = 8.7$
5. $A = 48.5°, C = 57.3°, b = 47.3$
6. $A = 67°, a = 100, c = 125$
7. $A = 65.2°, a = 21.3, b = 19$
8. $C = 53°, a = 140, c = 115$
9. In triangle ABC, let $c = 20$ and $B = 60°$. Find a value of b such that C has **(i)** two possible values, **(ii)** exactly one value, **(iii)** no value.
10. Repeat Exercise 9 for $c = 20$ and $B = 150°$.

In Exercises 11–18, use the Law of Cosines to solve each triangle ABC. Round each answer to the nearest tenth.

11. $a = 60, b = 90, c = 125$
12. $a = 15, b = 9, C = 120°$
13. $a = 40, c = 38, B = 80°$
14. $a = 10, b = 20, c = 22$
15. $a = 2.6, b = 3.7, c = 4.8$
16. $a = 15, c = 26, B = 115°$
17. $a = 12, b = 7, C = 130°$
18. $b = 75, c = 100, A = 80°$

In Exercises 19–24, find the area of each triangle ABC with the given information. Round each answer to the nearest square unit.

19. $a = 5$ meters, $b = 7$ meters, $c = 10$ meters
20. $a = 2.4$ meters, $b = 3.4$ meters, $c = 4.4$ meters
21. $A = 65°, b = 6$ feet, $c = 4$ feet
22. $A = 115°, b = 20$ inches, $a = 30$ inches
23. $A = 67°, B = 38°, b = 12$ m
24. $B = 46.7°, C = 29.4°, a = 7.3$ ft

In Exercises 25–28, plot each point in polar coordinates and find its rectangular coordinates.

25. $(24, 30°)$
26. $(12, -60°)$
27. $\left(-2, -\dfrac{\pi}{4}\right)$
28. $\left(-3, -\dfrac{3\pi}{4}\right)$

In Exercises 29–32, convert the rectangular coordinates of each point to polar coordinates (r, θ), with $r > 0$ and $0 \leq \theta < 2\pi$.

29. $(-2, 2)$
30. $(\sqrt{3}, 1)$
31. $(2\sqrt{3}, -2)$
32. $(-2, -2\sqrt{3})$

In Exercises 33–36, convert each rectangular equation to a polar equation.

33. $3x + 2y = 12$
34. $x^2 + y^2 = 36$
35. $x^2 + y^2 = 8x$
36. $x^2 + y^2 = 6y$

In Exercises 37–42, convert each polar equation to a rectangular equation.

37. $r = -3$
38. $\theta = \dfrac{5\pi}{6}$
39. $r = 3 \csc \theta$
40. $r = 2 \sec \theta$
41. $r = 1 - 2 \sin \theta$
42. $r = 3 \cos \theta$

In Exercises 43–50, eliminate the parameter to obtain a Cartesian equation of the curve. Graph the curve and indicate its orientation.

43. $x = t, y = 1 - t, 0 \leq t \leq 1$
44. $x = 3 \cos t, y = 3 \sin t, 0 \leq t \leq 2\pi$
45. $x = 3 \sin^2 t, y = 3 \cos^2 t, 0 \leq t \leq 2\pi$
46. $x = 3 \cos \theta, y = -3 \sin \theta, 0 \leq \theta \leq \pi$
47. $x = \sin t, y = \cos 2t, 0 \leq t \leq 2\pi$
48. $x = 3 \sec \theta, y = 3 \tan \theta, -\dfrac{\pi}{2} < \theta < \dfrac{\pi}{2}$
49. $x = t^3, y = t^2, -\infty < t < \infty$
50. $x = t^2 - 2t, y = t, -\infty < t < \infty$

Applying the Concepts

51. **Geometry.** A side of a parallelogram is 60 inches long and makes angles of 32° and 46° with the two diagonals. Find the lengths of the diagonals.

52. **Height of a flagpole.** Two friends, Carmen and Latisha, stand 300 feet apart with a flagpole situated between them. Their angles of elevation of the top of the flagpole are 34° and 27°. How tall is the flagpole? How far is each girl from it?

53. **Angle of elevation of a hill.** A tower 120 feet tall is located at the top of a hill. At a point 460 feet down the hill, the angle between the surface of the hill and the line of sight to the top of the tower is 12°. Find the angle of elevation of the hill to a horizontal plane.

54. **Distance between cars.** Two cars start from a point A along two straight roads. The angle between the two roads is 72°. The speeds of the two cars are 55 miles per hour and 65 miles per hour. How far apart are the two cars after 80 minutes? Round to the nearest tenth of a mile.

55. **Length of a pond.** A surveyor wants to find the length of a pond, but can do so only indirectly. She stands at a point A on one side of the pond and locates points B and C at opposite ends of the pond. The angle BAC is 73°, $AB = 560$ yd, and $AC = 720$ yd. Find the length of the pond.

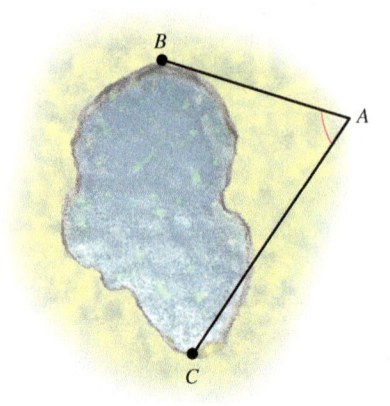

56. **Geometry.** A parallelogram has adjacent sides 120 and 140 feet long, and the angle between them is 48°. Find the lengths of the two diagonals of the parallelogram.

57. **Tangential circles.** Three circles of radii 60, 80, and 125 ft are tangent to each other externally. Find the angles of the triangle formed by joining the centers of the circles.

58. **Distance between planets.** The angle subtended at Earth by the lines joining Earth to Venus and the sun is 28°. If Venus is 67 million miles from the sun and Earth is 93 million miles from the sun, what is the distance (to the nearest million) between Earth and Venus?

59. In Washington, D.C., Constitution Avenue and Pennsylvania Avenue intersect at an angle of 19°. The White House is on Pennsylvania Avenue 5600 feet from this point of intersection. The National Academy of Sciences is on Constitution Avenue 8600 feet from this intersection. Both buildings lie on the same side of 4th Street. How far (to the nearest foot) is the White House from the National Academy of Sciences?

60. The Lincoln Memorial in Washington, D.C., is 4300 feet due west of the Washington Monument. The Pentagon is 7100 feet S 15° W from the Lincoln Memorial. How far is the Pentagon from the Washington Monument?

61. **Triangular plot.** A triangular plot of land has sides of length 310 feet, 415 feet, and 175 feet. Find the largest angle between the sides.

62. **Area.** Find the area of the triangular plot of Exercise 61.

EXERCISES FOR CALCULUS

1. If $f(t) \to L$ as $t \to a$, we say that the limit of $f(t)$ as $t \to a$ is L and write $\lim_{t \to a} f(t) = L$. Use a calculator. By choosing values of t close to 0, show that
$$\lim_{t \to a} \cos(t) = 1, \lim_{t \to 0} \sin(t) = 0, \text{ and } \lim_{t \to 0} \tan(t) = 0.$$

2. Let $f(t) = \dfrac{\sin t}{t}$. Note that $f(0)$ is undefined.
   a. For $t \neq 0$, show that $f(t)$ is an even function.
   b. Use a calculator to find $f(0,1), f(0.01), f(0.001)$, and $f(0.0001)$.
   c. What is the behavior of $f(t)$ as $t \to 0$?

3. In the figure, observe that in the unit circle
$$\text{area } \Delta AOB < \text{area sector } AOB < \text{area } \Delta AOQ.$$

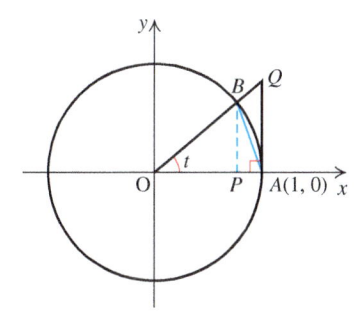

   a. Show that $\sin t < t < \tan t$.
   b. Deduce from part a. that $\cos t < \dfrac{\sin t}{t} < 1$.
   c. From part b. and Exercise 1, conclude that $\lim_{t \to 0} \dfrac{\sin t}{t} = 1$.

4. a. Show that $\lim_{t \to 0} \dfrac{\sin 2t}{t} = 2$.
   [*Hint:* Let $2t = x$; then if $t \to 0, x \to 0$ also.]
   b. Show that $\lim_{t \to 0} \dfrac{\sin ct}{t} = c$ for $c \neq 0$.
   c. Show that $\lim_{t \to 0} \dfrac{\sin^2 t}{t} = 0$.
   d. Show that $\lim_{t \to 0} \dfrac{1 - \cos t}{t} = 0$.

5. Show that of all of the parallelograms with adjacent sides a and b, the parallelogram with maximum area is a rectangle. [*Hint:* Area of a parallelogram $= ab \sin \theta$]

6. The rectangle formed by the vertices $(2, 0), (2, 5), (4, 5)$, and $(4, 0)$ is revolved about the x-axis to form a right circular cylinder. Find the volume of the generated cylinder.

7. Find the volume of the cylinder generated by revolving the rectangle with vertices $(a, 0), (a, b), (a + h, b)$, and $(a + h, 0)$ about the x-axis.

8. The rectangle of Exercise 6 is revolved about the y-axis to form a cylindrical shell. Find the volume of the shell.

9. Find the volume of the shell generated by revolving the rectangle of Exercise 7 about the y-axis.

10. Find the volume of the shell generated by revolving the rectangle of Exercise 6 about the line $y = 6$.

PRACTICE TEST A

1. In Problems 1–4, solve triangle *ABC*. Round each answer to the nearest tenth.

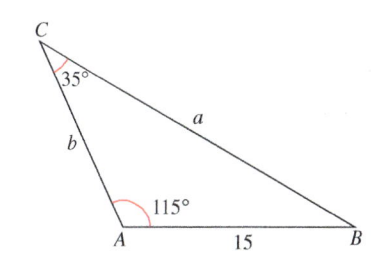

2. $A = 42°, B = 37°, a = 50$ meters.

3.

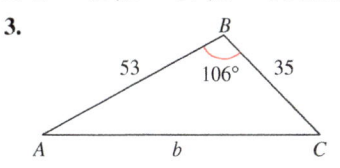

4. $a = 30$ feet, $b = 20$ feet, $c = 25$ feet.

5. Find the measure of the central angle of a circle of radius 8 feet that intercepts a chord of length 4.5 feet. Round to the nearest tenth of a degree.

6. Starting from point *A*, a surveyor walks 580 feet in the direction N 70.0° E. From that point, she walks 725 feet in the direction N 35.0° W. To the nearest foot, how far is she from her starting point?

In Problems 7–10, find the area of each triangle *ABC*. Round each answer to the nearest tenth square unit.

7. $a = 8, b = 12, C = 120°$
8. $C = 50°, a = 50, b = 60$
9. $A = 47°, B = 62°, c = 16$
10. $a = 2, b = 3, c = 4$
11. A parallelogram has adjacent sides 70 and 90 feet long, and the angle between them is 37°. Find the lengths of the two diagonals.
12. Three circles of radii 80, 100, and 150 feet are tangent to each other externally. Find the area of the triangle formed by joining the centers of the circles.
13. Convert $(-2, -45°)$ to rectangular coordinates.
14. Convert $(-\sqrt{3}, -1)$ to polar coordinates with $r > 0$ and $0 \leq \theta < 2\pi$.

446 Chapter 5 Applications of Trigonometric Functions

15. Convert the polar equation $r = -3$ to rectangular form. Identify the curve having this equation.
16. Convert $x^2 + y^2 - 4y = 0$ to polar form.
17. Check $r = \dfrac{3}{1 + \sin \theta}$ for symmetry.

In Problem 18–20, eliminate the parameter to obtain a Cartesian equation of the curve.

18. $x = 2 \sin^2 t, y = 3 \cos^2 t, 0 \le t \le 2\pi$
19. $x = 3 \cos t, y = 4 \sin t, 0 \le t \le 2\pi$
20. $x = 4 \sec t, y = 3 \tan t, -\infty < t < \infty$

PRACTICE TEST B

1. In triangle ABC, which is closest to the correct value for b?
   a. $b = 13.4$
   b. $b = 20.8$
   c. $b = 20.2$
   d. $b = 0.05$

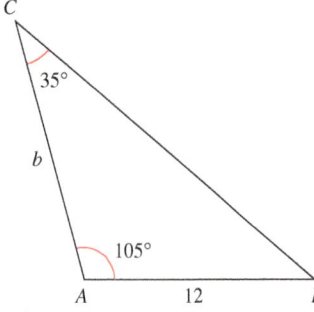

2. In triangle ABC, if $A = 27°, B = 50°$, and $a = 25$ meters, which is closest to the correct value for c?
   a. $c = 0.03$ meter
   b. $c = 42.2$ meters
   c. $c = 53.7$ meters
   d. $c = 0.02$ meter

3. Which statement is true for the measurements $A = 25°, a = 25$ feet, $b = 30$ feet? [The angle measurements in choices (c) and (d) are given to the nearest tenth.]
   a. No triangle has these measurements.
   b. One triangle has these measurements.
   c. Two triangles, with $c_1 = 48.7$ and $c_2 = 5.6$, have these measurements (in feet).
   d. Two triangles, with $c_1 = 46$ and $c_2 = 6.4$, have these measurements (in feet).

4. In triangle ABC, which is closest to the correct value for b?
   a. $b = 35.8$
   b. $b = 57.2$
   c. $b = 18.7$
   d. $b = 48.4$

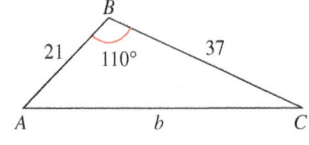

5. In triangle ABC, with $a = 12$ feet, $b = 13$ feet, and $c = 7$ feet, which is closest to the correct value for A?
   a. 24°
   b. 66°
   c. 81.8°
   d. 32.2°

6. Find the measure of the central angle of a circle of radius 9 feet that intercepts a chord of length 3 feet. Round to the nearest tenth of a degree.
   a. 70.8°
   b. 63.6°
   c. 26.4°
   d. 19.2°

7. A hunter leaves a straight east–west road at a point A, walking 2 miles at a bearing N 40° W to a point B. From point B, he walks at a bearing of S 20° E and arrives back on the road at a point C. How far, to the nearest tenth of a mile, is point B from point C?
   a. 4.5 miles
   b. 2.9 miles
   c. 1.6 miles
   d. 5.3 miles

8. How many triangles are possible with the measurements $A = 43°, a = 185$ feet, $b = 248$ feet?
   a. No triangle
   b. One triangle
   c. Two triangles
   d. Infinitely many triangles

In Problems 9–11, find the area of each triangle ABC. Answers are rounded to the nearest square unit.

9. $b = 5.1$ ft, $c = 3.8$ ft, and $A = 48°$
   a. 6.5 ft²
   b. 7.2 ft²
   c. 7.9 ft²
   d. 4.7 ft²

10. $A = 30°, B = 45°$, and $c = 2.7$ m
    a. 2.4 m²
    b. 1.6 m²
    c. 1.3 m²
    d. 3.2 m²

11. $a = 40$ in., $b = 30$ in., and $c = 20$ in.
    a. 512 in.²
    b. 312 in.²
    c. 290 in.²
    d. 262 in.²

12. The area of an equilateral triangle whose perimeter is 60 units is
    a. $250\sqrt{3}$ unit².
    b. $200\sqrt{3}$ unit².
    c. 150 unit².
    d. $100\sqrt{3}$ unit².

13. Convert $(-4, -30°)$ to rectangular coordinates.
    a. $(-2\sqrt{3}, -2)$
    b. $(2, -2\sqrt{3})$
    c. $(-2\sqrt{3}, 2)$
    d. $(2\sqrt{3}, -2)$

14. Convert $(3, -\sqrt{3})$ to polar coordinates, with $r > 0$ and $0 \le \theta < 360°$.
    a. $(12, 330°)$
    b. $(2\sqrt{3}, 330°)$
    c. $(2\sqrt{3}, 150°)$
    d. $(\sqrt{3}, 150°)$

15. Which polar equation has a circle for its graph?
    a. $r = -4 \cos \theta + 1$
    b. $r = -2$
    c. $r \sin \theta = 2$
    d. $r \cos \theta = 1$

16. Convert $x^2 + y^2 - 2x = 0$ to polar form.
    a. $r = 2 \sin \theta$
    b. $r^2 = 1 + \sin \theta$
    c. $r = 2 \cos \theta$
    d. $r = 1 + 2 \cos \theta$

17. Convert $2x + 3y = 6$ to polar form.
    a. $r = \dfrac{6}{2 \cos \theta + 3 \sin \theta}$
    b. $r = \dfrac{6}{2 \sin \theta + 3 \cos \theta}$
    c. $r = 6[2 \cos\theta + 3 \sin \theta]$
    d. $r = \dfrac{2 \sin \theta + 3 \cos \theta}{6}$

18. The curve $r = 3 + 2 \cos \theta$ is symmetric about
    a. the x-axis.
    b. the y-axis.
    c. the origin.
    d. no symmetry.

In Exercises 19 and 20, find the Cartesian equation of the given parametric equation.

19. $x = 2 \cos^2 t, y = 5 \sin^2 t, -\infty < t < \infty$
    a. $2x + 5y = 10$
    b. $5x + 2y = 10$
    c. $5x + 2y = 1$
    d. $2x + 5y = 1$

20. $x = 4 \cos t, y = 3 \sin t, -\infty < t < \infty$
    a. $x^2 + y^2 = 5$
    b. $16x^2 + 9y^2 = 25$
    c. $9x^2 + 16y^2 = 144$
    d. $16x^2 + 9y^2 = 144$

CHAPTER 6

Further Topics in Algebra

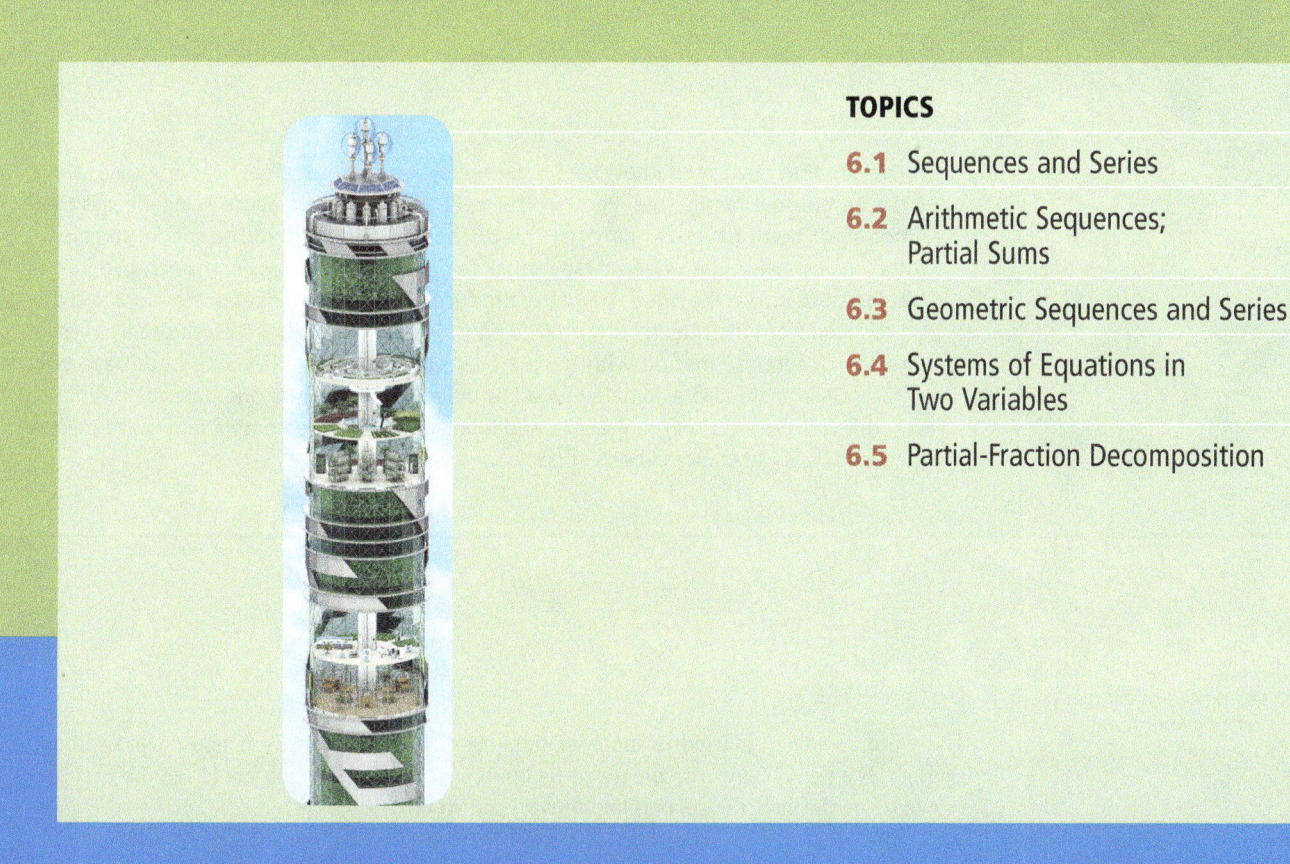

TOPICS

6.1 Sequences and Series

6.2 Arithmetic Sequences; Partial Sums

6.3 Geometric Sequences and Series

6.4 Systems of Equations in Two Variables

6.5 Partial-Fraction Decomposition

Sequences and series find applications in a wide range of areas, from finance to medicine to forensic science.

Systems of equations are frequently the tools used to describe relationships among variables representing different interrelated quantities. These same systems of equations are used to predict results associated with changes in the quantities.

The quantities involved can come from the study of medicine, taxes, elections, business, agriculture, city planning, or any discipline in which multiple quantities interact.

SECTION 6.1 Sequences and Series

BEFORE STARTING THIS SECTION, REVIEW

1. Finding function values (Section 1.3, page 80)

OBJECTIVES

1. Use sequence notation and find specific and general terms in a sequence.
2. Use factorial notation.
3. Use summation notation to write partial sums of a series.

◆ The Family Tree of the Honeybee

The bee that most of us know best is the honeybee, an insect that lives in a colony and has an unusual family tree. Perhaps the most surprising fact about honeybees is that not all of them have two parents. This phenomenon begins with a special female in the colony called the *queen*. Many other female bees, called *worker bees*, live in the colony, but unlike the queen bee, they produce no eggs. Male bees do no work and are produced from the queen's unfertilized eggs. The female bees are produced as a result of the queen bee mating with a male bee. Consequently, all female bees have two parents—a male and a female—whereas male bees have just one parent—a female. In this section, we study sequences, and in Example 4, we see how a famous sequence accurately counts a honeybee's ancestors.

1 Use sequence notation and find specific and general terms in a sequence.

Sequences

The word *sequence* is used in mathematics in much the same way it is used in ordinary English. If a person saw a *sequence* of bad movies, then that person could list the first bad movie he or she saw, the second bad movie, and so on.

Sequence

An **infinite sequence** is a function whose domain is the set of positive integers. The function values, written as

$$a_1, a_2, a_3, a_4, \ldots, a_n, \ldots,$$

are called the **terms** of the sequence. The **nth term**, a_n, is called the **general term** of the sequence.

If the domain of a function consists of only the first n positive integers, the sequence is called a **finite sequence**.

In computer science, finite sequences of the form $a_1, a_2, \ldots, a_n$ are called *strings*. Sometimes for convenience, we include 0 in the domain of the function that defines a sequence, and we write the sequence terms as $a_0, a_1, a_2, a_3, \ldots$. The *n*th term of this sequence is a_{n-1}.

We can use subscripts on variables other than a to represent the terms of a sequence. We use the variable b in Example 1**c**.

SIDE NOTE

Notice in Example 1 that converting $a_n = 5n - 1$ to standard function notation gives us $f(n) = 5n - 1$, a linear function.

TECHNOLOGY CONNECTION

A graphing calculator can write the terms of a sequence and graph the sequence. You must first select *sequence* mode.

The first four terms of the sequence $a_n = \dfrac{1}{n+1}$ and the graph of this sequence are shown here. Use the right arrow key ▶ to scroll to the right to display the sequence values corresponding to $n = 3$ and $n = 4$.

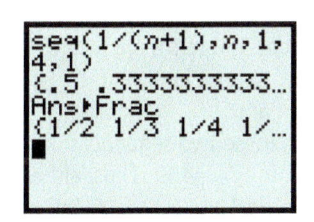

```
seq(1/(n+1),n,1,
4,1)
{.5 .3333333333...
Ans▶Frac
{1/2 1/3 1/4 1/...
```

Now press GRAPH.

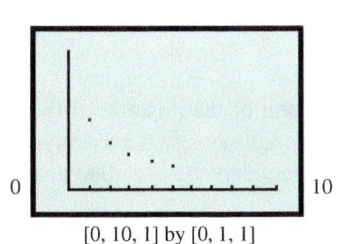

[0, 10, 1] by [0, 1, 1]

EXAMPLE 1   Writing the Terms of a Sequence from the General Term

Write the first four terms of each sequence.

a. $a_n = 5n - 1$ **b.** $a_n = \dfrac{1}{n+1}$ **c.** $b_n = (-1)^{n+1}\left(\dfrac{1}{n}\right)$

Solution

The first four terms of each sequence are found by replacing n with the integers 1, 2, 3, and 4 in the equation defining a_n.

a. $a_n = 5n - 1$ General term of the sequence
$a_1 = 5(1) - 1 = 4$ 1st term: Replace n with 1.
$a_2 = 5(2) - 1 = 9$ 2nd term: Replace n with 2.
$a_3 = 5(3) - 1 = 14$ 3rd term: Replace n with 3.
$a_4 = 5(4) - 1 = 19$ 4th term: Replace n with 4.

The first four terms of the sequence are: 4, 9, 14, and 19.

b. $a_n = \dfrac{1}{n+1}$

$a_1 = \dfrac{1}{1+1} = \dfrac{1}{2}$

$a_2 = \dfrac{1}{2+1} = \dfrac{1}{3}$

$a_3 = \dfrac{1}{3+1} = \dfrac{1}{4}$

$a_4 = \dfrac{1}{4+1} = \dfrac{1}{5}$

The first four terms of the sequence are: $\dfrac{1}{2}, \dfrac{1}{3}, \dfrac{1}{4},$ and $\dfrac{1}{5}$.

c. $b_n = (-1)^{n+1}\left(\dfrac{1}{n}\right)$

$b_1 = (-1)^{1+1}\left(\dfrac{1}{1}\right) = (-1)^2(1) = 1$

$b_2 = (-1)^{2+1}\left(\dfrac{1}{2}\right) = (-1)^3\left(\dfrac{1}{2}\right) = -\dfrac{1}{2}$

$b_3 = (-1)^{3+1}\left(\dfrac{1}{3}\right) = (-1)^4\left(\dfrac{1}{3}\right) = \dfrac{1}{3}$

$b_4 = (-1)^{4+1}\left(\dfrac{1}{4}\right) = (-1)^5\left(\dfrac{1}{4}\right) = -\dfrac{1}{4}$

The first four terms of the sequence are: $1, -\dfrac{1}{2}, \dfrac{1}{3},$ and $-\dfrac{1}{4}$.

Practice Problem 1 Write the first four terms of the sequence with general term $a_n = -2^n$.

Frequently, the first few terms of a sequence exhibit a pattern that *suggests* a natural choice for the general term of the sequence.

EXAMPLE 2   Finding a General Term of a Sequence from a Pattern

Write the general term a_n for a sequence whose first five terms are given.

a. $1, 4, 9, 16, 25, \ldots$ **b.** $0, \dfrac{1}{2}, -\dfrac{2}{3}, \dfrac{3}{4}, -\dfrac{4}{5}, \ldots$

Solution

Write the position number of the term above each term of the sequence and look for a pattern that connects the term to the position number of the term.

a.
$$n: \quad 1, \quad 2, \quad 3, \quad 4, \quad 5, \quad \ldots, \quad n$$
$$\text{term:} \quad 1, \quad 4, \quad 9, \quad 16, \quad 25, \quad \ldots, \quad a_n$$

Apparent pattern: Here $1 = 1^2, 4 = 2^2, 9 = 3^2, 16 = 4^2$, and $25 = 5^2$. Each term is the square of the position number of that term. This suggests $a_n = n^2$.

b.
$$n: \quad 1, \quad 2, \quad 3, \quad 4, \quad 5, \quad \ldots, \quad n$$
$$\text{term:} \quad 0, \quad \frac{1}{2}, \quad -\frac{2}{3}, \quad \frac{3}{4}, \quad -\frac{4}{5}, \quad \ldots, \quad a_n$$

Apparent pattern: When the terms alternate in sign and $n = 1$, we use factors such as $(-1)^n$ if we want to begin with the factor -1 or we use factors such as $(-1)^{n+1}$ if we want to begin with the factor 1. Notice that each term can be written as a quotient with denominator equal to the position number and numerator equal to one less than the position number, suggesting the general term $a_n = (-1)^n \left(\dfrac{n-1}{n} \right)$.

Practice Problem 2 Write a general term for a sequence whose first five terms are $0, -\dfrac{1}{2}, \dfrac{2}{3}, -\dfrac{3}{4}, \dfrac{4}{5}, \ldots$.

WARNING

In Example 2, we found the general term of a sequence by finding a pattern in the first few terms of the sequence. However, when only a finite number of successive terms are given without a rule that defines the general term, a *unique* general term cannot be found.

To indicate why this is so, consider the two sequences $a_n = n^2$ and $b_n = n^2 + (n-1)(n-2)(n-3)(n-4)$. The first four terms of these two sequences are identical: 1, 4, 9, 16. However, the fifth term is different: $a_5 = 25$, but $b_5 = 49$. Thus, either a_n or b_n could correctly describe a sequence whose first four terms are 1, 4, 9, and 16. So you can *never* find a unique general term from a finite number of terms in a sequence.

HISTORICAL NOTE

Leonardo of Pisa (Fibonacci)

(1175–1250) Leonardo, often known today by the name Fibonacci (son of Bonaccio) was born around 1175. He traveled extensively through North Africa and the Mediterranean, probably on business with his father. At each location, he met with Islamic scholars and absorbed the mathematical knowledge of the Islamic world. After returning to Pisa around 1200, he started writing books on mathematics. He was one of the earliest European writers on algebra. In his most famous book *Liber abaci*, or *Book of Calculations*, he introduced the Western world to the rules of computing with the new Hindu–Arabic numerals.

Recursive Formulas

Up to this point, we have expressed the general term as a function of the position of the term in the sequence. A second approach is to define a sequence *recursively*. A **recursive formula** requires that one or more of the first few terms of the sequence be specified and all other terms be defined in relation to previously defined terms.

For example, the **Fibonacci sequence** is a famous sequence that is defined recursively and shows up often in nature. In this sequence, we specify the first two terms as $a_0 = 1$ and $a_1 = 1$; each subsequent term is the sum of the two terms immediately preceding it. So we have

$a_0 = 1, a_1 = 1, a_2 = a_0 + a_1 = 1 + 1 = 2$	Add the two given terms.
$a_1 = 1, a_2 = 2, a_3 = a_1 + a_2 = 1 + 2 = 3$	Add the two previous terms.
$a_2 = 2, a_3 = 3, a_4 = a_2 + a_3 = 2 + 3 = 5$	Add the two previous terms.
$a_3 = 3, a_4 = 5, a_5 = a_3 + a_4 = 3 + 5 = 8$	Add the two previous terms.

The first six terms of the sequence are then

$$1, 1, 2, 3, 5, 8.$$

This sequence can also be defined in subscript notation:

$$a_0 = 1, a_1 = 1, a_k = a_{k-2} + a_{k-1}, \text{ for all } k \geq 2$$

Replacing k with 2, 3, 4, and 5, respectively, in the equation $a_k = a_{k-2} + a_{k-1}$ will again produce the values $a_2 = 2, a_3 = 3, a_4 = 5$, and $a_5 = 8$.

EXAMPLE 3 Finding Terms of a Recursively Defined Sequence

Write the first five terms of the recursively defined sequence

$$a_1 = 4, a_{n+1} = 2a_n - 9.$$

Solution

We are given the first term of the sequence: $a_1 = 4$.

$a_2 = 2a_1 - 9$	Replace n with 1 in $a_{n+1} = 2a_n - 9$.
$a_2 = 2(4) - 9 = -1$	Replace a_1 with 4.
$a_3 = 2a_2 - 9$	Replace n with 2 in $a_{n+1} = 2a_n - 9$.
$a_3 = 2(-1) - 9 = -11$	Replace a_2 with -1.
$a_4 = 2a_3 - 9$	Replace n with 3 in $a_{n+1} = 2a_n - 9$.
$a_4 = 2(-11) - 9 = -31$	Replace a_3 with -11.
$a_5 = 2a_4 - 9$	Replace n with 4 in $a_{n+1} = 2a_n - 9$.
$a_5 = 2(-31) - 9 = -71$	Replace a_4 with -31.

Thus, the first five terms of the sequence are

$$4, -1, -11, -31, -71.$$

Practice Problem 3 Write the first five terms of the recursively defined sequence $a_1 = -3, a_{n+1} = 2a_n + 5$.

EXAMPLE 4 Diagramming the Family Tree of a Honeybee

Find a sequence that accurately counts the ancestors of a male honeybee. The first term of the sequence should give the number of parents of a male honeybee; the second term, the number of grandparents; the third term, the number of great-grandparents; and so on.

Solution

Recall from the introduction to this section that female honeybees have two parents—a male and a female—but that male honeybees have just one parent—a female.

First, we consider the family tree of a male honeybee. A male bee has one parent: a female. Because his mother had two parents, he has two grandparents. Because his grandmother had two parents and his grandfather had only one parent, he has three great-grandparents.

Now to count the bee's ancestors, we count backward using Figure 6.1. You should recognize the first six terms of the Fibonacci sequence: 1, 1, 2, 3, 5, 8. This sequence is defined by $a_0 = 1, a_1 = 1$, and $a_k = a_{k-2} + a_{k-1}$, for all $k \geq 2$.

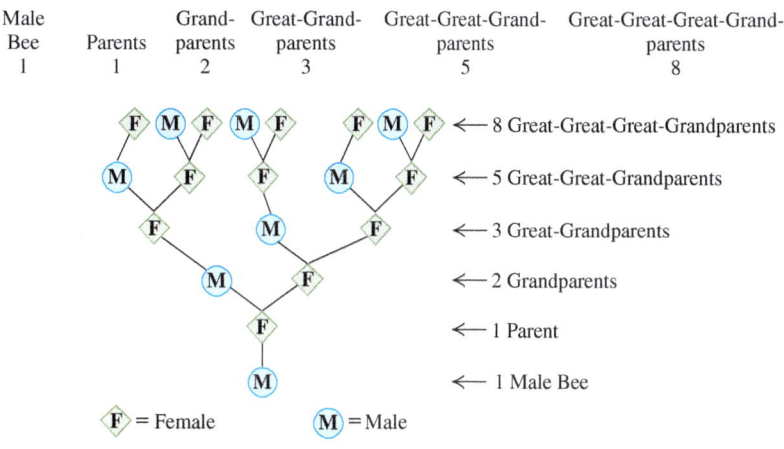

FIGURE 6.1

Practice Problem 4 Find a sequence that accurately counts the ancestors of a female honeybee.

2 Use factorial notation.

Factorial Notation

Special types of products, called factorials, appear in some very important sequences.

DO YOU KNOW?

The symbol n! was first introduced by Christian Kramp of Strasbourg in 1808 as a convenience for the printer. Alternative symbols ⌊n and $\pi(n)$ were used until late in the nineteenth century.

A typical calculator will compute approximate values of n! up to $n = 69$.

$$69! \approx 1.7112 \times 10^{98}$$

Factorial

For any positive integer n, **n factorial** (written as n!) is defined by

$$n! = n \cdot (n-1) \cdots 4 \cdot 3 \cdot 2 \cdot 1.$$

As a special case, **zero factorial** (0!) is defined by

$$0! = 1.$$

Here are the first eight values of n!

$$0! = 1 \qquad\qquad 4! = 4 \cdot 3 \cdot 2 \cdot 1 = 24$$
$$1! = 1 \qquad\qquad 5! = 5 \cdot 4 \cdot 3 \cdot 2 \cdot 1 = 120$$
$$2! = 2 \cdot 1 = 2 \qquad 6! = 6 \cdot 5 \cdot 4 \cdot 3 \cdot 2 \cdot 1 = 720$$
$$3! = 3 \cdot 2 \cdot 1 = 6 \qquad 7! = 7 \cdot 6 \cdot 5 \cdot 4 \cdot 3 \cdot 2 \cdot 1 = 5040$$

The values of n! get large very quickly. For example, $10! = 3,628,800$ and $12! = 479,001,600$.

In simplifying some expressions involving factorials, it is helpful to note that

$$n! = n \cdot (n-1)!$$

TECHNOLOGY CONNECTION

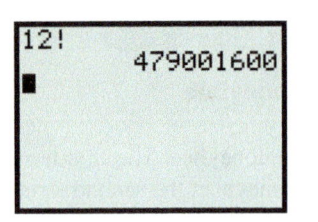

```
12!
        479001600
■
```

EXAMPLE 5 Simplifying a Factorial Expression

Simplify.

a. $\dfrac{16!}{14!}$ b. $\dfrac{(n+1)!}{(n-1)!}$

Solution

a. $\dfrac{16!}{14!} = \dfrac{16 \cdot 15 \cdot 14!}{14!}$ $16! = 16 \cdot 15! = 16 \cdot 15 \cdot 14!$

$= 16 \cdot 15 = 240$ Divide out the factor 14!

b. $\dfrac{(n+1)!}{(n-1)!} = \dfrac{(n+1) \cdot n \cdot (n-1)!}{(n-1)!}$ $(n+1)! = (n+1) \cdot n! = (n+1) \cdot n \cdot (n-1)!$

$= (n+1)n$ Divide out the common factor $(n-1)!$

Practice Problem 5 Simplify.

a. $\dfrac{13!}{12!}$ b. $\dfrac{n!}{(n-3)!}$

EXAMPLE 6 Writing Terms of a Sequence Involving Factorials

Write the first five terms of the sequence whose general term is $a_n = \dfrac{(-1)^{n+1}}{n!}$.

Solution

Replace n in the formula for the general term with each positive integer from 1 through 5.

$$a_1 = \frac{(-1)^{1+1}}{1!} = \frac{(-1)^2}{1} = 1 \qquad a_4 = \frac{(-1)^{4+1}}{4!} = \frac{(-1)^5}{24} = -\frac{1}{24}$$

$$a_2 = \frac{(-1)^{2+1}}{2!} = \frac{(-1)^3}{2} = -\frac{1}{2} \qquad a_5 = \frac{(-1)^{5+1}}{5!} = \frac{(-1)^6}{120} = \frac{1}{120}$$

$$a_3 = \frac{(-1)^{3+1}}{3!} = \frac{(-1)^4}{6} = \frac{1}{6}$$

Using these five terms, we could write this sequence as

$$1, -\frac{1}{2}, \frac{1}{6}, -\frac{1}{24}, \frac{1}{120}, \ldots.$$

Practice Problem 6 Write the first five terms of the sequence whose general term is

$$a_n = \frac{(-1)^n 2^n}{n!}.$$

Summation Notation

3 Use summation notation to write partial sums of a series.

If we know the general term of a sequence, we can represent a *sum* of terms of the sequence by using *summation* (or *sigma*) *notation*. In this notation, the Greek letter Σ (capital sigma) indicates that we are to *add* the given terms. The letter Σ corresponds to the English letter S, the first letter of the word *sum*.

SUMMATION NOTATION

The sum of the first n terms of a sequence $a_1, a_2, a_3, \ldots, a_n, \ldots$ is denoted by

$$\sum_{i=1}^{n} a_i = a_1 + a_2 + a_3 + \cdots + a_n.$$

The letter i in the summation notation is called the **index of summation**, n is called the **upper limit**, and 1 is called the **lower limit** of the summation.

TECHNOLOGY CONNECTION

A graphing calculator in *sequence* mode can compute a finite sum of sequence terms.

The viewing window shows $\sum_{j=4}^{7} (2j^2 - 1)$. Note that the calculator replaces the variable j with n.

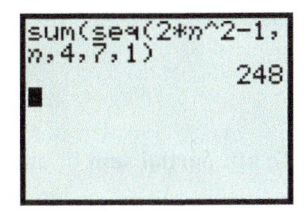

```
sum(seq(2*n^2-1,
n,4,7,1)
              248
```

In summation notation, the index need not start at $i = 1$. Also, any letter may be used in place of the index i. In general, we have

$$\sum_{j=k}^{n} a_j = a_k + a_{k+1} + \cdots + a_n, \quad 0 \leq k \leq n.$$

EXAMPLE 7 Evaluating Sums Given in Summation Notation

Find each sum.

a. $\sum_{i=1}^{9} i$ **b.** $\sum_{j=4}^{7} (2j^2 - 1)$ **c.** $\sum_{k=0}^{4} \frac{2^k}{k!}$

Solution

a. Replace i with successive integers from 1 to 9 inclusive; then add.

$$\sum_{i=1}^{9} i = 1 + 2 + 3 + 4 + 5 + 6 + 7 + 8 + 9 = 45$$

b. The index of summation is j. Evaluate $(2j^2 - 1)$ for all consecutive integers from 4 through 7 inclusive; then add.

$$\sum_{j=4}^{7} (2j^2 - 1) = [2(4)^2 - 1] + [2(5)^2 - 1] + [2(6)^2 - 1] + [2(7)^2 - 1]$$
$$= 31 + 49 + 71 + 97 = 248$$

c. The index of summation is k. Evaluate $\dfrac{2^k}{k!}$ for consecutive integers 0 through 4, inclusive; then add.

$$\sum_{k=0}^{4} \frac{2^k}{k!} = \frac{2^0}{0!} + \frac{2^1}{1!} + \frac{2^2}{2!} + \frac{2^3}{3!} + \frac{2^4}{4!}$$
$$= \frac{1}{1} + \frac{2}{1} + \frac{4}{2} + \frac{8}{6} + \frac{16}{24} = 1 + 2 + 2 + \frac{4}{3} + \frac{2}{3} = 7$$

Practice Problem 7 Find the following sum: $\sum_{k=0}^{3} (-1)^k k!$

The familiar properties of real numbers, such as the distributive, commutative, and associative properties, can be used to prove the summation properties listed next.

SUMMATION PROPERTIES

Let a_k and b_k represent the general terms of two sequences and let c represent any real number. Then

1. $\sum_{k=1}^{n} c = c \cdot n$

2. $\sum_{k=1}^{n} ca_k = c \sum_{k=1}^{n} a_k$

3. $\sum_{k=1}^{n} (a_k + b_k) = \sum_{k=1}^{n} a_k + \sum_{k=1}^{n} b_k$

4. $\sum_{k=1}^{n} (a_k - b_k) = \sum_{k=1}^{n} a_k - \sum_{k=1}^{n} b_k$

5. $\sum_{k=1}^{n} a_k = \sum_{k=1}^{j} a_k + \sum_{k=j+1}^{n} a_k$, for $1 \leq j < n$

Series

We all know how to add numbers, but do you think it is possible to add infinitely many numbers? It is sometimes possible to add infinitely many numbers, although a complete explanation of this kind of addition must be put off until a calculus course. For now, we need some definitions.

Series

Let $a_1, a_2, a_3, \ldots, a_k, \ldots$ be an infinite sequence. Then

1. The sum of all terms of the sequence is called a **series** and is denoted by

$$a_1 + a_2 + a_3 + \cdots = \sum_{i=1}^{\infty} a_i.$$

2. The sum of the first n terms of the sequence is called the **nth partial sum** of the series and is denoted by

$$a_1 + a_2 + a_3 + \cdots + a_n = \sum_{i=1}^{n} a_i.$$

EXAMPLE 8 Writing a Partial Sum in Summation Notation

Write each sum in summation notation.

a. $3 + 5 + 7 + \cdots + 21$ **b.** $\dfrac{1}{4} + \dfrac{1}{9} + \cdots + \dfrac{1}{49}$

Solution

a. The finite series $3 + 5 + 7 + \cdots + 21$ is a sum of consecutive odd numbers from 3 to 21. Each of these numbers can be expressed in the form $2k + 1$, starting with $k = 1$ and ending with $k = 10$. With k as the index of summation, 1 as the lower limit, and 10 as the upper limit, we write

$$3 + 5 + 7 + \cdots + 21 = \sum_{k=1}^{10} (2k + 1).$$

b. The finite series $\dfrac{1}{4} + \dfrac{1}{9} + \cdots + \dfrac{1}{49}$ is a sum of fractions, each of which has numerator 1 and denominator k^2, starting with $k = 2$ and ending with $k = 7$. We write

$$\dfrac{1}{4} + \dfrac{1}{9} + \cdots + \dfrac{1}{49} = \sum_{k=2}^{7} \dfrac{1}{k^2}.$$

Practice Problem 8 Write the following in summation notation: $2 - 4 + 6 - 8 + 10 - 12 + 14$.

SECTION 6.1 Exercises

Basic Concepts and Skills

In Exercises 1–20, write the first four terms of each sequence.

1. $a_n = 3n - 2$
2. $a_n = 2n + 1$
3. $a_n = 1 - \dfrac{1}{n}$
4. $a_n = 1 + \dfrac{1}{n}$
5. $a_n = -n^2$
6. $a_n = n^3$
7. $a_n = \dfrac{2n}{n + 1}$
8. $a_n = \dfrac{3n}{n^2 + 1}$
9. $a_n = (-1)^{n+1}$
10. $a_n = (-3)^{n-1}$
11. $a_n = 3 - \dfrac{1}{2^n}$
12. $a_n = \left(\dfrac{3}{2}\right)^n$
13. $a_n = 0.6$
14. $a_n = -0.4$
15. $a_n = \dfrac{(-1)^n}{n!}$
16. $a_n = \dfrac{n}{n!}$
17. $a_n = (-1)^n 3^{-n}$
18. $a_n = (-1)^n 3^{n-1}$
19. $a_n = \dfrac{e^n}{2n}$
20. $a_n = \dfrac{2^n}{e^n}$

In Exercises 21–34, write a general term a_n for each sequence. Assume that n begins with 1.

21. $1, 4, 7, 10, \ldots$
22. $7, 9, 11, 13, \ldots$
23. $\dfrac{1}{2}, \dfrac{1}{3}, \dfrac{1}{4}, \dfrac{1}{5}, \ldots$
24. $\dfrac{2}{3}, \dfrac{3}{4}, \dfrac{4}{5}, \dfrac{5}{6}, \ldots$
25. $2, -2, 2, -2, \ldots$
26. $-3, 6, -9, 12, \ldots$
27. $\dfrac{1}{2}, \dfrac{3}{4}, \dfrac{9}{8}, \dfrac{27}{16}, \ldots$
28. $\dfrac{-1}{2}, \dfrac{1}{4}, \dfrac{-1}{8}, \dfrac{1}{16}, \ldots$
29. $1 \cdot 2, 2 \cdot 3, 3 \cdot 4, 4 \cdot 5, \ldots$
30. $\dfrac{-1}{1 \cdot 2}, \dfrac{1}{2 \cdot 3}, \dfrac{-1}{3 \cdot 4}, \dfrac{1}{4 \cdot 5}, \ldots$
31. $2 + \dfrac{1}{2}, 2 - \dfrac{1}{3}, 2 + \dfrac{1}{4}, 2 - \dfrac{1}{5}, \ldots$
32. $1 + \dfrac{1}{2}, 1 + \dfrac{1}{3}, 1 + \dfrac{1}{4}, 1 + \dfrac{1}{5}, \ldots$
33. $\dfrac{3^2}{2}, \dfrac{3^3}{3}, \dfrac{3^4}{4}, \dfrac{3^5}{5}, \ldots$
34. $\dfrac{e}{2}, \dfrac{e^2}{4}, \dfrac{e^3}{8}, \dfrac{e^4}{16}, \ldots$

In Exercises 35–44, write the first five terms of each recursively defined sequence.

35. $a_1 = 2, a_{n+1} = a_n + 3$
36. $a_1 = 5, a_{n+1} = a_n - 1$
37. $a_1 = 3, a_{n+1} = 2a_n$
38. $a_1 = 1, a_{n+1} = \dfrac{1}{2} a_n$
39. $a_1 = 7, a_{n+1} = -2a_n + 3$
40. $a_1 = -4, a_{n+1} = -3a_n - 5$

41. $a_1 = 2, a_{n+1} = \dfrac{1}{a_n}$ **42.** $a_1 = -1, a_{n+1} = \dfrac{-1}{a_n}$

43. $a_1 = 25, a_{n+1} = \dfrac{(-1)^n}{5a_n}$ **44.** $a_1 = 12, a_{n+1} = \dfrac{(-1)^n}{3a_n}$

In Exercises 45–50, use a graphing calculator to (a) find the first ten terms of the sequence and (b) graph the first ten terms of the sequence.

45. $a_n = 3n^2 - 1$ **46.** $a_n = 4 - \dfrac{3}{n}$

47. $a_n = n\left(1 - \dfrac{1}{n}\right)$ **48.** $a_n = n^3 - n^2$

49. $a_n = 1 - \dfrac{1}{a_{n-1}}, a_1 = \dfrac{1}{2}$ **50.** $a_n = a_{n-1}^2, a_1 = 1$

In Exercises 51–58, simplify the factorial expression.

51. $\dfrac{3!}{5!}$ **52.** $\dfrac{8!}{10!}$

53. $\dfrac{12!}{11!}$ **54.** $\dfrac{20!}{18!}$

55. $\dfrac{n!}{(n+1)!}$ **56.** $\dfrac{(n-1)!}{(n-2)!}$

57. $\dfrac{(2n+1)!}{(2n)!}$ **58.** $\dfrac{(2n+1)!}{(2n-1)!}$

In Exercises 59–70, find each sum.

59. $\sum\limits_{k=1}^{7} 5$ **60.** $\sum\limits_{j=1}^{4} 12$

61. $\sum\limits_{j=0}^{5} j^2$ **62.** $\sum\limits_{k=0}^{4} k^3$

63. $\sum\limits_{i=1}^{5} (2i - 1)$ **64.** $\sum\limits_{k=0}^{6} (1 - 3k)$

65. $\sum\limits_{j=3}^{7} \dfrac{j+1}{j}$ **66.** $\sum\limits_{k=3}^{8} \dfrac{1}{k+1}$

67. $\sum\limits_{i=2}^{6} (-1)^i 3^{i-1}$ **68.** $\sum\limits_{k=2}^{4} (-1)^{k+1} k$

69. $\sum\limits_{k=4}^{7} (2 - k^2)$ **70.** $\sum\limits_{j=4}^{9} (j^3 + 1)$

In Exercises 71–78, write each sum in summation notation.

71. $1 + 3 + 5 + 7 + \cdots + 101$

72. $2 + 4 + 6 + 8 + \cdots + 102$

73. $\dfrac{1}{5(1)} + \dfrac{1}{5(2)} + \dfrac{1}{5(3)} + \dfrac{1}{5(4)} + \cdots + \dfrac{1}{5(11)}$

74. $1 + \dfrac{2}{2 \cdot 3} + \dfrac{2}{3 \cdot 4} + \dfrac{2}{4 \cdot 5} + \cdots + \dfrac{2}{9 \cdot 10}$

75. $1 - \dfrac{1}{2} + \dfrac{1}{3} - \dfrac{1}{4} + \cdots - \dfrac{1}{50}$

76. $1 - 2 + 3 - 4 + \cdots + (-50)$

77. $\dfrac{1}{2} + \dfrac{2}{3} + \dfrac{3}{4} + \dfrac{4}{5} + \cdots + \dfrac{10}{11}$

78. $2 + \dfrac{2^2}{2} + \dfrac{2^3}{3} + \dfrac{2^4}{4} + \cdots + \dfrac{2^{10}}{10}$

In Exercises 79–84, use a graphing calculator to find each sum.

79. $\sum\limits_{i=1}^{10} 12i^2$ **80.** $\sum\limits_{i=1}^{50} (2i + 7)$

81. $\sum\limits_{k=5}^{30} \dfrac{7}{1 - k^2}$

82. $\sum\limits_{k=10}^{25} \dfrac{(k+1)^2}{k}$

83. $\sum\limits_{j=1}^{100} (-1)^j j$

84. $\sum\limits_{j=8}^{50} \left(7 + \dfrac{(-1)^j}{j}\right)$

Applying the Concepts

85. Free fall. In the absence of friction, a freely falling body will fall about 16 feet the first second, 48 feet the next second, 80 feet the third second, 112 feet the fourth second, and so on. How far has it fallen during
   a. the seventh second?
   b. the nth second?

86. Bacterial growth. A colony of 1000 bacteria doubles in size every hour. How many bacteria will there be in
   a. 2 hours?
   b. 5 hours?
   c. n hours?

87. Workplace fines. A contractor who quits work on a house was told she would be fined $50 if she did not resume work on Monday, $75 if she failed to resume work on Tuesday, $100 if she did not resume work on Wednesday, and so on (including weekends). How much will her fine be on the ninth day she fails to show up for work?

88. Appreciating value. A painting valued at $30,000 is expected to appreciate $1280 the first year, $1240 the second year, $1200 the third year, and so on. How much will the painting appreciate in
   a. the seventh year?
   b. the tenth year?

89. Cell phone use. At the end of the first six months after a company began providing cell phones to its sales force, it averaged 600 cell minutes per month. For the next four years, its monthly cell phone use doubled every six months. How many cell minutes per month were being used at the end of three years?

90. Motorcycle acceleration. A motorcycle travels 10 yards the first second and then increases its speed by 20 yards per second in each succeeding second. At this rate, how far will the motorcycle travel during
   a. the eighth second?
   b. the nth second?

91. Compound interest. Suppose $10,000 is deposited into an account that earns 6% interest compounded semiannually. The balance in the account after n compounding periods is given by the sequence

$$A_n = 10{,}000\left(1 + \dfrac{0.06}{2}\right)^n, n = 1, 2, 3, \ldots.$$

   a. Find the first six terms of this sequence.
   b. Find the balance in the account after eight years.

92. Compound interest. Suppose $100 is deposited at the beginning of a year into an account that earns 8% interest compounded quarterly. The balance in the account after n compounding periods is

$$A_n = 100\left(1 + \frac{0.08}{4}\right)^n, n = 1, 2, 3, \ldots.$$

a. Find the first six terms of this sequence.
b. Find the balance in the account after ten years.

93. Real estate value. Analysts estimate that a $100,000 condominium will increase 5% in value each year for the next seven years. Find the value of the condominium in each of those years. Write a formula for a sequence whose first seven terms give these values.

94. Diminishing savings. Laura withdraws 10% of her $50,000 savings each year. Find the amount remaining in her account for each of the next ten years. Write a formula for a sequence whose first ten terms give these values.

Critical Thinking / Discussion / Writing

95. The triangular tiles used in the figures shown have white interiors and gold edges. A sequence of figures is obtained by adding one triangle to the previous figure.

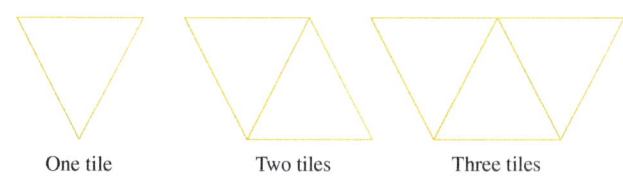

One tile   Two tiles   Three tiles

a. Write a recursive sequence whose nth term, a_n, gives the number of gold edges in the nth figure.
b. Write a recursive sequence whose nth term, b_n, gives the number of gold edges that lie on the perimeter of the nth figure.

96. The figure shows a sequence of successively smaller squares formed by connecting the midpoints of the sides of the preceding larger square. The area of the largest square is 1.
a. Write the first five terms of the sequence whose general term, a_n, gives the area of the nth square.
b. Write the general term, a_n, of the sequence.

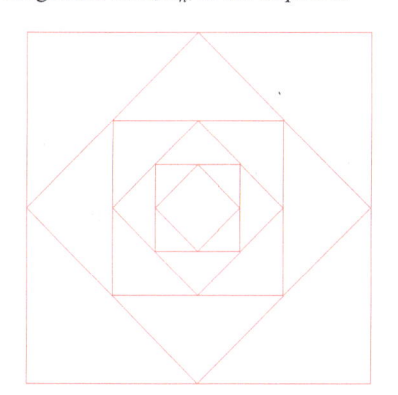

97. Find the largest integer value for the upper limit k if

$$\sum_{n=1}^{k} (n^2 + n)$$

$$= \sum_{n=1}^{k} [(n-1)(n-2)(n-3)(n-4)(n-5) + n^2 + n].$$

98. The Ulam conjecture. Define the sequence a_n recursively by

$$a_n = \begin{cases} \dfrac{a_{n-1}}{2} & \text{if } a_{n-1} \text{ is even} \\ 3a_{n-1} + 1 & \text{if } a_{n-1} \text{ is odd}. \end{cases}$$

The Ulam conjecture says that if your first term a_0 is any positive integer, the sequence a_n will eventually reach the integer 1. Verify the conjecture for $a_0 = 13$.

Maintaining Skills

In Exercises 99–104, simplify each expression.

99. $\left(\dfrac{1}{2}\right)^{-1}$ **100.** $\left(\dfrac{9}{4}\right)^{\frac{1}{2}}$

101. $\left(\dfrac{1}{27}\right)^{-\frac{1}{3}}$ **102.** $\left(\dfrac{-217}{1462}\right)^0$

103. $\dfrac{5}{9} \div \dfrac{1}{3} \cdot \dfrac{6}{5}$ **104.** $\dfrac{3}{7} \div \dfrac{4}{21} \cdot \dfrac{-1}{9}$

In Exercises 105–108, expand each expression.
105. $(x - 1)^2$ **106.** $(2x + 3)^2$
107. $(2 - x)^3$
108. $(x + 3)^3$

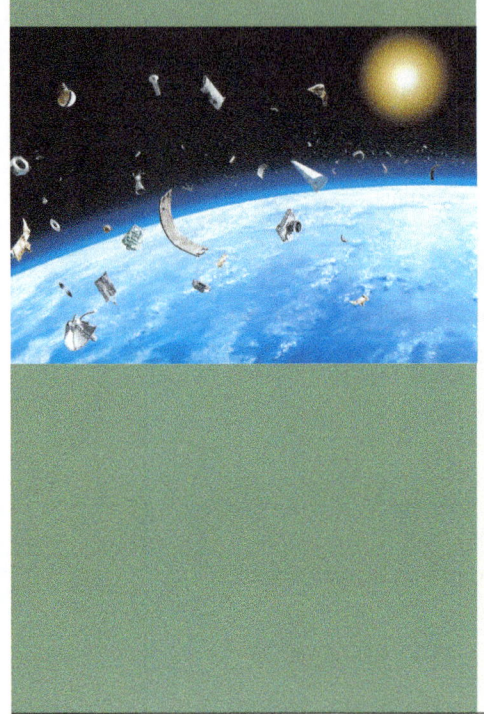

SECTION 6.2 Arithmetic Sequences; Partial Sums

BEFORE STARTING THIS SECTION, REVIEW

1. The general term of a sequence (Section 6.1, page 448)
2. Partial sums (Section 6.1, page 454)

OBJECTIVES

1. Identify an arithmetic sequence and find its common difference.
2. Find the sum of the first n terms of an arithmetic sequence.

◆ Falling Space Junk

Once more than 300 kilograms (700 pounds) of "space junk" crashed to the ground in South Africa. The material was eventually identified as part of a Delta 2 rocket used to launch a global positioning satellite. Once an object begins a free fall toward Earth, it falls (in the absence of friction) 16 feet in the first second, 48 feet in the second second, 80 feet in the third second, and so on. The number of feet traveled in succeeding seconds is the sequence

$$16, 48, 80, 112, \ldots.$$

This is an example of an *arithmetic sequence*. We investigate such a sequence in Example 5.

1 Identify an arithmetic sequence and find its common difference.

Arithmetic Sequence

When the difference $(a_{n+1} - a_n)$ between any two consecutive terms of a sequence is always the same number, the sequence is called an *arithmetic sequence*. In other words, the sequence $a_1, a_2, a_3, a_4, \ldots$ is an arithmetic sequence if $a_2 - a_1 = a_3 - a_2 = a_4 - a_3 = \cdots$.

Arithmetic Sequence

The sequence

$$a_1, a_2, a_3, a_4, \ldots, a_n, \ldots$$

is an **arithmetic sequence**, or an **arithmetic progression**, if there is a number d such that each term in the sequence except the first is obtained from the preceding term by adding d to it. The number d is called the **common difference** of the arithmetic sequence. We have

$$d = a_{n+1} - a_n, \quad n \geq 1.$$

EXAMPLE 1 Finding the Common Difference

Find the common difference of each arithmetic sequence.

a. $3, 23, 43, 63, 83, \ldots$ **b.** $29, 19, 9, -1, -11, \ldots$

Solution

a. The common difference is $23 - 3 = 20$ (or $43 - 23, 63 - 43$, or $83 - 63$).
b. The common difference is $19 - 29 = -10$ (or $9 - 19, -1 - 9$, or $-11 - (-1)$).

Practice Problem 1 Find the common difference of the arithmetic sequence

$$3, -2, -7, -12, -17, \ldots.$$

An arithmetic sequence can be completely specified by giving the first term a_1 and the common difference d. Let's see why. Suppose we are told that the sequence $a_1, a_2, a_3, a_4, \ldots$ is an arithmetic sequence and that $a_1 = 5$ and $d = 4$. Then

$$a_2 - a_1 = 4, \quad \text{so} \quad a_2 = a_1 + 4 = 5 + 4 = 9;$$
$$a_3 - a_2 = 4, \quad \text{so} \quad a_3 = a_2 + 4 = 9 + 4 = 13;$$
$$a_4 - a_3 = 4, \quad \text{so} \quad a_4 = a_3 + 4 = 13 + 4 = 17; \text{ and so on.}$$

Rewriting $d = a_{n+1} - a_n$ as $a_{n+1} = a_n + d$, we have a recursive definition of an arithmetic sequence.

Recursive Definition of an Arithmetic Sequence

An arithmetic sequence $a_1, a_2, a_3, a_4, \ldots, a_n, \ldots$ can be defined recursively. The recursive formula

$$a_{n+1} = a_n + d \text{ for } n \geq 1$$

defines an arithmetic sequence with first term a_1 and common difference d.

Consider an arithmetic sequence with first term a_1 and common difference d. Using the recursive definition $a_{n+1} = a_n + d$, for $n \geq 1$, we write

$a_1 = a_1$	The first term is a_1.
$a_2 = a_1 + d$	Replace n with 1 in $a_{n+1} = a_n + d$
$a_3 = a_2 + d = (a_1 + d) + d = a_1 + 2d$	Replace a_2 with $a_1 + d$; simplify.
$a_4 = a_3 + d = (a_1 + 2d) + d = a_1 + 3d$	Replace a_3 with $a_1 + 2d$; simplify.
$a_5 = a_4 + d = (a_1 + 3d) + d = a_1 + 4d$	Replace a_4 with $a_1 + 3d$; simplify.
$\vdots$	
$a_n = a_{n-1} + d$	
$\quad = [a_1 + (n-2)d] + d$	Replace a_{n-1} with $a_1 + (n-2)d$.
$\quad = a_1 + (n-1)d.$	Simplify.

Thus, we can find the general term a_n from the values a_1 and d.

*n*TH TERM OF AN ARITHMETIC SEQUENCE

If a sequence $a_1, a_2, a_3, \ldots$ is an arithmetic sequence, then its nth term, a_n, is given by

$$a_n = a_1 + (n-1)d,$$

where a_1 is the first term and d is the common difference.

EXAMPLE 2 Finding the *n*th Term of an Arithmetic Sequence

Find an expression for the nth term of the arithmetic sequence

$$3, 7, 11, 15, \ldots.$$

Solution

Because $a_1 = 3$ and $d = 7 - 3 = 4$, we have

$$a_n = a_1 + (n-1)d \quad \text{Formula for the } n\text{th term}$$
$$= 3 + (n-1)4 \quad \text{Replace } a_1 \text{ with 3 and } d \text{ with 4.}$$
$$= 3 + 4n - 4 = 4n - 1 \quad \text{Simplify.}$$

Note that in the expression $a_n = 4n - 1$, we get the sequence $3, 7, 11, 15, \ldots$ by substituting $n = 1, 2, 3, 4, \ldots$.

Practice Problem 2 Find an expression for the nth term of the arithmetic sequence

$$-3, 1, 5, 9, 13, 17, \ldots.$$

EXAMPLE 3 Finding the Common Difference of an Arithmetic Sequence

Find the common difference d and the nth term a_n of the arithmetic sequence whose 5th term is 15 and whose 20th term is 45.

Solution

$$a_n = a_1 + (n-1)d \quad \text{Formula for the } n\text{th term}$$
$$45 = a_1 + (20-1)d \quad \text{Replace } n \text{ with 20; } a_n = a_{20} = 45.$$
$$(1) \quad 45 = a_1 + 19d \quad \text{Simplify.}$$

Also,

$$15 = a_1 + (5-1)d \quad \text{Replace } n \text{ with 5; } a_n = a_5 = 15.$$
$$(2) \quad 15 = a_1 + 4d \quad \text{Simplify.}$$

Solving the system of equations

$$\begin{cases} a_1 + 19d = 45 & (1) \\ a_1 + 4d = 15 & (2) \end{cases}$$

gives $d = 2$ and $a_1 = 7$. We substitute $a_1 = 7$ and $d = 2$ in the formula for the nth term.

$$a_n = a_1 + (n-1)d \quad \text{Formula for the } n\text{th term}$$
$$a_n = 7 + (n-1)2 \quad \text{Replace } a_1 \text{ with 7 and } d \text{ with 2.}$$
$$= 7 + 2n - 2 = 2n + 5 \quad \text{Simplify.}$$

The nth term of this sequence is given by

$$a_n = 2n + 5, \quad n \geq 1.$$

Practice Problem 3 Find the common difference d and the nth term a_n of the arithmetic sequence whose 4th term is 41 and whose 15th term is 8.

2 Find the sum of the first n terms of an arithmetic sequence.

Sum of the First n Terms of an Arithmetic Sequence

We can use the following formula in many applications involving the sum of the first n terms of an arithmetic sequence.

$$1 + 2 + 3 \cdots + n = \frac{n(n+1)}{2}$$

HISTORICAL NOTE
Karl Friedrich Gauss (1777–1855)

Karl Friedrich Gauss discovered this formula in his arithmetic class at the age of 8. One day, the story goes, the teacher became so incensed with the class that he assigned the students the task of adding up all of the numbers from 1 to 100. As Gauss's classmates dutifully began to add, Gauss walked up to the teacher and presented the answer, 5050. The story goes that the teacher was neither impressed nor amused. Here are Gauss's calculations:

$S = 1 + 2 + 3 + \cdots + 100$
$S = 100 + 99 + 98 + \cdots + 1$
$2S = 101 + 101 + 101 + \cdots + 101$
  (100 terms)
$2S = 100 \times 101$
$S = \dfrac{100 \times 101}{2} = 5050$

It is said that Karl Friedrich Gauss (1777–1855) discovered this formula when he realized that any sum of numbers added in reverse order produces the same sum. Therefore, if S denotes the sum of the first n natural numbers, then

$S = 1 + 2 + 3 + \cdots + n$
$S = n + (n-1) + (n-2) + \cdots + 1$
$2S = (n+1) + (n+1) + (n+1) + \cdots + (n+1)$ Add $S + S = 2S$.
  (n terms)

$2S = n(n+1)$
$S = \dfrac{n(n+1)}{2}$ Divide both sides by 2.

For example, if $n = 100$, we have

$S = 1 + 2 + 3 + \cdots + 100 = \dfrac{100(100+1)}{2}$
$= \dfrac{100(101)}{2} = 5050.$

We can use this method to calculate the sum S_n of the first n terms of any arithmetic sequence.

First, we write the sum S_n of the first n terms:

$S_n = a_1 + (a_1 + d) + (a_1 + 2d) + (a_1 + 3d) + \cdots + a_n$

We can also write S_n (with the terms in reverse order) by starting with a_n and *subtracting* the common difference d:

$S_n = a_n + (a_n - d) + (a_n - 2d) + (a_n - 3d) + \cdots + a_1$

Adding the two equations for S_n, we find that the d's in the sums "drop out" [for example, $(a_1 + d) + (a_n - d) = a_1 + a_n$] and we have

$2S_n = (a_1 + a_n) + (a_1 + a_n) + \cdots + (a_1 + a_n)$
  (n terms)

$2S_n = n(a_1 + a_n)$
$S_n = n\left(\dfrac{a_1 + a_n}{2}\right)$ Divide both sides by 2.

SUM OF THE FIRST n TERMS OF AN ARITHMETIC SEQUENCE

Let $a_1, a_2, a_3, \ldots a_n$ be the first n terms of an arithmetic sequence with common difference d. The sum S_n of these n terms is given by

$$S_n = n\left(\dfrac{a_1 + a_n}{2}\right),$$

where $a_n = a_1 + (n-1)d$.

EXAMPLE 4 Finding the Sum of Terms of a Finite Arithmetic Sequence

Find the sum of the following arithmetic sequence of numbers:

$$1 + 4 + 7 + \cdots + 25$$

Solution

In this arithmetic sequence, $a_1 = 1$ and $d = 3$. Let's first find the number of terms.

$$a_n = a_1 + (n-1)d \quad \text{Formula for } n\text{th term}$$
$$25 = 1 + (n-1)3 \quad \text{Replace } a_n \text{ with 25, } a_1 \text{ with 1, and } d \text{ with 3.}$$
$$24 = (n-1)3 \quad \text{Subtract 1 from both sides.}$$
$$8 = n - 1 \quad \text{Divide both sides by 3.}$$
$$n = 9 \quad \text{Solve for } n.$$
$$S_n = n\left(\frac{a_1 + a_n}{2}\right) \quad \text{Formula for } S_n$$
$$S_9 = 9\left(\frac{1 + 25}{2}\right) \quad \text{Replace } n \text{ with 9, } a_1 \text{ with 1, and } a_n(=a_9) \text{ with 25.}$$
$$= 9(13) = 117 \quad \text{Simplify.}$$

So $1 + 4 + 7 + \cdots + 25 = 117$.

Practice Problem 4 Find the sum $\dfrac{2}{3} + \dfrac{5}{6} + 1 + \dfrac{7}{6} + \dfrac{4}{3} + \dfrac{3}{2} + \dfrac{5}{3} + \dfrac{11}{6} + 2 + \dfrac{13}{6}$.

EXAMPLE 5   Calculating the Distance Traveled by a Freely Falling Object

In the introduction to this section, we described the arithmetic sequence 16, 48, 80, 112, ... whose terms gave the number of feet that freely falling space junk falls in successive seconds. For this sequence, find the following:

a. The common difference d
b. The nth term a_n
c. The distance a piece of space junk travels in 10 seconds

Solution

a. The common difference $d = a_2 - a_1 = 48 - 16 = 32$.
b. $a_n = a_1 + (n-1)d \quad \text{Formula for the } n\text{th term}$
$ = 16 + (n-1)32 \quad \text{Replace } a_1 \text{ with 16 and } d \text{ with 32.}$
$ = 32n - 16 \quad \text{Simplify.}$
c. The sum of the first ten terms of this sequence gives the distance the piece travels in ten seconds. We find a_{10} from the formula for a_n.

$$a_n = 32n - 16 \quad \text{From part b}$$
$$a_{10} = 32(10) - 16 \quad \text{Replace } n \text{ with 10.}$$
$$= 304$$
$$S_n = n\left(\frac{a_1 + a_n}{2}\right) \quad \text{Formula for } S_n$$
$$S_{10} = 10\left(\frac{16 + 304}{2}\right) \quad \text{Replace } n \text{ with 10, } a_1 \text{ with 16, and } a_n(=a_{10}) \text{ with 304.}$$
$$= 1600 \quad \text{Simplify.}$$

The piece falls 1600 feet in ten seconds.

Practice Problem 5 In Example 5, find the distance the piece travels during the eleventh through fifteenth seconds.

SECTION 6.2 Exercises

Basic Concepts and Skills

In Exercises 1–14, determine whether each sequence is arithmetic. For those that are, find the first term a_1 and the common difference d.

1. $1, 2, 3, 4, 5, \ldots$
2. $1, 3, 5, 7, 9, \ldots$
3. $2, 5, 8, 11, 14, \ldots$
4. $10, 7, 4, 1, -2, \ldots$
5. $1, \frac{1}{2}, 0, -\frac{1}{2}, -\frac{1}{4}, \ldots$
6. $2, 4, 8, 16, 32, \ldots$
7. $1, -1, 2, -2, 3, \ldots$
8. $-\frac{1}{4}, \frac{1}{4}, \frac{3}{4}, \frac{5}{4}, \frac{7}{4}, \ldots$
9. $0.6, 0.2, -0.2, -0.6, -1, \ldots$
10. $2.3, 2.7, 3.1, 3.5, 3.9, \ldots$
11. $a_n = 2n + 6$
12. $a_n = 1 - 5n$
13. $a_n = 1 - n^2$
14. $a_n = 2n^2 - 3$

In Exercises 15–24, find an expression for the nth term of the arithmetic sequence.

15. $5, 8, 11, 14, 17, \ldots$
16. $4, 7, 10, 13, 16, \ldots$
17. $11, 6, 1, -4, -9, \ldots$
18. $9, 5, 1, -3, -7, \ldots$
19. $\frac{1}{2}, \frac{1}{4}, 0, -\frac{1}{4}, -\frac{1}{2}, \ldots$
20. $\frac{2}{3}, \frac{5}{6}, 1, \frac{7}{6}, \frac{4}{3}, \ldots$
21. $-\frac{3}{5}, -1, -\frac{7}{5}, -\frac{9}{5}, -\frac{11}{5}, \ldots$
22. $\frac{1}{2}, 2, \frac{7}{2}, 5, \frac{13}{2}, \ldots$
23. $e, 3 + e, 6, + e, 9 + e, 12 + e, \ldots$
24. $2\pi, 2(\pi + 2), 2(\pi + 4), 2(\pi + 6), 2(\pi + 8) \ldots,$

In Exercises 25–30, find the common difference d and the nth term a_n of the arithmetic sequence with the specified terms.

25. 4th term 21; 10th term 60
26. 3rd term 15; 21st term 87
27. 7th term 8; 15th term -8
28. 5th term 12; 18th term -1
29. 3rd term 7; 23rd term 17
30. 11th term -1; 31st term 5

In Exercises 31–40, find the sum of each arithmetic sequence.

31. $1 + 2 + 3 + \cdots + 50$
32. $2 + 4 + 6 + \cdots + 102$
33. $1 + 3 + 5 + \cdots + 99$
34. $5 + 10 + 15 + \cdots + 200$
35. $3 + 6 + 9 + \cdots + 300$
36. $4 + 7 + 10 + \cdots + 301$
37. $2 - 1 - 4 - \cdots - 34$
38. $-3 - 8 - 13 - \cdots - 48$
39. $\frac{1}{3} + 1 + \frac{5}{3} + \cdots + 7$
40. $\frac{3}{5} + 2 + \frac{17}{5} + \cdots + \frac{101}{5}$

In Exercises 41–46, find the sum of the first n terms of the given arithmetic sequence.

41. $2, 7, 12, \ldots; n = 50$
42. $8, 10, 12, \ldots; n = 40$
43. $-15, -11, -7, \ldots; n = 20$
44. $-20, -13, -6, \ldots; n = 25$
45. $3.5, 3.7, 3.9, \ldots; n = 100$
46. $-7, -6.5, -6, \ldots; n = 80$

In Exercises 47–52, find n for the given value of a_n in each arithmetic sequence.

47. $a_n = 75; 1, 3, 5, \ldots$
48. $a_n = 120; 2, 4, 6, \ldots$
49. $a_n = 95; -1, 3, 7, \ldots$
50. $a_n = 83; -5, -3, -1, \ldots$
51. $a_n = 50\sqrt{3}; 2\sqrt{3}, 4\sqrt{3}, 6\sqrt{3}, \ldots$
52. $a_n = 73\pi; 3\pi, 5\pi, 7\pi, \ldots$

Applying the Concepts

In Exercises 53–59, assume that the indicated sequence is arithmetic.

53. **Orange picking.** When Eric started work as an orange picker, he picked 10 oranges in the first minute, 12 in the second minute, 14 in the third minute, and so on. How many oranges did Eric pick in the first half hour?

54. **Exercise.** Walking up a steep hill, Jan walks 60 feet in the first minute, 57 feet in the second minute, 54 feet in the third minute, and so on.
    a. How far will Jan walk in the nth minute?
    b. How far will Jan walk in the first 15 minutes?

55. **Contest winner.** A contest winner will receive money each month for three years. The winner receives $50 the first month, $75 the second month, $100 the third month, and so on. How much money will the winner have collected after 30 months?

56. **Salary.** Darren took a 12-month temporary job with a monthly salary that increased a fixed amount each month. He can't recall the starting salary, but does remember that he was paid $820 at the end of the third month and $910 for his last month's work. How much was Darren's total pay for the entire 12 months?

464 Chapter 6 Further Topics in Algebra

57. **Hourly wage.** Antonio's new weekend job started at an hourly wage of $12.75. He is guaranteed a raise of 25¢ an hour every three months for the next four years. What will Antonio's hourly wage be at the end of four years?

58. **Competing job offers.** Denzel is considering offers from two companies. A marketing company pays $32,500 the first year and guarantees a raise of $1300 each year; an exporting company pays $36,000 the first year, with a guaranteed raise of $400 each year. Over a five-year period, which company will pay more? How much more?

59. **Theater seating.** A theater has 25 rows of seats. The first row has 20 seats, the second row has 22 seats, the third row has 24 seats, and so on. How many seats are in the theater?

60. **Bricks in a driveway.** A brick driveway has 50 rows of bricks. The first row has 16 bricks, and the fiftieth row has 65 bricks. Assuming that the sequence of numbers giving the number of bricks in rows 1 through 50 is arithmetic, how many bricks does the driveway contain?

61. **Stacked boxes.** A stack of boxes has 17 rows. The bottom row has 40 boxes, and the top row has 8 boxes. Assuming the sequence of numbers that gives the number of boxes in succeeding rows is arithmetic, how many boxes does the stack contain?

Critical Thinking / Discussion / Writing

62. Only 3 customers showed up for the opening day of a flea market. However, 9 came the second day, and an additional 6 customers came on each subsequent day. After several days, there was a total of 192 customers. How many days had the flea market been open?

63. A sequence is **harmonic** if the reciprocals of the terms of the sequence form an arithmetic sequence. Is the sequence $\frac{1}{2}, \frac{3}{5}, \frac{3}{4}, 1, \frac{3}{2}, 3 \ldots$ harmonic? Explain your reasoning.

64. Find the nth term of an arithmetic sequence whose first term is 10 and whose 21st term is 0.

65. If a_1 and d are the first term and common difference, respectively, of an arithmetic sequence, find the first term and difference of an arithmetic sequence whose terms are the negatives of the terms in the given sequence.

66. How many terms are in the series $\sum_{i=22}^{100} 17i^3$?

67. The sum of the first n counting numbers is 12,403. Find the value of n.

Maintaining Skills

In Exercises 68–73, simplify each expression.

68. $\dfrac{-16}{625} \div \dfrac{-8}{125}$

69. $\dfrac{27}{512} \div \dfrac{9}{64}$

70. $\dfrac{(-1)^{n+1}}{(-1)^n}$

71. $\dfrac{3^{n+1}}{3^n}$

72. $\dfrac{-5}{3^{n+1}} \div \dfrac{-5}{3^n}$

73. $\dfrac{6}{11^{n+1}} \div \dfrac{6}{11^n}$

In Exercises 74 and 75, solve each equation for x.

74. $\dfrac{x}{7^n} = 7$

75. $\dfrac{x}{\left(\dfrac{1}{2}\right)^n} = \dfrac{1}{2}$

In Exercises 76 and 77, simplify each expression.

76. $5 + 5x + 5x^2 - x(5 + 5x + 5x^2)$

77. $a + ax + ax^2 + ax^3 - x(a + ax + ax^2 + ax^3)$

SECTION 6.3 Geometric Sequences and Series

BEFORE STARTING THIS SECTION, REVIEW

1. The general term of a sequence (Section 6.1, page 448)
2. Partial sums (Section 6.1, page 454)
3. Compound interest (Section 3.1, page 228)

OBJECTIVES

1. Identify a geometric sequence and find its common ratio.
2. Find the sum of a finite geometric sequence.
3. Solve annuity problems.
4. Find the sum of an infinite geometric sequence.

◆ Spreading the Wealth

Cities across the nation compete to host a Super Bowl. Cities across the world compete to host the Olympic Games. What is the incentive? The *multiplier effect* is often listed among the benefits for hosting one of these events. The effect refers to the phenomenon that occurs when money spent directly on an event has a ripple effect on the economy that is many times the amount spent directly on the event.

Suppose, for example, that $10 million is spent directly by tourists for hotels, meals, tickets, cabs, and so on. Some portion of that amount will be respent by its recipients on groceries, clothes, gas, and so on.

This process is repeated over and over, affecting the economy much more than the initial $10 million spent. Under certain assumptions, the spending just described results in an *infinite geometric series*, whose sum is called the *multiplier*. Example 8 illustrates this effect.

Geometric Sequence

1 Identify a geometric sequence and find its common ratio.

When the *ratio* $\dfrac{a_{n+1}}{a_n}$ of any two consecutive terms of a sequence is always the same number, the sequence is called a *geometric sequence*. In other words, the sequence $a_1, a_2, a_3, a_4, \ldots$ is a geometric sequence if

$$\frac{a_2}{a_1} = \frac{a_3}{a_2} = \frac{a_4}{a_3} = \cdots .$$

Geometric Sequence

The sequence

$$a_1, a_2, a_3, a_4, \ldots, a_n, \ldots$$

is a **geometric sequence**, or a **geometric progression**, if there is a number r such that each term except the first in the sequence is obtained by multiplying the previous term by r. The number r is called the **common ratio** of the geometric sequence. We have

$$\frac{a_{n+1}}{a_n} = r, \quad n \geq 1$$

466 Chapter 6 Further Topics in Algebra

EXAMPLE 1   Finding the Common Ratio

Find the common ratio for each geometric sequence.

a. 2, 10, 50, 250, 1250, ...   **b.** $-162, -54, -18, -6, -2, ...$

Solution

a. The common ratio is $\frac{10}{2} = 5 \left(\text{or } \frac{50}{10}, \frac{250}{50}, \text{ or } \frac{1250}{250} \right)$.

b. The common ratio is $\frac{-54}{-162} = \frac{1}{3} \left(\text{or } \frac{-18}{-54}, \frac{-6}{-18}, \text{ or } \frac{-2}{-6} \right)$.

Practice Problem 1 Find the common ratio for the geometric sequence 6, 18, 54, 162, 486, ...

A geometric sequence can be completely specified by giving the first term a_1 and the common ratio r. Suppose, for example, that we are told that the sequence $a_1, a_2, a_3, a_4, ...$ is geometric and that $a_1 = 3$ and $r = 4$. Then we can find a_2:

$$\frac{a_2}{a_1} = 4, \text{ so } a_2 = 4a_1 = 4 \cdot 3 = 12$$

Now we continue:

$$\frac{a_3}{a_2} = 4, \text{ so } a_3 = 4a_2 = 4 \cdot 12 = 48;$$

$$\frac{a_4}{a_3} = 4, \text{ so } a_4 = 4a_3 = 4 \cdot 48 = 192; \text{ and so on.}$$

By rewriting $\frac{a_{n+1}}{a_n} = r$ as $a_{n+1} = ra_n$, we have a recursive definition of a geometric sequence.

Recursive Definition of a Geometric Sequence

A geometric sequence $a_1, a_2, a_3, a_4, ..., a_n, ...$ can be defined recursively. The recursive formula

$$a_{n+1} = ra_n, \quad n \geq 1$$

defines a geometric sequence with the first term a_1 and the common ratio r.

EXAMPLE 2   Determining Whether a Sequence Is Geometric

Determine whether each sequence is geometric. If it is, find the first term and the common ratio.

a. $a_n = 4^n$ **b.** $b_n = \frac{3}{2^n}$ **c.** $c_n = 1 - 5^n$

Solution

a. The sequence 4, 16, 64, 256, ..., 4^n, ... appears to be geometric, with $a_1 = 4$ and $r = 4$. To verify this, we check that $\frac{a_{n+1}}{a_n} = 4$ for all $n \geq 1$.

$a_{n+1} = 4^{n+1}$ Replace n with $n+1$ in the general term $a_n = 4^n$.

$\frac{a_{n+1}}{a_n} = \frac{4^{n+1}}{4^n} = 4$ $\frac{4^{n+1}}{4^n} = \frac{4^n \cdot 4}{4^n} = 4$

Consequently, $\frac{a_{n+1}}{a_n} = 4$ for all $n \geq 1$, and the sequence is geometric.

b. The sequence $\frac{3}{2}, \frac{3}{4}, \frac{3}{8}, \frac{3}{16}, \ldots$ appears to be geometric, with $b_1 = \frac{3}{2}$ and $r = \frac{1}{2}$.

To verify this, we check that $\frac{b_{n+1}}{b_n} = \frac{1}{2}$ for all $n \geq 1$.

$$b_{n+1} = \frac{3}{2^{n+1}}$$ Replace n with $n+1$ in the general term $b_n = \frac{3}{2^n}$.

$$\frac{b_{n+1}}{b_n} = \frac{3}{2^{n+1}} \div \frac{3}{2^n} \qquad \frac{b_{n+1}}{b_n} = b_{n+1} \div b_n$$

$$= \frac{3}{2^{n+1}} \cdot \frac{2^n}{3} = \frac{1}{2} \qquad 2^{n+1} = 2 \cdot 2^n$$

Consequently, $\frac{b_{n+1}}{b_n} = \frac{1}{2}$ for all $n \geq 1$, and the sequence is geometric.

c. The sequence $-4, -24, -124, -624, \ldots, 1 - 5^n, \ldots$ is *not* a geometric sequence because not all pairs of consecutive terms have identical quotients $\frac{c_{n+1}}{c_n}$. The quotient of the first two terms is $\frac{-24}{-4} = 6$, but the quotient of the third and second terms is $\frac{-124}{-24} = \frac{31}{6}$.

Practice Problem 2 Determine whether the sequence $a_n = 2\left(\frac{3}{2}\right)^n$ is geometric. If so, find the first term and the common ratio.

From the equation $a_{n+1} = a_n r$ used in the recursive definition of a geometric sequence, we see that each term after the first is obtained by multiplying the preceding term by the common ratio r.

The General Term of a Geometric Sequence

Every geometric sequence can be written in the form

$$a_1, a_1 r, a_1 r^2, a_1 r^3, \ldots, a_1 r^{n-1}, \ldots,$$

where r is the common ratio. Because $a_1 = a_1(1) = a_1 r^0$, the **nth term of the geometric sequence** is

$$a_n = a_1 r^{n-1}, \text{ for } n \geq 1.$$

EXAMPLE 3 Finding Terms in a Geometric Sequence

For the geometric sequence $1, 3, 9, 27, \ldots$, find each of the following:

a. a_1 **b.** r **c.** a_n

Solution

a. The first term of the sequence is given: $a_1 = 1$.

b. The sequence is geometric, so we can take the ratio of any two consecutive terms $\frac{a_{n+1}}{a_n}$. The ratio of the first two terms gives us

$$r = \frac{3}{1} = 3.$$

c. $a_n = a_1 r^{n-1}$ Formula for the nth term
$ = (1)(3^{n-1})$ Replace a_1 with 1 and r with 3.
$ = 3^{n-1}$

Practice Problem 3 For the geometric sequence $2, \dfrac{6}{5}, \dfrac{18}{25}, \dfrac{54}{125}, \ldots$, find the following:

a. a_1 **b.** r **c.** a_n

Notice in Example 3 that converting $a_n = 3^{n-1}$ to standard function notation gives us $f(n) = 3^{n-1}$, an exponential function. A geometric sequence with first term a_1 and common ratio r can be viewed as an exponential function $f(n) = a_1 r^{n-1}$ with the set of natural numbers as its domain.

EXAMPLE 4   Finding a Particular Term in a Geometric Sequence

Find the 23rd term of a geometric sequence whose first term is 10 and whose common ratio is 1.2.

Solution

$a_n = a_1 r^{n-1}$ Formula for the nth term
$a_{23} = 10(1.2)^{23-1}$ Replace n with 23, a_1 with 10, and r with 1.2.
$\phantom{a_{23}} = 10(1.2)^{22}$
$\phantom{a_{23}} \approx 552.06$ Use a calculator.

Practice Problem 4 Find the 18th term of a geometric sequence whose first term is 7 and whose common ratio is 1.5.

2 Find the sum of a finite geometric sequence.

Finding the Sum of a Finite Geometric Sequence

We can find a formula for the sum S_n of the first n terms of a geometric sequence. (We have already found a similar formula for arithmetic sequences.)

$S_n = a_1 + a_1 r + a_1 r^2 + a_1 r^3 + \cdots + a_1 r^{n-1}$ Definition of S_n
$r S_n = a_1 r + a_1 r^2 + a_1 r^3 + a_1 r^4 + \cdots + a_1 r^n$ Multiply both sides by r.
$S_n - r S_n = a_1 - a_1 r^n$ Subtract the second equation from the first.
$(1 - r) S_n = a_1(1 - r^n)$ Factor both sides.
$S_n = \dfrac{a_1(1 - r^n)}{1 - r}, r \neq 1$ Divide both sides by $1 - r$.

SUM OF THE FIRST n TERMS OF A GEOMETRIC SEQUENCE

Let $a_1, a_2, a_3, \ldots, a_n$ be the first n terms of a geometric sequence with first term a_1 and common ratio r. The sum S_n of these terms is

$$S_n = \sum_{i=1}^{n} a_1 r^{i-1} = \dfrac{a_1(1 - r^n)}{1 - r}, r \neq 1.$$

A geometric sequence with $r = 1$ is a sequence in which all terms are identical; that is, $a_n = a_1(1)^{n-1} = a_1$. Consequently, the sum of the first n terms, S_n, is

$$\underbrace{a_1 + a_1 + \cdots + a_1}_{n \text{ terms}} = n a_1.$$

Section 6.3 ■ Geometric Sequences and Series   469

TECHNOLOGY CONNECTION

The following screen shows the result of Example 5a obtained using a graphing calculator.

```
sum(seq(5*0.7^(n
-1),n,1,15,1)
       16.58754064
```

EXAMPLE 5   Finding the Sum of Terms of a Finite Geometric Sequence

Find each sum.

a. $\sum_{i=1}^{15} 5(0.7)^{i-1}$ **b.** $\sum_{i=1}^{15} 5(0.7)^{i}$

Solution

a. We can evaluate $\sum_{i=1}^{15} 5(0.7)^{i-1}$ by using the formula for $S_n = \sum_{i=1}^{n} a_1 r^{i-1} = \frac{a_1(1 - r^n)}{1 - r}$,

where $a_1 = 5$, $r = 0.7$, and $n = 15$.

$S_{15} = \sum_{i=1}^{15} a_1 r^{i-1} = 5\left[\frac{1 - (0.7)^{15}}{1 - 0.7}\right]$ Replace a_1 with 5, $r = 0.7$, and $n = 15$.

≈ 16.5875 Use a calculator.

b. $\sum_{i=1}^{15} 5(0.7)^{i} = (0.7)\sum_{i=1}^{15} 5(0.7)^{i-1}$ Factor out 0.7 from each term.

$\approx (0.7)(16.5875)$ From part **a**

$= 11.6113$

Practice Problem 5 Find the sum $\sum_{i=1}^{17} 3(0.4)^{i}$.

3 Solve annuity problems.

Annuities

One of the most important applications of finite geometric series is the computation of the value of an *annuity*. An **annuity** is a sequence of equal periodic payments. When a fixed rate of compound interest applies to all payments, the sum of all of the payments made plus all interest can be found by using the formula for the sum of a finite geometric sequence. The **value of an annuity** (also called the **future value of an annuity**) is the sum of all payments and interest. To develop a formula for the value of an annuity, we suppose $P is deposited at the end of each year at an annual interest rate i compounded annually.

The compound interest formula $A = P(1 + i)^t$ gives the total value after t years when $P earns an annual interest rate i (in decimal form) compounded once a year. At the end of the first year, the initial payment of $P is made and the annuity's value is $P. At the end of the second year, $P is again deposited. At this time, the first deposit has earned interest during the second year. The value of the annuity after two years is

$$P \quad + \quad P(1 + i).$$

↑ Payment made at the end of year 2

↑ First payment plus interest earned for one year

The value of the annuity after three years is

$$P \quad + P(1 + i) \quad + P(1 + i)^2$$

↑ Payment made at the end of year 3

↑ Second payment plus interest earned for one year

↑ First payment plus interest earned for two years

The value of the annuity after t years is

$$P + P(1 + i) + P(1 + i)^2 + P(1 + i)^3 + \cdots + P(1 + i)^{t-1}.$$

↑ Payment made at the end of year t

↑ First payment plus interest earned for $t - 1$ years

RECALL

The sum S_n of the first n terms of a geometric sequence with first term a_1 and common ratio r is given by the formula

$$S_n = \frac{a_1(1 - r^n)}{1 - r}.$$

The value of the annuity after t years is the sum of a geometric sequence with first term $a_1 = P$ and common ratio $r = 1 + i$. We compute this sum A as follows:

$$A = S_n = \frac{a_1(1 - r^n)}{1 - r} \quad \text{Formula for } S_n$$

$$A = \frac{P[1 - (1 + i)^t]}{1 - (1 + i)} \quad \text{Replace } n \text{ with } t, a_1 \text{ with } P, \text{ and } r \text{ with } 1 + i.$$

$$= \frac{P[1 - (1 + i)^t]}{-i} \quad \text{Simplify.}$$

$$= P\frac{[(1 + i)^t - 1]}{i} \quad \text{Multiply numerator and denominator by } -1.$$

If interest is compounded n times per year and equal payments are made at the end of each compounding period, we adjust the formula as we did in Section 3.1 to account for the more frequent compounding.

VALUE OF AN ANNUITY

Let P represent the payment in dollars made at the end of each of n compounding periods per year and let i be the annual interest rate. Then the value A of the annuity after t years is as follows:

$$A = P\left[\frac{\left(1 + \frac{i}{n}\right)^{nt} - 1}{\frac{i}{n}}\right]$$

EXAMPLE 6 Finding the Value of an Annuity

An individual retirement account (IRA) is a common way to save money to provide funds after retirement. Suppose you make payments of $1200 into an IRA at the end of each year at an annual interest rate of 4.5% per year, compounded annually. What is the value of this annuity after 35 years?

Solution

Each annuity payment is $P = \$1200$. The annual interest rate is 4.5% (so $i = 0.045$), and the number of years is $t = 35$. Because interest is compounded annually, $n = 1$. The value of the annuity is

$$A = 1200\left[\frac{\left(1 + \frac{0.045}{1}\right)^{(1)35} - 1}{\frac{0.045}{1}}\right]$$

$$= \$97{,}795.94 \quad \text{Use a calculator.}$$

The value of the IRA after 35 years is $97,795.94.

Practice Problem 6 If in Example 6 the end-of-year payments are $1500, the annual interest rate remains 4.5%, compounded annually, and the payments are made for 30 years, what is the value of the annuity?

4 Find the sum of an infinite geometric sequence.

Infinite Geometric Series

The sum S_n of the first n terms of a geometric series is given by the formula $S_n = \dfrac{a_1(1 - r^n)}{1 - r}$.

If this finite sum S_n approaches a number S as $n \to \infty$ (that is, as n gets larger and larger), we say that S is the **sum of the infinite geometric series** and we write

$$S = \sum_{i=1}^{\infty} a_1 r^{i-1}.$$

If r is any real number with $-1 < r < 1$ (equivalently, $|r| < 1$), then the value of r^n, like the value of $\left(\dfrac{1}{2}\right)^n$, gets closer to 0 as n gets larger. We indicate that the values of r^n approach 0 by writing

$$\lim_{n \to \infty} r^n = 0 \text{ if } |r| < 1.$$

The expression $\lim_{n \to \infty} r^n$ is read "the limit, as n approaches infinity, of r to the n" or "the limit, as n approaches infinity, of r to the nth power."

When $|r| < 1$,

$$S_n = \dfrac{a_1(1 - r^n)}{1 - r} \to \dfrac{a_1(1 - 0)}{1 - r} = \dfrac{a_1}{1 - r} \quad \text{as } n \to \infty.$$

SUM OF THE TERMS OF AN INFINITE GEOMETRIC SEQUENCE

If $|r| < 1$, the infinite sum

$$a_1 + a_1 r + a_1 r^2 + a_1 r^3 + \cdots + a_1 r^{n-1} + \cdots$$

is given by

$$S = \sum_{i=1}^{\infty} a_1 r^{i-1} = \dfrac{a_1}{1 - r}.$$

When $|r| \geq 1$, the infinite geometric series does not have a sum. This is because the value of r^n does not approach 0 as $n \to \infty$. For example, if $r = 2$, we have

$$2^1 = 2, 2^2 = 4, 2^3 = 8, 2^4 = 16, 2^5 = 32, 2^6 = 64, \ldots.$$

EXAMPLE 7   Finding the Sum of an Infinite Geometric Series

Find the sum $2 + \dfrac{3}{2} + \dfrac{9}{8} + \dfrac{27}{32} + \cdots$.

Solution

The first term is $a_1 = 2$, and the common ratio is

$$r = \dfrac{\frac{3}{2}}{2} = \dfrac{3}{4}.$$

472 Chapter 6 Further Topics in Algebra

Because $|r| = \dfrac{3}{4} < 1$, we can use the formula for the sum of an infinite geometric series.

$$S = \dfrac{a_1}{1-r} \quad \text{Formula for } S$$

$$= \dfrac{2}{1-\dfrac{3}{4}} \quad \text{Replace } a_1 \text{ with 2 and } r \text{ with } \dfrac{3}{4}.$$

$$= 8 \quad \text{Simplify.}$$

Practice Problem 7 Find the sum $3 + \dfrac{6}{3} + \dfrac{12}{9} + \dfrac{24}{27} + \cdots$.

EXAMPLE 8 Calculating the Multiplier Effect

The host city for the Super Bowl expects that tourists will spend $10,000,000. Assume that 80% of this money is spent again in the city, then 80% of this second round of spending is spent again, and so on. In the introduction to this section, we said that such a spending pattern results in a geometric series whose sum is called the *multiplier*. Find this series and its sum.

Solution

We start our series with the $10,000,000 brought into the city and add the subsequent amounts spent.

$$10{,}000{,}000 + 10{,}000{,}000\,(0.80) + 10{,}000{,}000\,(0.80)^2 + 10{,}000{,}000\,(0.80)^3 + \cdots$$

- original 10,000,000
- 80% of 10,000,000
- 80% of previous amount
- 80% of previous amount

Using the formula $\sum_{i=1}^{\infty} ar^{i-1} = \dfrac{a}{1-r}$ for the sum of an infinite geometric series, we have

$$\sum_{i=1}^{\infty} (10{,}000{,}000)(0.80)^{i-1} = \dfrac{\$10{,}000{,}000}{1-0.80} = \$50{,}000{,}000.$$

This amount should shed some light on why there is so much competition to serve as the host city for a major sports event.

Practice Problem 8 Find the series and the sum that results in Example 8 assuming that 85%, rather than 80%, occurs in the repeated spending.

SECTION 6.3 Exercises

Basic Concepts and Skills

In Exercises 1–20, determine whether each sequence is geometric. If it is, find the first term and the common ratio.

1. $3, 6, 12, 24, \ldots$
2. $2, 4, 8, 16, \ldots$
3. $1, 5, 10, 20, \ldots$
4. $1, 1, 3, 3, 9, 9, \ldots$
5. $1, -3, 9, -27, \ldots$
6. $-1, 2, -4, 8, \ldots$
7. $7, -7, 7, -7, \ldots$
8. $1, -2, 4, -8, \ldots$
9. $9, 3, 1, \dfrac{1}{3}, \ldots$
10. $5, 2, \dfrac{4}{5}, \dfrac{8}{25}, \ldots$
11. $a_n = \left(-\dfrac{1}{2}\right)^n$
12. $a_n = 5\left(\dfrac{2}{3}\right)^n$

13. $a_n = 2^{n-1}$
14. $a_n = -(1.06)^{n-1}$
15. $a_n = 7n^2 + 1$
16. $a_n = 1 - (2n)^2$
17. $a_n = 3^{-n}$
18. $a_n = 50(0.1)^{-n}$
19. $a_n = 5^{n/2}$
20. $a_n = 7^{\sqrt{n}}$

In Exercises 21–28, find the first term a_1, the common ratio r, and the nth term a_n for each geometric sequence.

21. $2, 10, 50, 250, \ldots$
22. $-3, -6, -12, -24, \ldots$
23. $5, \dfrac{10}{3}, \dfrac{20}{9}, \dfrac{40}{27}, \ldots$
24. $1, \sqrt{3}, 3, 3\sqrt{3}, \ldots$
25. $0.2, -0.6, 1.8, -5.4, \ldots$
26. $1.3, -0.26, 0.052, -0.0104, \ldots$
27. $\pi^4, \pi^6, \pi^8, \pi^{10}, \ldots$
28. $e^2, 1, e^{-2}, e^{-4}, \ldots$

In Exercises 29–38, find the indicated term of each geometric sequence.

29. a_7 when $a_1 = 5$ and $r = 2$
30. a_7 when $a_1 = 8$ and $r = 3$
31. a_{10} when $a_1 = 3$ and $r = -2$
32. a_{10} when $a_1 = 7$ and $r = -2$
33. a_6 when $a_1 = \dfrac{1}{16}$ and $r = 3$
34. a_6 when $a_1 = \dfrac{1}{81}$ and $r = 3$
35. a_9 when $a_1 = -1$ and $r = \dfrac{5}{2}$
36. a_9 when $a_1 = -4$ and $r = \dfrac{3}{4}$
37. a_{20} when $a_1 = 500$ and $r = -\dfrac{1}{2}$
38. a_{20} when $a_1 = 1000$ and $r = -\dfrac{1}{10}$

In Exercises 39–44, find the sum S_n of the first n terms of each geometric sequence.

39. $\dfrac{1}{10}, \dfrac{1}{2}, \dfrac{5}{2}, \dfrac{25}{2}, \ldots; n = 10$
40. $6, 2, \dfrac{2}{3}, \dfrac{2}{9}, \ldots; n = 10$
41. $\dfrac{1}{25}, -\dfrac{1}{5}, 1, -5, \ldots; n = 12$
42. $-10, \dfrac{1}{10}, \dfrac{-1}{1000}, \dfrac{1}{100,000}, \ldots; n = 12$
43. $5, \dfrac{5}{4}, \dfrac{5}{4^2}, \dfrac{5}{4^3}, \ldots; n = 8$
44. $2, \dfrac{2}{5}, \dfrac{2}{5^2}, \dfrac{2}{5^3}, \ldots; n = 8$

In Exercises 45–52, find each sum.

45. $\displaystyle\sum_{i=1}^{5} \left(\dfrac{1}{2}\right)^{i-1}$
46. $\displaystyle\sum_{i=1}^{5} \left(\dfrac{1}{5}\right)^{i-1}$
47. $\displaystyle\sum_{i=1}^{8} 3\left(\dfrac{2}{3}\right)^{i-1}$
48. $\displaystyle\sum_{i=1}^{8} 2\left(\dfrac{3}{5}\right)^{i-1}$
49. $\displaystyle\sum_{i=3}^{10} \dfrac{2^{i-1}}{4}$
50. $\displaystyle\sum_{i=3}^{10} \dfrac{5^{i-1}}{2}$
51. $\displaystyle\sum_{i=1}^{20} \left(-\dfrac{3}{5}\right)\left(-\dfrac{5}{2}\right)^{i-1}$
52. $\displaystyle\sum_{i=1}^{20} \left(-\dfrac{1}{4}\right)(3^{2-i})$

In Exercises 53–58, use a graphing calculator to find each sum.

53. $\dfrac{1}{5} + \dfrac{1}{10} + \dfrac{1}{20} + \cdots + \dfrac{1}{320}$
54. $2 + 6 + 18 + \cdots + 1458$
55. $\displaystyle\sum_{n=1}^{10} 3^{n-2}$
56. $\displaystyle\sum_{n=1}^{15} \left(\dfrac{1}{7}\right)^{n-1}$
57. $\displaystyle\sum_{n=1}^{10} (-1)^{n-1} 3^{2-n}$
58. $\displaystyle\sum_{n=1}^{20} (-4)^{3-n}$

In Exercises 59–68, find each sum.

59. $\dfrac{1}{3} + \dfrac{1}{9} + \dfrac{1}{27} + \dfrac{1}{81} + \cdots$
60. $\dfrac{5}{2} + \dfrac{5}{4} + \dfrac{5}{8} + \dfrac{5}{16} + \cdots$
61. $-\dfrac{1}{2} + \dfrac{1}{4} - \dfrac{1}{8} + \dfrac{1}{16} - \cdots$
62. $-\dfrac{3}{2} + \dfrac{3}{4} - \dfrac{3}{8} + \dfrac{3}{16} + \cdots$
63. $8 - 2 + \dfrac{1}{2} - \dfrac{1}{8} + \cdots$
64. $1 - \dfrac{3}{5} + \dfrac{9}{25} - \dfrac{27}{125} + \cdots$
65. $\displaystyle\sum_{n=0}^{\infty} 5\left(\dfrac{1}{3}\right)^n$
66. $\displaystyle\sum_{n=0}^{\infty} 3\left(\dfrac{1}{4}\right)^n$
67. $\displaystyle\sum_{n=0}^{\infty} \left(-\dfrac{1}{4}\right)^n$
68. $\displaystyle\sum_{n=0}^{\infty} \left(-\dfrac{1}{3}\right)^n$

474 Chapter 6 Further Topics in Algebra

In Exercises 69–74, find n for the given value of a_n in each geometric sequence.

69. $a_n = 512; 2, 4, 8, \ldots$

70. $a_n = \dfrac{3}{1024}; 3, \dfrac{3}{2}, \dfrac{3}{4}, \ldots$

71. $a_n = -\dfrac{1}{4096}; 1, -\dfrac{1}{2}, \dfrac{1}{4}, \ldots$

72. $a_n = \dfrac{1}{2187}; 3, -1, \dfrac{1}{3}, \ldots$

73. $a_n = \dfrac{5}{5{,}764{,}801}; 5, -\dfrac{5}{7}, \dfrac{5}{49}, \ldots$

74. $a_n = \dfrac{1}{1{,}000{,}000}; 1000, 100, 10, \ldots$

Applying the Concepts

75. **Population growth.** The population in a small town is increasing at the rate of 3% per year. If the present population is 20,000, what will the population be at the end of five years?

76. **Growth of a zoo collection.** The butterfly collection at a local zoo started with only 25 butterflies. If the number of butterflies doubled each month, how many butterflies were in the collection after 12 months?

77. **Savings growth.** Ramón deposits $100 on the last day of each month into a savings account that pays 6% annually, compounded monthly. What is the balance in the account after 36 compounding periods?

78. **Savings growth.** Kat deposits $300 semiannually into a savings account for her son. If the account pays 8% annually, compounded semiannually, what is the balance in the account after 20 compounding periods?

79. **Ancestors.** Every person has two parents, four grandparents, eight great-grandparents, and so on. Find the number of ancestors a person has in the tenth generation back.

80. **Bacterial growth.** A colony of bacteria doubles in number every day. If there are 1000 bacteria now, how many will there be
    a. on the 7th day?
    b. on the nth day?

81. **Rebounding ball.** A particular ball always rebounds $\dfrac{3}{5}$ the distance it falls. If the ball is dropped from a height of 5 meters, how far will it travel before coming to a stop?

82. **Rebounding ball.** Repeat Exercise 81 assuming that the ball is dropped from a height of 9 meters.

Critical Thinking / Discussion / Writing

83. The accompanying figure shows the first six squares in an infinite sequence of successively smaller squares formed by connecting the midpoints of the sides of the preceding larger square. The area of the largest square is 1. Find the total area of the indicated sections that are shaded in every other square as the number of squares increases to infinity.

84. The accompanying figure shows the first three equilateral triangles in an infinite sequence of successively smaller equilateral triangles formed by connecting the midpoints of the sides of the preceding larger triangle. The largest triangle has sides of length 4. Find the total perimeter of all of the triangles.

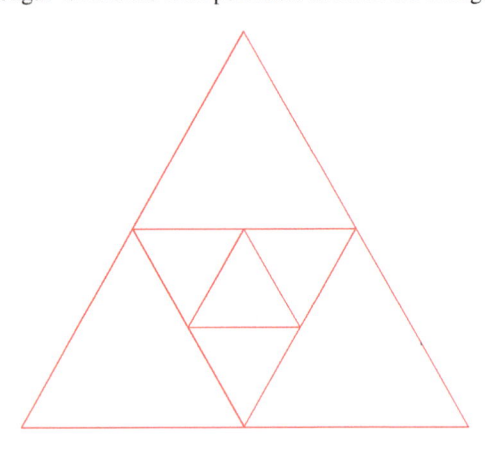

85. If a_n denotes the nth term of an arithmetic sequence with $a_1 = 10$ and $d = 2.7$ and if b_n denotes the nth term of a geometric sequence with $b_1 = 10$ and $r = 2$, which number is larger, a_{1001} or b_{1001}?

86. If the sum $\dfrac{a(1 - r^n)}{1 - r}$ of the first n terms of a geometric sequence is 1023, find a when $n = 10$ and $r = \dfrac{1}{2}$.

87. Find n and k so that the sum $5 + 5 \cdot 2 + 5 \cdot 2^2 + \cdots + 5 \cdot 2^{15}$ can be written as $\displaystyle\sum_{i=0}^{n} 5 \cdot 2^i$ or as $\displaystyle\sum_{i=1}^{k} 5 \cdot 2^{i-1}$.

88. The sum of three consecutive terms of a geometric sequence is 35, and their product is 1000. Find the numbers.
    [*Hint:* Let $\dfrac{a}{r}$, a, and ar be the three terms of geometric sequence.]

89. The sum of three consecutive terms of an arithmetic sequence is 15. If 1, 4, and 19 are respectively added to the three terms, the resulting numbers form three consecutive terms of a geometric sequence. Find the numbers.

Maintaining Skills

In Exercises 90–93, find the value of a variable that satisfies the given condition.

90. Find x if $3x + 2y = 7$ and $y = \dfrac{1}{2}$.

91. Find x if $-2x + 9y = 5$ and $y = 3$.

92. Find y if $-4x + 7y = 7$ and $x = 0$.

93. Find y if $23x - 14y = -5$ and $x = 1$.

94. Find the slope of the line with equation $10x - 2y = 28$.

95. Find the slope of the line with equation $15x + 5y = 2$.

96. Find an equation of the line through the point $(5, -1)$ parallel to the line with equation $6x - 3y = 7$.

97. Find an equation of the line through the point $(3, 3)$ parallel to the line with equation $x + 2y = 1$.

98. Find values for a and b so that the equation $ax + by = 3$ has the same graph as the equation $2x - 4y = 12$.

99. Find values for a and b so that the equation $ax + by = 2$ has the same graph as the equation $4x - y = -1$.

SECTION 6.4

Systems of Equations in Two Variables

BEFORE STARTING THIS SECTION, REVIEW

1. Graphs of equations (Section 1.1, page 55)
2. Graphs of linear equations (Section 1.2, page 69)

OBJECTIVES

1. Verify a solution to a system of equations.
2. Solve a system of equations by the graphical method.
3. Solve a system of equations by the substitution method.
4. Solve a system of equations by the elimination method.
5. Solve applied problems by solving systems of equations.

Al-Khwarizmi (780–850)

Al-Khwarizmi came to Baghdad from the town of Khwarizm, which is now *Khivga*, Uzbekistan. The term *algorithm* is a corruption of the name al-khwarizmi. Originally, the word *algorism* was used for the rules for performing arithmetic by using decimal notation. *Algorism* evolved into the word *algorithm* by the eighteenth century. The word *algorithm* now means a finite set of precise instructions for performing a computation or for solving a problem. (The picture shown is a Soviet postage stamp from 1983 depicting Al-Khwarizmi.)

◆ Algebra–Iraq Connection

The most important contributions of medieval Islamic mathematicians lie in the area of algebra. One of the greatest Islamic scholars was Muhammad ibn Musa al-Khwarizmi (A.D. 780–850). Al-Khwarizmi was one of the first scholars in the House of Wisdom, an academy of scientists, established by caliph al-Mamun in the city of Baghdad in what is now Iraq. Jews, Christians, and Muslims worked together in scholarly pursuits in the academy during this period. The eminent scholar al-Khwarizmi wrote several books on astronomy and mathematics. Western Europeans first learned about algebra from his books. The word *algebra* comes from the Arabic *al-jabr*, part of the title of his book *Kitab al-jabr wal-muqabala*.

This book was translated into Latin and was a widely used text. The Arabic word *al-jabr* means "restoration," as in restoring broken parts. (At one time, it was not unusual to see the sign "Algebrista y Sangrador," meaning "bone setter and blood letter," at the entrance of a Spanish barber's shop. The sign informed customers of the barber's side business.) Al-Khwarizmi used the term in the mathematical sense of removing a negative quantity on one side of an equation and restoring it as a positive quantity on the other side. This procedure is among the most widely used in solving equations in both theoretical and practical situations. See Example 10.

1 Verify a solution to a system of equations.

System of Equations

A set of equations with common variables is called a **system of equations**. If each equation in a system of equations is linear, then the set of equations is called a **system of linear equations** or a **linear system of equations**. However, if at least one equation in a system of equations is nonlinear, then the set of equations is called a **system of nonlinear equations**. In this section, we will solve systems of two equations in two variables, such as:

$$\begin{cases} 2x - y = 5 \\ x + 2y = 5 \end{cases}$$

Notice that we designate a system of equations by using a left brace. The brace is placed to remind us that the equations must be dealt with simultaneously. A system of equations is sometimes referred to as a set of *simultaneous equations*. A **solution of a system of equations in two variables** x and y is an ordered pair of numbers (a, b) such that when x is replaced with a and y is replaced with b, the resulting equations are true. The **solution set of a system of equations** is the set of all solutions of the system.

EXAMPLE 1   Verifying a Solution

Verify that the ordered pair $(3, 1)$ is a solution of the system of linear equations

$$\begin{cases} 2x - y = 5 & (1) \\ x + 2y = 5 & (2) \end{cases}$$

Solution

To verify that $(3, 1)$ is a solution, replace x with 3 and y with 1 in both equations.

Equation (1)	Equation (2)
$2x - y = 5$	$x + 2y = 5$
$2(3) - 1 \stackrel{?}{=} 5$	$3 + 2(1) \stackrel{?}{=} 5$
$5 = 5$ ✓	$5 = 5$ ✓

Because the ordered pair $(3, 1)$ satisfies both equations, it is a solution of the system.

Practice Problem 1 Verify that $(1, 3)$ is a solution of the system

$$\begin{cases} x + y = 4 \\ 3x - y = 0 \end{cases}$$

In this section, you will learn three methods of solving a system of two equations in two variables: the *graphical method,* the *substitution method,* and the *elimination method.*

2 Solve a system of equations by the graphical method.

Graphical Method

Recall that the graph of a linear equation

$$ax + by = c \quad (a \text{ and } b \text{ not both zero})$$

is a line. Consider a system of linear equations:

$$\begin{cases} a_1 x + b_1 y = c_1 \\ a_2 x + b_2 y = c_2 \end{cases}$$

A solution of this system (if any) is a point in the xy-plane whose coordinates satisfy both equations. To find or estimate the solution set of this system, therefore, we graph both equations on the same coordinate axes and find the coordinates of any points of intersection. This procedure is called the **graphical method**.

EXAMPLE 2   Solving a System by the Graphical Method

Use the graphical method to solve the system of equations.

$$\begin{cases} 2x - y = 4 & (1) \\ 2x + 3y = 12 & (2) \end{cases}$$

Solution

Step 1   **Graph both equations on the same coordinate axes.**
  (i) We first graph equation (1) by finding its intercepts.
    a. To find the y-intercept, set $x = 0$ and solve for y.
       We have $2(0) - y = 4$, or $y = -4$; so the y-intercept is -4.

478 Chapter 6 Further Topics in Algebra

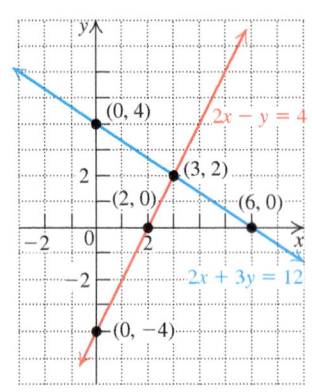

FIGURE 6.2 Graphical solution

TECHNOLOGY CONNECTION

In Example 2, solve each equation for y. Graph

$Y_1 = 2x - 4$ and

$Y_2 = -\dfrac{2}{3}x + 4$

on the same screen.

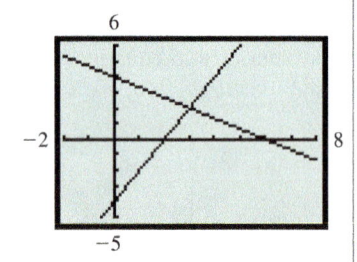

Use the **intersect** feature on your calculator to find the point of intersection.

b. To find the x-intercept, set $y = 0$ and solve for x. We have
$2x - 0 = 4$, or $x = 2$; so the x-intercept is 2.
The points $(0, -4)$ and $(2, 0)$ are on the graph of equation (1), which is sketched in red in Figure 6.2.

(ii) We now graph Equation (2) using intercepts.
Here the x-intercept is 6 and the y-intercept is 4; so joining the points $(0, 4)$ and $(6, 0)$ produces the line sketched in blue in Figure 6.2.

Step 2 **Find the point(s) of intersection of the two graphs.** We observe in Figure 6.2 that the point of intersection of the two graphs is $(3, 2)$.

Step 3 **Check your solution(s).** Replace x with 3 and y with 2 in equations (1) and (2).

Equation (1)	Equation (2)
$2x - y = 4$	$2x + 3y = 12$
$2(3) - 2 \stackrel{?}{=} 4$	$2(3) + 3(2) \stackrel{?}{=} 12$
$4 = 4$ ✓	$12 = 12$ ✓

Step 4 **Write the solution set for the system.**
The solution set is $\{(3, 2)\}$.

Practice Problem 2 Solve the system of equations graphically.

$$\begin{cases} x + y = 2 \\ 4x + y = -1 \end{cases}$$

If a system of equations has at least one solution (as in Example 2), the system is **consistent**. A system of equations with no solution is **inconsistent**, and its solution set is the empty set, ∅.

A system of two linear equations in two variables must have one of the following types of solution sets.

1. One solution (the lines intersect; see Figure 6.3(a)): the system is consistent, and the equations in the system are said to be **independent**.
2. No solution (the lines are parallel; see Figure 6.3(b)): the system is inconsistent.
3. Infinitely many solutions (the lines coincide; see Figure 6.3(c)): the system is consistent, and the equations in the system are said to be **dependent**.

One solution

(a) Consistent system; independent equations

No solution

(b) Inconsistent system

Infinitely many solutions

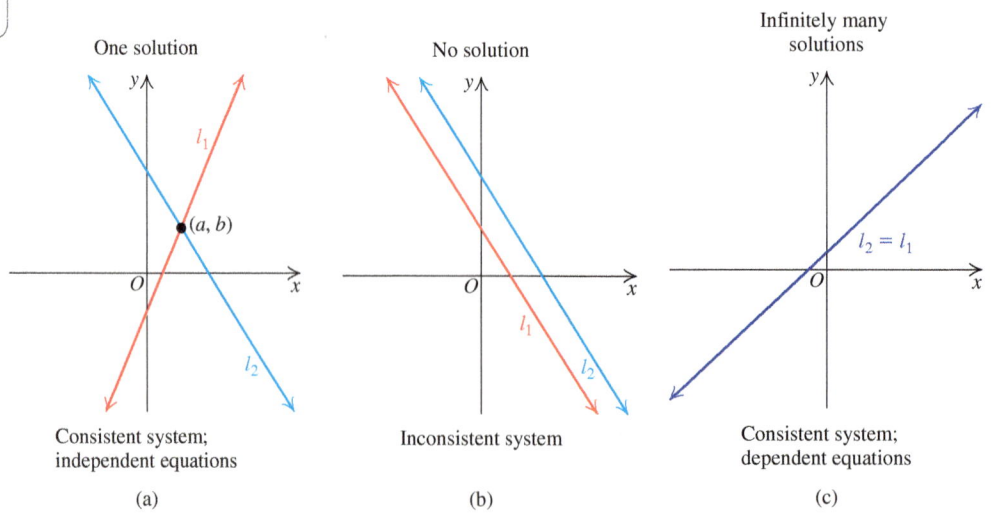

(c) Consistent system; dependent equations

FIGURE 6.3 Possible solution sets of a system of two linear equations

3 Solve a system of equations by the substitution method.

Substitution Method

Another way to solve a linear system of equations is called the **substitution method**.

PROCEDURE IN ACTION

EXAMPLE 3 The Substitution Method

OBJECTIVE
Reduce the solution of the system to the solution of one equation in one variable by substitution.

EXAMPLE
Solve the system.
$$\begin{cases} 2x - 5y = 3 & (1) \\ y - 2x = 9 & (2) \end{cases}$$

Step 1 Solve for one variable. Choose one of the equations and express one of its variables in terms of the other variable.

1. We choose equation (2) and express y in terms of x.
   $y = 2x + 9$ (3) Solve equation (2) for y.

Step 2 Substitute. Substitute the expression found in Step 1 into the other equation to obtain an equation in one variable.

2. $2x - 5y = 3$ Equation (1)
   $2x - 5(2x + 9) = 3$ Substitute $2x + 9$ for y.
   $2x - 10x - 45 = 3$ Distributive property

Step 3 Solve the equation obtained in Step 2.

3. $-8x - 45 = 3$ Combine like terms.
   $-8x = 3 + 45$ Add 45 to both sides.
   $x = -6$ Solve for x.

Step 4 Back-substitute. Substitute the value(s) you found in Step 3 back into the expression you found in Step 1. The result is the solution(s).

4. $y = 2x + 9$ Equation (3) from Step 1
   $y = 2(-6) + 9 = -3$ Substitute $x = -6$.
   Because $x = -6$ and $y = -3$, the solution set is $\{(-6, -3)\}$.

Step 5 Check. Check your answer(s) in the original equations.

5. Check: $x = -6$ and $y = -3$.

   Equation (1) | Equation (2)
   $2(-6) - 5(-3) \stackrel{?}{=} 3$ | $-3 - 2(-6) \stackrel{?}{=} 9$
   $-12 + 15 = 3 \checkmark$ | $-3 + 12 = 9 \checkmark$

Practice Problem 3 Solve: $\begin{cases} x - y = 5 \\ 2x + y = 7 \end{cases}$

SIDE NOTE

It is easy to make arithmetic errors in the many steps involved in the process of solving a system of equations. You should check your solution by substituting it into each equation in the original system.

It is possible that in the process of solving the equation in Step 3, you obtain an equation equivalent to the equation $0 = k$, where k is a *nonzero* constant. In such cases, the false statement, equivalent to $0 = k$, indicates that the system is *inconsistent*.

EXAMPLE 4 Attempting to Solve an Inconsistent System of Equations

Solve the system of equations.
$$\begin{cases} x + y = 3 & (1) \\ 2x + 2y = 9 & (2) \end{cases}$$

Solution
Step 1 Solve equation (1) for y in terms of x.

$y = 3 - x$ Add $-x$ to both sides.

480 Chapter 6 Further Topics in Algebra

Step 2 Substitute this expression into equation (2).

$$2x + 2y = 9 \quad \text{Equation (2)}$$
$$2x + 2(3 - x) = 9 \quad \text{Replace } y \text{ with } 3 - x \text{ (from Step 1).}$$

Step 3 Solve for x.

$$2x + 6 - 2x = 9 \quad \text{Distributive property}$$
$$\underbrace{6 = 9}_{\text{False}} \quad \text{Simplify.}$$

Because the equation $6 = 9$ is false, the system is inconsistent. Figure 6.4 shows that the graphs of the two equations in this system are parallel lines. Because the lines do not intersect, the system has no solution. The solution set is $\varnothing$.

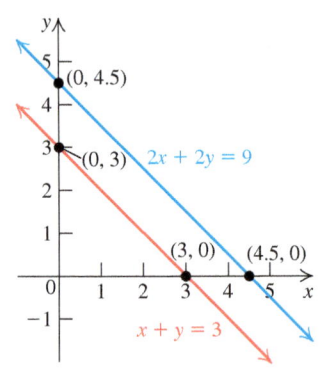

FIGURE 6.4 Inconsistent system

Practice Problem 4 Solve the system of equations.

$$\begin{cases} x - 3y = 1 \\ -2x + 6y = 3 \end{cases}$$

In Step 3, it is also possible to end up with an equation that is equivalent to the equation $0 = 0$. In such cases, the equations are *dependent* and the system has infinitely many solutions.

EXAMPLE 5 Solving a Dependent System

Solve the system of equations.

$$\begin{cases} 4x + 2y = 12 & (1) \\ -2x - y = -6 & (2) \end{cases}$$

Solution

Step 1 Solve equation (2) for y in terms of x.

$$-2x - y = -6 \quad \text{Equation (2)}$$
$$-y = -6 + 2x \quad \text{Add } 2x \text{ to both sides.}$$
$$y = 6 - 2x \quad \text{Multiply both sides by } -1.$$

Step 2 Substitute $(6 - 2x)$ for y in equation (1).

$$4x + 2y = 12 \quad \text{Equation (1)}$$
$$4x + 2(6 - 2x) = 12 \quad \text{Replace } y \text{ with } 6 - 2x \text{ (from Step 1).}$$

Step 3 Solve for x.

$$4x + 12 - 4x = 12 \quad \text{Distributive property}$$
$$\underbrace{12 = 12}_{\text{True}} \quad \text{Simplify.}$$

The equation is true for *every* value of x. Thus, *any* value of x can be used in the equation $y = 6 - 2x$ for back-substitution.

The solutions of the system are of the form $(x, 6 - 2x)$, and the solution set is

$$\{(x, 6 - 2x) \mid x \text{ any real number}\}.$$

In other words, the solution set consists of the infinite number of ordered pairs (x, y) lying on the line with equation $4x + 2y = 12$, as shown in Figure 6.5. You can find particular solutions by replacing x with any particular real number. For example, if we let $x = 0$ in $(x, 6 - 2x)$, we find that $(0, 6 - 2(0)) = (0, 6)$ is a solution of the system. Similarly, letting $x = 1$, we find that $(1, 4)$ is a solution.

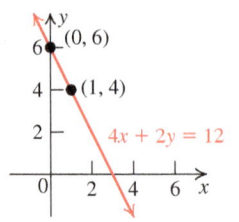

FIGURE 6.5 Dependent equations

Practice Problem 5 Solve the system of equations.

$$\begin{cases} -2x + y = -3 \\ 4x - 2y = 6 \end{cases}$$

EXAMPLE 6 Using Substitution to Solve a Nonlinear System

Solve the system of equations by the substitution method.

$$\begin{cases} 4x + y = -3 & (1) \\ -x^2 + y = 1 & (2) \end{cases}$$

Solution

We follow the same steps as those used in the substitution method for solving a system of linear equations.

Step 1 **Solve for one variable.** In equation (2), you can express y in terms of x:

$$y = x^2 + 1 \quad (3) \qquad \text{Solve equation (2) for } y.$$

Step 2 **Substitute.** Substitute $x^2 + 1$ for y in equation (1).

$$\begin{aligned} 4x + y &= -3 & &\text{Equation (1)} \\ 4x + (x^2 + 1) &= -3 & &\text{Replace } y \text{ with } (x^2 + 1). \\ 4x + x^2 + 4 &= 0 & &\text{Add 3 to both sides.} \\ x^2 + 4x + 4 &= 0 & &\text{Rewrite in descending powers of } x. \end{aligned}$$

Step 3 **Solve** the equation resulting from Step 2.

$$\begin{aligned} (x + 2)(x + 2) &= 0 & &\text{Factor.} \\ x + 2 &= 0 & &\text{Zero-product property} \\ x &= -2 & &\text{Solve for } x. \end{aligned}$$

Step 4 **Back-substitution.** Substitute $x = -2$ in equation (3) to find the corresponding y-value.

$$\begin{aligned} y &= x^2 + 1 & &\text{Equation (3)} \\ y &= (-2)^2 + 1 & &\text{Replace } x \text{ with } -2. \\ y &= 5 & &\text{Simplify.} \end{aligned}$$

Because $x = -2$ and $y = 5$, the apparent solution set of the system is $\{(-2, 5)\}$.

Step 5 **Check.** Replace x with -2 and y with 5 in equation (1) and equation (2).

Equation (1)	Equation (2)
$4x + y = -3$	$-x^2 + y = 1$
$4(-2) + 5 \stackrel{?}{=} -3$	$-(-2)^2 + 5 \stackrel{?}{=} 1$
$-8 + 5 = -3$ ✓	$-4 + 5 = 1$ ✓

The graphs of the line $4x + y = -3$ and the parabola $y = x^2 + 1$ confirm that the solution set is $\{(-2, 5)\}$. See Figure 6.6.

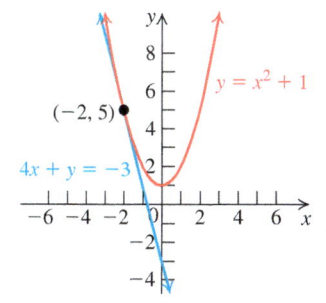

FIGURE 6.6

Practice Problem 6 Solve the system of equations by the substitution method.

$$\begin{cases} x^2 + y = 2 \\ 2x + y = -1 \end{cases}$$

Elimination Method

4 Solve a system of equations by the elimination method.

The **elimination method** of solving a system of equations is also called the **addition method**.

PROCEDURE IN ACTION

EXAMPLE 7   The Elimination Method

OBJECTIVE

Solve a system of two linear equations by first eliminating one variable.

EXAMPLE

Solve the system.

$$\begin{cases} 2x + 3y = 21 & (1) \\ 3x - 4y = 23 & (2) \end{cases}$$

Step 1 Adjust the coefficients. If necessary, multiply both equations by appropriate numbers to get two new equations in which the coefficients of the variable to be eliminated are opposites.

1. We arbitrarily select y as the variable to be eliminated.

$$\begin{cases} 8x + 12y = 84 & \text{Multiply equation (1) by 4.} \\ 9x - 12y = 69 & \text{Multiply equation (2) by 3.} \end{cases}$$

Coefficients are opposites of each other.

Step 2 Add the equations. Add the resulting equations to get an equation in one variable.

2. $8x + 12y = 84$
    $9x - 12y = 69$
    _____
    $17x = 153$

Adding the two equations from Step 1 eliminates the y-terms.

Step 3 Solve the resulting equation.

3. $17x = 153$ Equation from Step 2

    $x = \dfrac{153}{17}$ Divide both sides by 17.

    $x = 9$ Simplify.

Step 4 Back-substitute the value you found into one of the original equations to solve for the other variable.

4. $2x + 3y = 21$ Equation (1)
    $2(9) + 3y = 21$ Replace x with 9.
    $18 + 3y = 21$ Simplify.
    $3y = 3$ Subtract 18 from both sides.
    $y = 1$ Solve for y.

Step 5 Write the solution set from Steps 3 and 4.

5. The solution set is $\{(9, 1)\}$.

Step 6 Check your solution(s) in the original equations (1) and (2).

6. Check $x = 9$ and $y = 1$.

    Equation (1)   Equation (2)
    $2(9) + 3(1) \stackrel{?}{=} 21$ $3(9) - 4(1) = 23$
    $18 + 3 = 21$ ✓ $27 - 4 = 23$ ✓

Practice Problem 7 Solve the system.

$$\begin{cases} 3x + 2y = 3 \\ 9x - 4y = 4 \end{cases}$$

Just as in the substitution method, if you add the equations in Step 2 and the resulting equation becomes $0 = k$, where $k \neq 0$, then the system is inconsistent: it has no solutions. As an illustration, if you solve the system of equations in Example 4 by the elimination method, you will obtain the equation $0 = 3$. You now conclude that the system is inconsistent. Similarly, in Step 2, if you obtain $0 = 0$, then the equations are dependent. See Example 5.

SIDE NOTE

You should use the elimination method when the terms involving one of the variables can easily be eliminated by adding multiples of the equations. Otherwise, use substitution to solve the system. The graphing method is usually used to confirm or visually interpret the result obtained from the other two methods.

The next example illustrates how some nonlinear systems can be solved by making a substitution to create a linear system and then solving the new system by the elimination method. We then find the solution(s) to the original equations by the substitution method.

EXAMPLE 8 Solving a Nonlinear System by a Linearizing Substitution

Solve the system.

$$\begin{cases} \dfrac{2}{x} + \dfrac{5}{y} = -5 & (1) \\ \dfrac{3}{x} - \dfrac{2}{y} = -17 & (2) \end{cases}$$

Solution

Replace $\dfrac{1}{x}$ with u and $\dfrac{1}{y}$ with v. [Note that $\dfrac{2}{x} = 2\left(\dfrac{1}{x}\right) = 2u$, and so on.] Equations (1) and (2) become

$$\begin{cases} 2u + 5v = -5 & (3) \\ 3u - 2v = -17 & (4) \end{cases}$$

Now we solve the new system for u and v using the elimination method.

Step 1 We choose to eliminate the variable u.

$6u + 15v = -15$	(5)	Multiply equation (3) by 3.
$-6u + 4v = 34$	(6)	Multiply equation (4) by -2.

Step 2 $\quad 19v = 19 \quad$ (7) $\quad$ Add equations (5) and (6).

Step 3 $\quad v = 1 \quad\quad\quad\quad\quad$ Solve equation (7) for v.

Step 4 $\quad 2u + 5v = -5 \quad\quad$ Equation (3)

$\quad\quad\quad\quad 2u + 5(1) = -5 \quad\quad$ Back-substitute $v = 1$.

$\quad\quad\quad\quad u = -5 \quad\quad\quad\quad\quad$ Solve for u.

Step 5 Now solve for x and y, the variables in the original system.

$u = \dfrac{1}{x} \quad$ and $\quad v = \dfrac{1}{y} \quad\quad$ Original substitution

$-5 = \dfrac{1}{x} \quad\quad\quad 1 = \dfrac{1}{y} \quad\quad$ Replace u with -5 and v with 1.

$x = -\dfrac{1}{5} \quad\quad\quad y = 1 \quad\quad$ Solve for x and y.

The solution set of the original system is $\left\{\left(-\dfrac{1}{5}, 1\right)\right\}$.

Step 6 Verify that the ordered pair $\left(-\dfrac{1}{5}, 1\right)$ is the solution of the original system of equations (1) and (2).

Practice Problem 8 Solve the system.

$$\begin{cases} \dfrac{4}{x} + \dfrac{3}{y} = 1 \\ \dfrac{2}{x} - \dfrac{6}{y} = 3 \end{cases}$$

EXAMPLE 9 Using Elimination to Solve a Nonlinear System

Solve the system of equations by the elimination method.

$$\begin{cases} x^2 + y^2 = 25 & (1) \\ x^2 - y = 5 & (2) \end{cases}$$

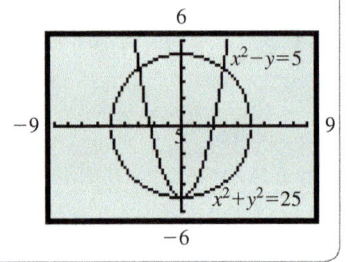

Solution

Step 1 **Adjust the coefficients.** Because x has the same power in both equations, we choose the variable x for elimination. We multiply equation (2) by -1 and obtain the equivalent system.

$$\begin{cases} x^2 + y^2 = 25 & (1) \\ -x^2 + y = -5 & (3) \end{cases} \quad \text{Multiply equation (2) by } -1.$$

Step 2 $\quad y^2 + y = 20 \quad (4) \qquad$ Add equations (1) and (3).

Step 3 Solve the equation found in Step 2: $y^2 + y = 20$.

$$y^2 + y - 20 = 0 \qquad \text{Subtract 20 from both sides.}$$
$$(y + 5)(y - 4) = 0 \qquad \text{Factor.}$$
$$y + 5 = 0 \quad \text{or} \quad y - 4 = 0 \qquad \text{Zero-product property}$$
$$y = -5 \quad \text{or} \quad y = 4 \qquad \text{Solve for } y.$$

Step 4 **Back-substitute** the values in one of the original equations to solve for the other variable.

(i) Substitute $y = -5$ in equation (2) and solve for x.

$$x^2 - y = 5 \qquad \text{Equation (2)}$$
$$x^2 - (-5) = 5 \qquad \text{Replace } y \text{ with } -5.$$
$$x^2 = 0 \qquad \text{Simplify.}$$
$$x = 0 \qquad \text{Solve for } x.$$

So, $(0, -5)$ is a solution of the system.

(ii) Substitute $y = 4$ in equation (2) and solve for x.

$$x^2 - y = 5 \qquad \text{Equation (2)}$$
$$x^2 - 4 = 5 \qquad \text{Replace } y \text{ with } 4.$$
$$x^2 = 9 \qquad \text{Add 4 to both sides.}$$
$$x = \pm 3 \qquad \text{Square root property}$$

So, $(3, 4)$ and $(-3, 4)$ are solutions of the system.
From (i) and (ii), the apparent solution set for the system is $\{(0, -5), (3, 4), (-3, 4)\}$.

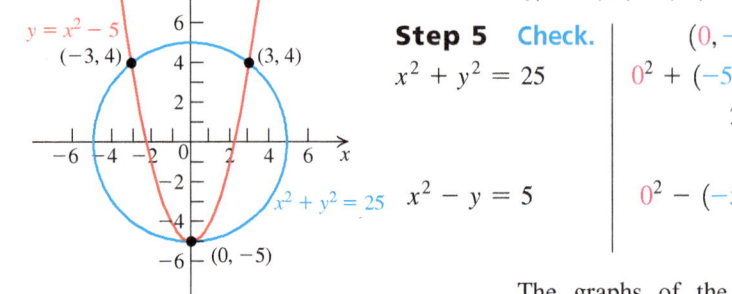

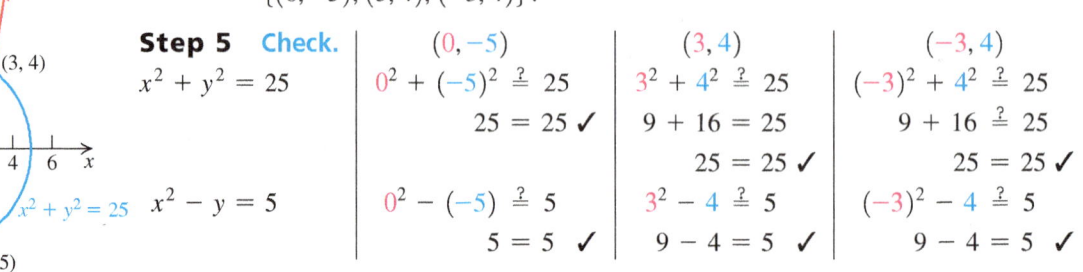

Step 5 **Check.**

	$(0, -5)$	$(3, 4)$	$(-3, 4)$
$x^2 + y^2 = 25$	$0^2 + (-5)^2 \stackrel{?}{=} 25$	$3^2 + 4^2 \stackrel{?}{=} 25$	$(-3)^2 + 4^2 \stackrel{?}{=} 25$
	$25 = 25 \checkmark$	$9 + 16 = 25$	$9 + 16 \stackrel{?}{=} 25$
		$25 = 25 \checkmark$	$25 = 25 \checkmark$
$x^2 - y = 5$	$0^2 - (-5) \stackrel{?}{=} 5$	$3^2 - 4 \stackrel{?}{=} 5$	$(-3)^2 - 4 \stackrel{?}{=} 5$
	$5 = 5 \checkmark$	$9 - 4 = 5 \checkmark$	$9 - 4 = 5 \checkmark$

The graphs of the circle $x^2 + y^2 = 25$ and the parabola $y = x^2 - 5$ sketched in Figure 6.7 confirm the three solutions.

TECHNOLOGY CONNECTION

To use a graphing calculator to graph the system in Example 9, you must solve $x^2 + y^2 = 25$ for y and enter the two resulting equations.

$$Y_1 = \sqrt{25 - x^2}$$
$$Y_2 = -\sqrt{25 - x^2}$$

Then enter

$$Y_3 = x^2 - 5$$

and use the ZSquare feature.

FIGURE 6.7

Practice Problem 9 Solve the system of equations.

$$\begin{cases} x^2 + 2y^2 = 34 \\ x^2 - y^2 = 7 \end{cases}$$

Applications of Systems of Equations

5 Solve applied problems by solving systems of equations.

It is well known that as the price of a product increases, demand for it decreases, and as the price increases, the supply of the product also increases. The **equilibrium point** is the ordered pair (x, p) such that the number of units, x, and the price per unit, p, satisfy *both* the demand and supply equations.

EXAMPLE 10 Finding the Equilibrium Point

Find the equilibrium point if the supply and demand functions for a new brand of digital video recorder (DVR) are given by the system

$p = 60 + 0.0012x$ (1)   Supply equation
$p = 80 - 0.0008x$ (2)   Demand equation

where p is the price in dollars and x is the number of units.

> **SIDE NOTE**
> The graph of a system of equations allows you to use geometric intuition to understand algebra. If the geometric and algebraic representations of a solution do not agree, you must go back and find the error(s).

Solution

We substitute the value of p from equation (1) into equation (2) and solve the resulting equation.

$$p = 80 - 0.0008x \quad \text{Demand equation}$$
$$60 + 0.0012x = 80 - 0.0008x \quad \text{Replace } p \text{ with } 60 + 0.0012x.$$
$$0.002x = 20 \quad \text{Collect like terms and simplify.}$$
$$x = \frac{20}{0.002} = 10{,}000 \quad \text{Solve for } x.$$

The equilibrium point occurs when the supply and demand for DVRs is 10,000 units. To find the price p, we back-substitute $x = 10{,}000$ into either of the original equations (1) or (2).

$$p = 60 + 0.0012x \quad \text{Equation (1)}$$
$$= 60 + 0.0012(10{,}000) \quad \text{Replace } x \text{ with } 10{,}000.$$
$$= 72 \quad \text{Simplify.}$$

The equilibrium point is (10,000, 72). See Figure 6.8.

Check. Verify that the ordered pair (10,000, 72) satisfies both equations (1) and (2).

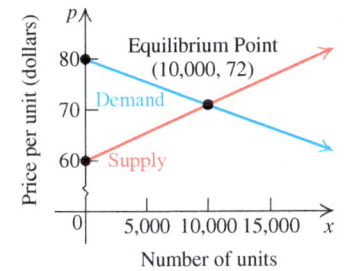

FIGURE 6.8 Price of DVRs

Practice Problem 10 Find the equilibrium point.

$$\begin{cases} p = 20 + 0.002x & \text{Supply equation} \\ p = 77 - 0.008x & \text{Demand equation} \end{cases}$$

In general, to solve an applied problem involving two unknowns, we need to set up two equations using two variables to represent the two unknowns. The general rule is that *the number of equations formed must be equal to the number of unknown quantities involved.*

SECTION 6.4 Exercises

Basic Concepts and Skills

In Exercises 1–6, determine which ordered pairs are solutions of each system of equations.

1. $\begin{cases} 2x + 3y = 3 \\ 3x - 4y = 13 \end{cases}$ $(1, -3), (3, -1), (6, 3), \left(5, \frac{1}{2}\right)$

2. $\begin{cases} x + 2y = 6 \\ 3x + 6y = 18 \end{cases}$ $(2, 2), (-2, 4), (0, 3), (1, 2)$

3. $\begin{cases} 5x - 2y = 7 \\ -10x + 4y = 11 \end{cases}$ $\left(\frac{5}{4}, 1\right), \left(0, \frac{11}{4}\right), (1, -1), (3, 4)$

4. $\begin{cases} x - 2y = -5 \\ 3x - y = 5 \end{cases}$ $(1, 3), (-5, 0), (3, 4), (3, -4)$

486 Chapter 6 Further Topics in Algebra

5. $\begin{cases} x + y = 1 \\ \frac{1}{2}x + \frac{1}{3}y = 2 \end{cases}$ $(0, 1), (1, 0), \left(\frac{2}{3}, \frac{3}{2}\right), (10, -9)$

6. $\begin{cases} \frac{2}{x} + \frac{3}{y} = 2 \\ \frac{6}{x} + \frac{18}{y} = 9 \end{cases}$ $(3, 2), (2, 3), (4, 3), (3, 4)$

In Exercises 7–16, estimate the solution(s) (if any) of each system by the graphical method. Check your solution(s). For any dependent equations, write your answer with x being arbitrary.

7. $\begin{cases} x + y = 3 \\ x - y = 1 \end{cases}$
8. $\begin{cases} x + y = 10 \\ x - y = 2 \end{cases}$
9. $\begin{cases} x + 2y = 6 \\ 2x + y = 6 \end{cases}$
10. $\begin{cases} 2x - y = 4 \\ x - y = 3 \end{cases}$
11. $\begin{cases} 3x - y = -9 \\ y = 3x + 6 \end{cases}$
12. $\begin{cases} 5x + 2y = 10 \\ y = -\frac{5}{2}x - 5 \end{cases}$
13. $\begin{cases} x + y = 7 \\ y = 2x \end{cases}$
14. $\begin{cases} y - x = 2 \\ y + x = 9 \end{cases}$
15. $\begin{cases} 3x + y = 12 \\ y = -3x + 12 \end{cases}$
16. $\begin{cases} 2x + 3y = 6 \\ 6y = -4x + 12 \end{cases}$

In Exercises 17–30, determine whether each system is *consistent* or *inconsistent*. If the system is consistent, determine whether the equations are dependent or independent. Do not solve the system.

17. $\begin{cases} y = -2x + 3 \\ y = 3x + 5 \end{cases}$
18. $\begin{cases} 3x + y = 5 \\ 2x + y = 4 \end{cases}$
19. $\begin{cases} 2x + 3y = 5 \\ 3x + 2y = 7 \end{cases}$
20. $\begin{cases} 2x - 4y = 5 \\ 3x + 5y = -6 \end{cases}$
21. $\begin{cases} 3x + 5y = 7 \\ 6x + 10y = 14 \end{cases}$
22. $\begin{cases} 3x - y = 2 \\ 9x - 3y = 6 \end{cases}$
23. $\begin{cases} x + 2y = -5 \\ 2x - y = 4 \end{cases}$
24. $\begin{cases} x + 2y = -2 \\ 2x - 3y = 5 \end{cases}$
25. $\begin{cases} 2x - 3y = 5 \\ 6x - 9y = 10 \end{cases}$
26. $\begin{cases} 3x + y = 2 \\ 15x + 5y = 15 \end{cases}$
27. $\begin{cases} -3x + 4y = 5 \\ \frac{9}{2}x - 6y = \frac{15}{2} \end{cases}$
28. $\begin{cases} 6x + 5y = 11 \\ 9x + \frac{15}{2}y = 21 \end{cases}$
29. $\begin{cases} 7x - 2y = 3 \\ 11x - \frac{3}{2}y = 8 \end{cases}$
30. $\begin{cases} 4x + 7y = 10 \\ 10x + \frac{35}{2}y = 25 \end{cases}$

In Exercises 31–40, solve each system of equations by the substitution method. Check your solutions. For linear systems, if the equations are dependent, write your answer in the ordered pair form given in Example 5.

31. $\begin{cases} y = 2x + 1 \\ 5x + 2y = 9 \end{cases}$
32. $\begin{cases} x = 3y - 1 \\ 2x - 3y = 7 \end{cases}$
33. $\begin{cases} 3x - y = 5 \\ x + y = 7 \end{cases}$
34. $\begin{cases} 2x + y = 2 \\ 3x - y = -7 \end{cases}$
35. $\begin{cases} 2x - y = 5 \\ -4x + 2y = 7 \end{cases}$
36. $\begin{cases} 3x + 2y = 5 \\ -9x - 6y = 15 \end{cases}$
37. $\begin{cases} x - y = 2 \\ x^2 - 4x + y^2 = -2 \end{cases}$
38. $\begin{cases} 2x - y = 1 \\ x^2 - 8y + y^2 = -6 \end{cases}$
39. $\begin{cases} x - 2y = 5 \\ -3x + 6y = -15 \end{cases}$
40. $\begin{cases} x + y = 3 \\ 2x + 2y = 6 \end{cases}$

In Exercises 41–50, solve each system of equations by the elimination method. Check your solutions. For any linear systems, if the equations are dependent, write your answer as in Example 5.

41. $\begin{cases} x - y = 1 \\ x + y = 5 \end{cases}$
42. $\begin{cases} 2x - 3y = 5 \\ 3x + 2y = 14 \end{cases}$
43. $\begin{cases} x + y = 0 \\ 2x + 3y = 3 \end{cases}$
44. $\begin{cases} x + y = 3 \\ 3x + y = 1 \end{cases}$
45. $\begin{cases} x^2 + 2y^2 = 12 \\ -5x^2 + 7y^2 = 8 \end{cases}$
46. $\begin{cases} x^2 - 6y^2 = 19 \\ 3x^2 + 2y^2 = 77 \end{cases}$
47. $\begin{cases} x - y = 2 \\ -2x + 2y = 5 \end{cases}$
48. $\begin{cases} x + y = 5 \\ 2x + 2y = -10 \end{cases}$
49. $\begin{cases} 4x + 6y = 12 \\ 2x + 3y = 6 \end{cases}$
50. $\begin{cases} 4x + 7y = -3 \\ -8x - 14y = 6 \end{cases}$

In Exercises 51–70, use any method to solve each system of equations. For any dependent equations, write your answer as in Example 5.

51. $\begin{cases} 2x + y = 9 \\ 2x - 3y = 5 \end{cases}$
52. $\begin{cases} x + 2y = 10 \\ x - 2y = -6 \end{cases}$
53. $\begin{cases} 2x + 5y = 2 \\ x + 3y = 2 \end{cases}$
54. $\begin{cases} 4x - y = 6 \\ 3x - 4y = 11 \end{cases}$
55. $\begin{cases} 2x + 3y = 7 \\ 3x + y = 7 \end{cases}$
56. $\begin{cases} x = 3y + 4 \\ x = 5y + 10 \end{cases}$
57. $\begin{cases} 2x + 3y = 9 \\ 3x + 2y = 11 \end{cases}$
58. $\begin{cases} 3x - 4y = 0 \\ y = \frac{2x + 1}{3} \end{cases}$
59. $\begin{cases} \frac{x}{4} + \frac{y}{6} = 1 \\ x + 2(x - y) = 7 \end{cases}$
60. $\begin{cases} \frac{x}{3} + \frac{y}{5} = 12 \\ x - y = 4 \end{cases}$
61. $\begin{cases} x - y^2 = 2 \\ 2x + 3y = 3 \end{cases}$
62. $\begin{cases} x + 2y = 6 \\ y - x^2 = 0 \end{cases}$
63. $\begin{cases} 5x - 2y = 7 \\ x^2 + y^2 = 2 \end{cases}$
64. $\begin{cases} x - 2y = -5 \\ x^2 + y^2 = 25 \end{cases}$
65. $\begin{cases} x^2 + y^2 = 20 \\ x^2 - y^2 = 12 \end{cases}$
66. $\begin{cases} x^2 - y^2 = 9 \\ 4x^2 + 5y^2 = 180 \end{cases}$
67. $\begin{cases} \frac{3}{x} + \frac{1}{y} = 4 \\ \frac{6}{x} - \frac{1}{y} = 2 \end{cases}$
68. $\begin{cases} \frac{6}{x} + \frac{3}{y} = 0 \\ \frac{4}{x} + \frac{9}{y} = -1 \end{cases}$

69. $\begin{cases} \dfrac{5}{x} + \dfrac{10}{y} = 3 \\ \dfrac{2}{x} - \dfrac{12}{y} = -2 \end{cases}$
70. $\begin{cases} \dfrac{3}{x} + \dfrac{4}{y} = 1 \\ \dfrac{6}{x} + \dfrac{4}{y} = 3 \end{cases}$

Applying the Concepts

In Exercises 71–74, the demand and supply functions of a product are given. In each case, p represents the price in dollars per unit and x represents the number of units in hundreds. Find the equilibrium point.

71. $\begin{cases} 2p + x = 140 & \text{Demand equation} \\ 12p - x = 280 & \text{Supply equation} \end{cases}$

72. $\begin{cases} 7p + x = 150 & \text{Demand equation} \\ 10p - x = 20 & \text{Supply equation} \end{cases}$

73. $\begin{cases} 2p + x = 25 & \text{Demand equation} \\ x - p = 13 & \text{Supply equation} \end{cases}$

74. $\begin{cases} p + 2x = 96 & \text{Demand equation} \\ p - x = 39 & \text{Supply equation} \end{cases}$

75. **Investment.** Last year Mrs. García invested $50,000. She put part of the money in a real estate venture that paid 7.5% for the year and the rest in a small-business venture that returned 12% for the year. The combined income from the two investments for the year totaled $5190. How much did she invest at each rate?

76. **Investment.** Mr. Sharma invested a total of $30,000 in two ventures for a year. The annual return from one of them was 8%, and the other paid 10.5% for the year. He received a total income of $2550 from both investments. How much did he invest at each rate?

77. **Mixture problem.** An herb that sells for $5.50 per pound is mixed with tea that sells for $3.20 per pound to produce a 100-pound mix that is worth $3.66 per pound. How many pounds of each ingredient does the mix contain?

78. **Mixture problem.** A chemist had a solution of 60% acid. She added some distilled water, reducing the acid concentration to 40%. She then added 1 more liter of water to further reduce the acid concentration to 30%. How much of the 30% solution did she then have?

79. **Airplane speed.** With the help of a tail wind, a plane travels 3000 kilometers in 5 hours. The return trip against the wind requires 6 hours. Assume that the direction and the wind speed are constant. Find both the speed of the plane in still air and the wind speed. [*Hint:* Remember that the wind helps the plane in one direction and hinders it in the other.]

80. **Motorboat speed.** A motorboat travels up a stream a distance of 12 miles in 2 hours. If the current had been twice as strong, the trip would have taken 3 hours. How long should it take for the return trip down the stream?

Critical Thinking / Discussion / Writing

81. If $(x - 2)$ is a factor of both $f(x) = x^3 - 4x^2 + ax + b$ and $g(x) = x^3 - ax^2 + bx + 8$, find the values of a and b.

82. Consider the system of equations
$$\begin{cases} a_1 x + b_1 y = c_1 \\ a_2 x + b_2 y = c_2, \end{cases}$$
where $a_1, b_1, c_1, a_2, b_2,$ and c_2 are constants. Find a relationship (an equation) among these constants such that the system of equations has
  a. only one solution. Find the solution in terms of the constants.
  b. no solution.
  c. infinitely many solutions.

83. Let $3x + 4y - 12 = 0$ be the equation of a line l_1 and let $P(5, 8)$ be a point.
  a. Find an equation of a line l_2 passing through the point P and perpendicular to l_1.
  b. Find the point Q of the intersection of the two lines l_1 and l_2.
  c. Find the distance $d(P, Q)$. This is the distance from the point P to the line l_1.

84. Use the procedure outlined in Exercise 83 to show that the (perpendicular) distance from a point $P(x_1, y_1)$ to the line $ax + by + c = 0$ is given by $\dfrac{|ax_1 + by_1 + c|}{\sqrt{a^2 + b^2}}$.

85. Use the formula from Exercise 84 to find the distance from the given point P to the given line l.
  a. $P(2, 3), l: x + y - 7 = 0$
  b. $P(-2, 5), l: 2x - y + 3 = 0$
  c. $P(3, 4), l: 5x - 2y - 7 = 0$
  d. $P(0, 0), l: ax + by + c = 0$

Maintaining Skills

In Exercises 86–91, solve each equation for the specified variable.

86. $\dfrac{1}{2 \times 3} = \dfrac{1}{2} + \dfrac{B}{3}$, for B

87. $\dfrac{1}{n(n+1)} = \dfrac{1}{n} + \dfrac{B}{(n+1)}$, for B

88. $2y - 6x = 12$, for y

89. $5x - y = 9$, for y

90. $3x - 12y = 15$, for x

91. $4y - x = 19$, for x

In Exercises 92–97, factor each expression.

92. $x^2 + 5x + 6$
93. $x^2 - 3x - 10$
94. $2x^2 + 5x - 3$
95. $3x^2 + x - 2$
96. $6x^2 - 17x + 5$
97. $12x^2 - 11x + 2$

SECTION 6.5

Partial-Fraction Decomposition

BEFORE STARTING THIS SECTION, REVIEW

1. Division of polynomials (Section 2.3, page 179)

OBJECTIVES

1. Become familiar with partial-fraction decomposition.
2. Decompose $\dfrac{P(x)}{Q(x)}$ when $Q(x)$ has only distinct linear factors.
3. Decompose $\dfrac{P(x)}{Q(x)}$ when $Q(x)$ has repeated linear factors.

Georg Simon Ohm (1787–1854)

Georg Ohm was born in Erlangen, Germany. His father, Johann, a mechanic and a self-educated man, was interested in philosophy and mathematics and gave his children an excellent education in physics, chemistry, mathematics, and philosophy through his own teachings. The Ohm's law, named in Georg Ohm's honor, appeared in his famous pamphlet *Die galvanische Kette, mathematisch bearbeitet* (1827). Ohm's work was not appreciated in Germany, where it was first published. Eventually, it was recognized by the British Royal Society, which awarded Ohm the Copley Medal in 1841. In 1849, Ohm became a curator of the Bavarian Academy's science museum and began to lecture at the University of Munich. Finally, in 1852 (two years before his death), Ohm achieved his lifelong ambition of becoming chair of physics at the University of Munich.

◆ Ohm's Law

If a voltage is applied across a resistor (such as a wire), a current will flow. Ohm's law states that in a circuit at constant temperature,

$$I = \frac{V}{R},$$

where V is the voltage (measured in volts), I is the current (measured in amperes), and R is the resistance (measured in ohms). A **parallel circuit** consists of two or more resistors connected as shown in Figure 6.9. In the figure, the current I from the source divides into two currents, I_1 and I_2, with I_1 going through one resistor and I_2 going through the other, so that $I = I_1 + I_2$. Suppose we want to find the total resistance R in the circuit due to the two resistors R_1 and R_2 connected in parallel. Ohm's law can be used to show that in the parallel circuit in Figure 6.9, $\dfrac{1}{R} = \dfrac{1}{R_1} + \dfrac{1}{R_2}$. This result is called the *parallel property of resistors*. In Example 4, we examine the parallel property of resistors by using partial fractions.

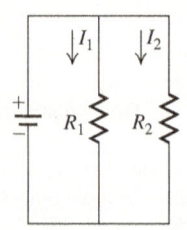

FIGURE 6.9 Parallel resistors

488 Chapter 6 Further Topics in Algebra

1 Become familiar with partial-fraction decomposition.

Partial Fractions

Recall that to add two rational expressions such as $\dfrac{2}{x+3}$ and $\dfrac{3}{x-1}$ required you to find the least common denominator, rewrite expressions with the same denominator, and then add the corresponding numerators. This procedure gives

$$\frac{2}{x+3}+\frac{3}{x-1}=\frac{2(x-1)}{(x+3)(x-1)}+\frac{3(x+3)}{(x-1)(x+3)}=\frac{5x+7}{(x+3)(x-1)}.$$

In some applications of algebra to more advanced mathematics, you need a *reverse* procedure for *splitting* a fraction such as $\dfrac{5x+7}{(x+3)(x-1)}$ into the simpler fractions to obtain

$$\frac{5x+7}{(x+3)(x-1)}=\frac{2}{x+3}+\frac{3}{x-1}.$$

Each of the two fractions on the right is called a **partial fraction**. Their sum is called the **partial-fraction decomposition** of the rational expression on the left.

A rational expression $\dfrac{P(x)}{Q(x)}$ is called **improper** if the degree $P(x) \geq$ degree $Q(x)$ and is called **proper** otherwise. Examples of improper rational expressions are

$$\frac{x^3}{x^2-1},\quad \frac{(x+2)(x-1)}{(x-2)(x+3)},\quad \text{and}\quad \frac{x^2+x+1}{2x+3}.$$

Using long division, we can express an improper rational fraction $\dfrac{P(x)}{Q(x)}$ as the sum of a polynomial and a proper rational fraction:

$$\underset{\underset{\text{Fraction}}{\text{Improper}}}{\dfrac{P(x)}{Q(x)}} \;=\; \underset{\text{Polynomial}}{S(x)} \;+\; \underset{\underset{\text{Fraction}}{\text{Proper}}}{\dfrac{R(x)}{Q(x)}}$$

We therefore restrict our discussion of the decomposition of $\dfrac{P(x)}{Q(x)}$ into partial fractions to cases involving proper rational fraction. We also assume that $P(x)$ and $Q(x)$ have no common factor.

RECALL

Any polynomial $Q(x)$ with real coefficients can be factored so that each factor is linear or is an irreducible quadratic factor.

The problem of decomposing a proper fraction $\dfrac{P(x)}{Q(x)}$ into partial fractions depends on the type of factors in the denominator $Q(x)$. In this section, we consider two cases for $Q(x)$, where $Q(x)$ has only linear factors.

We remark that if $Q(x)$ has an irreducible factor ax^2+bx+c of multiplicity m, then the corresponding group in the partial fraction decomposition has the form:

$$\frac{A_1x+B_1}{ax^2+bx+c}+\frac{A_2x+B_2}{(ax^2+bx+c)^2}+\cdots+\frac{A_mx+B_m}{(ax^2+bx+c)^m}.$$

2 Decompose $\frac{P(x)}{Q(x)}$ when $Q(x)$ has only distinct linear factors.

$Q(x)$ Has Only Distinct Linear Factors

> **CASE 1: THE DENOMINATOR IS THE PRODUCT OF DISTINCT (NONREPEATED) LINEAR FACTORS**
>
> Suppose $Q(x)$ can be factored as
>
> $$Q(x) = c(x - a_1)(x - a_2) \cdots (x - a_n),$$
>
> with no factor repeated. The partial-fraction decomposition of $\frac{P(x)}{Q(x)}$ is of the form
>
> $$\frac{P(x)}{Q(x)} = \frac{A_1}{x - a_1} + \frac{A_2}{x - a_2} + \cdots + \frac{A_n}{x - a_n}$$
>
> where $A_1, A_2, \ldots, A_n$ are constants to be determined.

We can find the constants $A_1, A_2, \ldots, A_n$ by using the following procedure. Note that if the number of constants is small, we use the letters $A, B, C, \ldots$, instead of $A_1, A_2, A_3, \ldots$.

PROCEDURE IN ACTION

EXAMPLE 1 Partial-Fraction Decomposition

OBJECTIVE
Find the partial-fraction decomposition of a rational expression.

EXAMPLE
Find the partial-fraction decomposition of $\frac{3x + 26}{(x - 3)(x + 4)}$.

Step 1 Write the form of the partial-fraction decomposition with the unknown constants $A, B, C, \ldots$ in the numerators of the decomposition.

1. $\dfrac{3x + 26}{(x - 3)(x + 4)} = \dfrac{A}{x - 3} + \dfrac{B}{x + 4}$

Step 2 Multiply both sides of the equation in Step 1 by the original denominator. Use the distributive property and eliminate common factors. Simplify.

2. Multiply both sides by $(x - 3)(x + 4)$ and simplify to obtain
$$3x + 26 = (x + 4)A + (x - 3)B$$

Step 3 Write both sides of the equation in Step 2 in descending powers of x and equate the coefficients of like powers of x.

3. $3x + 26 = (A + B)x + (4A - 3B)$
$$\begin{cases} 3 = A + B & \text{Equate coefficients of } x. \\ 26 = 4A - 3B & \text{Equate constant coefficients.} \end{cases}$$

Step 4 Solve the linear system resulting from Step 3 for the constants $A, B, C, \ldots$.

4. Solving the system of equations in Step 3, we obtain $A = 5$ and $B = -2$.

Step 5 Substitute the values you found for $A, B, C, \ldots$ into the equation in Step 1 and write the partial-fraction decomposition.

5. $\dfrac{3x + 26}{(x - 3)(x + 4)} = \dfrac{5}{x - 3} + \dfrac{-2}{x + 4}$ Replace A with 5 and B with -2 in Step 1.

$\dfrac{3x + 26}{(x - 3)(x + 4)} = \dfrac{5}{x - 3} - \dfrac{2}{x + 4}$

Practice Problem 1 Find the partial-fraction decomposition of $\dfrac{2x - 7}{(x + 1)(x - 2)}$.

EXAMPLE 2 — Finding the Partial-Fraction Decomposition When the Denominator Has Only Distinct Linear Factors

Find the partial-fraction decomposition of the expression:

$$\frac{10x - 4}{x^3 - 4x}$$

Solution

First, factor the denominator.

$$\begin{aligned} x^3 - 4x &= x(x^2 - 4) & &\text{Distributive property} \\ &= x(x - 2)(x + 2) & &x^2 - 4 = (x - 2)(x + 2) \end{aligned}$$

Each of the factors x, $x - 2$, and $x + 2$ becomes a denominator for a partial fraction.

Step 1 The partial-fraction decomposition is given by

$$\frac{10x - 4}{x(x - 2)(x + 2)} = \frac{A}{x} + \frac{B}{x - 2} + \frac{C}{x + 2} \qquad \text{Use } A, B, \text{ and } C \text{ instead of } A_1, A_2, \text{ and } A_3.$$

Step 2 Multiply both sides by the common denominator $x(x - 2)(x + 2)$.

$$x(x-2)(x+2)\left[\frac{10x - 4}{x(x-2)(x+2)}\right] = x(x - 2)(x + 2)\left[\frac{A}{x} + \frac{B}{x - 2} + \frac{C}{x + 2}\right]$$

$$\begin{aligned} 10x - 4 &= A(x - 2)(x + 2) + Bx(x + 2) + Cx(x - 2) \quad (1) & &\text{Distributive property; simplify.} \\ &= A(x^2 - 4) + B(x^2 + 2x) + C(x^2 - 2x) & &\text{Multiply.} \\ &= Ax^2 - 4A + Bx^2 + 2Bx + Cx^2 - 2Cx & &\text{Distributive property} \\ &= (A + B + C)x^2 + (2B - 2C)x - 4A & &\text{Combine like terms.} \end{aligned}$$

Step 3 Now use the fact that *two equal polynomials have equal corresponding coefficients*. Writing $10x - 4 = 0x^2 + 10x - 4$, we have

$$0x^2 + 10x - 4 = (A + B + C)x^2 + (2B - 2C)x - 4A.$$

Equating corresponding coefficients leads to the system of equations.

$$\begin{cases} A + B + C = 0 & \text{Equate coefficients of } x^2. \\ 2B - 2C = 10 & \text{Equate coefficients of } x. \\ -4A = -4 & \text{Equate constant coefficients.} \end{cases}$$

Step 4 Solve the system of equations in Step 3 to obtain $A = 1$, $B = 2$, and $C = -3$. (See Exercise 39.)

Step 5 The partial-fraction decomposition is

$$\frac{10x - 4}{x^3 - 4x} = \frac{1}{x} + \frac{2}{x - 2} + \frac{-3}{x + 2} = \frac{1}{x} + \frac{2}{x - 2} - \frac{3}{x + 2}.$$

Alternative Solution of Example 2

In Example 2, to find the constants A, B, and C, we *equated coefficients* of like powers of x and solved the resulting system. An alternative (and sometimes quicker) method is to *substitute* well-chosen values for x in equation (1) found in Step 2:

$$10x - 4 = A(x - 2)(x + 2) + Bx(x + 2) + Cx(x - 2) \qquad (1)$$

TECHNOLOGY CONNECTION

You can reinforce your partial-fraction connection graphically: The graph of

$$Y_1 = \frac{10x - 4}{x^3 - 4x}$$

appears to be identical to the graph of

$$Y_2 = \frac{1}{x} + \frac{2}{x - 2} - \frac{3}{x + 2}.$$

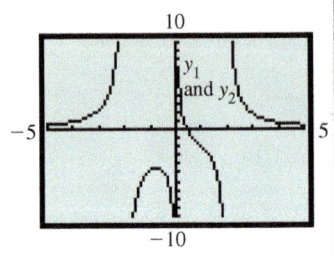

[Graph showing y_1 and y_2 on window -5 to 5 horizontally and -10 to 10 vertically]

The graphs of Y_1 and Y_2 are the same.

Substitute $x = 2$ in equation (1) to cause the terms containing A and C to be 0.

$10(2) - 4 = A(2 - 2)(2 + 2) + B(2)(2 + 2) + C(2)(2 - 2)$

$16 = 8B$ *Simplify.*

$2 = B$ *Solve for B.*

Now substitute $x = -2$ in equation (1) to get

$10(-2) - 4 = A(-2 - 2)(-2 + 2) + B(-2)(-2 + 2) + C(-2)(-2 - 2)$

$-24 = 8C$ *Simplify.*

$-3 = C$ *Solve for C.*

Finally, substitute $x = 0$ in equation (1) to get

$10(0) - 4 = A(0 - 2)(0 + 2) + B(0)(0 + 2) + C(0)(0 - 2)$

$-4 = -4A$ *Simplify.*

$1 = A$ *Solve for A.*

Because we found the same values for A, B, and C, the partial-fraction decomposition will also be the same.

Practice Problem 2 Find the partial-fraction decomposition of:

$$\frac{3x^2 + 4x + 3}{x^3 - x}$$

---

3 Decompose $\frac{P(x)}{Q(x)}$ when $Q(x)$ has repeated linear factors.

Q(x) Has Repeated Linear Factors

CASE 2: THE DENOMINATOR HAS A REPEATED LINEAR FACTOR

Let $(x - a)^m$ be the linear factor $(x - a)$ that is repeated m times in $Q(x)$. Then the portion of the partial-fraction decomposition of $\frac{P(x)}{Q(x)}$ that corresponds to the factor $(x - a)^m$ is

$$\frac{A_1}{x - a} + \frac{A_2}{(x - a)^2} + \cdots + \frac{A_m}{(x - a)^m},$$

where $A_1, A_2, \ldots, A_m$ are constants.

EXAMPLE 3 Finding the Partial-Fraction Decomposition When the Denominator Has Repeated Linear Factors

Find the partial-fraction decomposition of: $\dfrac{x + 4}{(x + 3)(x - 1)^2}$

Solution

Step 1 The linear factor $(x - 1)$ is repeated twice, and the factor $(x + 3)$ is nonrepeating. So the partial-fraction decomposition has the form

$$\frac{x + 4}{(x + 3)(x - 1)^2} = \frac{A}{x + 3} + \frac{B}{x - 1} + \frac{C}{(x - 1)^2}.$$

Steps 2–4 Multiply both sides of the equation in Step 1 by the original denominator, $(x + 3)(x - 1)^2$; then use the distributive property on the right side and simplify to get

$$x + 4 = A(x - 1)^2 + B(x + 3)(x - 1) + C(x + 3).$$

Substitute $x = 1$ in the previous equation to obtain
$$1 + 4 = A(1 - 1)^2 + B(1 + 3)(1 - 1) + C(1 + 3).$$
This simplifies to $5 = 4C$, or $C = \dfrac{5}{4}$.

To find A, we substitute $x = -3$ to get
$$-3 + 4 = A(-3 - 1)^2 + B(-3 + 3)(-3 - 1) + C(-3 + 3).$$
This simplifies to $1 = 16A$, or $A = \dfrac{1}{16}$.

To find B, we replace x with any convenient number (say, 0). We have
$$0 + 4 = A(0 - 1)^2 + B(0 + 3)(0 - 1) + C(0 + 3)$$
$$4 = A - 3B + 3C \qquad\qquad\qquad\qquad \text{Simplify.}$$

Now replace A with $\dfrac{1}{16}$ and C with $\dfrac{5}{4}$ in the previous equation.

$$4 = \dfrac{1}{16} - 3B + 3\left(\dfrac{5}{4}\right)$$
$$64 = 1 - 48B + 60 \qquad \text{Multiply both sides by 16.}$$
$$B = -\dfrac{1}{16} \qquad\qquad\qquad \text{Solve for } B.$$

We have $A = \dfrac{1}{16}$, $B = -\dfrac{1}{16}$, and $C = \dfrac{5}{4}$.

Step 5 Substitute $A = \dfrac{1}{16}$, $B = -\dfrac{1}{16}$, and $C = \dfrac{5}{4}$ in the decomposition in Step 1 to get

$$\dfrac{x + 4}{(x + 3)(x - 1)^2} = \dfrac{\tfrac{1}{16}}{x + 3} + \dfrac{-\tfrac{1}{16}}{x - 1} + \dfrac{\tfrac{5}{4}}{(x - 1)^2}$$

or

$$\dfrac{x + 4}{(x + 3)(x - 1)^2} = \dfrac{1}{16(x + 3)} - \dfrac{1}{16(x - 1)} + \dfrac{5}{4(x - 1)^2}.$$

Practice Problem 3 Find the partial fraction decomposition of
$$\dfrac{x + 5}{x(x - 1)^2}.$$

EXAMPLE 4 Calculating the Resistance in a Circuit

The total resistance R due to two resistors connected in parallel is given by
$$R = \dfrac{x(x + 5)}{3x + 10}.$$

Use partial-fraction decomposition to write $\dfrac{1}{R}$ in terms of partial fractions. What does each term represent?

Solution

From $R = \dfrac{x(x+5)}{3x+10}$, we get the following:

$\dfrac{1}{R} = \dfrac{3x+10}{x(x+5)}$ Take the reciprocal of both sides.

$\dfrac{3x+10}{x(x+5)} = \dfrac{A}{x} + \dfrac{B}{x+5}$ Form of partial-fraction decomposition

$3x + 10 = A(x+5) + Bx$ Multiply both sides by $x(x+5)$ and simplify.

Substitute $x = 0$ in the preceding equation to obtain $5A = 10$, or $A = 2$. Now substitute $x = -5$ in the same equation to get $-5 = -5B$, or $B = 1$. Thus, we have

$\dfrac{1}{R} = \dfrac{3x+10}{x(x+5)} = \dfrac{2}{x} + \dfrac{1}{x+5}$

$\dfrac{1}{R} = \dfrac{1}{\dfrac{x}{2}} + \dfrac{1}{x+5}$ Rewrite the right side.

The previous equation says that if two resistors with resistances $R_1 = \dfrac{x}{2}$ and $R_2 = x+5$ are connected in parallel, they will produce a total resistance R, where $R = \dfrac{x(x+5)}{3x+10}$.

Practice Problem 4 Repeat Example 4 assuming that $R = \dfrac{(x+1)(x+2)}{4x+7}$.

RECALL

The total resistance R in a circuit due to two resistors R_1 and R_2 connected in parallel satisfies the equation

$\dfrac{1}{R} = \dfrac{1}{R_1} + \dfrac{1}{R_2}$.

SECTION 6.5 Exercises

Basic Concepts and Skills

In Exercises 1–6, write the form of the partial-fraction decomposition of each rational expression. You do not need to solve for the constants.

1. $\dfrac{1}{(x-1)(x+2)}$

2. $\dfrac{x}{(x+1)(x-3)}$

3. $\dfrac{1}{x^2+7x+6}$

4. $\dfrac{3}{x^2-6x+8}$

5. $\dfrac{2}{x^3-x^2}$

6. $\dfrac{x-1}{(x+2)^2(x-3)}$

In Exercises 7–30, find the partial-fraction decomposition of each rational expression.

7. $\dfrac{2x+1}{(x+1)(x+2)}$

8. $\dfrac{7}{(x-2)(x+5)}$

9. $\dfrac{1}{x^2+4x+3}$

10. $\dfrac{x}{x^2+5x+6}$

11. $\dfrac{2}{x^2+2x}$

12. $\dfrac{x+9}{x^2-9}$

13. $\dfrac{x}{(x+1)(x+2)(x+3)}$

14. $\dfrac{x^2}{(x-1)(x^2+5x+4)}$

15. $\dfrac{x-1}{(x+1)^2}$

16. $\dfrac{3x+2}{x^2(x+1)}$

17. $\dfrac{2x^2+x}{(x+1)^3}$

18. $\dfrac{x}{(x-1)(x+1)}$

19. $\dfrac{-x^2+3x+1}{x^3+2x^2+x}$

20. $\dfrac{5x^2-8x+2}{x^3-2x^2+x}$

21. $\dfrac{1}{x^2(x+1)^2}$

22. $\dfrac{2}{(x-1)^2(x+3)^2}$

23. $\dfrac{1}{(x^2-1)^2}$

24. $\dfrac{3}{(x^2-4)^2}$

25. $\dfrac{x-1}{(2x-3)^2}$

26. $\dfrac{6x+1}{(3x+1)^2}$

27. $\dfrac{6x+7}{4x^2+12x+9}$

28. $\dfrac{2x+3}{9x^2+30x+25}$

29. $\dfrac{x-3}{x^3+x^2}$

30. $\dfrac{x^2+2x+4}{x^3+x^2}$

Applying the Concepts

In Exercises 31–34, express each sum as a fraction of whole numbers in lowest terms.

$$\left[\text{Hint: } \frac{1}{n(n+1)} = \frac{1}{n} - \frac{1}{n+1} \right]$$

31. $\dfrac{1}{1 \cdot 2} + \dfrac{1}{2 \cdot 3} + \dfrac{1}{3 \cdot 4} + \cdots + \dfrac{1}{n(n+1)}$

32. $\dfrac{2}{1 \cdot 3} + \dfrac{2}{2 \cdot 4} + \dfrac{2}{3 \cdot 5} + \cdots + \dfrac{2}{100 \cdot 102}$

33. $\dfrac{2}{1 \cdot 3} + \dfrac{2}{3 \cdot 5} + \dfrac{2}{5 \cdot 7} + \cdots + \dfrac{2}{(2n-1)(2n+1)}$

34. $\dfrac{2}{1 \cdot 2 \cdot 3} + \dfrac{2}{2 \cdot 3 \cdot 4} + \dfrac{2}{3 \cdot 4 \cdot 5} + \cdots + \dfrac{2}{100 \cdot 101 \cdot 102}$

$$\left[\text{Hint: Write} \right.$$
$$\frac{2}{k(k+1)(k+2)} = \frac{1}{k} - \frac{2}{k+1} + \frac{1}{k+2}$$
$$\left. = \frac{1}{k} - \frac{1}{k+1} - \frac{1}{k+1} + \frac{1}{k+2} \right]$$

In Exercises 35–38, the total resistance R in a circuit is given. Use partial-fraction decomposition to write $\dfrac{1}{R}$ in terms of partial fractions. Interpret each term in the partial fraction.

35. Circuit resistance. $R = \dfrac{(x+1)(x+3)}{2x+4}$

36. Circuit resistance. $R = \dfrac{(x+2)(x+4)}{7x+20}$

37. Circuit resistance. $R = \dfrac{R_1 R_2 R_3}{R_1 R_2 + R_2 R_3 + R_3 R_1}$

38. Circuit resistance. $R = \dfrac{x(x+2)(x+4)}{3x^2 + 12x + 8}$

Critical Thinking / Discussion / Writing

39. Solve the system of equations in Example 2 to verify the values of the constants A, B, and C.

40. Find the partial-fraction decomposition of

$$\frac{4x}{x^4 + 4}.$$

[Hint: $x^4 + 4 = x^4 + 4 + 4x^2 - 4x^2 = (x^2+2)^2 - (2x)^2$
$= (x^2 - 2x + 2)(x^2 + 2x + 2)$]

41. Find the partial-fraction decomposition of

$$\frac{x^2 - 8x + 18}{(x-5)^3}.$$

In Exercises 42–45, find the partial-fraction decomposition by first using long division.

42. $\dfrac{x^2 + 4x + 5}{x^2 + 3x + 2}$

43. $\dfrac{(x-1)(x+2)}{(x+3)(x-4)}$

44. $\dfrac{2x^4 + x^3 + 2x^2 - 2x - 1}{(x-1)(x^2+1)}$

45. $\dfrac{x^4}{(x-1)(x-2)(x-3)}$

Maintaining Skills

In Exercises 46–57, simplify each expression.

46. $(3x^2)^3$

47. $(2x^3)(-x^4)^7$

48. $(-(2x)^4)(13x^{25})^0$

49. $((-5x)^2)((-27x)^0)^{14}$

50. $\dfrac{15}{2x^4} \cdot \dfrac{6x^2}{25}$

51. $\dfrac{-5x^2}{8y^2} \cdot \dfrac{64y^4}{25x^3}$

52. $\dfrac{2yz^2}{7x^5} \cdot \dfrac{-21x^3}{8y^2 z^3}$

53. $\dfrac{(-xy^2)^3 z}{x(yz^2)^2} \cdot \dfrac{(-2y^2 z)^2 x}{(x^2 y)^3 z}$

54. $\dfrac{3x}{8x+4} \cdot \dfrac{12x+6}{27x^2}$

55. $\dfrac{1-x}{y-1} \cdot \dfrac{1-y}{x-1}$

56. $\dfrac{x^2 - 5x + 6}{x^3 - 3x^2 + 2x} \cdot \dfrac{x^2 - 5x + 4}{x^2 - 7x + 12}$

57. $\dfrac{x^2 - 6x + 8}{x^2 - 5x + 4} \cdot \dfrac{3x^2 - 12x + 9}{x^3 - 5x^2 + 6x}$

SUMMARY Definitions, Concepts, and Formulas

6.1 Sequences and Series

i. An infinite sequence is a function whose domain is the set of positive integers. The function values $a_1, a_2, a_3, \ldots, a_n, \ldots$ are called the **terms** of the sequence. The term a_n is the **nth term**, or **general term**, of the sequence.

ii. *Recursive formula*
A sequence may be defined recursively, with the nth term of the sequence defined in relation to previous terms.

iii. *Factorial notation*
$$n! = n(n-1) \cdots 3 \cdot 2 \cdot 1 \text{ and } 0! = 1$$

iv. *Summation notation*
$$\sum_{i=1}^{n} a_i = a_1 + a_2 + a_3 + \cdots + a_n$$

v. *Series*
Given an infinite sequence $a_1, a_2, a_3, \ldots, a_n, \ldots$, the sum $\sum_{i=1}^{n} a_i$ is called an **nth partial sum** and the sum $\sum_{i=1}^{\infty} a_i = a_1 + a_2 + a_3 + \cdots$ is called the **sum of the series**.

6.2 Arithmetic Sequences; Partial Sums

i. A sequence is an **arithmetic sequence** if each term after the first differs from the preceding term by a constant. The constant difference d between the consecutive terms is called the **common difference**. We have $d = a_n - a_{n-1}$ for all $n \geq 2$.

ii. The **nth term** a_n of the arithmetic sequence $a_1, a_2, a_3, \ldots, a_n, \ldots$ is given by $a_n = a_1 + (n-1)d, n \geq 1$, where d is the common difference.

iii. The **sum** S_n of the first n terms of the arithmetic sequence $a_1, a_2, \ldots, a_n, \ldots$ is given by the formula
$$S_n = n\left(\frac{a_1 + a_n}{2}\right).$$

6.3 Geometric Sequences and Series

i. A sequence is a **geometric sequence** if each term after the first is a constant multiple of the preceding term. The constant ratio r between the consecutive terms is called the **common ratio**. We have $\dfrac{a_{n+1}}{a_n} = r, n \geq 1$.

ii. The **nth term** a_n of the geometric sequence $a_1, a_2, a_3, \ldots, a_n, \ldots$ is given by $a_n = a_1 r^{n-1}, n \geq 1$, where r is the common ratio.

iii. The **sum** S_n of the first n terms of the geometric sequence $a_1, a_2, \ldots, a_n, \ldots$ with common ratio $r \neq 1$ is given by the formula
$$S_n = \frac{a_1(1 - r^n)}{1 - r}.$$

The **sum S of an infinite geometric sequence** is given by
$$S = \sum_{i=1}^{\infty} a_1 r^{i-1} = \frac{a_1}{1-r} \text{ if } |r| < 1.$$

iv. An **annuity** is a sequence of equal payments made at equal time. Suppose $\$P$ is the payment made at the end of each of n compounding periods per year and i is the annual interest rate. The value A of the annuity after t years is
$$A = P\left[\frac{\left(1 + \dfrac{i}{n}\right)^{nt} - 1}{\dfrac{i}{n}}\right]$$

v. If $|r| < 1$, the infinite geometric series $a_1 + a_1 r + a_1 r^2 + \cdots + a_1 r^{n-1} + \cdots$ has the sum $S = \dfrac{a_1}{1-r}$.
When $|r| \geq 1$, the series does not have a finite sum.

6.4 Systems of Equations in Two Variables

A **system of equations** is a set of equations with common variables.

A **solution** of a system of equations in two variables x and y is an ordered pair of numbers (a, b) that satisfies all equations in the system. The solution set of a system of two linear equations of the form $ax + by = c$ is the point(s) of intersection of the graphs of the equations.

Systems with at least one solution are **consistent** and systems with no solution are called **inconsistent**. A system of two linear equations in two variables may have one solution (**independent equations**), no solution (**inconsistent system**), or infinitely many solutions (**dependent equations**).

Three methods of solving a system of equations are (1) the graphical method; (2) the substitution method; and (3) the elimination, or addition, method.

6.5 Partial-Fraction Decomposition

In partial-fraction decomposition, we reverse the addition of rational expressions. A rational expression $\dfrac{P(x)}{Q(x)}$ is called a proper fraction if deg $P(x) <$ deg $Q(x)$.

We consider two cases in decomposing $\dfrac{P(x)}{Q(x)}$ into partial fractions:

1. $Q(x)$ has only distinct linear factors. (See page 490.)
2. $Q(x)$ has repeated linear factors. (See page 492.)

REVIEW EXERCISES

In Exercises 1–4, write the first five terms of each sequence.

1. $a_n = 2n - 3$
2. $a_n = \dfrac{n(n-2)}{2}$
3. $a_n = \dfrac{n}{2n+1}$
4. $a_n = (-2)^{n-1}$

In Exercises 5 and 6, write a general term a_n for each sequence.

5. $30, 28, 26, 24, \ldots$
6. $-1, 2, -4, 8, \ldots$

In Exercises 7–10, simplify each expression.

7. $\dfrac{9!}{8!}$
8. $\dfrac{10!}{7!}$
9. $\dfrac{(n+1)!}{n!}$
10. $\dfrac{n!}{(n-2)!}$

In Exercises 11–14, find each sum.

11. $\sum_{k=1}^{4} k^3$
12. $\sum_{j=1}^{5} \dfrac{1}{2j}$
13. $\sum_{k=1}^{7} \dfrac{k+1}{k}$
14. $\sum_{k=1}^{5} (-1)^k 3^{k+1}$

In Exercises 15 and 16, write each sum in summation notation.

15. $1 + \dfrac{1}{2} + \dfrac{1}{3} + \dfrac{1}{4} + \cdots + \dfrac{1}{50}$
16. $2 - 4 + 8 - 16 + 32$

In Exercises 17–20, determine whether each sequence is arithmetic. If a sequence is arithmetic, find the first term a_1 and the common difference d.

17. $11, 6, 1, -4, \ldots$
18. $\dfrac{2}{3}, \dfrac{5}{6}, 1, \dfrac{7}{6}, \ldots$
19. $a_n = n^2 + 4$
20. $a_n = 3n - 5$

In Exercises 21–24, find an expression for the nth term of each arithmetic sequence.

21. $3, 6, 9, 12, 15, \ldots$
22. $5, 9, 13, 17, 21, \ldots$
23. $x, x+1, x+2, x+3, x+4, \ldots$
24. $3x, 5x, 7x, 9x, 11x, \ldots$

In Exercises 25 and 26, find the common difference d and the nth term a_n for each arithmetic sequence.

25. 3rd term: 7; 8th term: 17
26. 5th term: -16; 20th term: -46

In Exercises 27 and 28, find each sum.

27. $7 + 9 + 11 + 13 + \cdots + 37$
28. $\dfrac{1}{4} + \dfrac{1}{2} + \dfrac{3}{4} + 1 + \cdots + 15$

In Exercises 29 and 30, find the sum of the first n terms of each arithmetic sequence.

29. $3, 8, 13, \ldots; n = 40$
30. $-6, -5.5, -5, \ldots; n = 60$

In Exercises 31–34, determine whether each sequence is geometric. If a sequence is geometric, find the first term a_1 and the common ratio r.

31. $4, -8, 16, -32, \ldots$
32. $\dfrac{1}{5}, \dfrac{1}{10}, \dfrac{1}{20}, \dfrac{1}{40}, \ldots$
33. $\dfrac{1}{3}, 1, \dfrac{5}{3}, 9, \ldots$
34. $a_n = 2^{-n}$

In Exercises 35 and 36, for each geometric sequence, find the first term a_1, the common ratio r, and the nth term a_n.

35. $16, -4, 1, -\dfrac{1}{4}, \ldots$
36. $-\dfrac{5}{6}, -\dfrac{1}{3}, -\dfrac{2}{15}, -\dfrac{4}{75}, \ldots$

In Exercises 37 and 38, find the indicated term of each geometric sequence.

37. a_{10} when $a_1 = 2, r = 3$
38. a_{12} when $a_1 = -2, r = \dfrac{3}{2}$

In Exercises 39 and 40, find the sum S_n of the first n terms of each geometric sequence.

39. $\dfrac{1}{10}, \dfrac{1}{5}, \dfrac{2}{5}, \dfrac{4}{5}, \ldots; n = 12$
40. $2, -1, \dfrac{1}{2}, -\dfrac{1}{4}, \ldots; n = 10$

In Exercises 41–44, find each sum.

41. $\dfrac{1}{2} + \dfrac{1}{6} + \dfrac{1}{18} + \dfrac{1}{54} + \cdots$
42. $-5 - 2 - \dfrac{4}{5} - \dfrac{8}{25} - \cdots$
43. $\sum_{i=1}^{\infty} \left(\dfrac{3}{5}\right)^i$
44. $\sum_{i=1}^{\infty} 7\left(-\dfrac{1}{4}\right)^i$

In Exercises 45–62, solve each system of equations by using the method of your choice. Identify systems with no solution and systems with infinitely many solutions.

45. $\begin{cases} 3x - y = -5 \\ x + 2y = 3 \end{cases}$
46. $\begin{cases} x + 3y + 6 = 0 \\ y = 4x - 2 \end{cases}$
47. $\begin{cases} 2x + 4y = 3 \\ 3x + 6y = 10 \end{cases}$
48. $\begin{cases} x - y = 2 \\ 2x - 2y = 9 \end{cases}$
49. $\begin{cases} 3x - y = 3 \\ \dfrac{1}{2}x + \dfrac{1}{3}y = 2 \end{cases}$
50. $\begin{cases} 0.02y - 0.03x = -0.04 \\ 1.5x - y = 3 \end{cases}$
51. $\begin{array}{l} x + 3y = 13 \\ x - 2y = 3 \end{array}$
52. $\begin{array}{l} 2x - 3y = 1 \\ 3x - 4y = 3 \end{array}$
53. $\begin{array}{l} 6x + y = 12 \\ -7x + 2y = 5 \end{array}$
54. $\begin{array}{l} 5x - y = 5 \\ 3x + 2y = -10 \end{array}$
55. $\begin{array}{l} 3x + 8y = 0 \\ 5x + 2y = 0 \end{array}$
56. $\begin{array}{l} 3x - 2y = 1 \\ -7x + 3y = 1 \end{array}$

57. $3x + y = 3$
    $6x + 2y = 11$

58. $x - y = 2$
    $-2x + 2y = 5$

59. $\dfrac{2}{u} + \dfrac{1}{v} = 3$
    $\dfrac{4}{u} - \dfrac{2}{v} = 0$

60. $\dfrac{3}{u} + \dfrac{1}{v} = 4$
    $\dfrac{6}{u} - \dfrac{1}{v} = 2$

61. $\dfrac{6}{u} + \dfrac{3}{v} = 0$
    $\dfrac{4}{u} + \dfrac{9}{v} = -1$

62. $\dfrac{5}{u} + \dfrac{10}{v} = 3$
    $\dfrac{2}{u} - \dfrac{12}{v} = -2$

In Exercises 63–68, find the partial-fraction decomposition of each rational expression.

63. $\dfrac{x + 4}{x^2 + 5x + 6}$

64. $\dfrac{x + 14}{x^2 + 3x - 4}$

65. $\dfrac{3x^2 + x + 1}{x(x - 1)^2}$

66. $\dfrac{3x^2 + 2x + 3}{(x^2 - 1)(x + 1)}$

67. $\dfrac{x^2 + 2x + 3}{(x - 2)^2(x + 2)}$

68. $\dfrac{2x}{x^4 - 1}$

$\left[\text{Hint: for Exercise 68} \\ = \dfrac{A}{x - 1} + \dfrac{B}{x + 1} + \dfrac{Cx + D}{x^2 + 1}\right]$

Applying the Concepts

69. **Investment.** A speculator invested part of $15,000 in a high-risk venture and received a return of 12% at the end of the year. The rest of the $15,000 was invested at 4% annual interest. The combined annual income from the two sources was $1300. How much was invested at each rate?

70. **Agriculture.** A farmer earns a profit of $525 per acre of tomatoes and $475 per acre of soybeans. His soybean acreage is 5 acres more than twice his tomato acreage. If his total profit from the two crops is $24,500, how many acres of tomatoes and how many acres of soybeans does he have?

71. **Geometry.** The area of a rectangle is 63 square feet, and its perimeter is 33 feet. What are the dimensions of the rectangle?

72. **Numbers.** Twice the sum of the reciprocals of two numbers is 13, and the product of the numbers is $\dfrac{1}{9}$. Find the numbers.

73. **Geometry.** The hypotenuse of a right triangle is 17. If one leg of the triangle is increased by 1 and the other leg is increased by 4, the hypotenuse becomes 20. Find the sides of the triangle.

74. **Paper route.** Chris covers her paper route, which is 21 miles long, by 7:30 A.M. each day. If her average rate of travel were 1 mile faster each hour, she would cover the route by 7 A.M. What time does she start in the morning?

75. **Agriculture.** A rectangular pasture with an area of 6400 square meters is divided into three smaller pastures by two fences parallel to the shorter sides. The width of two of the smaller pastures is the same, and the width of the third is twice that of the others. Find the dimensions of the original pasture if the perimeter of the larger of the subdivisions is 240 meters.

76. **Leasing.** Budget Rentals leases its compact cars for $23.00 per day plus $0.17 per mile. Dollar Rentals leases the same car for $24.00 per day plus $0.22 per mile.
    a. Find the cost functions describing the daily cost of leasing from each company.
    b. Graph the two functions in part (a) on the same coordinate axes.
    c. From which company should you lease the car if you plan to drive (i) 50 miles per day; (ii) 60 miles per day; (iii) 70 miles per day?

77. **Apartment lease.** Alisha and Sunita signed a lease on an apartment for nine months. At the end of six months, Alisha got married and moved out. She paid the landlady an amount equal to the difference between double-occupancy rental and single-occupancy rental for the remaining three months, and Sunita paid the single-occupancy rate for the same three months. If the nine-month rental cost Alisha $2340 and Sunita $4140, what were the single and double monthly rates?

78. **Seating arrangement.** The 600 graduating seniors at Central State College are seated in rows, each of which contains the same number of chairs, and every chair is occupied. If five more chairs were in each row, everyone could be seated in four fewer rows. How many chairs are in each row?

EXERCISES FOR CALCULUS

1. A colony of a certain type of bacteria doubles every day. If there are 1000 bacteria now, how many will there be after seven days?

2. A ball that rebounds half way back from any height from which it falls is initially dropped from a height of 1 foot. How far does it travel?

3. A ball that rebounds a fraction r back from any height from which it falls ($0 < r < 1$) is initially dropped from a height of a feet. How far does it travel?

4. If 80 cents of every dollar spent in a certain town is spent again in that town and an extra million dollars is suddenly put into the town's economy, how much spending results from the million dollars?

In Exercises 5 and 6, show that each sequence term is larger than the preceding one (that is, that $a_n < a_{n+1}$ for each positive integer n).

5. $a_n = \dfrac{3n}{n + 1}$

6. $a_n = \dfrac{n^2}{2 + n}$

In Exercises 7 and 8, express each repeating decimal as a geometric series and express the sum of the series as a fraction.

7. $0.252525\ldots$
8. $0.612612612\ldots$

9. A two-digit number equals eight times the sum of its digits. When three times the tens digit is added to twice the units digit, the result is 25. Find the number.

10. The tens digit of a two-digit number is 2 less than the units digit. If 1 is subtracted from the number, the result is 8 times the units digit. Find the number.

11. Estimate the area under the graph of $f(x) = x^2$ and above the interval $[1, 4]$ by
    a. adding the areas of the rectangles with base $[n, n + 1]$ and height $f(n)$ for $1 \le n \le 3$ to obtain an underestimate of the area. (See the figure.)
    b. adding the areas of the rectangles with base $[n, n + 1]$ and height $f(n + 1)$ for $1 \le n \le 3$ to obtain an overestimate of the area. (See the figure.)

    The actual area is between the estimates found in parts (a) and (b).

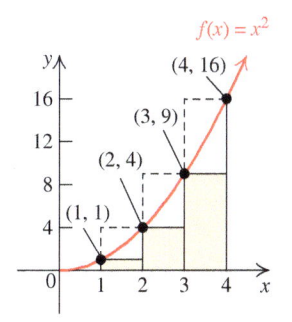

12. Estimate the area under the graph of $f(x) = \dfrac{1}{x}$ and above the interval $[1, 4]$ by
    a. adding the areas of the rectangles with base $[n, n + 1]$ and height $f(n)$ for $1 \le n \le 3$ to obtain an overestimate of the area. (See the figure.)
    b. adding the areas of the rectangles with base $[n, n + 1]$ and height $f(n + 1)$ for $1 \le n \le 3$ to obtain an underestimate of the area. (See the figure.)

    The actual area is between the estimates found in parts (a) and (b).

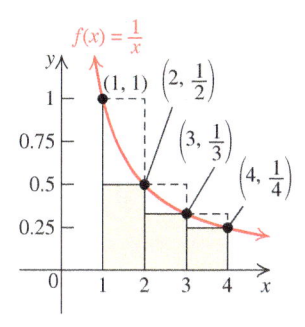

PRACTICE TEST A

In Problems 1 and 2, write the first five terms of each sequence. State whether the sequence is arithmetic or geometric.

1. $a_n = 3(5 - 4n)$
2. $a_n = -3(2^n)$

3. Write the first five terms of the sequence defined by $a_1 = -2$ and $a_n = 3a_{n-1} + 5$ for $n \ge 2$.

4. Evaluate $\dfrac{(n - 1)!}{n!}$.

5. Find a_7 for the arithmetic sequence with $a_1 = -3$ and $d = 4$.

6. Find a_8 for the geometric sequence with $a_1 = 13$ and $r = -\dfrac{1}{2}$.

In Problems 7–9, find each sum.

7. $\displaystyle\sum_{k=1}^{20} (3k - 2)$
8. $\displaystyle\sum_{k=1}^{5} \left(\dfrac{3}{4}\right)(2^{-k})$
9. $\displaystyle\sum_{k=1}^{\infty} 18\left(\dfrac{1}{100}\right)^k$; write the answer as a fraction in lowest terms.

In Problems 10–16, solve each system of equations.

10. $\begin{cases} 2x - y = 4 \\ 2x + y = 4 \end{cases}$

11. $\begin{cases} x + 2y = 8 \\ 3x + 6y = 24 \end{cases}$

12. $\begin{cases} -2x + y = 4 \\ 4x - 2y = 4 \end{cases}$

13. $\begin{cases} 3x + 3y = -15 \\ 2x - 2y = -10 \end{cases}$

14. $\begin{cases} \dfrac{5}{3}x + \dfrac{y}{2} = 14 \\ \dfrac{2}{3}x - \dfrac{y}{8} = 3 \end{cases}$

15. $\begin{cases} y = x^2 \\ 3x - y + 4 = 0 \end{cases}$

16. $\begin{cases} x - 3y = -4 \\ 2x^2 + 3x - 3y = 8 \end{cases}$

17. Two gold bars together weigh a total of 485 pounds. One bar weighs 15 pounds more than the other.
    a. Write a system of equations that describes these relationships.
    b. How much does each bar weigh?

For Problems 18 and 19, write the form of the partial-fraction decomposition of the given rational expression. You do not need to solve for the constants.

18. $\dfrac{2x}{(x-5)(x+1)}$

19. $\dfrac{-5x^2 + x - 8}{(x-2)(x+1)^2}$

20. Find the partial-fraction decomposition of the rational expression

$$\dfrac{x+3}{(x+4)^2(x-7)}.$$

PRACTICE TEST B

In Problems 1 and 2, write the first five terms of each sequence. State whether the sequence is arithmetic or geometric.

1. $a_n = 4(2n - 3)$
   a. $-1, 1, 3, 5, 7$; arithmetic
   b. $-1, 1, 3, 5, 7$; geometric
   c. $-4, 4, 12, 20, 28$; arithmetic
   d. $-4, 4, 12, 20, 28$; geometric

2. $a_n = 2(4^n)$
   a. $2, 8, 32, 128, 512$; arithmetic
   b. $2, 8, 32, 128, 512$; geometric
   c. $8, 32, 128, 512, 2048$; arithmetic
   d. $8, 32, 128, 512, 2048$; geometric

3. Write the first five terms of the sequence defined by $a_1 = 5$ and $a_n = 2a_{n-1} + 4$, for $n \geq 2$.
   a. $5, 14, 32, 68, 140$
   b. $5, 14, 19, 24, 29$
   c. $5, 9, 13, 17, 21$
   d. $5, 8, 10, 12, 14$

4. Evaluate $\dfrac{(n+2)!}{n+2}$.
   a. $2!$
   b. 1
   c. $(n+1)!$
   d. $n+1$

5. Find a_8 for the arithmetic sequence with $a_1 = -6$ and $d = 3$.
   a. 15
   b. -271
   c. 30
   d. 18

6. Find a_{10} for the geometric sequence with $a_1 = 247$ and $r = \dfrac{1}{3}$.
   a. 250
   b. $\dfrac{247}{19,683}$
   c. $\dfrac{247}{59,049}$
   d. $\dfrac{247}{177,147}$

In Problems 7 and 8, find each sum.

7. $\displaystyle\sum_{k=1}^{45} (4k - 7)$
   a. 4185
   b. 3735
   c. 4027.5
   d. 3825

8. $\displaystyle\sum_{k=1}^{5} \left(\dfrac{4}{3}\right)(2^k)$
   a. $\dfrac{248}{3}$
   b. $\dfrac{251}{3}$
   c. $\dfrac{242}{3}$
   d. $\dfrac{287}{3}$

9. Evaluate $\displaystyle\sum_{k=1}^{\infty} 8(-0.3)^{k-1}$.
   a. 6
   b. 6.15
   c. -6.15
   d. -4.15

In Problems 10–16, solve each system of equations.

10. $\begin{cases} 2x - 3y = 16 \\ x - y = 7 \end{cases}$
    a. $\{(5, -2)\}$
    b. $\{(8, 0)\}$
    c. $\{(0, -7)\}$
    d. $\{(4, -3)\}$

11. $\begin{cases} 6x - 9y = -2 \\ 3x - 5y = -6 \end{cases}$
    a. $\{(-3, 0)\}$
    b. $\left\{\left(0, \dfrac{5}{6}\right)\right\}$
    c. $\left\{\left(\dfrac{44}{3}, 10\right)\right\}$
    d. $\{(1, 1)\}$

12. $\begin{cases} 2x - 5y = 9 \\ 4x - 10y = 18 \end{cases}$
    a. $\left\{\left(0, \dfrac{9}{2}\right)\right\}$
    b. $\left\{\left(\dfrac{5}{2}y + \dfrac{9}{2}, y\right)\right\}$
    c. $\left\{\left(x, \dfrac{5}{2}x + \dfrac{9}{2}\right)\right\}$
    d. $\varnothing$

13. $\begin{cases} 3x + 5y = 1 \\ -6x - 10y = 2 \end{cases}$
    a. $\left\{\left(0, -\dfrac{1}{5}\right)\right\}$
    b. $\{(2, -1)\}$
    c. $\varnothing$
    d. $\left\{\left(-\dfrac{5}{3}y + \dfrac{1}{3}, y\right)\right\}$

14. $\begin{cases} \dfrac{1}{5}x + \dfrac{2}{5}y = 1 \\ \dfrac{1}{4}x - \dfrac{1}{3}y = \dfrac{-5}{12} \end{cases}$
    a. $\{(-15, 10)\}$
    b. $\{(1, 2)\}$
    c. $\{(-10, -15)\}$
    d. $\{(1, 0)\}$

15. $\begin{cases} x - 3y = \dfrac{1}{2} \\ -2x + 6y = -1 \end{cases}$
    a. $\left\{\left(\dfrac{1}{2}, 0\right)\right\}$
    b. $\left\{\left(2, \dfrac{1}{2}\right)\right\}$
    c. $\varnothing$
    d. $\left\{\left(3y + \dfrac{1}{2}, y\right)\right\}$

16. $\begin{cases} 3x + 4y = 12 \\ 3x^2 + 16y^2 = 48 \end{cases}$
   a. $\{(0,4),(3,0)\}$
   b. $\{(0,-4)\}$
   c. $\{(0,-4),(3,0)\}$
   d. $\{(4,0),(2,\frac{3}{2})\}$

17. Student tickets for a dance cost $2, and nonstudent tickets cost $5. Three hundred tickets were sold, and the total ticket receipts were $975. How many of each type of ticket were sold?
   a. 150 student tickets
      150 nonstudent tickets
   b. 200 student tickets
      100 nonstudent tickets
   c. 210 student tickets
      90 nonstudent tickets
   d. 175 student tickets
      125 nonstudent tickets

For Problems 18 and 19, write the form of the partial-fraction decomposition of the given rational functions. You do not need to solve for the constants.

18. $\dfrac{x}{(x+2)(x-7)}$
   a. $\dfrac{A}{x+2} + \dfrac{Bx}{x-7}$
   b. $\dfrac{Ax}{x+2} + \dfrac{Bx}{x-7}$
   c. $\dfrac{A}{x+2} + \dfrac{B}{x-7}$
   d. $\dfrac{A}{x+2} + \dfrac{Bx+C}{x-7}$

19. $\dfrac{7-x}{(x-3)(x+5)^2}$
   a. $\dfrac{A}{x-3} + \dfrac{B}{x+5} + \dfrac{C}{(x+5)^2}$
   b. $\dfrac{A}{x-3} + \dfrac{B}{x+5} + \dfrac{Cx+D}{(x+5)^2}$
   c. $\dfrac{A}{x+3} + \dfrac{B}{(x+5)^2}$
   d. $\dfrac{A}{x+3} + \dfrac{Bx+C}{(x+5)^2}$

20. Find the partial-fraction decomposition of the rational expression
$$\dfrac{x^2+15x+18}{x^3-9x}.$$
   a. $\dfrac{-2}{x} + \dfrac{26}{x-9}$
   b. $\dfrac{-2}{x} + \dfrac{4}{x+3} - \dfrac{1}{x-3}$
   c. $\dfrac{-2}{x} + \dfrac{4}{x-3} - \dfrac{1}{x+3}$
   d. $\dfrac{-2}{x} + \dfrac{3x-15}{x^2-9}$

Answers to Practice Problems

CHAPTER P

Section P.1

Answers to Practice Problems

1. $\dfrac{710}{333}$ **3. a.** $(-\infty, \infty)$ **b.** $[2, 5)$ **4. a.** 10 **b.** 1 **c.** 1
5. 9 **6. a.** $\dfrac{1}{2}$ **b.** 1 **c.** $\dfrac{4}{9}$ **7. a.** $\dfrac{1}{4x^8}$ **b.** $-\dfrac{1}{xy^3}$

Section P.2

Answers to Practice Problems

1. a. -2, **b.** 2, **c.** 3, **d.** Not a real number **2. a.** $2\sqrt[3]{9}$,
b. $2\sqrt[4]{3a^2}$, **c.** a^2 **3. a.** $13\sqrt{3}$ **b.** 0 **4.** $\sqrt[10]{972}$
5. a. $\dfrac{7\sqrt{2}}{4}$, **b.** $\dfrac{\sqrt[3]{2ab^2}}{2b^2}$ **6.** $\dfrac{1}{(x+3)(\sqrt{x}+\sqrt{3})}$
7. a. $12x^{7/10}$, **b.** $\dfrac{5}{x^{7/12}}$, **c.** $\dfrac{1}{x^{2/15}}$ **8.** $\dfrac{3(x+2)}{2(x+3)^{1/2}}$
9. $\dfrac{1}{3}x^{1/3}(7x+20)$

Section P.3

Answers to Practice Problems

1. -1 **2.** $\varnothing$ **3.** $\{-4, -21\}$ **4.** $\{3 - \sqrt{2}, 3 + \sqrt{2}\}$
5. $\left\{-\dfrac{1}{2}, \dfrac{2}{3}\right\}$ **6. a.** $\{-2, 0, 2\}$ **b.** $\{-2, 2, 5\}$ **7.** $\{-10, 1\}$
8. $\{10\}$ **9.** $\{13\}$ **10.** $\{4, 16\}$ **11.** $\{-1, 2\}$

Section P.4

Answers to Practice Problems

1. a. $(2, \infty)$
b. $(-\infty, 5]$
2. a. $(-\infty, \infty)$ **b.** $\varnothing$
3. $(-\infty, 2] \cup [4, \infty)$ **4.** $\left[-1, \dfrac{3}{2}\right)$
5. $a = -11, b = 10$ **6.** $(-1, 4)$ **7.** $[-6, 1)$ **8.** $[-3, 1]$
9. $\left(-\infty, -\dfrac{9}{2}\right] \cup \left[\dfrac{3}{2}, \infty\right)$ **10. a.** $(-\infty, \infty)$ **b.** $\varnothing$
11. $(-\infty, -6) \cup \left(-\dfrac{14}{5}, \infty\right)$ **12.** $\delta = 0.05$

Section P.5

Answers to Practice Problems

1. a. $4 - 2i$ **b.** $-1 + 4i$ **2. a.** $26 + 2i$ **b.** $-15 - 21i$
3. 37 **4. a.** $1 + i$ **b.** $-\dfrac{15}{41} - \dfrac{12}{41}i$

CHAPTER 1

Section 1.1

Answers to Practice Problems

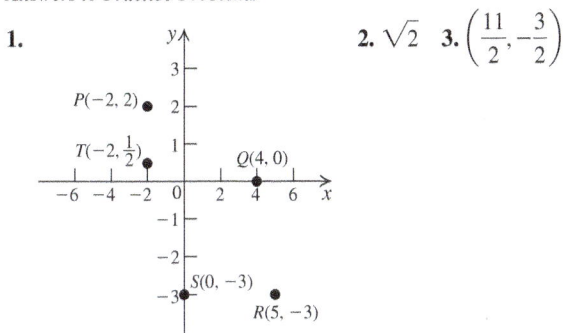

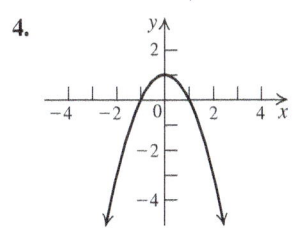

1. [graph showing points $P(-2, 2)$, $T(-2, \frac{1}{2})$, $Q(4, 0)$, $S(0, -3)$, $R(5, -3)$]

2. $\sqrt{2}$ **3.** $\left(\dfrac{11}{2}, -\dfrac{3}{2}\right)$

4. [parabola graph]

5. x-intercepts: $-2, \dfrac{1}{2}$; y-intercept: -2 **6.** Symmetric

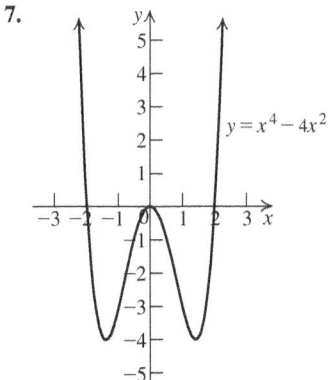

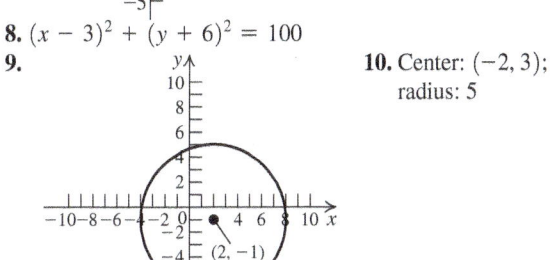

7. [graph of $y = x^4 - 4x^2$]

8. $(x - 3)^2 + (y + 6)^2 = 100$

9. [graph of circle centered at $(2, -1)$]

10. Center: $(-2, 3)$; radius: 5

P-1

Section 1.2

Answers to Practice Problems

1. $-\frac{8}{13}$, for every 13 units increase in x, the y-value decreases by 8 units. **2.** $y + 3 = -\frac{2}{3}(x + 2); y = -\frac{2}{3}x - \frac{13}{3}$

3. $y + 4 = 5(x + 3); y = 5x + 11$ **4.** $y + 3 = 2x; y = 2x - 3$

5.

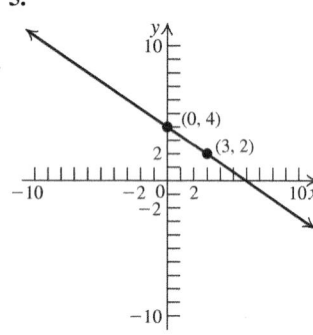

6.

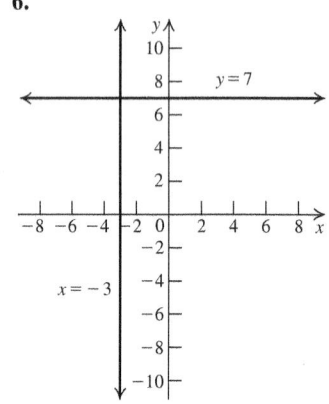

7.

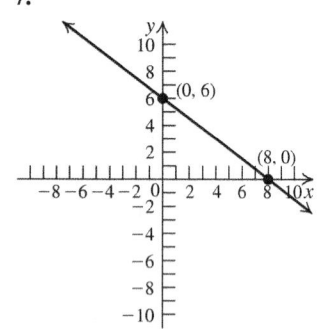

8. Between 176.8 and 179.4 cm
9. a. $4x - 3y + 23 = 0$
b. $5x - 4y - 31 = 0$

Section 1.3

Answers to Practice Problems
1. a. Yes **b.** No **2. a.** No **b.** Yes **3. a.** 0 **b.** -7
c. $-2x^2 - 4hx + 5x - 2h^2 + 5h$ **4.** 44 sq. units
5. a. $(-\infty, 1)$ **b.** $(-\infty, -2] \cup (3, \infty)$ **6. a.** No **b.** Yes
c. $[0, 9]$ **7.** No **8. a.** Yes **b.** $-3, -1$ **c.** -5 **d.** $-5, 1$
9. Domain: $(-3, \infty)$; range: $(-2, 2] \cup \{3\}$
10. Domain: $[1, \infty)$; range: $[6, 12]$; $C(11) = 11.994$
11. If $c =$ cost per foot on land, then
$C = c[1.3\sqrt{(500)^2 + x^2} + 1200 - x]$.

Section 1.4

Answers to Practice Problems
1. $g(x) = 2x + 6$ **2.** Decreasing on $(-\infty, -3)$; constant on $(-3, 2)$; increasing on $(2, \infty)$. **3.** Contraction to 7.33 mm
4. a. Even **b.** Odd **c.** Neither
5.

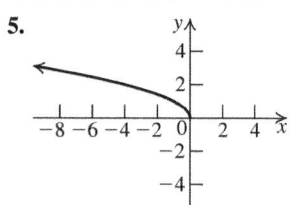

Domain: $(-\infty, 0]$; range: $[0, \infty)$

6.

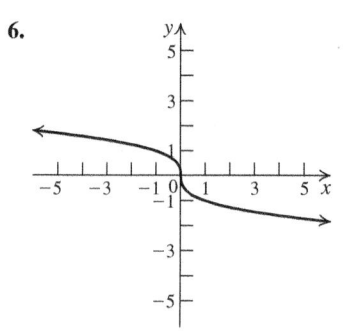

Domain: $(-\infty, \infty)$; range: (∞, ∞)
7. $f(-2) = (-2)^2 = 4; f(3) = 2(3) = 6$
8.

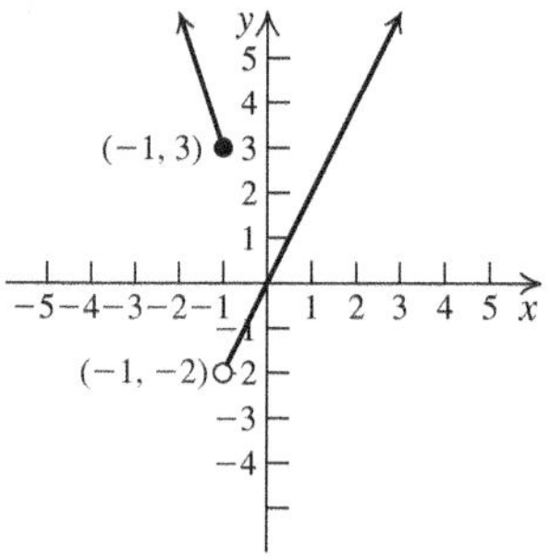

9. $f(-3.4) = -4; f(4.7) = 4$ **10.** -6 **11.** -1
12. $-2x + 1 - h$

Section 1.5

Answers to Practice Problems
1. The graph of g is the graph of f shifted one unit up; the graph of h is the graph of f shifted two units down.

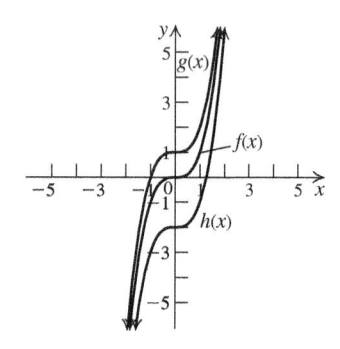

2. The graph of g is the graph of f shifted one unit to the right; the graph of h is the graph of f shifted two units to the left.

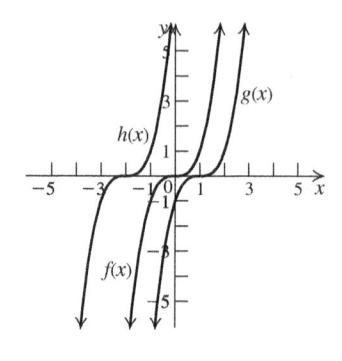

3.

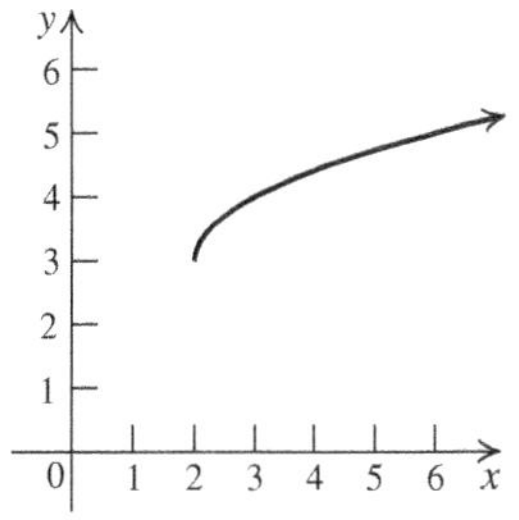

4. Shift the graph of $y = x^2$ one unit to the right, reflect the resulting graph about the x-axis, and shift it two units up.

5.

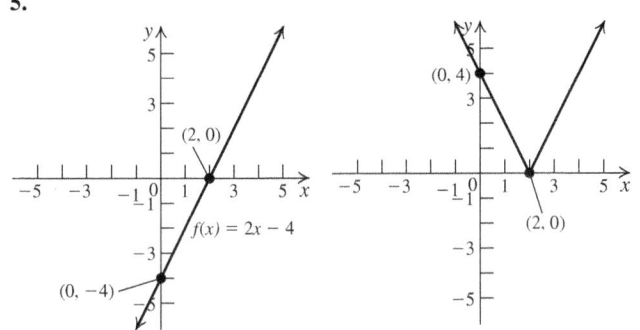

6. The graph of g is the graph of f vertically stretched by multiplying each of its y-coordinates by 2.

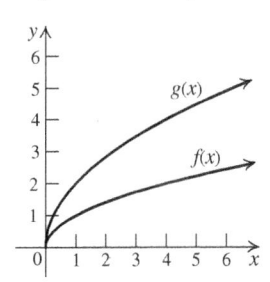

7. a. 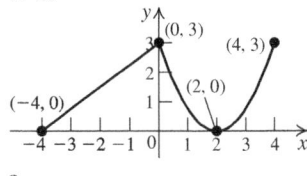   **b.**

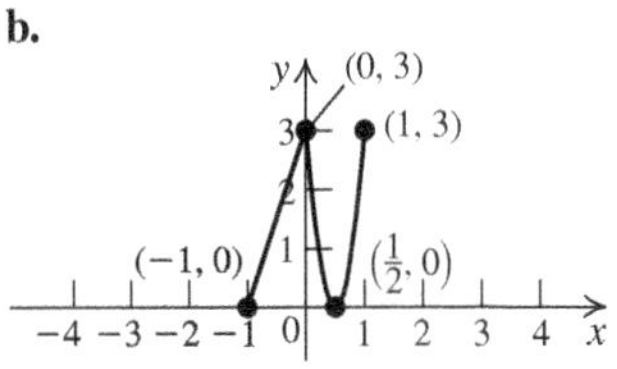

8. 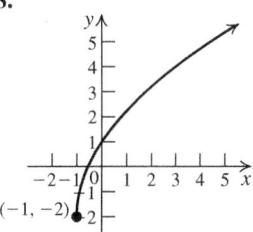   **9.**

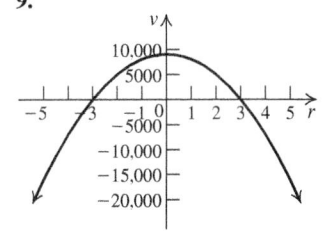

Section 1.6

Answers to Practice Problems

1. $(f + g)(x) = x^2 + 3x + 1$
$(f - g)(x) = -x^2 + 3x - 3$
$(fg)(x) = 3x^3 - x^2 + 6x - 2$
$\left(\dfrac{f}{g}\right)(x) = \dfrac{3x - 1}{x^2 + 2}$

2. a. $(f \circ g)(0) = -5$ **b.** $(g \circ f)(0) = 1$

3. a. $(g \circ f)(x) = 2x^2 - 8x + 9$
b. $(f \circ g)(x) = 1 - 2x^2$
c. $(g \circ g)(x) = 8x^4 + 8x^2 + 3$

4. $(g \circ f)(x) = 2\sqrt{x} + 5$; domain: $[0, \infty)$

5. $(h \circ f)(x) = x - 2$; domain: $[2, \infty)$ **6.** $f(x) = \dfrac{1}{x}$

7. a. $A = 9\pi t^2$ **b.** $A = 324\pi$ square miles
8. a. $r(x) = x - 4500$ **b.** $d(x) = 0.94x$
c. (i) $(r \circ d)(x) = 0.94x - 4500$
(ii) $(d \circ r)(x) = 0.94x - 4230$
d. $(d \circ r)(x) - (r \circ d)(x) = 270$

Section 1.7

Answers to Practice Problems

1. Not one-to-one; the horizontal line $y = 1$ intersects the graph at two different points.

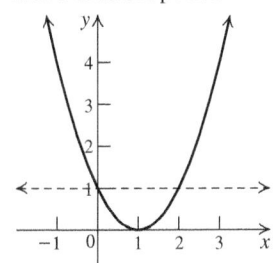

2. a. -3 **b.** 4

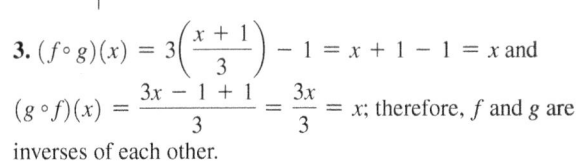

3. $(f \circ g)(x) = 3\left(\dfrac{x + 1}{3}\right) - 1 = x + 1 - 1 = x$ and
$(g \circ f)(x) = \dfrac{3x - 1 + 1}{3} = \dfrac{3x}{3} = x$; therefore, f and g are inverses of each other.

4.   **5.** $f^{-1}(x) = \dfrac{3 - x}{2}$

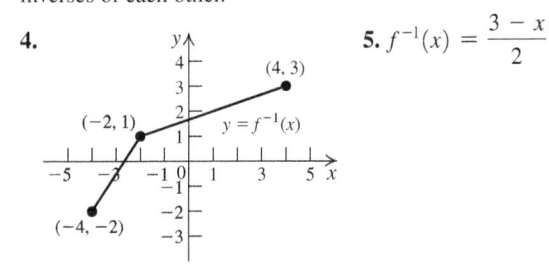

6. $f^{-1}(x) = \dfrac{3x}{1 - x}, x \neq 1$

7. Domain: $(-\infty, -3) \cup (-3, \infty)$; range: $(-\infty, 1) \cup (1, \infty)$
8. $G^{-1}(x) = -\sqrt{x + 1}$ **9.** 3597 ft

CHAPTER 2

Section 2.1

Answers to Practice Problems

1. $y = 3(x - 1)^2 - 5$; f has minimum value -5 **2.** The graph is a parabola $a = -2$, $h = -1, k = 3$. It opens downward. Vertex: $(-1, 3)$. It has maximum value 3; x-intercepts: $\pm\sqrt{\dfrac{3}{2}} - 1$; y-intercept: 1

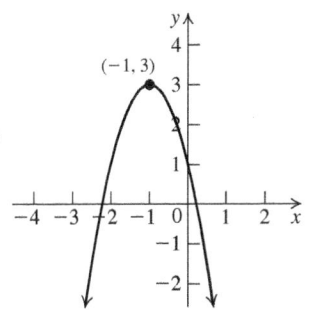

3. The graph is a parabola. It opens upward. Vertex $\left(\frac{1}{2}, \frac{-27}{4}\right)$; it has a minimum value $-\frac{27}{4}$; x-intercepts: $-1, 2$; y-intercept: -6

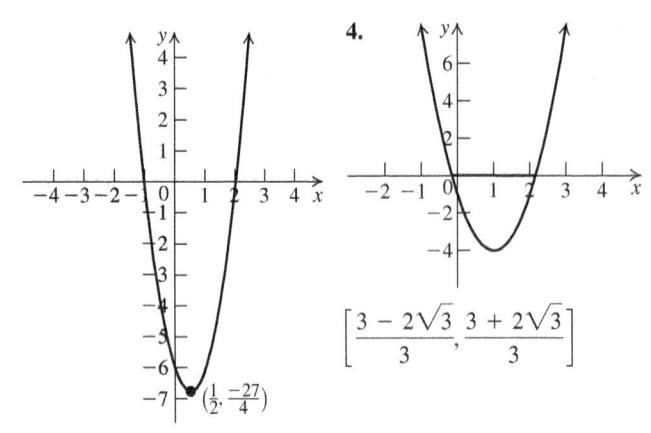

4. $\left[\frac{3 - 2\sqrt{3}}{3}, \frac{3 + 2\sqrt{3}}{3}\right]$

5. 1800 m²

Section 2.2

Answers to Practice Problems

1. a. Not a polynomial function **b.** A polynomial function; the degree is 7, the leading term is $2x^7$, and the leading coefficient is 2.

2. $P(x) = 4x^3 + 2x^2 + 5x - 17 = x^3\left(4 + \frac{2}{x} + \frac{5}{x^2} - \frac{17}{x^3}\right)$

When $|x|$ is large, the terms $\frac{2}{x}, \frac{5}{x^2}$, and $-\frac{17}{x^3}$ are close to 0. Therefore, $P(x) \approx x^3(4 + 0 + 0 - 0) = 4x^3$.

3. $y \to -\infty$ as $x \to -\infty$ and $y \to -\infty$ as $x \to \infty$ **4.** $\frac{3}{2}$

5. $f(1) = -7, f(2) = 4$. Because $f(1)$ and $f(2)$ have opposite signs, by this Intermediate Value Theorem, f has a real zero between 1 and 2. **6.** $-5, -1$, and 3 **7.** -5 and -3 with multiplicity 1 and 1 with multiplicity 2 **8.** 3

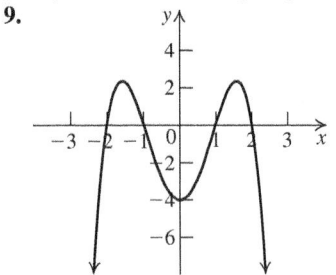

9. **10.** 207.378 L

Section 2.3

Answers to Practice Problems

1. The quotient is $2x^2 + 7x + 3$, and the remainder is 2.

2. The remainder is 4. **3.** 32 **4.** $\left\{-2, -\frac{2}{3}, 3\right\}$ **5.** 18

6. $\{-2\}$

Section 2.4

Answers to Practice Problems

1. Number of positive zeros: 1; number of negative zeros: 0 or 2
2. Upper bound: 1; lower bound: -2
3. a. $P(x) = 3(x + 2)(x - 1)(x - 1 - i)(x - 1 + i)$
b. $P(x) = 3x^4 - 3x^3 - 6x^2 + 18x - 12$

4. $-3, -3, 2 + 3i, 2 + 3i, 2 - 3i, 2 - 3i, i, -i$
5. The zeros are 1, 2, 2i, and $-2i$.
6. The zeros are 2, 4, $1 + i$, and $1 - i$.

Section 2.5

Answers to Practice Problems

1. Domain: $(-\infty, -1) \cup (-1, 5) \cup (5, \infty)$
2. a. Vertical asymptote: $x = 2$;
Horizontal asymptote: $y = 0$
Domain: $(-\infty, 2) \cup (2, \infty)$
Range: $(-\infty, 0) \cup (0, \infty)$

b. Vertical asymptote: $x = -1$;
Horizontal asymptote: $y = 2$
Domain: $(-\infty, -1) \cup (-1, \infty)$
Range: $(-\infty, 2) \cup (2, \infty)$

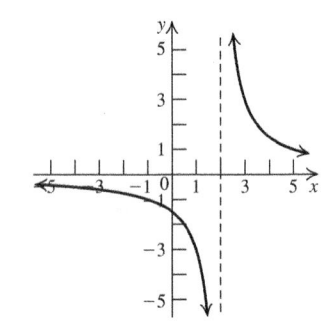

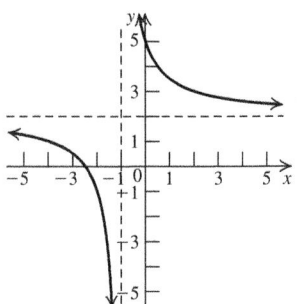

3. $x = -5$ and $x = 2$ **4.** $x = -3$ **5. a.** $y = \frac{2}{3}$

b. No horizontal asymptote
6. x-intercept: 0; y-intercept: 0; vertical asymptotes: $x = -1$ and $x = 1$; horizontal asymptote: x-axis. The graph is symmetric with respect to the origin. The graph is above the x-axis on $(-1, 0) \cup (1, \infty)$ and below the x-axis on $(-\infty, -1) \cup (0, 1)$.

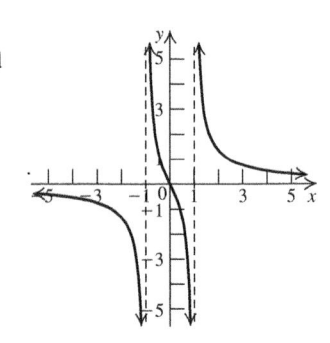

7. x-intercepts: $\pm\frac{\sqrt{2}}{2}$;
y-intercept: $\frac{1}{3}$; vertical asymptotes: $x = -\frac{3}{2}$ and $x = 1$;
horizontal asymptote: $y = 1$;
symmetry: none. The graph is above the line $y = 1$ on $\left(-\infty, -\frac{3}{2}\right) \cup (1, 2)$ and below it on $\left(-\frac{3}{2}, 1\right) \cup (2, \infty)$. The graph crosses the horizontal asymptote at $(2, 1)$.

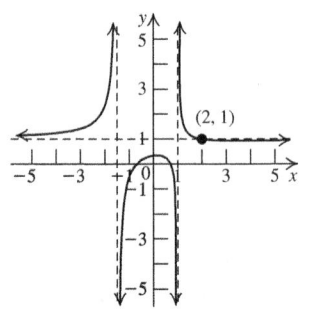

8. No x-intercept; y-intercept: $\frac{1}{2}$; no vertical asymptote; horizontal asymptote: $y = 1$. The graph is symmetric with respect to the y-axis. The graph is below the line $y = 1$ on $(-\infty, \infty)$.

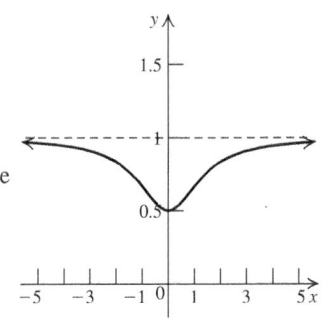

9. No x-intercept; y-intercept: -2; vertical asymptote: $x = 1$; no horizontal asymptote; $y = x + 1$ is an oblique asymptote; symmetry: none. The graph is above the line $y = x + 1$ on $(1, \infty)$ and below the line $y = x + 1$ on $(-\infty, 1)$.

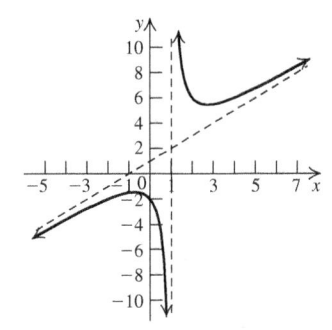

10. a. $R(10) = \$30$ billion
$R(20) = \$40$ billion
$R(30) = \$42$ billion
$R(40) = \$40$ billion
$R(50) = \$35.7$ billion
$R(60) = \$30$ billion
c. 29%

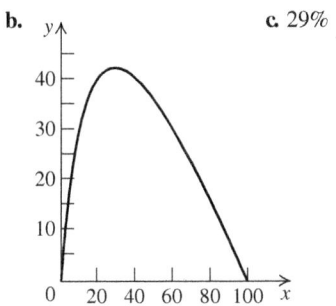

CHAPTER 3

Section 3.1

Answers to Practice Problems

1. $f(2) = \frac{1}{16}$
$f(0) = 1$
$f(-1) = 4$
$f\left(\frac{5}{2}\right) = \frac{1}{32}$
$f\left(-\frac{3}{2}\right) = 8$

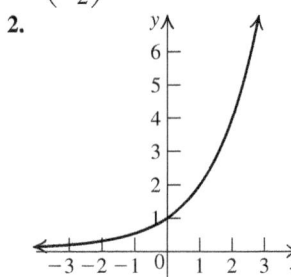

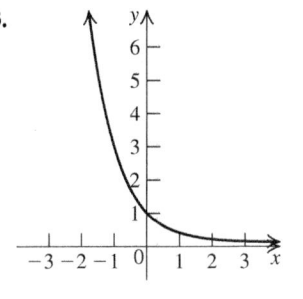

4. a. 7^x **b.** $4\left(\frac{1}{2}\right)^x$ **5.** $\$11,500$ **6. a.** $A = \$11,485.03$
b. $I = \$3,485.03$ **7. (i)** $\$5325.00$ **(ii)** $\$5330.28$ **(iii)** $\$5333.01$
(iv) $\$5334.86$ **(v)** $\$5335.76$ **8.** 10.023% **9.** $\$14,764.48$
10. $y = -e^{x-1} - 2$

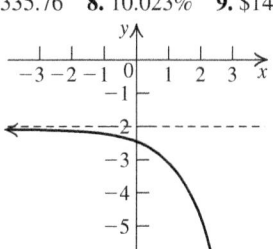

11. a. 11.96229 **b.** 4.76720 **12.** $\$7471.10$

Section 3.2

Answers to Practice Problems

1. a. $\log_2 1024 = 10$ **b.** $\log_9\left(\frac{1}{3}\right) = -\frac{1}{2}$ **c.** $\log_a p = q$
2. a. $2^6 = 64$ **b.** $v^w = u$ **3. a.** 2 **b.** $-1/2$ **c.** -5
4. a. 8 **b.** 5 **c.** $x = -2$ or $x = 3$ **5.** $(-\infty, 1)$
6.

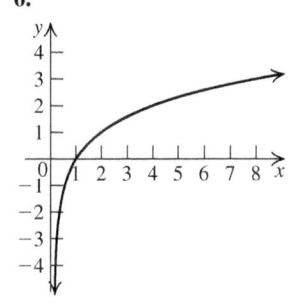

7. Shift the graph of $y = \log_2 x$ three units right and reflect the graph about the x-axis.

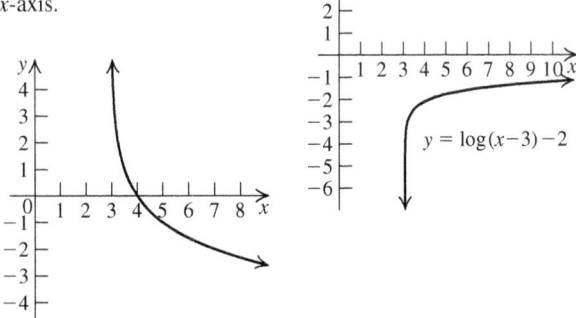

8. $y = \log(x-3) - 2$

9. a. -1 **b.** ≈ 0.693 **10. a.** 17 years **b.** 21.97%
11. 5.423 min **12. a.** 28,935 m³ **b.** 12.6 yr

Section 3.3

Answers to Practice Problems

1. a. -1 **b.** 13 **2. a.** $\ln(2x - 1) - \ln(x + 4)$
b. $\log 2 + \frac{1}{2}\log x + \frac{1}{2}\log y - \frac{1}{2}\log z$ **3.** $\log\sqrt{x^2 - 1}$
4. $2.215088 \times 10^{1343}$ **5.** $\frac{\log 15}{\log 3} = \frac{\ln 15}{\ln 3} \approx 2.46497$

6. a. $y = 5 - \frac{2}{\log 3}\log x$ **b.** $y = 5 - 2\log_3 x$
7. 25 yr **8.** $\approx 65.34\%$

Section 3.4

Answers to Practice Problems

1. a. $x = 5$ **b.** $x = \dfrac{2}{3}$ **2.** $x = \dfrac{\ln\left(\dfrac{11}{7}\right)}{\ln 3} - 1 \approx -0.59$

3. $x = \dfrac{\ln 3}{2\ln 2 - \ln 3} \approx 3.82$ **4.** $x = \ln 5 \approx 1.609$

5. a. United States: 343.61 million; Pakistan: 255.96 million
b. Sometime in 2022 (after 11.69 yr) **c.** In 23.68 yr (sometime in 2034)   **6.** $x = e^{3/2}$ **7.** $\{3\}$ **8.** $x = 9$
9. The average growth rate was approximately 1.74%.
10. $(-\infty, -2)$ **11.** $\left(-\infty, \dfrac{1-e^2}{3}\right)$

CHAPTER 4

Section 4.1

Answers to Practice Problems

1.

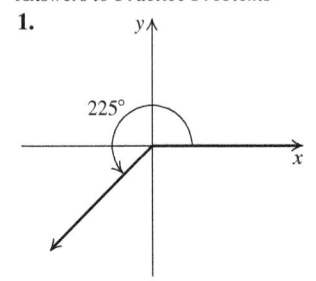

2. $-\dfrac{\pi}{4}$ radians   **3.** $270°$ **4.** Complement $= 23°$; supplement $= 113°$ **5.** $\dfrac{5\pi}{2} \approx 7.85$ meters   **6.** ≈ 790 miles

Section 4.2

Answers to Practice Problems

1. $\left(-\dfrac{2}{3}, \dfrac{\sqrt{5}}{3}\right)$ and $\left(-\dfrac{2}{3}, -\dfrac{\sqrt{5}}{3}\right)$

2. a. $\sin t = \dfrac{3}{5}, \cos t = -\dfrac{4}{5}, \tan t = -\dfrac{3}{4}$

$\csc t = \dfrac{5}{3}, \sec t = -\dfrac{5}{4}, \cot t = -\dfrac{4}{3}$

b. $\sin t = \dfrac{2\sqrt{2}}{3}, \cos t = \dfrac{1}{3}, \tan t = 2\sqrt{2}$

$\csc t = \dfrac{3\sqrt{2}}{4}, \sec t = 3, \cot t = \dfrac{\sqrt{2}}{4}$

3. $\sin \dfrac{5\pi}{2} = 1, \cos \dfrac{5\pi}{2} = 0, \tan \dfrac{5\pi}{2}$ is undefined

$\csc \dfrac{5\pi}{2} = 1, \sec \dfrac{5\pi}{2}$ is undefined, $\cot \dfrac{5\pi}{2} = 0$

4. a. $\left(-\dfrac{\sqrt{3}}{2}, -\dfrac{1}{2}\right)$, **b.** $\left(-\dfrac{1}{2}, \dfrac{\sqrt{3}}{2}\right)$ **c.** $\left(-\dfrac{\sqrt{3}}{2}, -\dfrac{1}{2}\right)$

5. a. $P\left(\dfrac{31\pi}{6}\right) = \left(-\dfrac{\sqrt{3}}{2}, -\dfrac{1}{2}\right)$; $\cos \dfrac{31\pi}{6} = -\dfrac{\sqrt{3}}{2}, \sin \dfrac{31\pi}{6} = -\dfrac{1}{2}$,

$\tan \dfrac{31\pi}{6} = \dfrac{1}{\sqrt{3}} = \dfrac{\sqrt{3}}{3}$ **b.** $P\left(-\dfrac{28\pi}{3}\right) = \left(-\dfrac{1}{2}, \dfrac{\sqrt{3}}{2}\right)$;

$\cos\left(-\dfrac{28\pi}{3}\right) = -\dfrac{1}{2}, \sin\left(-\dfrac{28\pi}{3}\right) = \dfrac{\sqrt{3}}{2}, \tan\left(-\dfrac{28\pi}{3}\right) = -\sqrt{3}$

6. $\sin(-270°) = 1, \cos(-270°) = 0, \tan(-270°)$ is undefined $\csc(-270°) = 1, \sec(-270°)$ is undefined, $\cot(-270°) = 0$

7. $\sin \theta = -\dfrac{5\sqrt{29}}{29}, \cos \theta = \dfrac{2\sqrt{29}}{29}, \tan \theta = -\dfrac{5}{2}$

$\csc \theta = -\dfrac{\sqrt{29}}{5}, \sec \theta = \dfrac{\sqrt{29}}{2}, \cot \theta = -\dfrac{2}{5}$

8. a. -0.41 **b.** 2.03 **9.** $\sin \theta = -\dfrac{4\sqrt{41}}{41}, \sec \theta = \dfrac{\sqrt{41}}{5}$

Section 4.3

Answers to Practice Problems

1.

2.

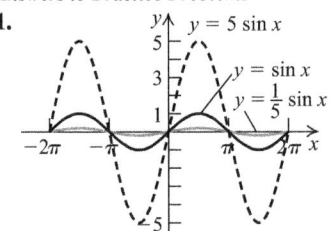

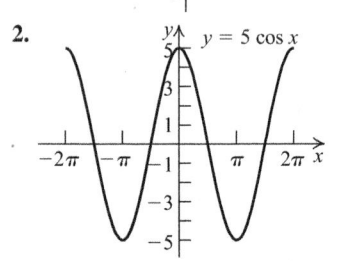

Amplitude $= 5$; range $= [-5, 5]$

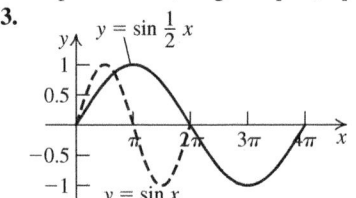

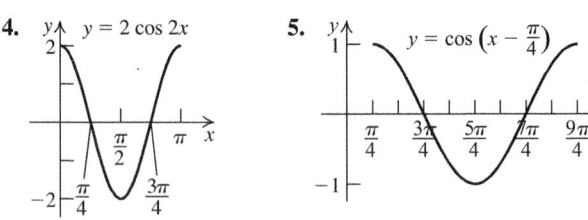

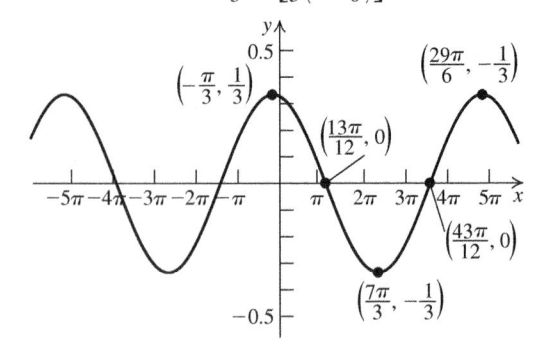

3.

4.

5.

6. Amplitude $= \dfrac{1}{3}$; period $= 5\pi$; phase shift $= -\dfrac{\pi}{6}$

$y = \dfrac{1}{3}\cos\left[\dfrac{2}{5}\left(x + \dfrac{\pi}{6}\right)\right]$

7. $y = -2\cos[3(x + \pi)]$

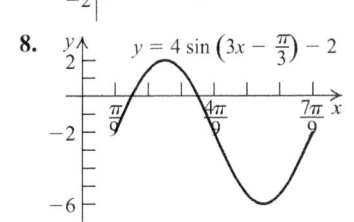

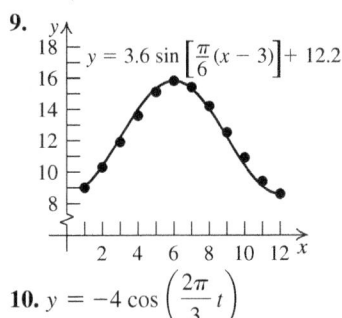

8. $y = 4\sin\left(3x - \dfrac{\pi}{3}\right) - 2$

9. $y = 3.6\sin\left[\dfrac{\pi}{6}(x-3)\right] + 12.2$

10. $y = -4\cos\left(\dfrac{2\pi}{3}t\right)$

Section 4.4

Answers to Practice Problems

1. $y = \dfrac{\sqrt{3}}{3}x + \left(3 - \dfrac{2\sqrt{3}}{3}\right)$

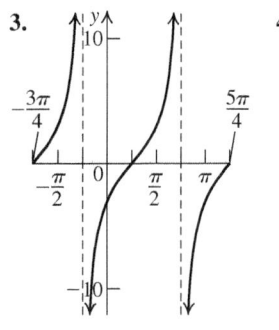

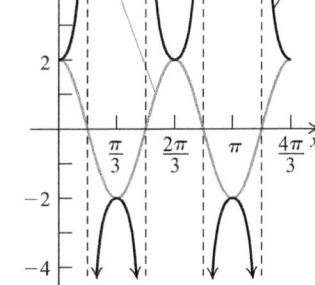

2. [graph with points $(-\tfrac{\pi}{2}, 3)$ and $(0, -2)$]

3. [graph] **4.** [graph of $y = 2\cos 3x$ and $y = 2\sec 3x$]

5. $[2.6, 5.1]$

Section 4.5

Answers to Practice Problems

1. a. $-\dfrac{\pi}{3}$ **b.** $-\dfrac{\pi}{2}$ **2. a.** $\dfrac{3\pi}{4}$ **b.** $\dfrac{\pi}{3}$ **3.** $\dfrac{\pi}{6}$ **4.** $\dfrac{\pi}{3}$
5. b. 1 **6. a.** 1.3490 **b.** 0.2898 **c.** 2.9320
7. a. 53.1301° **b.** 4.4117° **c.** −85.2364° **8.** $-\dfrac{\pi}{2}$
9. $\dfrac{2\sqrt{2}}{3}$ **10.** $\dfrac{\sqrt{9-x^2}}{x}$ **11.** $\cot\theta$ **12.** 80°

Section 4.6

Answers to Practice Problems

1. $\sin\theta = \dfrac{4}{5}$ $\csc\theta = \dfrac{5}{4}$ **2.** $\sin\theta = \dfrac{2\sqrt{2}}{3}$ $\csc\theta = \dfrac{3\sqrt{2}}{4}$
$\cos\theta = \dfrac{3}{5}$ $\sec\theta = \dfrac{5}{3}$ $\cos\theta = \dfrac{1}{3}$ $\sec\theta = 3$
$\tan\theta = \dfrac{4}{3}$ $\cot\theta = \dfrac{3}{4}$ $\tan\theta = 2\sqrt{2}$ $\cot\theta = \dfrac{\sqrt{2}}{4}$

3. a. $\sec 69° \approx 2.7904$ **b.** $\cot 15° \approx 3.7321$
4. $B = 34°, a \approx 21.4, b \approx 14.4$ **5.** $A \approx 48.4°, B \approx 41.6°,$
$a \approx 27.6$ **6.** ≈ 1.096 mi **7.** $\approx 20{,}320$ ft
8. ≈ 43 ft **9.** 80°

Section 4.7

Answers to Practice Problems

1. $\sin x = \dfrac{2\sqrt{5}}{5}, \cos x = -\dfrac{\sqrt{5}}{5}, \tan x = -\dfrac{1}{2}$
$\csc x = \dfrac{\sqrt{5}}{2}, \sec x = -\sqrt{5}, \cot x = -2$

2. For $x = \pi$, $\cos\pi = -1 \neq 1 = 1 - \sin\pi$ **8.** $3\cos\theta$

Section 4.8

Answers to Practice Problems

1. $\dfrac{\sqrt{6}+\sqrt{2}}{4}$ **2.** $\dfrac{\sqrt{2}-\sqrt{6}}{4}$ **5.** $\dfrac{1}{2}$
7. a. $-\dfrac{120}{169}$ **b.** $-\dfrac{119}{169}$ **c.** $\dfrac{120}{119}$ **8. a.** $-\dfrac{\sqrt{3}}{2}$ **b.** $-\dfrac{\sqrt{2}}{2}$
9. a. $\dfrac{3}{8} - \dfrac{1}{2}\cos 2x + \dfrac{1}{8}\cos 4x$ **10.** $-\dfrac{\sqrt{26}}{26}$

CHAPTER 5

Section 5.1

Answers to Practice Problem

1. $C = 45°, b \approx 15.4$ in., $c \approx 12.0$ in. **2.** 10,576 ft **3.** None.
4. None. **5.** Solution 1: $A \approx 99.2°, B \approx 45.8°, a \approx 20.7$ ft
Solution 2: $A \approx 10.8°, B \approx 134.2°, a \approx 3.9$ ft **6.** No triangle
exists. **7.** $A \approx 31.3°, B \approx 88.7°, b \approx 57.7$ ft **8. a.** ≈ 31.5 mi
b. ≈ 15.6 mi **9.** $b \approx 21.8, A \approx 36.6°, C \approx 83.4°$
10. $A \approx 41.9°, B \approx 86.0°, C \approx 52.1°$ **11.** No triangle exists.

Section 5.2

Answers to Practice Problems

1. a. 124 ft **b.** 672 ft² **2.** 375.2 ft² **3.** 90° **4.** 203.4 in²
5. 93.5 m² **6.** 14,801 gal

Section 5.3

Answers to Practice Problems

1. a. [figure: ray with $P(2, -150°)$] **b.** $(-2, 30°)$
c. $(2, 210°)$ **d.** $(-2, -330°)$
2. a. $\left(-\dfrac{3}{2}, -\dfrac{3\sqrt{3}}{2}\right)$ **b.** $(\sqrt{2}, -\sqrt{2})$ **3.** $\left(\sqrt{2}, \dfrac{5\pi}{4}\right)$
4. $(24.8, 36.0°)$ **5.** $r^2 - 2r\sin\theta + 3 = 0$
6. a. $x^2 + y^2 = 25$; a circle centered at the origin with radius 5
b. $y = -x\sqrt{3}$; a line through the origin with slope $-\sqrt{3}$
c. $x = 1$; a vertical line through $x = 1$ **d.** $x^2 + (y-1)^2 = 1$; a circle with center $(0, 1)$ and radius 1

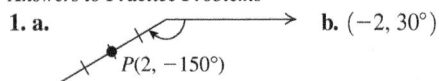

7. [spiral graph with points at $\pi, 2\pi, 3\pi, 4\pi, 5\pi, 6\pi, 7\pi, 8\pi, 9\pi$]

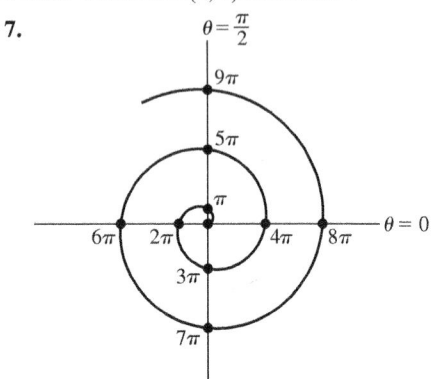

8. [circle graph with labeled points $(2+\sqrt{3}, \tfrac{2\pi}{3})$, $(4, \tfrac{\pi}{2})$, $(2+\sqrt{3}, \tfrac{\pi}{3})$, $(3, \tfrac{5\pi}{6})$, $(3, \tfrac{\pi}{6})$, $(2, \pi)$, $(2, 0)$, $(0, \tfrac{3\pi}{2})$]

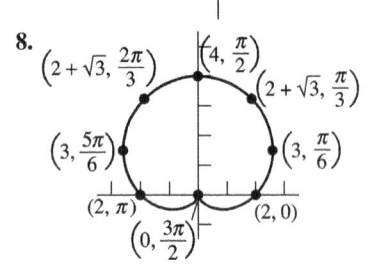

9. [four-petaled rose graph]

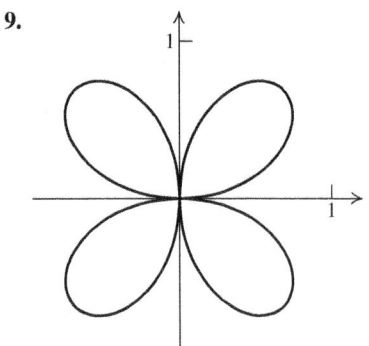

10. [limaçon with inner loop]

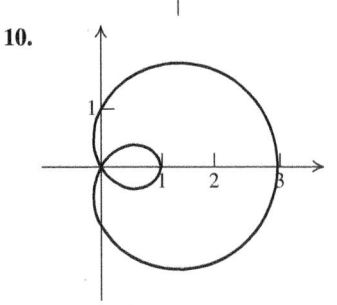

11. [ellipse graph]

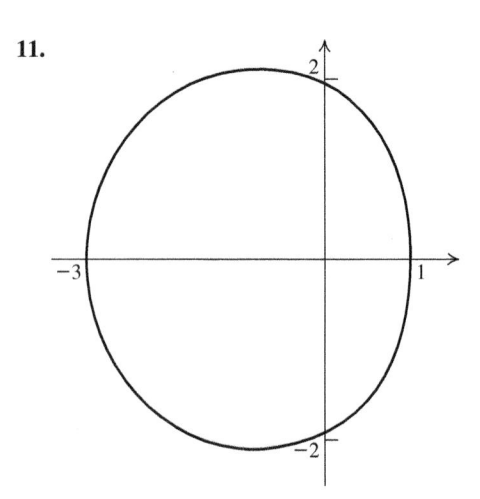

Section 5.4

Answers to Practice Problems

1. [parabola graph]

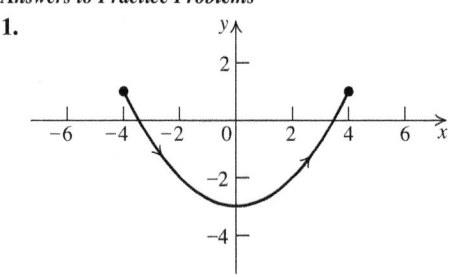

2. The graph is $y = (x-1)^2 + 3$ starting at the vertex $(1, 3)$ and continuing to the left for $x < 1$.

3. $\dfrac{x^2}{16} - \dfrac{y^2}{9} = 1$

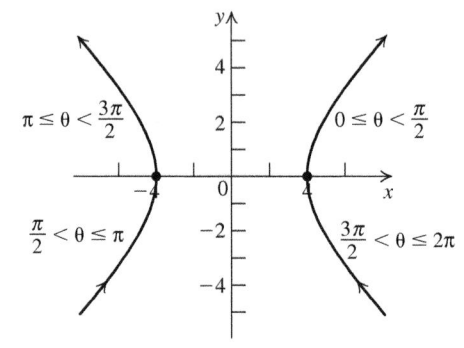

[hyperbola graph with $\pi \le \theta < \tfrac{3\pi}{2}$, $0 \le \theta < \tfrac{\pi}{2}$, $\tfrac{\pi}{2} < \theta \le \pi$, $\tfrac{3\pi}{2} < \theta \le 2\pi$]

4. a. A semicircle centered at the origin, starting at $(1, 0)$ and continuing counterclockwise to $(-1, 0)$.
b. Circles of radius 1 around the origin, starting at $(1, 0)$ and continuing counterclockwise for infinitely many rotations.
5. a. $x = \cos t, y = \sin t$; the particle moves counterclockwise starting at $(1, 0)$ and finishing at $(-1, 0)$.
b. $x = \cos 2t, y = \sin 2t$; the particle moves counterclockwise starting at $(1, 0)$ and finishing at $(-1, 0)$.
c. $x = -\cos t, y = \sin t$; the particle moves clockwise starting at $(-1, 0)$ and finishing at $(1, 0)$.
6. $x = at - b\sin t, y = a - b\cos t$

CHAPTER 6

Section 6.1

Answers to Practice Problems

1. $a_1 = -2, a_2 = -4, a_3 = -8, a_4 = -16$
2. $a_n = (-1)^{n+1}\left(1 - \dfrac{1}{n}\right)$ **3.** $a_1 = -3, a_2 = -1, a_3 = 3,$ $a_4 = 11, a_5 = 27$ **4.** $a_0 = 1, a_1 = 2$ and $a_k = a_{k-2} + a_{k-1}$, for all $k \geq 2$ **5. a.** 13 **b.** $n(n-1)(n-2)$
6. $a_1 = -2, a_2 = 2, a_3 = -\dfrac{4}{3}, a_4 = \dfrac{2}{3}, a_5 = -\dfrac{4}{15}$ **7.** -4
8. $\displaystyle\sum_{k=1}^{7}(-1)^{k+1}(2k)$

Section 6.2

Answers to Practice Problems

1. -5 **2.** $a_n = 4n - 7$ **3.** $d = -3, a_n = 53 - 3n$
4. $\dfrac{85}{6}$ **5.** 2000 ft

Section 6.3

Answers to Practice Problems

1. 3 **2.** It is geometric; $a_1 = 3, r = \dfrac{3}{2}$.
3. a. $a_1 = 2$ **b.** $r = \dfrac{3}{5}$ **c.** $a_n = 2\left(\dfrac{3}{5}\right)^{n-1}$
4. $7(1.5)^{17} \approx 6896.8288$ **5.** ≈ 2 **6.** \$91,510.60
7. 9 **8.** $\displaystyle\sum_{n=0}^{\infty}(10{,}000{,}000)(0.85)^n \approx \$66{,}666{,}666.67$

Section 6.4

Answers to Practice Problems

1. Equation (1): $(1) + (3) = 4$; equation (2): $3(1) - (3) = 0$
2. $\{(-1, 3)\}$ **3.** $\{(4, -1)\}$ **4.** $\varnothing$
5. $\{(x, 2x - 3) \mid x \text{ any real number}\}$
6. $\{(-1, 1), (3, -7)\}$ **7.** $\left\{\left(\dfrac{2}{3}, \dfrac{1}{2}\right)\right\}$ **8.** $\{(2, -3)\}$
9. $\{(4, 3), (4, -3), (-4, 3), (-4, -3)\}$ **10.** $(5700, 31.4)$

Section 6.5

Answers to Practice Problems

1. $\dfrac{3}{x + 1} - \dfrac{1}{x - 2}$ **2.** $\dfrac{5}{x - 1} - \dfrac{3}{x} + \dfrac{1}{x + 1}$
3. $\dfrac{5}{x} + \dfrac{6}{(x - 1)^2} - \dfrac{5}{x - 1}$
4. $\dfrac{1}{R} = \dfrac{3}{x + 1} + \dfrac{1}{x + 2} = \dfrac{1}{\dfrac{x + 1}{3}} + \dfrac{1}{x + 2}$;
$R_1 = \dfrac{x + 1}{3}, R_2 = x + 2$

Answers

CHAPTER P

Section P.1

Basic Concepts and Skills:
1. -0.8 terminating 3. $0.\overline{27}$ repeating 5. Rational 7. Irrational
9. Rational 11. $\frac{15}{4}$ 13. $\frac{211}{99}$ 15. $\frac{419}{99}$ 17. $(-2,5), [1,3]$
19. $(-6,10), \varnothing$ 21. $(-\infty, 7), (-\infty, 3)$ 23. -4 25. $5 - \sqrt{2}$
27. 12 29. 3 31. $|8 - 3| = 5$ 33. $|-6 - (-20)| = 14$
35. 11 37. -1 39. -6 41. -13 43. $\frac{62}{15}$ 45. x^4 47. $-\frac{8}{x}$
49. $\frac{1}{xy^2}$ 51. x^{33} 53. $\frac{4x^2}{y^2}$ 55. $\frac{1}{x^4}$ 57. $-243x^5y^5$ 59. $\frac{x^5}{y^4}$

Applying the Concepts:
61. $119.5 \le x \le 134.5$
63. a. $|124 - 120| = 4$ b. $|137 - 120| = 17$ c. $|114 - 120| = 6$
65. a. $(2x)(2x) = 4x^2 = 2^2 A$ b. $(3x)(3x) = 3x^2 = 3^2 A$

Maintaining Skills:
67. a. $\frac{14}{21}$ b. $\frac{12}{20}$ 68. a. $\frac{3}{7}$ b. $\frac{19}{24}$ c. $\frac{19}{18}$ d. $\frac{1}{6}$ e. $\frac{5}{16}$ f. $\frac{18}{65}$ 69. 54 70. 3

Section P.2

Basic Concepts and Skills:
1. 8 3. 4 5. -3 7. $-\frac{1}{2}$ 9. 3 11. Not a real number 13. 1
15. -7 17. $7\sqrt{3}$ 19. $9\sqrt{5}$ 21. $3\sqrt{2x}$ 23. $-\sqrt[3]{3}$ 25. $2\sqrt[3]{3x}$
27. $\sqrt{2x}(x^2 + 3x - 20)$ 29. $(2x - y)^2\sqrt{3xy}$ 31. $\frac{2\sqrt{3}}{3}$
33. $\frac{\sqrt{2} - x}{2 - x^2}$ 35. $\sqrt{3} - \sqrt{2}$ 37. $\frac{7 - 2\sqrt{10}}{3}$
39. $\frac{2x + h - 2\sqrt{x(x+h)}}{h}$ 41. $\frac{1}{\sqrt{4+h}+2}$ 43. $\frac{1}{2+\sqrt{4-x}}$
45. $\frac{1}{\sqrt{x}+2}$ 47. $\frac{2x}{\sqrt{x^2+4x}+x}$ 49. 5 51. -2 53. 4
55. $-\frac{1}{125}$ 57. $\frac{125}{27}$ 59. $x^{9/10}$ 61. $x^{1/10}$ 63. $4x^4$ 65. $\frac{1}{9x^4y^2}$
67. $5x^{5/4}$ 69. $\frac{x^3}{y^8}$ 71. $3^{1/2}$ 73. x^3 75. x^2y^3 77. 3
79. $\frac{2}{3}x^{1/3}(7x - 6)$ 81. $2(3x + 1)^{1/3}(7x - 1)$
83. $(x^2 + 1)^{1/2}(14x^3 - 5x^2 + 8x - 2)$

Applying the Concepts:
85. 40 cm 87. 2 cm/sec

Maintaining Skills:
89. -2 90. 51 91. -1 92. $-\frac{1}{4}$ 93. $x^2 - 6x + 9$
94. $x^2 + 7x + 12$ 95. $4x^2 - 9$ 96. $9x^2 - \frac{1}{4}$

Section P.3

Basic Concepts and Skills:
1. $\{-1\}$ 3. Identity 5. $\{28\}$ 7. $\{11\}$ 9. $\left\{\frac{4}{5}\right\}$ 11. $\varnothing$
13. Identity 15. $\{0, 5\}$ 17. $\{-1, 6\}$ 19. $\{-1 \pm \sqrt{6}\}$
21. $\{2 \pm 2\sqrt{2}\}$ 23. $\{0, 2\}$ 25. $\{-1, 0, 1\}$ 27. $\left\{-\frac{3}{2}, 2\right\}$
29. $\varnothing$ 31. $\{1\}$ 33. $\{-2, 1\}$ 35. $\{6\}$ 37. $\{3, 7\}$
39. $\{8, 64\}$ 41. $\{1\}$ 43. $\{\pm 2\sqrt{2}, \pm \sqrt{3}\}$ 45. $\left\{-\frac{4}{3}, -\frac{2}{3}\right\}$
47. $\{-5, -1\}$ 49. $\{\varnothing\}$ 51. $\{-5b, 2b\}$ 53. $\left\{\frac{1 + \sqrt{5}}{2}\right\}$
55. 0, 8 59. 2

Applying the Concepts:
63. 2 in. 65. 40 km 67. 25 liters

Maintaining Skills:
69. $x + 10$ 70. $-6x + 7$ 71. $-3x + 1$ 72. $1.6x + 1.3$
73. $-5x + 3$ 74. -3

Section P.4

Basic Concepts and Skills:
1. $(-\infty, -1)$
3. $(-\infty, -6]$
5. $(-\infty, \infty)$
7. $(-\infty, -2) \cup (2, \infty)$ 9. $\left(-\infty, \frac{11}{2}\right]$ 11. $(-\infty, -2] \cup (6, \infty)$
13. $[1, 5)$ 15. $\varnothing$ 17. $(-\infty, 5]$ 19. $(-2, -1)$ 21. $(-3, 3]$
23. $\left(-1, \frac{1}{2}\right]$ 25. $a = 5, b = 6$ 27. $a = 1, b = 3$
29. $a = -1, b = 19$ 31. $[-6, 2]$ 33. $(-2, 1) \cup (2, \infty)$
35. $(-\sqrt{2}, 0) \cup (0, \sqrt{2})$ 37. $(-2, -1] \cup [3, 4)$ 39. $(-3, 6]$
41. $\left(-1, \frac{1}{7}\right] \cup (3, \infty)$ 43. $(-4, 2)$ 45. $\left(-\frac{1}{2}, \frac{7}{2}\right)$
47. $(-\infty, 1) \cup (4, \infty)$ 49. $\varnothing$ 51. $(-\infty, \infty)$ 53. $\left(-\frac{1}{2}, \infty\right)$
55. $(-\infty, -6] \cup [0, \infty)$ 57. $[-5, -2) \cup (-2, -1]$
59. $\left(\frac{1}{2}, 1\right) \cup \left(1, \frac{5}{2}\right)$ 61. 0.05 63. 0.03

Applying the Concepts:
65. $2012.50 to $2100.00 67. 86 to 100 69. 21 to 29 hr

Maintaining Skills:
71. $2x^3$ 72. $\frac{y^5}{x^3}$ 73. $\frac{27}{8}$ 74. $3y^{10}$ 75. $9x^{6z}$ 76. a^{5m+3}

Section P.5

Basic Concepts and Skills:
1. $x = 3, y = 2$ 3. $x = 2, y = -4$ 5. $8 + 3i$ 7. $-7i$ 9. $15 + 6i$
11. $-3 + 15i$ 13. $3 + 11i$ 15. 13 17. $24 + 7i$ 19. $-141 - 24\sqrt{3}i$
21. $-26 - 18i$ 23. $\bar{z} = 2 + 3i; z\bar{z} = 13$ 25. $z = 4 - 5i; z\bar{z} = 41$
27. $\bar{z} = \sqrt{2} + 3i; z\bar{z} = 11$ 29. $5i$ 31. $1 + 2i$ 33. $\frac{43}{65} - \frac{6}{65}i$
35. i 37. i 43. $14 - 8i$ 45. $-2 - 16i$

Applying the Concepts:
47. $9 + i$ 49. $Z = \frac{595}{74} + \frac{315}{74}i$ 51. $V = 45 + 11i$
53. $I = \frac{17}{15} + \frac{4}{15}i$

A-1

Maintaining Skills:
55. $h = \dfrac{3V}{\pi r^2}$
56. (number line from 2 to 5, open at both ends)
57. $4 + \sqrt{15}$
58. $t = 3, t = 7$

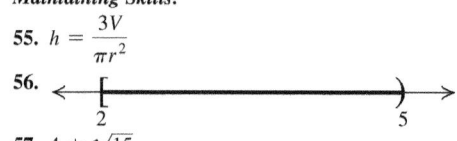

CHAPTER 1

Section 1.1
Basic Concepts and Skills:
1. (graph with plotted points)
$(2, 2)$ Quadrant I
$(3, -1)$ Quadrant IV
$(-1, 0)$, x-axis
$(-2, -5)$, Quadrant III
$(0, 0)$, origin
$(-7, 4)$, Quadrant II
$(0, 3)$, y-axis
$(-4, 2)$, Quadrant II

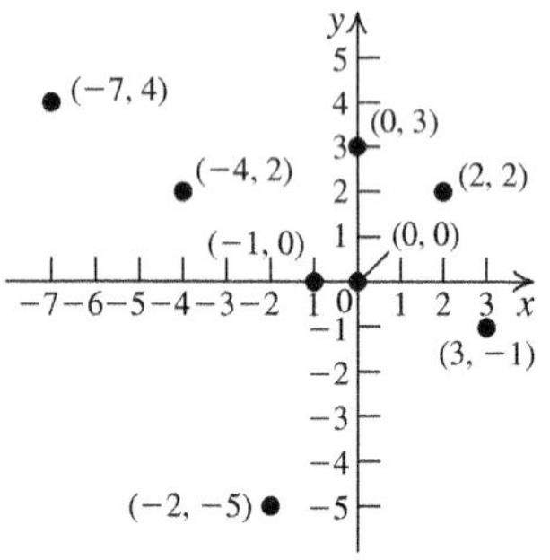

3. a. On the y-axis **b.** Vertical line intersecting the x-axis at -1

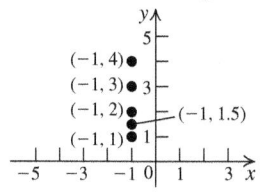

5. a. 5 **b.** $(0.5, 5)$ **7. a.** $\sqrt{(x+2)^2 + (y-3)^2}$ **b.** $\left(\dfrac{x-2}{2}, \dfrac{y+3}{2}\right)$
9. $\left(-\dfrac{1}{2}, 7\right)$ **11.** Yes **13.** No; $9\sqrt{2} + 5 + \sqrt{61}$ **15.** Isosceles
17. Scalene **19.** Equilateral
21. (graph) **23.** (graph)
25. (graph) **27.** (graph)
29. (graph) **31.** x-intercepts: $-3, 0, 3$; y-intercepts: $-2, 0, 2$

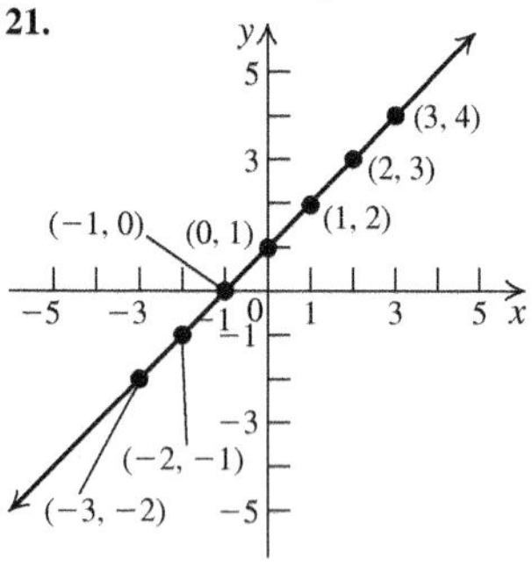

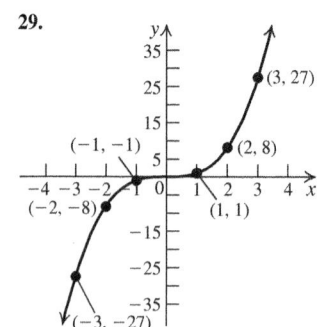

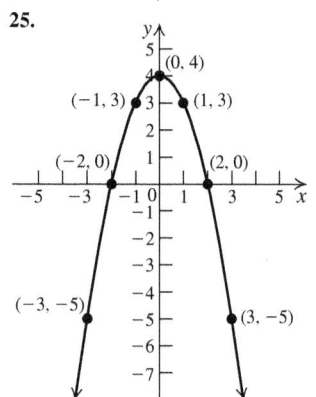

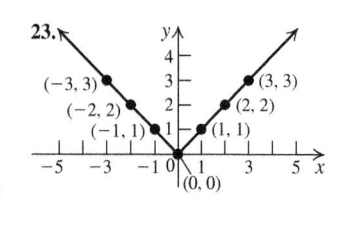

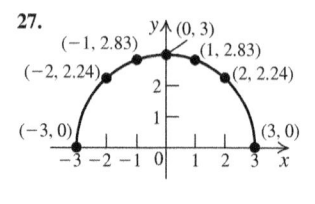

33. x-intercept: 4; y-intercept: 3 **35.** x-intercepts: 2, 4; y-intercept: 8
37. x-intercepts: $-2, 2$; y-intercepts: $-2, 2$ **39.** x-intercepts: $-1, 1$;
no y-intercept **41.** Not symmetric with respect to the x-axis, symmetric
with respect to the y-axis, not symmetric with respect to the origin
43. Not symmetric with respect to the x-axis, not symmetric with respect
to the y-axis, symmetric with respect to the origin **45.** Not symmetric
with respect to the x-axis, symmetric with respect to the y-axis, not
symmetric with respect to the origin **47.** Not symmetric with respect
to the x-axis, not symmetric with respect to the y-axis, symmetric with
respect to the origin **49.** Center: $(-1, 3)$; radius: 4
51. $(x - 3)^2 + (y - 1)^2 = 9$ **53.** $(x + 3)^2 + (y + 2)^2 = 9$
55. $(x - 5)^2 + (y - 1)^2 = 25$ **57. a.** Circle; center: $(1, 1)$;
radius: $\sqrt{6}$ **b.** x-intercepts: $1 \pm \sqrt{5}$; y-intercepts: $1 \pm \sqrt{5}$
59. a. Circle; center: $(0, -1)$; radius: 1 **b.** x-intercept: 0;
y-intercepts: $-2, 0$ **61.** point $(1, -2)$

Applying the Concepts:
63. $|x| = |y|$ **65.** $x = \dfrac{y^2}{4} + 1$ **67. a.** $12 million **b.** $5.5 million
c. (graph)

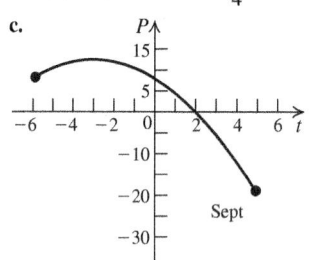

d. -8 and 2. They represent the months when there is neither profit nor
loss. **e.** 8; it represents the profit in July 2012.
69. a. After 0 second: 320 ft; after
1 second: 432 ft; after 2 seconds:
512 ft; after 3 seconds: 560 ft; after
4 seconds: 576 ft; after 5 seconds:
560 ft; after 6 seconds: 512 ft
b. (graph)

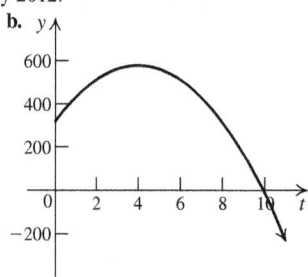

c. $0 \le t \le 10$ **d.** t-intercept: 10; it represents the time when the object
hits the ground. y-intercept: 320; it shows the height of the building.

Critical Thinking / Discussion / Writing:
71. The graph is the union of the graphs of $y = \sqrt{2x}$ and $y = -\sqrt{2x}$.
(graph)

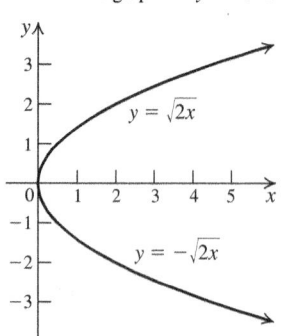

72. See the graph of $y = x$. **73.** False because it gives the y-intercepts **74.** For example, $y = -(x + 2)(x - 3)$ **75.** Answers may vary slightly.

Maintaining Skills:

76. $\frac{19}{24}$ **77.** $\frac{1}{12}$ **78.** $\frac{2}{3}$ **79.** $\frac{2}{3}$ **80.** $x^2 + 3x + 2$
81. $2x^2 + 7x + 6$ **82.** $4x^2 - 9$

Section 1.2

Basic Concepts and Skills:

1. $\frac{4}{3}$, rising **3.** 0, horizontal **5. a.** ℓ_3 **b.** ℓ_2 **c.** ℓ_4 **d.** ℓ_1
7. a. $y = 0$ **b.** $x = 0$
9. $y = \frac{1}{2}x + 4$ **11.** $y = -\frac{3}{2}x + 4$

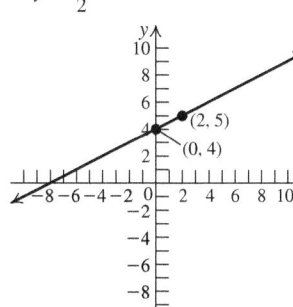

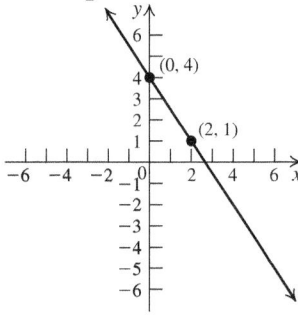

13. $y = -4$ 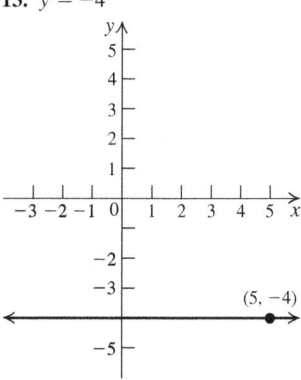   **15.** $y = 2x + 1$

17. $y = -\frac{6}{7}x - \frac{27}{7}$ **19.** $x = 5$ **21.** $y = 0$ **23.** $y = 2x + 5$
25. $y = \frac{4}{3}x + 4$ **27.** $y = 7$ **29.** $y = -5$ **31. a.** Parallel
b. Neither **c.** Perpendicular
33. $m = -\frac{1}{2}$; y-intercept = 2;   **35.** $m = \frac{1}{3}$; y-intercept = 3;
x-intercept = 4   x-intercept = -9

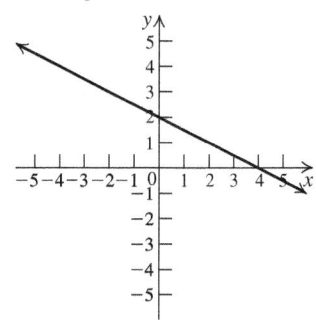

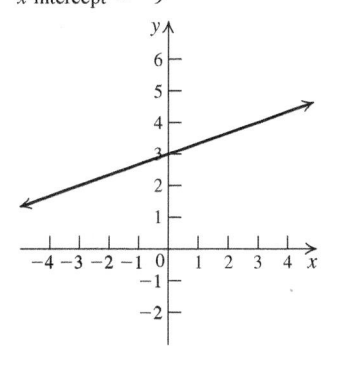

37. m is undefined; no y-intercept; x-intercept = 5

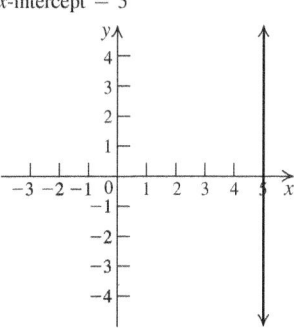

39. m is undefined; y-intercepts = y-axis; x-intercept = 0

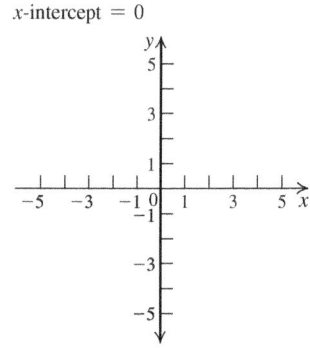

43. $\frac{x}{3} + \frac{y}{2} = 1$; x-intercept = 3; y-intercept = 2 **45.** $y = -x - 2$
47. $y = x + 2$; because the coordinates of the point $(-1, 1)$ satisfy this equation, $(-1, 1)$ also lies on this line. **49.** Parallel **51.** Neither
53. Perpendicular **55.** Neither **57.** $y = -x + 2$ **59.** $y = -3x - 2$
61. $y = -\frac{1}{6}x + 4$

Applying the Concepts:

63. a. x: time in weeks; y: amount of money in the account; $y = 7x + 130$
b. Slope: weekly deposit in the account; y-intercept: initial deposit
65. a. $F = \frac{9}{5}C + 32$ **b.** One degree Celsius change in the temperature is equal to $\frac{9}{5}$ degrees change in Fahrenheit. **c.** 104°F, 77°F, 23°F, 14°F
d. 37.78°C, 32.22°C, 23.89°C, −23.33°C, −28.89°C **e.** 36.44°C to 37.56°C
f. at −40°
67. a. $y = 23.1x + 117.8$ **b.** Slope: cost of producing one modem; y-intercept: fixed overhead cost **c.** $395

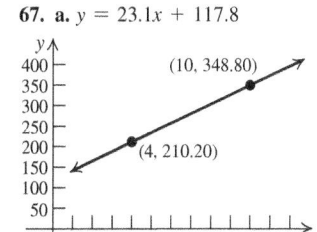

Critical Thinking / Discussion / Writing:

70. $y = \frac{5}{12}x + \frac{169}{12}$ **71.** The slope of the line that passes through $(1, -1)$ and $(-2, 5)$ is -2. The slope of the line that passes through $(1, -1)$ and $(3, -5)$ is -2. The slope of the line that passes through $(-2, 5)$ and $(3, -5)$ is -2.
72. a. This is a family of lines parallel to the line $y = -2x$, and they all have slope equal to -2.   **b.** This is a family of lines that pass through the point $(0, -4)$ and they all have y-intercept -4

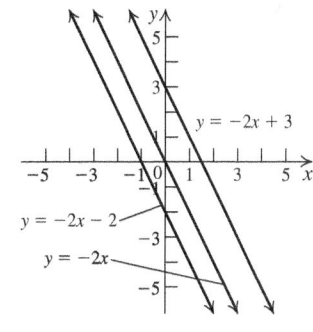

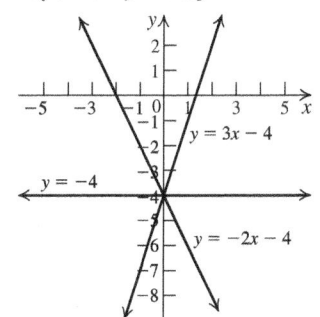

Maintaining Skills:

73. $\dfrac{x-3}{x-2}$ **74.** $\dfrac{2x+1}{(x+3)^2}$ **75.** $\dfrac{x+2}{x^2-1}$ **76.** $\dfrac{5x-8}{x^2-6x+8}$

77. $\dfrac{1}{3}$ **78.** $-x$

Section 1.3

Basic Concepts and Skills:

1. Domain: $\{a, b, c\}$; range: $\{d, e\}$; function

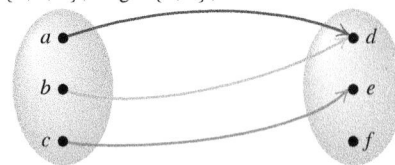

3. Domain; $\{a, b, c\}$; range: $\{1, 2\}$; function

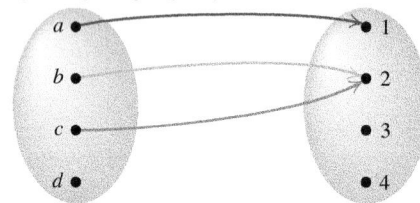

5. Domain: $\{0, 3, 8\}$; range: $\{-3, -2, -1, 1, 2\}$; not a function **7.** Yes
9. Yes **11.** No **13.** Yes **15.** No **17.** Yes **19. a.** 0 **b.** $\dfrac{2\sqrt{3}}{3}$
c. Not defined **d.** Not defined **e.** $\dfrac{-2x}{\sqrt{4-x^2}}$ **21.** 5
23. $(-\infty, 9) \cup (9, \infty)$ **25.** $(-\infty, -1) \cup (-1, 1) \cup (1, \infty)$ **27.** $[3, \infty)$
29. $(-\infty, -2) \cup (-2, -1) \cup (-1, \infty)$ **31.** $[3, 11]$ **33.** $[-7, 2]$
35. a. yes **b.** No **c.** $[5, 11]$ **37.** Yes **39.** Yes **41.** No
43. $x = -2$ or 3 **45. a.** No **b.** $-1 \pm \sqrt{3}$ **c.** 5 **d.** $-1 \pm \dfrac{\sqrt{14}}{2}$
47. $[-9, \infty)$ **49.** $-3, 4, 7, 9$ **51.** $f(-7) = 4, f(1) = 5, f(5) = 2$
53. $\{-3.75, -2.25, 3\} \cup [12, \infty)$ **55.** $[-9, \infty)$
57. $g(-4) = -1, g(1) = 3, g(3) = 4$ **59.** $[-9, -5)$

Applying the Concepts:

61. $A(x) = x^2$; 16; area of the square **63. a.** $C(x) = 5.5x + 75{,}000$ **b.** $R(x) = 9x$ **c.** $P(x) = 3.5x - 75{,}000$ **d.** 21,429 **e.** \$86,000
65. After 4 hours: 12 ml; after 8 hours: 21 ml; after 12 hours: 27.75 ml; after 16 hours: 32.81 ml; after 20 hours: 36.61 ml

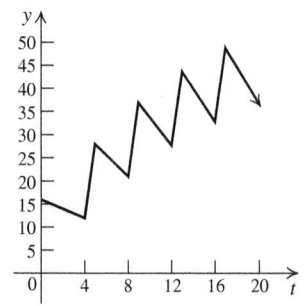

67. $P = 28x - x^2$ **69.** $S = 2x^2 + \dfrac{256}{x}$ **71.** $A = \dfrac{x^2}{4\pi} + \dfrac{1}{16}(20-x)^2$
73. $C = 2x^2 + \dfrac{6912}{x}$ **75.** $d = [(x-2)^2 + (x^3 - 3x + 5)^2]^{1/2}$

Critical Thinking / Discussion / Writing:

78. a. $y = \sqrt{x-2}$ **b.** $y = \dfrac{1}{\sqrt{x-2}}$ **c.** $y = \sqrt{2-x}$ **d.** $y = \dfrac{1}{\sqrt{2-x}}$
79. a. $ax^2 + bx + c = 0$ **b.** $y = c$ **c.** $b^2 - 4ac < 0$ **d.** Not possible
80. a. 9 **b.** 8 **81.** n^m

Maintaining Skills:

82. $(3x+2)(x+4)$ **83.** $(x-2)(x-4)$ **84.** $x(x+3)(x+4)$
85. $(5x+4)(5x-4)$ **86.** $(x-1)(x+1)(x+5)$
87. $(3x-4)(2x+3)$

Section 1.4

Basic Concepts and Skills:

1. $f(x) = x + 1$ **3.** $f(x) = 2x + 3$

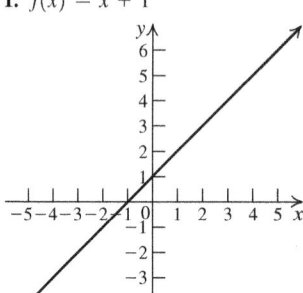

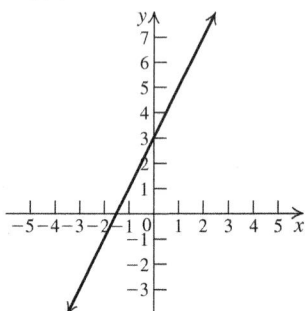

5. $f(x) = -3x + 4$ 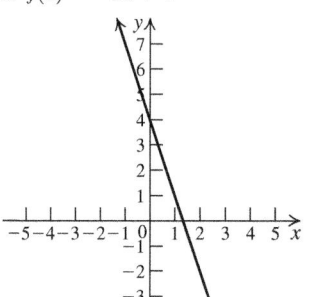 **7.** $f(x) = \dfrac{1}{2}x + 3$

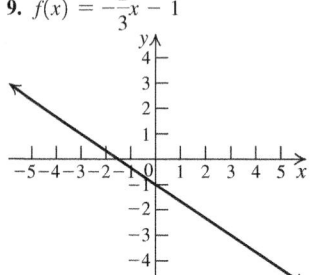

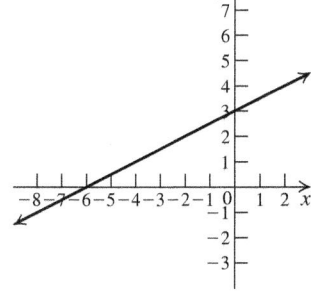

9. $f(x) = -\dfrac{2}{3}x - 1$ **11.** Constant on $(-\infty, \infty)$

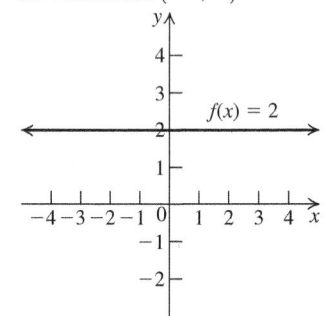

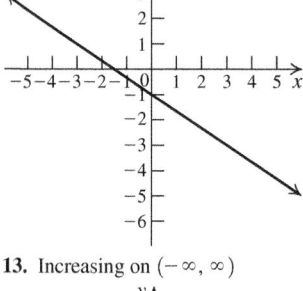

13. Increasing on 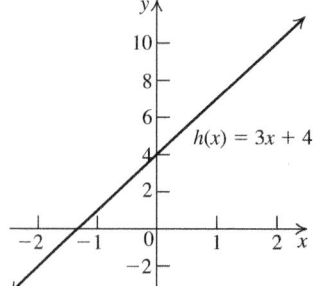 $(-\infty, \infty)$ **15.** Decreasing on $(-\infty, 0)$; increasing on $(0, \infty)$

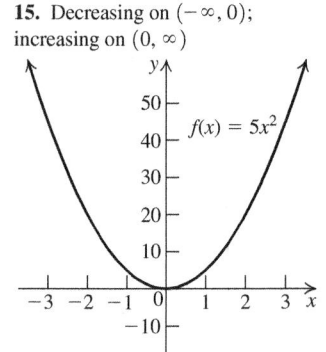

17. Decreasing on $(-\infty, 0)$; increasing on $(0, \infty)$

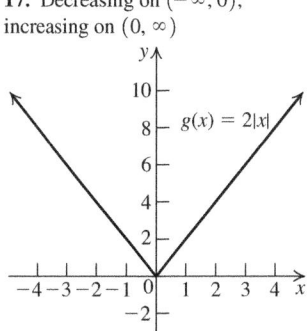

19. Decreasing on $(-\infty, \infty)$

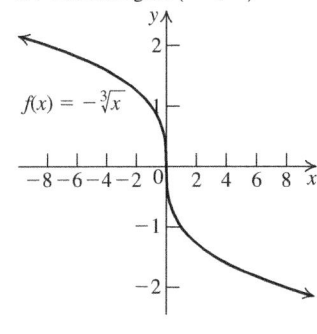

21. Even **23.** Neither **25.** Even **27.** Odd **29.** Even **31.** Odd
33. Neither **35. a.** Domain: $[-2, 6]$; range: $[-3, 2]$ **b.** x-intercept: $\frac{2}{5}$; y-intercept: $-\frac{1}{2}$ **c.** Increasing on $(-2, 2)$, constant on $(2, 6)$ **d.** Neither even nor odd **37. a.** Domain: $[-2, 4]$; range: $[-1, 2]$ **b.** x-intercepts: $-2, 0$; y-intercept: 0 **c.** Decreasing on $(-2, -1)$ and $(2, 4)$, increasing on $(-1, 2)$ **d.** Neither even nor odd **39. a.** Domain: $(0, \infty)$; range: $(0, \infty)$ **b.** No x- or y-intercept **c.** Decreasing on $(0, \infty)$ **d.** Neither even nor odd
41. a. Domain: $(-\infty, \infty)$; range: $(0, \infty)$ **b.** No x-intercept; y-intercept: 1 **c.** Increasing on $(-\infty, \infty)$ **d.** Neither even nor odd
43. a. $f(1) = 2, f(2) = 2, f(3) = 3$
b.

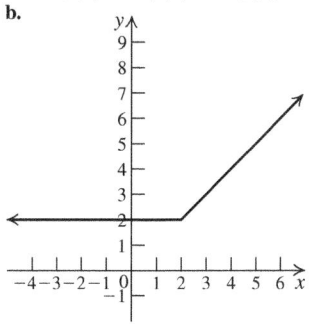

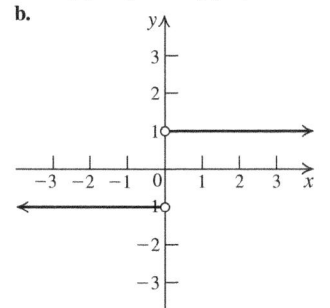

45. a. $f(-15) = -1, f(12) = 1$
b.
c. Domain: $(-\infty, 0) \cup (0, \infty)$; range: $\{-1, 1\}$

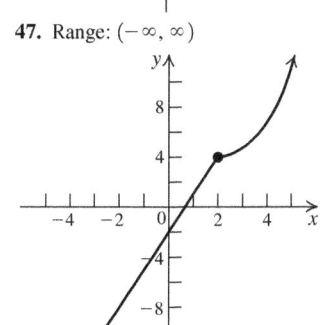

47. Range: $(-\infty, \infty)$

49. Range: $(-\infty, 0] \cup [2, \infty)$

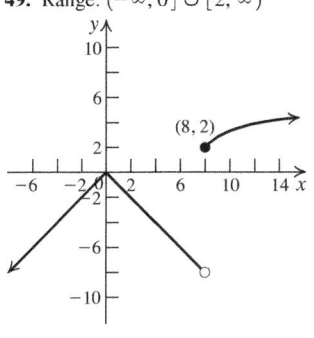

51. Range: $(-\infty, 0] \cup \{1, 2\}$ **53.** -2 **55.** 10 **57.** $3x + 7$

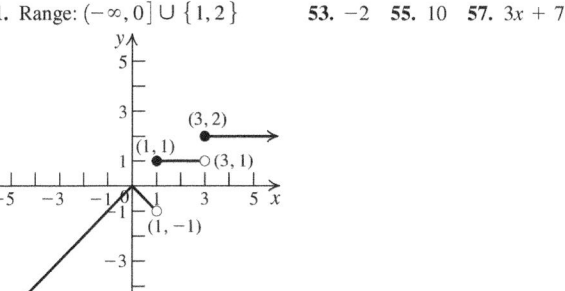

59. $-\dfrac{4}{x}$ **61.** $4x + 2h + 3$ **63.** 0

Applying the Concepts:
65. $[0, \infty)$; the particle's motion is tracked indefinitely from time $t = 0$.
67. The graph is above the t-axis on the intervals $(0, 9)$ and $(21, 24)$; the particle is moving forward between 0 and 9 seconds and again between 21 and 24 seconds. **69.** $|v|$ is increasing on $(0, 3), (5, 6), (11, 15)$, and $(21, 23)$; the particle is moving forward on $(0, 9)$ and $(21, 23)$, and moving backwards on; $(11, 19)$. **71.** Maximum speed is attained between times $t = 15$ and $t = 16$. **73.** The particle is moving forward with increasing velocity.
75. a.

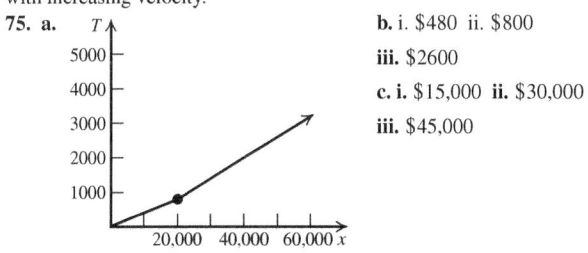

b. i. $480 **ii.** $800
iii. $2600
c. i. $15,000 **ii.** $30,000
iii. $45,000

Critical Thinking / Discussion / Writing:
77. a. $f(x) = |x|$ **b.** $f(x) = 0$ **c.** $f(x) = x$ **d.** $f(x) = \sqrt{-x^2}$
e. $f(x) = 1$ **f.** Not possible **78.** $f(x) = 0$ **79.** c is an endpoint maximum of f if c is an endpoint of an interval in the domain of f and there is an interval $(x_1, c]$ or $[c, x_2)$ in the domain of f such that $f(c) \geq f(x)$ for every x in the interval. **80.** c is an endpoint minimum of f if c is an endpoint of an interval in the domain of f and there is an interval $(x_1, c]$ or $[c, x_2)$ in the domain of f such that $f(c) \leq f(x)$ for every x in the interval. **81. a.** $f(x) = x$ on the interval $[-1, 1]$
b. $f(x) = x$ on the interval $[-1, \infty)$ **c.** $f(x) = x$ on the interval $(-\infty, 1]$ **d.** $f(x) = x$

Maintaining Skills:
82. $5\sqrt{2}$ **83.** $5\sqrt[3]{2}$ **84.** $1 \pm \sqrt{2}$ **85.** $\dfrac{2 \pm \sqrt{3}}{3}$
86. $-x^2$ **87.** x^2

Section 1.5

Basic Concepts and Skills:
1. a. Shift two units up **b.** Shift one unit down **3. a.** Shift one unit left
b. Shift two units right **5. a.** Shift one unit left and two units down
b. Shift one unit right and three units up **7. a.** Reflect about x-axis
b. Reflect about y-axis **9. a.** Stretch vertically by a factor of 2
b. Compress horizontally by a factor of 2 **11. a.** Reflect about x-axis, shift one unit up **b.** Reflect about y-axis, shift one unit up
13. a. Shift one unit up **b.** Shift one unit left

A-6 Answers ■ Section 1.5

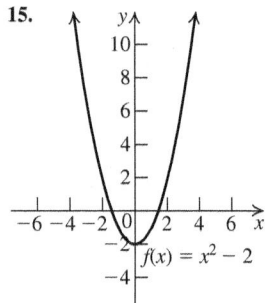

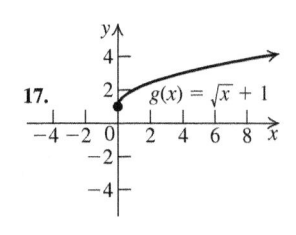

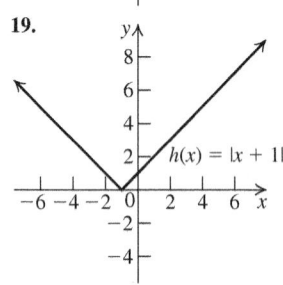

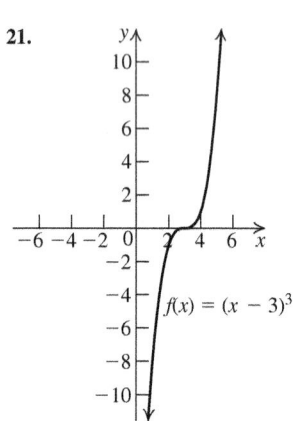

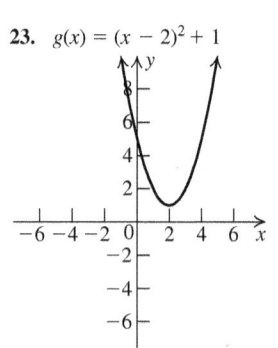

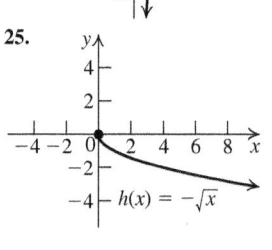

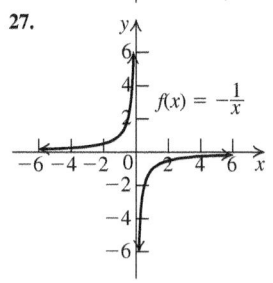

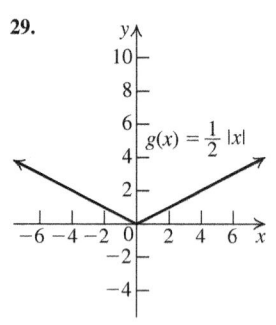

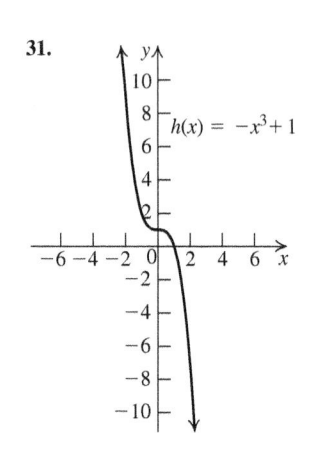

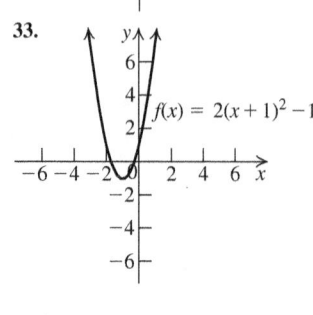

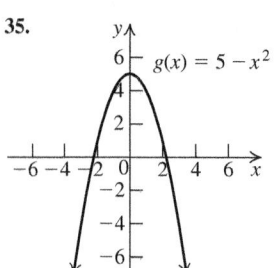

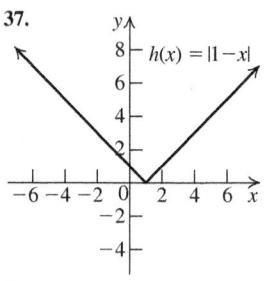

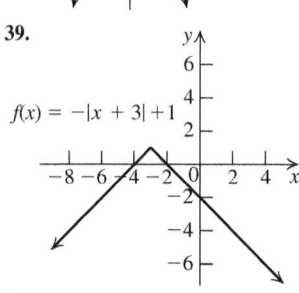

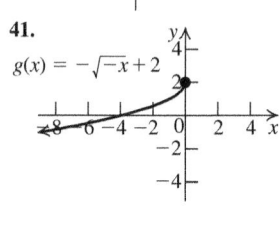

15. [graph of $f(x) = x^2 - 2$]

17. [graph of $g(x) = \sqrt{x} + 1$]

19. [graph of $h(x) = |x + 1|$]

21. [graph of $f(x) = (x - 3)^3$]

23. $g(x) = (x - 2)^2 + 1$ [graph]

25. [graph of $h(x) = -\sqrt{x}$]

27. [graph of $f(x) = -\dfrac{1}{x}$]

29. [graph of $g(x) = \dfrac{1}{2}|x|$]

31. [graph of $h(x) = -x^3 + 1$]

33. [graph of $f(x) = 2(x + 1)^2 - 1$]

35. [graph of $g(x) = 5 - x^2$]

37. [graph of $h(x) = |1 - x|$]

39. [graph of $f(x) = -|x + 3| + 1$]

41. [graph of $g(x) = -\sqrt{-x} + 2$]

43. [graph of $h(x) = 2[[x + 1]]$]

45. $y = x^3 + 2$ **47.** $y = -|x|$

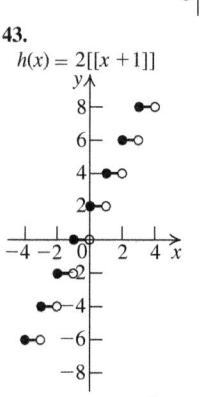

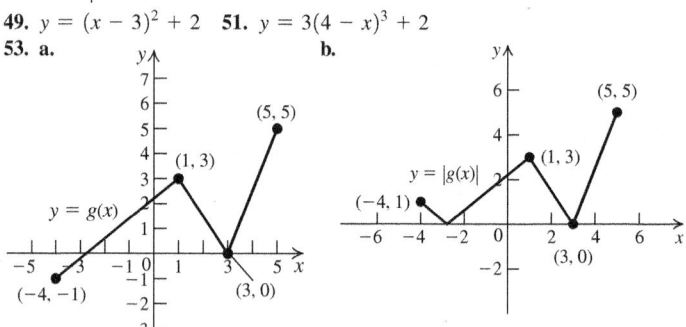

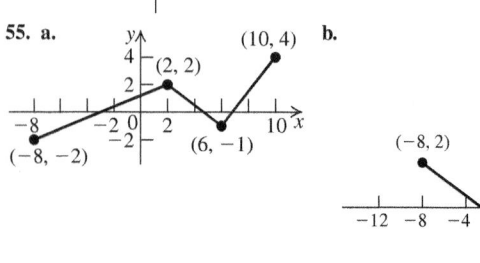

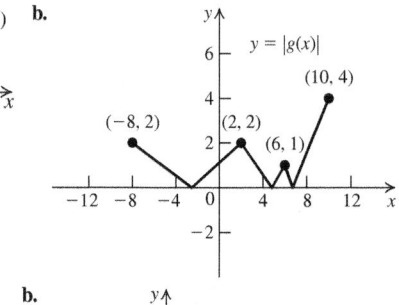

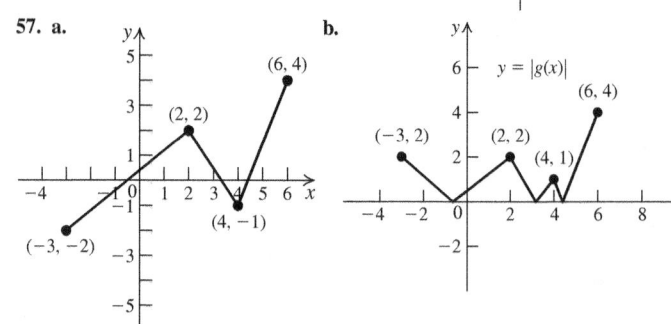

49. $y = (x - 3)^2 + 2$ **51.** $y = 3(4 - x)^3 + 2$

53. a. [graph of $y = g(x)$ with points $(-4, -1)$, $(1, 3)$, $(3, 0)$, $(5, 5)$]   **b.** [graph of $y = |g(x)|$ with points $(-4, 1)$, $(1, 3)$, $(3, 0)$, $(5, 5)$]

55. a. [graph with points $(-8, -2)$, $(2, 2)$, $(6, -1)$, $(10, 4)$]   **b.** [graph of $y = |g(x)|$ with points $(-8, 2)$, $(2, 2)$, $(6, 1)$, $(10, 4)$]

57. a. [graph with points $(-3, -2)$, $(2, 2)$, $(4, -1)$, $(6, 4)$]   **b.** [graph of $y = |g(x)|$ with points $(-3, 2)$, $(2, 2)$, $(4, 1)$, $(6, 4)$]

59. a.

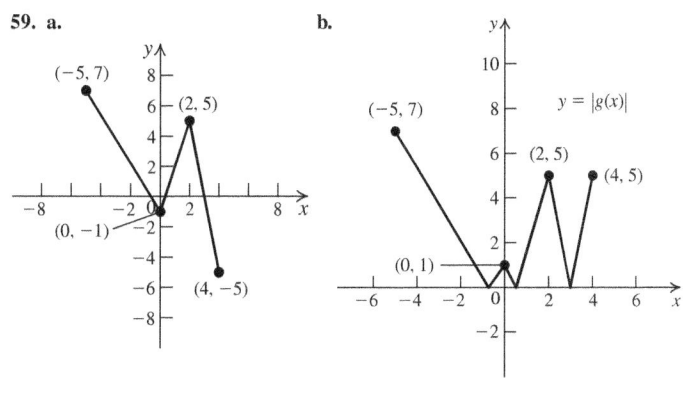

b.

Applying the Concepts:
61. $g(x) = f(x) + 800$ **63.** $p(x) = 1.02(f()x + 500)$
65. The first coordinate gives the month. The second coordinate gives the hours of daylight.

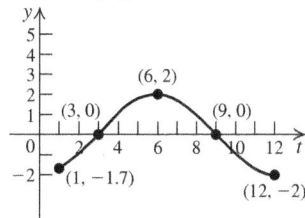

Critical Thinking / Discussion / Writing:
67. a. x-intercept 1 **b.** x-intercept −2; y-intercept 15 **c.** x-intercept −2; y-intercept −3 **d.** x-intercept 2; y-intercept 3 **68. a.** x-intercept 2 **b.** x-intercept 4; y-intercept −2 **c.** x-intercept 4; y-intercept 1 **d.** x-intercept −4; y-intercept −1 **69. a.** The graph of g is the graph of h shifted three units to the right and three units up. **b.** The graph of g is the graph of h shifted one unit to the left and one unit down. **c.** The graph of g is the graph of h stretched horizontally and vertically by a factor of 2. **d.** The graph of g is the graph of h stretched horizontally by a factor of 3, reflected about the y-axis, stretched vertically by a factor of 3, and reflected about the x-axis. **70.** Stretch the graph of $y = f(-4x)$ horizontally by a factor of 4 and reflect the resulting graph about the y-axis.

Maintaining Skills:
71. $\dfrac{3 - \sqrt{5}}{4}$ **72.** $\dfrac{\sqrt{5} + \sqrt{3}}{2}$ **73.** $-(5 + 2\sqrt{6})$
74. $\dfrac{2x - 2\sqrt{x(x+h)} + h}{h}$ **75.** $\dfrac{1}{\sqrt{x+h} + \sqrt{z}}$
76. $\dfrac{x}{2(\sqrt{x^2 + x + 1} + \sqrt{x^2 + 1})}$

Section 1.6

Basic Concepts and Skills:
1. a. $(f + g)(-1) = -1$ **b.** $(f - g)(0) = 0$ **c.** $(f \cdot g)(2) = -8$
d. $\left(\dfrac{f}{g}\right)(1) = -2$ **3. a.** $(f + g)(-1) = 0$
b. $(f - g)(0) = \dfrac{1}{2}(\sqrt{2} - 2)$ **c.** $(f \cdot g)(2) = \dfrac{5}{2}$ **d.** $\left(\dfrac{f}{g}\right)(1) = \dfrac{1}{3\sqrt{3}}$
5. a. $(f + g)(x) = x^2 + x - 3$; domain: $(-\infty, \infty)$
b. $(f - g)(x) = -x^2 + x - 3$; domain: $(-\infty, \infty)$
c. $(f \cdot g)(x) = x^3 - 3x^2$; domain: $(-\infty, \infty)$
d. $\left(\dfrac{f}{g}\right)(x) = \dfrac{x - 3}{x^2}$; domain: $(-\infty, 0) \cup (0, \infty)$
e. $\left(\dfrac{g}{f}\right)(x) = \dfrac{x^2}{x - 3}$; domain: $(-\infty, 3) \cup (3, \infty)$

7. a. $(f + g)(x) = x^3 + 2x^2 + 4$; domain: $(-\infty, \infty)$
b. $(f - g)(x) = x^3 - 2x^2 - 6$; domain: $(-\infty, \infty)$
c. $(f \cdot g)(x) = 2x^5 + 5x^3 - 2x^2 - 5$; domain: $(-\infty, \infty)$
d. $\left(\dfrac{f}{g}\right)(x) = \dfrac{x^3 - 1}{2x^2 + 5}$; domain: $(-\infty, \infty)$
e. $\left(\dfrac{g}{f}\right)(x) = \dfrac{2x^2 + 5}{x^3 - 1}$; domain: $(-\infty, 1) \cup (1, \infty)$
9. a. $(f + g)(x) = 2x + \sqrt{x} - 1$; domain: $[0, \infty)$
b. $(f - g)(x) = 2x - \sqrt{x} - 1$; domain: $[0, \infty)$
c. $(f \cdot g)(x) = 2x\sqrt{x} - \sqrt{x}$; domain: $[0, \infty)$
d. $\left(\dfrac{f}{g}\right)(x) = \dfrac{2x - 1}{\sqrt{x}}$; domain: $(0, \infty)$
e. $\left(\dfrac{g}{f}\right)(x) = \dfrac{\sqrt{x}}{2x - 1}$; domain: $\left[0, \dfrac{1}{2}\right) \cup \left(\dfrac{1}{2}, \infty\right)$
11. a. $(f + g)(x) = \dfrac{2 + x}{x + 1}$; domain: $(-\infty, -1) \cup (-1, \infty)$
b. $(f - g)(x) = \dfrac{2 - x}{x + 1}$; domain: $(-\infty, -1) \cup (-1, \infty)$
c. $(f \cdot g)(x) = \dfrac{2x}{(x + 1)^2}$; domain: $(-\infty, -1) \cup (-1, \infty)$
d. $\left(\dfrac{f}{g}\right)(x) = \dfrac{2}{x}$; domain: $(-\infty, -1) \cup (-1, 0) \cup (0, \infty)$
e. $\left(\dfrac{g}{f}\right)(x) = \dfrac{x}{2}$; domain: $(-\infty, -1) \cup (-1, \infty)$
13. $g(f(x)) = 2x^2 + 1$; $g(f(2)) = 9$; $g(f(-3)) = 19$ **15.** 11 **17.** 31
19. −5 **21.** $4c^2 - 5$ **23.** $8a^2 + 8a - 1$ **25.** 7
27. $\dfrac{2x}{x + 1}$; $x \neq 0, x \neq -1$ **29.** $\sqrt{-3x - 1}$; $\left(-\infty, -\dfrac{1}{3}\right]$
31. $(f \circ g)(x) = \sqrt{x^2 - 1}$; domain: $(-\infty, -1] \cup [1, \infty)$
33. a. $(f \circ g)(x) = 8x^2 - 2x - 1$; domain: $(-\infty, \infty)$
b. $(g \circ f)(x) = 4x^2 + 6x - 1$; domain: $(-\infty, \infty)$
c. $(f \circ f)(x) = 8x^4 + 24x^3 + 24x^2 + 9x$; domain: $(-\infty, \infty)$
d. $(g \circ g)(x) = 4x - 3$; domain: $(-\infty, \infty)$
35. a. $(f \circ g)(x) = x$; domain: $[0, \infty)$
b. $(g \circ f)(x) = |x|$; domain: $(-\infty, \infty)$
c. $(f \circ f)(x) = x^4$; domain: $(-\infty, \infty)$
d. $(g \circ g)(x) = \sqrt[4]{x}$; domain: $[0, \infty)$
37. a. $(f \circ g)(x) = \sqrt{\sqrt{4 - x} - 1}$; domain: $(-\infty, 3]$
b. $(g \circ f)(x) = \sqrt{4 - \sqrt{x - 1}}$; domain: $[1, 17]$
c. $(f \circ f)(x) = \sqrt{\sqrt{x - 1} - 1}$; domain: $[2, \infty)$
d. $(g \circ g)(x) = \sqrt{4 - \sqrt{4 - x}}$; domain: $[-12, 4]$
39. a. $(f \circ g)(x) = \dfrac{3}{|2x + 3|}$; domain: $\left(-\infty, -\dfrac{3}{2}\right) \cup \left(-\dfrac{3}{2}, \infty\right)$
b. $(g \circ f)(x) = \dfrac{6}{|x|} + 3$; domain: $(-\infty, 0) \cup (0, \infty)$
c. $(f \circ g)(x) = |x|$; domain: $(-\infty, 0) \cup (0, \infty)$
d. $(g \circ g)(x) = 4x + 9$; domain: $(-\infty, \infty)$
41. $f(x) = \sqrt{x}$; $g(x) = x + 2$ **43.** $f(x) = x^{10}$; $g(x) = x^2 - 3$
45. $f(x) = \dfrac{1}{x}$; $g(x) = 3x - 5$ **47.** $f(x) = \sqrt[3]{x}$; $g(x) = x^2 - 7$
49. $f(x) = \dfrac{1}{|x|}$; $g(x) = x^3 - 1$

Applying the Concepts:
51. a. Cost function **b.** Revenue function **c.** Selling price of x shirts after sales tax **d.** Profit function **53. a.** $f(x) = 1.1x$; $g(x) = x + 8$
b. $(f \circ g)(x) = 1.1x + 8.8$; a final test score, which originally was x, after adding 8 points and then increasing the score by 10%
c. $(g \circ f)(x) = 1.1x + 8$; a final test score, which originally was x, after increasing the score by 10% and then adding 8 points **d.** 85.8 and 85.0
e. No **f. (i)** 73.82 **(ii)** 74.55
55. a. $f(x) = \pi x^2$ **b.** $g(x) = \pi(x + 30)^2$ **c.** The area between the fountain and the fence **d.** $16,052 **57. a.** $(f \circ g)(t) = \pi(2t + 1)^2$
b. $A(t) = \pi(2t + 1)^2$ **c.** They are the same.

Beyond the Basics:
59. a. $f + g = \{(-2, 3), (1, 0), (3, 2)\}$
b. $fg = \{(-2, 0), (1, -4), (3, 0)\}$
c. $\dfrac{f}{g} = \{(1, -1), (3, 0)\}$ **d.** $f \circ g = \{(0, 1), (1, 3), (3, 1)\}$

Critical Thinking / Discussion / Writing:
61. c. $h(x) = \dfrac{h(x) + h(-x)}{2} + \dfrac{h(x) - h(-x)}{2}$
62. a. $h(x) = o(x) + e(x)$, where $e(x) = x^2 + 3$ and $o(x) = -2x$
b. $h(x) = o(x) + e(x)$, where $e(x) = \dfrac{[\![x]\!] + [\![-x]\!]}{2}$ and
$o(x) = x + \dfrac{[\![x]\!] - [\![-x]\!]}{2}$
63. a. Domain: $(-\infty, 0) \cup [1, \infty)$ **b.** Domain: $[0, 2]$ **c.** Domain: $[1, 2]$
d. Domain: $[1, 2)$ **64. a.** Domain: $\emptyset$ **b.** Domain: $(-\infty, 0)$
65. a. Even function **b.** Odd function **c.** Neither function **d.** Even function **e.** Even function **f.** Odd function **66. a.** Odd function **b.** Even function **c.** Even function **d.** Even function

Maintaining Skills:
67. $\{1\}$ **68.** $\{-6\}$ **69.** $\emptyset$ **70.** $\{-1 \pm \sqrt{6}\}$ **71.** $\left\{-\dfrac{3}{2}\right\}$ **72.** $\{4\}$

Section 1.7

Basic Concepts and Skills:
1. One-to-one **3.** Not one-to-one **5.** Not one-to-one **7.** One-to-one
9. 2 **11.** -1 **13.** a **15.** 337 **17.** -1580 **19. a.** 3 **b.** 3 **c.** 19 **d.** 5
21. a. 2 **b.** 1 **c.** 269
29. 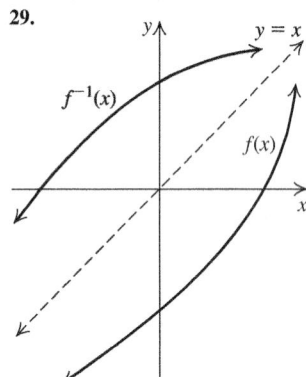
31. 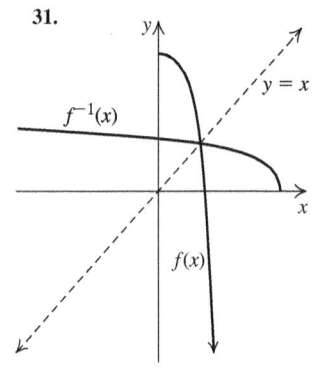
33. 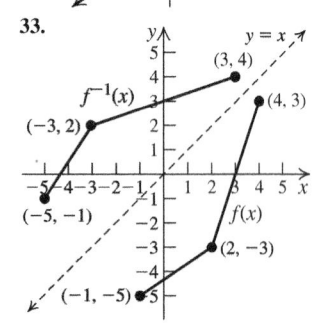
35. a. One-to-one **b.** $f^{-1}(x) = 5 - \dfrac{1}{3}x$
c. **d.** Domain f: $(-\infty, \infty)$
x-intercept: 5
y-intercept: 15
Domain f^{-1}: $(-\infty, \infty)$
x-intercept: 15
y-intercept: 5

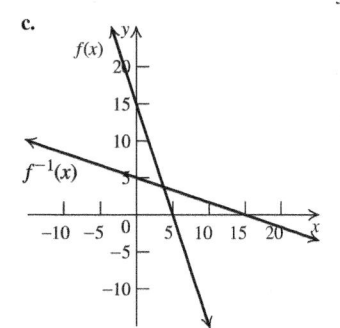

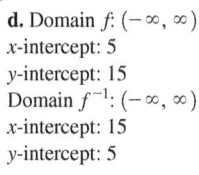

37. Not one-to-one
39. a. One-to-one **b.** $f^{-1}(x) = (x - 3)^2$
c. **d.** Domain f: $[0, \infty)$
No x-intercept
y-intercept: 3
Domain f^{-1}: $[3, \infty)$
x-intercept: 3
No y-intercept

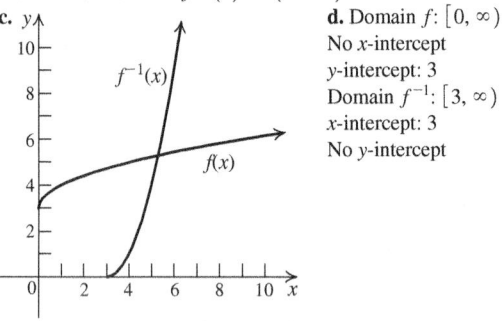

41. a. One-to-one **b.** $g^{-1}(x) = (x^3 - 1)$
c. **d.** Domain g: $(-\infty, \infty)$
x-intercept: -1
y-intercept: 1
Domain g^{-1}: $(-\infty, \infty)$
x-intercept: 1
y-intercept: -1

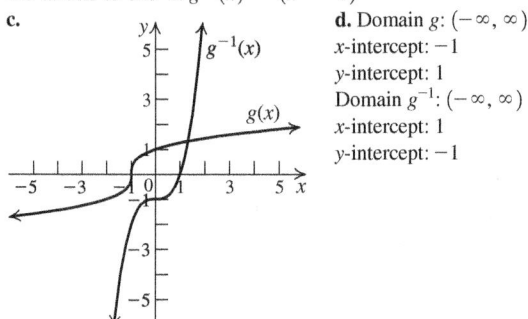

43. a. One-to-one **b.** $f^{-1}(x) = \dfrac{x + 1}{x}$
c. **d.** Domain f: $(-\infty, 1) \cup (1, \infty)$
No x-intercept
y-intercept: -1
Domain f^{-1}: $(-\infty, 0) \cup (0, \infty)$
x-intercept: -1
No y-intercept

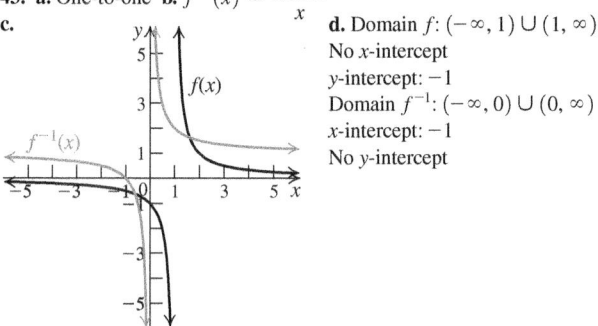

45. a. One-to-one **b.** $f^{-1}(x) = x^2 - 4x + 3$
c. **d.** Domain f: $[-1, \infty)$
y-intercept: 3
No x-intercept
Domain f^{-1}: $[2, \infty)$
x-intercept is 3
No y-intercept

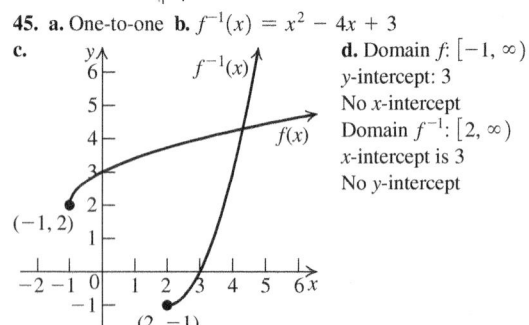

47. $f^{-1}(x) = \dfrac{2x + 1}{x - 1}$
Domain: $(-\infty, 2) \cup (2, \infty)$; range: $(-\infty, 1) \cup (1, \infty)$
49. $f^{-1}(x) = \dfrac{1 - x}{x + 2}$
Domain: $(-\infty, -1) \cup (-1, \infty)$; range: $(-\infty, -2) \cup (-2, \infty)$

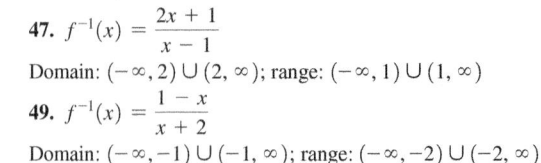

51. $y = \sqrt{-x}$, $x \leq 0$ **53.** $y = x$, $x \geq 0$

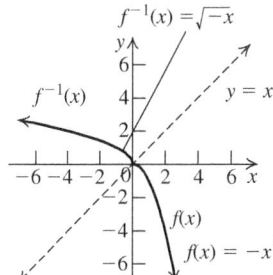

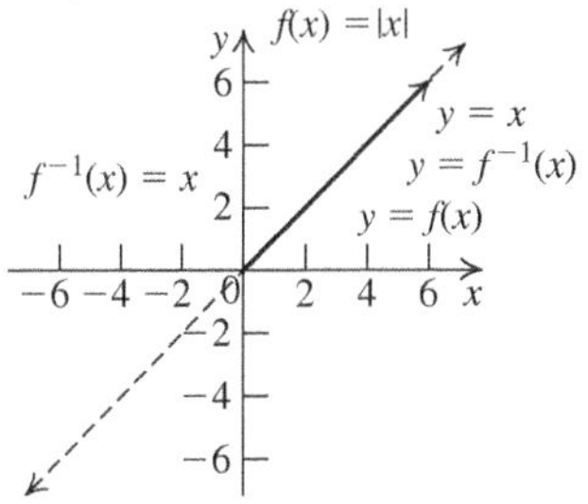

55. $y = -\sqrt{x-1}$, $x \geq 1$ 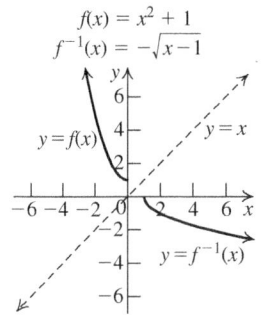 **57.** $y = -\sqrt{2-x}$, $x \leq 2$

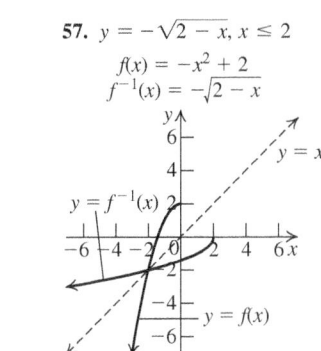

Applying the Concepts:
59. a. Dollars to euros: $f(x) = 0.75x$ (x is the number of dollars; $f(x)$ is the number of euros). Euros to dollars: $g(x) = 1.25x$ (x is the number of euros; $g(x)$ is the number of dollars) **b.** $g(f(x)) = 0.9375x \neq x$; therefore, g and f are not inverse functions. **c.** She loses money
61. a. $w = \begin{cases} 4 + 0.05x, & \text{if } x > 60 \\ 7, & \text{if } x \leq 60 \end{cases}$
b. It does not have an inverse because it is constant on $(0, 60)$ and hence is not one-to-one. **c.** Restriction of the domain $[60, \infty)$
63. a. $x = \dfrac{1}{64}V^2$. It gives the height of the water in terms of the velocity.
b. (i) 14.0625 ft **(ii)** 6.25 ft

Critical Thinking / Discussion / Writing:
65. No. $f(x) = x^3 - x$ is odd, but it does not have an inverse because $f(0) = f(1)$. So it is not one-to-one. **66.** Yes. The function $f = \{(0, 1)\}$ is even, and it has an inverse: $f^{-1} = \{(1, 0)\}$.
67. Yes because increasing and decreasing functions are one-to-one
68. a. $R = \{(-1, 1), (0, 0), (1, 1)\}$ **b.** $R = \{(-1, 1), (0, 0), (1, 2)\}$

Maintaining Skills:
69. $(-\infty, \infty)$; $(-\infty, \infty)$ **70.** $(-\infty, \infty)$; $[-5, \infty)$ **71.** $(-\infty, 1]$; $[0, \infty)$
72. $[-2, 2]$; $[0, 2]$ **73.** $(-\infty, -2) \cup (-2, \infty)$; $(-\infty, 1) \cup (1, \infty)$
74. $(-\infty, -1] \cup (2, \infty)$; $[0, 1) \cup (1, \infty)$

Review Exercises

Basic Concepts and Skills:
1. False **3.** True **5.** False **7.** True **9. a.** $2\sqrt{5}$ **b.** $(1, 4)$ **c.** $\dfrac{1}{2}$
11. a. $5\sqrt{2}$ **b.** $\left(\dfrac{13}{2}, -\dfrac{11}{2}\right)$ **c.** -1 **13. a.** $\sqrt{34}$ **b.** $\left(\dfrac{7}{2}, -\dfrac{9}{2}\right)$ **c.** $\dfrac{5}{3}$
17. $(4, 5)$ **19.** $\left(\dfrac{31}{18}, 0\right)$ **21.** Not symmetric with respect to the x-axis, symmetric with respect to the y-axis, not symmetric with respect to the origin
23. Symmetric with respect to the x-axis, not symmetric with respect to the y-axis, not symmetric with respect to the origin

25. x-intercept: 4; y-intercept: 2; not symmetric with respect to the x-axis, not symmetric with respect to the y-axis, not symmetric with respect to the origin

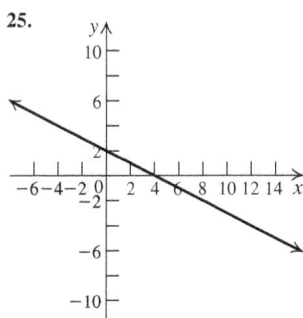

27. x-intercept: 0; y-intercept: 0; not symmetric with respect to the x-axis, not symmetric with respect to the y-axis, not symmetric with respect to the origin

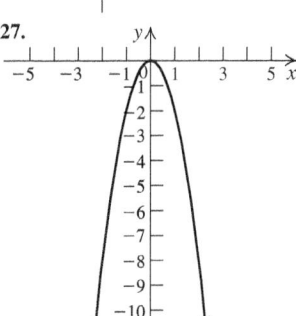

29. x-intercept: 0; y-intercept: 0; not symmetric with respect to the x-axis, not symmetric with respect to the y-axis, symmetric with respect to the origin

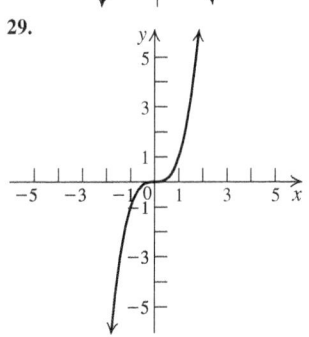

31. No x-intercept; y-intercept: 2; not symmetric with respect to the x-axis, symmetric with respect to the y-axis, not symmetric with respect to the origin

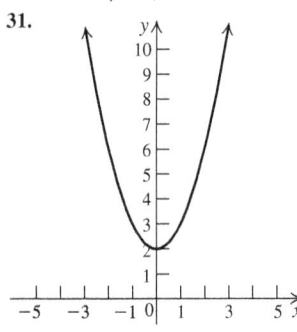

33. x-intercepts: -4 and 4; y-intercepts: -4 and 4; symmetric with respect to the x-axis, symmetric with respect to the y-axis, symmetric with respect to the origin

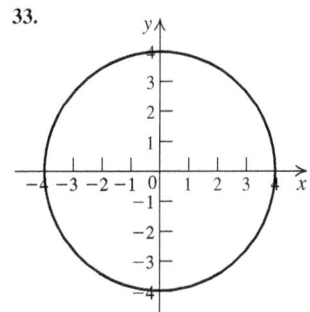

35. $(x-2)^2 + (y+3)^2 = 25$ **37.** $(x+2)^2 + (y+5)^2 = 4$

39. $\dfrac{x}{2} - \dfrac{y}{5} = 1 \Rightarrow y = \dfrac{5}{2}x - 5$;

line with slope $\dfrac{5}{2}$; y-intercept: -5,

x-intercept: 2

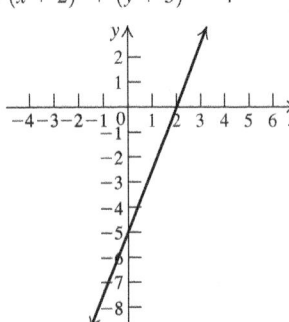

41. $x^2 + y^2 - 2x + 4y - 4 = 0 \Rightarrow (x-1)^2 + (y+2)^2 = 9$; circle centered at $(1, -2)$ and with radius 3

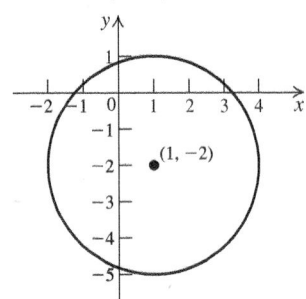

x-intercepts: $1 \pm \sqrt{5}$
y-intercepts: $-2 \pm 2\sqrt{2}$

43. $y = -2x + 4$ **45.** $y = -2x + 5$ **47.** $y = -\dfrac{4}{3}x + \dfrac{13}{3}$

49. Not a function **51.**

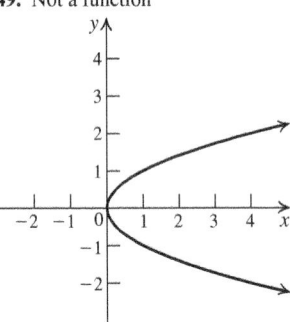

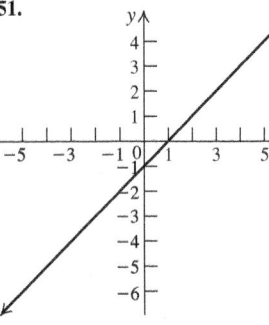

Domain: $(-\infty, \infty)$; range: $(-\infty, \infty)$; function

53. Not a function

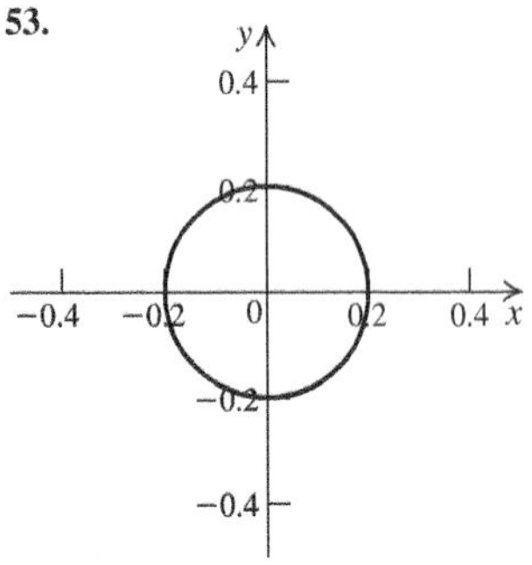

55. Domain: $\{1\}$; range: $(-\infty, \infty)$; not a function

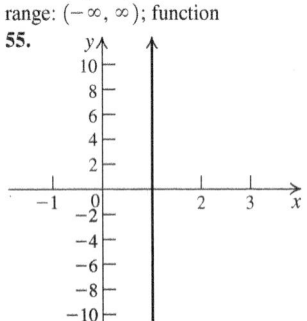

57. $y = |x+1|$

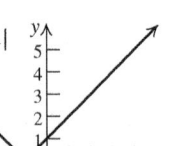

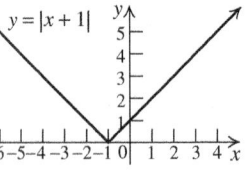

Domain: $(-\infty, \infty)$; range: $[0, \infty)$; function

59. -5 **61.** $x = 1$ **63.** 3
65. -10 **67.** 22 **69.** $3x^2 - 5$ **71.** $9x + 4$ **73.** 3
77.

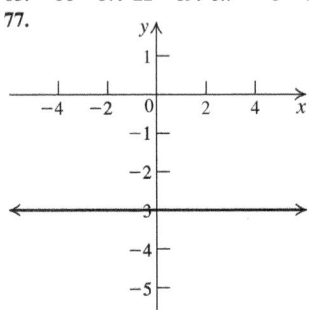

Domain: $(-\infty, \infty)$; range: $\{-3\}$; constant on $(-\infty, \infty)$

79.

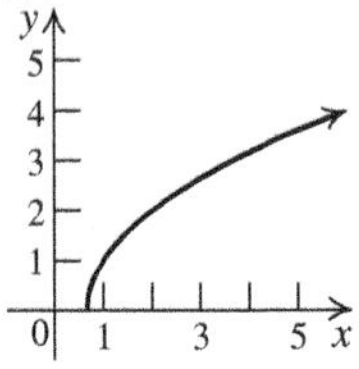

Domain: $\left[\dfrac{2}{3}, \infty\right)$; range: $[0, \infty)$;

increasing on $\left(\dfrac{2}{3}, \infty\right)$

81.

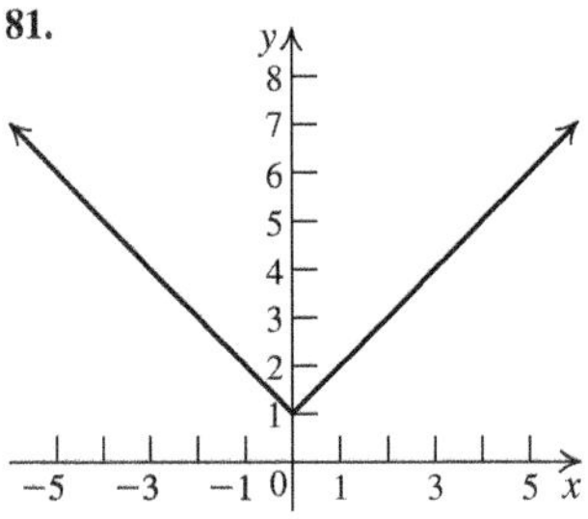

Domain: $(-\infty, \infty)$; range: $[1, \infty)$; decreasing on $(-\infty, 0)$, increasing on $(0, \infty)$

83. The graph of g is the graph of f shifted one unit to the left.

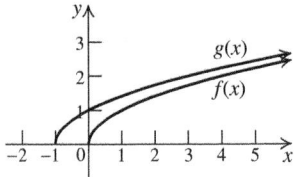

85. The graph of g is the graph of f shifted two units to the right and reflected in the x-axis.

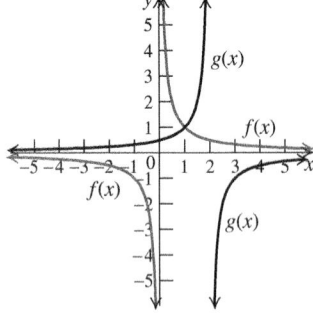

87. Even; not symmetric with respect to the x-axis, symmetric with respect to the y-axis, not symmetric with respect to the origin

89. Even; not symmetric with respect to the x-axis, symmetric with respect to the y-axis, not symmetric with respect to the origin
91. Neither even nor odd; not symmetric with respect to the x-axis, not symmetric with respect to the y-axis, not symmetric with respect to the origin **93.** $f(x) = (g \circ h)(x)$, where $g(x) = \sqrt{x}$ and $h(x) = x^2 - 4$
95. $h(x) = (f \circ g)(x)$, where $f(x) = \sqrt{x}$ and $g(x) = \dfrac{x-3}{2x+5}$

97. One-to-one; $f^{-1}(x) = x - 2$

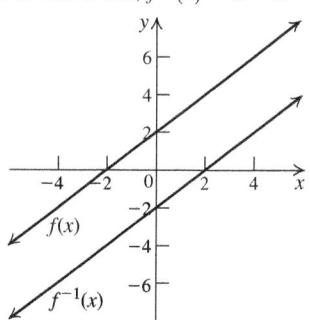

99. One-to-one; $f^{-1}(x) = x^3 + 2$

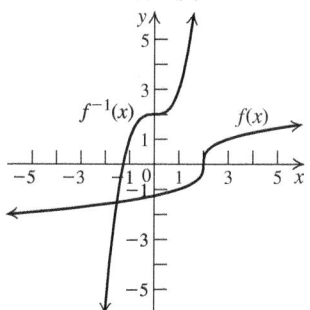

101. $f^{-1}(x) = x^2 - 8x + 17; x \geq 4$
Domain f: $[1, \infty)$; Range f: $[1, \infty)$

103. a. $f(x) = \begin{cases} 3x + 6, & \text{if } -3 \leq x \leq -2 \\ \frac{1}{2}x + 1, & \text{if } -2 < x < 0 \\ x + 1, & \text{if } 0 \leq x \leq 3 \end{cases}$

b. Domain: $[-3, 3]$; range: $[-3, 4]$ **c.** x-intercept: -2; y-intercept: 1

d. graph with points $(-3, 4)$, $(0, 1)$, $(2, 0)$, $(3, -3)$

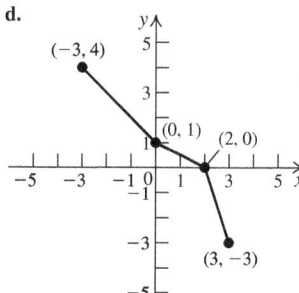

e. graph with points $(-3, 3)$, $(-2, 0)$, $(0, -1)$, $(3, -4)$

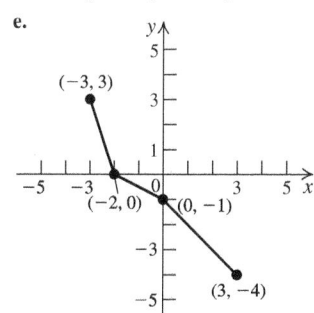

f. graph with points $(-3, -2)$, $(-2, 1)$, $(0, 2)$, $(3, 5)$

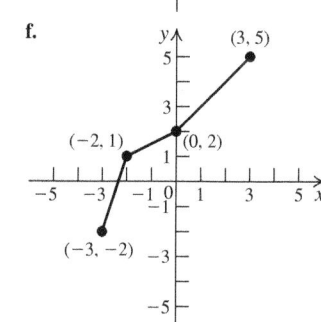

g. graph with points $(-4, -3)$, $(-3, 0)$, $(-1, 1)$, $(2, 4)$

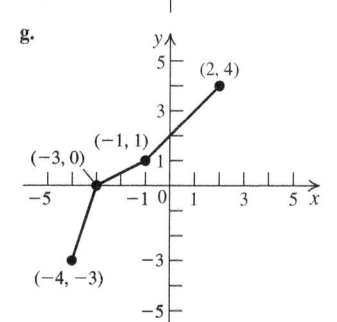

h. graph with points $(-3, -6)$, $(-2, 0)$, $(0, 2)$, $(3, 8)$

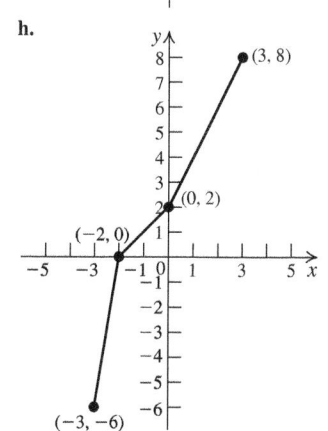

i. graph with points $(-1.5, -3)$, $(-1, 0)$, $(0, 1)$, $(1.5, 4)$

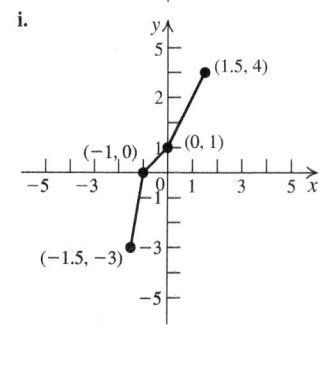

j. graph with points $(-6, -3)$, $(-4, 0)$, $(0, 1)$, $(6, 3)$

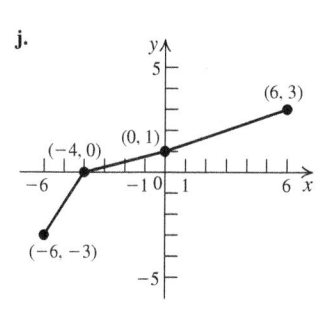

k. It satisfies the horizontal-line test.

l. graph with points $(-3, -3)$, $(0, -2)$, $(1, 0)$, $(4, 3)$

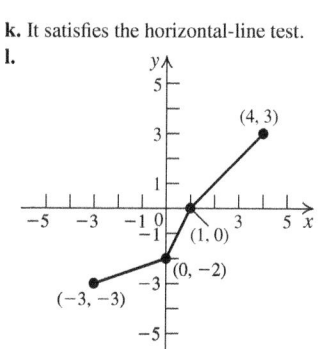

105. a. $C = 0.875w - 22{,}125$ **b.** Slope: the cost of disposing 1 pound of waste; x-intercept: the amount of waste that can be disposed with no cost; y-intercept: the fixed cost **c.** \$510,750 **d.** 1,168,142.86 lb
107. a. She won \$98. **b.** She was winning at a rate of \$49/h.
c. Yes. After 20 hours **d.** \$5/hr
109. a. $0.5\sqrt{0.000004t^4 + 0.004t^2 + 5}$ **b.** 1.13

Exercises for Calculus
1. $V = x(24 - 2x)(40 - 2x)$; domain: $(0, 12)$
2. For sales of x dollars
$I = \begin{cases} 25{,}000 + 0.06x & \text{if } 0 \leq x \leq 100{,}000 \\ 31{,}000 + 0.08(x - 100{,}000) & \text{if } x > 100{,}000 \end{cases}$

3. a. $\dfrac{1}{\sqrt{x+2}}$ **b.** $\dfrac{1}{\sqrt{x^2+8}+3}$ **c.** $\dfrac{1}{\sqrt{x+h}+\sqrt{x}}$ **d.** $\dfrac{x}{x+\sqrt{x}}$

4. 5 **5.** m **6.** $-\dfrac{1}{x(x+h)}$ **7. a.** $\dfrac{2}{3}x^{-4}$ **b.** $\dfrac{6}{5}x^{-1/2}$ **c.** $2x^{-3/4}$ **d.** $x^{1/12}$

8. $2x^{1/2} + x^{-2} - 5x^{-3}$
9. a. positive on $(-1, 0) \cup (2, \infty)$; negative on $(-\infty, -1) \cup (0, 2)$
b. > 0 on $(-\infty, -2) \cup (3, \infty)$; < 0 on $(-2, 1) \cup (1, 3)$
10. a. step graph   **b.** step graph

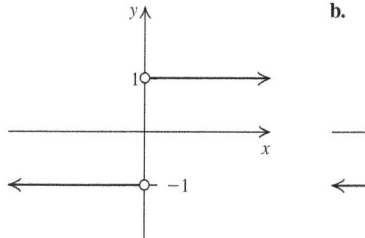

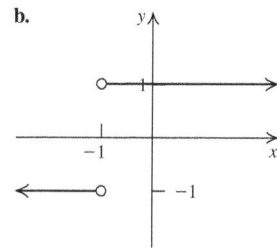

Practice Test A
1. Symmetric about the x-axis **2.** x-intercepts: $-1, 0, 3$; y-intercept: 0
3. circle centered at $C(1,1)$; x-intercepts: $1 \pm \sqrt{3}$; y-intercepts: $1 \pm \sqrt{3}$

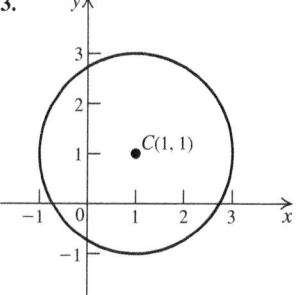

4. $y = -x + 9$ **5.** $y = 4x - 9$ **6.** -36 **7.** -1
8. $x^4 - 4x^3 + 2x^2 + 4x$ **9. a.** $f(-1) = -3$ **b.** $f(0) = -2$
c. $f(1) = -1$ **10.** $[0, 1)$ **11.** $(-\infty, -3] \cup [2, \infty)$ **12.** 2 **13.** Even
14. Increasing on $(-\infty, 0) \cup (2, \infty)$; decreasing on $(0, 2)$ **15.** Shift the graph of f three units to the right. **16.** 2.5 sec **17.** 2
18. $f^{-1}(x) = \dfrac{1}{x-1}$ **19.** $A(x) = 100x + 1000$
20. a. \$87.50 **b.** 110 mi

Practice Test B
1. d **2.** b **3.** d **4.** d **5.** c **6.** b **7.** d **8.** b **9.** a **10.** c **11.** a
12. b **13.** a **14.** a **15.** b **16.** d **17.** c **18.** c **19.** b **20.** a

CHAPTER 2

Section 2.1

Basic Concepts and Skills:
1. f **3.** a **5.** h **7.** g **9.** $y = -2x^2$ **11.** $y = 5x^2$
13. $y = 2x^2$ **15.** $y = -(x + 3)^2$ **17.** $y = 2(x - 2)^2 + 5$
19. $y = \dfrac{11}{49}(x - 2)^2 - 3$ **21.** $y = -12\left(x - \dfrac{1}{2}\right)^2 + \dfrac{1}{2}$
23. $y = \dfrac{3}{4}(x + 2)^2$ **25.** $y = \dfrac{3}{4}(x - 3)^2 - 1$
27. $y = (x + 2)^2 - 4$. The graph is the graph of $y = x^2$ shifted two units to the left and four units down. Vertex: $(-2, -4)$; axis: $x = -2$; x-intercepts: -4 and 0; y-intercept: 0

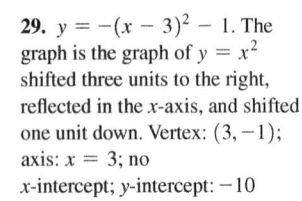

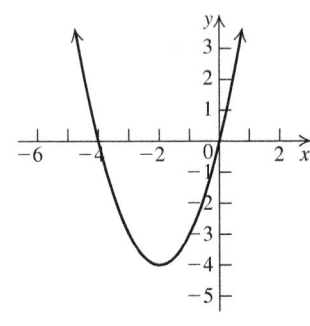

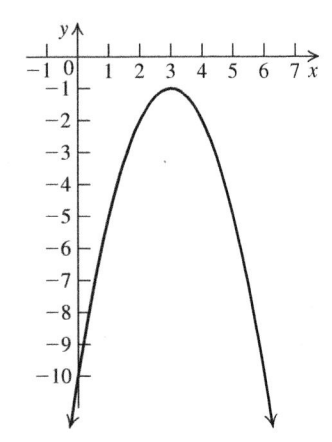

29. $y = -(x - 3)^2 - 1$. The graph is the graph of $y = x^2$ shifted three units to the right, reflected in the x-axis, and shifted one unit down. Vertex: $(3, -1)$; axis: $x = 3$; no x-intercept; y-intercept: -10

31. $y = 2(x - 2)^2 + 1$. The graph is the graph of $y = x^2$ shifted two units to the right, stretched vertically by a factor of 2, and shifted one unit up.

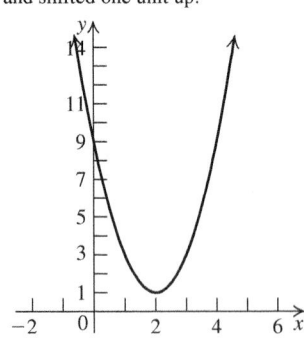

Vertex: $(2, 1)$; axis: $x = 2$; no x-intercept; y-intercept: 9

33. $y = -3(x - 3)^2 + 16$. The graph is the graph of $y = x^2$ shifted 3 units to the right, stretched vertically by a factor of 3, reflected about the x-axis, and shifted 16 units up.

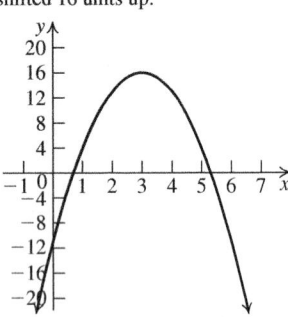

Vertex: $(3, 16)$; axis: $x = 3$; x-intercepts: $3 \pm \dfrac{4}{3}\sqrt{3}$; y-intercept: -11

35. a. Opens up **b.** $(4, -1)$ **c.** $x = 4$ **d.** x-intercepts: 3 and 5; y-intercept: 15
e.

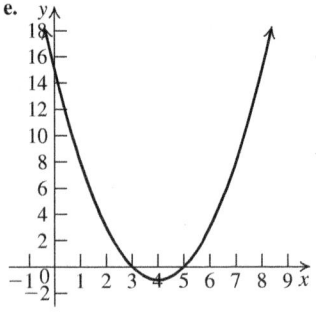

37. a. Opens up **b.** $\left(\dfrac{1}{2}, -\dfrac{25}{4}\right)$ **c.** $x = \dfrac{1}{2}$ **d.** x-intercepts: -2 and 3; y-intercept: -6
e.

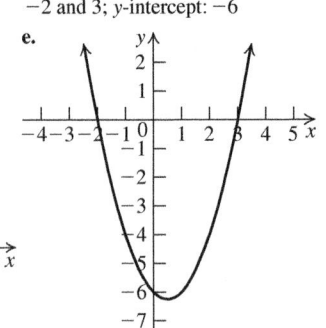

39. a. Opens up **b.** $(1, 3)$ **c.** $x = 1$ **d.** x-intercepts: none; y-intercept: 4
e.

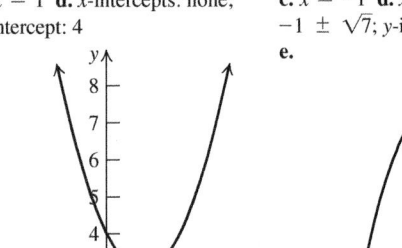

41. a. Opens down **b.** $(-1, 7)$ **c.** $x = -1$ **d.** x-intercepts: $-1 \pm \sqrt{7}$; y-intercept: 6
e.

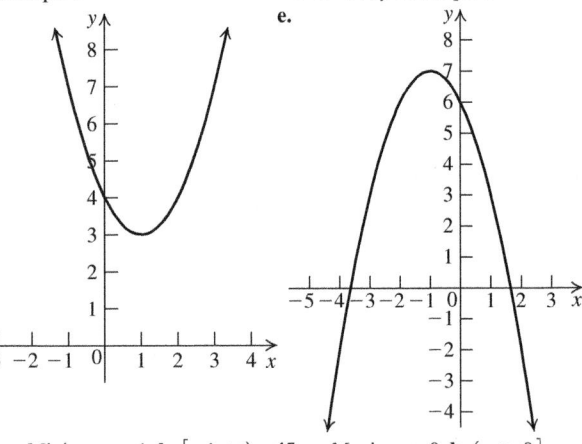

43. a. Minimum: -1 **b.** $[-1, \infty)$
45. a. Maximum: 0 **b.** $(-\infty, 0]$
47. a. Minimum: -5 **b.** $[-5, \infty)$
49. a. Maximum: 16 **b.** $(-\infty, 16]$
51.

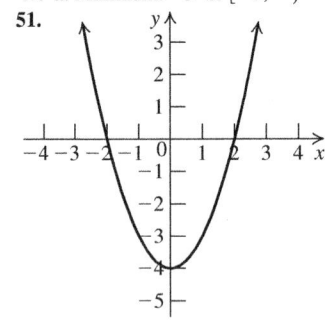

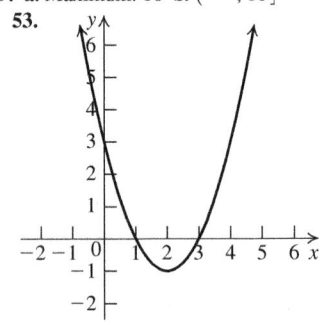

Solution: $[-2, 2]$

53.

Solution: $(-\infty, 1) \cup (3, \infty)$

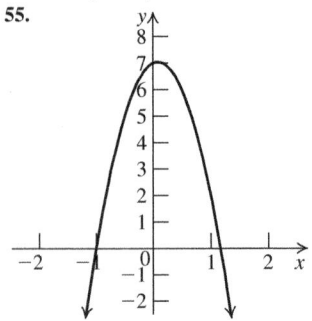

55.

Solution: $\left(-1, \dfrac{7}{6}\right)$

Applying the Concepts:
57. Dimensions: 75 m × 100 m; area: 7500 m² **59. a.** 64 ft
b. After 4 sec **61.** $3 - \sqrt{2}, 3 + \sqrt{2}$

Critical Thinking / Discussion / Writing:
64. (5, 19)

Maintaining Skills:
65. -1 **66.** $\dfrac{1}{16}$ **67.** $-\dfrac{1}{16}$ **68.** $6x^6$ **69.** 1 **70.** $-6x^6$ **71.** $\dfrac{9x^2}{4}$
72. $x^4 + 3x^2 - 5x$

Section 2.2

Basic Concepts and Skills:
1. Polynomial function; degree: 5; leading term: $2x^5$; leading coefficient: 2 **3.** Polynomial function; degree: 3; leading term: $\dfrac{2}{3}x^3$; leading coefficient: $\dfrac{2}{3}$ **5.** Polynomial function; degree: 4; leading term: πx^4; leading coefficient: π **7.** Presence of $|x|$ **9.** Domain is not $(-\infty, \infty)$ **11.** Presence of $\sqrt{x}$ **13.** Domain is not $(-\infty, \infty)$ **15.** Graph is not continuous **17.** Graph not continuous **19.** Not a function **21.** c **23.** a **25.** d **27.** Zeros: $x = -5$, multiplicity: 1, the graph crosses the x-axis; $x = 1$, multiplicity: 1, the graph crosses the x-axis; maximum number of turning points: 1 **29.** Zeros: $x = -1$, multiplicity: 2, the graph touches but does not cross the x-axis; $x = 1$, multiplicity: 3, the graph crosses the x-axis; maximum number of turning points: 4 **31.** Zeros: $x = -\dfrac{2}{3}$, multiplicity: 1, the graph crosses the x-axis; $x = \dfrac{1}{2}$, multiplicity: 2, the graph touches but does not cross the x-axis; maximum number of turning points: 2 **33.** Zeros: $x = 0$, multiplicity: 2, the graph touches but does not cross the x-axis; $x = 3$, multiplicity: 2, the graph touches but does not cross the x-axis; maximum number of turning points: 3 **35.** 2.09 **37.** 2.28
39. a. $y \to \infty$ as $x \to -\infty$ and $y \to \infty$ as $x \to \infty$ **b.** No zeros **c.** The graph is above the x-axis on $(-\infty, \infty)$. **d.** 3 **e.** y-axis symmetry. **f.** 1
g.

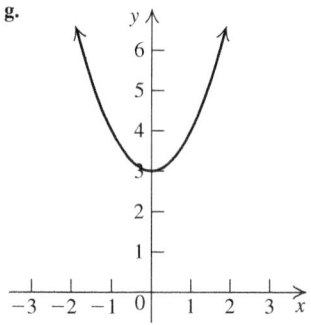

41. a. $y \to \infty$ as $x \to -\infty$ and $y \to \infty$ as $x \to \infty$
b. Zeros: $x = -7$, the graph crosses the x-axis; $x = 3$, the graph crosses the x-axis.
c. The graph is above the x-axis on $(-\infty, -7)$ and $(3, \infty)$ and below the x-axis on $(-7, 3)$.
d. -21 **e.** no symmetries **f.** 1
g.

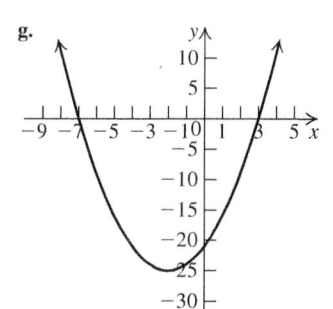

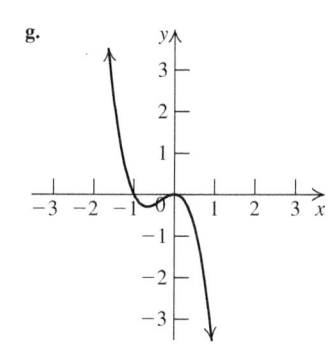

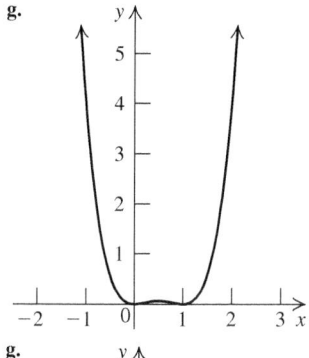

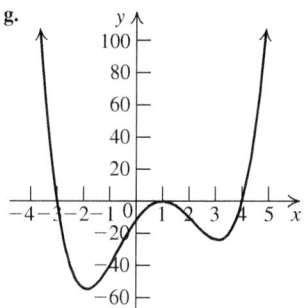

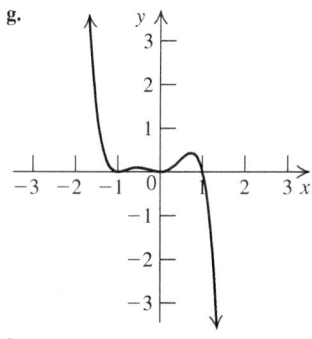

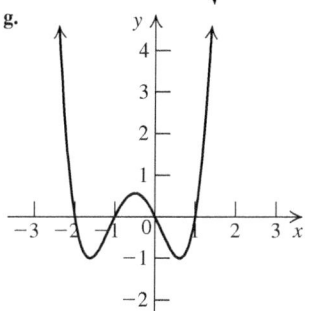

43. a. $y \to \infty$ as $x \to -\infty$ and $y \to -\infty$ as $x \to \infty$ **b.** Zeros: $x = -1$, the graph crosses the x-axis; $x = 0$, the graph touches but does not cross the x-axis.
c. The graph is above the x-axis on $(-\infty, -1)$ and below the x-axis on $(-1, 0) \cup (0, \infty)$. **d.** 0
e. No symmetries. **f.** 2
g.

45. a. $y \to \infty$ as $x \to -\infty$ and $y \to \infty$ as $x \to \infty$ **b.** Zeros: $x = 0$, the graph touches but does not cross the x-axis; $x = 1$, the graph touches but does not cross the x-axis.
c. The graph is above the x-axis on $(-\infty, 0) \cup (0, 1) \cup (1, \infty)$. **d.** 0
e. No symmetries **f.** 3
g.

47. a. $y \to \infty$ as $x \to -\infty$ and $y \to \infty$ as $x \to \infty$
b. Zeros: $x = -3$, the graph crosses the x-axis; $x = 1$, the graph touches but does not cross the x-axis; $x = 4$, the graph crosses the x-axis. **c.** The graph is above the x-axis on $(-\infty, -3)$ and $(4, \infty)$ and below the x-axis on $(-3, 1) \cup (1, 4)$. **d.** -12
e. No symmetries **f.** 3
g.

49. a. $y \to \infty$ as $x \to -\infty$ and $y \to -\infty$ as $x \to \infty$
b. Zeros: $x = -1$, the graph touches but does not cross the x-axis; $x = 0$, the graph touches but does not cross the x-axis; $x = 1$, the graph crosses the x-axis.
c. The graph is above the x-axis on $(-\infty, -1) \cup (-1, 0) \cup (0, 1)$ and below the x-axis on $(1, \infty)$. **d.** 0
e. No symmetries. **f.** 4
g.

51. a. $y \to \infty$ as $x \to -\infty$ and $y \to \infty$ as $x \to \infty$
b. Zeros: $x = -2$, the graph crosses the x-axis; $x = -1$, the graph crosses the x-axis; $x = 0$, the graph crosses the x-axis; $x = 1$, the graph crosses the x-axis.
c. The graph is above the x-axis on $(-\infty, -2) \cup (-1, 0) \cup (1, \infty)$ and below the x-axis on $(-2, -1) \cup (0, 1)$. **d.** 0 **e.** No symmetries. **f.** 3
g.

Answers — Section 2.5

Applying the Concepts:
53. a. Zeros: $x = 0$, multiplicity 2; $x = 4$, multiplicity 1
c. 2 **d.** Domain: $[0, 4]$. The portion between the x-intercepts constitutes the graph of $R(x)$.
b.

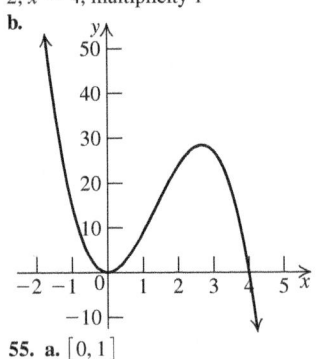

55. a. $[0, 1]$

b.

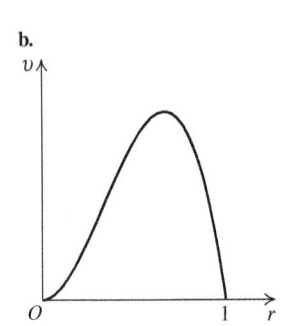

Critical Thinking / Discussion / Writing:
57. degree 5, because the graph has five x-intercepts **58.** degree 5 because the graph has four turning points **59.** degree 6, because the graph has five turning points **60.** degree 6 because the graph has five turning points **61.** No, because the domain of any polynomial function is $(-\infty, \infty)$, which includes the point $x = 0$. **62.** Yes. The graph of $f(x) = x^2 + 1$ has no x-intercept. **63.** Not possible, this is not compatible with its end behavior. **64.** Not possible, this is not compatible with its end behavior.

Maintaining Skills:
65. $-4x^2$ **66.** $\dfrac{1}{x^2}$ **67.** $\dfrac{3x^2}{2}$ **68.** $x^3 - 3$ **69.** $5x^2 + 2x - 1$
70. a. $7 = 1 \cdot 7$ **b.** $7 = 7 \cdot 1$ **c.** $7 = (-1)(-7)$ **d.** $7 = (-7)(-1)$
71. $-7, -1, 1, 7$ **72.** $\pm 1, \pm 2, \pm 3, \pm 4, \pm 5, \pm 6, \pm 8, \pm 10, \pm 12,$ $\pm 15, \pm 20, \pm 24, \pm 30, \pm 40, \pm 60, \pm 120$.

Section 2.3

Basic Concepts and Skills:
1. The quotient is $x^2 - 7$, and the remainder is -5. **3.** The quotient is $x^2 + 2x - 11$, and the remainder is 12. **5.** The quotient is $x^3 - x^2 + 4$, and the remainder is 13. **7.** The quotient is $2x^2 + 5x - \frac{1}{2}$, and the remainder is $\frac{3}{4}$. **9.** The quotient is $2x^2 - 6x + 6$, and the remainder is -1. **11. a.** 5 **b.** 3 **c.** $\dfrac{15}{8}$ **d.** 1301 **13. a.** -17 **b.** -27
c. -56 **d.** 24 **23.** $k = -1$ **25.** $k = -7$ **27.** Remainder: 1
29. Remainder: 6 **31.** $f(x) = 2(x - 2)(x + 1)(x + 2)$
33. $f(x) = -2(x - 1)^2(x - 2)(x + 2)$ **35.** $\{\pm\frac{1}{3}, \pm 1, \pm\frac{5}{3}, \pm 5\}$
37. $\{\pm\frac{1}{4}, \pm\frac{1}{2}, \pm\frac{3}{4}, \pm 1, \pm\frac{3}{2}, \pm 2, \pm 3, \pm 6\}$ **39.** $\{-2, 1, 2\}$ **41.** $\{\frac{2}{3}\}$
43. $\{-\frac{3}{2}\}$ **45.** $\{-1, 2\}$ **47.** $\{-3, -1, 1, 4\}$ **49.** No rational zeros
51. $\{1, -3 \pm \sqrt{11}\}$ **53.** $\left\{1, -1, \dfrac{3 \pm \sqrt{5}}{2}\right\}$

Applying the Concepts:
55. $2x^2 + 1$ cm **57.** $(2, 4)$

Critical Thinking / Discussion / Writing:
59. a. False **b.** False **60. a.** n is an odd integer. **b.** n is an even integer. **c.** There is no such n. **d.** n is any positive integer.
63. $\left\{-\dfrac{3}{2}, \dfrac{1}{2}\right\}$

Maintaining Skills:
64. $x^3 + 2x + 3; 0$ **65.** $x^2 - x + 1; 0$ **66.** $-x^5 + 2x; 3$
67. $x^3 + 2x^2 + x + 1; 6$ **68.** $\pm 1, \pm 11$
69. $\pm 1, \pm 2, \pm 3, \pm 6, \pm 9, \pm 18$ **70.** $\pm 1, \pm 3, \pm 17, \pm 51$
71. $\pm 1, \pm 2, \pm 3, \pm 4, \pm 6, \pm 8, \pm 9, \pm 12, \pm 18, \pm 24, \pm 36, \pm 72$

Section 2.4

Basic Concepts and Skills:
1. Number of positive zeros: 0 or 2; number of negative zeros: 1
3. Number of positive zeros: 0 or 2; number of negative zeros: 1
5. Number of positive zeros: 1 or 3; number of negative zeros: 0 or 2
7. Number of positive zeros: 1 or 3; number of negative zeros: 1
9. Upper bound: $\frac{1}{3}$; lower bound: $-\frac{1}{3}$ **11.** Upper bound: $\frac{1}{3}$; lower bound: $-\frac{7}{3}$ **13.** Upper bound: 31; lower bound: -31 **15.** Upper bound: $\dfrac{7}{2}$; lower bound: $-\dfrac{7}{2}$ **17.** $x = \pm 5i$ **19.** $x = 2 \pm 3i$
21. $x = -2 \pm 3i$ **23.** $\{2, -1 + i\sqrt{3}, -1 - i\sqrt{3}\}$
25. $\{2, 3i, -3i\}$ **27.** $3 - i$ **29.** $2 + i$ **31.** $-i, -3i$
33. $P(x) = 2x^4 - 20x^3 + 70x^2 - 180x + 468$
35. $P(x) = 7x^5 - 119x^4 + 777x^3 - 2415x^2 + 3500x - 1750$
37. $-1, 0, 3i, -3i$ **39.** $-2, 0, 1, 3 + i, 3 - i$ **41.** $1, 4 + i, 4 - i$
43. $-\dfrac{4}{3}, 1 + 3i, 1 - 3i$ **45.** $1, \dfrac{3}{2} + \dfrac{3}{2}i, \dfrac{3}{2} - \dfrac{3}{2}i$ **47.** $-3, 1, 3 + i, 3 - i$
49. $\dfrac{1}{2}, 2, 3, i, -i$ **51.** $f(x) = \dfrac{1}{3}(x^2 + 9)$
53. $f(x) = \dfrac{1}{2}(2 - x)(x^2 + 1)(x^2 + 4)$
55. There are three cube roots: $1, -\dfrac{1}{2} + \dfrac{1}{2}\sqrt{3}i$, and $-\dfrac{1}{2} - \dfrac{1}{2}\sqrt{3}i$.
61. $f(x) = -4(x - 2)(x^2 - 2x + 5)$
$y \to \infty$ as $x \to -\infty$, $y \to -\infty$ as $x \to \infty$.
63. $f(x) = -2(x^2 - 1)(x^2 - 6x + 10)$
$y \to -\infty$ as $x \to -\infty$, $y \to -\infty$ as $x \to \infty$.

Critical Thinking / Discussion / Writing:
66. a. According to Descartes's Rule of Signs, the polynomial $x^3 + 6x - 20$ has one positive zero and no negative zero. Therefore, there is exactly one real solution.
b. Substituting $v - u$ for x, we obtain
$(v - u)^3 + 6(v - u) = v^3 - 3v^2u + 3vu^2 - u^3 + 6(v - u)$
$= (v^3 - u^3) + (6 - 3uv)(v - u)$
$= 20 - (6 - 3(2))(v - u) = 20.$
Therefore, $x = v - u$ is the solution.
c. $u = \sqrt[3]{-10 \pm 6\sqrt{3}}$, $v = \dfrac{2}{\sqrt[3]{-10 \pm 6\sqrt{3}}}$, $v = \sqrt[3]{10 \pm 6\sqrt{3}}$
67. $\{-4, -1 + i, -1 - i\}$

Maintaining Skills:
68. $(f + g)(x) = 3x^2 + 3x + 2$
69. $(g + h)(x) = x^3 + 3x^2 + 2x - 1$
70. $(f \cdot g)(x) = 6x^3 + 11x^2 + x - 3$
71. $g \cdot h(x) = 3x^5 + x^4 + 2x^3 + x^2 - x$
72. $(h - s)(x) = x$
73. $(s - g)(x) = x^3 - 3x^2 - x + 1$
74. $(f \circ s)(x) = 2x^3 + 3$
75. $(h \circ s)(x) = x^9 + x^3$

Section 2.5

Basic Concepts and Skills:
1. $(-\infty, -4) \cup (-4, \infty)$ **3.** $(-\infty, \infty)$ **5.** $(-\infty, -2) \cup (-2, 3) \cup (3, \infty)$
7. $(-\infty, 2) \cup (2, 4) \cup (4, \infty)$ **9.** ∞ **11.** ∞ **13.** 1
15. $(-\infty, -2) \cup (-2, 1) \cup (1, \infty)$ **17.** $x = -2\ x = 1$

19. Vertical asymptote: $x = 4$
Horizontal asymptote: $y = 0$
Domain: $(-\infty, 4) \cup (4, \infty)$
Range: $(-\infty, 0) \cup (0, \infty)$

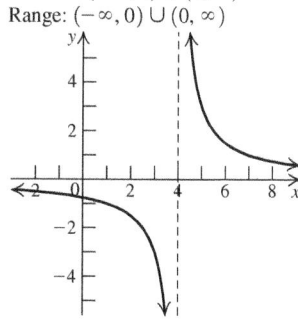

21. Vertical asymptote: $x = -\dfrac{1}{3}$
Horizontal asymptote: $y = -\dfrac{1}{3}$
Domain: $\left(-\infty, -\dfrac{1}{3}\right) \cup \left(-\dfrac{1}{3}, \infty\right)$
Range: $\left(-\infty, -\dfrac{1}{3}\right) \cup \left(-\dfrac{1}{3}, \infty\right)$

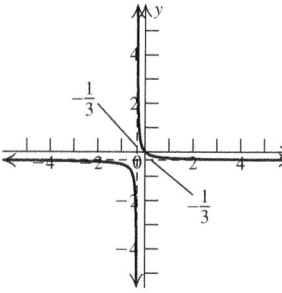

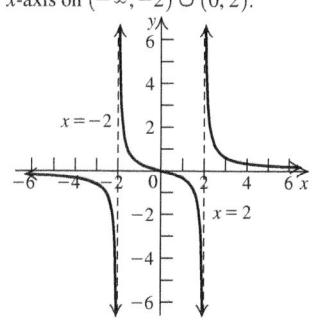

55. x-intercept: 0; y-intercept: 0; vertical asymptotes: $x = -3$ and $x = 3$; horizontal asymptote: $y = -2$. The graph is symmetric in the y-axis. The graph is above the line $y = -2$ on $(-3, 3)$ and below it on $(-\infty, -3) \cup (3, \infty)$.

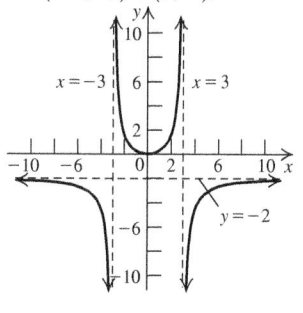

57. No x-intercept; y-intercept: -1; vertical asymptotes: $x = -\sqrt{2}$ and $x = \sqrt{2}$; horizontal asymptote: x-axis. The graph is symmetric with respect to the y-axis. The graph is above the x-axis on $(-\infty, -\sqrt{2}) \cup (\sqrt{2}, \infty)$ and below the x-axis on $(-\sqrt{2}, \sqrt{2})$.

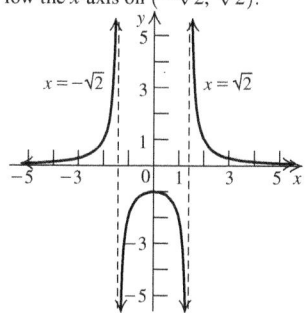

23. Vertical asymptote: $x = -2$
Horizontal asymptote: $y = -3$
Domain: $(-\infty, -2) \cup (-2, \infty)$
Range: $(-\infty, -3) \cup (-3, \infty)$

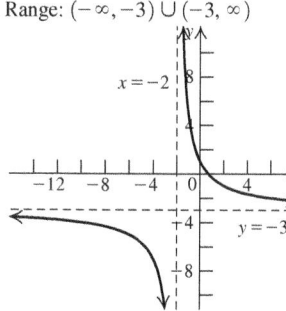

25. Vertical asymptote: $x = 4$
Horizontal asymptote: $y = 5$
Domain: $(-\infty, 4) \cup (4, \infty)$
Range: $(-\infty, 5) \cup (5, \infty)$

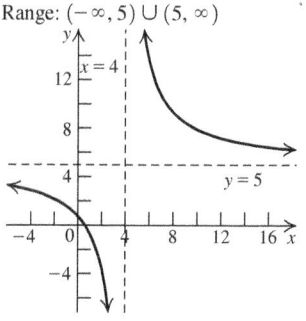

59. x-intercept: -1; y-intercept: $-\dfrac{1}{6}$; vertical asymptotes: $x = -3$ and $x = 2$; horizontal asymptote: x-axis; symmetry: none. The graph is above the x-axis on $(-3, -1) \cup (2, \infty)$ and below the x-axis on $(-\infty, -3) \cup (-1, 2)$.

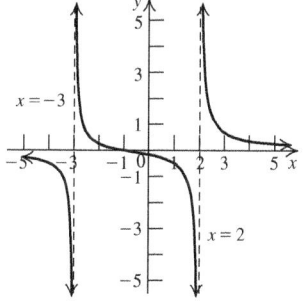

61. x-intercept: 0; y-intercept: 0; no vertical asymptote; horizontal asymptote: $y = 1$. The graph is symmetric with respect to the y-axis. The graph is above the x-axis on $(-\infty, \infty)$ and below the line $y = 1$ on $(-\infty, \infty)$.

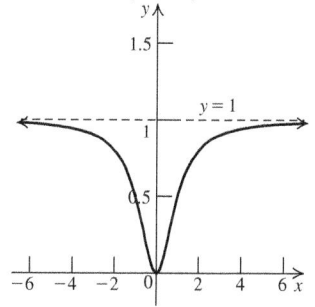

27. $x = 1$ **29.** $x = -4$ and $x = 3$ **31.** $x = -3$ and $x = 2$
33. $x = -3$ **35.** No vertical asymptote **37.** $y = 0$ **39.** $y = \dfrac{2}{3}$
41. No horizontal asymptote **43.** $y = 0$ **45.** d **47.** e **49.** a
51. x-intercept: 0; y-intercept: 0; vertical asymptote: $x = 3$; horizontal asymptote: $y = 2$; symmetry: none. The graph is above the line $y = 2$ on $(3, \infty)$ and below it on $(-\infty, 3)$.

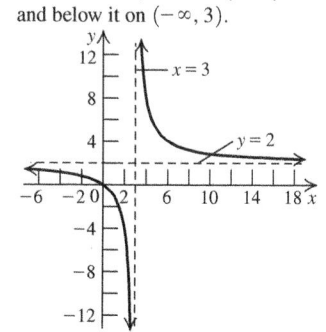

53. x-intercept: 0; y-intercept: 0; vertical asymptotes: $x = -2$ and $x = 2$; horizontal asymptote: x-axis. The graph is symmetric with respect to the origin. The graph is above the x-axis on $(-2, 0) \cup (2, \infty)$ and below the x-axis on $(-\infty, -2) \cup (0, 2)$.

63. x-intercept: ± 2; No y-intercept; hole at $\left(0, \dfrac{4}{9}\right)$ vertical asymptotes: $x = -3$ and $x = 3$; horizontal asymptote: $y = 1$. The graph is symmetric with respect to the y-axis. The graph is above the line $y = 1$ on $(-\infty, -3) \cup (3, \infty)$ and below it on $(-3, 0) \cup (0, 3)$.

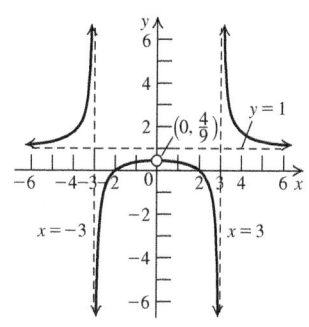

65. No x-intercept; y-intercept: -2; hole at $(2, 0)$ no vertical asymptote; no horizontal asymptote; symmetry: none. The graph is above the x-axis on $(2, \infty)$ and below the x-axis on $(-\infty, 2)$.

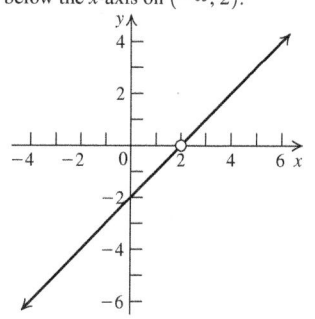

A-16 Answers ■ Review Exercises

67. $f(x) = \dfrac{-2(x-1)}{x-2}$ **69.** $f(x) = \dfrac{(x-1)(x-3)}{x(x-2)}$

71. Oblique asymptote: $y = 2x$ **73.** Oblique asymptote: $y = x$

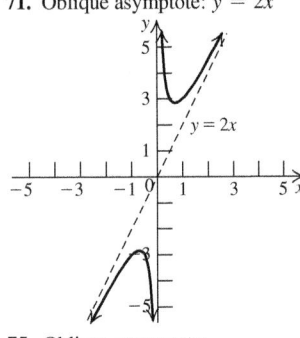

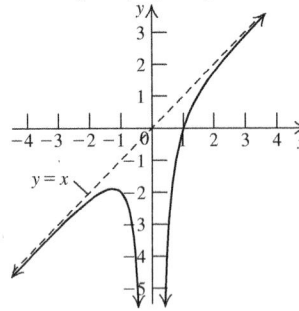

75. Oblique asymptote: $y = x - 2$

77. Oblique asymptote: $y = x - 2$

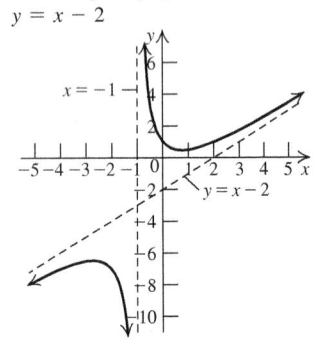

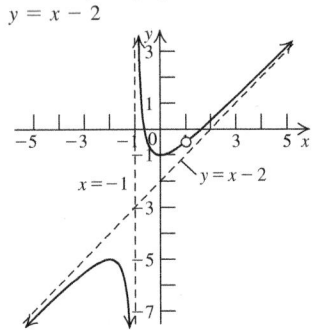

Applying the Concepts:
79. a.

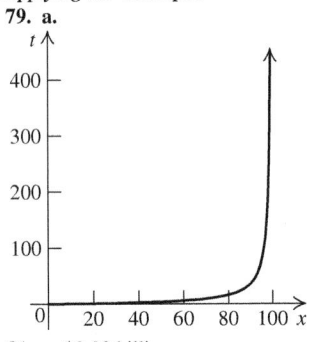

b. i. about 4 min **ii.** about 12 min **iii.** about 76 min **iv.** 397 min
c. i. ∞ **ii.** Not applicable; the domain is $x < 100$. **d.** No

81. a. $3.02 billion   **c.** 90.89%
b.

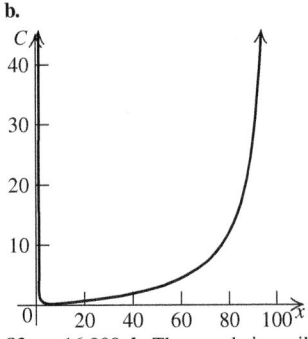

83. a. 16,000 **b.** The population will stabilize at 4000.

Critical Thinking / Discussion / Writing:
84. $f(x) = \dfrac{2-x}{x-3}$ **85.** $f(x) = \dfrac{x^2}{x^2-1}$ **86.** $f(x) = \dfrac{-x^2+1}{x}$

87. $f(x) = \dfrac{x^2+x-6}{x-3}$ **88.** $f(x) = \dfrac{x+4}{x-2}$

89. $f(x) = \dfrac{x-4}{x^2-x-2}$ **90.** $f(x) = \dfrac{4x^2-1}{(x-1)^2}$

91. $f(x) = \dfrac{2x^2}{x^2+1}$ **92.** $f(x) = \dfrac{3x^2-x+1}{x-1}$

93. No, because in that case, there would be values of x with two different corresponding function values; so it would not be a function.

Maintaining Skills:
94. $y = -\dfrac{2}{3}x - \dfrac{1}{3}$ **95.** $y = \dfrac{5}{4}x - \dfrac{1}{2}$ **96.** $[-3, -1] \cup [3, \infty)$
97. $\left\{-\dfrac{2}{3}, 4\right\}$

Review Exercises

Basic Concepts and Skills:
1. i. Opens up **ii.** Vertex: (1, 2)
iii. Axis: $x = 1$ **iv.** No x-intercept
v. y-intercept: 3 **vi.** Decreasing on $(-\infty, 1)$, increasing on $(1, \infty)$

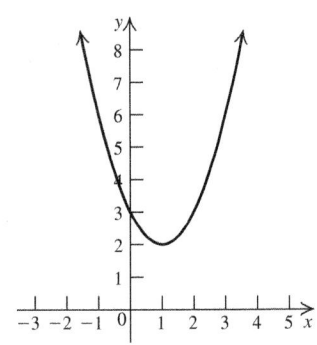

3. i. Opens down **ii.** Vertex: (3, 4)
iii. Axis: $x = 3$ **iv.** x-intercepts: $3 \pm \sqrt{2}$ **v.** y-intercept: -14
vi. Increasing on $(-\infty, 3)$, decreasing on $(3, \infty)$

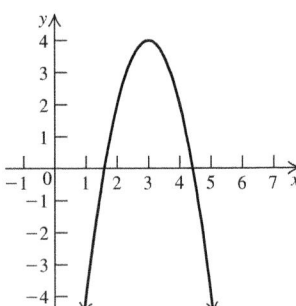

5. i. Opens down **ii.** Vertex: (0, 3)
iii. Axis: y-axis **iv.** x-intercepts: $\pm \dfrac{\sqrt{6}}{2}$ **v.** y-intercept: 3
vi. Increasing on $(-\infty, 0)$, decreasing on $(0, \infty)$

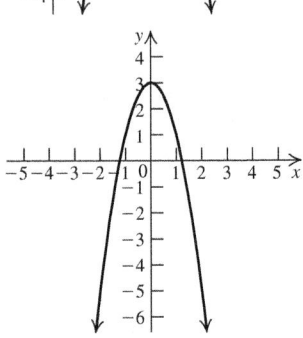

7. i. Opens up **ii.** Vertex: (1, 1)
iii. Axis: $x = 1$ **iv.** No x-intercept
v. y-intercept: 3 **vi.** Decreasing on $(-\infty, 1)$, increasing on $(1, \infty)$

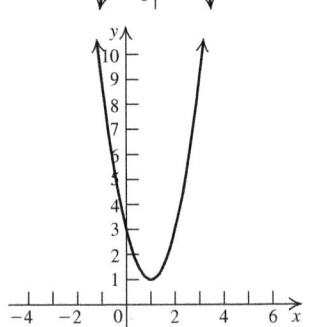

9. i. Opens up
ii. Vertex: $\left(\dfrac{1}{3}, \dfrac{2}{3}\right)$
iii. Axis: $x = \dfrac{1}{3}$
iv. No x-intercept
v. y-intercept: 1
vi. Decreasing on $\left(-\infty, \dfrac{1}{3}\right)$, increasing on $\left(\dfrac{1}{3}, \infty\right)$

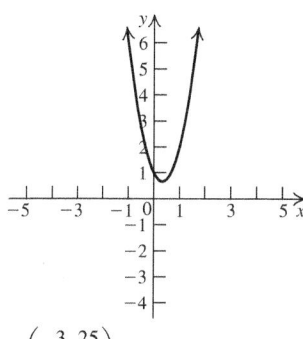

11. Minimum: $(2, -1)$ **13.** Maximum: $\left(-\dfrac{3}{4}, \dfrac{25}{8}\right)$

15. Shift the graph of $y = x^3$ one unit to the left and two units down.

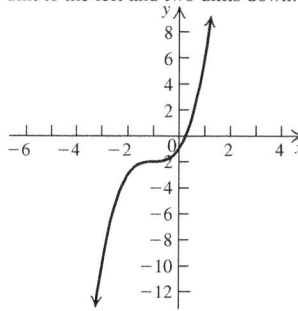

17. Shift the graph of $y = x^3$ one unit right, reflect the resulting graph about the x-axis, and shift it one unit up.

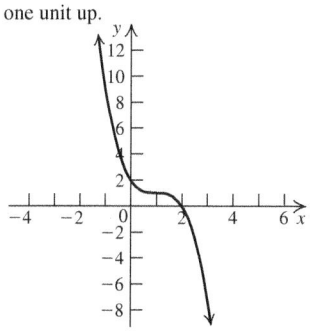

19. i. $f(x) \to -\infty$ as $x \to -\infty$ and $f(x) \to \infty$ as $x \to \infty$.
ii. Zeros: $x = -2$, multiplicity: 1, the graph crosses the x-axis; $x = 0$, multiplicity: 1, the graph crosses the x-axis; $x = 1$, multiplicity: 1, the graph crosses the x-axis. iii. x-intercepts: $-2, 0, 1$; y-intercept: 0 iv. The graph is above the x-axis on $(-2, 0) \cup (1, \infty)$ and below the x-axis on $(-\infty, -2) \cup (0, 1)$.
v. Symmetry: none

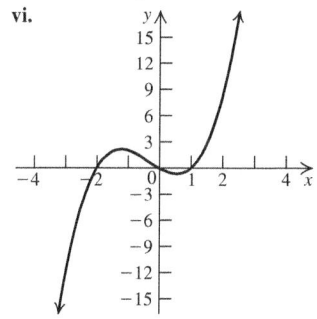

21. i. $f(x) \to -\infty$ as $x \to -\infty$ and $f(x) \to -\infty$ as $x \to \infty$.
ii. Zeros: $x = 0$, multiplicity: 2, the graph touches but does not cross the x-axis; $x = 1$, multiplicity: 2, the graph touches but does not cross the x-axis.
iii. x-intercepts: 0, 1; y-intercept: 0
iv. The graph is below the x-axis on $(-\infty, 0) \cup (0, 1) \cup (1, \infty)$.
v. Symmetry: none

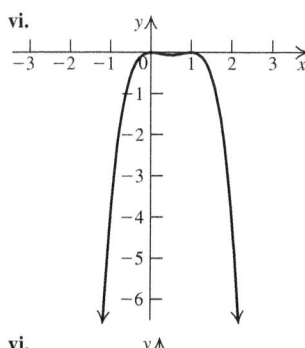

23. i. $f(x) \to -\infty$ as $x \to -\infty$ and $f(x) \to -\infty$ as $x \to \infty$.
ii. Zeros: $x = -1$, multiplicity: 1, the graph crosses the x-axis; $x = 0$; multiplicity: 2, the graph touches but does not cross the x-axis; $x = 1$ multiplicity: 1, the graph crosses the x-axis.
iii. x-intercepts: $-1, 0, 1$; y-intercept: 0 iv. The graph is above the x-axis on $(-1, 0) \cup (0, 1)$ and below the x-axis on $(-\infty, -1) \cup (1, \infty)$.
v. The graph is symmetric with respect to the y-axis.

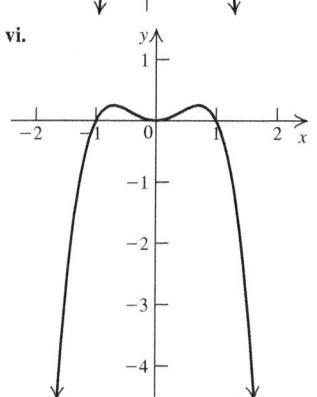

25. Quotient: $2x + 3$; remainder: -7
27. Quotient: $8x^3 - 12x^2 + 14x - 21$; remainder: 186
29. Quotient: $x^2 + 3x - 3$; remainder: -6
31. Quotient: $2x^3 - 5x^2 + 10x - 17$; remainder: 182
33. -11 **35.** 88 **37.** 1, 2, 4 **39.** $-3, -2, \dfrac{1}{3}$
41. $-6, -3, -2, -1, 1, 2, 3, 6$ **43.** $-2, -\dfrac{1}{5}, 0$ **45.** $-3, -2, 2$
47. $-3, 1, 2$ **49.** $-1, 3, 3i, -3i$ **51.** $2i, -2i, -1 + 2i, -1 - 2i$
53. $\{-2, 1, 2\}$ **55.** $\left\{-1, -\dfrac{1}{2}, \dfrac{3}{2}\right\}$ **57.** $\{-1, 2, i, -i\}$ **59.** The only possible rational roots are $-2, -1, 1$, and 2. None of them satisfies the equation. **61.** 1.88
63. x-intercept: -1; no y-intercept; vertical asymptote: y-axis; horizontal asymptote: $y = 1$; symmetry: none. The graph is above the $y = 1$ on $(0, \infty)$ and below it on $(-\infty, 0)$.

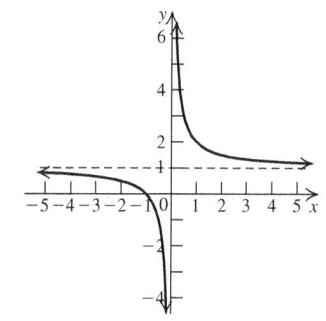

65. x-intercept: 0; y-intercept: 0; vertical asymptotes: $x = -1$ and $x = 1$; horizontal asymptote: x-axis; symmetry: none. The graph is above the x-axis on $(-1, 0) \cup (1, \infty)$ and below the x-axis on $(-\infty, -1) \cup (0, 1)$.

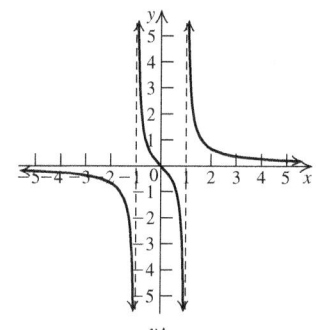

67. x-intercept: 0; y-intercept: 0; vertical asymptotes: $x = -3$ and $x = 3$; no horizontal asymptote; symmetry about origin. Oblique asymptote: $y = x$. The graph is above the line $y = x$ on $(-3, 0) \cup (3, \infty)$ and below it on $(-\infty, -3) \cup (0, 3)$.

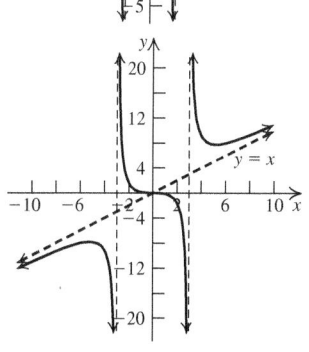

69. x-intercept: 0; y-intercept: 0; vertical asymptotes: $x = -2$ and $x = 2$; no horizontal asymptote. The graph is symmetric with respect to the y-axis. The graph is above the x-axis on $(-\infty, -2) \cup (2, \infty)$ and below the x-axis on $(-2, 0) \cup (0, 2)$.

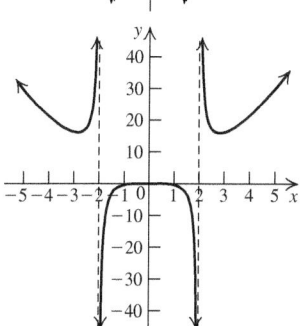

A-18 Answers ■ Exercises for Calculus

Applying the Concepts:
71. Maximum height: 1000. The missile hits the ground at $x = 200$.

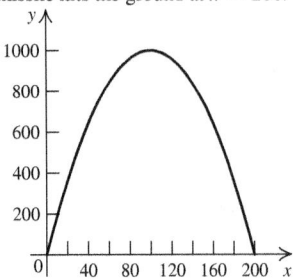

73. $x = 10\sqrt{6} \approx 24.5$ ft, $y = \dfrac{20\sqrt{6}}{3} \approx 16.3$ ft

75. a.   **b.** $x = 50$

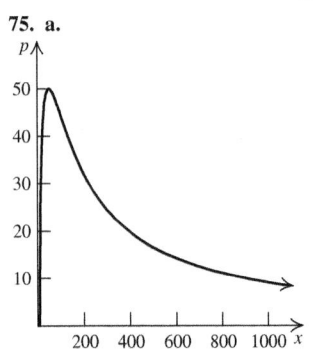

77. a. 11.59 in. **b.** No

Exercises for Calculus

1. $24x^2$ **2.** $\dfrac{x-5}{x-1}$ **3.** $-1, 2; f(x) < g(x)$ on $(-1, 2)$
4. $-2, 2; f(x) < g(x)$ on $(-2, 2)$ **5.** $0, 1, 3; f(x) > g(x)$ on $(0, 1), f(x) < g(x)$ on $(1, 3)$ **6.** $-2, 1, 2; f(x) > g(x)$ on $(-2, 1), f(x) < g(x)$ on $(1, 2)$
7. a. $f(x) \to 0$ as $x \to -\infty$, $f(x) \to 0$ as $x \to \infty$, $f(x) \to -\infty$ as $x \to 0^-$, and $f(x) \to \infty$ as $x \to 0^+$; no x-intercept; no y-intercept; vertical asymptote: y-axis; horizontal asymptote: x-axis. The graph is symmetric with respect to the origin. The graph is above the x-axis on $(0, \infty)$ and below the x-axis on $(-\infty, 0)$.

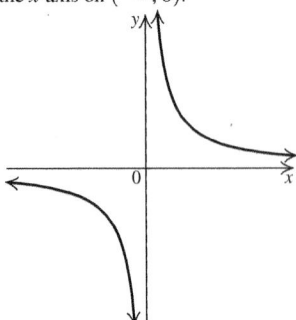

b. $f(x) \to 0$ as $x \to -\infty$, $f(x) \to 0$ as $x \to \infty$, $f(x) \to \infty$ as $x \to 0^-$, and $f(x) \to -\infty$ as $x \to 0^+$; no x-intercept; no y-intercept; vertical asymptote: y-axis; horizontal asymptote: x-axis. The graph is symmetric with respect to the origin. The graph is above the x-axis on $(-\infty, 0)$ and below the x-axis on $(0, \infty)$.

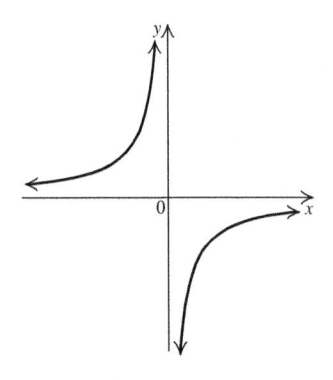

c. $f(x) \to 0$ as $x \to \infty$, $f(x) \to 0$ as $x \to -\infty$, $f(x) \to \infty$ as $x \to 0^-$, and $f(x) \to \infty$ as $x \to 0^+$; no x-intercept; no y-intercept; vertical asymptote: y-axis; horizontal asymptote: x-axis. The graph is symmetric with respect to the y-axis. The graph is above the x-axis on $(-\infty, 0) \cup (0, \infty)$.

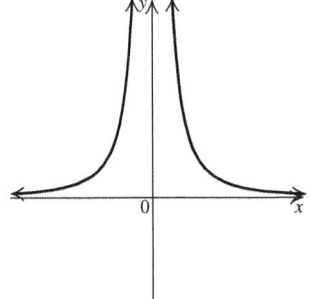

d. $f(x) \to 0$ as $x \to -\infty$, $f(x) \to 0$ as $x \to \infty$, $f(x) \to -\infty$ as $x \to 0^-$, and $f(x) \to -\infty$ as $x \to 0^+$; no x-intercept; no y-intercept; vertical asymptote: y-axis; horizontal asymptote: x-axis. The graph is symmetric with respect to the y-axis. The graph is below the x-axis on $(-\infty, 0) \cup (0, \infty)$.

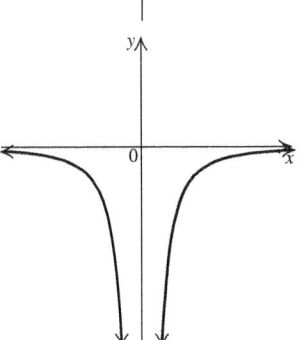

8. a. If c is a zero of $f(x)$, then because c is not a factor of the numerator, $x = c$ is a vertical asymptote. **b.** Because the numerator is positive, the sign of $f(x)$ and $g(x)$ is the same. **c.** If the graphs intersect for some value x, then $f(x) = g(x)$. It implies that $f(x) = \dfrac{1}{f(x)}$, from which it follows that $f(x) = \pm 1$. **d.** If $f(x)$ increases (decreases, remains constant), the denominator of $g(x)$ increases (decreases, remains constant). Therefore, $g(x)$ decreases (increases, remains constant).

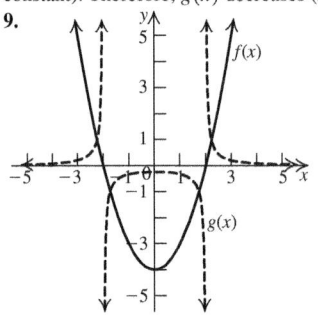

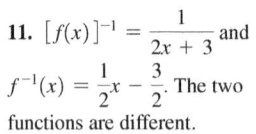

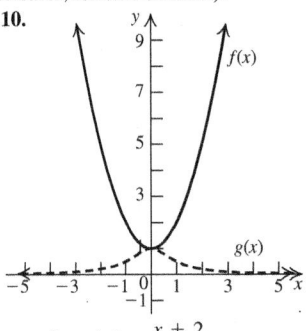

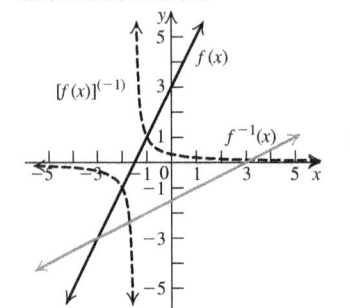

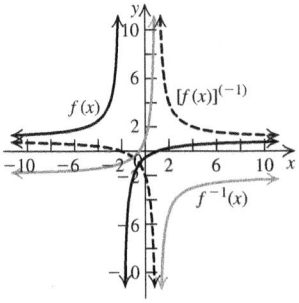

9.   **10.**

11. $[f(x)]^{-1} = \dfrac{1}{2x+3}$ and $f^{-1}(x) = \dfrac{1}{2}x - \dfrac{3}{2}$. The two functions are different.

12. $[f(x)]^{-1} = \dfrac{x+2}{x-1}$ and $f^{-1}(x) = \dfrac{2x+1}{1-x}$.

13. $A(x) = \dfrac{2V}{x} + 2\pi x^2 \text{ in}^2$ **14. a.** $\dfrac{3}{4}$ **b.** $\dfrac{4 - \sqrt{25 - (3+h)^2}}{h}$

c. $\dfrac{h(h+6)}{h(4 + \sqrt{25-(3+h)^2})} = \dfrac{h+6}{4 + \sqrt{25-(3+h)^2}}, h \neq 0$

d. The slopes of the secant lines are very close to the slope of the tangent line when h is close to zero.

Practice Test A
1. x-intercepts: $3 \pm \sqrt{7}$
2. [graph] **3.** $(1, 10)$
4. $(-\infty, -4) \cup (-4, 1) \cup (1, \infty)$.
5. Quotient: $x^2 - 4x + 3$; remainder: 0
6. [graph]
7. $-2, 1, 2$ **8.** Quotient: $-3x^2 + 5x + 1$ **9.** $P(-2) = -53$
10. $-2, 2, 5$ **11.** $0, -\dfrac{1}{2} + \dfrac{\sqrt{61}}{2}, -\dfrac{1}{2} - \dfrac{\sqrt{61}}{2}$
12. $-9, -\dfrac{9}{2}, -3, -\dfrac{3}{2}, -1, -\dfrac{1}{2}, \dfrac{1}{2}, 1, \dfrac{3}{2}, 3, \dfrac{9}{2}, 9$
13. $f(x) \to -\infty$ as $x \to -\infty$ and $f(x) \to \infty$ as $x \to \infty$.
14. $x = -2$, multiplicity: 3; $x = 2$, multiplicity: 1
15. Number of positive zeros: 0 or 2; number of negative zeros: 1
16. Horizontal asymptote: $y = 2$; vertical asymptotes: $x = -4$ and $x = 5$
17. $y = x - 6$ **18.** At $x = 3$ **19.** $x = 15$ (15,000 units)
20. $V(x) = x(8 - 2x)(17 - 2x)$

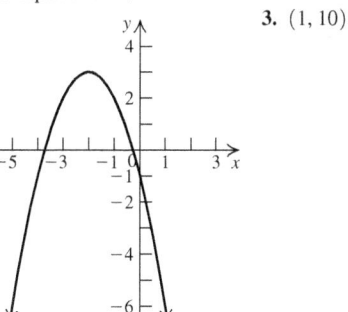

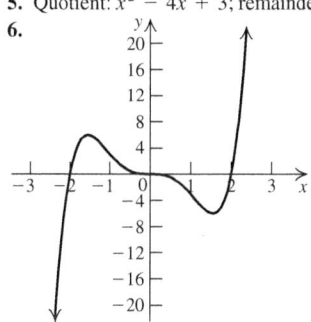

Practice Test B
1. b **2.** d **3.** a **4.** b **5.** d **6.** c **7.** c **8.** b **9.** c **10.** a
11. a **12.** d **13.** c **14.** b **15.** c **16.** c **17.** d **18.** d **19.** b
20. d

CHAPTER 3

Section 3.1

Basic Concepts and Skills:
1. $25, \dfrac{1}{5}$ **3.** $0.0892, 11.2116$ **5.** $\dfrac{4}{9}, \dfrac{729}{64}$ **7. a.** 4^x **b.** 3^x
9. a. $\dfrac{1}{5} \cdot 5^x$ **b.** $5 \cdot \left(\dfrac{1}{5}\right)^x$

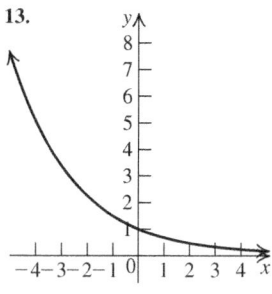

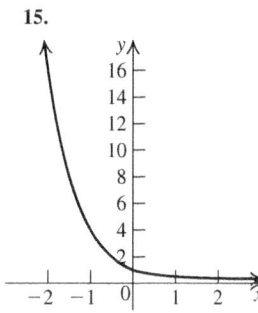

11. [graph] **13.** [graph]
15. [graph] **17.** [graph]

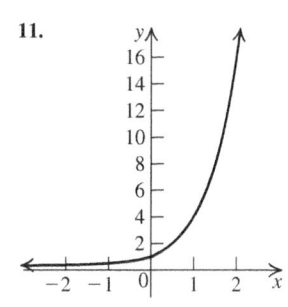

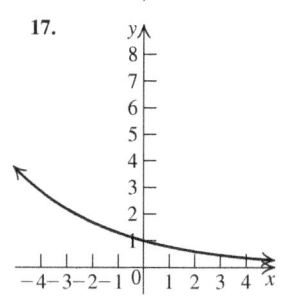

19. Yes, reflection in the y-axis **21.** c **23.** a
25. Domain: $(-\infty, \infty)$; Range: $(0, \infty)$ Horizontal asymptote: $y = 0$ 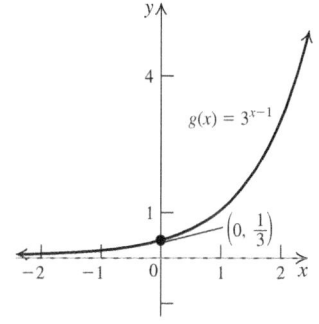
27. Domain: $(-\infty, \infty)$; Range: $(0, \infty)$ Horizontal asymptote: $y = 0$

[graph: $g(x) = 3^{x-1}$, point $(0, \tfrac{1}{3})$]
[graph: $g(x) = 4^{-x}$]

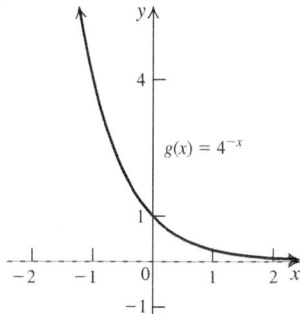

29. Domain: $(-\infty, \infty)$; Range: $(-\infty, 4)$ Horizontal asymptote: $y = 4$
31. Domain: $(-\infty, \infty)$; Range: $(-\infty, 3)$ Horizontal asymptote: $y = 3$

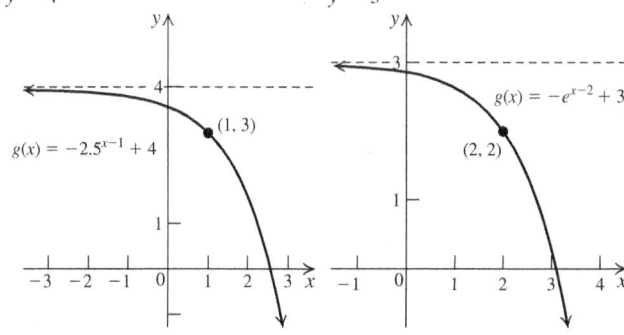

[graph: $g(x) = -2.5^{x-1} + 4$, point $(1, 3)$]
[graph: $g(x) = -e^{x-2} + 3$, point $(2, 2)$]

A-20 Answers ■ Section 3.1

33. Domain: $(-\infty, \infty)$; Range: $[1, \infty)$ No Horizontal asymptote.

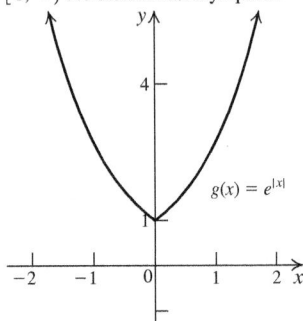

$g(x) = e^{|x|}$

35. $f(x) = (1.5)^x + 2$; $f(2) = 4.25$ **37.** $f(x) = \left(\dfrac{1}{3}\right)^x - 2$; $f(2) = -\dfrac{17}{9}$ **39.** $y = 2^{x+2} + 5$ **41.** $y = 2\left(\dfrac{1}{2}\right)^x - 5$ **43.** $2500
45. $5764.69 **47. a.** $7936.21 **b.** $4436.21 **49. a.** $12,365.41 **b.** $4865.41 **51.** $4631.93 **53.** $4493.29
55. $f(x) = e^{-x}; y = 0$ **57.** $f(x) = e^{x-2}; y = 0$

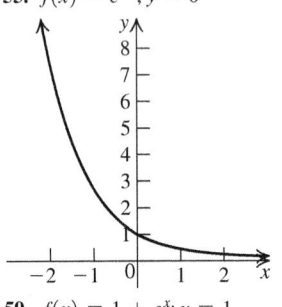

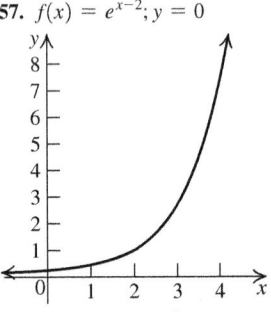

59. $f(x) = 1 + e^x; y = 1$ 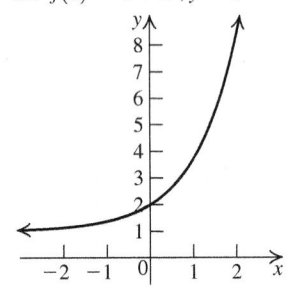 **61.** $f(x) = -e^{x-2} + 3; y = 3$

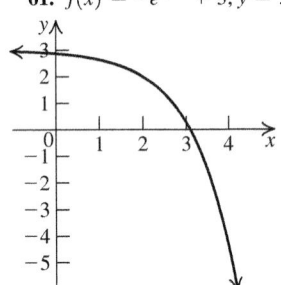

Applying the Concepts:
63. a. i. 176.6°C **ii.** 148.1°C **b.** $t = 5$ hours **c.** 25°C **65.** $220,262
67. $35,496.43 **69.** 16.81% **71.** 0.1357 mm² **73.** 3.9%
75. a. $0.015 \times 2^{30} \approx 16{,}106{,}127$ cm
b. $0.015 \times 2^{40} \approx 16{,}492{,}674{,}420$ cm
c. $0.015 \times 2^{50} \approx 1.689 \times 10^{13}$ cm

Critical Thinking / Discussion / Writing:
77. For $a = 1$, f is a constant function. If a is 0 or negative, then the domain is not $(-\infty, \infty)$ **78.** $x = \dfrac{m}{n}$, where n is odd
79. For $a > 1$, $y \to 0$ as $x \to \infty$ and $y \to b$ as $x \to -\infty$; for $0 < a < 1$, $y \to b$ as $x \to \infty$ and $y \to 0$ as $x \to -\infty$ **82.** $x = 1, x = 2$

Maintaining Skills:
83. 1 **84.** 0.1 **85.** -2 **86.** 5 **87.** 49 **88.** 3 **89.** $3\sqrt{2}$ **90.** $2\sqrt{3}$

91. Domain: $(-\infty, \infty)$
$f(x) = x^2 - 4$

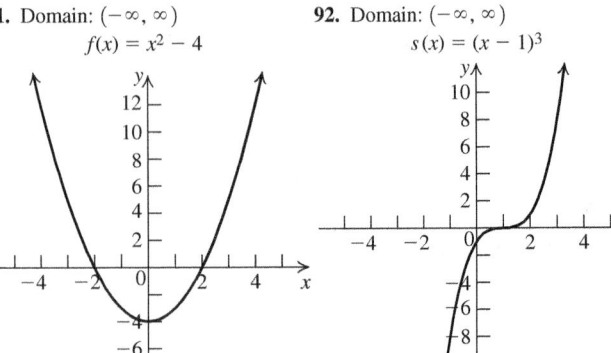

92. Domain: $(-\infty, \infty)$
$s(x) = (x - 1)^3$

93. Domain: $[0, \infty)$
$h(x) = \sqrt{x} + 3$

94. Domain: $[-1, \infty)$
$g(x) = \sqrt{x + 1}$

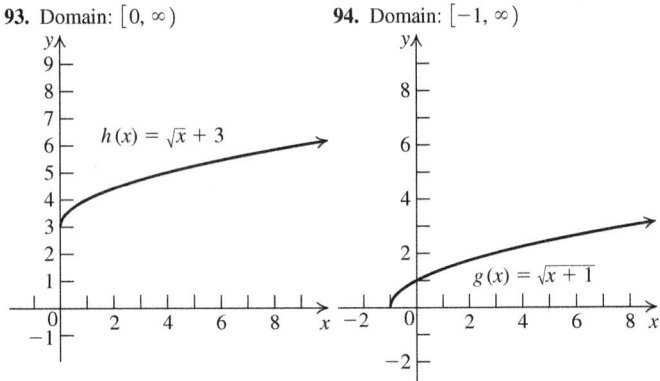

95. Domain: $(-\infty, 1]$
$f(x) = 2\sqrt{1 - x}$

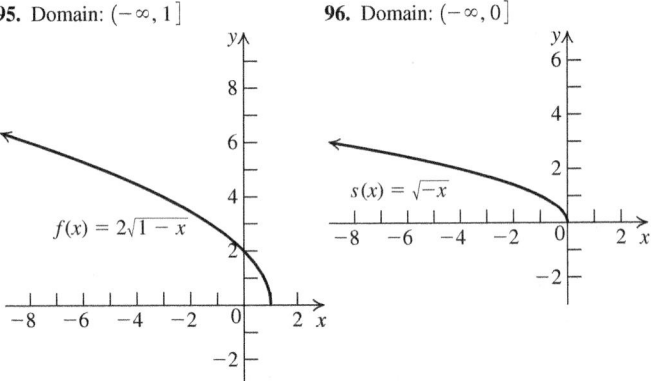

96. Domain: $(-\infty, 0]$
$s(x) = \sqrt{-x}$

97. Domain: $(-\infty, -1) \cup (-1, \infty)$
$h(x) = \dfrac{1}{x + 1}$

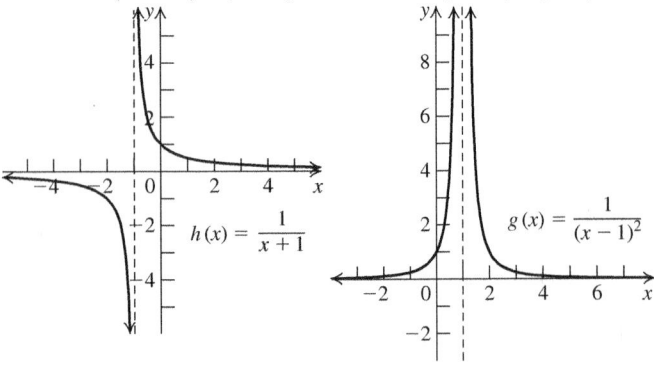

98. Domain: $(-\infty, 1) \cup (1, \infty)$
$g(x) = \dfrac{1}{(x - 1)^2}$

Section 3.2

Basic Concepts and Skills:

1. $\log_5 25 = 2$ **3.** $\log_{49}\left(\frac{1}{7}\right) = -\frac{1}{2}$ **5.** $\log_{1/16} 4 = -\frac{1}{2}$ **7.** $\log_{10} 1 = 0$
9. $\log_{10} 0.1 = -1$ **11.** $\log_a 5 = 2$ **13.** $2^5 = 32$ **15.** $10^2 = 100$
17. $10^0 = 1$ **19.** $10^{-2} = 0.01$ **21.** $8^{1/3} = 2$ **23.** $e^x = 2$ **25.** 3
27. 4 **29.** -3 **31.** $\frac{3}{2}$ **33.** $\frac{1}{4}$ **35.** $x = 25$ **37.** $x = 127$
39. $x = -4$ **41.** $x = 5$ **43.** $x = 2$ **45.** $x = 3, 4$ **47.** $(-1, \infty)$
49. $(5, \infty)$ **51.** $(1, \infty)$ **53.** or $(2, \infty)$ **55.** $(1, 2)$
57. a. f **b.** a **c.** d **d.** b **e.** e **f.** c
59. Domain $= (-3, \infty)$ **61.** Domain $= (0, \infty)$
Range $= (-\infty, \infty)$ Range $= (-\infty, \infty)$
Asymptote: $x = -3$ Asymptote: $x = 0$

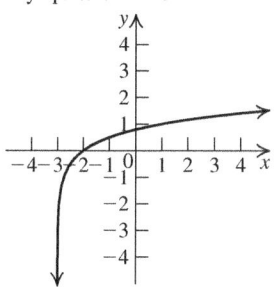

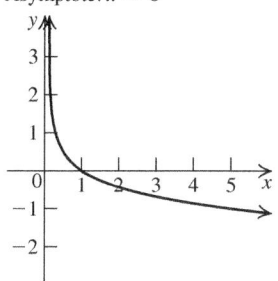

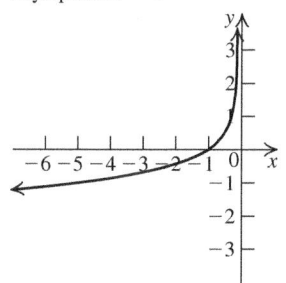

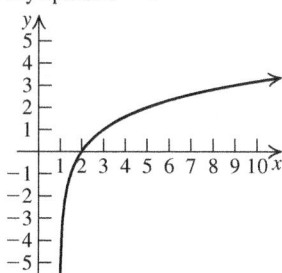

63. Domain $= (-\infty, 0)$ **65.** Domain $= (0, \infty)$
Range $= (-\infty, \infty)$ Range $= [0, \infty)$
Asymptote: $x = 0$ Asymptote: $x = 0$

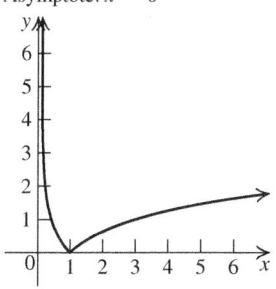

67. Domain $= (1, \infty)$ **69.** Domain $= (-\infty, 3)$
Range $= (-\infty, \infty)$ Range $= (-\infty, \infty)$
Asymptote: $x = 1$ Asymptote: $x = 3$

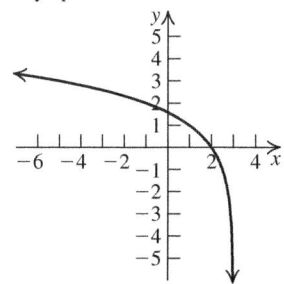

71. Domain $= (-\infty, 3)$ **73.** Domain $= (-\infty, 0) \cup (0, \infty)$
Range $= (-\infty, \infty)$ Range $= (-\infty, \infty)$
Asymptote: $x = 3$ Asymptote: $x = 0$

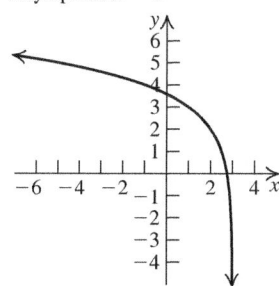

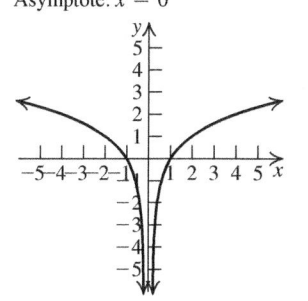

75. 1 **77.** 7 **79.** 5 **81.** $x = 100$ **83.** $x = e$
85. **87.**

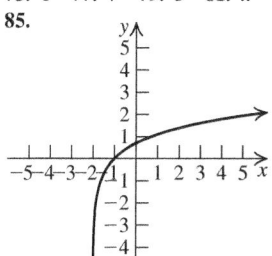

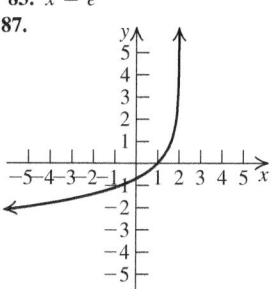

89.

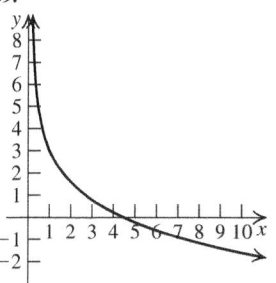

Applying the Concepts:
91. 8.66 yr **93. a.** ≈ 31.2 million **b.** 1.16%
c. ≈ 59.8 years or 2070 **d.** ≈ 31 yrs or 2041 **95.** At 8:17 P.M. (353 min earlier than 2:10 A.M., assuming that the temperature of the live body was 98.6°F) **97.** 76 min

Critical Thinking / Discussion / Writing:
99. a. $24{,}659.70 **b.** 4.05% **100. a.** $286{,}504.80 **b.** $84{,}584.55
c. $41{,}042.50 **101.** 1 **102.** 16 **103.** 16 **105. a.** Yes **b.** The one-to-one property **106.** To avoid expressions that are not real numbers, such as $\sqrt{-1}$

Maintaining Skills:
107. a^9 **108.** a^6 **109.** a^4 **110.** a^2 **111.** $\frac{-10}{7}$ **112.** $\left\{\frac{3}{2}, -1\right\}$
113. 3 **114.** 0 **115.** $y = \frac{1}{2}x + \frac{9}{2}$

Section 3.3

Basic Concepts and Skills:
1. 0.78 **3.** 0.7 **5.** -1.7 **7.** 7.3 **9.** $\frac{16}{3}$ **11.** 0.56
13. $\ln x + \ln(x - 1)$ **15.** $\frac{1}{2}\log(x^2 + 1) - \log(x + 3)$
17. $2\log_b x + 3\log_b y + \log_b z$
19. $\ln x + \frac{1}{2}\ln(x - 1) - \ln(x^2 + 2)$
21. $2\ln(x + 1) - \ln(x - 3) - \frac{1}{2}\ln(x + 4)$
23. $\ln(x + 1) + \frac{1}{2}[\ln(x^2 + 2) - \ln(x^2 + 5)]$
25. $3\ln x + 4\ln(3x + 1) - \frac{1}{2}\ln(x^2 + 1) + 5\ln(x + 2) - 2\ln(x - 3)$
27. $\log_2(7x)$ **29.** $\log\left(\frac{z\sqrt{x}}{y}\right)$ **31.** $\log_2(zy^2)^{1/5}$ **33.** $\ln(xy^2z^3)$
35. $\ln\left(\frac{x^2}{\sqrt{x^2 + 1}}\right)$ **37.** $1.4036 \cdot 10^{217}$ **39.** 234^{567} **41.** 362
43. 2.322 **45.** -1.585 **47.** 1.760 **49.** 3.6 **51.** $\frac{1}{2}$ **53.** 1 **55.** 18
57. 2 **59.** $y = 2 - \log x$ **61.** $y = 2 - \ln x$ **63.** $y = 1 + 3\log_5 x$
65. $y = -1 + 5\log_2 x$ **67.** ≈ 10.7 yrs **69.** ≈ 10.4 hrs

Applying the Concepts:
71. $K \approx 0.012775$ **73.** 0.010498 **75.** 12.53 yr **77.** 12.60 yr

Critical Thinking / Discussion / Writing:
79. Because $\log \frac{1}{2}$ is a negative number, $3 < 4$ implies that $3\log\frac{1}{2} > 4\log\frac{1}{2}$, so the second step is not correct. **80.** The domain of $2\log x$ is $(0, \infty)$, while the domain of $\log(x^2)$ is $(-\infty, 0) \cup (0, \infty)$.
81. 12978189 digits

Maintaining Skills:
82. 11 **83.** -4 **84.** 2 **85.** 1 **86.** $t^2 - t + 1 = 0$
87. $3t^2 - 2t - 7 = 0$ **88.** $t^2 + 4 = 0$ **89.** $3t^2 + 12 = 0$
90. -2 **91.** $\frac{1}{3}$ **92.** $-4, 1$ **93.** $-3, 8$ **94.** $(-\infty, 7)$ **95.** $(-\infty, 5]$
96. $(2, \infty)$ **97.** $[3, \infty)$

Section 3.4

Basic Concepts and Skills:
1. $x = 4$ **3.** $x = \frac{5}{3}$ **5.** $x = \pm\frac{7}{2}$ **7.** $\emptyset$ **9.** $x = 1$ **11.** $x = \frac{1}{2}$
13. $x = \pm 9$ **15.** $x = 1.585$ **17.** $x = 0.453$ **19.** $x = 1.766$
21. $x = 0.934$ **23.** $x = -1.159$ **25.** $x = -3.944$ **27.** $t = 11.007$
29. $x = 2.807$ **31.** $x = 0.631, 1.262$ **33.** $x = 1.262$ **35.** $x = 0.232$
37. $x = \frac{\ln 2}{\ln 3} \approx 0.631$ **39.** $x = \frac{\ln 18 - \ln 7}{\ln 3} \approx 0.860$ **41.** $\emptyset$
43. $x = -\frac{49}{20}$ **45.** $x = 3, -2$ **47.** $x = 5, 2$ **49.** $x = 8$ **51.** $x = 1$
53. $x = 3$ **55.** $x = 4$ **57.** $\emptyset$ **59.** $x = \frac{2}{3}, \frac{5}{2}$ **61.** $a = 30, k = 2; 500$
63. $a = 5, k = \frac{1}{2}; 31$ **65.** $a = 2, k = \ln\left(\frac{17}{2}\right); 0.068$
67. $a = -2, k = \ln\left(\frac{11}{18}\right); -7.902$ **69.** $x \geq \frac{\ln 2}{\ln 0.3}$ **71.** $x \leq 0$
73. $(-3, 17)$ **75.** $x \geq e + 5$ **77.** $\left(\frac{7}{3}, 5\right)$

Applying the Concepts:
79. a. 10.087 yr **b.** 9.870 yr **c.** 9.821 yr **d.** 9.797 yr **e.** 9.796 yr
81. a. $\frac{\ln 2}{\ln(1+r)}$ **b.** $t = \frac{\ln 2}{r}$ yr **83. a.** 20 **b.** 4490
85. a. $I = 10^{8.6}$ W/m² **b.** $E = 10^{17.3}$ joules **87.** $\log 150 \approx 2.2$
89. ≈ 74 dB **91.** $10^{-4.7}$ W/m²

Critical Thinking / Discussion / Writing:
93. $4, \frac{1}{32}$ **94.** $\frac{1}{9}, 3$ **95.** $x = 1.10$ **96.** $x = 16$ **97.** $x = 2$
98. $t = \frac{\ln a}{k}$ **99. a.** $-7, 9$ **b.** 9 **c.** $-7, 9$

Maintaining Skills:
100. $(0, 0); 1$ **101.** $(-2, 3); 4$ **102.** $(1, 0); 2$ **103.** $(0, -2); 2$
104. $(1, -2); 3$

Review Exercises

Basic Concepts and Skills:
1. False **3.** False **5.** True **7.** False **9.** True **11.** h **13.** f
15. d or e **17.** a
19. Domain $= (-\infty, \infty)$ **21.** Domain $= (-\infty, \infty)$
Range $= (0, \infty)$ Range $= (3, \infty)$
Asymptote: $y = 0$ Asymptote: $y = 3$

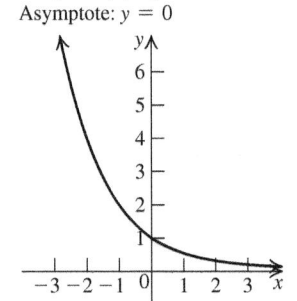

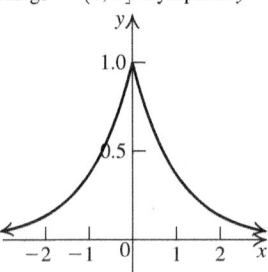

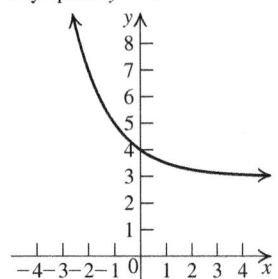

23. Domain $= (-\infty, \infty)$ **25.** Domain $= (-\infty, 0)$
Range $= (0, 1]$ Asymptote: $y = 0$ Range $= (-\infty, \infty)$
                                        Asymptote: $x = 0$

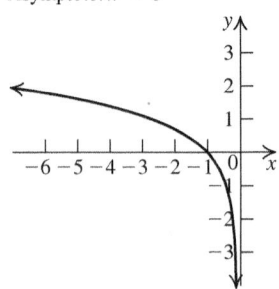

27. Domain $= (1, \infty)$ **29.** Domain $= (-\infty, 0)$
Range $= (-\infty, \infty)$ Range $= (-\infty, \infty)$
Asymptote: $x = 1$ Asymptote: $x = 0$

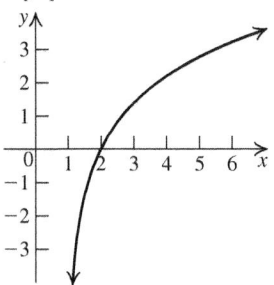

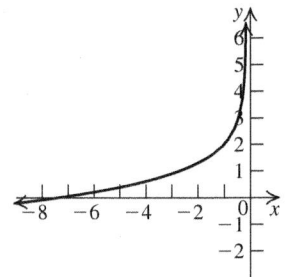

31. a. x-intercept $= \ln\left(\frac{2}{3}\right)$;
y-intercept $= 1$
b. as $x \to \infty, y \to 3$
as $x \to -\infty, y \to -\infty$

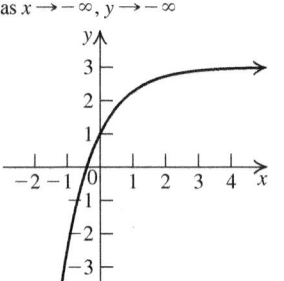

33. a. x-intercept: none;
y-intercept $= 1$
b. as $x \to \infty, y \to 0$
as $x \to -\infty, y \to 0$

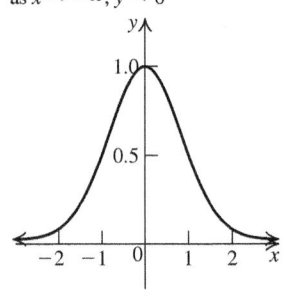

35. $a = 10, k = 2, f(2) = 160$
37. $a = 3, k = -0.2824, f(4) = 0.38898$ **39.** $y = 3 \cdot 2^x$
41. $y = \log_5(x - 1)$ **43. a.** $y = -2^{(x-1)} - 3$ **b.** $y = -2^{(x-1)} + 3$
45. $\ln x + 2\ln y + 3\ln z$ **47.** $\ln x + \frac{1}{2}\ln(x^2 + 1) - 2\ln(x^2 + 3)$
49. $y = 3x$ **51.** $y = -1 \pm \sqrt{x - 2}$ **53.** $y = \frac{\sqrt{x^2 - 1}}{x^2 + 1}$ **55.** 4
57. $-1 \pm \sqrt{5}$ **59.** $\frac{\ln 23}{\ln 3} \approx 2.854$ **61.** $\frac{\ln 19}{\ln 273} \approx 0.525$
63. 0 **65.** $\frac{-\ln 3}{\ln\left(\frac{(1.7)^3}{9}\right)} \approx 1.815$ **67.** $\frac{5}{2}$ **69.** 0 **71.** 8 **73.** 5
75. 9 **77.** 3 **79.** $x \geq -1$ **81.** $\left(-\frac{7}{2}, \frac{93}{2}\right)$

Applying the Concepts:
83. 11.4 yr **85.** 40 hr **87.** At 5% compounded yearly, $A = \$9849.70$.
At 4.75% compounded monthly, $A = \$9754.74$. The 5% investment will provide the greater return. **89. a.** 34 million **b.** In 2206 (after 199.3 yr)

91. a.

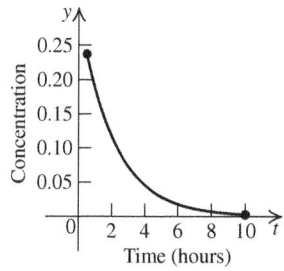

b. ≈ 5 hr

93. $t = 6.57$ min   **95. a.** 5000 people/square mile
b. 7459 people/square mile   **c.** ≈ 13.73 miles   **97.** $I = 0.2322I_0$
99. a. 10.98 ft/sec   **b.** 5.29 ft/sec   **c.** 201   **101. a.** 1 g
b. Remains stationary at 6 g   **c.** 4.6 days

Exercises for Calculus
1. $k = \ln a$ **2. a.** 0.22313, −1.499, 0.22335; 0.36788, −0.999, 0.36825;
0.60653, −0.499, 0.60714; 1.00000, 0.001, 1.00100; 1.64872, 0.501,
1.65037; 2.71828, 1.001, 2.72100; 4.48169, 1.501, 4.48617
b. The slope of this secant line is approximately e^x.
3. $f(t) = 1000 \cdot 8^{t/2}$
4. ≈ 1.37

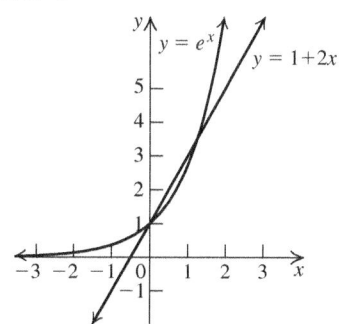

5. $a = 10{,}000$
$x > \ln(10); -x < -\ln 10; -x < \ln\frac{1}{10}; e^{-x} < e^{\ln(\frac{1}{10})}; e^{-x} < \frac{1}{10}$
$x > \ln(100); -x < -\ln(100); -x < \ln\left(\frac{1}{100}\right); e^{-x} < e^{\ln(\frac{1}{100})}; e^{-x} < \frac{1}{100}$
6. $2\log_2 x + \log_2(1-x)$
7. $4\log x + 3\log(2x-3) + 5\log(x+7)$
8. $2\ln x + 5\ln(x-1) - \frac{1}{2}\ln(x^2+1)$ **9.** ≈ 6.7 months
10. Using the continuous compounding formula with $r = 1$ and $t = 1$,
we have $A = 1e^{1 \cdot 1} = e$.

Practice Test A
1. $x = -3$ **2.** $x = 32$
3. Range $= (-\infty, 1)$; horizontal asymptote: $y = 1$ **4.** −3   **5.** $x = 3$
6. −3   **7.** $\ln 3x^5$ **8.** $\dfrac{\ln\left(\frac{5}{2}\right)}{\ln 2}$ **9.** $\ln 2$
10. $\ln 2 + 3\ln x - 5\ln(x+1)$ **11.** −5   **12.** $y = \ln(x-1) + 3$
13.

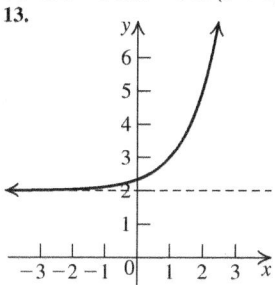

14. $(-\infty, 0)$ **15.** $\ln\dfrac{x^3(x^3+2)}{\sqrt{3x^2+2}}$ **16.** $x = 3$ **17.** $x = 3$
18. $\frac{1}{2}[\ln 2 + 3\ln x + 2\ln y]$ **19.** $A(t) = 15{,}000(1.0175)^{4t}$
20. In 1983 (after 23.03 yr)

Practice Test B
1. b   **2.** b   **3.** b   **4.** d   **5.** d   **6.** b   **7.** b   **8.** c   **9.** d   **10.** d   **11.** b
12. d   **13.** b   **14.** a   **15.** a   **16.** d   **17.** b   **18.** c   **19.** d   **20.** b

CHAPTER 4

Section 4.1

Basic Concepts and Skills:
1.   **3.**

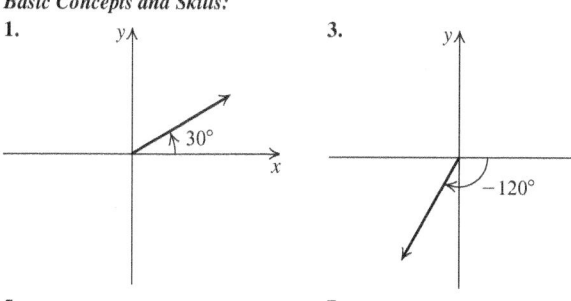

5.   **7.**

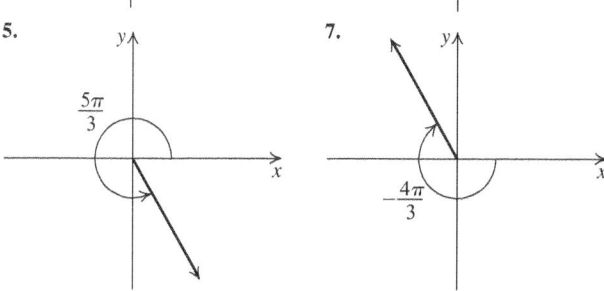

9. $\dfrac{\pi}{9}$ **11.** $-\pi$ **13.** $\dfrac{7\pi}{4}$ **15.** $\dfrac{8\pi}{3}$ **17.** $-\dfrac{17\pi}{6}$ **19.** 15°   **21.** −100°
23. 300°   **25.** 450°   **27.** −495°   **29.** 0.21 radian   **31.** −1.47 radians
33. 53.86°   **35.** −470.40°   **37.** Complement: 43°; supplement: 133°
39. Complement: none because the measure of the angle is greater than
90°; supplement: 60°   **41.** Complement: none because the measure of
the angle is greater than 90°; supplement: none because the measure of the
angle is greater than 180°   **43.** 0.28 radian   **45.** 2.095 radians
47. 1.309 m   **49.** 78 m

Applying the Concepts:
51. 70.69 in.   **53.** $\dfrac{2\pi}{3}$ radians   **55.** 2.62 ft   **57.** ≈ 508 mi
59. ≈ 462 mi   **61.** ≈ 6.366°   **63.** ≈ 1037 mph   **65.** 132,000 radians
per hour   **69.** 69°

Critical Thinking / Discussion / Writing:
72. 4   **73.** The line of latitude through N 40°30′13″

Maintaining Skills:
75. Yes   **76.** Yes   **77.** Yes   **78.** Yes   **79.** No   **80.** No

Section 4.2

Basic Concepts and Skills:
1. $\pm\dfrac{\sqrt{3}}{2}$ **3.** $\pm\dfrac{\sqrt{7}}{4}$ **5.** None
7. $\sin t = \dfrac{1}{3}, \cos t = \dfrac{2\sqrt{2}}{3}, \tan t = \dfrac{\sqrt{2}}{4}$
   $\csc t = 3, \sec t = \dfrac{3\sqrt{2}}{4}, \cot t = 2\sqrt{2}$
9. $\sin t = \dfrac{2\sqrt{2}}{3}, \cos t = \dfrac{1}{3}, \tan t = 2\sqrt{2}$
   $\csc t = \dfrac{3\sqrt{2}}{4}, \sec t = -3, \cot t = -\dfrac{\sqrt{2}}{4}$
11. $\sin t = -\dfrac{\sqrt{3}}{2}, \cos t = \dfrac{1}{2}, \tan t = -\sqrt{3}$
   $\csc t = \dfrac{2\sqrt{3}}{3}, \sec t = 2, \cot t = -\dfrac{\sqrt{3}}{3}$ **13.** 0   **15.** −1   **17.** 0

A-24 Answers — Section 4.3

19. $\sin t = -\dfrac{1}{2}, \cos t = \dfrac{\sqrt{3}}{2}, \tan t = -\dfrac{\sqrt{13}}{3}$

21. $\sin t = -\dfrac{\sqrt{3}}{2}, \cos t = -\dfrac{1}{2}, \tan t = \sqrt{3}$

23. $\sin t = -\dfrac{1}{2}, \cos t = -\dfrac{\sqrt{3}}{2}, \tan t = \dfrac{\sqrt{3}}{3}$

25. $\sin t = \dfrac{1}{2}, \cos t = -\dfrac{\sqrt{3}}{2}, \tan t = -\dfrac{\sqrt{3}}{3}$

27. $\sin t = -\dfrac{1}{2}, \cos t = \dfrac{\sqrt{3}}{2}, \tan t = -\dfrac{\sqrt{3}}{3}$

29. $\sin t = \dfrac{1}{2}, \cos t = \dfrac{\sqrt{3}}{2}, \tan t = \dfrac{\sqrt{3}}{3}$

31. $\sin t = -\dfrac{\sqrt{3}}{2}, \cos t = \dfrac{1}{2}, \tan t = -\sqrt{3}$

33. 0 35. $1 + \dfrac{\sqrt{2}}{2}$ 37. -1

39. $\sin\theta = \dfrac{3}{5}\quad \csc\theta = \dfrac{5}{3}$
$\cos\theta = -\dfrac{4}{5}\quad \sec\theta = -\dfrac{5}{4}$
$\tan\theta = -\dfrac{3}{4}\quad \cot\theta = -\dfrac{4}{3}$

41. $\sin\theta = -\dfrac{1}{2}\quad \csc\theta = -2$
$\cos\theta = -\dfrac{\sqrt{3}}{2}\quad \sec\theta = -\dfrac{2\sqrt{3}}{3}$
$\tan\theta = \dfrac{\sqrt{3}}{3}\quad \cot\theta = \sqrt{3}$

43. $\sin\theta = \dfrac{\sqrt{2}}{2}\quad \csc\theta = \sqrt{2}$
$\cos\theta = \dfrac{\sqrt{2}}{2}\quad \sec\theta = \sqrt{2}$
$\tan\theta = 1\quad \cot\theta = 1$

45. $\sin\theta = -\dfrac{5}{13}\quad \csc\theta = -\dfrac{13}{5}$
$\cos\theta = \dfrac{12}{13}\quad \sec\theta = \dfrac{13}{12}$
$\tan\theta = -\dfrac{5}{12}\quad \cot\theta = -\dfrac{12}{5}$

47. 0.97 49. 1.21 51. 0.84 53. -0.66 55. 1.00

57. $\sin\theta = -\dfrac{12}{13}\quad \csc\theta = -\dfrac{13}{12}$
$\cos\theta = -\dfrac{5}{13}\quad \sec\theta = -\dfrac{13}{5}$
$\tan\theta = \dfrac{12}{5}\quad \cot\theta = \dfrac{5}{12}$

59. $\sin\theta = \dfrac{4}{5}\quad \csc\theta = \dfrac{5}{4}$
$\cos\theta = -\dfrac{3}{5}\quad \sec\theta = -\dfrac{5}{3}$
$\tan\theta = -\dfrac{4}{3}\quad \cot\theta = -\dfrac{3}{4}$

61. $\sin\theta = \dfrac{3}{5}\quad \csc\theta = \dfrac{5}{3}$
$\cos\theta = -\dfrac{4}{5}\quad \sec\theta = -\dfrac{5}{4}$
$\tan\theta = -\dfrac{3}{4}\quad \cot\theta = -\dfrac{4}{3}$

63. $60°$ 65. $40°$ 67. $\dfrac{\pi}{4}$

69. $\dfrac{\pi}{6}$ 71. $x = \dfrac{\sqrt{3}}{2}, y = \dfrac{1}{2}$

Applying the Concepts:
73. **a.** 100 **b.** 75   75. **a.** 13 ft **b.** 7 ft

Critical Thinking / Discussion / Writing:
77. **a.** $\tan\left(\theta + \dfrac{\pi}{2}\right) = -\cot\theta$ **b.** $\cot\left(\theta + \dfrac{\pi}{2}\right) = -\tan\theta$
c. $\sec\left(\theta + \dfrac{\pi}{2}\right) = -\csc\theta$ **d.** $\csc\left(\theta + \dfrac{\pi}{2}\right) = \sec\theta$
81. $180° + \theta$ 82. **a.** $e^{\pi/2}$ **b.** $\ln \pi$

Maintaining Skills:
83. Odd   84. Even   85. Neither   86. Odd   87. Even   88. Neither

Section 4.3
Basic Concepts and Skills:

1. $y = 2\sin x$ 3. $y = -\dfrac{1}{2}\sin x$

5. $y = \cos 2x$ 7. $y = \cos \dfrac{2}{3}x$

9. $y = \cos\left(x + \dfrac{\pi}{2}\right)$ 11. $y = \cos\left(x - \dfrac{\pi}{3}\right)$

13. $y = 2\cos\left(x - \dfrac{\pi}{2}\right)$ 15. $y = (\sin x) + 1$

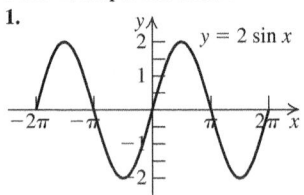

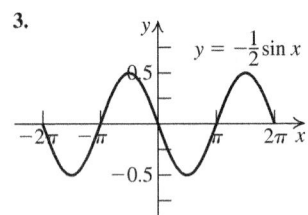

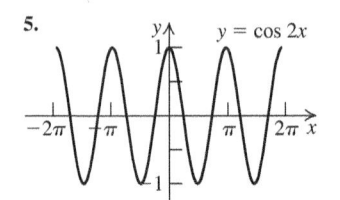

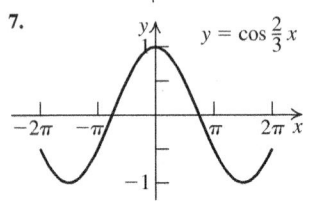

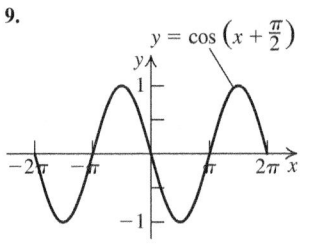

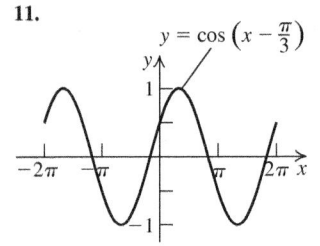

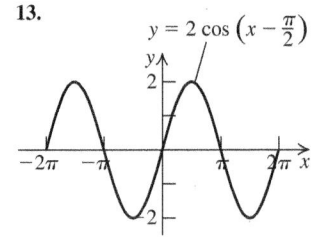

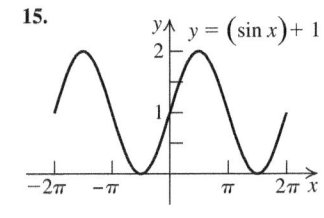

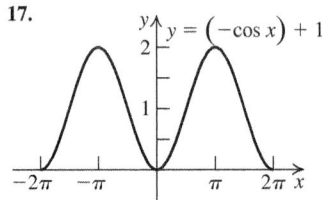

17. $y = (-\cos x) + 1$

19. Amplitude $= 5$; period $= 2\pi$; phase shift $= \pi$
21. Amplitude $= 7$; period $= \dfrac{2\pi}{9}$; phase shift $= -\dfrac{\pi}{6}$
23. Amplitude $= 6$; period $= 4\pi$; phase shift $= -2$
25. Amplitude $= 0.9$; period $= 8\pi$; phase shift $= \dfrac{\pi}{4}$

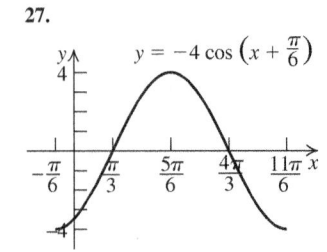

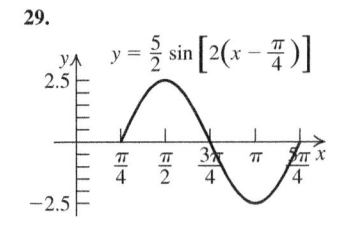

27. $y = -4\cos\left(x + \dfrac{\pi}{6}\right)$ 29. $y = \dfrac{5}{2}\sin\left[2\left(x - \dfrac{\pi}{4}\right)\right]$

31. [graph: $y = -5\cos\left[4\left(x - \frac{\pi}{6}\right)\right]$]

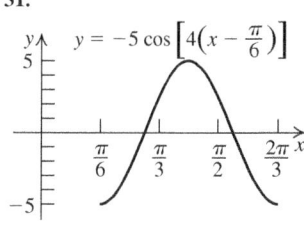

33. [graph: $y = \frac{1}{2}\sin\left[4\left(x + \frac{\pi}{4}\right)\right] + 2$]

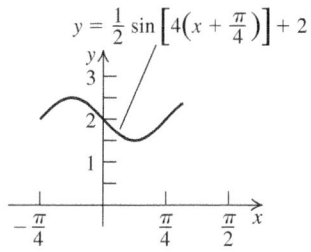

35. $y = 4\cos\left[2\left(x + \frac{\pi}{6}\right)\right]$; period $= \pi$; phase shift $= -\frac{\pi}{6}$

37. $y = -\frac{3}{2}\sin\left[2\left(x - \frac{\pi}{2}\right)\right]$; period $= \pi$; phase shift $= \frac{\pi}{2}$

39. $y = 3\cos \pi x$; period $= 2$; phase shift $= 0$

41. $y = \frac{1}{2}\cos\left[\frac{\pi}{4}(x + 1)\right]$; period $= 8$; phase shift $= -1$

43. $y = 2\sin\left[\pi\left(x + \frac{3}{\pi}\right)\right]$; period $= 2$; phase shift $= -\frac{3}{\pi}$

Applying the Concepts:

45. a. Period $= \frac{1}{70}$. The pulse is the frequency of the function; it says how many times the heart beats in one minute.

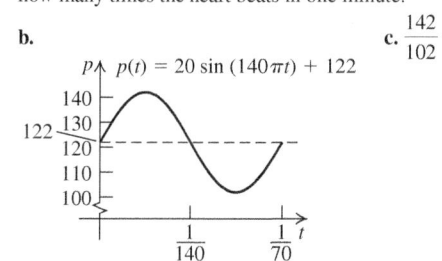

b. [graph: $p(t) = 20\sin(140\pi t) + 122$]

c. $\frac{142}{102}$

47. a. 800 kangaroos **b.** 500 kangaroos **c.** $\frac{\pi}{2}$ yr ≈ 1.57 yr

Critical Thinking / Discussion / Writing:

49. a. 10 units **b.** 5 units **50.** 10 **51.** $y = 3\cos 2x$

52. $y = -2\sin 2x$ **53.** $y = 3\sin\left[2\left(x - \frac{\pi}{4}\right)\right]$

54. $y = 2\cos\left[2\left(x + \frac{\pi}{4}\right)\right]$

Maintaining Skills:

56. $y = \frac{1}{2}x - 4$ **57.** Domain: $(-\infty, \infty)$; range: $[0, \infty)$

58. Zeros: $-7, 2$; horizontal asymptotes: $y = \frac{1}{2}$; vertical asymptotes: $x = -3, x = \frac{1}{2}$ **59.** $\frac{2\pi}{3}$

Section 4.4

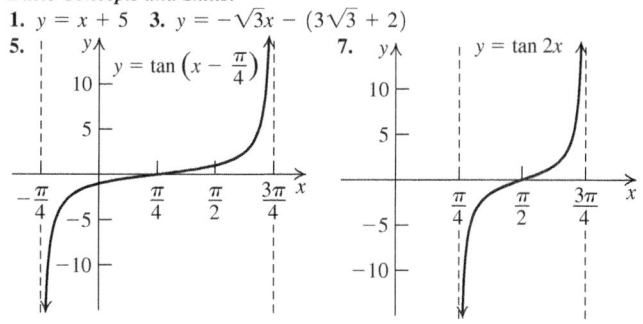

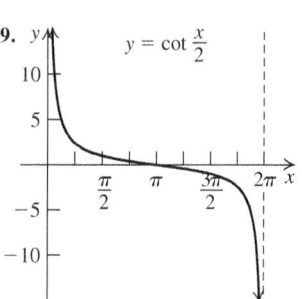

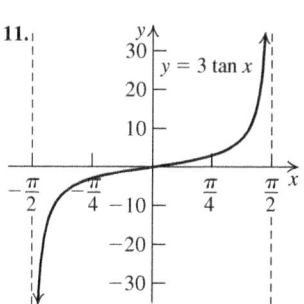

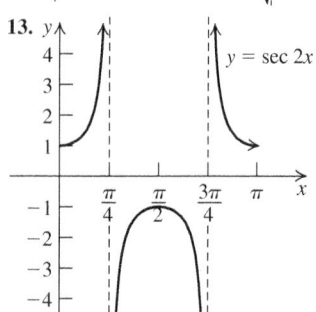

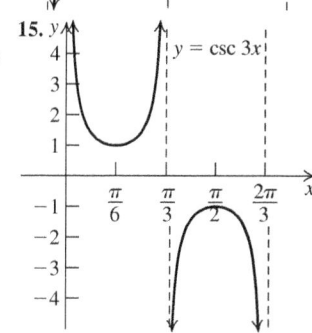

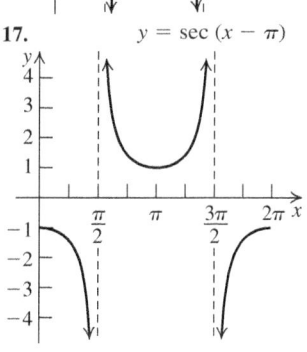

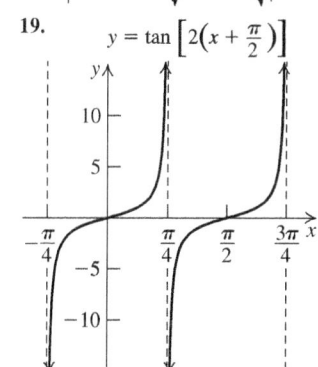

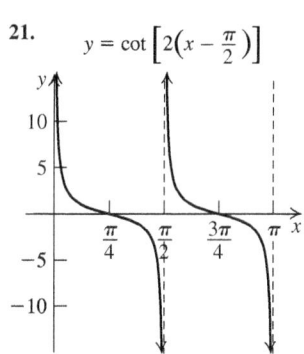

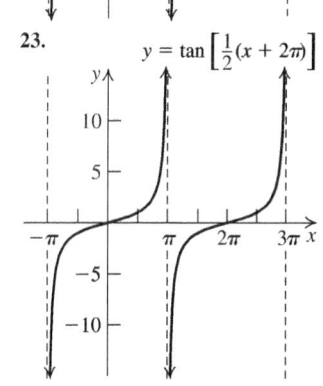

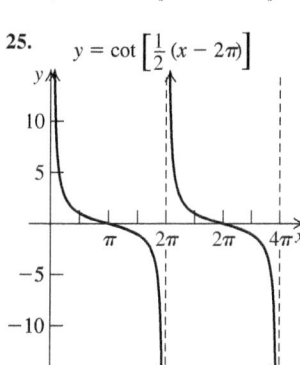

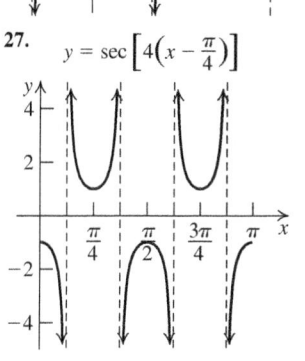

Basic Concepts and Skills:

1. $y = x + 5$ **3.** $y = -\sqrt{3}x - (3\sqrt{3} + 2)$

5. [graph: $y = \tan\left(x - \frac{\pi}{4}\right)$]

7. [graph: $y = \tan 2x$]

9. [graph: $y = \cot \frac{x}{2}$]

11. [graph: $y = 3\tan x$]

13. [graph: $y = \sec 2x$]

15. [graph: $y = \csc 3x$]

17. [graph: $y = \sec(x - \pi)$]

19. [graph: $y = \tan\left[2\left(x + \frac{\pi}{2}\right)\right]$]

21. [graph: $y = \cot\left[2\left(x - \frac{\pi}{2}\right)\right]$]

23. [graph: $y = \tan\left[\frac{1}{2}(x + 2\pi)\right]$]

25. [graph: $y = \cot\left[\frac{1}{2}(x - 2\pi)\right]$]

27. [graph: $y = \sec\left[4\left(x - \frac{\pi}{4}\right)\right]$]

29. $y = 3\csc\left(x + \frac{\pi}{2}\right)$

31. $y = \tan\left[\frac{2}{3}\left(x - \frac{\pi}{2}\right)\right]$

33. $y = -5\tan\left[2\left(x + \frac{\pi}{3}\right)\right]$

35. $y = \frac{1}{3}\cot\left[2(x - \pi)\right]$

Applying the Concepts:
37. a. $d(t) = 20\tan\frac{\pi t}{5}$

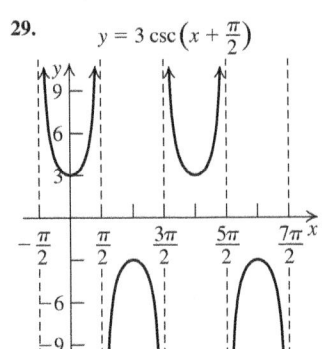

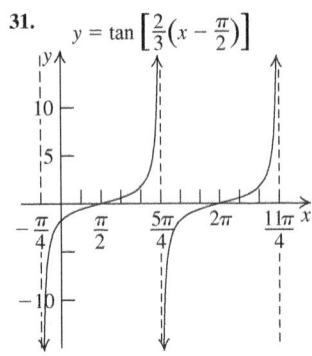

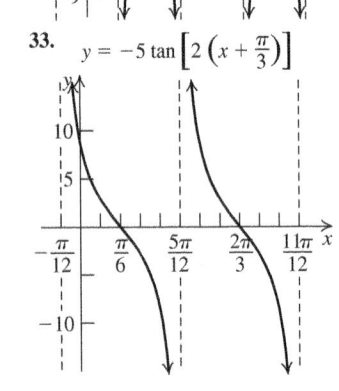

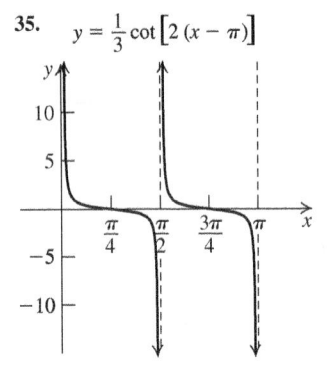

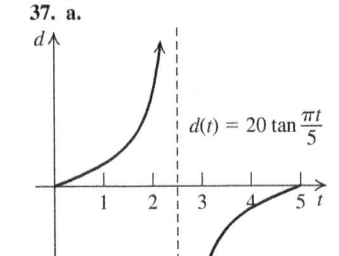

b. The light is pointing parallel to the wall.

Critical Thinking / Discussion / Writing:
39. $k = -\frac{7\pi}{4}$ **40.** 6 **41.** $y = -2\csc\left(x + \frac{\pi}{2}\right)$
42. $y = 5\cot\left(\frac{x}{2} - \frac{\pi}{2}\right)$ **43.** $y = 2\tan\left(x + \frac{\pi}{2}\right)$ **44.** $y = 2\csc 2x$
45. $y = -3\sec\left(\frac{x}{2} - \pi\right)$ **46.** $y = 7\tan\left[\frac{\pi}{8}(x + 1)\right]; -1.39, 10.48$
47. $y = -6\tan\left[\frac{\pi}{10}(x - 2)\right]$ 11.78, 0.95

Maintaining Skills:
48. one-to-one; $f^{-1}(x) = \frac{x + 3}{2}$ **49.** Not one-to-one
50. one-to-one; $f^{-1}(x) = \sqrt{\frac{x}{2}}$ **51.** True **52.** True

Section 4.5

Basic Concepts and Skills:
1. 0 **3.** $-\frac{\pi}{6}$ **5.** π **7.** Undefined **9.** $\frac{\pi}{3}$ **11.** $-\frac{\pi}{4}$ **13.** $\frac{3\pi}{4}$
15. Undefined **17.** $\frac{2\pi}{3}$ **19.** $\frac{1}{8}$ **21.** $\frac{\pi}{7}$ **23.** 247 **25.** $-\frac{\pi}{3}$ **27.** $-\frac{\pi}{3}$
29. $\frac{\pi}{3}$ **31.** $\frac{\pi}{6}$ **33.** $\frac{2\pi}{3}$ **35.** $-\frac{\pi}{3}$ **37.** $\frac{2\pi}{3}$ **39.** 1 **41.** -1 **43.** 1
45. 53.13° **47.** $-43.63°$ **49.** 73.40° **51.** 85.91° **53.** $-88.64°$
55. $\frac{\sqrt{5}}{3}$ **57.** $\frac{3}{5}$ **59.** $\frac{2\sqrt{29}}{29}$ **61.** $\frac{3}{4}$ **63.** $\frac{4\sqrt{17}}{17}$ **65.** $\sqrt{3}$
67. $\frac{x}{\sqrt{x + x^2}}$ **69.** $\frac{\sqrt{25 - 9x^2}}{3x}$ **71.** $\frac{x}{\sqrt{4 - x^2}}$ **73.** $\cot\theta$
75. $\sin\theta$ **77.** $\csc\theta$

Applying the Concepts:
79. 62° **81.** 159°

Maintaining Skills:
85. $\frac{2}{x^2 - 1}$ **86.** $\frac{x}{x^2 - 1}$ **87.** $x + a$ **88.** 0

Section 4.6

Basic Concepts and Skills:
1. $\sin\theta = \frac{2\sqrt{5}}{25}$ $\csc\theta = \frac{5\sqrt{5}}{2}$ **3.** $\sin\theta = \frac{3}{5}$ $\csc\theta = \frac{5}{3}$
$\cos\theta = \frac{11\sqrt{5}}{25}$ $\sec\theta = \frac{5\sqrt{5}}{11}$ $\cos\theta = \frac{4}{5}$ $\sec\theta = \frac{5}{4}$
$\tan\theta = \frac{2}{11}$ $\cot\theta = \frac{11}{2}$ $\tan\theta = \frac{3}{4}$ $\cot\theta = \frac{4}{3}$

5. $\sin\theta = \frac{\sqrt{2}}{10}$ $\csc\theta = 5\sqrt{2}$ **7.** $\sin\theta = \frac{\sqrt{5}}{3}$ $\csc\theta = \frac{3\sqrt{5}}{5}$
$\cos\theta = \frac{7\sqrt{2}}{10}$ $\sec\theta = \frac{5\sqrt{2}}{7}$ $\cos\theta = \frac{2}{3}$ $\sec\theta = \frac{3}{2}$
$\tan\theta = \frac{1}{7}$ $\cot\theta = 7$ $\tan\theta = \frac{\sqrt{5}}{2}$ $\cot\theta = \frac{2\sqrt{5}}{5}$

9. $\sin\theta = \frac{5\sqrt{34}}{34}$ $\csc\theta = \frac{\sqrt{34}}{5}$ **11.** $\sin\theta = \frac{5}{13}$ $\csc\theta = \frac{13}{5}$
$\cos\theta = \frac{3\sqrt{34}}{34}$ $\sec\theta = \frac{\sqrt{34}}{3}$ $\cos\theta = \frac{12}{13}$ $\sec\theta = \frac{13}{12}$
$\tan\theta = \frac{5}{3}$ $\cot\theta = \frac{3}{5}$ $\tan\theta = \frac{5}{12}$ $\cot\theta = \frac{12}{5}$

13. 0.8480 **15.** 0.5095 **17.** 2.3662 **19.** $B = 40°, a \approx 7.0, b \approx 5.9$
21. $B = 54°, c \approx 20.4, b \approx 16.5$
23. $c \approx 14.0, A \approx 63.6°, B \approx 26.4°$
25. $a \approx 11, A \approx 49.6°, B \approx 40.4°$
27. $c \approx 12.806, \sin\theta \approx 0.625, \tan\theta \approx 0.800$
29. $\cos\theta \approx 0.291, \tan\theta \approx 3.286$
31. $\sin\theta = 0.5, b \approx 15.588, c = 18$
33. $\alpha = 30°, a \approx 12.124, c = 14$

Applying the Concepts:
35. ≈ 11 ft **37.** ≈ 1029 m **39.** 2500 ft **41.** ≈ 29 ft
43. ≈ 19 ft **45.** ≈ 34 ft **47.** $\approx 12{,}994$ ft **49.** 62°
51. ≈ 2545 mi **53.** 159°

Critical Thinking / Discussion / Writing:
56. 1 **57.** $\tan\beta > \tan\alpha$

Maintaining Skills:
59. $(2, 8)$ **60.** $(-5, 2)$ **61.** $(-\infty, -2] \cup [4, \infty)$
62. $(-\infty, -2] \cup \left[-\frac{2}{3}, \infty\right)$ **63.** $(-\infty, 1)$ **64.** $\left(\frac{2}{3}, 4\right)$

Section 4.7

Basic Concepts and Skills:

1. $\tan x = -\frac{4}{3}, \cot x = -\frac{3}{4}, \sec x = -\frac{5}{3}, \csc x = \frac{5}{4}$
3. $\tan x = \frac{\sqrt{2}}{2}, \cot x = \sqrt{2}, \sec x = \frac{\sqrt{6}}{2}, \csc x = \sqrt{3}$
5. $\sin x = -\frac{\sqrt{5}}{5}, \cos x = -\frac{2\sqrt{5}}{5}, \cot x = 2, \csc x = -\sqrt{5}$
7. $\sin x = \frac{1}{3}, \cos x = \frac{2\sqrt{2}}{3}, \tan x = \frac{\sqrt{2}}{4}, \sec x = \frac{3\sqrt{2}}{4}$
9. $\cos x = -\frac{12}{13}, \tan x = \frac{5}{12}, \sec x = -\frac{13}{12}, \csc x = -\frac{13}{5}$
11. $\sin x = -\frac{2\sqrt{2}}{3}, \cos x = \frac{1}{3}, \tan x = -2\sqrt{2},$
    $\cot x = -\frac{\sqrt{2}}{4}, \csc x = -\frac{3\sqrt{2}}{4}$
13. 2 15. 1 17. $\csc^2 x + \cot^2 x$
19. 1 21. $\tan x - 3$
23. $x = \pi; 0 \neq 2$ 25. $x = \pi; -1 \neq 1$
27. $x = \frac{\pi}{3}; \frac{3}{4} \neq \frac{1}{4}$ 65. $\sec \theta$ 67. $\tan \theta$ 69. $\cot \theta$

Applying the Concepts:

71. $20 \csc \theta$ 73. $\tan \theta = \frac{m_1 - m_2}{1 + m_1 m_2}$

Critical Thinking / Discussion / Writing:

76. $t = 0$ 77. $a = b$, for $a, b \neq 0$ 78. $a = b$, and $a, b \neq 0$

Maintaining Skills:

80. $2\sqrt{5}$ 81. 2 82. x

Section 4.8

Basic Concepts and Skills:

1. $\frac{\sqrt{6} + \sqrt{2}}{4}$ 3. $2 - \sqrt{3}$ 5. $-\sqrt{6} - \sqrt{2}$ 7. $\frac{\sqrt{6} - \sqrt{2}}{4}$
9. $-2 - \sqrt{3}$ 11. $2 + \sqrt{3}$ 21. $\frac{\sqrt{3}}{2}$ 23. 1 25. $\frac{56}{65}$ 27. $\frac{16}{63}$
29. a. $-\frac{24}{25}$ b. $\frac{7}{25}$ c. $-\frac{24}{7}$
31. a. $\frac{8}{17}$ b. $-\frac{15}{17}$ c. $-\frac{8}{15}$
33. a. $-\frac{4}{5}$ b. $-\frac{3}{5}$ c. $\frac{4}{3}$
35. $-\frac{\sqrt{3}}{2}$ 37. $-\frac{\sqrt{3}}{3}$ 39. $\frac{\sqrt{2}}{2}$
45. $\frac{\sqrt{2} - \sqrt{3}}{2}$ 47. $\frac{\sqrt{2} + \sqrt{2}}{2}$ 49. $-\sqrt{2} - 1$
51. a. $\frac{2\sqrt{5}}{5}$ b. $\frac{\sqrt{5}}{5}$ c. 2
53. a. $\sqrt{\frac{13 + 3\sqrt{13}}{26}}$ b. $\sqrt{\frac{13 - 3\sqrt{13}}{26}}$ c. $\sqrt{\frac{13 + 3\sqrt{13}}{13 - 3\sqrt{13}}}$
55. a. $\sqrt{\frac{5 - \sqrt{5}}{10}}$ b. $\sqrt{\frac{5 + \sqrt{5}}{10}}$ c. $\sqrt{\frac{5 - \sqrt{5}}{5 + \sqrt{5}}}$

Applying the Concepts:

57. 100 watts 59. 1200 watts 63. 0 65. $\left\{\frac{\pi}{4}\right\}$

Critical Thinking / Discussion / Writing:

67. a. $\frac{4}{5}$ b. $\frac{3}{5}$ c. $\frac{24}{25}$ d. $-\frac{7}{25}$ e. $-\frac{24}{7}$ f. $\frac{\sqrt{5}}{5}$ g. $\frac{2\sqrt{5}}{5}$ h. $\frac{1}{2}$
69. $\frac{1 + \cot x \cot y}{\cot y - \cot x}$

Maintaining Skills:

72. Odd 73. Neither 74. Even 75. Even 76. Even

Review Exercises

Basic Concepts and Skills:

1. $\frac{\pi}{9}$ 3. $-\frac{\pi}{3}$ 5. $50°$ 7. 1.257 m
9. $\sin \theta = \frac{4\sqrt{17}}{17}, \cos \theta = \frac{\sqrt{17}}{17}, \tan \theta = 4, \cot \theta = \frac{1}{4},$
   $\sec \theta = \sqrt{17}, \csc \theta = \frac{\sqrt{17}}{4}$
11. $\sin \theta = \frac{2}{3}, \cos \theta = -\frac{\sqrt{5}}{3}, \tan \theta = -\frac{2\sqrt{5}}{5}, \cot \theta = -\frac{\sqrt{5}}{2},$
    $\sec \theta = -\frac{3\sqrt{5}}{5}, \csc \theta = \frac{3}{2}$
13. $\sin \theta = -\frac{3}{5}, \tan \theta = \frac{3}{4}, \cot \theta = \frac{4}{3}, \sec \theta = -\frac{5}{4}, \csc \theta = -\frac{5}{3}$
15. $\cos \theta = -\frac{4}{5}, \tan \theta = -\frac{3}{4}, \cot \theta = -\frac{4}{3}, \sec \theta = -\frac{5}{4}, \csc \theta = \frac{5}{3}$
17. Amplitude = 14; period = π; phase shift = $-\frac{\pi}{14}$
19. Vertical stretch = 6; period = $\frac{\pi}{2}$; phase shift = $-\frac{\pi}{10}$

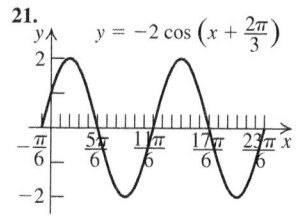

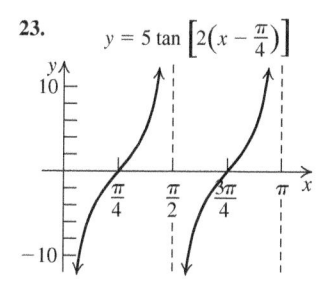

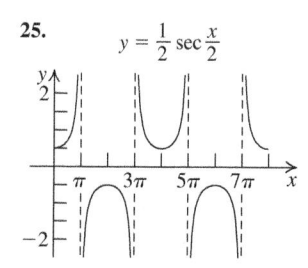

21.

23.

25.

27. $\frac{5\pi}{8}$ 29. $\frac{\pi}{3}$ 31. $\frac{\sqrt{2}}{2}$
33. $-\sqrt{3}$ 35. $B = 60°, b \approx 10.4, c = 12$
37. $A \approx 31.0°, B \approx 59.0°, c \approx 5.8$ 51. $\frac{\sqrt{2} + \sqrt{3}}{2}$
53. $\sqrt{6} - \sqrt{2}$ 55. 1 57. -1 59. $-\frac{16}{65}$ 61. $\frac{63}{65}$

Applying the Concepts:

77. ≈ 44 in 79. ≈ 866 mi 81. 1.5 yr

Exercises for Calculus

3. b. $12r \tan \dfrac{\pi}{6}$ **4. b.** $6r^2 \tan \dfrac{\pi}{6}$

5. b. $20\pi \approx 62.8319$ **c.** $80, 64.9839, 62.9147, 62.8525$; approach 20π = circumference of the circle.

6. b. $100\pi \approx 314.1593$ **c.** $400, 324.9197, 314.5733, 314.2627$. Approach the area of the circle

7. a. $A_2 = \left(r \cos \dfrac{\pi}{4}, r \sin \dfrac{\pi}{4} \right)$,

$A_3 = \left(r \cos \dfrac{\pi}{2}, r \sin \dfrac{\pi}{2} \right)$,

$A_4 = \left(r \cos \dfrac{3\pi}{4}, r \sin \dfrac{3\pi}{4} \right)$

$A_5 = (r \cos \pi, r \sin \pi)$,

$A_6 = \left(r \cos \dfrac{5\pi}{4}, r \sin \dfrac{5\pi}{4} \right)$,

$A_7 = \left(r \cos \dfrac{3\pi}{2}, r \sin \dfrac{3\pi}{2} \right)$,

$A_8 = \left(r \cos \dfrac{7\pi}{4}, r \sin \dfrac{7\pi}{4} \right)$

b. $r\sqrt{2 - \sqrt{2}}$ **c.** $8r\sqrt{2 - \sqrt{2}}$

8. a. $\dfrac{r^2}{2\sqrt{2}}$ **b.** $2\sqrt{2}\, r^2$

9. b. $10\pi \approx 31.4159$
c. $28.2843, 30.9017, 31.3953, 31.4108$; approach 10π
10. b. $25\pi \approx 78.5398$ **b.** $50, 73.4732, 78.3333, 78.4881$; approach 25π

Practice Test A

1. 2.4435 **2.** $252°$ **3.** $\dfrac{2\sqrt{29}}{29}$ **4.** Quadrant II **5.** $\dfrac{13}{5}$ **6.** $37°$

7. 0.223 **8.** Amplitude $= 19$; period $= \dfrac{\pi}{6}$; phase shift $= -3\pi$

9. $-\dfrac{\pi}{3}$ **10.** $\dfrac{2\pi}{3}$ **11.** $-\dfrac{\sqrt{2}}{4}$ **15.** $10(\sqrt{3} + 1)$ **16.** 18 ft **17.** $\dfrac{24}{25}$

18. Odd **19.** No; let $x = y = \dfrac{\pi}{2}$. **20.** $\theta = 45°$

Practice Test B

1. b **2.** d **3.** a **4.** d **5.** b **6.** c **7.** a **8.** b **9.** c **10.** d
11. d **12.** d **13.** b **14.** c **15.** a **16.** c **17.** b **18.** a
19. c **20.** b

CHAPTER 5

Section 5.1

Basic Concepts and Skills:
1. $C = 63°, a \approx 98.2 \text{ ft}, b \approx 93.0 \text{ ft}$
3. $B = 27°, a \approx 64.8 \text{ m}, b \approx 31.3 \text{ m}$ **5.** $B = 65°, C = 50°, c \approx 16.9 \text{ ft}$
7. $B \approx 23.7°, C \approx 41.3°, c \approx 51.0 \text{ m}$ **9.** $b \approx 20.2, A \approx 44°, C \approx 30°$
11. $A \approx 130.5°, B \approx 27.1°, C \approx 22.4°$
13. $C = 105°, b \approx 89.2 \text{ m}, c \approx 150.3 \text{ m}$
15. $B = 79°, b \approx 102.3 \text{ cm}, c \approx 85.4 \text{ cm}$
17. $B = 98°, a \approx 47.1 \text{ ft}, b \approx 81.2 \text{ ft}$
19. $A = 70°, a \approx 55.1 \text{ in.}, c \approx 54.0 \text{ in.}$
21. $A = 24°, a \approx 13.3 \text{ ft}, b \approx 30.7 \text{ ft}$
23. $C = 98.5°, a \approx 17.7 \text{ m}, b \approx 21.7 \text{ m}$
25. $B \approx 34°, C \approx 106°, c \approx 34.4$ **27.** $C = 60°, B = 90°, c = 25\sqrt{3}$
29. No triangle exists. **31.** No triangle exists.
33. $B \approx 56.8°, C \approx 23.2°, c \approx 16.0$ **35.** No triangle exists.
37. Solution 1: $C \approx 55.3°, A \approx 78.7°, a \approx 47.7$
  Solution 2: $C \approx 124.7°, A \approx 9.3°, a \approx 7.9$ **39.** No triangle exists.
41. Solution 1: $C \approx 49.0°, B \approx 89.0°, b \approx 82.2$
  Solution 2: $C \approx 131.0°, B \approx 7.0°, b \approx 10.0$
43. No triangle exists. **45.** $c \approx 21, A \approx 38.2°, B \approx 21.8°$
47. $a \approx 11.5, B \approx 50.4°, C \approx 67.6°$

49. $A \approx 81.3°, B \approx 46.4°, C \approx 52.3°$
51. $A \approx 28.3°, B \approx 43.3°, C \approx 108.4°$
53. $A \approx 23.7°, B \approx 36.4°, C \approx 119.9°$
55. $a \approx 5.7, B \approx 33.8°, C \approx 48.5°$
57. $b \approx 5.0, A \approx 46.3°, C \approx 65.4°$
59. $A \approx 38.4°, B \approx 49.1°, C \approx 92.5°$

Applying the Concepts:
61. $\approx 399 \text{ ft}$ **63. a.** 25.7 mi **b.** 14.5 mi **65.** 33 or 127 million miles
67. 9.1 miles **69. a.** 3689 m **b.** 1634.5 m

Critical Thinking / Discussion / Writing:
74. $90°$ **75.** $a = \sqrt{13}, b = 5\sqrt{2}, c = \sqrt{53}, A \approx 29.1°, B \approx 72.2°, C \approx 78.7°$
76. $a \approx 4.3, A \approx 12.5°, C \approx 17.5°$
77. $c \approx 11.98, C \approx 93.6°, B \approx 56.4°, c \approx 5.34, C \approx 26.4°, B \approx 123.6°$
78. No triangle can be formed.

Maintaining Skills:
79. 15×45 **80.** 14 in. **81.** $\dfrac{9}{10}\pi$ **82.** $\dfrac{7}{4}\pi \text{ ft}^2$

Section 5.2

Basic Concepts and Skills:
1. 654.2 sq in. **3.** 118.7 sq km **5.** 62.5 sq mm **7.** 25,136.6 sq ft
9. 84.3 sq ft **11.** 167.1 sq yd **13.** 178.6 sq cm **15.** 1621.9 sq ft
17. 2.9 **19.** 1240.2 **21.** 13.5 **23.** 7.7 **25.** $30°$ or $150°$ **27.** $90°$
31. 187.35 cm^2

Applying the Concepts:
33. 493,941 sq mi **35.** $\approx \$775{,}668$

Critical Thinking / Discussion / Writing:
40. b. $(6\pi - 9\sqrt{3}) \text{ in}^2$

Maintaining Skills:
47. Only the x-axis **48.** Only the origin
49. The x-axis, y-axis, and origin
50. x-intercepts: $-3 \pm \sqrt{21}$ y-intercepts: $-2, 6$

Section 5.3

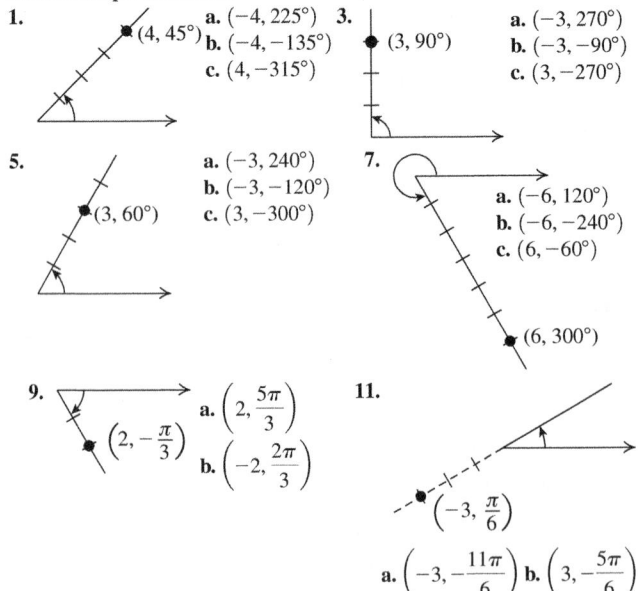

Basic Concepts and Skills:
1. (4, 45°) **a.** $(-4, 225°)$ **3.** (3, 90°) **a.** $(-3, 270°)$
 b. $(-4, -135°)$ **b.** $(-3, -90°)$
 c. $(4, -315°)$ **c.** $(3, -270°)$

5. (3, 60°) **a.** $(-3, 240°)$ **7.** **a.** $(-6, 120°)$
 b. $(-3, -120°)$ **b.** $(-6, -240°)$
 c. $(3, -300°)$ **c.** $(6, -60°)$

(6, 300°)

9. $\left(2, -\dfrac{\pi}{3}\right)$ **a.** $\left(2, \dfrac{5\pi}{3}\right)$ **11.**
 b. $\left(-2, \dfrac{2\pi}{3}\right)$ $\left(-3, \dfrac{\pi}{6}\right)$

a. $\left(-3, -\dfrac{11\pi}{6}\right)$ **b.** $\left(3, -\dfrac{5\pi}{6}\right)$

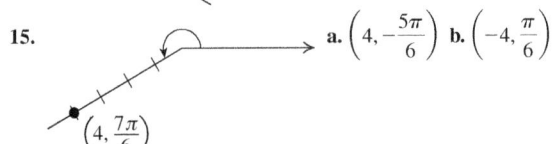

13. $\left(-2, -\frac{\pi}{6}\right)$ **a.** $\left(-2, \frac{11\pi}{6}\right)$ **b.** $\left(2, \frac{5\pi}{6}\right)$

15.   **a.** $\left(4, -\frac{5\pi}{6}\right)$ **b.** $\left(-4, \frac{\pi}{6}\right)$ $\left(4, \frac{7\pi}{6}\right)$

17. $\left(\frac{3}{2}, \frac{3\sqrt{3}}{2}\right)$ **19.** $\left(\frac{5}{2}, -\frac{5\sqrt{3}}{2}\right)$ **21.** $(-3, 0)$ **23.** $(\sqrt{3}, 1)$
25. $\left(\sqrt{2}, \frac{7\pi}{4}\right)$ **27.** $\left(3\sqrt{2}, \frac{\pi}{4}\right)$ **29.** $\left(3\sqrt{2}, \frac{7\pi}{4}\right)$ **31.** $\left(2, \frac{2\pi}{3}\right)$
33. $r = 4$ **35.** $r = 4 \cot\theta \csc\theta$ **37.** $r = 6\sin\theta$ **39.** $r^2 = \dfrac{1}{\sin\theta\cos\theta}$
41. $x^2 + y^2 = 4$; circle centered at $(0, 0)$ with radius 2
43. $y = -x$; a line with slope -1 and y-intercept 0
45. $(x-2)^2 + y^2 = 4$; a circle centered at $(2, 0)$ with radius 2
47. $x^2 + (y+1)^2 = 1$; a circle centered at $(0, -1)$ with radius 1

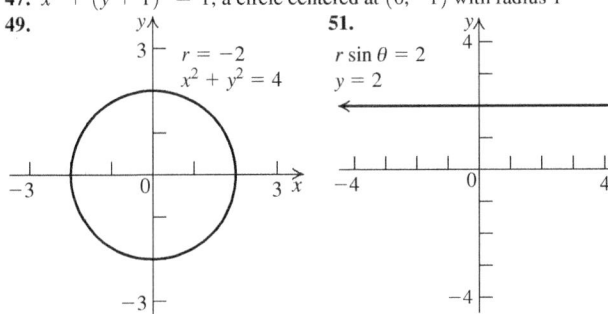

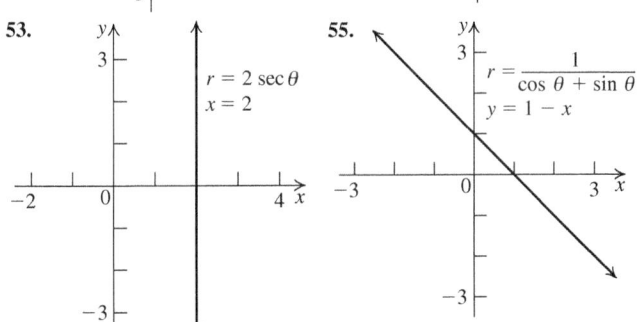

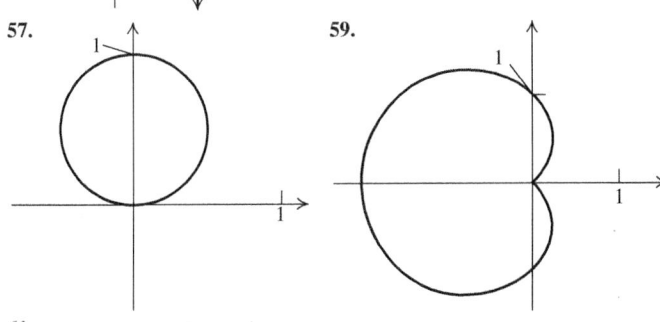

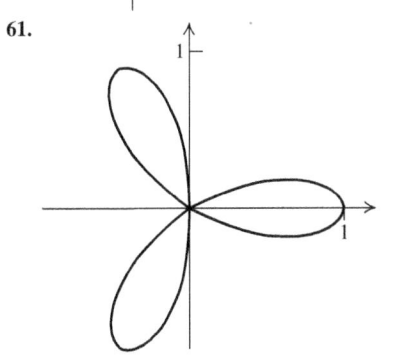

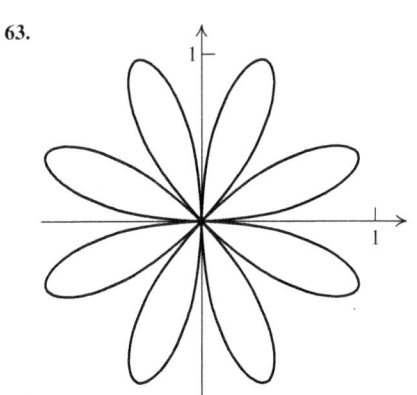

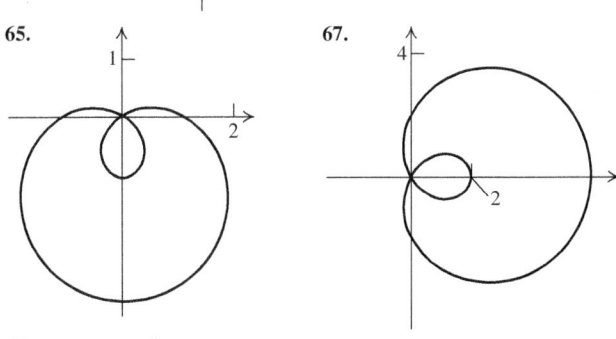

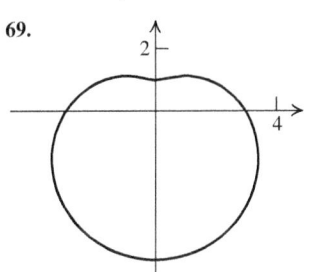

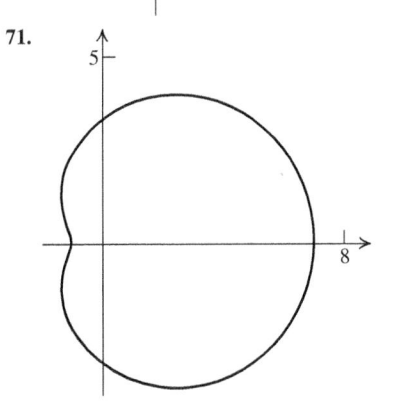

49. $r = -2$; $x^2 + y^2 = 4$

51. $r\sin\theta = 2$; $y = 2$

53. $r = 2\sec\theta$; $x = 2$

55. $r = \dfrac{1}{\cos\theta + \sin\theta}$; $y = 1 - x$

57.

59.

61.

63.

65.

67.

69.

71.

Applying the Concepts:
73. $(21.8, 39.6°)$ **75.** $(18.3, -45.8°)$

Critical Thinking / Discussion / Writing:
77. False   **78.** True   **79.** True   **80.** True
82. Center $\left(\dfrac{b}{2}, \dfrac{a}{2}\right)$; radius $\sqrt{\dfrac{b^2}{4} + \dfrac{a^2}{4}}$

Maintaining Skills:
85. $y = -\dfrac{2}{5}x + \dfrac{29}{5}$ **86.** $y = -\sqrt{3}x + (2 - 3\sqrt{3})$ **87.** $45°$
88. $x^2 + y^2 + 10x - 4y + 20 = 0$ **89.** $C(2, -1), r = \sqrt{5}$

Section 5.4

Basic Concepts and Skills:

1.

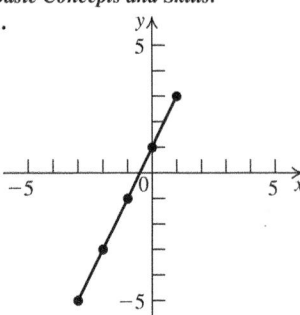

3.

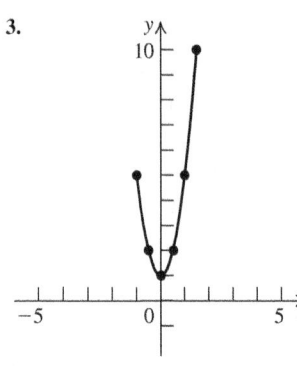

5.

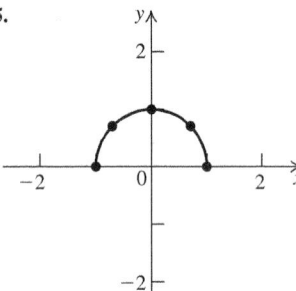

7.

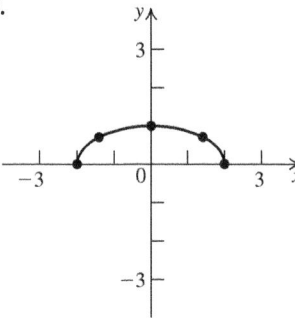

9. $y = 2x$

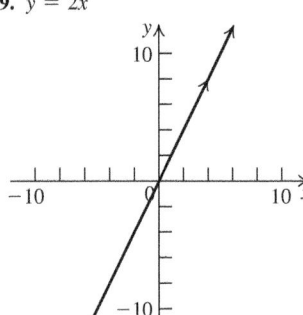

11. $y = 3x, 0 \leq x \leq 8$

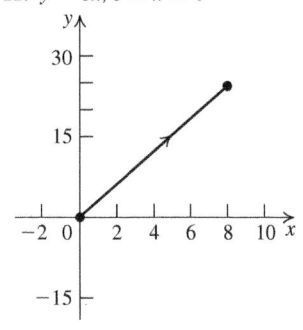

13. $y = 3 - x^2, 0 \leq x \leq 2$

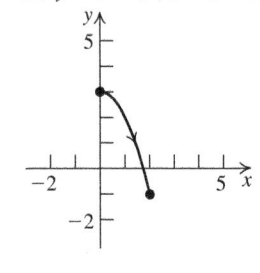

15. $y = \sqrt{4 - x^2}, 0 \leq x \leq 2$

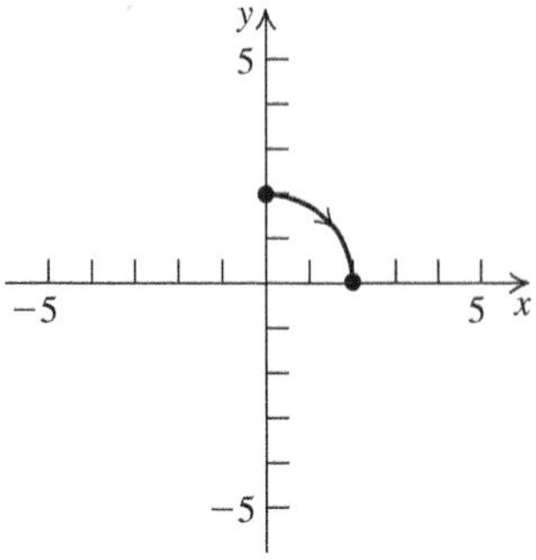

17. $y = 3\sqrt{1 - \dfrac{x^2}{4}}, 0 \leq x \leq 2$

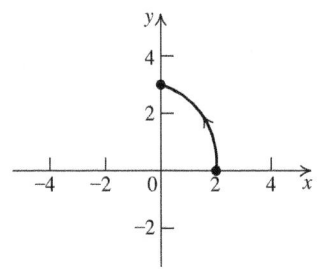

19. $y = \sqrt{x^2 - 4}, x \geq 2$

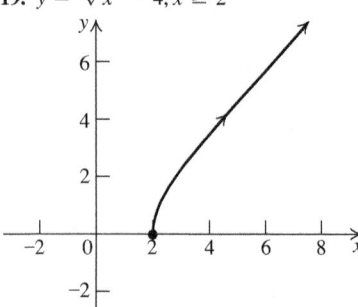

21. $x^2 + y^2 = 1$

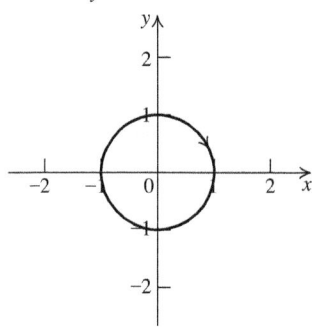

23. $y = 1 - x, 0 \leq x \leq 1$

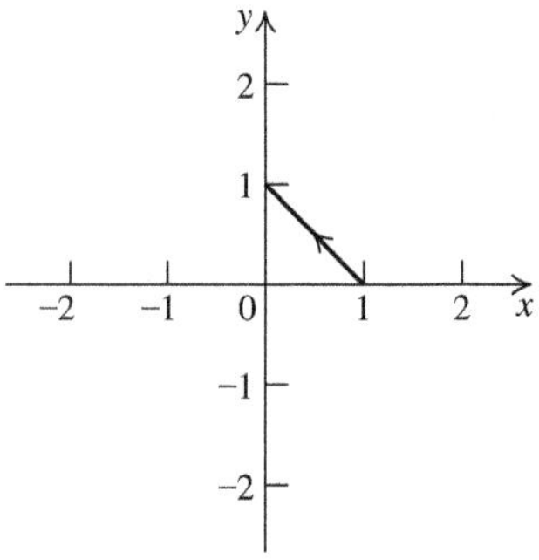

25. $\dfrac{x^2}{9} + \dfrac{y^2}{4} = 1$, traversed counterclockwise

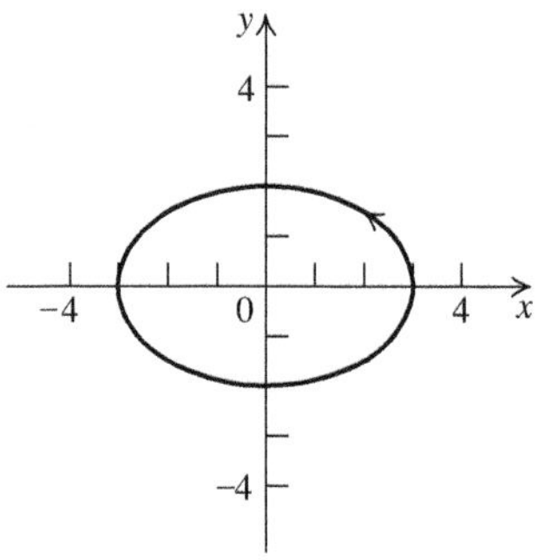

27. $y = (2x^2 - 1)^2, -1 \leq x \leq 1$, traversed infinitely often in both directions

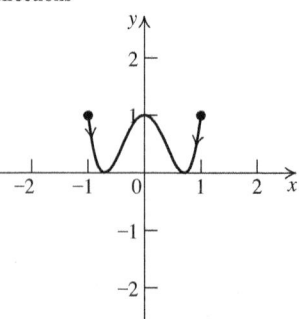

29. $(x - 1)^2 + (y - 2)^2 = 1$, traversed counterclockwise

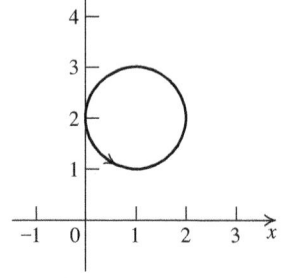

31. $\dfrac{(y-2)^2}{9} - \dfrac{(x+1)^2}{4} = 1, x \geq -1, y \geq 5$

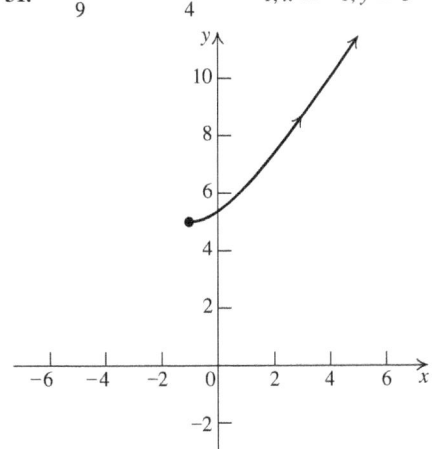

33. $y = x^2, x \geq 1$ 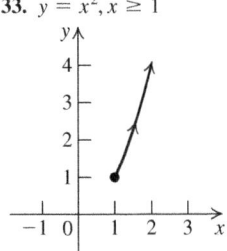   **35.** $y = \ln x, x \geq 1$

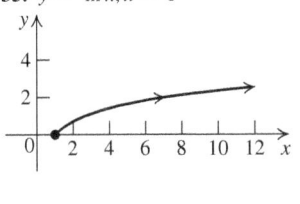

37. $x = t, y = 3t + 2$; the particle moves along the entire line from left to right.   **39.** $x = e^t, y = 3e^t + 2$; the particle moves to the right on the x-interval $0 < x < \infty$.   **41.** $x = -t, y = t^2 - 4$; the particle moves along the entire parabola from right to left.   **43.** $x = \cos t, y = \cos^2 t - 4$; the particle oscillates on the parabola between the points $(-1, -3)$ and $(1, -3)$.   **45.** $x = \sin t, y = \cos t, -\dfrac{\pi}{2} \leq t \leq \dfrac{\pi}{2}$; the motion is counterclockwise from $(0, -1)$ to $(0, 1)$.

47. $x = -\sin t, y = \cos t, -\dfrac{\pi}{2} \leq t \leq \dfrac{\pi}{2}$; the motion is clockwise from $(0, 1)$ to $(0, -1)$.

Applying the Concepts:
49. $x = 32\sqrt{3}t, y = 32t - 16t^2$ **51.** After 2 sec
53. $x = 800t, y = 800\sqrt{3}t - 16t^2 + 96$ **55.** Approximately 13.13 mi

Critical Thinking / Discussion / Writing:
57. b. $x = 2 - 3t, y = -3 + 7t, 0 \leq t \leq 1$
58. b. $x = 1 + 2\cos t, y = 2\sin t, 0 \leq t \leq 2\pi$
59. (a) represents the line; (b) and (d) represent rays included in the line but not the entire line, and (c) represents a different line.

60. a. The equation is $y = x \tan \theta - \dfrac{g}{2}\left(\dfrac{x}{v_0 \cos \theta}\right)^2$.   **b.** The vertex is $\left(\dfrac{v_0^2}{g}\sin \theta \cos \theta, \dfrac{v_0^2}{2g}\sin^2 \theta\right)$.   **c.** The flight time is $\dfrac{2v_0}{g}\sin \theta$.
d. The range is $\dfrac{2v_0^2}{g}\sin \theta \cos \theta$.   **e.** The range is a maximum when $2 \sin \theta \cos \theta = \sin 2\theta$ is a maximum, which happens for $\theta = 45°$. The maximum range is $\dfrac{v_0^2}{g}$.   **f.** Both the range and height at the vertex depend on the square of the speed.

Maintaining Skills:
61. $-1, 2, 5, 20$ **62.** $-3, -1, -\dfrac{1}{3}, \dfrac{1}{2}$ **63.** 15   **64.** 46

Review Exercises

Basic Concepts and Skills:
1. $C = 105°, a \approx 66.5, c \approx 59.4$ **3.** No triangle exists.
5. $B = 74.2°, a \approx 36.8, c \approx 41.4$ **7.** $B \approx 54.1°, C \approx 60.7°, c \approx 20.5$

9. Answers will vary. Sample answers: **i.** 18, **ii.** $10\sqrt{3}$, **iii.** 10
11. $A \approx 26.6°, B \approx 42.1°, C \approx 111.3°$
13. $A \approx 51.7°, C \approx 48.3°, b \approx 50.2$
15. $A \approx 32.5°, B \approx 49.8°, C \approx 97.7°$
17. $A \approx 32.0°, B \approx 18.0°, c \approx 17.3$
19. 16 sq. m   **21.** 11 sq. ft   **23.** 104 sq. m
25. $(24, 30°)$, $(12\sqrt{3}, 12)$

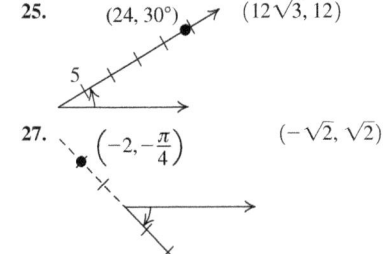

27. $\left(-2, -\dfrac{\pi}{4}\right)$, $(-\sqrt{2}, \sqrt{2})$

29. $\left(2\sqrt{2}, \dfrac{3\pi}{4}\right)$ **31.** $\left(4, \dfrac{11\pi}{6}\right)$ **33.** $r = \dfrac{12}{3\cos\theta + 2\sin\theta}$
35. $r = 8\cos\theta$ **37.** $x^2 + y^2 = 9$ **39.** $y = 3$
41. $(x^2 + y^2 + 2y)^2 = x^2 + y^2$
43. $y = 1 - x, 0 \leq x \leq 1$ **45.** $y = 3 - x, 0 \leq x \leq 3$, traversed twice in each direction

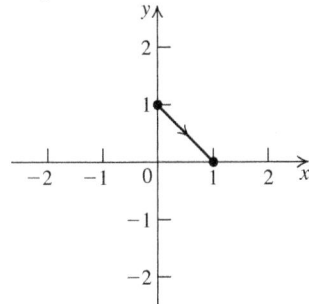

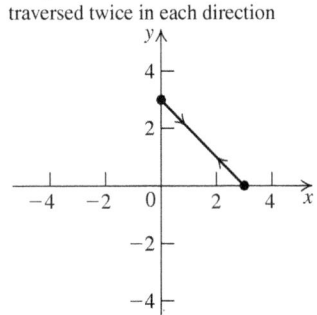

47. $y = -2x^2 + 1, -1 \leq x \leq 1$, traversed once in each direction   **49.** $y = x^{2/3}, -\infty < x < \infty$

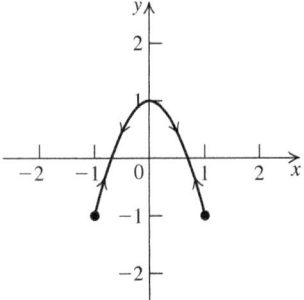

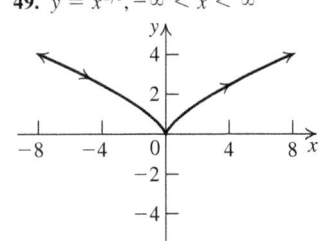

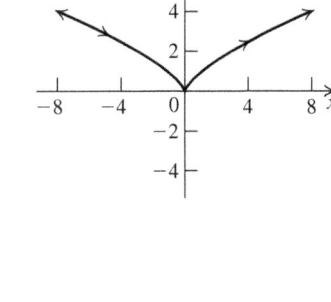

Applying the Concepts:
51. 88.2 in., 65.0 in.   **53.** 25.2°   **55.** 772.2 yd
57. $A \approx 76.8°, B \approx 61.5°, C \approx 41.7°$ **59.** 3775 ft   **61.** 114.8°

Exercises for Calculus
2. b. 0.99833, 0.999983, 0.999999, 0.999999   **c.** $f(t) \to 1$ as $t \to 0$
6. 50π **7.** $b^2 h\pi$ **8.** 60π **9.** $(2a + h)hb\pi$ **10.** 70π

Practice Test A
1. $B = 30°, a \approx 23.7, b \approx 13.1$ **2.** $C = 101°, b \approx 45.0$ m, $c \approx 73.4$ m   **3.** $A \approx 28.2°, C \approx 45.8°, b \approx 71.1$
4. $A \approx 82.8°, B \approx 41.4°, C \approx 55.8°$ **5.** 32.7°   **6.** 803 ft
7. 41.6 sq. units   **8.** 1149.1 sq. units   **9.** 87.4 sq. units
10. 2.9 sq. units   **11.** 54.2 ft, 151.9 ft   **12.** 19,899.7 sq. ft
13. $(-\sqrt{2}, \sqrt{2})$ **14.** $\left(2, \dfrac{7\pi}{6}\right)$ **15.** $x^2 + y^2 = 9$, circle with center $(0, 0)$ and radius 3   **16.** $r = 4\sin\theta$ **17.** Symmetric about the line $\theta = \pi/2$
18. $\dfrac{x}{2} + \dfrac{y}{3} = 1$ **19.** $\dfrac{x^2}{9} + \dfrac{y^2}{16} = 1$ **20.** $\dfrac{x^2}{16} - \dfrac{y^2}{9} = 1$

Practice Test B
1. a   **2.** c   **3.** c   **4.** d   **5.** b   **6.** d   **7.** c   **8.** c   **9.** b   **10.** c   **11.** c
12. d   **13.** c   **14.** b   **15.** b   **16.** c   **17.** a   **18.** a   **19.** b   **20.** c

CHAPTER 6

Section 6.1

Basic Concepts and Skills:
1. $a_1 = 1, a_2 = 4, a_3 = 7, a_4 = 10$
3. $a_1 = 0, a_2 = \frac{1}{2}, a_3 = \frac{2}{3}, a_4 = \frac{3}{4}$
5. $a_1 = -1, a_2 = -4, a_3 = -9, a_4 = -16$
7. $a_1 = 1, a_2 = \frac{4}{3}, a_3 = \frac{3}{2}, a_4 = \frac{8}{5}$
9. $a_1 = 1, a_2 = -1, a_3 = 1, a_4 = -1$
11. $a_1 = \frac{5}{2}, a_2 = \frac{11}{4}, a_3 = \frac{23}{8}, a_4 = \frac{47}{16}$
13. $a_1 = 0.6, a_2 = 0.6, a_3 = 0.6, a_4 = 0.6$
15. $a_1 = -1, a_2 = \frac{1}{2}, a_3 = -\frac{1}{6}, a_4 = \frac{1}{24}$
17. $a_1 = -\frac{1}{3}, a_2 = \frac{1}{9}, a_3 = -\frac{1}{27}, a_4 = \frac{1}{81}$
19. $a_1 = \frac{e}{2}, a_2 = \frac{e^2}{4}, a_3 = \frac{e^3}{6}, a_4 = \frac{e^4}{8}$
21. $a_n = 3n - 2$ 23. $a_n = \frac{1}{n+1}$ 25. $a_n = (-1)^{n+1}(2)$
27. $a_n = \frac{3^{n-1}}{2^n}$ 29. $a_n = n(n+1)$ 31. $a_n = 2 - \frac{(-1)^n}{n+1}$
33. $a_n = \frac{3^{n+1}}{n+1}$ 35. $a_1 = 2, a_2 = 5, a_3 = 8, a_4 = 11, a_5 = 14$
37. $a_1 = 3, a_2 = 6, a_3 = 12, a_4 = 24, a_5 = 48$
39. $a_1 = 7, a_2 = -11, a_3 = 25, a_4 = -47, a_5 = 97$
41. $a_1 = 2, a_2 = \frac{1}{2}, a_3 = 2, a_4 = \frac{1}{2}, a_5 = 2$
43. $a_1 = 25, a_2 = -\frac{1}{125}, a_3 = -25, a_4 = \frac{1}{125}, a_5 = 25$
45. a. 2, 11, 26, 47, 74, 107, 146, 191, 242, 299
b. 400

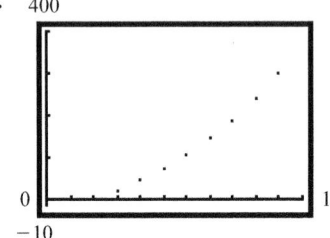

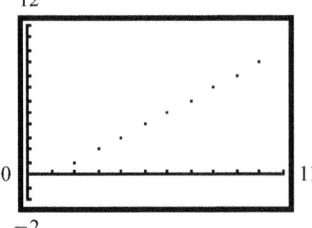

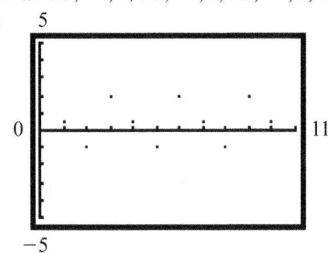

47. a. 0, 1, 2, 3, 4, 5, 6, 7, 8, 9
b. 12

49. a. 0.5, −1, 2, 0.5, −1, 2, 0.5, −1, 2, 0.5
b. 5

51. $\frac{1}{20}$ 53. 12 55. $\frac{1}{n+1}$ 57. $2n + 1$ 59. 35 61. 55 63. 25
65. $\frac{853}{140}$ 67. 183 69. -118 71. $\sum_{k=1}^{51}(2k-1)$ 73. $\sum_{k=1}^{11}\frac{1}{5k}$
75. $\sum_{k=1}^{50}\frac{(-1)^{k+1}}{k}$ 77. $\sum_{k=1}^{10}\left(\frac{k}{k+1}\right)$ 79. 4620
81. -1.345 (rounded to three decimal places) 83. 50

Applying the Concepts:
85. a. 208 ft b. $(16 + (n-1)32)$ ft 87. $250 89. 19,200 min
91. a. $A_1 = 10{,}300, A_2 = 10{,}609, A_3 = 10{,}927.27, A_4 = 11{,}255.09$,
$A_5 = 11{,}592.74, A_6 = 11{,}940.52$ b. $16{,}047.10
93. 1st year: $105,000; 2nd year: $110,250; 3rd year: $115,762.50;
4th year: $121,550.63; 5th year: $127,628.16; 6th year: $134,009.56;
7th year: $140,710.04; formula: $a_n = (100{,}000)(1.05^n)$

Critical Thinking / Discussion / Writing:
95. a. $a_1 = 3, a_{n+1} = a_n + 2$ b. $b_1 = 3, b_{n+1} = b_n + 1$
96. a. $a_1 = 1, a_2 = \frac{1}{2}, a_3 = \frac{1}{4}, a_4 = \frac{1}{8}, a_5 = \frac{1}{16}$ b. $a_n = \frac{1}{2^{n-1}}$
97. $k = 5$ 98. The terms leading to 1 are 13, 40, 20, 10, 5, 16, 8, 4, 2, and 1.

Maintaining Skills:
99. 2 100. $\frac{3}{2}$ 101. 3 102. 1 103. 2 104. $-\frac{1}{4}$
105. $x^2 - 2x + 1$ 106. $4x^2 + 12x + 9$
107. $-x^3 + 6x^2 - 12x + 8$ 108. $x^3 + 9x^2 + 27x + 27$

Section 6.2

Basic Concepts and Skills:
1. Arithmetic; $a_1 = 1, d = 1$ 3. Arithmetic; $a_1 = 2, d = 3$
5. Not Arithmetic 7. Not Arithmetic 9. Arithmetic;
$a_1 = 0.6, d = -0.4$ 11. Arithmetic; $a_1 = 8, d = 2$
13. Not arithmetic 15. $a_n = 3n + 2$ 17. $a_n = 16 - 5n$
19. $a_n = \frac{3-n}{4}$ 21. $a_n = -\frac{2n+1}{5}$ 23. $a_n = 3n + e - 3$
25. $d = \frac{13}{2}, a_n = \frac{13}{2}n - 5$ 27. $d = -2, a_n = 22 - 2n$
29. $d = \frac{1}{2}, a_n = \frac{n+11}{2}$ 31. 1275 33. 2500 35. 15,150
37. -208 39. $\frac{121}{3}$ 41. 6225 43. 460 45. 1340
47. $n = 38$ 49. $n = 25$ 51. $n = 25$

Applying the Concepts:
53. 1170 55. $12,375 57. $16.75 59. 1100 61. 408

Critical Thinking / Discussion / Writing:
62. 8 63. Yes, it is harmonic because the reciprocals form an arithmetic sequence with $d = -\frac{1}{3}$. 64. $a_n = \frac{21-n}{2}$ 65. First term $-a_1$; difference: $-d$ 66. 79 67. $n = 157$

Maintaining Skills:
68. $\frac{2}{5}$ 69. $\frac{3}{8}$ 70. -1 71. 3 72. $\frac{1}{3}$ 73. $\frac{1}{11}$
74. 7^{n+1} 75. $\left(\frac{1}{2}\right)^{n+1}$ 76. $5 - 5x^3$ 77. $a - ax^4$

Section 6.3

Basic Concepts and Skills:
1. Geometric; $a_1 = 3, r = 2$ 3. Not geometric
5. Geometric; $a_1 = 1, r = -3$ 7. Geometric; $a_1 = 7, r = -1$
9. Geometric; $a_1 = 9, r = \frac{1}{3}$ 11. Geometric; $a_1 = -\frac{1}{2}, r = -\frac{1}{2}$
13. Geometric; $a_1 = 1, r = 2$ 15. Not geometric

17. Geometric; $a_1 = \frac{1}{3}, r = \frac{1}{3}$ **19.** Geometric; $a_1 = \sqrt{5}, r = \sqrt{5}$
21. $a_1 = 2, r = 5, a_n = (2)(5)^{n-1}$ **23.** $a_1 = 5, r = \frac{2}{3}$, $a_n = (5)\left(\frac{2}{3}\right)^{n-1}$ **25.** $a_1 = 0.2, r = -3, a_n = 0.2(-3)^{n-1}$
27. $a_1 = \pi^4, r = \pi^2, a_n = \pi^{2n+2}$ **29.** $a_7 = 320$
31. $a_{10} = -1536$ **33.** $a_6 = \frac{243}{16}$ **35.** $a_9 = -\frac{390{,}625}{256}$
37. $a_{20} = -\frac{125}{131{,}072}$ **39.** $S_{10} = \frac{1{,}220{,}703}{5}$ **41.** $S_{12} = -\frac{40{,}690{,}104}{25}$
43. $S_8 = \frac{109{,}225}{16{,}384}$ **45.** $\frac{31}{16}$ **47.** $\frac{6305}{729}$ **49.** 255 **51.** $\frac{3(5^{20} - 2^{20})}{35(2)^{19}}$
53. $\frac{127}{320}$ **55.** .9841333 (rounded to three decimal places)
57. 2.250 (rounded to three decimal places) **59.** $\frac{1}{2}$ **61.** $-\frac{1}{3}$
63. $\frac{32}{5}$ **65.** $\frac{15}{2}$ **67.** $\frac{4}{5}$ **69.** $n = 9$ **71.** $n = 13$ **73.** $n = 9$

Applying the Concepts:
75. 23,185 **77.** $3933.61 **79.** 1024 **81.** 20 m

Critical Thinking / Discussion / Writing:
83. $\frac{2}{3}$ **84.** 24 **85.** b_{1001} is larger **86.** $a = 512$
87. $n = 15, k = 16$ **88.** 5, 10, and 20
89. 2, 5, and 8; or 26, 5, and -16

Maintaining Skills:
90. 2 **91.** 11 **92.** 1 **93.** 2 **94.** 5 **95.** -3 **96.** $2x - y = 11$
97. $x + 2y = 9$ **98.** $a = \frac{1}{2}, b = -1$ **99.** $a = -8, b = 2$

Section 6.4

Basic Concepts and Skills:
1. $(3, -1)$ **3.** None of them **5.** $(10, -9)$
7. $(2, 1)$ **9.** $(2, 2)$

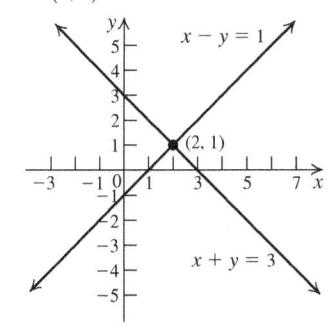

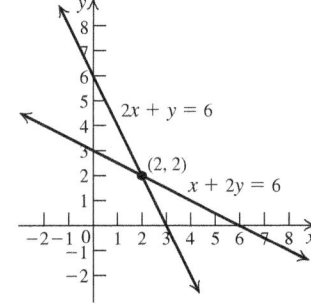

11. Inconsistent **13.** $\left(\frac{7}{3}, \frac{14}{3}\right)$

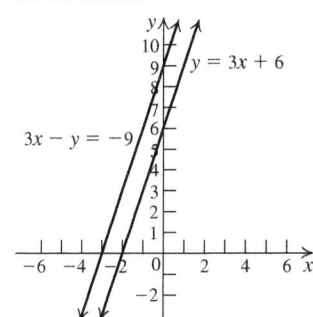

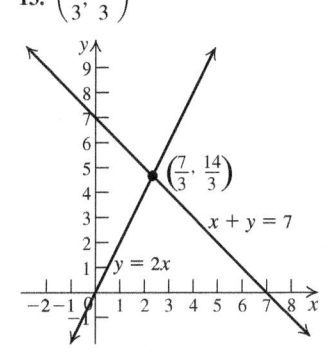

15. $\{(x, 12 - 3x)\}$

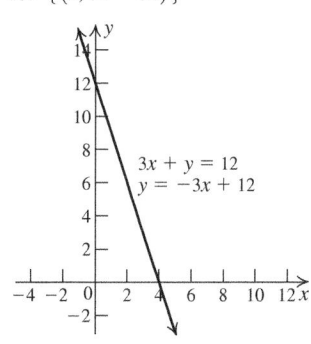

17. Independent **19.** Independent **21.** Dependent **23.** Independent
25. Inconsistent **27.** Inconsistent **29.** Independent **31.** $\left\{\left(\frac{7}{9}, \frac{23}{9}\right)\right\}$
33. $\{(3, 4)\}$ **35.** $\varnothing$ **37.** $\{(1, -1)(3, 1)\}$ **39.** $\{(x, \frac{1}{2}(x - 5))\}$
41. $\{(3, 2)\}$ **43.** $\{(-3, 3)\}$ **45.** $\{(2, 2), (2, -2), (-2, 2), (-2, -2)\}$
47. $\varnothing$ **49.** $\left\{\left(x, 2 - \frac{2}{3}x\right)\right\}$ **51.** $\{(4, 1)\}$ **53.** $\{(-4, 2)\}$
55. $\{(2, 1)\}$ **57.** $\{(3, 1)\}$ **59.** $\left\{\left(\frac{19}{6}, \frac{5}{4}\right)\right\}$
61. $\left\{(3, -1), \left(\frac{9}{4}, -\frac{1}{2}\right)\right\}$ **63.** $\left\{(1, -1), \left(\frac{41}{29}, \frac{1}{29}\right)\right\}$
65. $\{(4, 2), (4, -2), (-4, 2), (-4, -2)\}$
67. $\left\{\left(\frac{3}{2}, \frac{1}{2}\right)\right\}$ **69.** $\{(5, 5)\}$

Applying the Concepts:
71. $(80, 30)$ **73.** $(17, 4)$ **75.** $18,000 at 7.5% and $32,000 at 12%
77. 20 lb of herb and 80 lb of tea **79.** The speed of the plane is 550 km/hr, and the wind speed is 50 km/hr.

Critical Thinking / Discussion / Writing:
81. $a = 4, b = 0$ **82. a.** $x = \frac{b_2c_1 - b_1c_2}{a_1b_2 - a_2b_1}, y = \frac{a_1c_2 - a_2c_1}{a_1b_2 - a_2b_1}$, where $a_1b_2 - a_2b_1 \neq 0$ **b.** $\frac{c_1}{b_1} \neq \frac{c_2}{b_2}, \frac{a_1}{a_2} = \frac{b_1}{b_2}$ **c.** $\frac{a_1}{a_2} = \frac{b_1}{b_2} = \frac{c_1}{c_2}$
83. a. $3y - 4x = 4$ **b.** $\left(\frac{4}{5}, \frac{12}{5}\right)$ **c.** 7
84. a. $b(x - x_1) - a(y - y_1) = 0$
b. $\left(\frac{b^2x_1 - aby_1 - ac}{a^2 + b^2}, \frac{-abx_1 + a^2y_1 - bc}{a^2 + b^2}\right)$ **85. a.** $\sqrt{2}$
b. $\frac{6\sqrt{5}}{5}$ **c.** 0 **d.** $\frac{|c|}{\sqrt{a^2 + b^2}}$ if $a \neq 0$ or $b \neq 0$

Maintaining Skills:
86. -1 **87.** -1 **88.** $y = 3x + 6$ **89.** $y = 5x - 9$
90. $x = 4y + 5$ **91.** $x = 4y - 19$ **92.** $(x + 3)(x + 2)$
93. $(x + 2)(x - 5)$ **94.** $(2x - 1)(x + 3)$ **95.** $(3x - 2)(x + 1)$
96. $(3x - 1)(2x - 5)$ **97.** $(4x - 1)(3x - 2)$

Section 6.5

Basic Concepts and Skills:
1. $\frac{A}{x - 1} + \frac{B}{x + 2}$ **3.** $\frac{A}{x + 6} + \frac{B}{x + 1}$ **5.** $\frac{A}{x^2} + \frac{B}{x} + \frac{C}{x - 1}$
7. $\frac{3}{x + 2} - \frac{1}{x + 1}$ **9.** $\frac{-1}{2(x + 3)} + \frac{1}{2(x + 1)}$ **11.** $\frac{-1}{x + 2} + \frac{1}{x}$
13. $\frac{2}{x + 2} - \frac{3}{2(x + 3)} - \frac{1}{2(x + 1)}$ **15.** $\frac{-2}{(x + 1)^2} + \frac{1}{x + 1}$
17. $\frac{2}{x + 1} - \frac{3}{(x + 1)^2} + \frac{1}{(x + 1)^3}$ **19.** $\frac{1}{x} + \frac{3}{(x + 1)^2} - \frac{2}{x + 1}$

21. $\dfrac{-2}{x} + \dfrac{1}{x^2} + \dfrac{1}{(x+1)^2} + \dfrac{2}{x+1}$

23. $\dfrac{1}{4(x-1)^2} + \dfrac{1}{4(x+1)^2} + \dfrac{1}{4(x+1)} - \dfrac{1}{4(x-1)}$

25. $\dfrac{1}{2(2x-3)} + \dfrac{1}{2(2x-3)^2}$ 27. $\dfrac{-2}{(2x+3)^2} + \dfrac{3}{2x+3}$

29. $\dfrac{4}{x} - \dfrac{3}{x^2} - \dfrac{4}{x+1}$

Applying the Concepts:

31. $\dfrac{n}{n+1}$ 33. $\dfrac{2n}{2n+1}$ 35. $\dfrac{1}{R} = \dfrac{1}{x+1} + \dfrac{1}{x+3}$

37. $\dfrac{1}{R} = \dfrac{1}{R_1} + \dfrac{1}{R_2} + \dfrac{1}{R_3}$

Critical Thinking / Discussion / Writings:

40. $\dfrac{1}{x^2-2x+2} - \dfrac{1}{x^2+2x+2}$ 41. $\dfrac{1}{x-5} + \dfrac{2}{(x-5)^2} + \dfrac{3}{(x-5)^3}$

42. $1 + \dfrac{2}{x+1} - \dfrac{1}{x+2}$ 43. $1 - \dfrac{4}{7(x+3)} + \dfrac{18}{7(x-4)}$

44. $2x + 3 + \dfrac{2x-1}{x^2+1} + \dfrac{1}{x-1}$

45. $x + 6 - \dfrac{16}{x-2} + \dfrac{81}{2(x-3)} + \dfrac{1}{2(x-1)}$

Maintaining Skills:

46. $27x^6$ 47. $-2x^{31}$ 48. $-16x^4$ 49. $25x^2$ 50. $\dfrac{9}{5x^2}$ 51. $\dfrac{-8y^2}{5x}$

52. $\dfrac{-3}{4x^2yz}$ 53. $\dfrac{-4y^5}{x^3z^2}$ 54. $\dfrac{1}{6x}$ 55. 1 56. $\dfrac{1}{x}$ 57. $\dfrac{3}{x}$

Review Exercises

1. $a_1 = -1, a_2 = 1, a_3 = 3, a_4 = 5, a_5 = 7$
3. $a_1 = \dfrac{1}{3}, a_2 = \dfrac{2}{5}, a_3 = \dfrac{3}{7}, a_4 = \dfrac{4}{9}, a_5 = \dfrac{5}{11}$
5. $a_n = 32 - 2n$ 7. 9 9. $n+1$ 11. 100
13. $\dfrac{1343}{140}$ 15. $\sum_{k=1}^{50} \dfrac{1}{k}$ 17. Arithmetic; $a_1 = 11, d = -5$
19. Not arithmetic 21. $a_n = 3n$ 23. $a_n = x + n - 1$
25. $d = 2, a_n = 2n + 1$ 27. 352 29. 4020
31. Geometric: $a_1 = 4, r = -2$
33. Not geometric 35. $a_1 = 16, r = -\dfrac{1}{4}, a_n = \dfrac{(-1)^{n-1}}{4^{n-3}}$
37. $a_{10} = 39{,}366$ 39. $S_{12} = \dfrac{819}{2}$ 41. $\dfrac{3}{4}$ 43. $\dfrac{3}{2}$

45. $\{(-1, 2)\}$ 47. $\varnothing$ 49. $\{(2, 3)\}$ 51. $\{(7, 2)\}$
53. $\{(1, 6)\}$ 55. $\{(0, 0)\}$ 57. $\varnothing$ 59. $\left\{\left(\dfrac{4}{3}, \dfrac{2}{3}\right)\right\}$
61. $\{(14, -7)\}$ 63. $\dfrac{2}{x+2} - \dfrac{1}{x+3}$ 65. $\dfrac{2}{x-1} + \dfrac{1}{x} + \dfrac{5}{(x-1)^2}$
67. $\dfrac{13}{16(x-2)} + \dfrac{11}{4(x-2)^2} + \dfrac{3}{16(x+2)}$

Applying the Concepts:
69. $8750 was invested at 12%, and $6250 was invested at 4%.
71. 10.5 ft by 6 ft 73. 15 and 8 75. The original width is 40 m, and the original length is 160 m. 77. $450 for the single and $600 for the double

Exercises for Calculus

1. $128{,}000$ 2. 3 ft 3. $\dfrac{a(1+r)}{1-r}$ 4. $5 million 7. $\dfrac{25}{99}$
8. $\dfrac{68}{111}$ 9. 72 10. 57 11. a. 14 sq. units b. 29 sq. units
12. a. ≈ 1.833 sq. units b. ≈ 1.083 sq. units

Practice Test A

1. $a_1 = 3, a_2 = -9, a_3 = -21, a_4 = -33, a_5 = -45$; arithmetic sequence 2. $a_1 = -6, a_2 = -12, a_3 = -24, a_4 = -48, a_5 = -96$; geometric sequence 3. $a_1 = -2, a_2 = -1, a_3 = 2, a_4 = 11, a_5 = 38$
4. $\dfrac{1}{n}$ 5. $a_7 = 21$ 6. $a_8 = -\dfrac{13}{128}$ 7. 590 8. $\dfrac{93}{128}$ 9. $\dfrac{2}{11}$
10. $\{(2, 0)\}$ 11. $\{(8 - 2y, y)\}$ 12. $\varnothing$ 13. $\{(-5, 0)\}$
14. $\{(6, 8)\}$ 15. $\{(-1, 1), (4, 16)\}$ 16. $\left\{\left(-3, \dfrac{1}{3}\right), (2, 2)\right\}$
17. a. $\begin{cases} x + y = 485 \\ x - y = 15 \end{cases}$ b. $x = 250, y = 235$ 18. $\dfrac{A}{(x-5)} + \dfrac{B}{(x+1)}$
19. $\dfrac{A}{x-2} + \dfrac{B}{x+1} + \dfrac{C}{(x+1)^2}$
20. $-\dfrac{10}{121(x+4)} + \dfrac{1}{11(x+4)^2} + \dfrac{10}{121(x-7)}$

Practice Test B

1. c 2. d 3. a 4. c 5. a 6. b 7. d 8. a 9. b 10. a 11. c
12. b 13. c 14. b 15. d 16. d 17. d 18. c 19. a 20. c

CREDITS

Text Credits

Page 1–32 "The ups and downs of drug levels" by Bob Munk from POSITIVELY AWARE, May/June 2001. Copyright © 2001 by Test Positive Aware Network. Reprinted with permission. **379** DODD, ANNABEL Z., THE ESSENTIAL GUIDE TO TELECOMMUNICATIONS, 3rd Ed., ©2002. Reprinted and Electronically reproduced by permission of Pearson Education, Inc., Upper Saddle River, New Jersey.

Photo Credits

COVER Brian Stablyk/Photographer's Choice/Getty Images **COVER** Bloomberg/Getty Images **51** Pearson Education, Inc. **52** Mark Yuill/Fotolia **53** HO/AFP/Newscom **65** George Doyle/Stockbyte/Getty Images **88** iStockphoto **94** Mykhaylo Palinchak/Shutterstock **001** Steve Cukrov/Shutterstock **077** Piotr Marcinski/Fotolia **109** Hinrich Bäsemann/picture alliance/Hinrich Bäsemann/Newscom **124** Chris Ridley/Travel Seychelles/Alamy **136** Image Source/Alamy **154** MasterLu/Fotolia **155** Nickolae/Fotolia **164** David Carillet/Shutterstock **179** Geo Stock/Getty Images **184** Mary Evans Picture Library/Alamy **188** Eldon Doty/Getty Images **198** Pitris/Fotolia **221** Dmitrijs Gerciks/Fotolia **222** Georgios Kollidas/Fotolia **231** G. Smith/Fotolia **239** Richard Nowitz/Getty Images **253** Science Source/Photo Researchers, Inc. **261** Tom Reichner/Shutterstock **265** Pearson Education, Inc. **272** Art Directors & TRIP/Alamy **282** Rob Marmion/Shutterstock **293** Galyna Andrushko/Shutterstock **310** Everett Collection **327** Anton Gvozdikov/Shutterstock **337** Robert Hardholt/Shutterstock **350** North Wind Picture Archives **362** David De Lossy/Getty Images **373** Jose Manuel Revuelta Luna/Alamy **394** Jagdish Agarwal/Corbis **395** Oliver Sved/Shutterstock **419** Mmdi/Digital Vision/Getty Images **447** Georgette Douwma/Getty Images **448** Science Source/Photo Researchers, Inc. **450** Chris Butler/Science Source/Photo Researchers, Inc. **458** A. Wittmann/Goettingen University **461** ZUMA Wire Service/Alamy **465** Juulijs/Fotolia **476** Mary Evans Picture Library/Alamy **488**

INDEX

A

AAA triangles, 396
AAS triangles
   defined, 395–396
   finding area of, 414–415
   solving, 397–398
Absolute values
   defined, 30
   distance on number line and, 8
   in equations, 30–31
   graph of, 101, 103
   in inequalities, 40–43
   as piecewise function, 101
   of real numbers, 7–8
Acute angles, 284
Addition
   of complex numbers, 46
   of functions, 125–126
   of radicals, 14
Addition method, 482–484
Addition property, 33
Algebra. *See also specific topics*
   Fundamental Theorem of, 191
   word origin, 476
Algebraic expressions, trigonometric and, 346–347
Algebraic numbers/functions, transcendental vs., 233
Algorithms
   defined, 180
   division, 179–181
   word origin, 476
al-Khwarizmi, Muhammad ibn Musa, 476
Altitude of triangle, 396
Ambiguous case, defined, 399
Amperes, 373
Amplitudes, finding and changing, 315–319
Angles
   application, 290
   arc length formula, 289–290
   definitions, 283–284
   degree-radian conversions, 286–288
   of depression and elevation, 355
   finding complements and supplements, 288–289
   measuring, 284–286
   reference, 306–307
   in standard position, 284, 285, 303–304
   trigonometric functions of (*See under* Trigonometric functions)
Annuity problems, 469–470
Apollonius, 52
Applied problems. *See also specific topics (e.g., bioavailability)*
   angles, 290
   cosecant function, 334
   cosine function, 324–325
   exponential equations, 268–269
   exponential functions, 227–233
   functions in general, 88–89
   geometric sequences, 469–470
   Heron's formula, 416
   inverse functions, 143–144
   inverse trigonometric functions, 347
   Law of Sines, 398, 403–404
   logarithmic functions, 247–250
   logarithms, 260–262
   partial-fraction decomposition, 493–494
   polar coordinates, 423–424
   polynomial functions, 174–175
   polynomials, 184
   quadratic functions, 160–161
   rational functions, 210
   right triangles, 355–358
   sequences
      arithmetic, 462
      Fibonacci, 451
      geometric, 472
   sine function, 322–323
   systems of linear equations in two variables, 485
Arc length formula, 289–290
Arccosine function, 339
Arcsine function, 337–338
Arctangent function, 340
Area
   list of formulas, 411
   of rectangles, 82, 411–412
   of semicircles, 412
   of triangles, 411–416
Arithmetic sequences, 458–462
Ars Magna, 188
ASA triangles
   defined, 395–396
   finding area of, 414–415
   solving, 397–398
Asymptotes
   horizontal and vertical, 200–205
   oblique, 208–210
Average rate of change, 104–106
Axes
   reflections about, 112–113
   of symmetry, 57, 156

B

Ball attached to spring, 324–325
Base(s)
   change-of-, 258–259
   exponential equations with different, 267
   of exponential functions, defined, 223
Bearings, 403–404
Bees, 448, 451
Bermuda Triangle, 411
Binary operations, defined, 2
Bioavailability, 77, 88–89
Blanket, electric, 373
Blood pressure, 109, 119, 293
Book of Calculations, 450
Bounds on real zeros, 189–190
British thermal units (BTU), 184

C

Calculators. *See* Graphing calculator usage
Cameras, rotation angle of, 337, 357–358
Car sales, applying composition to, 131–132
Carbon dating, 261–262
Cardano, Gerolamo, 188
Carnarvon, Lord, 253
Carter, Howard, 253
Cartesian equations
   conversion between polar and, 424–425
   parametric equations vs., 439–440
Cartesian plane, 52–53
Celsius, conversion to Fahrenheit, 37–38
Center of circles, 60
Central angles, radian measure of, 286
Change-of-base formula, 258–259
Chapman, Helen, 65
Chemical toxins in lake, 249–250
Chessboard, grains of wheat on, 222
Cholesterol-reducing drugs, 88–89
Circles
   arc length formula, 289–290
   definitions, 60
   equation of, 60–62
   geometry formulas for, 411
   graphing, 61
   polar equations and, 431
Circular functions. *See* Trigonometric functions
Circumference
   of circle, 411
   of semicircles, 412
   of unit circle, 294
Cities, distance between, 290
Closed interval, 6
Clouds, finding height of, 356
Coffee, McDonald's, 239, 248–249
Cofunction identities, 375–376
Commentarii Academia Petropolitanae, 78
Common difference, 458–460
Common logarithms, 245–246. *See also* Logarithms
Common ratio, 465–466
Complementary angles
   finding, 288–289
   trigonometric functions for, 353
Completing the square
   converting quadratic functions to standard form by, 158
   solving quadratic equations by, 24–26
Complex conjugate product theorem, 47
Complex numbers
   adding and subtracting, 46
   defined, 191
   definitions, 45
   multiplying and dividing, 46–48
   powers of i, 48–49
Complex polynomials and zeros of polynomials, 191–192
Composite functions
   applications, 131–132
   domain of, 128–130
   main discussion, 126–128
   trigonometric, 344–347
Compound inequalities, 36–38
Compound interest, 228–233
Compressing functions, 115–117
Conditional equations, 22, 23
Congruent triangles, 395
Conjugate Pairs Theorem, 192–195
Conjugates
   of complex numbers, 47–48, 192
   radical expressions, 16
   verifying identities using, 368–369
Consistent systems, 478
Constant functions, 94, 95–96, 103
Constant term, 164
Constants, defined, 2
Continuous compound interest, 231–233
Continuous curves, 165
Conway, John, 69

I-1

Coordinate axes, 52
Coordinate line, 5. *See also* Real numbers
Coordinate plane, 52–53. *See also* Graphs and graphing
Correspondence diagrams, 78
Cosecant function. *See also* Trigonometric functions
   application, 334
   defined, 295
   graph and properties of, 331–334
   inverse, 341
Cosine function. *See also* Law of Cosines; Trigonometric functions
   amplitude and period of, 315–319
   application, 324–325
   defined, 295
   phase shifts, 319–323
   properties of, 310–312
   simple harmonic motion, 323–325
   sum and difference formulas for, 373–375
Cotangent function. *See also* Trigonometric functions
   defined, 295
   graph and properties of, 331–334
   inverse, 341
Coterminal angles, 284, 306
Coughing, 94, 97
Counterexamples, 365
Cube root functions, 99–100, 103
Cube roots, 13
Cubes, 411
Cubic polynomials, 165
Cubing function, 103
Current (electric), 373, 488
Curtate cycloids, 440
Curves
   plane, 434–440
   rose, 431
Cycle of sine/cosine graphs, 311–312
Cycloids, 440

D

De Motu Cordis (On the Motion of the Heart), 293
De Propria Vita, 188
Decay, 226, 234–235, 260–262
Decimals, 3
Decomposition
   of functions, 130
   partial-fraction (*See* Partial-fraction decomposition)
Decreasing functions, 95–96
Degrees. *See also* Angles
   measuring angles using, 284–285
   radians and, 286–288
   real zeros, turning points, and, 170–173
Delta 2 rocket, 458
Demand and supply equations, 485
Denominator, rationalizing, 15–16
Dependent systems
   defined, 478
   using substitution method for, 480
Dependent variables, defined, 55, 77
Depressed equations, 183
Descartes, René
   background information on, 53
   coordinate plane discovery by, 52
   Rule of Signs, 188–189
   slope and, 69
Devil's Triangle, 411
Diameter of circles, 60
Difference formulas. *See* Sum and difference formulas
Difference quotients, 106
Digits, common logarithms and number of, 258
Direction of increasing parameter, 435
Distance formula
   in coordinate plane, 53–54
   on number line, 8
Division
   of complex numbers, 48
   of functions, 125–126
   of polynomials, 179–181
Domain
   of combined functions, 125–126
   of composite functions, 128–130
   of exponential functions, 223, 244
   finding from graph, 86–87
   of functions in general, 78, 82–83
   of inverse functions, 138, 142–143
   of linear functions, 94
   of logarithmic functions, 242–243, 244
   of polynomial functions, 165
   of quadratic functions, 156
   of rational functions, 199–200
   of relations, 79
   of sine and cosine functions, 310–311
   of tangent function, 328
   of variable in equation, 21
Double-angle formulas, 379–381
Drug's bioavailability, 77, 88–89

E

e (Euler number), 232
Electric blanket, 373
Elements of sets, defined, 2
Elimination method, 482–484
Empty set, 2
End behavior of polynomial functions, 166–168
Endpoint, defined, 283
Energy units, 184
Equal ordered pairs, defined, 52
Equations
   absolute values in, 30–31
   of circles, 60–62
   converting between polar and rectangular forms, 424–425
   equivalent, 21–22
   exponential, 266–269
   functions defined by, 80–81
   graphing (*See also* Graphs and graphing)
      by plotting points, 55–56
      polar, 425–431
   inconsistent, 21
   linear (*See* Lines and linear equations)
   logarithmic, 269–272
   in one variable, defined, 21
   polar, graphing, 425–431
   polynomial, 27–28
   quadratic, 23–27
   quadratic in form, 30
   radical, 29–30
   rational, 28–29
   trigonometric, 364–370
Equilibrium point, 485
Equivalent equations, 21–22
Equivalent inequalities, 34
Estimating large numbers, 257–258
Euler, Leonhard, 78, 231
Euler number, 232
Even degree, power functions of, 166
Even functions, 98
Even-odd identities, 363
Even-odd properties
   of cosecant, secant, and cotangent functions, 331–332
   of sine and cosine functions, 311
   of tangent function, 328
Everest, Mount, 395
Exponential equations, 266–269
Exponential functions
   applications
      continuous compound interest, 231–233
      growth and decay, 234–235
      simple and compound interest, 227–231
   converting from/to logarithmic form, 240–241
   defined, 223
   evaluating, 223–224
   finding, 226–227
   graphing, 224–226, 233–234
   natural, 233–235
   properties of, 226, 244
Exponential inequalities, 272–273
Exponents
   integer, 8–9
   rational, 17–19
   rules of, 9–10
Expressions
   algebraic and trigonometric, 346–347
   radical, 13–19
Extraneous solutions of rational equations, 28
Extreme values, 97
Exxon Valdez, 124

F

Factor Theorem, 183
Factorials, 452–453
Factoring
   solving polynomial equations by, 27–28
   solving quadratic equations by, 24
Factorization Theorem for Polynomials, 191–192
Factors, defined, 2
Factors of polynomials, defined, 180
Fahrenheit, conversion from Celsius, 37–38
Federal taxes and revenues, 198, 210
Femur, inferring height from, 72
Fiber optics cables, cost of, 89
Fibonacci sequence, 450–451
Filene's Basement store, 337
Finite sequences
   defined, 448
   geometric, 468–469
Fiore, Antonio Maria, 188
FOIL method, for complex numbers, 46–47
Formulas. *See also* Area; Sum and difference formulas
   annuity value, 470
   arc length, 289–290
   change-of-base, 258–259
   compound interest, 230
   continuous compound interest, 232

distance
  in coordinate plane, 53–54
  on number line, 8
double-angle, 379–381
geometry, 411–412
half-angle, 382–384
half-life, 260
Heron's, 415–416
midpoint, 54–55
population growth, 265
power-reducing, 381–382
simple interest, 228
velocity of air when coughing, 94, 97
velocity of blood flow, 119
volume, 411
water pressure, 136, 143
Fractions, partial. *See* Partial-fraction decomposition
Free-falling objects, 155, 458, 462
Frequency in simple harmonic motion, 324
Functions
  absolute value (*See* Absolute values)
  applications, 88–89
  average rate of change of, 104–106
  combining, 124–126
  composite (*See* Composite functions)
  constant, 94, 95–96, 103
  cube root, 99–100, 103
  cubing, 103
  decomposing, 130
  decreasing, 95–96
  definitions, 77–78
  domain of, 82–83
  evaluating, 81–82
  even, 98
  exponential (*See* Exponential functions)
  graphs of
    basic, 102–104
    getting information from, 86–87
    identifying, 84–86
  greatest integer, 102, 104
  identity, 94, 103
  increasing, 95–96
  inverse (*See* Inverse functions)
  linear, 94–95
  list of basic, 102–104
  logarithmic (*See* Logarithmic functions)
  notation, 78–79
  odd, 98
  one-to-one, 136–137, 226
  piecewise, 100–102
  polynomial (*See* Polynomial functions)
  power, 166
  properties of, 95–98
  quadratic (*See* Quadratic functions)
  range of, 83–84
  rational (*See* Rational functions)
  rational power, 104
  reciprocal, 104
  representations of, 79–81
  square root, 98–99, 103
  squaring, 103
  step, 102
  transformations of (*See under* Transformations)
  trigonometric (*See* Trigonometric functions)
Fundamental Theorem of Algebra, 191

G

Galilei, Galileo, 155
Gauss, Karl Friedrich, 461
General form
  of equation of circle, 61–62
  of equation of line, 70–72
General term of sequences, 448–450
Geometric sequences
  applications
    annuities, 469–470
    multiplier effect, 472
  overview, 465–468
  sum of finite, 468–469
  sum of infinite, 471–472
Geometry formulas, 411–412
Global warming, 179
Grains of wheat on chessboard, 222
Graphical method, 477–478
Graphing calculator usage
  angle measurement and conversion, 285, 287, 288
  complex numbers, entering, 45
  conversion between polar and rectangular coordinates, 422, 423
  cosines, 318, 321
  exponential functions
    Euler number, 233
    evaluating, 223
  exponents, integer, 9
  factorials, 452
  geometric sequences, 469
  horizontal lines, 70
  inverse functions, 143
  inverse trigonometric functions, 342, 343–344
  lines in slope-intercept form, 71
  logarithms
    common, 246
    natural, 247
    other, 259
  parametric equations, 440
  partial fractions, 492
  polar equations, 427
  polynomial functions, 167
    turning points, 172
    zeros, 169
  rational functions
    connected vs. dot, 204
    with hole, 203
    revenue curve, 210
  right triangles, 356
  sequences, 449
  square roots, 12
  summation, 453, 469
  systems of linear equations in two variables, 478
  systems of nonlinear equations in two variables, 484
  trigonometric functions
    angles, 304–305
    tangent, cotangent, secant, cosecant, 329
  trigonometric identities, 366
Graphs and graphing
  circles, 61
  compound inequalities, 36–37
  cosecant, secant, and cotangent functions, 331–334
  cosine function, 312–315 (*See also* Cosine function; Sine function)
  cube root functions, 99
  exponential functions, 224–226
    natural, 233–234
  functions in general, 80, 84–87
  horizontal shifts, 111–112
  inequalities using test points, 38–40
  intercepts of, finding, 56–57
  inverse functions, 140–141 (*See also* Inverse trigonometric functions)
  of linear functions, 95
  linear inequalities in one variable, 34–35
  logarithmic functions, 243–247
  ordered pairs, 53
  piecewise functions, 101–102
  plane curves, 435, 437
  by plotting points, 55–56 (*See also under* Points)
  points, 52–53
  polar equations, 425–431
  polynomial functions, 173–174
  quadratic functions, 155–160
  rational functions, 199–200, 205–210
    using translations, 201–202
  real numbers, 5
  sinusoidal, 312–315 (*See also* Cosine function; Sine function)
  square root functions, 99
  step functions, 102
  symmetries of, 57–59
  systems of linear equations in two variables, 477–478
  tangent function, 329–331
  vertical shifts, 109–110, 112
Great Trigonometric Survey (GTS), 395
Greatest integer function, 102, 104
Greenhouse effect, 179
Growth, population, 265, 268–269, 271–272
Growth and decay, 226, 234–235, 260–262

H

Hales, Stephen, 293
Half-angle formulas, 382–384
Half-life, 260–262
Harvey, William, 293
Height
  of clouds, finding, 356
  inferring from femur, 72
  of Kilimanjaro, finding, 356–357
  of mountains, estimating, 398
Heron's formula, 415–416
Hipparchus, 362
Honeybees, 448, 451
Horizontal asymptotes, 200–205
Horizontal lines, 69–70
Horizontal shifts
  of functions in general, 109–112
  of sinusoidal graphs, 319–321
Horizontal stretching/compressing, 116–117
Horizontal-line test, 137–138

I

Identities
  cofunction, 375–376
  equations that are, 21, 23
  trigonometric (*See* Trigonometric identities)

Identity function, 94, 103
Imaginary part of complex numbers, 45
Improper expressions, 180, 489
Inconsistent equations, 21, 23
Inconsistent systems, 478, 479–480
Increasing functions, 95–96
Independent systems, 478
Independent variables, defined, 55, 77
Index in radical expressions, 13
Index of summation, 453
India, 395
Individual retirement account (IRA), 470
Inequalities
   absolute values in, 40–43
   compound, 36–38
   definitions and properties, 33–34
   equivalent, 34
   exponential, 272–273
   linear, 34–35
   logarithmic, 272–273
   rational, 39–40
   real numbers, 5
   symbols for, 5
   test-point method, 38–40
Infinite intervals, 6
Infinite sequences
   defined, 448
   geometric, 471–472
Initial point of plane curves, 434
Initial side, defined, 283
Integers
   defined, 3
   as exponents, 8–9
Intercepts, finding, 56–57. *See also* Lines and linear equations
Interest, 227–233
Intermediate Value Theorem, 170
Intersection of intervals, 7
Intervals, 5–7
Inverse functions. *See also* Inverse trigonometric functions
   applications, 143–144
   defined, 138–139
   finding, 140–142, 143
   finding domain and range of, 142–143
   graphing logarithmic functions using, 243–244
   one-to-one, 136–137, 226
   verifying, 139–140
Inverse trigonometric functions
   application, 347
   arccosine, 339
   arcsine, 337–338
   arctangent, 340
   composition of, 344–347
   of cosecant, secant, and cotangent functions, 341
   evaluating, 341–344
   identities, 342–343
Investments
   annuities, 469–470
   doubling, 247–248
   interest, 227–233
IRA, 470
Iraq, algebra and, 476
Irrational numbers, defined, 3–4

K

Kepler, Johannes
   background information on, 164
   volume of wine barrels, 164, 174–175
Key points in sine/cosine graphs, 314–315
Kilimanjaro, 350, 356–357
King Tut, 253, 261–262
Kitab al-jabr wal-muqabala, 476
Kramp, Christian, 452
Krypton-85, 260

L

Lake, chemical toxins in, 249–250
Lambton, William, 395
Large numbers, estimating, 257–258
Latitudes, 283, 290
Law of Cooling, 248–249
Law of Cosines
   deriving, 404–405
   solving SAS triangles, 405–406
   solving SSS triangles, 406–407
Law of Sines
   applications, 398, 403–404
   bearings, 403–404
   deriving, 396–397
   solving AAS and ASA triangles, 397–398
   solving SSA triangles, 399–403
Leading coefficient, defined, 164
Leading term, defined, 164
Leading-term test, 168
Lemniscates, 431
Leonardo of Pisa, 450
Libby, Willard Frank, 261
Liber abaci, 450
Liebeck, Stella, 239
Like radicals, 14
Limaçons, 430–431
Line segments
   finding midpoint of, 54–55
   parameterizing, 439
Linear functions, 94–95
Linear inequalities in one variable, 34–35
Linear systems of equations. *See* Systems of linear equations in two variables
Lines and linear equations
   defined, 65
   general form of, 70–72
   horizontal and vertical, 69–70
   in one variable, 22–23
   parallel and perpendicular, 72–73
   slopes of, 65–67
Lines of symmetry, 57, 156
Logarithmic equations, 269–272
Logarithmic functions. *See also* Logarithms
   applications, 247–250
   converting from/to exponential form, 240–241
   defined, 239–240
   domain of, 242–243
   graphing, 243–247
   modeling with, 259–260
   properties of, 244
Logarithmic inequalities, 272–273
Logarithms. *See also* Logarithmic functions
   applications, 260–262
   basic properties of, 243, 270–271
   changing base of, 258–259
   common, 245–246
   estimating large numbers, 257–258
   evaluating, 241–242
   exponential equations, solving using, 266–268
   logistic growth model, 265, 271–272
   natural, 247
   rules of, 253–257
Logistic growth model, 265, 271–272
Long division of polynomials, 180–181
Longitudes, 283, 290
Longley, Wild Bill, 65
Lower bound on zeros, 189–190
Lower limit of summation, 453
Lowest terms, 200

M

Mach numbers, 334
Magnitude of cosecant, secant, and cotangent functions, 331
Maximum value of quadratic functions, 156–157
McDonald's coffee, 239, 248–249
Members of sets, defined, 2
Midpoint formula, 54–55
Minimum value of quadratic functions, 156–157
Mount Everest, 395
Mount Kilimanjaro, 350, 356–357
Mountains, estimating height of, 398
Multiplication
   of complex numbers, 46–47
   of functions, 125–126
   of radicals, 14–15
Multiplication property, 33
Multiplicity of zeros, 171–172
Multiplier effect, 465, 472
Munk, Bob, 77

N

Natural exponential functions, 233–235
Natural logarithms, 247. *See also* Logarithms
Natural numbers, defined, 2
Navigation, using bearings, 403–404
Negative angles, defined, 283–284
Negative numbers, 5, 9
Newton, Sir Isaac, 248–249
Nominal rate, defined, 229
Nonlinear systems of equations, 476, 481, 483–484
Nonrigid transformations, defined, 115
Nonterminating repeating decimals, 3
Notations
   factorial, 452–453
   functional, 78–79
   set-builder, 5–6
   summation, 453–455
Nova Stereometria Doliorum Vinariorum, 164
Nth partial sum, 454–455
Nth roots, 12–13
Nth term
   of arithmetic sequences, 459–460
   of geometric sequences, 467–468
   of sequences in general, 448
Number line, 5. *See also* Real numbers
Number of Zeros Theorem, 192
Numerator, rationalizing, 15–16

O

Oblique asymptotes, 208–210
Oblique triangles, solving, 395–396. *See also* Law of Sines
Obtuse angles, defined, 284
Odd, even- properties. *See* Even-odd properties
Odd degree, power functions of, 166
Odd functions, 98
Odd-degree polynomials with real zeros, 193
Ohm's Law, 488
Oil spill, 124, 131
On the Motion of the Heart, 293
One-to-one functions, 136–137, 226. *See also* Inverse functions
One-to-one property of logarithms, 270
Open interval, 6
Ordered pairs
  defined, 52
  functions defined by, 79–80
  graphing, 53
Orientation of plane curves, 435
Origin
  in coordinate plane, 52
  on number line, 5

P

Parabolas, parts of, 156
Parallel circuits, 488, 493–494
Parallel lines, 72–73
Parametric equations of plane curves, 434–440
Partial-fraction decomposition
  application, 493–494
  defined, 489
  when denominator has
    only distinct linear factors, 490–492
    repeated linear factors, 492–494
Peak, defined, 88
Perfect squares, defined, 4
Perfect-square trinomials, defined, 25
Perimeter formulas, 411–412
Periodic functions, 311–312
Periods
  of cosecant, secant, and cotangent functions, 331–332
  of sine and cosine functions
    defined, 311–312
    finding and changing, 315–319
  of tangent function, 328
Perpendicular lines, 72–73
Petroleum consumption, 184
Pharmacokinetics, 32, 88–89
Phase shifts, 319–323
Pi (π), defined, 3
Piecewise functions, 100–102
Pitiscus, 350
Plane curves
  definitions, 434–435
  eliminating parameter, 436–438
  finding parametric equations of, 438–440
  graphing, 435, 437
Plane figures, geometry formulas for, 411
Points
  converting between polar and rectangular forms, 421–424
  determining whether on graph, 86
  finding distance between two, 53–54
  graphing equations by plotting, 55–56
  graphing logarithmic functions by plotting, 243
  graphing polar equations by plotting, 426
  on number line (*See* Number line)
  plotting in coordinate plane, 52–53
  plotting using polar coordinates, 419–421
  on unit circle, finding, 294
Point-slope form, 67–68
Poiseuille, Jean Louis Marie, 109, 119
Polar coordinates
  application, 423–424
  converting equations between polar and rectangular forms, 424–425
  converting points between polar and rectangular forms, 421–424
  definitions, 419
  graphing equations, 425–431
  plotting points using, 419–421
  symmetry, tests for, 426
Polygons, geometry formulas for, 411
Polynomial equations, 27–28
Polynomial functions. *See also* Polynomials; Quadratic functions
  application, 174–175
  defined, 155
  definitions, 164–165
  end behavior of, 166–168
  graphing, 173–174
  properties of, 165–166
  relationship between degrees, real zeros, and turning points, 170–173
  zeros of (*See* Zeros of polynomial functions)
Polynomials. *See also* Polynomial functions
  application, 184
  defined, 179
  dividing, 179–181
  Factorization Theorem for, 191–192
  Rational Zeros Test, 184–185
  Remainder and Factor Theorems, 181–184
Population growth, 265, 268–269, 271–272
Positive angles, defined, 283
Positive numbers, 5, 8
Power (in watts), 373
Power functions, 166
Power of a product rule, 9
Power of a quotient rule, 9
Power rule
  for exponents, 9
  for logarithms, 254–257
Power-reducing formulas, 381–382
Powers of i, 48–49
Practical problems. *See* Applied problems
Principal, defined, 227
Principal square and nth roots, 12–13, 45
Product rule
  for exponents, 9
  for logarithms, 254–257, 271
Proper expressions, 180, 489
Pure imaginary numbers, 45
Pythagorean identities, 363
Pythagorean Theorem
  distance formula and, 53–54
  generalizing (*See also* Law of Cosines)

Q

Quadrantal angles, 284, 301–302
Quadrants, 52
Quadratic equations, 23–27
Quadratic form, exponential equations in, 267–268
Quadratic formula, 26–27
Quadratic functions
  applications, 160–161
  characteristics from graph, 160
  converting to standard form, 158
  graphing
    in any form, 158–160
    in standard form, 155–158
Quadratic in form, 30
Quadratic inequalities, test-point method for solving, 38–40
Quartic polynomials, 165
Quetelet, Lambert, 272
Quintic polynomials, 165
Quotient identities, 363
Quotient rule
  for exponents, 9
  for logarithms, 254–257, 271
Quotients, converting decimal rationals to, 3

R

Radians, 286–288. *See also* Angles
Radicals
  definitions, 12–13
  equations involving, 29–30
  expressions, simplifying, 13–14
  operations on, 14–15
  properties of, 13
  rational exponents, 17–19
  rationalizing denominator/numerator, 15–16
Radiocarbon dating, 261–262
Radius of circles, 60
Range
  of exponential functions, 244
  finding from graph, 86–87
  of functions in general, 78, 83–84
  of inverse functions, 138, 142–143
  of linear functions, 94
  of logarithmic functions, 242, 244
  of relations, 79
  of sine and cosine functions, 310–311
  of tangent function, 328
Rational approximation, 4
Rational equations, 28–29
Rational exponents, 17–19
Rational functions
  application, 210
  defined, 198–200
  graphing, 205–210
    using translations, 201–202
  with oblique asymptotes, 208–210
  vertical and horizontal asymptotes, 200–205
Rational inequalities, 39–40
Rational numbers, defined, 3
Rational power functions, 104
Rational Zeros Test, 184–185
Rationalizing, 15–16
Rays, defined, 283

Real numbers
  absolute values, 7–8
  defined, 4
  integer exponents, 8–9
  interval notation, 5–7
  on number line, 5
  sets of, 2–5
  trigonometric functions of (See under Trigonometric functions)
Real part of complex numbers, 45
Real zeros of polynomial functions, 170–173, 189–190. See also Zeros of polynomial functions
Reciprocal functions
  properties and graph of, 104
  trigonometric, 331–334
Reciprocal identities, 363
Reciprocal square function, 104
Rectangles
  area of, 82, 411–412
  maximum area of, 160–161
  perimeter of, 411–412
Rectangular coordinates, converting between polar and, 421–424
Rectangular equations, converting between polar and, 424–425
Rectangular solids, geometry formulas for, 411
Recursive definitions/formulas, 88, 450–451, 459, 466
Reference angles, 306–307
Reflections, 112–115
Relations, defined, 79–80
Relative maxima/minima of functions, 97
Remainder Theorem, 181–183
Resistance (ohms), 488, 493–494
Retail theft, 337, 347, 357–358
Retirement, IRA, 470
Revenue curves, 198, 210
Right angles, defined, 284
Right circular cones, 411
Right circular cylinders, 411
Right triangles. See also Pythagorean Theorem
  applications, 355–358
  evaluating trigonometric functions in, 351–353
  expressing trigonometric functions using, 350–351
  solving, 354–355
Rigid transformations, defined, 109
Rise, defined, 65
Rivers, finding width of, 357
Roots, 12–13. See also Radicals; Solutions and solution sets
Rose curves, 431
Roster method, 5
Rotation angle of cameras, 337, 357–358
Rule of Signs, Descartes's, 188–189
Run, defined, 65

S

SAS triangles
  defined, 395–396
  finding area of, 411–414
  solving, 405–406
Secant function. See also Trigonometric functions
  defined, 295
  graph and properties of, 331–334
  inverse, 341

Security cameras, rotation angle of, 337, 357–358
Semicircles, 412
Semiperimeter, 415
Sequences
  applications
    arithmetic, 462
    Fibonacci, 451
    geometric, 469–470, 472
  arithmetic, 458–462
  defined, 448
  factorial notation, 452–453
  geometric, 465–472
  recursively defined, 450–451
  summation notation, 453–455
  writing and finding terms of, 449–451
Series, 454–455. See also Sequences
Set-builder notation, 5–6
Sets of numbers, 2–3
Shifts. See Horizontal shifts; Vertical shifts
Sickdhar, Radhanath, 395
Sign-chart method, 38–40
Signs, variation of, 188–189
Similar triangles, 351, 395
Simple harmonic motion, 323–325
Simple interest, 227–228
Sine function. See also Law of Sines; Trigonometric functions
  amplitude and period of, 315–319
  application, 322–323
  defined, 295
  graphing, 312–315
  phase shifts, 319–323
  properties of, 310–312
  simple harmonic motion, 323–325
  sum and difference formulas for, 377
Sinusoidal curves/graphs, 315. See also Cosine function; Sine function
Slope-intercept form, 68–69
Slopes. See also Lines and linear equations
  defined, 65–66
  finding and interpreting, 66–67
  main facts about, 67
  tangents as, 327
  use of m to represent, 69
Smooth curves, 165
Solids, geometry formulas for, 411
Solutions and solution sets
  of equations, 21, 55
  extraneous, 28
  of inequalities, 33
  of systems of equations, 477
Sound waves, 334
Space junk, 458, 462
Spheres, 411
Spirals, 431
Square root functions, 98–99, 103
Square roots, 12
Squares, geometry formulas for, 411
Squaring function, 103
SSA triangles
  defined, 395–396
  solving, 399–403
SSS triangles
  defined, 395–396
  finding area of, 415–416
  solving, 406–407

Standard form
  for complex numbers, 45
  for equation of circle, 60–62
  for linear equations, 22
  for quadratic equations, 24
  for quadratic functions, 156–158
Standard position, angles in, 284, 285, 303–304
Step functions, 102
Straight angles, defined, 284
Stretching functions, 115–117
Substitution method, 479–481
Subtraction
  of complex numbers, 46
  of functions, 125–126
  of radicals, 14
Sum and difference formulas
  for cosine, 373–375
  for sine, 377
  for tangent, 378
Summation, 453–455. See also Sequences
Super Bowl, 472
Supplementary angles, finding, 288–289
Supply and demand equations, 485
Surface areas, 411
Symmetries
  defined, 57–58
  finding terminal points by, 299–300
  graphing using, 59
  inverse functions and, 140–141
  in polar coordinates, 426
  tests for, 58–59
  on unit circle, 294
Synthetic division of polynomials, 180–181
Systems of equations, defined, 476
Systems of linear equations in two variables
  application, 485
  elimination method, 482–484
  graphical method, 477–478
  substitution method, 479–481
  verifying solutions to, 476–477
Systems of nonlinear equations, 476, 481, 483–484

T

Tables, functions defined by, 80
Tangent function. See also Trigonometric functions
  defined, 295
  graphing, 329–331
  properties of, 327–328
  sum and difference formulas for, 378
Tartaglia, Nicolo, 188
Tax rates and revenues, 198, 210
Technology connection. See Graphing calculator usage
Terminal points. See also Trigonometric functions
  defined, 294–296
  finding, 299–300
  of plane curves, 434
Terminal side, defined, 283
Terminating decimals, 3
Terms of sequences, 448–451. See also Sequences
Test-point method, 38–40
Theft, retail, 337, 347, 357–358
Theorems
  complex conjugate product, 47
  Conjugate Pairs, 192–195

Factor, 183
Factorization Theorem for Polynomials, 191–192
Fundamental Theorem of Algebra, 191
Intermediate Value, 170
Number of Zeros, 192
Pythagorean, 53–54
Remainder, 181–183
Toxins in lake, 249–250
Transcendental numbers/functions, 233
Transformations
  of exponential functions, 233–234
  of functions in general
    combining, 114–115, 118–119
    defined, 109
    reflections, 112–115
    stretching or compressing, 115–117
    vertical and horizontal shifts, 109–112
  of logarithmic functions, 244–246
Transitive property, 33
Translations of rational functions, 201–202
Trapezoids, 411
Triangles. *See also* Pythagorean Theorem
  AAS and ASA
    finding area of, 414–415
    solving, 397–398
  area of, 411–416
  congruent, 395
  geometry formulas for, 411
  oblique, solving, 395–396 (*See also* Law of Sines)
  right (*See* Right triangles)
  SAS
    finding area of, 411–414
    solving, 405–406
  similar, 351, 395
  SSA, solving, 399–403
  SSS
    finding area of, 415–416
    solving, 406–407
Triangulation, 395
Trichotomy property, 33
Trigonometric equations, verifying identities, 364–370
Trigonometric expressions, algebraic and, 346–347
Trigonometric functions. *See also specific functions (e.g., sine function)*
  of angles
    common angles, 302
    coterminal angles, 306
    defined, 301–302
    finding values, 303–304
    reference angles, 306–307
    signs of, 305–306
    using calculator, 304–305

composition of, 344–347
evaluating using identities, 363–364
inverse (*See* Inverse trigonometric functions)
of real numbers
  defined, 294–296
  finding exact values, 296–300
right triangles and (*See* Right triangles)
visualizing, on unit circle, 362
Trigonometric identities. *See also* Sum and difference formulas
  basic/fundamental, 362–364
  cofunction, 375–376
  double-angle formulas, 379–381
  half-angle formulas, 382–384
  inverse, 342–343
  power-reducing formulas, 381–382
  verifying, 364–370
Trigonometry. *See also specific concepts*
  founder of, 362
  right-triangle (*See* Right triangles)
  word origin, 350
Trinomials, perfect-square, defined, 25
Trough, defined, 88
Turning points, 97, 172–173
Tutankhamun, King Nebkheperure, 253, 261–262

U

Unbounded intervals, 6
Union of intervals, 7
Unit circle, 293–294, 362. *See also* Trigonometric functions
Upper bound on zeros, 189–190
Upper limit of summation, 453
The Ups and Downs of Drug Levels, 77

V

Variables, defined, 2
Variation of sign, 188–189
Velocity
  of air when coughing, 94, 97
  of blood flow, 119
Verdi, John, 411
Verhulst, Pierre François, 265, 272
Vertex. *See* Vertices
Vertical asymptotes, 200–205
Vertical lines, 69–70
Vertical shifts
  of functions in general, 109–112
  of sinusoidal graphs, 322–323
Vertical stretching/compressing, 115–116
Vertical-line test, 84–85

Vertices
  of angles, 283
  of parabolas
    defined, 156
    finding, 159
Voltage, 373, 488
Volume formulas, 411

W

Water pressure, 136, 143–144
Waugh, Andrew, 395
Whole numbers, defined, 2
Width of rivers, finding, 357
Wine barrels, volume of, 164, 174–175

X

x-axis. *See* Axes
x-coordinate, defined, 52
x-intercepts, finding, 56–57
xy-plane, defined, 52

Y

y-axis. *See* Axes
y-coordinate, defined, 52–53
y-intercepts, finding, 56–57

Z

Zero-product property, solving quadratic equations using, 24
Zero(s)
  of cosecant, secant, and cotangent functions, 331
  division by, 2
  as exponent, 9
  factorial, 452
  multiplicity of, 171–172
  Number of, Theorem, 192
  of polynomial functions (*See* Zeros of polynomial functions)
  Rational Zeros Test, 184–185
  of sine and cosine functions, 311
  of tangent function, 328
Zeros of polynomial functions
  complex, 191–192
  Conjugate Pairs Theorem, 192–195
  degree and, 170–173
  finding, 168–170, 194–195
  finding bounds on real, 189–190
  finding possible number of positive and negative, 188–189

Algebra—Formulas and Definitions

Real-Number Properties
$a + b = b + a$
$(a + b) + c = a + (b + c)$
$a(b + c) = ab + ac$
$a + 0 = 0 + a = a$
$a + (-a) = 0$
$a \cdot 0 = 0 \cdot a = 0$
If $a \cdot b = 0$, then $a = 0$ or $b = 0$.

$ab = ba$
$(ab)c = a(bc)$
$(a + b)c = ac + bc$
$a \cdot 1 = 1 \cdot a = a$
$a\left(\dfrac{1}{a}\right) = 1, a \neq 0$

Absolute Value
$|a| = a$ if $a \geq 0$ and $|a| = -a$ if $a < 0$.

Assume that $a > 0$ and u is an algebraic expression:
$|u| = a$ is equivalent to $u = a$ or $u = -a$.
$|u| < a$ is equivalent to $-a < u < a$.
$|u| > a$ is equivalent to $u < -a$ or $u > a$.

Exponents
$a^n = \underbrace{a \cdot a \cdots a}_{n \text{ factors}}$
$a^0 = 1, \; a \neq 0$
$a^{-n} = \dfrac{1}{a^n}, \; a \neq 0$
$a^m a^n = a^{m+n}$
$\dfrac{a^m}{a^n} = a^{m-n}$
$(a^m)^n = a^{mn}$
$(a \cdot b)^n = a^n \cdot b^n$
$\left(\dfrac{a}{b}\right)^n = \dfrac{a^n}{b^n}$
$\left(\dfrac{a}{b}\right)^{-n} = \left(\dfrac{b}{a}\right)^n = \dfrac{b^n}{a^n}$

Logarithms
$y = \log_a x$ means $a^y = x$
$\log_a a^x = x$
$\log_a 1 = 0$
$\log x = \log_{10} x$
$\log_a MN = \log_a M + \log_a N$
$\log_a \dfrac{M}{N} = \log_a M - \log_a N$
$\log_a M^r = r \log_a M$
$\log_b x = \dfrac{\log_a x}{\log_a b} = \dfrac{\log x}{\log b} = \dfrac{\ln x}{\ln b}$ (change-of-base formula)

$a^{\log_a x} = x$
$\log_a a = 1$
$\ln x = \log_e x$

Radicals
$\sqrt{a^2} = |a|$
$\sqrt[n]{a^n} = |a|$ if n is even
$\sqrt[n]{a^n} = a$ if n is odd
$a^{\frac{1}{n}} = \sqrt[n]{a}$
$\sqrt[n]{ab} = \sqrt[n]{a} \cdot \sqrt[n]{b}$
$a^{\frac{m}{n}} = (\sqrt[n]{a})^m = \sqrt[n]{a^m}$
$\sqrt[n]{\dfrac{a}{b}} = \dfrac{\sqrt[n]{a}}{\sqrt[n]{b}}$

Rational Expressions
(Assume all denominators not to be zero.)
$\dfrac{A}{B} \cdot \dfrac{C}{D} = \dfrac{A \cdot C}{B \cdot D}$

$\dfrac{\frac{A}{B}}{\frac{C}{D}} = \dfrac{A}{B} \div \dfrac{C}{D} = \dfrac{A}{B} \cdot \dfrac{D}{C} = \dfrac{A \cdot D}{B \cdot C}$

$\dfrac{A}{C} \pm \dfrac{B}{C} = \dfrac{A \pm B}{C}$
$\dfrac{A}{B} \pm \dfrac{C}{D} = \dfrac{A \cdot D \pm B \cdot C}{B \cdot D}$

Common Products and Factors
$A^2 - B^2 = (A + B)(A - B)$
$A^2 + 2AB + B^2 = (A + B)^2$
$A^2 - 2AB + B^2 = (A - B)^2$
$A^3 - B^3 = (A - B)(A^2 + AB + B^2)$
$A^3 + B^3 = (A + B)(A^2 - AB + B^2)$
$A^3 + 3A^2B + 3AB^2 + B^3 = (A + B)^3$
$A^3 - 3A^2B + 3AB^2 - B^3 = (A - B)^3$

Distance Formula
The distance between $P(x_1, y_1)$ and $Q(x_2, y_2)$ is
$$d(P, Q) = \sqrt{(x_2 - x_1)^2 + (y_2 - y_1)^2}.$$

Midpoint Formula
The midpoint $M(x, y)$ of the line segment with endpoints $P(x_1, y_1)$ and $Q(x_2, y_2)$ is
$$M(x, y) = \left(\dfrac{x_1 + x_2}{2}, \dfrac{y_1 + y_2}{2}\right).$$

Lines
Slope $m = \dfrac{\text{rise}}{\text{run}} = \dfrac{y_2 - y_1}{x_2 - x_1}$

Point-slope form	$y - y_1 = m(x - x_1)$
Slope-intercept form	$y = mx + b$
Horizontal line	$y = k$
Vertical line	$x = k$
General form	$ax + by + c = 0$

Quadratic Formula
The solutions to the equation $ax^2 + bx + c = 0$, $a \neq 0$, are
$$x = \dfrac{-b \pm \sqrt{b^2 - 4ac}}{2a}.$$

Standard Form of a Circle
$(x - h)^2 + (y - k)^2 = r^2$ center: (h, k)
                                 radius: r

Standard Form of a Parabola

$(y - k)^2 = \pm 4p(x - h)$ vertex: (h, k) and $(p > 0)$
$(x - h)^2 = \pm 4p(y - k)$

Functions/Graphs

Constant Function
$f(x) = c$

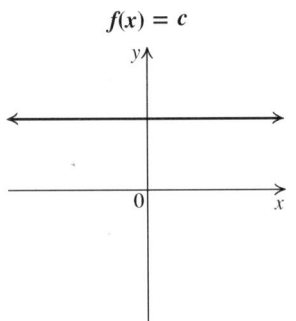

Identity Function
$f(x) = x$

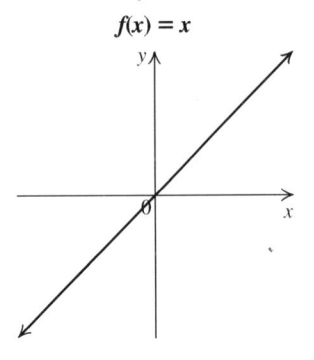

Squaring Function
$f(x) = x^2$

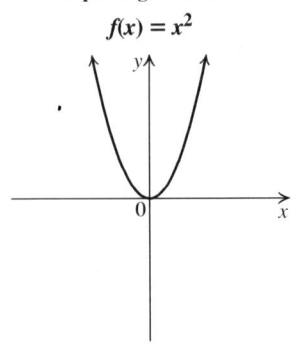

Cubing Function
$f(x) = x^3$

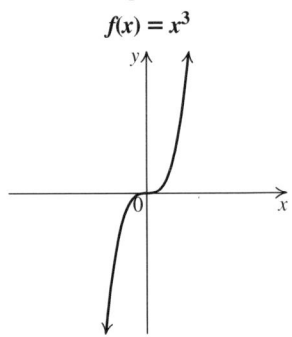

Absolute Value Function
$f(x) = |x|$

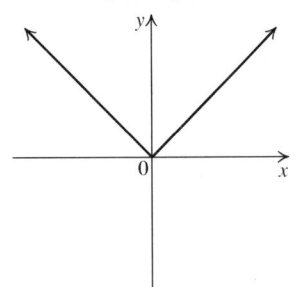

Square Root Function
$f(x) = \sqrt{x}$

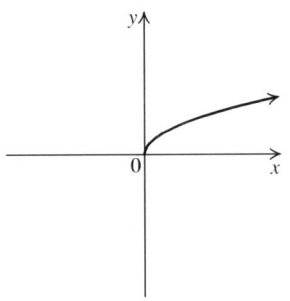

Cube Root Function
$f(x) = \sqrt[3]{x}$

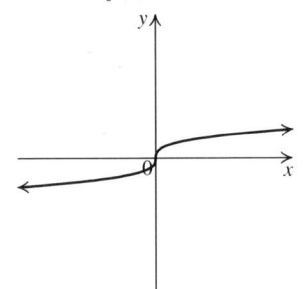

Greatest Integer Function
$f(x) = [\![x]\!]$

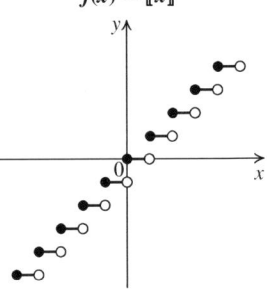

Reciprocal Function
$f(x) = \dfrac{1}{x}$

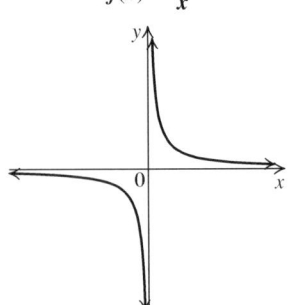

Reciprocal Square Function
$f(x) = \dfrac{1}{x^2}$

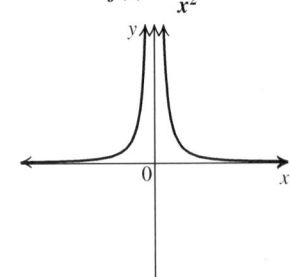

Rational Power Functions
$f(x) = x^{\frac{3}{2}} = \left(x^{\frac{1}{2}}\right)^3$

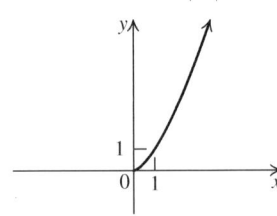

$f(x) = x^{\frac{2}{3}} = \left(x^{\frac{1}{3}}\right)^2$

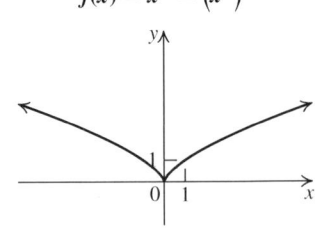